Ω IAS/PARK CITY
MATHEMATICS SERIES
Volume 26

Random Matrices

Alexei Borodin
Ivan Corwin
Alice Guionnet
Editors

American Mathematical Society
Institute for Advanced Study

Ian Morrison, Series Editor
Alexei Borodin, Ivan Corwin, and Alice Guionnet, Volume Editors.

IAS/Park City Mathematics Institute runs mathematics education programs that bring together high school mathematics teachers, researchers in mathematics and mathematics education, undergraduate mathematics faculty, graduate students, and undergraduates to participate in distinct but overlapping programs of research and education. This volume contains the lecture notes from the Graduate Summer School program

2010 *Mathematics Subject Classification.* Primary 15B52, 60B20, 82C22, 60H25, 82B44, 35Q53, 46L54.

Library of Congress Cataloging-in-Publication Data

Names: Borodin, Alexei, editor. | Corwin, Ivan, 1984– editor. | Guionnet, Alice, editor. | Institute for Advanced Study (Princeton, N.J.)

Title: Random matrices / Alexei Borodin, Ivan Corwin, Alice Guionnet, editors.

Description: Providence : American Mathematical Society, [2019] | Series: IAS/Park City mathematics series ; volume 26 | "Institute for Advanced Study." | Includes bibliographical references.

Identifiers: LCCN 2019017274 | ISBN 9781470452803 (alk. paper)

Subjects: LCSH: Random matrices. | Matrices. | AMS: Linear and multilinear algebra; matrix theory – Special matrices – Random matrices. msc | Probability theory and stochastic processes – Probability theory on algebraic and topological structures – Random matrices (probabilistic aspects; for algebraic aspects see 15B52). msc | Statistical mechanics, structure of matter – Time-dependent statistical mechanics (dynamic and nonequilibrium) – Interacting particle systems. msc | Probability theory and stochastic processes – Stochastic analysis – Random operators and equations. msc | Statistical mechanics, structure of matter – Equilibrium statistical mechanics – Disordered systems (random Ising models, random Schrödinger operators, etc.). msc | Partial differential equations – Equations of mathematical physics and other areas of application – KdV-like equations (Korteweg-de Vries). msc | Functional analysis – Selfadjoint operator algebras (C^*-algebras, von Neumann (W^*-) algebras, etc.) – Free probability and free operator algebras. msc

Classification: LCC QA196.5 .R3585 2019 | DDC 512.9/434–dc23

LC record available at https://lccn.loc.gov/2019017274

∞ The paper used in this book is acid-free and falls within the guidelines
established to ensure permanence and durability.
Visit the AMS home page at `https://www.ams.org/`

10 9 8 7 6 5 4 3 2 1   24 23 22 21 20 19

Contents

Preface v

Introduction vii

Riemann–Hilbert Problems
Percy Deift 1

The Semicircle Law and Beyond: The Shape of Spectra of Wigner Matrices
Ioana Dumitriu 41

The Matrix Dyson Equation and its Applications for Random Matrices
László Erdős 75

Counting equilibria in Complex Systems via Random Matrices
Yan V. Fyodorov 159

A Short Introduction to Operator Limits of Random Matrices
Diane Holcomb and Bálint Virág 213

From the Totally Asymmetric Simple Exclusion Process to the KPZ Fixed Point
Jeremy Quastel and Konstantin Matetski 251

Delocalization of Eigenvectors of Random Matrices
Mark Rudelson 303

Microscopic Description of Log and Coulomb Gases
Sylvia Serfaty 341

Random Matrices and Free Probability
Dimitri Shlyakhtenko 389

Least Singular Value, Circular Law, and Lindeberg Exchange
Terence Tao 461

IAS/Park City Mathematics Series
Volume 26, Pages 5–6
https://doi.org/10.1090/pcms/026/00840

Preface

The IAS/Park City Mathematics Institute (PCMI) was founded in 1991 as part of the Regional Geometry Institute initiative of the National Science Foundation. In mid-1993 the program found an institutional home at the Institute for Advanced Study (IAS) in Princeton, New Jersey.

The IAS/Park City Mathematics Institute encourages both research and education in mathematics and fosters interaction between the two. The three-week summer institute offers programs for researchers and postdoctoral scholars, graduate students, undergraduate students, high school students, undergraduate faculty, K-12 teachers, and international teachers and education researchers. The Teacher Leadership Program also includes weekend workshops and other activities during the academic year.

One of PCMI's main goals is to make all of the participants aware of the full range of activities that occur in research, mathematics training and mathematics education: the intention is to involve professional mathematicians in education and to bring current concepts in mathematics to the attention of educators. To that end, late afternoons during the summer institute are devoted to seminars and discussions of common interest to all participants, meant to encourage interaction among the various groups. Many deal with current issues in education: others treat mathematical topics at a level which encourages broad participation.

Each year the Research Program and Graduate Summer School focuses on a different mathematical area, chosen to represent some major thread of current mathematical interest. Activities in the Undergraduate Summer School and Undergraduate Faculty Program are also linked to this topic, the better to encourage interaction between participants at all levels. Lecture notes from the Graduate Summer School are published each year in this series. The prior volumes are:

- Volume 1: *Geometry and Quantum Field Theory* (1991)
- Volume 2: *Nonlinear Partial Differential Equations in Differential Geometry* (1992)
- Volume 3: *Complex Algebraic Geometry* (1993)
- Volume 4: *Gauge Theory and the Topology of Four-Manifolds* (1994)
- Volume 5: *Hyperbolic Equations and Frequency Interactions* (1995)
- Volume 6: *Probability Theory and Applications* (1996)
- Volume 7: *Symplectic Geometry and Topology* (1997)
- Volume 8: *Representation Theory of Lie Groups* (1998)
- Volume 9: *Arithmetic Algebraic Geometry* (1999)

- Volume 10: *Computational Complexity Theory* (2000)
- Volume 11: *Quantum Field Theory, Supersymmetry, and Enumerative Geometry* (2001)
- Volume 12: *Automorphic Forms and their Applications* (2002)
- Volume 13: *Geometric Combinatorics* (2004)
- Volume 14: *Mathematical Biology* (2005)
- Volume 15: *Low Dimensional Topology* (2006)
- Volume 16: *Statistical Mechanics* (2007)
- Volume 17: *Analytical and Algebraic Geometry* (2008)
- Volume 18: *Arithmetic of L-functions* (2009)
- Volume 19: *Mathematics in Image Processing* (2010)
- Volume 20: *Moduli Spaces of Riemann Surfaces* (2011)
- Volume 21: *Geometric Group Theory* (2012)
- Volume 22: *Geometric Analysis* (2013)
- Volume 23: *Mathematics and Materials* (2014)
- Volume 24: *Geometry of Moduli Spaces and Representation Theory* (2015)
- Volume 25: *The Mathematics of Data* (2016)

The American Mathematical Society publishes material from the Undergraduate Summer School in their Student Mathematical Library and from the Teacher Leadership Program in the series IAS/PCMI—The Teacher Program.

After more than 25 years, PCMI retains its intellectual vitality and continues to draw a remarkable group of participants each year from across the entire spectrum of mathematics, from Fields Medalists to elementary school teachers.

Rafe Mazzeo
PCMI Director
March 2019

IAS/Park City Mathematics Series
Volume 26, Pages 7–11
https://doi.org/10.1090/pcms/026/00841

Introduction

Alexei Borodin, Ivan Corwin, and Alice Guionnet

The decade leading up to the 2017 Park City Mathematics Institute saw tremendous breakthroughs in the domains of random matrix theory and interacting particle systems. Major long-standing problems were resolved via methods including local relaxation of Dyson Brownian motion, stochastic operator limits, Coulomb gas electrostatics, Lindeberg exchange, free probability, high dimensional geometry, quantum integrable systems and symmetric function theory. All of these developments have prompted new directions of research and brought into focus possibly avenues for further breakthrough.

The time was right for a large-scale summer school and Park City made for a spectacular venue. There were ten graduate courses with four hours of lectures as well as additional problem sessions. In addition, there was a highly active research program with thirty six lectures delivered on related topics to the courses.

In gathering together these lecture notes, we hope to do two things. First, and foremost, is to provide an introductory level text for students who wish to begin to explore the wide and exciting world of random matrix theory. These lecture notes offer very different perspectives and approaches, and thus will hopefully appeal to a wide variety of interests. The second goal of these notes is to serve as a piece of mathematical memorabilia—to capture some of the energy and enjoyment shared by the hundreds of participants who came to Park City in summer 2017. So, imagine the rarified air of the mountains and enjoy the read.

Random matrix theory has many roots and many branches, perhaps explaining why it has so successfully thrived as a research areas bridging mathematics and many other disciplines (such as statistics, physics, computer science, data science, numerical analysis, biology, ecology, engineering, operations research). In statistics, Wishart began the study of sample covariance matrices in the 1920s. Quite separately in nuclear physics, Wigner introduced and studied certain invariant ensembles in the 1950s. Goldstein and von Neumann came upon random matrix theory at a similar time from the perspective of numerical analysis and estimation of condition numbers. In number theory, in a surprising development in the 1970s, Montgomery recognized that random matrix statistics described the non-trivial zeros of the Riemann Zeta function. More recently, there have been a host of new motivations and sources for problems in random matrix theory, or uses of the tools which have been developed in its study. It is this constant growth and expansion of the field which has made it one of the most dynamic and exciting areas of mathematics.

While some applications of random matrix theory techniques come quite naturally, others (like the number theory ones mentioned above) come as a surprise and take a while to fully develop. In the late 1990s, such a mysterious link was discovered between random matrix ensembles and a few interacting particle systems, namely the longest increasing subsequence problem for random permutations and the closely related totally asymmetric simple exclusion process. This linked random matrix theory to a vibrant and growing area of probability and non-equilibrium statistical mechanics, and led to a bevy of new problems, methods and results. The origins of the link have been progressively exposed over time and have further connected these fields to asymptotic representation theory, quantum integrable systems and algebraic combinatorics.

Within random matrix theory, and more broadly probability and statistical mechanics, there are often two complementary themes—universality and integrability. Universality refers to the idea that randomness smooths out microscopic difference between systems and hence only certain key phenomenological properties of a system will control the large scale or long time behavior. The simplest instance of this concept at play is the central limit theorem for independent identically distributed random variables where, after fixing the mean and variance, all sums have the same universal Gaussian limit. Integrability (or sometimes exact solvability) refers to the search for models which enjoy enhanced algebraic structure which enables exact calculations and precise asymptotics. Indeed, with the central limit theorem example, coin flipping admits exact formulas in terms of binomial distributions which yielded for the first time (in 1738) the Gaussian distribution (long before it was proved universal around 1900). In a sense, universality says that many systems share a common limit, and integrability identifies precisely what that limit is.

The ten lecture courses at PCMI (as well as a host of research talks) addressed many of the themes described above. A natural starting point in the study of random matrices is to look at properties of ensembles with independent Gaussian entries (subject to symmetries such as being Hermitian). Such Gaussian ensembles (e.g. GOE, GUE, or GSE) can be studied through exact formulas and from their "integrable" structure one arrives at a host of precise "universality" predictions for more general classes of random matrices—for example, the Gaussian ensembles are both "Wigner matrices" and "invariant ensembles".

A "Wigner matrix" has independent identically distributed (iid) entries subject to symmetries (such as being Hermitian). Further generalizations away from the iid setting, such as to adjacency matrices for Erdös Reyni random graphs, or d-regular graphs. An "invariant ensemble" is a measure on matrices which is invariant under conjugation by the orthogonal, unitary, or symplectic group (each symmetry is a different class of invariant ensembles). Because of this invariance, the eigenvectors are Haar distributed on the corresponding symmetry group and the joint distribution of the eigenvalues takes a particularly nice form. Namely, they are distributed as point masses subject to a specific two-particle logarithmic (sometimes called Coulomb) pair potential along with a one-body potential. The Gaussian case corresponds to a quadratic one-particle potential.

For random matrix eigenvalues, there are two main scales—global and local. Global results involves the behavior of all of the eigenvalues—for instance, demonstrating a limit for their histogram, or for fluctuations of their linear statistics. Local results involve understanding individual eigenvalues—for instance, demonstrating that they converge to specific point processes (e.g. the Dyson Sine process in the bulk or Airy process around the edge).

Elements of global universality have a very long history (as opposed to local universality which has only seen progress in the past two decades). Dumitriu's lectures present the powerful and classical "moment method" for proving instances of global universality for Wigner ensembles. For instance, she demonstrates how the Wigner semi-circle law arises as the limit of the histogram of eigenvalues. She also describes more recent results related to the central limit theorem (and its connection to the Gaussian free field) for linear statistics of the eigenvalues. At the heart of these calculations are certain combinatorial problems which arise in computing these moments. Variants of such combinatorics also play an important role in Shlyakhtenko's lectures on free probability as we will describe further below.

Turning from global to local limits, Holcomb and Virág's lectures develop the method of "tridiagonal matrices" and their "stochastic operator" limits as a means to characterize the limiting point processes which arise on the local scale for Gaussian ensembles (in fact, the methods extend also to invariant ensembles and general β-ensembles). Gaussian ensembles can be tridiagonalized into a rather simple form. Such tridiagonal matrices can be thought of as discrete second order difference operators with noise. In the large N limit, these difference operators become differential operators and their (random) spectrum is exactly the desired limiting point process. This characterization is different than earlier descriptions of these point processes (e.g. via formulas for their correlation functions) and provides a useful tool in many calculations.

The past decade has seen tremendous progress in demonstrating local universality for Wigner ensembles (and further generalizations of them). Erdös, Rudelson and Tao each address aspects of this progress in their lectures. Erdös' lectures describes an approach to proving the universality of the local eigenvalue point process based on studying variants of Dyson's Brownian motion (DBM). Adding a small (in terms of variance) Gaussian ensemble to the Wigner matrix turns out to correspond to running a stochastic differential equation (DBM) on the eigenvalues. Universality comes from leveraging the fast convergence of DBM to "local equilibrium", i.e., to the local behavior of the Gaussian ensembles.

Rudelson's lectures focus on eigenvectors in addition to eigenvalues. For the Gaussian (or invariant) ensembles, the eigenvectors are independent of the eigenvalues and are distributed according to Haar measure on the relevant symmetry group. This shows that the eigenvectors are highly "delocalized" since there is no preferred direction. Rudelson addresses the challenge of showing that delocalization holds for a much wider class of matrices for which the eigenvalues and vectors do not simply decouple. At the heart of this chapter are methods from high dimensional geometry such as small ball probabilities for random projectors.

Tao's lectures addresses the behavior of singular values for non-Hermitian matrices, in particular how to bound the smallest such value. This problem relates to the earlier mentioned numerical analysis motivation to study random matrices, since the smallest singular value controls the speed of convergence for various related numerical schemes. He also leverages estimates from high dimensional geometry along with the "Lindeberg exchange" method to prove universality of various local eigenvalue properties for both non-Hermitian and Hermitian models. The "Lindeberg exchange" method studies the effect of swapping non-Gaussian entries with Gaussian entries.

Turning from Wigner to invariant ensembles, we come to Deift and Serfaty's lectures. For (unitary) invariant ensembles with general one-particle potentials, the "orthogonal polynomial" method expresses local eigenvalue statistics via the machinery of "determinantal point processes" involving an explicit correlation kernel. The kernel is written in terms of orthogonal polynomials related to the one-particle potential. For the Gaussian (quadratic) case, these are Hermite polynomials. However, for general potentials, they are not nearly as explicit and proving universality boils down to understanding the asymptotics of these orthogonal polynomials. This was achieve by studying asymptotics for an associated Riemann-Hilbert problem (RHP) via a nonlinear steepest descent method. This method is described in Deift's lectures, along with several other applications for RHP asymptotics coming from integrable systems and probability theory.

Serfaty's lectures also consider measures on point configurations given (in Gibbsian form) via a logarithmic (or Coulomb) two-particle pair potential and a general one-particle potential. In addition to working on the line, she also considers configurations on the plane (such as arise in the Ginibre ensemble of random matrices). The chapter studies these measures directly via electrostatics and shows how this method yields universality for fluctuations of linear statistics and a description of the large deviations for the empirical point process measure.

Random matrix methods and distributions have proved useful in a number of (not so obviously) related fields. Shlyakhtenko's lectures explains how notions from free probability (developed to study von Neumann algebras) closely relate to asymptotic behaviors of random matrix models. For instance, some applications for "free convolution", "free entropy" and free monotone transportation" are developed in relation to large N limits of random matrix eigenvalues.

Fyodorov's lectures turns to the role of random matrices as models for interactions in large systems of coupled differential equations (such as arise in ecology models). From this perspective, the stability of these systems proves to be closely related to the eigenvalues of the matrices. Such ideas and tools (such as the Kac-Rice formula for Gaussian processes) have also arisen in relation to performance of certain classes of algorithms in big data settings, and complexity of frustrated energy landscapes such as in spin glass models.

Quastel and Matetski's lectures study the Kardar-Parisi-Zhang (KPZ) universality class. Since the late 90s, this area of probability/theoretical physics has enjoyed immense and fruitful interactions with the superifically disjoint theory of random matrices. This is particularly true for the special KPZ class model known as the totally asymmetric simple exclusion process (TASEP). Through certain mappings, the study of TASEP can be related to the study of discrete variants of random matrix eigenvalues, and consequently many universal limit laws (like those named for Tracy and Widom) from random matrix theory also arise in this context. These lectures focus on describing asymptotic limits for transition probabilities for TASEP, which should govern the behavior of other models in the KPZ universality class. The key method involves novel representations and asymptotics for biorthogonal ensembles (related to non-intersecting paths).

This book exists due to the hard work of the lecturers and their coauthors in preparing and editing these notes, as well as the work of the anonymous referees in provided extensive helpful feedback through this process. Ian Morrison deserves special thanks as well for his tireless work as PMCI publisher in organizing and shepherding these notes into the form of the book that you now have.

We would also like to use this opportunity to thank and acknowledge some of the individuals who made PCMI 2017 a wonderful experience. Our involvement in the program is due to an invitation by Richard Hain. The role of program director soon after shifted to Rafe Mazzeo who guided us through every last intricacy of organizing this program. The amount of effort that Rafe put in to our program is immeasurable. The PCMI staff—Beth Brainard and Dena Vigil—likewise did extraordinary work in planning every piece of this program, and the steering committee, namely Bryna Kra and Michelle Wachs, were partners in this pursuit every step of the way.

One of the unique features of PCMI is that it combines the graduate/research program with several other programs involving undergraduate students, teachers of all levels, and even high-school students. The program brings everyone together under one big tent (quite literally for meals, as well as figuratively). We appreciate having had the opportunity for our research community (and ourselves) to participate in such an extraordinary and memorable event as this.

IAS/Park City Mathematics Series
Volume 26, Pages 1–40
https://doi.org/10.1090/pcms/026/00842

Riemann–Hilbert Problems

Percy Deift

Abstract. These lectures introduce the method of nonlinear steepest descent for Riemann-Hilbert problems. This method finds use in studying asymptotics associated to a variety of special functions such as the Painlevé equations and orthogonal polynomials, in solving the inverse scattering problem for certain integrable systems, and in proving universality for certain classes of random matrix ensembles. These lectures highlight a few such applications.

Contents

Lecture 1 1
Lecture 2 10
Lecture 3 20
Lecture 4 29

Lecture 1

These four lectures are an abridged version of 14 lectures that I gave at the Courant Institute on RHPs in 2015. These 14 lectures are freely available on the AMS website AMS Open Notes.

Basic references for RHPs are [8,12,28]. Basic references for complex function theory are [19,23,24]. Many more specific references will be given as the course proceeds.

Special functions are important because they provide *explicitly solvable models* for a vast array of phenomena in mathematics and physics. By "special functions" I mean Bessel functions, Airy functions, Legendre functions, and so on. If you have not yet met up with these functions, be assured, sooner or later, you surely will.

It works like this. Consider the *Airy equation* (see, e.g. [1,29])

$$y''(x) = xy(x), \quad -\infty < x < \infty. \tag{1.1}$$

Seek a solution of (1.1) in the form

$$y(x) = \int_\Sigma e^{xs} f(s)\, ds$$

for some functions $f(x)$ and some contours Σ in the complex plane $\mathbb{C}$. We have

$$y''(x) = \int_\Sigma s^2 e^{xs} f(s)\, ds$$

and

$$x\,y(x) = \int_\Sigma \left(\frac{d}{ds} e^{xs}\right) f(s)\, ds$$

$$= -\int e^{xs} f'(s)\, ds$$

provided we can drop the boundary terms. In order to solve (1.1) we need to have

$$-f'(s) = s^2 f$$

and so

$$f(s) = \text{const.}\; e^{-\frac{1}{3} s^3} .$$

Thus

$$y(x) = \text{const.} \int_\Sigma e^{xs - \frac{1}{3} s^3}\, ds$$

provides a solution of the Airy equation.

The particular choice

$$\text{const.} = \frac{1}{2\pi i}$$

and Σ in Figure 1.3 is known as *Airy's integral* $\mathrm{Ai}(x)$

(1.2) $$\mathrm{Ai}(x) = \frac{1}{2\pi i} \int_\Sigma e^{xz - \frac{1}{3} z^3}\, dz .$$

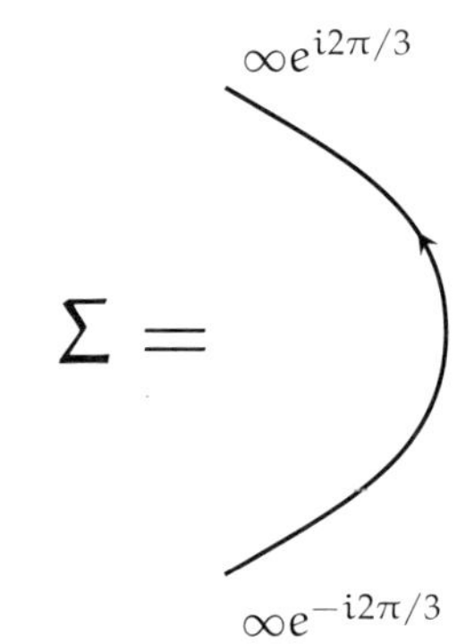

FIGURE 1.3. Σ for Airy's integral.

Other contours provide other, independent solutions of Airy's equation, such as $\mathrm{Bi}(x)$ (see [1]). Now the *basic fact of the matter* is that the integral representation (1.2) for $\mathrm{Ai}(x)$ enables us, using the classical method of *stationary phase/steepest descent*, to compute the asymptotics of $\mathrm{Ai}(x)$ as $x \to +\infty$ and $-\infty$ with *any desired accuracy*. We find, in particular [1, p. 448], that for $\zeta = \frac{2}{3}x^{3/2}$

$$\text{(1.4)} \qquad \mathrm{Ai}(x) \sim \frac{1}{2\sqrt{\pi}}\, x^{-\frac{1}{4}}\, e^{-\zeta} \sum_{k=0}^{\infty} (-1)^k c_k\, \zeta^{-k}$$

as $x \to +\infty$, where

$$c_0 = 1,$$

$$c_k = \frac{\Gamma\left(3k+\frac{1}{2}\right)}{54^k\, k!\, \Gamma\left(k+\frac{1}{2}\right)} = \frac{(2k+1)(2k+3)\dots(6k-1)}{(216)^k\, k!}, \qquad k \geqslant 1.$$

and that

$$\text{(1.5)} \qquad \begin{aligned} \mathrm{Ai}(-x) \sim \frac{1}{\sqrt{\pi}}\, x^{-1/4} \Bigg(& \sin\left(\zeta + \frac{\pi}{4}\right) \sum_{0}^{\infty} (-1)^k c_{2k}\, \zeta^{-2k} \\ & - \cos\left(\zeta + \frac{\pi}{4}\right) \sum_{0}^{\infty} (-1)^k c_{2k+1}\, \zeta^{-2k-1} \Bigg), \end{aligned}$$

as $x \to +\infty$.

Such results for solutions of general 2nd order equations are very rare. Formulae (1.4) and (1.5) solve the fundamental *connection problem* or *scattering problem* for solutions of the Airy equation. Thus, if we know that a solution $y(x)$ of the Airy equation behaves like

$$y(x) = \frac{1}{2\sqrt{\pi}}\, x^{-1/4}\, e^{-\zeta} \left(1 - \frac{c_1}{\zeta} + \dots\right)$$

as $x \to +\infty$, then we know *precisely* how it behaves as $x \to -\infty$, and vice versa, by (1.4) (1.5), see Figure 1.6.

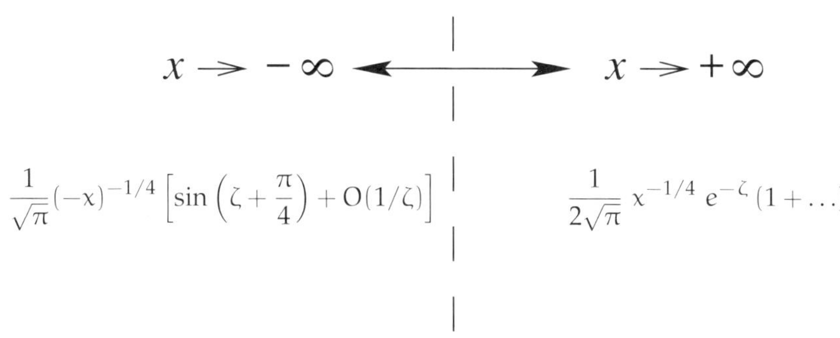

FIGURE 1.6. Asymptotics for Airy's integral.

Exercise 1.7. Use the classical steepest-descent method to verify (1.4) and (1.5). There are similar precise results for *all* the classical special functions. The diligent student should regard Abramowitz & Stegun [1] as an exercise book for the steepest descent method — verify all the asymptotic formulae!

Now in recent years it has become clear that a new and extremely broad class of problems in mathematics, engineering and physics is described by a *new* class of special functions, the so-called *Painlevé functions*. There are six Painlevé equations and we will say more about them later on. Whereas the classical special functions, such as Airy functions, Bessel functions, etc. typically arise in linear (or linearized problems) such as acoustics or electromagnetism, the Painlevé equations arise in nonlinear problems, and they are now recognized as forming the core of modern special function theory. Here are some examples of how Painlevé equations arise:

Example 1.8. Consider solutions of the modified Korteweg–de Vries equation (MKdV)

$$\begin{aligned} &u_t - 6u^2 u_x + u_{xxx} = 0, \quad -\infty < x < \infty, \quad t > 0, \\ &u(x,0) = u_0(x) \to 0 \quad \text{as} \quad |x| \to \infty. \end{aligned} \tag{1.9}$$

Then [16] as $t \to \infty$, in the region $|x| \leqslant c\, t^{1/3}$, $c < \infty$,

$$u(x,t) = \frac{1}{(3t)^{1/3}}\, p\left(\frac{x}{(3t)^{1/3}}\right) + O\left(\frac{1}{t^{2/3}}\right) \tag{1.10}$$

where $p(s)$ is a particular solution of the *Painlevé II (PII)* equation

$$p''(s) = s\, p(s) + 2\, p^3(s).$$

Example 1.11. Let $\pi := (\pi_1 \pi_2 \dots \pi_N) \in S_N$ be a permutation of the numbers $1, 2, \dots, N$. We say that $\pi_{i_1}, \pi_{i_2}, \dots \pi_{i_k}$ is an *increasing subsequence* of π of *length k* if

$$i_1 < i_2 < \cdots < i_k$$

and

$$\pi_{i_1} < \pi_{i_2} < \cdots < \pi_{i_k}.$$

Thus if $N = 6$ and $\pi = (413265)$, then 125 and 136 are increasing subsequences of π of length 3. Let $\ell_N(\pi)$ denote the length of a longest increasing subsequence of π, e.g., for $N = 6$ and π as above, $\ell_6(\pi) = 3$, which is the length of the longest increasing subsequences 125 and 136.

Now equip S_N with uniform measure. Thus

$$\text{Prob}\ (\ell_N \leqslant n) = \frac{\#\ \{\pi \in S_N : \ell_N(\pi) \leqslant n\}}{N!}.$$

Question. How does ℓ_N behave statistically as $N, n \to \infty$?

Theorem 1.12 ([2]). *Center and scale ℓ_N as follows:*

$$\ell_N \to X_N = \frac{\ell_N - 2\sqrt{N}}{N^{1/6}}$$

then

$$\lim_{N\to\infty} \text{Prob}\,(X_N \leqslant x) = e^{-\int_x^\infty (s-x)\, u^2(s)\, ds}$$

where $u(s)$ is the (unique) solution of Painlevé II (the so-called Hastings-McLeod solution) normalized such that

$$u(s) \sim \text{Ai}(s) \quad \textit{as} \quad s \to +\infty.$$

The distribution on the right in Theorem 1.12 is the famous Tracy-Widom distribution for the largest eigenvalue of a GUE matrix in the edge scaling limit.

Theorem 1 is one of a very large number of probabilistic problems in combinatorics and related areas, whose solution is expressed in terms of *Random Matrix Theory (RMT)* via Painlevé functions (see, e.g., [3]).

The *key* question is the following: Can we describe the solutions of the Painlevé equations as precisely as we can describe the solutions of the classical special functions such as Airy, Bessel, ... ? In particular, can we describe the solutions of the Painlevé equations asymptotically with arbitrary precision and solve the connection/scattering problem as in (1.4) and (1.5) for the Airy equation (or any other of the classical special functions):

$$\text{known behavior as } x \to +\infty \qquad \Rightarrow \qquad \text{known behavior as } x \to -\infty$$

and vice versa.

As we have indicated, at the technical level, connection formulae such as (1.4) and (1.5) can be obtained because of the existence of an integral representation such as (1.2) for the solution. Once we have such a representation the asymptotic behavior is obtained by applying the (classical) steepest descent method to the integral. There are, however, no known integral representations for solutions of the Painlevé equations and we are led to the following questions:

Question 1: Is there an analog of an integral representation for solutions of the Painlevé equations?

Question 2: Is there an analog of the classical steepest descent method which will enable us to extract precise asymptotic information about solutions of the Painlevé equations from this analog representation?

The answer to both questions is *yes*: In place of an integral representation such as (1.2), we have a *Riemann–Hilbert Problem (RHP)*, and in place of the classical steepest descent method we have the *nonlinear (or non-commutative) steepest descent method* for RHPs (introduced by P. Deift and X. Zhou [16]).

So what is a RHP? Let Σ be an oriented contour in the plane, see Figure 1.13.

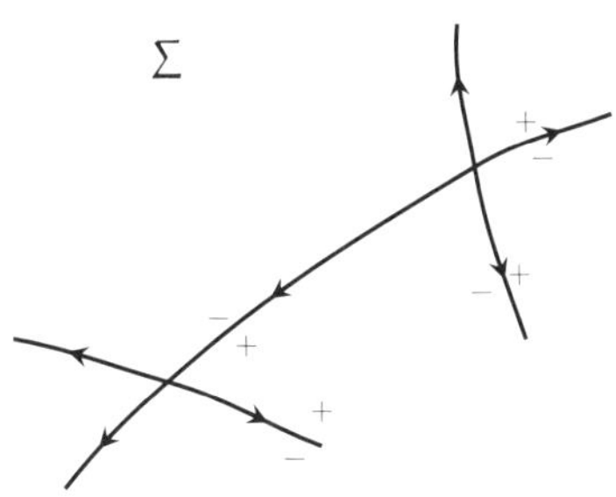

FIGURE 1.13. An oriented contour in the plane.

By convention, if we move along an arc in Σ in the direction of the orientation, the $(\pm)$-sides lie on the left (resp. right). Let $v : \Sigma \to GL(k, \mathbb{C})$, the *jump matrix*, be an invertible $k \times k$ matrix function defined on Σ with

$$v, v^{-1} \in L^\infty(\Sigma).$$

We say that an $n \times k$ matrix function $m(z)$ is a *solution* of the RHP (Σ, v) if

$$m(z) \text{ is analytic in } \mathbb{C}/\Sigma,$$

$$m_+(z) = m_-(z)v(z), z \in \Sigma,$$

$$\text{where } m_\pm(z) = \lim_{z' \to z_\pm} m(z').$$

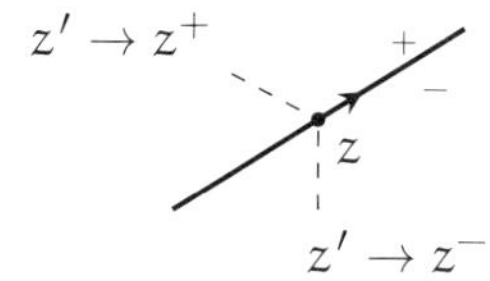

If, in addition, $n = k$ and

$$m(z) \to I_k \quad \text{as} \quad z \to \infty,$$

we say that $m(z)$ solves the *normalized* RHP (Σ, v).

RHPs involve a lot of technical issues. In particular

- How smooth should Σ be?
- What measure theory/function spaces are suitable for RHPs?
- What happens at points of self intersection (see Figure 1.14)?

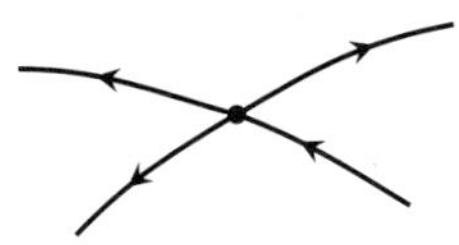

FIGURE 1.14. A point of self intersection.

- In what sense are the limits $m_\pm(z)$ achieved?
- In the case $n = k$, in what sense is the limit $m(z) \to I_k$ achieved?
- Does an $n \times k$ solution exist?
- In the normalized case, is the solution unique?

And most importantly

- at the analytical level, what kind of problem is a RHP? As we will see, the problem reduces to the analysis of singular integral equations on Σ.

There is not enough time in these 4 lectures to address all these issues systematically. Rather we will address specific issues as they arise.

As an example of how things work, we now show how PII is related to a RHP (see, e.g. [22]). Let Σ denote the union of six rays

$$\Sigma_k = e^{i(k-1)\pi/3}\rho, \quad \rho > 0, \quad 1 \leqslant k \leqslant 6$$

oriented outwards. Let p, q, r be complex numbers satisfying the relation

$$p + q + r + pqr = 0. \tag{1.15}$$

Let $v(z)$, $z \in \Sigma$, be constant on each ray as indicated in Figure 1.16 and for *fixed* $x \in \mathbb{C}$ set

$$v_x(z) = \begin{pmatrix} e^{-i\theta} & 0 \\ 0 & e^{i\theta} \end{pmatrix} v(z) \begin{pmatrix} e^{i\theta} & 0 \\ 0 & e^{-i\theta} \end{pmatrix}, \qquad z \in \Sigma$$

where

$$\theta = \theta_x(z) = \frac{4}{3}z^3 + xz.$$

Thus for $z \in \Sigma_3$

$$v_x(z) = \begin{pmatrix} 1 & r\,e^{-2i\theta} \\ 0 & 1 \end{pmatrix}$$

and so on.

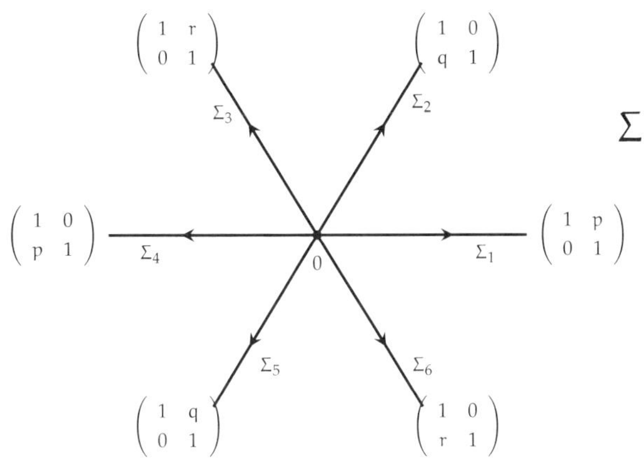

FIGURE 1.16. Six rays oriented outwards.

For *fixed* x, let $m_x(z)$ be the 2×2 matrix solution of the normalized RHP (Σ, v_x). Then

$$u(x) = 2i(m_1(x))_{12}$$

is a solution of the PII equation where

$$m_x(z) = I + \frac{m_1(x)}{z} + O\left(\frac{1}{z^2}\right)$$

as $z \to \infty$. (This result is due to Jimbo and Miwa [27], and independently to Flaschka and Newell [20].) The asymptotic behavior of $u(x)$ as $x \to \infty$ is then obtained from the RHP (Σ, v_x) by the nonlinear steepest descent method.

In the classical steepest descent method for integrals such as (1.2) above, the contour Σ is deformed so that the integral passes through a stationary phase point where the integrand is maximal and the main contribution to the integral then comes from a neighborhood of this point. The nonlinear (or non-commutative) steepest descent method for RHPs involves the same basic ideas as in the classical scalar case in that one deforms the RHP, $\Sigma \to \Sigma'$, in such a way that the exponential terms (see e.g. $e^{2i\theta}$ above) in the RHP have maximal modulus at points of the deformed contour Σ'. The situation is far more complicated than the scalar integral case, however, as the problem involves matrices that do not commute. In addition, terms of the form $e^{-2i\theta}$ also appear in the problem and must be separated algebraically from terms involving $e^{2i\theta}$, so that in the end the terms involving $e^{2i\theta}$ and $e^{-2i\theta}$ both have maximal modulus along Σ' (see [16–18]). A simple example of the nonlinear steepest descent method is given at the end of Lecture 4.

One finds, in particular, ([18], and also [22,25]) the following:
Let $-1 < q < 1$, $p = -q$, $r = 0$. Then as $x \to -\infty$,

$$u(x) = \frac{\sqrt{2\nu}}{(-x)^{1/4}} \cos\left(\frac{2}{3}(-x)^{3/2} - \frac{3}{2}\nu \log(-x) + \phi\right) + O\left(\frac{\log(-x)}{(-x)^{5/4}}\right) \tag{1.17}$$

where

$$\nu = \nu(q) = -\frac{1}{2\pi} \log\left(1 - q^2\right) \tag{1.18}$$

and

$$\phi = -3\nu \log 2 + \arg\Gamma(i\nu) + \frac{\pi}{2}\operatorname{sgn}(q) - \frac{\pi}{4}. \tag{1.19}$$

As $x \to +\infty$

$$u(x) = q \operatorname{Ai}(x) + O\left(\frac{e^{-4/3\, x^{3/2}}}{x^{1/4}}\right). \tag{1.20}$$

These asymptotics should be compared with (1.4), (1.5) for the Airy function. Note from (1.4) that as $x \to +\infty$

$$\operatorname{Ai}(x) \sim x^{-1/4} e^{-2/3\, x^{3/2}}.$$

Also observe that PII

$$u''(x) = x\, u(x) + 2u^3(x)$$

is a clearly a nonlinearization of the Airy equation

$$u''(x) = x\, u(x)$$

and so we expect similar solutions when the nonlinear term $2u^3(x)$ is small.

Also note that (1.17) and (1.18) solve the connection problem for PII. If we know the behavior of the solutions $u(x)$ of PII as $x \to +\infty$, then we certainly know q from (1.20). But then we know $\nu = \nu(q)$ and $\phi = \phi(q)$ in (1.18) and (1.19) and hence we know the asymptotics of $u(x)$ as $x \to -\infty$ from (1.17). Conversely, if we know the asymptotics of $u(x)$ as $x \to -\infty$, we certainly know $\nu > 0$ from (1.17) and hence we know q^2 from (1.18), $q^2 = 1 - e^{-2\pi\nu}$. But then again from (1.17), we know ϕ, and hence $\operatorname{sgn}(q)$ from (1.19). Thus we know q, and hence the asymptotics of the solution $u(x)$ as $x \to +\infty$ from (1.20). Finally note the similarity of the multiplier

$$e^{x z \quad \frac{1}{3} z^3} \tag{1.21}$$

for the Airy equation with the multiplier

$$e^{i\theta} = e^{i\left(x z + \frac{4}{3} z^3\right)} \tag{1.22}$$

in the RHP for PII. Setting $z \to i\,z$ in (1.21)

$$e^{x z - \frac{1}{3} z^3} \to e^{i\left(x z + \frac{1}{3} z^3\right)}$$

which agrees with (1.22) up to appropriate scalings.

Also note from (1.15) that PII is parameterized by parameters lying on a 2-dim variety: this corresponds to the fact that PII is second order.

The fortunate and remarkable fact is that the class of problems in physics, mathematics, and engineering expressible in terms of a RHP is very broad and growing. Here is one more, with more to come!

The RHP for the MKdV equation (1.9) is as follows (see e.g., [16]): Let $\Sigma = \mathbb{R}$, oriented from $-\infty$ to $+\infty$. For fixed $x,\ t \in \mathbb{R}$ let

$$v_{x,t}(z) = \begin{pmatrix} 1-|r(z)|^2 & -\overline{r(z)}\, e^{-2i\tau} \\ r(z)\, e^{2i\tau} & 1 \end{pmatrix}, \qquad z \in \mathbb{R} \tag{1.23}$$

where $\tau = \tau_{x,t}(z) = xz + 4tz^3$ and $r = r(z)$ is a given function in $L^\infty(\mathbb{R}) \cap L^2(\mathbb{R})$ with

$$\|r\|_\infty < 1$$

and

$$r(z) = -\overline{r(-z)}, \quad z \in \mathbb{R}.$$

There is a bijection from the initial data $u(x, t = 0) = u_0(x)$ for MKdV onto such functions $r(z)$ — see later. The function $r(z)$ is called the *reflection coefficient* for u_0, see (4.13).

Let $m = m_{x,t}(z)$ be the solution of the normalized RHP $(\Sigma, v_{x,t})$. Then

$$u(x,t) = 2\,(m_1(x,t))_{12}\ , \tag{1.24}$$

is the solution of MKdV with initial condition $u(x, t = 0) = u_0(x)$ corresponding to $r(z)$. Here

$$m_{x,t}(z) = I + \frac{m_1(x,t)}{z} + O\left(\frac{1}{z^2}\right)$$

as $z \to \infty$.

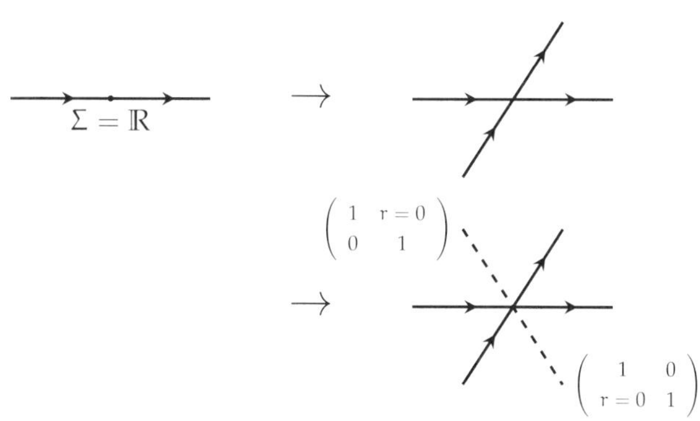

FIGURE 1.25. Obtaining six rays.

The asymptotic result (1.10) is obtained by applying the nonlinear steepest descent method to the RHP $(\Sigma, v_{x,t})$ in the region $|x| \leqslant c\ t^{1/3}$. In this case PII emerges as the RHP $(\Sigma, v_{x,t})$ is "deformed" into the RHP (Σ, v_x) in Figure 1.16.

As we will see, RHPs are useful not only for asymptotics, but also they can be used to determine symmetries and formulae/identities/equations, and also for analytical purposes.

Lecture 2

We now consider some of technical issues that arise for RHPs, which are listed with bullet points above.

A key role in RH theory is played by the *Cauchy operator*. We first consider the case when $\Sigma = \mathbb{R}$. Here the Cauchy operator $C = C_{\mathbb{R}}$ is given by

$$C\,f(z) = \int_{\mathbb{R}} \frac{f(s)}{s-z}\,đs, \qquad z \in \mathbb{C}/\mathbb{R}, \quad đs = \frac{ds}{2\pi i}$$

for suitable functions f on $\mathbb{R}$ (General refs for the case $\Sigma = \mathbb{R}$, and also when $\Sigma = \{|z| = 1\}$, are [19] and [23].) Assume first that $f \in \mathcal{S}(\mathbb{R})$, the Schwartz space of functions on $\mathbb{R}$. Let $z = x + i\epsilon$, $x \in \mathbb{R}$, $\epsilon > 0$. Then

$$\begin{aligned} C\,f(x+i\epsilon) &= \int_{\mathbb{R}} \frac{f(s)}{s-x-i\epsilon}\,đs \\ &= \int_{\mathbb{R}} f(s)\frac{i\epsilon}{(s-x)^2+\epsilon^2}\,đs + \int_{\mathbb{R}} f(s)\,\frac{s-x}{(s-x)^2+\epsilon^2}\,đs \\ &= \frac{1}{2}\int_{\mathbb{R}} f(s)\frac{1}{\pi}\,\frac{\epsilon}{(s-x)^2+\epsilon^2}\,ds + \frac{1}{2\pi i}\int_{\mathbb{R}} f(s)\,\frac{s-x}{(s-x)^2+\epsilon^2}\,ds \\ &:= I + II. \end{aligned}$$

Now

$$I = I_\epsilon = \frac{1}{2}\int_{\mathbb{R}} \frac{f(x+\epsilon u)}{\pi(u^2+1)}\,du.$$

Then, by dominated convergence,

$$\lim_{\epsilon\downarrow 0} I_\epsilon = f(x)\frac{1}{2\pi}\int_{\mathbb{R}} \frac{du}{u^2+1} = \frac{1}{2}\,f(x).$$

Write

$$II = II_\epsilon = II_{<\epsilon} + II_{>\epsilon}$$

where

$$II_{<\epsilon} = \frac{1}{2\pi i}\int_{|x-s|<\epsilon} f(s)\,\frac{s-x}{(s-x)^2+\epsilon^2}\,ds$$

and

$$II_{>\epsilon} = \frac{1}{2\pi i}\int_{|x-s|>\epsilon} f(s)\,\frac{s-x}{(s-x)^2+\epsilon^2}\,ds.$$

As

$$\frac{s-x}{(s-x)^2+\epsilon^2}$$

is an odd function about $s = x$, $II_{<\epsilon}$ can be written as

$$II_{<\epsilon} = \frac{1}{2\pi i}\int_{|s-x|<\epsilon} \frac{s-x}{(s-x)^2+\epsilon^2}\,(f(s)-f(x))\,ds$$

and so

$$\begin{aligned} |II_{<\epsilon}| &\leqslant \frac{1}{2\pi}\,\|f'\|_{L^\infty}\int_{|s-x|<\epsilon} \frac{(s-x)^2}{(s-x)^2+\epsilon^2}\,ds \\ &\leqslant \|f'\|_{L^\infty}\,\frac{2\epsilon}{2\pi} \end{aligned}$$

which goes to 0 as $\epsilon \downarrow 0$. Finally

$$\begin{aligned} II_{>\epsilon} &= \frac{1}{2\pi i}\int_{|s-x|>\epsilon}\left(\frac{s-x}{(s-x)^2+\epsilon^2}-\frac{1}{s-x}\right)f(s)\,ds \\ &+\frac{1}{2\pi i}\int_{|s-x|>\epsilon}\frac{1}{s-x}f(s)\,ds \\ &= III_\epsilon + IV_\epsilon . \end{aligned}$$

We have

$$\begin{aligned} |III_\epsilon| &= \frac{1}{2\pi}\left|\int_{|s-x|>\epsilon}\frac{\epsilon^2}{(s-x)^2+\epsilon^2}\,\frac{f(s)}{s-x}\,ds\right| \\ &= \frac{1}{2\pi}\left|\int_{|u|>1}\frac{1}{u^2+1}\,\frac{f(x+\epsilon u)}{u}\,du\right| \end{aligned}$$

and so as $\epsilon \downarrow 0$, again by dominated convergence,

$$|III_\epsilon| \to \frac{1}{2\pi}\,|f(x)|\left|\int_{|u|>1}\frac{du}{u(u^2+1)}\right| = 0$$

as the final integrand is odd.

Thus we see that for $\Sigma = \mathbb{R}$ and $f \in \mathcal{S}(\mathbb{R})$

$$C^+f(x) \equiv \lim_{\epsilon\downarrow 0} Cf\,(x+i\,\epsilon) = \frac{1}{2}\,f(x) + \frac{i}{2}Hf(x)$$

where

$$Hf(x) = \lim_{\epsilon\downarrow 0}\frac{1}{\pi}\int_{|s-x|>\epsilon}\frac{f(s)}{x-s}\,ds.$$

$Hf(x)$ is called the *Hilbert transform* of f. Note that

$$\begin{aligned} \frac{1}{\pi}\int_{|s-x|>\epsilon}\frac{f(s)}{x-s}\,ds &= \frac{1}{\pi}\int_{|s-x|>1}\frac{f(s)}{x-s}\,ds \\ &+\frac{1}{\pi}\int_{\epsilon<|s-x|<1}\frac{f(s)-f(x)}{x-s}\,ds \end{aligned}$$

which converges to

$$\frac{1}{\pi}\int_{|s-x|>1}\frac{f(s)}{x-s}\,ds + \frac{1}{\pi}\int_{|s-x|<1}\frac{f(s)-f(x)}{x-s}\,ds$$

as $\epsilon \downarrow 0$, so that $\lim_{\epsilon\downarrow 0}\frac{1}{\pi}\int_{|s-x|>\epsilon}\frac{f(s)}{x-s}\,ds$ indeed exists pointwise for $f \in \mathcal{S}$.

Similarly one finds

$$C^-f(x) \equiv \lim_{\epsilon\downarrow 0} Cf(x-i\epsilon) = -\frac{1}{2}\,f(x) + \frac{i}{2}\,Hf(x), \qquad x \in \mathbb{R}$$

and we obtain the fundamental relations for $f \in \mathcal{S}$

(2.1) $$C^+f - C^-f = f$$

and

$$C^+f + C^-f = iHf.$$

Exercise 2.2. Show that the limits $C^{\pm}f(x) = \lim_{\epsilon\downarrow 0} Cf(x \pm i\epsilon)$ are in fact *non-tangential limits* i.e. $C^+f(x) = \lim_{z'\to x} Cf(z')$ where z' lies in a cone of arbitrary opening angle $\alpha < \pi$ (see Figure 2.3), and similarly for $C^-f(x)$ (see refs. [5,6,8]).

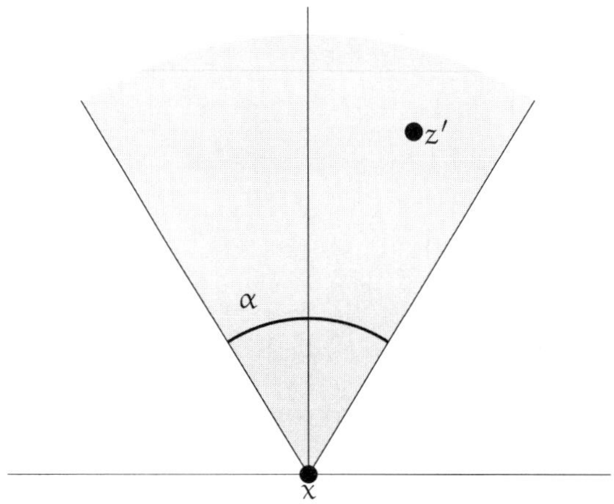

FIGURE 2.3. A cone of opening angle α.

A critical property of the singular integral operator H, and hence the operators $C^{\pm}$, is that, as we now show, H is a bounded operator from $L^p(\mathbb{R}) \to L^p(\mathbb{R})$ for all $1 < p < \infty$. To prove the result for L^2, recall that the Fourier transform

$$\hat{f}(z) = \mathcal{F}f(z) \equiv \lim_{R\to\infty} \int_{-R}^{R} e^{-izt} f(t) \frac{dt}{\sqrt{2\pi}}, \qquad z \in \mathbb{R}$$

and the inverse Fourier transform

$$\check{f}(x) = \mathcal{F}^{-1}f(x) \equiv \lim_{R\to\infty} \int_{-R}^{R} e^{izt} f(t) \frac{dt}{\sqrt{2\pi}}, \qquad x \in \mathbb{R}$$

are unitary maps

$$\|f\|_{L^2} = \|\hat{f}\|_{L^2} = \|\check{f}\|_{L^2}$$

from L^2 onto L^2. Moreover

$$\mathcal{F} \text{ and } \mathcal{F}^{-1} \text{ are indeed inverse to each other,}$$

$$\left(\hat{f}\right)^{\vee} = f = \left(\check{f}\right)^{\wedge}, \qquad f \in L^2(\mathbb{R}).$$

For $f \in \mathcal{S}(\mathbb{R})$, fix $\epsilon > 0$, and set

$$(C_\epsilon f)(x) \equiv Cf(x + i\epsilon).$$

Then

$$(2.4) \quad \begin{aligned} (\mathcal{F} C_\epsilon f)(z) &= \lim_{R\to\infty} \int_{-R}^{R} \left(\int_{\mathbb{R}} \frac{f(s)}{s - x - i\epsilon} đs \right) e^{-ixz} \frac{dx}{\sqrt{2\pi}} \\ &= \lim_{R\to\infty} \int_{\mathbb{R}} \frac{f(s)}{\sqrt{2\pi}} \left(\int_{-R}^{R} \frac{e^{-ixz}}{s - x - i\epsilon} đx \right) ds, \end{aligned}$$

by Fubini's Theorem.

Now for s fixed and R large, and $z > 0$

$$\int_{-R}^{R} \frac{e^{-ixz}}{s - x - i\epsilon} \,đx = -\int_{-R}^{R} \frac{e^{-ixz}}{x - (s - i\epsilon)} \,đx$$

$$= \int_{\substack{-R \qquad R \\ s - i\epsilon}} \frac{e^{-ixz}}{x - (s - i\epsilon)} \,đx - \int_{-R \quad \cdot \quad R} \frac{e^{-ixz}}{x - (s - i\epsilon)} \,đx$$

$$= e^{-i(s - i\epsilon)z} - \int_{-R \quad \cdot \quad R} \frac{e^{-ixz}}{x - (s - i\epsilon)} \,đx.$$

Exercise 2.5. Show that, for $z > 0$, we have

$$\lim_{R \to \infty} \int_{-R \quad \cdot \quad R} \frac{e^{-ixz}}{x - (s - i\epsilon)} \,đx = 0\,.$$

Hence for s fixed and $z > 0$

$$\lim_{R \to \infty} \int_{-R}^{R} \frac{e^{-ixz}}{s - x - i\epsilon} \,đx = e^{-isz}\, e^{-\epsilon z}.$$

But we also have

Exercise 2.6. For $z > 0$,

$$\int_{-R}^{R} \frac{e^{-ixz}}{s - x - i\epsilon} \,đx$$

is bounded in s uniformly for $R > 0$.

It follows that we may take the limit $R \to \infty$ in (2.4) in the s-integral and so for $z > 0$

$$\mathcal{F}\, C_\epsilon\, f(z) = \int_{\mathbb{R}} f(s)\, e^{-isz}\, e^{-\epsilon z} \frac{ds}{\sqrt{2\pi}} = e^{-\epsilon z}\, \mathcal{F} f(z).$$

Exercise 2.7. Show, by a similar argument, that

$$\mathcal{F}\, C_\epsilon\, f(z) = 0 \quad \text{for} \quad z < 0$$

Thus

$$C_\epsilon\, f = \mathcal{F}^{-1} \left(\chi_{>0}(\cdot)\, e^{-\epsilon\, \cdot} \right) \mathcal{F} f$$

where

$$\chi_{>0}(z)\, e^{-\epsilon z} = \begin{cases} e^{-\epsilon z} & \text{for } z > 0; \\ 0 & \text{for } z < 0. \end{cases}$$

Now as $\mathcal{S}(\mathbb{R})$ is dense in L^2, and as $\mathcal{F}^{-1}(\chi_{>0}(\cdot)\, e^{-\epsilon\, \cdot})\, \mathcal{F}$ is clearly bounded in L^2 it follows that

$$C_\epsilon\, f \quad \text{extends to a bounded operator on } L^2.$$

Moreover

$$\hat{C}\, f(x) \equiv \mathcal{F}^{-1} \chi_{>0}(\cdot)\, \mathcal{F}\, f$$

is clearly also a bounded operator in L^2 and for $f \in L^2$

$$\begin{aligned}\|C_\epsilon f - \hat{C} f\|_{L^2} &= \|\mathcal{F}^{-1} \chi_{>0}(\cdot)\left(e^{-\epsilon\,\cdot} - 1\right) \mathcal{F} f\|_{L^2} \\ &= \|\chi_{>0}(\cdot)\left(e^{-\epsilon\,\cdot} - 1\right) \mathcal{F} f\|_{L^2}\end{aligned}$$

which converges to 0 as $\epsilon \downarrow 0$, again by dominated convergence. In other words, for $f \in L^2$,

$$C f(x + i\epsilon) = \int_{\mathbb{R}} \frac{f(s)}{s - x - i\epsilon} đs \to \hat{C} f(x) \quad \text{in} \quad L^2(dx).$$

In particular, it follows by general measure theory, that for some sequence $\epsilon_n \downarrow 0$

$$C f(x + i\epsilon_n) \to \hat{C} f(x) \tag{2.8}$$

pointwise a.e. In particular (2.8) holds for $f \in \mathcal{S}(\mathbb{R})$. But then by our previous calculations, $C f(x + i\epsilon_n)$ converges pointwise for all x, and we conclude that for $f \in \mathcal{S}$ and a.e. x

$$C^+ f(x) = \frac{1}{2} f(x) + \frac{i}{2} H f(x) = \hat{C} f(x) \qquad \text{a.e. } x.$$

Thus $C^+ f$ and, hence $H f$, extend to bounded operators on $L^2(\mathbb{R})$ and

$$\frac{1}{2} \hat{f} + \frac{i}{2} \mathcal{F} H f = \mathcal{F} \hat{C} f = \chi_{>0} \hat{f}$$

and so

$$\begin{aligned}\mathcal{F} H f(z) &= \frac{2}{i}\left(\chi_{>0}(z) - \frac{1}{2}\right) \hat{f}(z) \\ &= -i \operatorname{sgn}(z)\, \hat{f}(z)\end{aligned}$$

where $\operatorname{sgn}(z) = +1$ if $z > 0$ and $\operatorname{sgn}(z) = -1$ if $z < 0$.

We have shown the following: For $f \in L^2$,

$$C^+ f = \frac{f}{2} + \frac{i}{2} H f = \frac{f}{2} + \frac{1}{2} \left(\hat{f} \operatorname{sgn}(\cdot)\right)^{\vee}$$

and similarly

$$C^- f = -\frac{f}{2} + \frac{i}{2} H f = -\frac{f}{2} + \frac{1}{2} \left(\hat{f} \operatorname{sgn}(\cdot)\right)^{\vee}.$$

$C_R =$

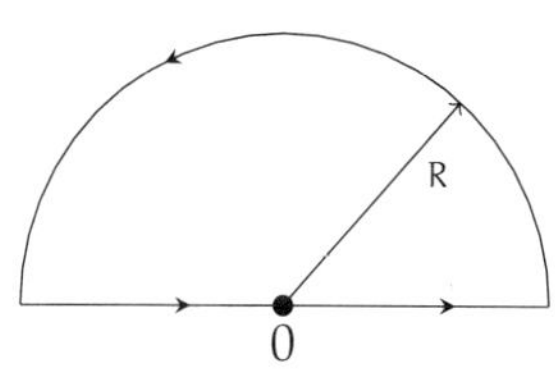

FIGURE 2.9. A semi-circle in the upper-half plane.

The following argument of Riesz shows that in fact $C^\pm$, and hence H, are bounded in $L^p(\mathbb{R})$, for all $1 < p < \infty$. Consider first the case $p = 4$. Suppose $f \in \mathcal{C}_0^\infty(\mathbb{R})$, the infinitely differentiable functions with compact support. Then as $z \to \infty$,

$$C f(z) = \int_{\mathbb{R}} \frac{f(s)}{s - z} đs = O\left(\frac{1}{z}\right)$$

and $C\,f(z)$ is continuous down to the axis. By Cauchy's theorem

$$\int_{C_R} (C\,f(z))^4 \; dz = 0$$

where C_R is given in Figure 2.9, and as

$$\int_{\overset{\frown}{-R\;\;R}} (C\,f(z))^4 \; dz \to 0 \text{ as } R \to \infty,$$

we conclude that

$$\int_{\mathbb{R}} \left(C^+ \, f(x)\right)^4 \; dx = 0.$$

But then as $C^+ \, f = \frac{f}{2} + \frac{i}{2} \, H\,f$ we obtain

$$0 = \int_{\mathbb{R}} \left(f^4 + 4f^3(Hf)\,i + 6f^2(Hf)^2\,i^2 + 4f(Hf)^3\,i^3 + (Hf)^4\,i^4\right) dx. \tag{2.10}$$

Now suppose that f is real. Then Hf is real and the real part of (2.10) yields

$$0 = \int_{\mathbb{R}} \left(f^4 - 6f^2(Hf)^2 + (Hf)^4\right) dx,$$

hence

$$\begin{aligned}\int_{\mathbb{R}} (Hf)^4 \, dx &= 6\int_{\mathbb{R}} f^2(Hf)^2 dx - \int_{\mathbb{R}} f^4 dx \\ &\leqslant 6\left(\int \frac{c}{2}\, f^4 dx + \int_{\mathbb{R}} \frac{1}{2c}\, (Hf)^4 dx\right) - \int_{\mathbb{R}} f^4 dx\end{aligned}$$

for any $c > 0$. Take $c = 6$. Then

$$\frac{1}{2}\int_{\mathbb{R}} (Hf)^4 dx \leqslant (18-1)\int_{\mathbb{R}} f^4 dx$$

or

$$\int_{\mathbb{R}} (Hf)^4 dx \leqslant 34 \int_{\mathbb{R}} f^4 dx.$$

The case when f is complex valued is handled by taking real and imaginary parts. Thus, by density, H maps L^4 boundedly to L^4.

Exercise 2.11. Show that H maps $L^p \to L^p$ for all $1 < p < \infty$. Hints:

(1) Show that the above argument works for all even integers p.
(2) Show that the result follows for all $p \geqslant 2$ by interpolation.
(3) Show that the result for $1 < p < 2$ now follows by duality.

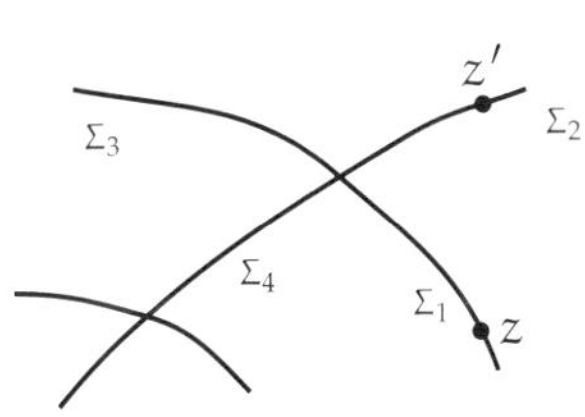

FIGURE 2.12. Contours that self-intersect.

Exercise 2.13. Show that H is *not* bounded from $L^1 \to L^1$. (However H maps $L^1 \to$ weak L^1.) As indicated in Lecture 1, RHPs take place on contours which self-intersect (see Figure 2.12).

We will need to know, for example, that if f is supported on Σ_1, say, and we consider

$$Cf(z') = \int_{\Sigma_1} \frac{f(z)}{z - z'} đz$$

for $z' \in \Sigma_2$, say, then $Cf(z') \in L^2(\Sigma_2)$ if $f \in L^2(\Sigma_1)$. Here is a prototype result which one can prove using the Mellin transform, which we recall is the Fourier transform for the multiplicative group $\{x > 0\}$. We have [5, p. 88] the following:

For $f \in L^2(0, \infty)$ and $r > 0$, set

$$C_\theta f(r) = \int_0^\infty \frac{f(s)}{s - \hat{z} r} đs, \qquad \hat{z} = e^{i\theta}$$

where $0 < \theta < 2\pi$. Then

$$\|C_\theta f\|_{L^2(dr)} \leqslant c_\theta \|f\|_{L^2(ds)} \tag{2.14}$$

where

$$c_\theta = \gamma^\gamma (1-\gamma)^{1-\gamma}, \qquad \gamma = \frac{\theta}{2\pi}.$$

One can also show that for any $1 < p < \infty$

$$\|C_\theta f\|_{L^p(dr)} \leqslant C_{\theta,p} \|f\|_{L^p(ds)}$$

for some $c_{\theta,p} < \infty$.

Results such as (2.14) are useful in many ways. For example, we have the following result.

Theorem 2.15. *Suppose* $f \in H^1(\mathbb{R}) = \{f \in L^2(\mathbb{R}) : f' \in L^2(\mathbb{R})\}$. *Then* $Cf(z)$ *is uniformly Hölder-*$\frac{1}{2}$ *in* $\overline{\mathbb{C}}_+$ *and in* $\overline{\mathbb{C}}_-$. *In particular,* Cf *is continuous down to the axis in* $\mathbb{C}_+$ *and in* $\mathbb{C}_-$.

Proof. For $z \in \mathbb{C} \setminus \mathbb{R}$

$$\frac{d}{dz} Cf(z) = \int_{\mathbb{R}} \frac{f(s)}{(s-z)^2} đs = -\int_{\mathbb{R}} \left(\frac{d}{ds}\left(\frac{1}{s-z}\right)\right) f(s) đs$$
$$= \int_{\mathbb{R}} \frac{f'(s)}{s-z} đs.$$

Now suppose $z', z'' \in \mathbb{C}_+$, and the straight line L through z', z'' intersects the line $\mathbb{R}$ at x at an angle θ as in Figure 2.16.

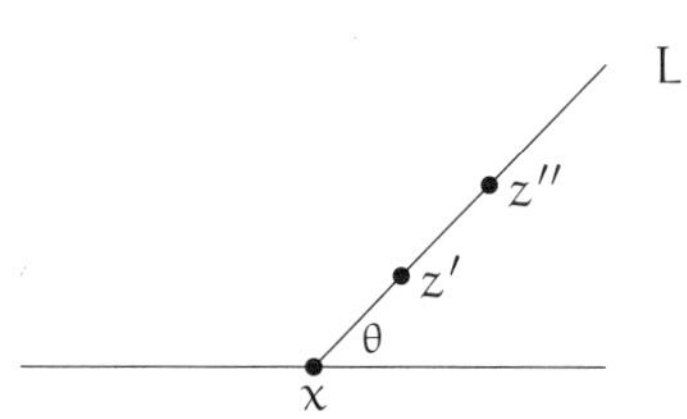

FIGURE 2.16. A line intersecting $\mathbb{R}$.

Then as $\int_{\mathbb{R}} \frac{f'(s)}{s-z} đs = \int_x^\infty \frac{f'(s)}{s-z} đs + \int_{-\infty}^x \frac{f'(s)}{s-z} đs$, and as $f' \in L^2(-\infty, x) \oplus L^2(x, \infty)$ it follows from (2.14) that

$$\int_0^\infty \left| \frac{d}{dz} Cf\left(re^{i\theta}\right) \right|^2 dr = \int_0^\infty \left| C f'(re^{i\theta}) \right|^2 dr \leqslant c\|f'\|_{L^2}.$$

But

$$\begin{aligned} \left|Cf(z'') - Cf(z')\right| &= \left| \int_{z' \to z'' \text{ in } L} \frac{d}{dz} Cf(z)\, dz \right| \\ &\leqslant \left|z'' - z'\right|^{\frac{1}{2}} \|\frac{d}{dz} Cf\|_{L^2(0,\infty)} \leqslant c|z'' - z'|^{\frac{1}{2}} \|f'\|_{L^2(\mathbb{R})}. \end{aligned}$$

□

We now consider general contours $\Sigma \subset \overline{\mathbb{C}} = \mathbb{C} \cup \{\infty\}$, which are *composed curves*: By definition a composed curve Σ is a finite union of arcs $\{\Sigma_i\}_{i=1}^n$ which can intersect only at their end points. Each arc Σ_i is homeomorphic to an interval $[a_i, b_i] \subset \mathbb{R}$:

$$\begin{cases} \varphi_i : [a_i, b_i] & \to \Sigma_i \subset \overline{\mathbb{C}}, \\ [a_i, b_i] \ni t & \to \varphi_i(t) \in \Sigma_i,\ \varphi_i(a_i) \neq \varphi_i(b_i). \end{cases}$$

Here $\overline{\mathbb{C}}$ has the natural topology generated by the sets $\{|z| < R_1\}$, $\{|z| > R_2\}$ where R_1, $R_2 > 0$. A loop, in particular the unit circle $T = \{|z| = 1\}$, is a composed curve on the understanding that it is a union of (at least) two arcs.

Although it is possible, and sometimes useful, to consider other function spaces (e.g. Hölder continuous functions), we will only consider RHPs in the sense of $L^p(\Sigma)$ for $1 < p < \infty$.

So the first question is "What is $L^p(\Sigma)$?". The natural measure theory for each arc Σ_i is generated by arc length measure μ as follows. If $z_0 = \varphi(t_0)$ and $z_n = \varphi(t_n)$ are the end-points of some arc $\Sigma \subset \mathbb{C}$, and $z_0, z_1, \dots, z_n$ is any partition of $[z_0, z_n] = \{\varphi(t) : t_0 \leqslant t \leqslant t_n\}$ (we assume z_{i+1} succeeds z_i in the ordering induced on Σ by φ, symbolically $z_i < z_{i+1}$, etc.) then

$$L = L_{[z_0, z_n]} \equiv \sup_{\text{all partitions } \{z_i\}} \sum_{i=0}^{n-1} |z_{i+1} - z_i|.$$

If $L < \infty$ we say that the arc $\Sigma = [z_0, z_n]$ is *rectifiable* and $L_{[z_0, z_n]}$ is its *arc length*. We will only consider composed curves Σ that are *locally rectifiable* i.e. for any $R > 0$, $\Sigma \cap \{|z| < R\}$ is rectifiable (note that the latter set is an at most countable union of simple arcs and rectifiability of the set means that the sum of the arc lengths of these arcs is finite. In particular, the unit circle $\mathbb{T}$ as a union of 2 rectifiable subarcs, is rectifiable, and $\mathbb{R}$ is locally rectifiable.) For any interval $[\alpha, \beta)$ on $\Sigma_i \subset \mathbb{C}$ (the case where Σ_i passes through ∞, must be treated separately — exercise!) define

$$\mu_i([\alpha, \beta)) = \text{ arc length } \alpha \to \beta.$$

Now the sets $\{[\alpha, \beta) : \alpha < \beta \text{ on } \Sigma_i\}$ form a semi-algebra (see [30]) and hence μ_i can be extended to a complete measure on a σ-algebra $\mathcal{A}$ containing the Borel sets on Σ_i. The restriction of the measure to the Borel sets is unique. For $1 \leqslant p < \infty$, we can define $L^p(\Sigma_i, d\mu_i)$ to be the set of f measurable with respect to $\mathcal{A}$ on Σ for which,

$$\int_{\Sigma_i} |f(z)|^p \, d\mu_i(z) < \infty,$$

and then all the "usual" properties go through. One usually writes $d\mu = |dz|$. For $\Sigma = \bigcup_{i=1}^n \Sigma_i$, $L^p(\Sigma, d\mu)$ is simply the *direct sum* of $L^p(\Sigma_i, d\mu_i)_{i=1}^n$.

Exercise 2.17. $|dz|$ is also equal to Hausdorff-1 measure on Σ_1.

Note that if $\Sigma_1 = \mathbb{R}$ and $\Sigma_2 = \left\{\left(x, x^3 \sin \frac{1}{x}\right) : x \in \mathbb{R}\right\}$ then $\Sigma = \Sigma_1 \cup \Sigma_2$ is not a composed curve, although Σ_1 and Σ_2 are both locally rectifiable.

For Σ as above we define the Cauchy operator for $h \in L^p(\Sigma, |dz|)$, $1 \leqslant p < \infty$, by

$$Cf(z) = C_\Sigma f(z) = \int_\Sigma \frac{f(\zeta)}{\zeta - z} đ\zeta, \qquad z \in \mathbb{C} \setminus \Sigma. \tag{2.18}$$

Given the homeomorphisms $\varphi_i : [a_i, b_i] \to \Sigma_i$, the contour Σ carries a natural orientation, and the integral here is a line integral following the orientation; if we parametrize the arcs Σ_i in Σ by arc length s,

$$0 \leqslant s \leqslant s_i, \qquad \zeta = \zeta(s), \qquad \text{then} \qquad \left|\frac{d\zeta(s)}{ds}(s)\right| = 1 \quad \text{(why?)}$$

and (2.18) is a sum over its subarcs Σ_i of integrals of the form

$$\int_0^{s_i} \frac{f(\zeta(s))}{\zeta(s) - z} \frac{d\zeta(s)}{ds} đs, \qquad z \in \mathbb{C} \setminus \Sigma$$

for each i, the integrand (clearly) lies in $L^p(ds : [0, s_i))$.

Now the fact of the matter is that many of the properties that were true for C_Σ when $\Sigma = \mathbb{R}$, go through for C_Σ in the general situation. (See, in particular, [24].) In particular for $f \in L^p(\Sigma, d\mu)$, the non-tangential limits

$$C_\Sigma^\pm f(z) = \lim_{z' \to z^\pm} C_\Sigma f(z') \tag{2.19}$$

exist pointwise a.e. on Σ. Figure 2.20 demonstrates non-tangential limits.

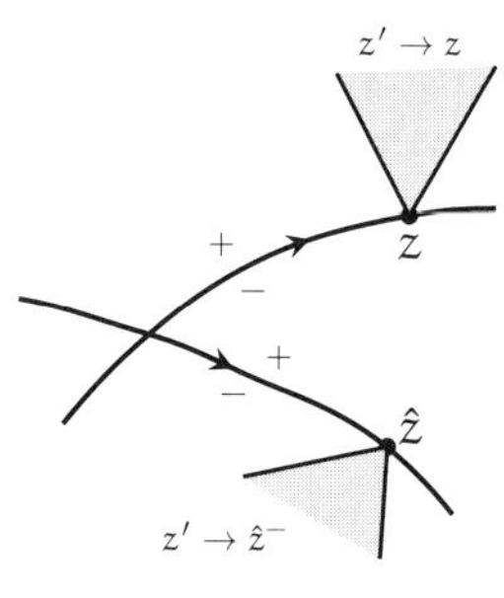

FIGURE 2.20. Non-tangential limits.

Note that as Σ_i is locally rectifiable, the tangent vector to the arc $\frac{d\zeta}{ds}$ exists at a.e. point $\zeta = \zeta(s)$: the normal to $\frac{d\zeta}{ds}$ bisects the cone.

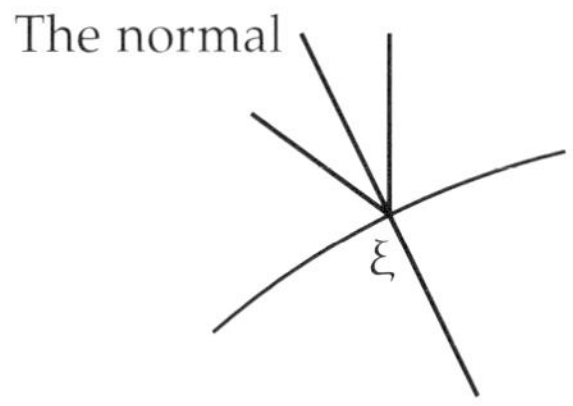

FIGURE 2.21. A contour and its normal.

Moreover,

$$C_{\Sigma}^{\pm} f(z) = \pm\frac{1}{2} f(z) + \frac{i}{2} Hf(z)$$

where the Hilbert transform is now given by

$$Hf(z) = \frac{1}{\pi} \lim_{\epsilon \downarrow 0} \int_{\substack{|s-z|>\epsilon \\ s\in\Sigma}} \frac{f(s)}{z-s}\, ds, \qquad z \in \Sigma \tag{2.22}$$

and the points $z \in \Sigma$ for which the non-tangential limits (2.19) exists are precisely the points for which the limit in (2.22) exists.

Again, for $f \in L^p(\Sigma, d\mu)$ with $1 \leqslant p < \infty$,

$$C^{+} f(z) - C^{-} f(z) = f(z)$$

and

$$C^{+} f(z) + C^{-} f(z) = i\, Hh(z).$$

The following issue is crucial for the analysis of RHPs:

Question. For which locally rectifiable contours Σ are the operators $C^{\pm}$ and H bounded in L^p, $1 < p < \infty$?

Quite remarkably, it turns out that there are *necessary* and *sufficient* conditions on a simple rectifiable curve for $C^{\pm}$, H to be bounded in $L^p(\Sigma)$, $1 < p < \infty$. The result is due to many authors, starting with Calderón [7], and then Coifman, Meyer and McIntosh [9], with Guy David [10] (see [6] for details and historical references) making the final decisive contribution.

Let Σ be a simple, rectifiable curve in $\mathbb{C}$. For any $z \in \Sigma$, and any $r > 0$, let

$$\ell_r(z) = \text{ arc length of } (\Sigma \cap D_r(z))$$

where $D_r(z)$ is the ball of the radius r centered at z, see Figure 2.23.

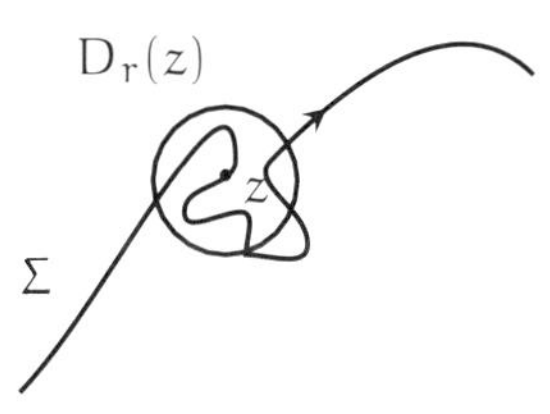

FIGURE 2.23. A ball $D_r(z)$ of radius r centered at z.

Set

$$\lambda = \lambda_\Sigma = \sup_{z\in\Sigma,\, r>0} \frac{\ell_r(z)}{r}.$$

Theorem 2.24. *Suppose $\lambda_\Sigma < \infty$. Then for any $1 < p < \infty$, the limit in* (2.22) *exists for a.e. $z \in \Sigma$ and defines a bounded operator for any $1 < p < \infty$*

$$\|Hf\|_{L^p} \leqslant c_p \|f\|_{L^p}, \quad f \in L^p, \quad c_p < \infty. \tag{2.25}$$

Conversely if the limit in (2.22) *exists a.e. and defines a bounded operator H in $L^p(\Sigma)$ for some $1 < p < \infty$, then H gives rise to a bounded operator for all p, $1 < p < \infty$, and $\lambda_\Sigma < \infty$.*

An excellent reference for the above Theorem, and more, is [6].

Remarks. Additional remarks:

(1) Locally rectifiable curves Σ for which $\lambda = \lambda_\Sigma < \infty$ are called Carleson curves,

(2) the constant c_p in (2.25) has the form $c_p = \phi_p(\lambda_\Sigma)$ for some continuous, increasing function, $\phi_p(t) \geqslant 0$, *independent* of Σ, such that $\phi_p(0) = 0$.

The fact that ϕ_p is independent of Σ, is very important for the nonlinear steepest descent method, where one deforms curves in a similar way to the classical steepest descent method for integrals.

Carleson curves are sometimes called AD-regular curves: the A and D denote Ahlfors and David. To get some sense of the subtlety of the above result, consider the following curve Σ with a cusp at the origin (see Figure 2.26):

$$\Sigma = \{0 \leqslant x \leqslant 1,\ y = 0\} \cup \{(x, x^2) : 0 \leqslant x \leqslant 1\}.$$

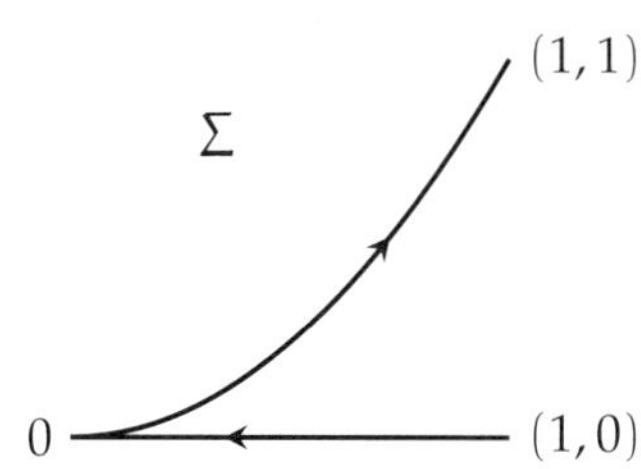

FIGURE 2.26. A cusp at the origin.

Clearly $\lambda_\Sigma < \infty$ so that the Hilbert transform H_Σ is bounded in L^p, $1 < p < \infty$.

Exercise 2.27. For Σ in Figure 2.26, prove directly that H_Σ is bounded in L^2. The presence of the cusp makes the proof surprisingly difficult.

Lecture 3

We now make the notion of a RHP precise (see [8,17,28]). Let Σ be a composite, oriented Carleson contour in $\mathbb{C}$ and let $v : \Sigma \to GL(n, \mathbb{C})$ be a jump matrix on Σ, with v, $v^{-1} \in L^\infty(\Sigma)$. Let $Ch(z) = C_\Sigma h(z)$, $C_\Sigma^\pm h$, $H_\Sigma h$ be the associated Cauchy and Hilbert operators.

We say that a pair of $L^p(\Sigma)$ function $f_\pm \in \partial C(L^p)$ if there exists a (unique) function $h \in L^p(\Sigma)$ such that

$$f_\pm(z) = (C^\pm h)(z), \qquad z \in \Sigma.$$

In turn we call $f(z) \equiv Ch(z)$, $z \in \mathbb{C}\backslash\Sigma$, the *extension* of $f_\pm = C^\pm h \in \partial C(L^p)$ *off* Σ.

Definition 3.1. *Fix $1 < p < \infty$. Given Σ, v and a measurable function f on Σ, we say that $m_\pm \in f + \partial C(L^p)$ solves an inhomogeneous RHP of the first kind* (IRHP1$_p$) *if*

$$m_+(z) = m_-(z)\, v(z), \qquad z \in \Sigma.$$

Definition 3.2. *Fix $1 < p < \infty$. Given Σ, v and a function $F \in L^p(\Sigma)$, we say that $M_\pm \in \partial\, C(L^p)$ solves an inhomogeneous RHP of the second kind* (IRHP2$_p$) *if*

$$M_+(z) = M_-(z)\, v(z) + F(z), \qquad z \in \Sigma.$$

Recall that m solves the normalized RHP (Σ, v) if, at least formally,

(3.3)
- $m(z)$ is a $n \times n$ analytic function in $\mathbb{C}\backslash\Sigma$,
- $m_+(z) = m_-(z)\, v(z), \qquad z \in \Sigma,$
- $m(z) \to I$ as $z \to \infty$.

More precisely, we make the following definition.

Definition 3.4. *Fix $1 < p < \infty$. We say that $m_\pm$ solves the normalized RHP $(\Sigma, v)_p$ if $m_\pm$ solves the* IRHP1$_p$ *with* $f \equiv I$.

In the above definition, if $m_\pm - I \in C^\pm h$, then clearly the extension

$$m(z) = I + Ch(z), \qquad z \in \mathbb{C}\backslash\Sigma,$$

off Σ solves the normalized RHP in the formal sense of (3.3).

IRHP1$_p$	$\equiv$	IRHP2$_p$
$\uparrow$		$\uparrow$
Invertibility of $1 - C_\omega$		Useful for deformations of RHP in $\mathbb{C}$

FIGURE 3.5. The uses of IRHP1$_p$ and IRHP2$_p$.

Let

$$v = (v^-)^{-1} v^+ = (I - \omega^-)^{-1} (I + \omega^+)$$
$$\omega^+ \equiv v^+ - I, \qquad \omega^- \equiv I - v^-,$$

be a *pointwise a.e.* factorization of v, i.e., $v(x) = (v^-(x))^{-1} v^+(x)$ for a.e. x, with $v^\pm$, $(v^\pm)^{-1} \in L^\infty$, and let $\omega = (\omega^-, \omega^+)$. Let C_ω denote the basic associated operator

$$C_\omega h \equiv C^+(h\,\omega^-) + C^-(h\,\omega^+)$$

acting on $L^p(\Sigma) - n \times n$ matrix valued functions h. As $\omega^\pm \in L^\infty$, $C_\omega \in \mathcal{L}(L^p)$, the bounded operators on L^p, for all $1 < p < \infty$. The utility of $\mathrm{IRHP1}_p$ and $\mathrm{IRHP2}_p$ will soon become clear, see Figure 3.5.

Theorem 3.6. *If* f *and* v *are such that* $f(v-I) \in L^p(\Sigma)$ *for some* $< p < \infty$, *then*

$$m_\pm = M_\pm + f$$

solves $\mathrm{IRHP1}_p$ *if* $M_\pm$ *solves* $\mathrm{IRPH2}_p$ *with* $F = f(v-I)$. *Conversely if* $F \in L^p(\Sigma)$, *then*

$$M_+ = m_+ + F, \qquad M_- = m_-$$

solves $\mathrm{IRHP2}_p$ *if* $m_\pm$ *solves* $\mathrm{IRHP1}_p$ *with* $f = C^- F$.

The first part of this result is straightforward: Suppose $M_\pm \in \partial C(L^p)$ solves $M_+ = M_- v + F$ on Σ with $F = f(v-I) \in L^p$. Then

$$M_+ = M_- v + f(v-I) = (M_- + f)v - f$$

or $m_+ = m_- v$ with $m_\pm = f + M_\pm \in f + \partial C(L^p)$. The converse is more subtle and is left as an exercise:

Exercise 3.7. Show $\mathrm{IRHP1}_p \Rightarrow \mathrm{IRHP2}_p$.

We now show that the RHPs $\mathrm{IRHP1}_p$ and $\mathrm{IRHP2}_p$, and, in particular, the normalized RHP $(\Sigma, v)_p$ are intimately connected with the singular integral operator $1 - C_\omega$.

Let $f \in L^p(\Sigma)$ and let $m_\pm = f + C^\pm h$ for some $h \in L^p(\Sigma)$. Also suppose $m_+ = m_- v = m_-(v^-)^{-1} v^+$. Set

$$\mu = m_- \left(v^-\right)^{-1} = m_+ \left(v^+\right)^{-1} \in L^p(\Sigma)$$

and define

$$H(z) = \left(C\left(\mu\left(\omega^+ + \omega^-\right)\right)\right)(z), \qquad z \in \mathbb{C}\backslash\Sigma.$$

Then we have on Σ, using $C^+ - C^- = 1$

$$\begin{aligned}
H_+ &= C^+\left(\mu\left(\omega^+ + \omega^-\right)\right) = C^+\mu\,\omega^+ + C^+\mu\,\omega^- \\
&= C^+\mu\,\omega^- + C^-\mu\,\omega^+ + \mu\,\omega^+ \\
&= C_\omega\,\mu + \mu\,\omega^+ = (C_\omega - 1)\,\mu + \mu\left(I + \omega^+\right) \\
&= (C_\omega - 1)\,\mu + \mu v^+ = (C_\omega - 1)\,\mu + m_+.
\end{aligned}$$

$$\text{i.e. } H_+ = (C_\omega - I)\,\mu + m_+.$$

Similarly

$$H_- = (C_\omega - 1)\,\mu + m_-.$$

Thus

$$m_\pm - f - H_\pm = (1 - C_\omega)\,\mu - f. \tag{3.8}$$

But $m_\pm - f - H_\pm \in \partial C(L^p)$; i.e. $m_\pm - f - H_\pm = C^\pm h$ for some $h \in L^p$. However, from (3.8)

$$C^+ h = C^- h \qquad \Rightarrow \qquad h = C^+ h - C^- h = 0.$$

We conclude that $(1 - C_\omega)\mu = f, \mu \in L^p$.

Conversely, if $\mu \in L^p(\Sigma)$ solves $(1 - C_\omega)\mu = f,$ then the above calculations show that $H \equiv C(\mu(\omega^+ + \omega^-))$ satisfies

$$H_\pm = -f + \mu v^\pm.$$

Thus setting $m_\pm = \mu v^\pm$, we see that $m_+ = m_- v$ and $m_\pm - f \in \partial C(L^p)$. In particular $\mu \in L^p$ solves $(1 - C_\omega)\mu = 0$ iff $m_\pm = \mu v^\pm$ solves the homogeneous RHP.

$$m_+ = m_- v, \qquad m_\pm \in \partial C(L^p). \tag{3.9}$$

We summarize the above calculations as follows:

Proposition 3.10. *Let* $1 < p < \infty$. *Then*

$$(1 - C_\omega) \quad \textit{is a bijection in} \quad L^p(\Sigma)$$

$$\Longleftrightarrow$$

$$\mathrm{IRHP1}_p \quad \textit{has a unique solution for all} \quad f \in L^p(\Sigma)$$

$$\Longleftrightarrow$$

$$\mathrm{IRHP2}_p \quad \textit{has a unique solution for all} \quad F \in L^p(\Sigma).$$

Moreover, if one, and hence all three of the above conditions, is satisfied, then for all $f \in L^p(\Sigma)$

$$\begin{aligned}(1 - C_\omega)^{-1} f &= m_+ (v^+)^{-1} = m_-(v^-)^{-1} \\ &= (M_+ + f)(v^+)^{-1} = (M_- + f)(v^-)^{-1}\end{aligned} \tag{3.11}$$

where $m_\pm$ *solves* $\mathrm{IRHP1}_p$ *with the given* f *and* $M_\pm$ *solves* $\mathrm{IRHP2}_p$ *with* $F = f(v - I)$ *(*$\in L^p$*!), and if* $M_\pm$ *solves* $\mathrm{IRHP2}_p$ *with* $F \in L^p(\Sigma)$*, then*

$$M_+ = \left((1 - C_\omega)^{-1}(C^- F)\right) v^+ + F \quad \textit{and}$$

$$M_- = \left((1 - C_\omega)^{-1}(C^- F)\right) v^-.$$

Finally, if $f \in L^\infty(\Sigma)$ and $v^\pm - I \in L^p(\Sigma)$, then (3.11) remains valid provided we interpret

$$(1 - C_\omega)^{-1} f \equiv f + (1 - C_\omega)^{-1} C_\omega f. \tag{3.12}$$

This is true, in particular, for the normalized RHP $(\Sigma, v)_p$ where $f \equiv I$.

Remark. If $1 - C_\omega$ is invertible, for one choice of $v^\pm$, then (**exercise**) it is invertible for all choices of $v^\pm$ such that

$$v = (v^-)^{-1} v^+, \qquad v^\pm, \quad (v^\pm)^{-1} \in L^\infty(\Sigma).$$

Note that if we take $v^+ = v, \quad v^- = I$, in particular, then

$$C_\omega h = C^-(h(v - I)).$$

The above Proposition implies, in particular, that if $\mu \in I + L^p$ solves

$$(1 - C_\omega)\,\mu = I \tag{3.13}$$

in the sense of (3.12) i.e. $\mu = I + \nu, \ \nu \in L^p$

$$(1 - C_\omega)\,\nu = C_\omega\, I = C^+\,\omega_- + C^-\,\omega_+ \in L^p \tag{3.14}$$

then $m_\pm = \mu v^+$ solves the normalized RHP $(\Sigma, v)_p$. *It is in this precise sense that the solution of the normalized RHP is equivalent to the solution of a singular integral equation* (3.13), (3.14) *on* Σ.

One very important consequence of the proof of Proposition 3.10 is given by the following

Corollary 3.15. *Let* $f \in L^p(\Sigma)$.

Let $m_\pm$ *solve* IRHP1$_p$ *with the given* f *and let* $M_\pm$ *solve* IRHP2$_p$ *with* $F = f(v - I)$. *Then*

$$\|(1 - C_\omega)^{-1} f\|_p \leqslant c\,\|m_\pm\|_p \tag{3.16}$$

and

$$\|(1 - C_\omega)^{-1} f\|_p \leqslant c'\,(\|M_\pm\|_p + \|f\|_p) \tag{3.17}$$

for some constants $c = c_p$, $c' = c'_p$. *In particular if we know, or can show, that* $\|m_\pm\|_p \leqslant$ *const* $\|f\|_p$, *or* $\|M_\pm\|_p \leqslant$ *const* $\|f\|_p$, *then we can conclude from* (3.16) *or* (3.17) *that* $(1 - C_\omega)^{-1}$ *is bounded in* L^p *with a corresponding bound. Conversely if we know that* $(1 - C_\omega)^{-1}$ *exists, then the above calculations show that* $\|m_\pm\|_p \leqslant \tilde{c}\|f\|_p$ *and* $\|M_\pm\|_p \leqslant \tilde{\tilde{c}}\|f\|_p$ *for corresponding constants* $\tilde{c}$, $\tilde{\tilde{c}}$.

Finally we consider uniqueness for the solution of the normalized RHP $(\Sigma, v)_p$ as given in Definition 3.4. Observe first that if $F(z) = (Cf)(z)$ for $f \in L^p(\Sigma)$ and $G(z) = (Cg)(z)$ for $g \in L^q(\Sigma)$, $\frac{1}{r} = \frac{1}{p} + \frac{1}{q} \leqslant 1$, $1 < p, q < \infty$, then a simple computation shows that

$$FG(z) = Ch(z) \tag{3.18}$$

where

$$h(s) = -\frac{1}{2i}\,(g(s)(Hf)(s) + f(s)(Hg)(s)) \tag{3.19}$$

where again $H f(s) =$ Hilbert transform $= \lim_{\epsilon \downarrow 0} \int_{|s'-s|>\epsilon} \frac{f(s')}{s-s'}\frac{ds'}{\pi}$, and similarly for $Hg(s)$. As h clearly lies in $L^r(\Sigma)$, $r \geqslant 1$, it follows that

$$F_+\,G_+(z) - F_-\,G_-(z) = h(z) \qquad \text{for a.e.} \quad z \in \Sigma.$$

(**Note:** $C^+\,h(z) - C^-\,h(z) = h(z)$ even if h is in L^1, even though $C^\pm$ is not bounded in L^1.)

Theorem 3.20. *Fix* $1 < p < \infty$. *Suppose* $m_\pm$ *solves the normalized RHP* $(\Sigma, v)_p$. *Suppose that* $m_\pm^{-1}$ *exists a.e. on* Σ *and* $m_\pm^{-1} \in I + \partial\, C(L^q)$, $1 < q < \infty$, $\frac{1}{r} = \frac{1}{p} + \frac{1}{q} \leqslant 1$. *Then the solution of the normalized RHP* $(\Sigma, v)_p$ *is unique.*

Proof. Suppose $\hat{m}_\pm = I + C^\pm \hat{h}$, $\hat{h} \in L^p(\Sigma)$ is a 2nd solution of the normalized RHP. We have, by assumption, $m_\pm^{-1} = I + C^\pm k$ for some $k \in L^q(\Sigma)$. (It is an **Exercise** to show that $I + (Ck)(z)$, the extension of $m_\pm^{-1}$ to $\mathbb{C}\backslash\Sigma$, is in fact $m(z)^{-1}$.).

Then arguing as above

$$\begin{aligned}\hat{m}_\pm m_\pm^{-1} - I &= (\hat{m}_\pm - I)\left(m_\pm^{-1} - I\right) + (\hat{m}_\pm - I) + \left(m_\pm^{-1} - I\right) \\ &= C^\pm h\end{aligned}$$

for some $h \in L^r(\Sigma) + L^p(\Sigma) + L^q(\Sigma)$.

Hence

$$\hat{m}_+ m_+^{-1} - \hat{m}_- m_-^{-1} = h.$$

But

$$\hat{m}_+ m_+^{-1} = (\hat{m}_- v)(m_- v)^{-1} = \hat{m}_- m_-^{-1}$$

and so $h = 0$. Thus $\hat{m}_\pm m_\pm^{-1} - I = 0$ or $\hat{m}_\pm = m_\pm$. □

Theorem 3.21. *If $n = 2$, $p = 2$ and $\det v(z) = 1$ a.e. on Σ, then the solution of the normalized RHP $(\Sigma, v)_2$ is unique.*

Proof. Because $n = 2$ and $p = 2$, (3.18), (3.19) $\Longrightarrow$ $(\det m(z))_\pm = 1 + C^\pm h$, where $h \in L^1(\Sigma) + L^2(\Sigma)$ and so $(\det m)_+ - (\det m)_- = h(z)$ a.e. But $\det m_+ = \det m_-$ as $\det v = 1$, and so $h \equiv 0$. But then $\det m(z)_+ = 1$. Hence, if

$$m_\pm = \begin{pmatrix} m_{11\pm} & m_{12\pm} \\ m_{21\pm} & m_{22\pm} \end{pmatrix}$$

we have

$$m_\pm^{-1} = \begin{pmatrix} m_{22\pm} & -m_{12\pm} \\ -m_{21\pm} & m_{11\pm} \end{pmatrix}$$

and so clearly $m_\pm^{-1} \in I + \partial C(L^2)$. The result now follows from Theorem 3.20. □

These results immediately imply that the normalized RHP $(\Sigma = \mathbb{R}, v_{x,t})$ for MKdV with $v_{x,t}$ given by (1.23) has a unique solution in $L^2(\mathbb{R})$. Indeed, factorize

$$\begin{aligned} v_{x,t}(z) = (v_{x,t}^-)^{-1} v_{x,t}^+ &= \left(I - w_{x,t}^-\right)^{-1}\left(I + w_{x,t}^+\right) \\ &= \begin{pmatrix} 1 & -\bar{r}e^{-2i\tau} \\ 0 & 1 \end{pmatrix}\begin{pmatrix} 1 & 0 \\ re^{2i\tau} & 1 \end{pmatrix}\end{aligned}$$

so that

$$w_{x,t} = \left(w_{x,t}^-, w_{x,t}^+\right) = \left(\begin{pmatrix} 0 & -\bar{r}e^{-2i\tau} \\ 0 & 0 \end{pmatrix}, \begin{pmatrix} 0 & 0 \\ re^{2i\tau} & 0 \end{pmatrix}\right)$$

But for $\Sigma = \mathbb{R}$, we have

Exercise 3.22. Both C^+ and $-C^-$ are orthogonal projections in $L^2(\mathbb{R})$ and so $\|C^\pm\|_{L^2} = 1$.

Using the Hilbert-Schmidt matrix norm $\|M\| = \left(\sum_{i,j} |M_{ij}|^2\right)^{\frac{1}{2}}$, we have

$$\begin{aligned}
\|C_{\omega_{x,t}} h\|_{L^2}^2 &= \left\| C^+ \begin{pmatrix} h_{11} & h_{12} \\ h_{21} & h_{22} \end{pmatrix} \begin{pmatrix} 0 & -\bar{r}\, e^{-2i\tau} \\ 0 & 0 \end{pmatrix} \right. \\
&\qquad \left. + C^- \left(\begin{pmatrix} h_{11} & h_{12} \\ h_{21} & h_{22} \end{pmatrix} \begin{pmatrix} 0 & 0 \\ re^{2i\tau} & 0 \end{pmatrix} \right) \right\|_{L^2}^2 \\
&= \left\| \begin{pmatrix} 0 & C^+ h_{11}(-\bar{r})\, e^{2i\tau} \\ 0 & C^+ h_{21}(-\bar{r})\, e^{-2i\tau} \end{pmatrix} \right. \\
&\qquad \left. + \begin{pmatrix} C^- h_{12}\, re^{2i\tau} & 0 \\ C^- h_{22}\, re^{2i\tau} & 0 \end{pmatrix} \right\|^2 \\
&= \|C^- h_{12}\, re^{2i\tau}\|_{L^2}^2 + \|C^- h_{22}\, re^{2i\tau}\|_{L^2}^2 \\
&\quad + \|C^+ h_{11}(-\bar{r})\, e^{-2i\tau}\|_{L^2}^2 + \|C^+ h_{21}(-\bar{r})\, e^{-2i\tau}\|_{L^2}^2 \\
&\leqslant \|r\|_\infty^2 \left(\|h_{12}\|_{L^2}^2 + \|h_{22}\|_{L^2}^2 + \|h_{11}\|_{L^2}^2 + \|h_{21}\|_{L^2}^2 \right) \\
&= \|r_\infty\|^2\, \|h\|_{L^2}^2
\end{aligned}$$

and so, as $\|r\|_\infty < 1$,

$$\|C_{\omega_{x,t}}\| < 1.$$

It follows that for each $x, t \in \mathbb{R}$, $(1 - C_{\omega_{x,t}})^{-1}$ exists in $L^2(\mathbb{R})$ and

$$\left\| (1 - C_{\omega_{x,t}})^{-1} \right\|_{L^2} \leqslant \frac{1}{1 - \|r\|_\infty} < \infty$$

and the proof of the existence and uniqueness for $(\Sigma, \nu_{x,t})$ follows from Proposition 3.10. On the other hand, just uniqueness alone follows from Theorem 3.21 as $\det \nu(z) \equiv 1$ on $\mathbb{R}$.

Now it turns out that a key role in the theory of RHPs is played by *Fredholm operators*. Recall that a bounded linear operator T from a Banach space X to a Banach space Y is *Fredholm* if

$$\dim \ker T < \infty$$

and

$$\dim\ \operatorname{coker}\ T < \infty \qquad \text{i.e. } Y/\operatorname{ran} T \text{ is a finite dimensional space.}$$

$$\text{If } T \text{ is Fredholm,} \quad \text{we define index } T \equiv \dim \ker T - \dim \operatorname{coker} T.$$

Exercise 3.23. If $T : X \to Y$ is Fredholm, then ran T is closed in Y.

Exercise 3.24. $T : X \to Y$ is Fredholm iff it has a *pseudo-inverse* $S \in \mathcal{L}(Y, X)$ such that $ST = 1_X + K$ and $TS = 1_Y + L$ where K is a compact operator in $\mathcal{L}(X)$ and L is a compact operator in $\mathcal{L}(Y)$.

We know that a normalized RHP $(\Sigma, v)_p$, say, has a (unique) solution if $(1 - C_\omega)^{-1}$ exists. The situation where we know, for example, that $\|C_\omega\|_{L^2} < 1$, as in the example $(\Sigma = \mathbb{R}, v_{x,t})$ above so that $(1 - C_\omega)^{-1}$ exists, is very rare. For example, for the KdV equation on $\mathbb{R}$

$$u_t + 6uu_x + u_{xxx} = 0,$$
$$u(x,0) = u_0(x) \to 0 \qquad \text{as} \qquad |x| \to \infty,$$

the associated RHP is exactly the same as $(\mathbb{R}, v_{x,t})$ for MKdV, except that now, generically,

(3.25) $$|r(z)| < 1 \qquad \text{for} \qquad |z| > 0$$

but

(3.26) $$|r(0)| = 1.$$

Thus $\|r\|_\infty = 1$ and the above proof of the existence and uniqueness for the RHP breaks down. A more general approach to proving the existence and uniqueness of solutions to normalized RHPs, is to attempt the following:

- Prove $1 - C_\omega$ is Fredholm.
- Prove $\operatorname{ind}(1 - C_\omega) = 0$.
- Prove $\dim \ker (1 - C_\omega) = 0$.

Then it follows that $1 - C_\omega$ is a bijection, and hence the normalized RHP (Σ, v) has a unique solution.

Let's see how this goes for KdV with normalized RHP $(\Sigma = \mathbb{R}, v_{x,t})$, but now r satisfies (3.25), (3.26). By our previous comments (see **Remark** above), it is enough to consider the special case $v^+ = v$, $v^- = I$ so that $\omega^+ = v - I$ and $\omega^- = 0$. Thus

$$C_\omega h = C^- h (v - I).$$

We assume $r(z)$ is continuous and $r(z) \to 0$ as $|z| \to \infty$. Let S be the operator

$$Sh = C^- h \left(v^{-1} - I\right).$$

Then

$$\begin{aligned} C_\omega Sh &= C^- \left(Sh(v - I)\right) \\ &= C^- \left[\left(C^- h \left(v^{-1} - I\right)\right)(v - I)\right] \\ &= C^- \left[\left(C^+ h \left(v^{-1} - I\right)\right)(v - I)\right] \\ &\quad - C^- \left[h \left(v^{-1} - I\right)(v - I)\right] \end{aligned}$$

as $C^+ - C^- = 1$. But $h\left(v^{-1} - I\right)(v - I) = h\left(2I - v - v^{-1}\right) = h(I - v) + h\left(I - v^{-1}\right)$.

Thus

$$\begin{aligned} C_\omega Sh = C^- &\left[\left(C^+ h \left(v^{-1} - I\right)\right)(v - I)\right] \\ &+ C^- \left(h \left(v^{-1} - I\right)\right) + C^- \left(h (v - I)\right) \end{aligned}$$

$$= C^-\left[\left(C^+h\left(v^{-1}-I\right)\right)(v-I)\right] + C_\omega h + Sh$$

and we see that

$$(1-C_\omega)(1-S)h = h + C^-\left[\left(C^+h\left(v^{-1}-I\right)\right)(v-I)\right].$$

But

Exercise 3.27. $Kh = C^-\left[\left(C^+h\left(v^{-1}-I\right)\right)(v-I)\right]$ is compact in $L^2(\mathbb{R})$.

Hint: $v-I$ is a continuous function which $\to 0$ as $|z|\to\infty$ and hence can be approximated in $L^\infty(\mathbb{R})$ by finite linear combinations of functions of the form $a/(z-z')$ for suitable constants a and points $z'\in\mathbb{C}\backslash\mathbb{R}$. Then use the following fact:

Exercise 3.28. If T_n, $n\geqslant 1$ are compact operators in $\mathcal{L}(X,Y)$ and $\|T_n - T\|\to 0$ as $n\to\infty$ for some operator $T\in\mathcal{L}(X,Y)$, then T is compact.

Similarly

$$(1-S)(1-C_\omega) = 1 + L, \qquad L \text{ compact.}$$

Thus $(1-C_\omega)$ is Fredholm.

Now we use the following fact:

Exercise 3.29. Suppose that for $\gamma\in[0,1]$, $T(\gamma)$ is a norm-continuous family of Fredholm operators. Then for $\gamma\in[0,1]$,

$$\operatorname{ind}T(\gamma) = \text{const } = \operatorname{ind}T(0) = \operatorname{ind}T(1).$$

Apply this fact to $C_{\omega(\gamma)}$, where we replace r by γr in $v_{x,t}$,

$$v_{x,t,\gamma} = \begin{pmatrix} 1-\gamma^2|r|^2 & -\gamma\bar{r}e^{-2i\tau} \\ \gamma re^{2i\tau} & 1\end{pmatrix}.$$

The proof above shows that $C_{\omega(\gamma)}$ is a norm continuous family of Fredholm operators and so $\operatorname{ind}(1-C_\omega) = \operatorname{ind}\left(1-C_{\omega(\gamma=1)}\right) = \operatorname{ind}\left(1-C_{\omega(\gamma=0)}\right) = 0$ as $C_{\omega(\gamma=0)} = 0$ and the index of the identity operator is clearly 0.

Finally suppose

$$(1-C_\omega)\,\mu = 0.$$

Then using (3.9), $m_+ = \mu v$ and $m_- = \mu$ solve $m_+ = m_- v$, $m_\pm\in\partial C(L^2)$.

Consider $P(z) = m(z)\,(m(\bar{z}))^*$ for $z\in\mathbb{C}^+$ where $m(z)$ is the extension of $m_\pm$ off $\mathbb{R}$ i.e. if $m_\pm = C^\pm h$, $h\in L^2$, then $m(z) = (Ch)(z)$. Then for a contour $\Gamma_{R,\epsilon}$, pictured in Figure 3.30, $\int_{\Gamma_{R,\epsilon}} P(z)\,dz = 0$ as $P(z)$ is analytic.

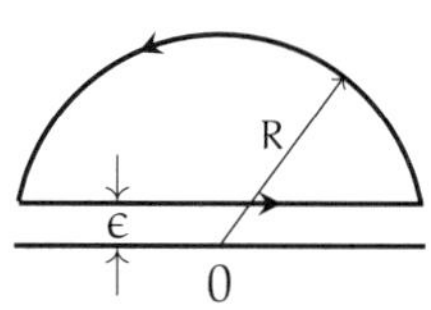

FIGURE 3.30. A semi-circle ϵ above the real axis.

Letting $\epsilon \downarrow 0$ and $R \to \infty$, we obtain (**exercise**) $\int_{-\infty}^{\infty} P_+(z)\, dz = 0$; i.e.

$$0 = \int_{\mathbb{R}} m_+(z)\, m_-(z)^*\, dz = \int_{\mathbb{R}} m_-(z)\, v(z)\, m_-(z)^*\, dz.$$

Taking adjoints and adding, we find

$$0 = \int_{\mathbb{R}} m_-(z)\, (v + v^*)\, (z)\, m_-(z)^*\, dz.$$

But a direct calculation shows that $v + v^*$ is diagonal and

$$(v + v^*)\,(z) = 2 \begin{pmatrix} 1 - |r(z)|^2 & 0 \\ 0 & 1 \end{pmatrix}.$$

Now since $|r(z)| < 1$ a.e. (in fact everywhere except $z = 0$), we conclude that $m_-(z) = 0$. But $\mu = m_-$ and so we see that $\ker(1 - C_\omega) = \{0\}$.

The result of the above chain of arguments is that the solution of the normalized RHP $(\Sigma, v_{x,t})$ for KdV exists and is unique. Such Fredholm arguments have wide applicability in Riemann–Hilbert Theory [22].

One last general remark. The scalar case $n = 1$ is special. This is because the RHP can be solved explicitly by formula. Indeed, if $m_+ = m_- v$, then it follows that $(\log m)_+ = (\log m_-) + \log v$ and hence $\log m(z)$ is given by Plemelj's formula, which provides the general solution of *additive* RHPs, via

$$\log m = C\,(\log v)\,(z) = \int_\Sigma \frac{\log v(s)}{s - z}\, đs$$

and so

$$m(z) = \exp\left(\int_\Sigma \frac{\log v(s)}{s - z}\, đs \right) \tag{3.31}$$

a formula which is easily checked directly. However, there is a hidden subtlety in the business: On $\mathbb{R}$, say, although $v(s)$ may go rapidly to 0 as $s \to \pm\infty$, $v(s)$ may wind around 0 and so $\log v(s)$ may not be integrable at both $\pm\infty$. Thus there is a *topological* obstacle to the existence of a solution of the RHP. If $n > 1$, there are many more such "hidden" obstacles.

Lecture 4

RHP's arise in many difference ways. For example, consider orthogonal polynomials: we are given a measure μ on $\mathbb{R}$ with finite moments,

$$\int_{\mathbb{R}} |x|^m\, d\mu(x) < \infty \qquad \text{for} \qquad m = 0, 1, 2, \ldots$$

Performing Gram-Schmidt on $1, x, x^2, \ldots$ with respect to $d\mu(x)$, we obtain (monic) orthogonal polynomials

$$\pi_n(x) = x^n + \ldots, \qquad , n \geqslant 0$$

such that

$$\int_{\mathbb{R}} \pi_n(x)\, \pi_m(x)\, d\mu u(x) = 0, \qquad n \neq m, \qquad n, m \geqslant 0.$$

(Here we assume that $d\mu$ has infinite support: otherwise there are only a finite number of such polynomials.)

Associated with the π_n's are the orthonormal polynomials

$$P_n(x) = \gamma_n \, \pi_n(x), \qquad \gamma_n > 0, \qquad n \geqslant 0 \tag{4.1}$$

such that

$$\int_{\mathbb{R}} P_n(x) \, P_m(x) \, d\mu(x) = \delta_{n,m}, \qquad n, m \geqslant 0.$$

Orthogonal polynomials are of great historical and continuing importance in many different areas of mathematics, from algebra, through combinatorics, to analysis. The classical orthogonal polynomials, such as the Hermite polynomials, the Legendre polynomials, the Krawchouk polynomials, are well known and much is known about their properties. In view of our earlier comments it should come as no surprise that much of this knowledge, particularly asymptotic properties, follows from the fact that these polynomials have integral representations analogous to the integral representation for the Airy function in the first lecture. For example, for the Hermite polynomials

$$\int_{\mathbb{R}} H_n(x) \, H_m(x) \, e^{-x^2} \, dx = 0 \qquad n \neq m, \qquad n, m \geqslant 0$$

one has the integral representation

$$H_n(x) = n! \int_{\mathcal{C}} \omega^{-n-1} \, e^{2x\omega - \omega^2} \, d\omega$$

where $\mathcal{C}$ is a (small) circle enclosing the origin, (Note: the H_n's are not monic, but are proportional to the π_n's, $H_n(x) = c_n \, \pi_n(x)$ where the c_n's are explicit) and the asymptotic behavior of the H_n's follow from the classical steepest descent method. For general weights, however, no such integral representations are known.

The Hermite polynomials play a key role in random matrix theory in the so-called Gaussian Unitary, Orthogonal and Symplectic Ensembles. However it was long surmised that local properties of random matrix ensembles were *universal*, i.e., independent of the underlying weights. In other words if one considers general weights such as $e^{-x^4} \, dx$, $e^{-(x^6+x^4)} \, dx$, etc., instead of the weight $e^{-x^2} \, dx$ for the Hermite polynomials, the local properties of the random matrices, at the technical level, boil down to analyzing the asymptotics of the polynomials orthogonal with respect to the weights $e^{-x^4} \, dx$, $e^{-(x^6+x^4)} \, dx$, etc., for which no integral representations are known. What to do?

It turns out however, that orthogonal polynomials with respect to an *arbitrary* weight can be expressed in terms of a RHP. Suppose $d\mu(x) = \omega(x) \, dx$, for some $\omega(x) \geqslant 0$ such that

$$\int_{\mathbb{R}} |x|^m \, \omega(x) \, dx < \infty, \qquad m = 0, 1, 2, \ldots.$$

and suppose for simplicity that

$$\omega \in H^1(\mathbb{R}) = \{f \in L^2 : f' \in L^2\}. \tag{4.2}$$

Fix $n \geqslant 0$ and let $Y^{(n)} = \{Y_{ij}^{(n)}(z)\}_{1\leqslant i,j\leqslant 2}$ solve the RHP $\left(\Sigma = \mathbb{R}, v = \left(\begin{smallmatrix} 1 & \omega \\ 0 & 1 \end{smallmatrix}\right)\right)$ normalized so that

$$Y^{(n)}(z)\begin{pmatrix} z^{-n} & 0 \\ 0 & z^{n} \end{pmatrix} \to I \quad \text{as} \quad z \to \infty.$$

Exercise 4.3. Show that we then have (see e.g. [12])

$$Y^{(n)}(z) = \begin{pmatrix} \pi_n(z) & C(\pi_n\, \omega) \\ -2\pi i\, \gamma_{n-1}^2\, \pi_{n-1}(z) & C\left(-2\pi i \gamma_{n-1}^2\, \pi_{n-1}\, \omega\right) \end{pmatrix}$$

where $C = C_{\mathbb{R}}$ is the Cauchy operator on $\mathbb{R}$, π_n, π_{n-1} are the monic orthogonal polynomials with respect to $\omega(x)\,dx$ and γ_{n-1} is the normalization coefficient for π_{n-1} as in (4.1). (Note that by (4.2) and Theorem 2.15, $Y^{(n)}(z)$ is continuous down to the axis for all z.) This discovery is due to Fokas, Its and Kitaev [21]. Moreover this is just exactly the kind of problem to which the nonlinear steepest descent method can be applied to obtain ([14, 15]) the asymptotics of the π_n's with comparable precision to the classical cases, Hermite, Legendre, ..., and so prove universality for unitary ensembles (and later, Deift and Gioev, Shcherbina, for Orthogonal & Symplectic Ensembles of random matrices, see [13] and the references therein).

As mentioned earlier, RHPs are useful not only for asymptotic analysis, but also to analyze analytical and algebraic issues. Here we show how RHPs give rise to difference equations, or differential equations, in other situations.

Consider the solution $Y^{(n)}$ for the orthogonal polynomial RHP $\left(\mathbb{R}, v = \left(\begin{smallmatrix} 1 & \omega \\ 0 & 1 \end{smallmatrix}\right)\right)$. The *key fact* is that the jump matrix $\left(\begin{smallmatrix} 1 & \omega \\ 0 & 1 \end{smallmatrix}\right)$ is *independent* of n: the dependence on n is only in the boundary condition

$$Y^{(n)}\begin{pmatrix} z^{-n} & 0 \\ 0 & z^{+n} \end{pmatrix} \to I\,.$$

So we have $Y_+^{(n+1)} = Y_-^{(n+1)}\, v$ and $Y_+^{(n)} = Y_-^{(n)}\, v$.

Let $R(z) = Y^{(n+1)}(z)\left(Y^{(n)}(z)\right)^{-1}$, $z \in \mathbb{C}\backslash\mathbb{R}$. Then

$$\begin{aligned} R_+(z) &= Y_+^{(n+1)}(z)\left(Y_+^{(n)}(z)\right)^{-1} \\ &= \left(Y_-^{(n+1)}(z)\, v(z)\right)\left(Y_-^{(n)}(z)\, v(z)\right)^{-1} \\ &= Y_-^{(n+1)}(z)\left(v(z)\, v(z)^{-1}\right)\left(Y_-^{(n)}(z)\right)^{-1} \\ &= R_-(z). \end{aligned}$$

Hence $R(z)$ has no jump across $\mathbb{R}$ and so, by an application of Morera's Theorem, $R(z)$ is in fact entire. But as $z \to \infty$

$$R(z) = \left[Y^{(n+1)}(z)\begin{pmatrix} z^{-n-1} & 0 \\ 0 & z^{n+1} \end{pmatrix}\right]\begin{pmatrix} z & 0 \\ 0 & z^{-1} \end{pmatrix}\left[Y^{(n)}(z)\begin{pmatrix} z^{-n} & 0 \\ 0 & z^{n} \end{pmatrix}\right]^{-1}$$

$$= \left(I + O\left(\frac{1}{z}\right)\right)\begin{pmatrix} z & 0 \\ 0 & z^{-1}\end{pmatrix}\left(I + O\left(\frac{1}{z}\right)\right)$$
$$= O(z).$$

Thus $R(z)$ must be a polynomial of order 1,

$$Y^{(n+1)}(z)\left(Y^{(n)}(z)\right)^{-1} = R(z) = Az + B$$

for suitable A and B, or,

$$Y^{(n+1)}(z) = (Az + B)\,Y^{(n)}(z) \tag{4.4}$$

which is a difference equation for orthogonal polynomials with respect to a fixed weight.

Exercise 4.5. Make the argument leading to (4.4) rigorous (why does $\left(Y^{(n)}\right)^{-1}$ exist, etc.)

Exercise 4.6. Show that (4.4) implies the familiar three term recurrence relation for orthogonal polynomials $p_n(z)$

$$b_n\, p_{n+1}(z) + (a_n - z)\, p_n(z) + b_{n-1}\, p_{n-1} = 0, \qquad n \geqslant 0$$

$a_n \in \mathbb{R}, \quad b_n > 0; \quad b_{-1} \equiv 0.$

Whereas the RHP for orthogonal polynomials comes "out of the blue", there are some systematic methods to produce RHP representations for certain problems of interest. This is true in particular for RHPs associated with ordinary differential equations. For example, consider the ZS–AKNS equation (Zakharov-Shabat, Ablowitz-Kaup-Newell-Segur)

$$\left[\partial_x - \left(iz\sigma + \begin{pmatrix} 0 & q(x) \\ \bar{q}(x) & 0 \end{pmatrix}\right)\right]\psi = 0, \qquad -\infty < x < \infty \tag{4.7}$$

(see e.g. [17]). Here $z \in \mathbb{C}$, $\sigma = \frac{1}{2}\begin{pmatrix} 1 & 0 \\ 0 & -1\end{pmatrix}$ and $q(x) \to 0$ at some sufficiently fast rate as $|x| \to \infty$. Equation (4.7) is intimately connected with the defocusing Nonlinear Schrödinger Equation (NLS) by virtue of the fact that the operator

$$L = (i\sigma)^{-1}\left(\partial_x - \begin{pmatrix} 0 & q \\ \bar{q} & 0\end{pmatrix}\right) \tag{4.8}$$

undergoes an *isospectral deformation* if $q = q(t) = q(x,t)$ solves NLS

$$\begin{aligned} iq_t + q_{xx} - 2|q|^2\, q &= 0 \\ q(x, t = 0) &= q_0(x). \end{aligned} \tag{4.9}$$

In other words, if $q = q(t)$ solves NLS then the *spectrum* of

$$L(t) = (i\sigma)^{-1}\left(\partial_x - \begin{pmatrix} 0 & q(x,t) \\ \overline{q(x,t)} & 0\end{pmatrix}\right)$$

is *constant:* Thus the spectrum of $L(t)$ provides constants of the motion for (4.9), and so NLS is "integrable". The key fact is that there is a RHP naturally associated with L which expresses the integrability of NLS in a form that is useful for analysis. Here we follow Beals and Coifman, see [4]. Let $q(x)$ in (4.8) be given with $q(x) \to 0$ as $|x| \to \infty$ sufficiently rapidly. Then for any $z \in \mathbb{C}\backslash\mathbb{R}$,

Exercise 4.10. The equation $(L - z)\psi = 0$ has a *unique* solution ψ such that $\psi(x,z)\,e^{-ixz\sigma} \to I$ as $x \to -\infty$ and is bounded $x \to \infty$. Such $\psi(x,z)$ are called *Beals-Coifman* solutions.

Remark 4.11. These solutions have the following properties:

(1) For fixed x, $\psi(x,z)$ is analytic in $\mathbb{C}\backslash\mathbb{R}$, and is continuous down to the axis. That is $\psi_\pm(x,z) = \lim_{\epsilon\downarrow 0}\psi(x, z \pm i\epsilon)$ exist for all $x, z \in \mathbb{R}$.

(2) For fixed x, $\psi(x,z)e^{-ixz\sigma} \to I$ as $z \to \infty$,

$$\psi(x,z)\,e^{-ixz\sigma} = I + \frac{m_1(x)}{z} + O\left(\frac{1}{z^2}\right), \qquad \text{as} \qquad z \to \infty \tag{4.12}$$

for some matrix residue term $m_1(x)$.

Now clearly $\psi_\pm(x,z)$, $z \in \mathbb{R}$, are two fundamental solutions of $(L - z)\psi = 0$ and so for $z \in \mathbb{R}$,

$$\psi_+(x,z) = \psi_-(x,z)\ v(z)$$

for all $x \in \mathbb{R}$, where $v(z)$ is *independent* of x. In other words, by (1) of Remark 4.11, $\psi(x,\cdot)$ *solves a RHP* $(\Sigma = \mathbb{R}, v)$, *normalized as in* (4.12). In this way differential equations give rise to RHPs in a systematic way.

One can calculate (**exercise**) the precise form of $v(z)$ and one finds

$$v(z) = \begin{pmatrix} 1 - |r(z)|^2 & r(z) \\ -\overline{r(z)} & 1 \end{pmatrix}, \qquad z \in \mathbb{R}$$

where, again (cf. (1.23) for MKdV) we have for r, the *reflection coefficient*,

$$\|r\|_\infty < 1.$$

Now the map

$$q \mapsto r = \mathcal{R}(q) \tag{4.13}$$

is a bijection between suitable spaces: $r = \mathcal{R}(q)$, the direct map, is constructed from q via the solutions $\psi(x,z)$ as above. The inverse map $r \mapsto \mathcal{R}^{-1}(r) = q$ is constructed by solving the RHP (Σ, v) normalized by (4.12) for any fixed x. One obtains

$$\psi(x,z)\,e^{-izx\sigma} = I + \frac{m_1(x;r)}{z} + O\left(\frac{1}{z^2}\right) \qquad \text{as} \qquad z \to \infty$$

and

$$q(x) = -i\,(m_1(x,r))_{12}$$

(cf (1.24) for MKdV).

Now if $q = q(t) = q(x,t)$ solves NLS then $r(t) = \mathcal{R}(q(t))$ evolves simply,

$$r(t) = r(t,z) = r(t=0,z)\,e^{-itz^2}, \qquad z \in \mathbb{R}$$

i.e. $t \to q(t) \to r(t) \to \log r(t) = \log r(t=0) - itz^2$ *linearizes* NLS. This leads to the following formula for the solution of NLS with initial data q_0

$$q(t) = \mathcal{R}^{-1}\left(e^{-it(\cdot)^2}\,\mathcal{R}(q_0)(\cdot)\right). \tag{4.14}$$

The effectiveness of this representation, which one should view as the RHP analog of NLS of the integral representation (1.2) for the Airy equation, depends on the effectiveness of the nonlinear steepest descent method for RHPs.

Question. Where in the representation (4.14) is the information encoded that $q(t)$ solves NLS?

The *answer* is as follows. Let $\psi(x,z,t)$ be the solution of the RHP with jump matrix

$$v_t(z) = \begin{pmatrix} 1-|r|^2 & re^{-itz^2} \\ -\bar{r}e^{itz^2} & 1 \end{pmatrix}$$

normalized as in (4.12). Set $H(x,z,t) = \psi(x,z,t)\,e^{-itz^2\sigma}$ and observe that

$$H_+ = H_- \begin{pmatrix} 1-|r|^2 & r \\ -\bar{r} & 1 \end{pmatrix} = H_-\,v \tag{4.15}$$

for which the jump matrix is *independent of* x *and* t. *This means that we can differentiate* (4.15) *with respect to* x *and* t, $H_{x+} = H_{x-}\,v$, $H_{t+} = H_{t-}\,v$ and conclude, as in the case of orthogonal polynomials, that $H_x\,H^{-1}$ and $H_t\,H^{-1}$ are entire, and evaluating these combinations as $z \to \infty$, we obtain two equations

$$H_x = D\,H \quad , \quad H_t = E\,H$$

for suitable polynomials matrix functions D and E. These functions constitute the famous Lax pair (D,E) for NLS. Compatibility of these two equations requires

$$\begin{aligned} & \partial_t\,\partial_x\,H = \partial_x\,\partial_t\,H \\ \Longrightarrow\quad & \partial_t(D\,H) = \partial_x\,(E\,H) \\ \Longrightarrow\quad & D_t\,H + D\,E\,H = E_x\,H + E\,D\,H \\ \Longrightarrow\quad & D_t + [D,\,E] = E_x \end{aligned}$$

which reduces directly to NLS. In this way RHP's lead to difference and differential equations.

Another systematic way that RHP's arise is through the distinguished class of so-called *integrable operators*. Let Σ be an oriented contour in $\mathbb{C}$ and let $f_1,\ldots,f_n$ and $g_1,\ldots,g_n$ be bounded measurable functions on Σ. We say that an operator K acting on $L^p(\Sigma)$, $1 < p < \infty$, is *integrable* if it has a kernel of the form

$$K(z,z') = \frac{\sum_{i=1}^n f_i(z)\,g_i(z')}{z-z'}, \qquad z,\,z' \in \Sigma; \qquad z \neq z'$$

for such L^∞ functions f_i, g_j,

$$(K\,h)(z) = \int_\Sigma K(z,z')\,h(z')\,dz'\,.$$

Integrable operators were first singled out as a distinguished class of operators by Sakhnovich [31] in the late 1960's, and their theory was developed fully by Its, Izergin, Korepin and Slavnov [26] in the early 1990's (see [11] for a full discussion). The famous sine kernel of random matrix theory

$$K_x(z,z') = \frac{\sin x(z-z')}{\pi(z-z')} = \frac{e^{ixz}\,e^{-ixz'} + \left(-e^{ixz'}\right)\cdot e^{ixz}}{2i\,\pi(z-z')}$$

is a prime example of such an operator, as is likewise the well-known Airy kernel operator.

Integrable operators form an algebra, but their *most remarkable property* is that their *inverses can be expressed* in terms of the solution of a naturally associated RHP. Indeed, let $m(z)$ be the solution of the normalized RHP (Σ, v) where

$$v(z) = I - 2\pi i f\,g^T, \quad f = (f_1, \dots, f_n)^T, g = (g_1, \dots, g_n)^T. \tag{4.16}$$

(Here we assume for simplicity that $\Sigma_{i=1}^n f_i(z)\,g_i(z) = 0$, for all $z \in \Sigma$ as in the sine-kernel: otherwise (4.16) must be slightly modified).

Then $(1-K)^{-1}$ has the form $1+L$ where L is an integrable operator

$$L(z,z') = \frac{\Sigma_{i=1}^n F_i(z)\,G_i(z')}{z-z'}, \qquad z, z' \in \Sigma, \qquad z \neq z'$$

and

$$\begin{cases} F = (F_1, \dots, F_n)^T = m_\pm f \\ G = (G_1, \dots, G_n)^T = (m_\pm^{-1})^T g. \end{cases} \tag{4.17}$$

This means that if, for example, K depends on parameters, as in the case of the sine kernel, asymptotic problems involving K as the parameters become large, are converted into asymptotic problems for a RHP, to which the nonlinear steepest descent method can be applied.

As an example, we show how to use RHP methods to give a proof of Szegő's celebrated Strong Limit Theorem. Let $\mathbb{T}$ be the unit circle.

Theorem 4.18 (Szegő Strong Limit Theorem). *Let* $\varphi(z) = e^{L(z)} \in L^1(\mathbb{T})$, $\varphi(z) > 0$, *where* $\sum_{k=1}^\infty k|L_k|^2 < \infty$ *and* $\{L_k\}$ *are Fourier coefficients of* $L(z)$. *Let* D_n *be the Toeplitz determinant generated by* φ, $D_n(\varphi) = \det\ X(\varphi)$ *where* $X(\varphi)$ *is the* $(n+1)\times(n+1)$ *matrix with entries* $\{\varphi_{i-j}\}_{0\leqslant i,\, j \leqslant n}$, *and* $\{\varphi_k\}$ *are the Fourier coefficients of* φ. *Then as* $n \to \infty$,

$$D_n = e^{(n+1)L_0 + \Sigma_{k=1}^\infty k|L_k|^2}\,(1+o(1)).$$

Sketch of proof. Let e_k, $0 \leqslant k \leqslant n$, be the standard basis in $\mathbb{C}^{n+1}$. Then the map $U_n : e_k \to z^k$, $0 \leqslant k \leqslant n$, $z \in \mathbb{T}$ takes $\mathbb{C}^{n+1}$ onto the trigonometric polynomials $\mathcal{P}_n = \left\{\sum_{j=0}^n a_j\, z^j\right\}$ of degree n and induces a map

$$\tau_n : \mathcal{P}_n \to \mathcal{P}_n$$

which is conjugate to $X(\varphi)$.

We then calculate

$$
\begin{aligned}
\tau_n z^k &= U_n X U_n^{-1} z^k \\
&= U_n X e_k \\
&= U_n \Big(\sum_{j=0}^{n} \varphi_{j-k} e_j\Big) \qquad (4.19)\\
&= \sum_{j=0}^{n} \varphi_{j-k} z^j, \qquad 0 \leqslant k \leqslant n.
\end{aligned}
$$

Now for any $p = \sum_{k=0}^{n} a_k z^k \in \mathcal{P}_n$

$$
\begin{aligned}
(\tau_n p)(z) &= \sum_{k=0}^{n} a_k \sum_{j=0}^{n} \varphi_{j-k} z^j \\
&= \sum_{k=0}^{n} a_k \sum_{j=0}^{n} \Big(\int_\Gamma (z')^{k-j-1} \varphi(z') đz'\Big) z^j \\
&= \sum_{k=0}^{n} a_k \int_\Gamma (z')^{k-1} \varphi(z') \frac{(z/z')^{n+1} - 1}{(z/z') - 1} đz' \\
&= \int_\Gamma \varphi(z') p(z') \frac{(z/z')^{n+1} - 1}{(z - z')} đz'.
\end{aligned}
$$

After some simple calculations (**Exercise**) one finds that

$$
\tau_n p = (1 - K_n) p, \qquad p \in \mathcal{P}_n \tag{4.20}
$$

where K_n is the integrable operator on $\mathbb{T}$ with kernel of the form

$$
K_n(z, z') = \frac{f_1(z) g_1(z') + f_2(z) g_2(z')}{z - z'}, \qquad z, z' \in \Gamma \tag{4.21}
$$

where

$$
\begin{aligned}
f &= (f_1, f_2)^T = \left(z^{n+1}, 1\right)^T \\
g &= (g_1, g_2)^T = \left(z^{-n-1} \frac{1 - \varphi(z)}{2\pi i}, -\frac{(1 - \varphi(z))}{2\pi i}\right)^T.
\end{aligned} \tag{4.22}
$$

We have, in particular, from (4.19) and (4.20), for $0 \leqslant k \leqslant n$,

$$
(1 - K_n) z^k = \sum_{j=0}^{n} \varphi_{j-k} z^j
$$

and for $k < 0$ and $k > n$ one easily shows that

$$
(1 - K_n) z^k = z^k + \sum_{j=0}^{n} \varphi_{j-k} z^i.
$$

Thus K_n is finite rank, and hence trace class, and $(1 - K_n)$ has block form with respect to the orthonormal basis $\{z^k\}_{-\infty}^{\infty}$ for $L^2(\Gamma)$ as given in Figure 4.23. And so

$$
D_n = \det \tau_n = \det X(\varphi) = \det(1 - K_n)
$$

I	0	0
$\cdots$	τ_n	$\cdots$
0	0	I

FIGURE 4.23. The block structure of $1-K_n$ in the basis $\{z^k\}_{-\infty}^{\infty}$.

Associated with the integrable operator K_n we have the normalized RHP $(\Sigma=\Gamma,v)$ where, by (4.16), (4.22)

$$v = I - 2\pi i\, f g^T = \begin{pmatrix} \varphi & -(\varphi-1)\, z^{n+1} \\ z^{-n-1}(\varphi-1) & 2-\varphi \end{pmatrix} \tag{4.24}$$

on $\mathbb{T}$. Now

$$\begin{aligned} \log D_n &= \log\det(1-K_n) \\ &= \operatorname{tr}\ \log(1-K_n) \\ &= \int_0^1 \frac{d}{dt}\ \operatorname{tr}\ \log(1-t\,K_n)\, dt \\ &= \int_0^1 \operatorname{tr}\ \left(\frac{1}{1-t\,K_n}\, K_n\right) dt. \end{aligned} \tag{4.25}$$

For $0\leqslant t\leqslant 1$, set

$$\varphi_t(z) = (1-t) + t\,\varphi(z), \qquad z\in\mathbb{T}.$$

Clearly $\varphi_t(z)>0$ and $\varphi_0(z)=1$, $\varphi_1(z)=\varphi(z)$. Now $\varphi_t - 1 = t(\varphi-1)$ and so we have from (4.21)

$$t\,K_n = K_{t,n} = \left[\left((z/z')^{n+1}-1\right)/(z-z')\right]\left[\left(1-\varphi_t(z')\right)/2\pi i\right]$$

and it follows that in (4.25)

$$\begin{aligned} \frac{1}{1-t\,K_n}\, t\,K_n &= \frac{1}{1-K_{t,n}}\, K_{t,n} \\ &= \frac{1}{1-K_{t,n}} - 1 \\ &= R_{t,n} \end{aligned}$$

where

$$R_{t,n}(z,z') = \frac{\sum_{j=1}^2 F_{t,j}(z)\, G_{t,j}(z')}{z-z'}$$

where by (4.17)

$$\begin{cases} F_t = (F_{t,1}, F_{t,2})^T = m_{t\,\pm}\, f_t, \\ G_t = (G_{t,1}, G_{t,2})^T = \left(m_{t\,\pm}^{-1}\right)^T g_t. \end{cases} \tag{4.26}$$

Here $m_{t\,\pm}$ refers to the solution of the RHP $(\mathbb{T}, v_t)$ where v_t involves φ_t rather than φ in (4.24), and similarly for f_t, g_t.

Hence (**Exercise**)

$$\log D_n = -\int_0^1 \left(\int_{\mathbb{T}} \left(\sum_{j=1}^{2} F'_{t,j}(z)\, G_{t,j}(z) \right) dz \right) \frac{dt}{t}. \tag{4.27}$$

So we see that in order to evaluate D_n as $n \to \infty$ we must evaluate the asymptotics of the solution m_t of the normalized RHP $(\mathbb{T}, v_t)$ as $n \to \infty$, for each $0 \leqslant t \leqslant 1$, and substitute this information into (4.27) using (4.26). This is precisely what can be accomplished [11] using the nonlinear steepest descent method.

Here we present the nonlinear steepest descent analysis in the case when $\varphi(z)$ is analytic in an annulus

$$A_\epsilon = \{z : 1-\epsilon < |z| < 1+\epsilon\}, \qquad \epsilon > 0$$

around $\mathbb{T}$. The idea of the proof, which is a common feature of all applications of the nonlinear steepest descent method, is to move the z^{n+1} term (or its analog in the general situation) in v_t into $|z| < 1$ and the z^{-n-1} term into $|z| > 1$: then as $n \to \infty$, these terms are exponentially small, and can be neglected.

But first we must separate the z^{n+1} and z^{-n-1} terms of v_t algebraically. This is done using the lower-upper pointwise factorization of v_t

$$v_t = \begin{pmatrix} 1 & 0 \\ z^{-n-1}\left(1-\varphi_t^{-1}\right) & 1 \end{pmatrix} \begin{pmatrix} \varphi_t & 0 \\ 0 & \varphi_t^{-1} \end{pmatrix} \begin{pmatrix} 1 & -\left(1-\varphi_t^{-1}\right) z^{n+1} \\ 0 & 1 \end{pmatrix} \tag{4.28}$$

which is easily verified.

Extend $\mathbb{T} = \Sigma \to \tilde{\Sigma} = \left\{|z| = \rho\right\} \cup \Sigma \cup \left\{|z| = \rho^{-1}\right\} = \Sigma_\rho \cup \Sigma \cup \Sigma_{\rho^{-1}}$ where we choose $1-\epsilon < \rho < 1 < \rho^{-1} < 1+\epsilon$. Now define a piecewise analytic function $\tilde{m}$ by the definitions in Figure 4.29.

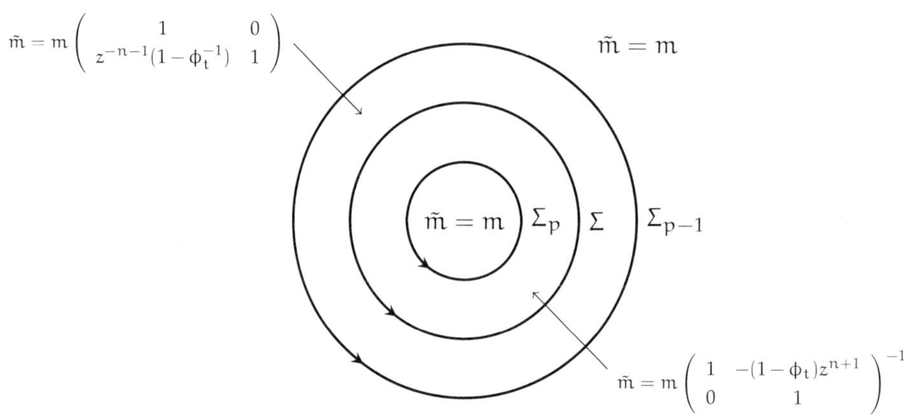

FIGURE 4.29. A piecewise definition of $\tilde{m}$.

This definition is motivated by the fact that

$$m_+ = m_- v_t = m_-(\cdot)(\cdot)(\cdot)$$

as in (4.28). It follows that $\tilde{m}(z)$ solves the normalized RHP $(\tilde{\Sigma}, \tilde{v})$ where

$$\tilde{v}(z) = \begin{pmatrix} 1 & 0 \\ z^{-n-1}\left(1 - \varphi_t^{-1}\right) & 1 \end{pmatrix} \quad \text{on} \quad \Sigma_{\rho^{-1}},$$

$$\tilde{v}(z) = \begin{pmatrix} \varphi_t(z) & 0 \\ 0 & \varphi_t(z)^{-1} \end{pmatrix} \quad \text{on} \quad \Sigma,$$

$$\tilde{v}(z) = \begin{pmatrix} 1 & -\left(1 - \varphi_t^{-1}\right) z^{n+1} \\ 0 & 1 \end{pmatrix} \quad \text{on} \quad \Sigma_\rho .$$

Now as $n \to \infty$, $\tilde{v}(z) \to I$ on Σ_ρ and on $\Sigma_{\rho-1}$. This means that $\tilde{m} \to m_\infty$ where m_∞ solves the normalized RHP (Σ, v_∞) where

$$v_\infty = v\big|_\Sigma = \begin{pmatrix} \varphi_t & 0 \\ 0 & \varphi_t^{-1} \end{pmatrix}.$$

But this RHP is a direct sum of *scalar* RHP's and hence can be solved explicitly, as noted earlier (cf. (3.31)). In this way we obtain the asymptotics of m as $n \to \infty$ and hence the asymptotics of the Toeplitz determinant D_n. □

Here is what, alas, I have not done and what I had hoped to do in these lectures (see AMS open notes):

- Show that in addition to the usefulness of RHP's for algebraic and asymptotic purposes, RHP's are also useful for analytic purposes. In particular, RHP's can be used to show that the Painlevé equations indeed have the Painlevé property.
- Show that in addition to RHP's arising "out of the blue" as in the case of orthogonal polynomials and systematically in the case of ODE's and also integrable operators, RHP's also arise in a systematic fashion in Wiener–Hopf Theory.
- Describe what happens to an RHP when the operator $1 - C_\omega$ is Fredholm, but not bijective, and
- Finally, I have not succeeded in showing you how the nonlinear steepest descent method works in general. All I have shown is one simple case.

References

[1] M Abramowitz and I A Stegun, *Handbook of mathematical functions*, National Bureau of Standards, Washington D.C., 1970. ↑1, 2, 3

[2] J Baik, P Deift, and K Johansson, *On the distribution of the length of the longest increasing subsequence of random permutations*, J. Am. Math. Soc. **12** (1999oct), no. 04, 1119–1179. ↑4

[3] J Baik, P Deift, and Suidan, *Combinatorics and Random Matrix Theory*, Amer. Math. Soc., Providence, RI, 2017. ↑5

[4] R Beals and R R Coifman, *Scattering and inverse scattering for first order systems*, Comm. Pure Appl. Math. **37** (1984), 39–90. ↑33

[5] R Beals, P Deift, and C Tomei, *Direct and inverse scattering on the line*, Mathematical Surveys and Monographs, vol. 28, American Mathematical Society, Providence, RI, 1988. ↑12, 16

[6] A Böttcher and Y I Karlovich, *Carleson Curves, Muckenhoupt Weights, and Toeplitz Operators*, Progress in Mathematics, vol. 154, Birkhäuser Verlag, Basel, 1997. ↑12, 19, 20

[7] A P Calderón, *Cauchy integrals on Lipschitz curves and related operators*, Proc. Nat. Acad. Sci. **74** (1977), 1324–1327. ↑19

[8] K F Clancey and I Gohberg, *Factorization of matrix functions and singular integral operators*, Operator Theory: Advances and Applications, vol. 3, Birkhäuser Verlag, Basel, 1981. ↑1, 12, 20

[9] R R Coifman, A McIntosh, and Y Meyer, *L'integrale de Cauchy définit un opérateur borné sur* L^2 *pour les courbes Lipschitziennes*, Ann. of Math. **116** (1982), 361 –388. ↑19

[10] G David, *L'integrale de Cauchy sur les courbes rectifiables*, Prepublication Univ. Paris-Sud, Dept. Math. **82T05** (1982). ↑19

[11] P Deift, *Integrable operators*, Amer. Math. Soc. Transl. **198** (1999), no. 2, 69–84. ↑35, 38

[12] P Deift, *Orthogonal Polynomials and Random Matrices: a Riemann–Hilbert Approach*, Amer. Math. Soc., Providence, RI, 2000. ↑1, 31

[13] P Deift and D Gioev, *Random Matrix Theory: Invariant Ensembles and Universality*, Courand Lecture Notes, vol. 18, Amer. Math. Soc., Providence, RI, 2009. ↑31

[14] P Deift, T Kriecherbauer, K T-R McLaughlin, S Venakides, and X Zhou, *Asymptotics for polynomials orthogonal with respect to varying exponential weights*, Internat. Math. Res. Not. **16** (1997), 759–782. ↑31

[15] P Deift, T Kriecherbauer, K T-R McLaughlin, S Venakides, and X Zhou, *Uniform asymptotics for polynomials orthogonal with respect to varying exponential weights and applications to universality questions in random matrix theory*, Commun. Pure Appl. Math. **52** (1999nov), no. 11, 1335–1425. ↑31

[16] P Deift and X Zhou, *A steepest descent method for oscillatory Riemann–Hilbert problems. Asymptotics for the MKdV Equation*, Ann. Math. **137** (1993), no. 2, 295–368. ↑4, 5, 7, 9

[17] P Deift and X Zhou, *Long-time asymptotics for solutions of the NLS equation with initial data in a weighted Sobolev space*, Comm. Pure Appl. Math. **56** (2003aug), 1029–1077, available at `0206222v2`. ↑7, 20, 32

[18] P Deift and X Zhou, *Asymptotics for the painlevé II equation*, Commun. Pure Appl. Math. **48** (1995), no. 3, 277–337. ↑7, 8

[19] P Duren, *Theory of* H^p *Spaces*, Academic Press, 1970. ↑1, 10

[20] H Flaschka and A C Newell, *Monodromy and spectrum preserving deformations, I*, Comm. Math. Phys. **76**, 67–116. ↑7

[21] A S Fokas, A R Its, and A V Kitaev, *The isomonodromy approach to matrix models in 2D quantum gravity*, Commun. Math. Phys. **147** (1992), 395–430. ↑31

[22] A S Fokas, A R Its, A A Kapaev, and V Y Novokshenov, *Painlevé Transcendents: the Riemann–Hilbert Approach*, Amer. Math. Soc., 2006. ↑6, 8, 29

[23] J Garnett, *Bounded Analytic Functions*, Graduate Texts in Mathematics, Springer New York, New York, NY, 2007. ↑1, 10

[24] G M Goluzin, *Geometric Theory of Functions of a Complex Variable*, Amer. Math. Soc., Providence, RI, 1969. ↑1, 18

[25] A R Its, A S Fokas, and A A Kapaev, *On the asymptotic analysis of the Painleve equations via the isomonodromy method*, Nonlinearity **7** (1994sep), no. 5, 1291–1325. ↑8

[26] A R Its, V E Izergin, V E Korepin, and N A Slavnov, *Differential equations for quantum correlation functions*, Int. J. Mod. Phys. B **4** (1990), 1003. ↑35

[27] M Jimbo and T Miwa, *Monodromy preserving deformations of linear ordinary differential equation with rational coefficients: II.*, Physica D **2** (1981), 407–448. ↑7

[28] G S Litvinchuk and I M Spitkovskii, *Factorization of Measurable Matrix Functions*, Operator Theory: Advances and Applications, vol. 25, Birkhäuser Basel, Basel, 1987. ↑1, 20

[29] F W J Olver, D W Lozier, R F Boisvert, and C W Clark, *NIST Handbook of Mathematical Functions*, Cambridge University Press, 2010. ↑1

[30] H Royden and P Fitzpatrick, *Real analysis*, 4th ed., Pearson, London, 2010. ↑17

[31] L A Sakhnovich, *Operators similar to the unitary operator with absolutely continuous spectrum*, Functional Anal. and Appl. **2** (1968), 48–60. ↑35

Department of Mathematics, Courant Institute of Mathematical Sciences, New York University
Email address: `deift@cims.nyu.edu`

IAS/Park City Mathematics Series
Volume 26, Pages 41–73
https://doi.org/10.1090/pcms/026/00843

The Semicircle Law and Beyond: The Shape of Spectra of Wigner Matrices

Ioana Dumitriu

Abstract. These lectures cover a celebrated basic result in random matrix theory: the semicircle law for Wigner matrices. We present it in its simplest form first, then remove most of the conditions, deepen the result by looking at fluctuations (aka the Central Limit Theorem), and show the generalization to overlapping Wigner submatrices and the connection to the Gaussian Free Field.

We do not assume here any significant acquaintance with the subject; just a good understanding of undergraduate probability and linear algebra.

1. Introduction

The purpose of these lecture notes is to introduce the reader to one of the central areas of Random Matrix Theory: empirical spectral distributions (ESDs) for random matrices (in this case, the Wigner model, defined below). The work of computing the limiting distributions of ESDs appeared first in the seminal 1955 paper that started the study of Wigner matrices [39]; the method was general enough to be applied to a much wider context. Wigner himself observed that the method extends [40], and the results were strengthened and deepened by Arnold [4], Grenander [21], and Trotter [38].

Since then, ESDs have been studied via their moments in most random matrix cases, from Wishart matrices (Marčenko-Pastur [26], Jonsson [25], Bai [5] and Bai and Silverstein [6, 31]), β-ensembles ([24], [14], [15]), banded Gaussian matrices ([30], [22]), more general banded matrices ([3]), graph-related matrices ([27], [13], [37]), etc. In most of these cases, the main feature of the matrix ensemble is that the distributions of the entries are (at least initially) considered to have moments of any size and which satisfy some growth conditions that allow for the method of moments to be applicable in the computation. The basic results thus obtained can then be strengthened via truncation and approximation for a wider class of entry distributions.

2010 *Mathematics Subject Classification.* Primary 15B52; Secondary 60F05.
Key words and phrases. spectra of random matrices, Wigner matrices, semicircle law, moment method, fluctuations, central limit theorem, Gaussian Free Field.

In these lecture notes, we set out to explain the method of moments and how it relates to finding the ESDs of Wigner matrices and enables us to prove the semicircle law (Section 2), and how one may extend the basic result to a stronger type of convergence or to a more general type of entry distribution (Section 3). Following this, one may examine (again, with the help of moments) the fluctuations from the semicircle law (Section 4) and even introduce a new dimension to the study of fluctuations and discover, in the context of overlapping submatrices, a connection to the Gaussian Free Field (Section 5). Most of what will be presented here (with the exception of the material in Section 5) is presented both more generally and more in depth in the seminal book ([2]); the author's approach here was to pick a particular thread of this very rich area and follow it ever deeper, while presenting it in a way consistent to a sequence of four lectures given to a wide audience.

2. First Results: the Weak Semicircle Law

To begin with, what is a random matrix? Depending on the perspective, one may think of it as a sample from a specific distribution over some set of matrices (symmetric, Hermitian, banded, etc.), or as a matrix, possibly with some structure (again, symmetric, Hermitian, banded, etc.) whose entries are sampled from certain distributions (with some level of generality). In some very particular cases[1], these two perspectives coincide. Here we will mostly take the second approach.

In a practical sense, random matrices originated as a model for uncertain, or noisy, or hard-to-quantify information; examples are the introduction of what is now known as the Wishart ensembles by John Wishart [41] and the Wigner ensembles in the '50s by Eugene Wigner ([39, 40]). The latter, in particular, were introduced as a substitute for a complicated Hamiltonian arising in many-particle interactions, connected to nuclear physics (for a good exposition of the subject, see [28]).

Throughout the lectures, we will prove the real version of each main theorem; similar results can be obtained in the complex and quaternion cases.

2.1. One name, many possible assumptions. The simplest way to define a real Wigner random matrix can be defined as below.

Definition 2.1.1. A $n \times n$ real Wigner matrix is defined as $W_n = (w_{ij})_{1 \leqslant i,j \leqslant n}$, with the following conditions:

- the matrix is symmetric, i.e., $w_{ij} = w_{ji}$ for all i, j;
- all variables w_{ij} are independent (subject to the condition of symmetry);
- the variables w_{ij} with $1 \leqslant i < j \leqslant n$ are identically distributed;
- the variables w_{ii} with $1 \leqslant i \leqslant n$ are identically distributed,
- all the variables w_{ij}, $1 \leqslant i, j \leqslant n$ are centered, i.e. $\mathbb{E}[w_{ij}] = 0$;

[1]For example, the Gaussian Orthogonal Ensemble, see Remark 2.1.3.

- the off-diagonal variables have variance 1 and the diagonal variables have finite variance: for all $1 \leqslant i < j \leqslant n$, $\mathbb{E}[w_{ij}^2] = 1)$ and for all $1 \leqslant i \leqslant n$, $\mathbb{E}[w_{ii}^2] = \delta < \infty)$.

In the beginning, the distributions of the entries will be required to satisfy other moment assumptions (see Assumption 1 below). These assumptions will then be removed for the most general proof of the convergence in probability for the semicircle law; later, for finer results like the CLT and the connection with the GFF, the assumptions will be reintroduced.

Assumption 2.1.2. Assume for all that $\geqslant 1$ that,

$$]\max_{i,j} \mathbb{E}(w_{i,j}^k) \leqslant m_k < \infty ,$$

for some fixed sequence $\{m_k\}_k$.

Remark 2.1.3. If the variables w_{ij}, $i \neq j$ are all distributed according to the centered, standard normal distribution $N(0,1)$ and all the variables w_{ii} are distributed according to the centered normal of variance 2, $N(0,2)$, then the matrix ensemble is known as the Gaussian Orthogonal Ensemble (GOE). The name reflects the ensemble's invariance under conjugation by (independent) random orthogonal matrices.

One of the main quantities of interest in the study of Wigner matrices is the spectrum, i.e., the set of eigenvalues of the matrix, which naturally will have a joint distribution. To place the spectrum on a compact set with high probability, it suffices to scale the matrix so as to make the total variance on each row bounded. Thus, we will actually work not with W_n, but with the scaling $\overline{W}_n = \frac{1}{\sqrt{n}} W_n$.

Below is the definition of the empirical spectral distribution of an arbitrary matrix.

Definition 2.1.4. Given an $n \times n$ real matrix A with eigenvalues (ordered by size) $\lambda_1(A) \leqslant \lambda_2(A) \leqslant \ldots \leqslant \lambda_n(A)$, the *empirical spectral distribution*, or ESD, for A is

$$f_A(x) = \frac{1}{n} \sum_{i=1}^{n} \delta_{\lambda_i(A)}(x) ,$$

that is, the distribution of an eigenvalue uniformly chosen from the spectrum.

The first question we will answer here regards the asymptotics of $f_{\overline{W}_n}$, as the size of the matrix goes to infinity. To this end, we can do a simple Matlab experiment and look at the resulting plot in Figure 2.1.6, which is a simple illustration of the theorem that follows.

Theorem 2.1.5. *Let $\{\overline{W}_n\}_n$ be a sequence of Wigner matrices with Assumption* 2.1.2, *of increasing sizes ($\overline{W}_n$ has size n). Then the sequence of ESDs corresponding to the Wigner matrices converges in probability to the semicircular distribution with density supported on $[-2,2]$:*

$$\sigma(x) = \frac{1}{2\pi} \sqrt{4 - x^2} .$$

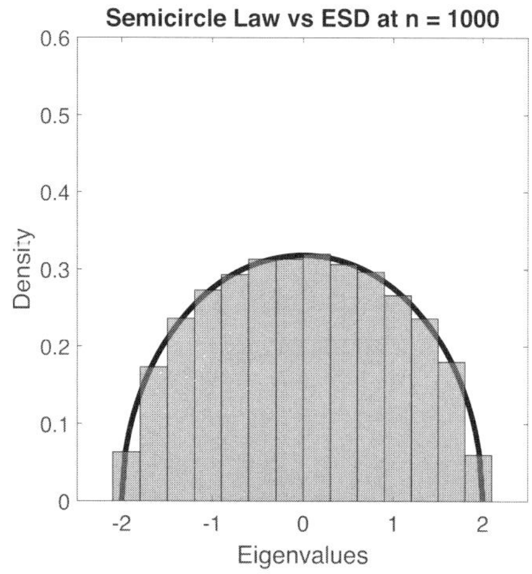

FIGURE 2.1.6. Scaled histogram of the ESD of a GOE matrix with $n = 1000$ (grey bins) plotted versus the semicircular law in black. We had to scale the histogram to cover an area of 1.

Now that we can see the "semicircle law" emerging, we can introduce the tool we will use for proving it, and construct our proof step-by-step.

2.2. Method of Moments. To show the semicircle law, we rely on the following well-known theorem, referred to in [2] and appearing in the Appendix of [8].

Theorem 2.2.1. *Under certain conditions (e.g., Carleman, Riesz, compact support) a distribution is uniquely defined by its set of moments,* $\left(m_k := \mathbb{E}_f(x^k)\right)_{k=1,2\ldots}$.

This method was first introduced by Chebyshev to show convergence to the normal distribution, and then used by Markov in his proof for the Central Limit Theorem.

If a distribution f is uniquely defined by its moments, then for a given sequence of distributions (f_n) for which $\mathbb{E}_{f_n}(x^k) \to \mathbb{E}_f(x^k)$ for all $k \geqslant 1$ integer, it follows that (f_n) converges weakly, in the sense of distributions, to f. This is the crux of the *moment method*, or the *method of moments*.

The semicircle law has odd and even moments $m_{2k+1} = 0$ for all $k \geqslant 0$, respectively, $m_{2k} = C_k$, where $C_k = \frac{1}{k+1}\binom{2k}{k}$, $k \geqslant 0$. The number C_k is known as the *Catalan number*, and a good reference for all the kinds of combinatorial structures counted by the Catalan numbers is [36]. (There are over 200 examples.)

The Carleman condition is that

$$\sum_{n=1}^{\infty} \frac{1}{m_{2k}^{1/2k}} = \infty\,;$$

with a little bit of work using the Stirling formula [1] for factorials, the Catalan number can be shown to be asymptotically proportional to $2^{2k}/k^{3/2}$, and so it follows that the semicircle distribution is unique with this set of moments.

When it comes to ESDs, things are a little more complex, since these are random distributions; some more care has to be taken to show the kind of convergence stated in Theorem 2.1.5; nevertheless, the very first step is to show moment convergence.

Theorem 2.2.2. *Let* $\{\overline{W}_n\}_n$ *be a sequence of Wigner matrices with Assumption* 2.1.2*, of increasing sizes* ($\overline{W}_n$ *has size* n)*. Then, for any fixed positive integer* k*, the sequence of moments of order* k *for the ESDs of the matrices* $\{\overline{W}_n\}_n$ *converges in probability to the* k*th moment of the semicircular distribution with density supported on* $[-2,2]$*:*

$$\sigma(x) = \frac{1}{2\pi}\sqrt{4-x^2}\,.$$

2.3. From moments to graphs. In this section we explain some notation and prove Theorem 2.2.2.

Proof of Theorem 2.2.2. The first step in the proof is to understand what the kth moment of the ESD of W_n is. Note that from the definition of the ESD it follows that

$$\mu_{n,k} := \mathbb{E}_{f_{\overline{W}_n}}(x^k) = \frac{1}{n}\sum_{i=1}^{n}(\lambda(\overline{W}_n))^k = \frac{1}{n}\mathrm{tr}(\overline{W}_n^k)\,.$$

Note that these are random variables which depend on W_n.

We will show that the moments $\mu_{n,k}$ converge to $C_{k/2}$ or to 0, depending on the parity of k, in two steps: first we will show that their expected values *taken over* $\overline{W}_n$ converge to $C_{k/2}$ or 0, and then we will show that their variances taken over $\overline{W}_n$ converge to 0. This means that for every k, the distributions of $\{\mu_{n,k}\}_n$ are highly concentrated around their expectations, and hence that the variables themselves converge to $C_{k/2}$ or 0, depending on the parity of k.

To prove the convergence in expectation of $\mu_{n,k}$, we will use a basic expression for the trace of the power of a matrix, and examine

$$\mathbb{E}_{\overline{W}_n}(\mu_{n,k}) = \frac{1}{n}\mathbb{E}_{\overline{W}_n}(\mathrm{tr}(\overline{W}_n^k)) \tag{2.3.1}$$

$$= \frac{1}{n^{k/2+1}}\mathbb{E}_{\overline{W}_n}\sum_{1\leqslant i_1,i_2,\ldots,i_k\leqslant n} w_{i_1i_2}w_{i_2i_3}\ldots w_{i_ki_1}\,. \tag{2.3.2}$$

Definition 2.3.3. Denote the k-tuple of numbers $I := (i_1, i_2, \ldots, i_k) \in \{1,2,\ldots,n\}^k$, and write $w_I = w_{i_1i_2}w_{i_2i_3}\ldots w_{i_ki_1}$. We will refer to w_I as a word.

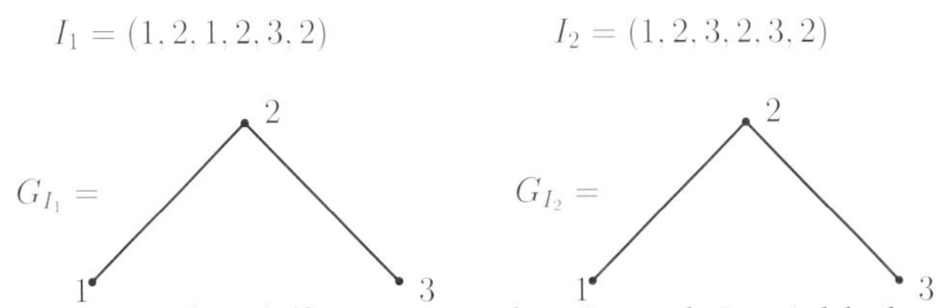

FIGURE 2.3.4. The different 6-tuples I_1 and I_2 yield the same graph $G_{I_1} = G_{I_2}$.

To each k-tuplet I we associate the graph G_I constructed over the set of distinct indices appearing in I (which may be fewer than k), and with edges given by the unordered pairs (i_j, i_{j+1}). G_I may have loops, but not multiple edges, and its edges are undirected. Note that this graph is not always unique (see Figure 2.3.4).

The ordered k-tuplet I defines a closed walk on G_I, since by convention we will think of $i_{k+1} = i_1$ (to reflect the fact that the last index in the word w_I is the same as its first). The sum over all possible I in the expression of the expected moment, then, becomes a sum over all possible closed walks of length k over complete graphs with at most k vertices, and over all possible labelings of these vertices with distinct numbers from $[n] := \{1, 2, \ldots, n\}$.

Consider w_I, and what it means to take its expectation over $\overline{W}_n$:

$$\mathbb{E}_{\overline{W}_n}(w_I) = \prod \mathbb{E}_{\overline{W}_n}(w_e^{f_e}), \tag{2.3.5}$$

where the index e (of the form (i_j, i_{j+1}) stands for edges of G_I appearing in I, and f_e is the number of times the walk uses the edges e.

By Assumption 2.1.2, the above expectation is always bounded in k.

If some edge e in G_I is used only once by the walk, the corresponding $f_e = 1$ and hence $\mathbb{E}_{\overline{W}_n}(w_I)$ is 0, since the variables $w_e = w_{ij}$ are centered and independent. Thus, in order for a walk to have nonzero contribution to the expectation of $\mu_{n,k}$, it must use every edge at least 2 times. **From now on, all walks we consider have this property.**

Let v be the number of (distinct) vertices in a graph G_I. Since each edge appears at least twice in the walk, and the total length of the walk is k, $v \leqslant k/2+1$. This is a consequence of the fact that the graph G_I is always connected (since it has a closed walk going through all vertices) and hence the number of its distinct edges is at least $v-1$; as every edge appears at least twice in a sequence of k edges (given by the walk itself), it follows that $k \geqslant 2e \geqslant 2(v-1)$ or $v \leqslant k/2+1$.

We argue that for $v < k/2+1$ fixed, the total contribution to the expectation of walks whose graphs have v vertices is negligible; since there is a bounded number of such vs, their contribution will still be negligible. It follows that, as $n \to \infty$, the only non-negligible contribution to the expectation will be made by walks with $v = k/2+1$.

Remark 2.3.6. Note that this is impossible when k is odd, and this is in fact why the expectation of odd moments converges to 0.

Suppose that $v < k/2+1$ and consider some fixed labels for vertices of the graph G_I; the total number of walks of length k of G_I is a function of k, thus bounded. The contribution of each walk to the expectation is also bounded, as per (2.3.5). Thus, the total contribution to the expectation from walks with a fixed set of v distinct indices is some explicit function of k, thus bounded. We have $\binom{n}{v} = O(n^v)$ ways of picking these indices from $[n]$; the factor $\frac{1}{n^{k/2+1}}$ in (2.3.2) contains a power of nis larger than v, as $v < k/2+1$, and hence even when we sum over all possible choices of these ν vertices, the contribution is negligible. (We have used $O(n^\nu)$ in the usual, asymptotic way.)

We are then left to deal with the case when k is even (see the Remark above) and the walk defined by I visits precisely $k/2+1$ distinct indices. In this case, every edge is walked on exactly twice, and in addition the connected graph G_I has exactly one less edge than vertices, i.e., it is a tree. Note then that the walk defines the tree, uniquely, and "reveals" it in a depth-first search.

The number of rooted unlabeled trees with $k/2+1$ vertices is $C_{k/2}$ (see, e.g., [36]). Since $\mathbb{E}(w_{ij}^2) = 1$ for any $i \neq j$, using (2.3.5) we see that the contribution to the expectation of $\mu_{n,k}$ of a closed walk on a tree whose vertices have $k/2+1$ ordered labels is $\frac{1}{n^{k/2+1}}$. The number of ways to choose these $k/2+1$ ordered labels from $[n]$ is $n^{k/2+1}(1+o(1))$.

Putting these three facts together with the above consideration, it follows that for any k, $\mathbb{E}_{\overline{W}_n}(\mu_{n,k}) \to 0$, when k is odd, respectively, $\mathbb{E}_{\overline{W}_n}(\mu_{n,k}) \to C_{k/2}$, when k is even.

So for each fixed k, the expectation of $\mu_{n,k}$ over $\overline{W}_n$ converges to a constant. To show that $\mu_{n,k}$ converges in probability to the same constant, we examine its variance:

$$Var\left[\frac{1}{n^{k/2+1}}\sum_I w_I\right] = \mathbb{E}\left(\frac{1}{n^{k/2+1}}\sum_I w_I - \mathbb{E}\left(\frac{1}{n^{k/2+1}}\sum_I w_I\right)\right)^2,$$

where the sums above are taken over all possible k-tuplets I with entries in $[n]$.

Expanding the right hand side in the equation above we get

$$Var\left[\frac{1}{n^{k/2+1}}\sum_I w_I\right] = \frac{1}{n^{k+2}}\sum_{I,J} Cov[w_I w_J], \tag{2.3.7}$$

where $Cov[w_I w_J]$ is the covariance of the two words taken with respect to $\overline{W}_n$. This is the quantity we will focus on now.

Consider the two graphs G_I and G_J. The total number of vertices in the union $G_{I\cup J}$ of the two graphs (defined as the graph whose vertex and edge sets are the union of the vertex, respectively, edge sets of G_I and G_J) is at most $2k$. Given some fixed labels on these $\nu \leqslant 2k$ vertices, the number of graphs one can define on them is n-independent, and so is the number of pairs of closed walks of length k on any such graph. Finally, the number of choices of vertex labels is $O(n^\nu)$. For an example of how this works, see Figure 2.3.8.

Moreover, if the two walks share no edges in common, independence of the matrix entries means that we must have $Cov[w_I w_J] = 0$. Thus, G_I and G_J must share edges.

The same considerations as before tell us that the only pairs of walks w_I, w_J with non-trivial contribution will have a union graph $G_{I\cup J}$ where each edge is walked on by the union of walks $w_{I\cup J}$ at least *twice*, and with the property that G_I and G_J overlap on at least one edge.

Since the total number of edges (with multiplicities) in the two walks is $2k$, and each actual edge in the union of graphs is walked on at least twice, the total number of edges in the union of graphs must be at most k. Moreover, since G_I and G_J are walk-defined (thus, connected) graphs which overlap on at least one edge, $G_{I \cup J}$ is connected, and so $v \leqslant k+1$.

Thus, for a fixed pair of graphs G_I and G_J with the desired properties, the number of all possible pairs of walks is a function of k, and the contribution to the covariance from each such pair of walks is bounded by Assumption 2.1.2 and the fact that k is fixed.

The only dependence on n is given by the choice of labels for the vertices.

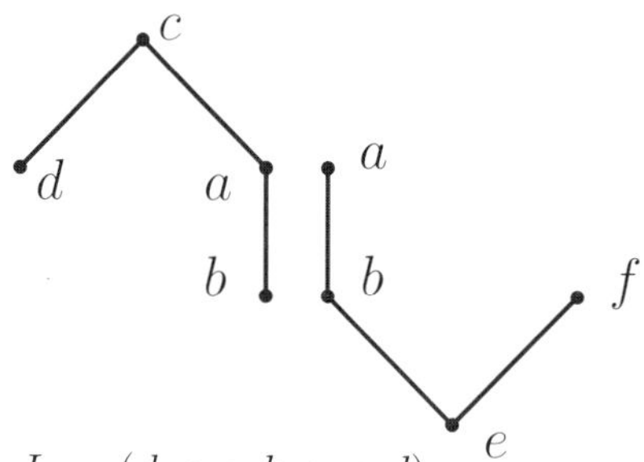

$I_1 = (d, c, a, b, a, c, d)$
$I_2 = (f, e, b, a, b, e, f)$
number of paris of walks of this type is $O(n^6)$

FIGURE 2.3.8. There are 6 vertices in the join of two graphs $G_{I_1 \cup I_2}$ above: a, b, c, d, e, f, hence the total number of choices for labels is $O(n^6)$.

As $v \leqslant k+1$, there are $O(n^v) = O(n^{k+1})$ such possible choices of labels from $\{1, 2, \ldots, n\}$ for the vertices involved, and thus the total contribution is $O(n^{k+1})$. But, as we can see from (2.3.7), the total sum is scaled by $O(n^{k+2})$; this means that the variance goes to 0 at least as fast as $1/n$. Since the variance of $\mu_{n,k}$ converges to 0 regardless of k, then by the Chebyshev inequality the moments themselves converge to the same values as their expectations: 0 if k is odd and $C_{k/2}$ if k is even. Theorem 2.2.2 is proved. □

Remark 2.3.9. Note that in the above we bounded, but did not actually find the speed with which the variances converges to 0. For now, this was enough. In reality, the variances of the moments decay not like $1/n$, but like $1/n^2$, as we will see in the main calculation in the next section.

2.4. Additional notes and context. The method of moments has been successfully used to calculate the asymptotics of the ESD of Wishart matrices [26], graphs [13,27], general β-ensembles [14,15,24], band matrices [3], etc. It has, notably, also been used to investigate the largest eigenvalues of matrix ensembles, since the seminal paper [19]; successive refinements lead to the first proof of universality at the edge of the spectrum, in a series of papers [32, 33, 35]. Recently, refinements of the method of moments have been used to investigate the spectral gap of adjacency and Laplacian matrices of random regular graphs [9], and a direct connection to orthogonal polynomials has been found [34]; this connection will be mentioned again in later sections.

The main benefit of the method is that it transforms a question about the spectrum of a matrix into a combinatorial question involving counting weighted closed walks on graphs. Sometimes local properties can be deduced by using the method of moments (as is the case with the largest eigenvalues) but in general they are not powerful enough to provide a refined local picture (like the Stieltjes transform method does).

We explore the connections to the combinatorics of Dyck paths and to tridiagonal matrices in the problems below.

2.5. Problems.

Problem 2.5.1. A *Dyck path* of size k is a lattice path from $(0,0)$ to $(2k,0)$ consisting of k up steps of the form $(1,1)$ and k down steps of the form $(-1,1)$ which never goes below the x-axis.

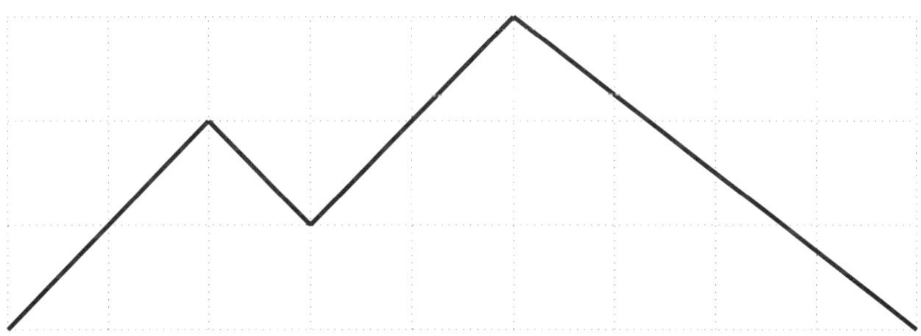

FIGURE 2.5.2. A Dyck path of size 4

a) Show that there is a bijection between rooted unlabelled trees with $k+1$ vertices and Dyck paths of size k.
b) Note that the k-th Catalan number $C_k = \frac{1}{k+1}\binom{2k}{k}$ satisfies

$$C_{k+1} = \sum_{j=0}^{k} C_j C_{k-j}, k \geqslant 1.$$

Let D_k denote the number of Dyck paths of size k. Prove that

$$D_k = C_k, k \geqslant 1.$$

Problem 2.5.3. Let P be a probability measure, and $\{P_n\}$ be a sequence of probability measures on $\mathbb{R}$ such that $\text{supp}(P) \subset [-M, M]$, $\text{supp}(P_n) \subset [-M, M]$ for all n and some $M > 0$. If $\int x^k dP_n \to \int x^k dP$ for all $k \in \mathbb{Z}^+$, as $n \to \infty$, show that $\int f dP_n \to \int f dP$ for all bounded continuous functions f as $n \to \infty$.

Problem 2.5.4. Let $M_n = \frac{G_n+G_n^T}{\sqrt{2}}$, and G_n is a $n \times n$ matrix with each entry an i.i.d. standard Gaussian random variable. It can be shown that M_n is similar to a symmetric tridiagonal matrix

$$T_n = \begin{pmatrix} N(0,2) & \chi_{n-1} & & & \\ \chi_{n-1} & N(0,2) & \chi_{n-2} & & \\ & \ddots & \ddots & \ddots & \\ & & \chi_2 & N(0,2) & \chi_1 \\ & & & \chi_1 & N(0,2) \end{pmatrix},$$

where $\chi_k = \sqrt{\sum_{i=1}^k X_i^2}$, and $\{X_i\}_{i=1}^k$ are independent and identically distributed standard Gaussian random variables. (All the random variables in the upper triangle and the diagonal line are independent.) Show that, for all $k \geqslant 1$,

$$\lim_{n\to\infty} \frac{1}{n}\mathbb{E}[\mathrm{tr}\left(\frac{T_n}{\sqrt{n}}\right)^{2k}] = C_k,$$

$$\lim_{n\to\infty} \frac{1}{n}\mathbb{E}[\mathrm{tr}\left(\frac{T_n}{\sqrt{n}}\right)^{2k+1}] = 0.$$

(Hint: by the Strong Law of Large Numbers, $\frac{\chi_n}{\sqrt{n}} \to 1$ almost surely as $n \to \infty$.)

3. Stronger results, weaker assumptions

In this section, we will strengthen Theorem 2.2.2 in two ways: we will show that the convergence in moments is in fact convergence in probability (thus proving Theorem 2.1.5), and we will remove the moment boundedness requirements of Assumption 2.1.2.

3.1. Convergence in Probability; Proof of Theorem 2.1.5.

Definition 3.1.1. For the sake of compactness of notation, in this section we will denote by $\langle f, g\rangle$ the value of the distribution f on the function g; more specifically,

$$\langle \sigma, g\rangle = \int_{-2}^{2} g(x)d\sigma(x)\,,$$

and

$$\langle f_{\overline{W}_n}, g\rangle = \frac{1}{n}\mathrm{tr}(g(\overline{W}_n))\,.$$

So far, in proving Theorem 2.2.2, we have shown that

1) $\mathbb{E}_{\overline{W}_n, f(\overline{W}_n)}(x^k) = \frac{1}{n}\mathbb{E}_{\overline{W}_n}\mathrm{tr}(\overline{W}_n^k) \to 0$, if k is odd.
2) $\mathbb{E}_{\overline{W}_n, f(\overline{W}_n)}(x^k) = \frac{1}{n}\mathbb{E}_{\overline{W}_n}\mathrm{tr}(\overline{W}_n^k) \to C_{k/2}$, if k is even.
3) $\mathrm{Var}_{\overline{W}_n, f(\overline{W}_n)}(x^k) = \frac{1}{n^2}\mathrm{Var}_{\overline{W}_n}\mathrm{tr}(\overline{W}_n^k) \to 0$ for all k.

This allowed us to conclude that the moments of the ESD themselves converge in probability:

$$\frac{1}{n}\,\mathrm{tr}(\overline{W}_n^k) = \mathbb{E}_{f_{\overline{W}_n}}(x^k) \to 0 \text{ or } C_{k/2}\,,\ \text{depending on the parity of } k.$$

We now show that the above, together with some tightness considerations, allows us to conclude that $f_{\overline{W}_n}$ converges to the semicircle distribution in probability. To do so, recall the Weierstrass approximation theorem.

Theorem 3.1.2. *(Weierstrass) Given any* $\epsilon > 0$, *compact interval* I, *and any function* $f \in C_c(I)$ *(continuous and compactly supported in* I*), there exists a polynomial* p_ϵ *such that* $|f(x) - p_\epsilon(x)| < \epsilon$ *uniformly on* I.

We will also use the following simple lemma, whose proof we provide below.

Lemma 3.1.3. *For any* $k \geqslant 1$, $C_k \leqslant 4^k$.

Proof. This is easily seen from the definition of $C_k = \frac{1}{k+1}\binom{2k}{k}$ and the fact that

$$\binom{2k}{k} \leqslant \sum_{i=0}^{2k} \binom{2k}{i} = 2^{2k} = 4^k . \qquad \square$$

For a matrix $\overline{W}_n$ with Assumption 2.1.2, one can identify its ESD $f_{\overline{W}_n}$ with the associated distribution $\langle f_{\overline{W}_n}, \cdot \rangle$, and note that this is defined by its effect on, e.g., compactly supported functions on $\mathbb{R}$.

To prove Theorem 2.1.5, we need to show that for a sequence $\{\overline{W}_n\}_n$ of Wigner matrices with Assumption 2.1.2, given $g \in C_b(\mathbb{R})$ (the set of all bounded, continuous functions) and $\epsilon > 0$,

$$\lim_{n \to \infty} P\left(\left| \langle f_{\overline{W}_n}, g \rangle - \langle \sigma, g \rangle \right| > \epsilon \right) = 0 . \tag{3.1.4}$$

Remark 3.1.5. One may take as definition for weak convergence (aka convergence in probability) that (3.1.4) occurs for any function $f \in C_b(\mathbb{R})$. It is a well known analysis fact that this can be weakened to $f \in C_c^\infty(\mathbb{R})$; hence, it is also sufficient to show it for $f \in C_c(\mathbb{R})$ or f Lipschitz. We are using the former above, and we will use the latter later in the proof.

Proof of Theorem 2.1.5. The most important idea in this proof is to show that the "spillover" from $[-2, 2]$ is negligible; in other words, that the ESDs' supports concentrate on $[-2.2]$.

Pick any $\delta > 0$. Choose R large, for example, larger than 5; by Markov's inequality,

$$P\left(\langle f_{\overline{W}_n}, |x|^k \mathbf{1}_{|x|>R} \rangle > \epsilon \right) \leqslant \frac{1}{\epsilon} \mathbb{E}_{\overline{W}_n} \left(\langle f_{\overline{W}_n}, |x|^k \mathbf{1}_{|x|>R} \rangle \right) .$$

It is easy to see that

$$\mathbb{E}_{\overline{W}_n} \left(\langle f_{\overline{W}_n}, |x|^k \mathbf{1}_{|x|>R} \rangle \right) \leqslant \frac{1}{R^k} \mathbb{E}_{\overline{W}_n} \left(\langle f_{\overline{W}_n}, x^{2k} \mathbf{1}_{|x|>R} \rangle \right) ,$$

due to the $|x| > R$ constraint. Thus we obtain

$$\limsup_{n \to \infty} P\left(\langle f_{\overline{W}_n}, |x|^k \mathbf{1}_{|x|>R} \rangle > \epsilon \right) \leqslant \frac{1}{\epsilon} \frac{C_k}{R^k} \leqslant \frac{1}{\epsilon} \frac{4^k}{R^k} . \tag{3.1.6}$$

As $k \to \infty$ and $R > 5$, the above upper bound decreases to 0; hence

$$\limsup_{n \to \infty} P\left(\langle f_{\overline{W}_n}, |x|^k \mathbf{1}_{|x|>R} \rangle > \epsilon \right) = 0 ;$$

thus, the "spillover" outside the interval $[-R, R]$ is insignificant.

In particular, this means that the sequence is *tight*, and we can use the following (well-known) lemma to change the condition that $g \in C_b(\mathbb{R})$ to $g \in C_c(\mathbb{R})$ (the set of continuous functions with compact support). The proof of the lemma below is not hard and we leave it as an exercise for the reader.

Lemma 3.1.7. *If the sequence of random measures $\{\mu_n\}_{n\geqslant 1}$ is* tight, *i.e., for all $\epsilon > 0$ there exists a compact $K = K_\epsilon$ such that $P(\mu_n(K^c) > \epsilon) < \epsilon$ for all n, then weak convergence is assured provided that for any function $g \in C_c(\mathbb{R})$,*

$$\lim_{n\to\infty} (\langle \mu_n, g\rangle - \langle \mu, g\rangle) = 0\,.$$

Moreover, it follows that proving the above for any class of functions which includes $C_c(\mathbb{R})$ (such as Lipschitz functions) will also suffice.

It is thus sufficient to examine the case when g is in the more constrained class $C_c(\mathbb{R})$, and in fact (following from the above) so that $\operatorname{supp} g \subseteq [-R, R]$. Let g be such a function.

Given ϵ, Weierstrass' theorem 3.1.2 says that there exists a polynomial $p = p_\epsilon$ such that $|p(x) - g(x)| < \epsilon/8$ uniformly on $[-R, R]$. Let

$$p_\epsilon = \sum_{i=0}^{m} c_i^\epsilon x^i\,.$$

Note that degree m does not depend on n.

Then

$$\begin{aligned} &|\langle f_{\overline{W}_n}, g\rangle - \langle \sigma, g\rangle| \\ &\leqslant |\langle f_{\overline{W}_n}, g\rangle - \langle f_{\overline{W}_n}, p\rangle| + |\langle \sigma, g\rangle - \langle \sigma, p\rangle| + |\langle f_{\overline{W}_n}, p\rangle - \langle \sigma, p\rangle| \\ &\leqslant \frac{\epsilon}{8} + |\langle f_{\overline{W}_n}, p\mathbf{1}_{|x|>R}\rangle| + \frac{\epsilon}{8} + |\langle f_{\overline{W}_n}, p\rangle - \langle \sigma, p\rangle| \\ &\leqslant \frac{\epsilon}{4} + |\langle f_{\overline{W}_n}, p\mathbf{1}_{|x|>R}\rangle| + |\langle f_{\overline{W}_n}, p\rangle - \langle \sigma, p\rangle|\,, \end{aligned} \tag{3.1.8}$$

since $g(x) = 0$ outside of $[-R, R]$ and

$$|\langle f_{\overline{W}_n}, g\rangle - \langle f_{\overline{W}_n}, p\rangle| \leqslant \sup_{x\in[-R,R]} |p(x) - g(x)| + |\langle f_{\overline{W}_n}, p\mathbf{1}_{|x|>R}\rangle|\,,$$

as both σ (the semicircle distribution) and g are identically 0 outside $[-R, R]$.

Since R was chosen to be large enough and p does not depend on n, it follows that we can pick n, R so that

$$\left(|\langle f_{\overline{W}_n}, \mathbf{1}_{|x|>R}\rangle|\right) < \frac{\epsilon}{100m}\,.$$

Since

$$|\langle f_{\overline{W}_n}, p\mathbf{1}_{|x|>R}\rangle| \leqslant \sum_{i=0}^{m} |c_i^\epsilon||\langle f_{\overline{W}_n}, |x|^i\mathbf{1}_{|x|>R}\rangle| < \frac{\epsilon}{100m} + \sum_{i=1}^{m} |c_i^\epsilon||\langle f_{\overline{W}_n}, |x|^i\mathbf{1}_{|x|>R}\rangle|\,,$$

it follows that for R and n large enough, using (3.1.6), one can make

$$P\left(|\langle f_{\overline{W}_n}, p\mathbf{1}_{|x|>R}\rangle| > \epsilon/4\right) < \frac{\delta}{2}\,.$$

We can now rewrite (3.1.8) as

$$
\begin{aligned}
&|\langle f_{\overline{W}_n}, g\rangle - \langle \sigma, g\rangle| \\
&\leqslant \frac{\epsilon}{4} + |\langle f_{W_n}, p\mathbf{1}_{|x|>R}\rangle| + |\langle \mathbb{E}_{\overline{W}_n}(f_{\overline{W}_n}), p\rangle - \langle \sigma, p\rangle| \\
&\qquad\qquad + |\langle f_{\overline{W}_n}, p\rangle - \langle \mathbb{E}_{\overline{W}_n}(f_{\overline{W}_n}), p\rangle| \\
&\leqslant \frac{\epsilon}{4} + |\langle f_{\overline{W}_n}, p\mathbf{1}_{|x|>R}\rangle| + \frac{\epsilon}{4} \qquad + |\langle f_{\overline{W}_n}, p\rangle - \langle \mathbb{E}_{\overline{W}_n}(f_{\overline{W}_n}), p\rangle| \\
&\leqslant \frac{\epsilon}{2} + |\langle f_{\overline{W}_n}, p\mathbf{1}_{|x|>R}\rangle| \qquad + |\langle f_{\overline{W}_n}, p\rangle - \langle \mathbb{E}_{\overline{W}_n}(f_{\overline{W}_n}), p\rangle|;
\end{aligned}
\tag{3.1.9}
$$

the last inequality is a consequence of Theorem 2.2.2; we simply pick n large enough to make

$$
|\langle \mathbb{E}_{\overline{W}_n}(f_{\overline{W}_n}), p\rangle - \langle \sigma, p\rangle| \leqslant \frac{\epsilon}{4}.
$$

Note that

$$
|\langle f_{\overline{W}_n}, p\rangle - \langle \mathbb{E}_{\overline{W}_n}(f_{\overline{W}_n}), p\rangle| \leqslant \sum_{i=0}^{m} c_i^{\epsilon} |\langle f_{\overline{W}_n}, x^i\rangle - \langle \mathbb{E}_{\overline{W}_n}(f_{\overline{W}_n}), x^i\rangle|,
$$

and since we established before that $Var(\langle f_{\overline{W}_n}, x^k\rangle) = Var(tr(\overline{W}_n^k)) = O(1/n)$, by applying the Chebyshev inequality to each summand, we obtain

$$
P\left(|\langle f_{\overline{W}_n}, p\rangle - \langle \mathbb{E}_{\overline{W}_n}(f_{\overline{W}_n}), p\rangle| > \epsilon/4\right) \leqslant \frac{1}{\epsilon^2} \sum_{i=0}^{m} c_i^{\epsilon} Var(\langle f_{\overline{W}_n}, x^k\rangle). \tag{3.1.10}
$$

Hence, we deduce that by choosing n large enough (recall that δ, ϵ, m are constant) we can make sure that

$$
P\left(|\langle f_{\overline{W}_n}, p\rangle - \langle \mathbb{E}_{\overline{W}_n}(f_{\overline{W}_n}), p\rangle| > \epsilon/4\right) \leqslant \delta/2. \tag{3.1.11}
$$

Finally, putting (3.1.9) together with (3.1.10) and (3.1.11), leads us to the conclusion that

$$
\begin{aligned}
&P\left(|\langle f_{\overline{W}_n}, g\rangle - \langle \sigma, g\rangle| \geqslant \epsilon\right) \\
&\leqslant P\left(|\langle f_{\overline{W}_n}, p\mathbf{1}_{|x|>R}\rangle| > \epsilon/4\right) + P\left(|\langle f_{\overline{W}_n}, p\rangle - \langle \mathbb{E}_{\overline{W}_n}(f_{\overline{W}_n}), p\rangle| > \epsilon/4\right) \leqslant \delta,
\end{aligned}
$$

for any n large enough.

The statement of the theorem now follows since we can choose δ to be arbitrarily small. □

3.2. Removal of Moment Assumptions. So far, we have relied on existence of moments of all orders for the variables w_{ij} (which is at the core of Assumption 2.1.2). The existence of these moments is not in fact needed, which is why we introduce the following.

Assumption 3.2.1. Assume that

(1) $\mathbb{E}(w_{ij}) = 0$ for all $1 \leqslant i, j \leqslant n$,
(2) $\mathbb{E}(w_{ij}^2) = 1$ for $1 \leqslant i < j \leqslant n$, and
(3) $\mathbb{E}(w_{ii}^2) = \sigma^2 < \infty$ for all $1 \leqslant i \leqslant n$ for some fixed $\sigma \geqslant 0$.

We will use a "truncation" procedure to make a smooth transition between Assumption 2.1.2 and the much less restrictive Assumption 3.2.1, by introducing a new set of variables which closely approximate the Assumption 3.2.1 ones, but which satisfy Assumption 2.1.2.

Given $\epsilon_1, \epsilon_2, \epsilon_3 > 0$ small, there exists a $K = K_{\epsilon_1,\epsilon_2,\epsilon_3}$ such that

$$\Pr[|w_{ij}| > K] \leqslant \epsilon_2 \,, \tag{3.2.2}$$

$$\mathbb{E}[w_{ij}^2 \mid |w_{ij}| > K] \leqslant \epsilon_3; \tag{3.2.3}$$

in addition, one may construct an array of independent variables $\hat{w}_{ij}$, $1 \leqslant i, j \leqslant n$ with $\hat{w}_{ij} = \hat{w}_{ij}$ for all i, j (symmetric) with

$$\hat{w}_{ij} = w_{ij} \text{ if } |w_{ij}| \leqslant K \,, \tag{3.2.4}$$

$$\operatorname{supp} \hat{w}_{ij} \subseteq [-K - \epsilon_1, K + \epsilon_1] \,, \tag{3.2.5}$$

while at the same time $E[\hat{w}_{ij}^2] = E[w_{ij}^2]$, $E[\hat{w}_{ij}] = 0$.

Remark 3.2.6. Note that (3.2.2) and (3.2.3) can be trivially fulfilled knowing that $\mathbb{E}(w_{ij} = 0)$ and $\mathbb{E}(w_{ij}^2) \leqslant \max\{1, \sigma^2\}$. Construction of variables $\hat{w}_{ij}$ with the desired properties is a simple exercise left to the reader. Also note that having $\mathbb{E}(w_{ij}^2) \leqslant \max\{1, \sigma^2\} < \infty$ means that in the above $K^2\epsilon_2$ (and consequently also $(K + \epsilon_1)^2\epsilon_2$) can be made as small as desired by taking ϵ_1, ϵ_2, ϵ_3 appropriately small.

Remark 3.2.7. The variables $\hat{w}_{ij}$ have moments of all orders, since they are compactly supported; therefore they satisfy Assumption 2.1.2. As such, defining new matrices $\widehat{W}_n$ with the new variables $\hat{w}_{ij}$, $1 \leqslant i, j \leqslant n$, we know by Theorem 2.1.5 that the ESDs of $\widehat{\overline{W}}_n$ converge to the semicircle law in probability.

All that remains to show is that the spectra of $\overline{W}_n$ and $\widehat{\overline{W}}_n$ are very close. For this, we shall use the Hoffman-Wielandt theorem below.

Theorem 3.2.8. *(Hoffman-Wielandt) Given symmetric $n \times n$ matrices A and B with eigenvalues $\lambda_1(A) \leqslant \lambda_2(A) \ldots \leqslant \lambda_n(A)$ and $\lambda_1(B) \leqslant \lambda_2(B) \ldots \leqslant \lambda_n(B)$,*

$$\sum_{i=1}^{n} (\lambda_i(A) - \lambda_i(B))^2 \leqslant tr(A - B)^2 \,.$$

Applying the above to the matrices $\overline{W}_n$ and $\widehat{\overline{W}}_n$ we obtain

$$\frac{1}{n}\mathrm{tr}(\overline{W}_n - \widehat{\overline{W}}_n)^2 = \frac{1}{n^2}\sum_{i,j} (w_{ij} - \hat{w}_{ij})^2 \mathbf{1}_{|w_{ij}| \geqslant K} \,.$$

Let $\epsilon > 0$ be small and use Markov's inequality to get

$$P\left[\frac{1}{n}\mathrm{tr}(\overline{W}_n - \widehat{\overline{W}}_n)^2 > \epsilon\right] \leqslant \frac{1}{\epsilon}\frac{1}{n^2}\sum_{i,j} \mathbb{E}\left((w_{ij} - \hat{w}_{ij})^2 \mathbf{1}_{|w_{ij}| \geqslant K}\right) \,.$$

If we first expand $(w_{ij} - \hat{w}_{ij})^2$, then choose K large enough to satisfy the additional requirement that $(K + \epsilon_1)^2\epsilon_2 < \epsilon^2$ and finally pick $\epsilon_3 < \epsilon^4/\max\{1, \sigma^2\}$, we obtain that

$$\begin{aligned}
&\mathbb{E}\left((w_{ij}-\hat{w}_{ij})^2\mathbf{1}_{|w_{ij}|\geqslant K}\right)\\
&\leqslant \mathbb{E}(w_{ij}^2\mathbf{1}_{|w_{ij}|\geqslant K})+\mathbb{E}(\hat{w}_{ij}^2\mathbf{1}_{|w_{ij}|\geqslant K})+2\left|\mathbb{E}\left(w_{ij}\hat{w}_{ij}\mathbf{1}_{|w_{ij}|\geqslant K}\right)\right|\\
&\leqslant \epsilon_3+(K+\epsilon_1)^2\epsilon_2+2\sqrt{\epsilon_3\max\{1,\sigma^2\}}\,,\\
&< 4\epsilon^2\,;
\end{aligned}$$

here the upper bound for the last (absolute value of) an expectation comes from the Cauchy-Schwartz inequality, together with the fact that $\mathbb{E}(\hat{w}_{ij}^2) \leqslant \max\{1, \sigma^2\}$.

This is then sufficient to conclude that

$$P\left[\frac{1}{n}\mathrm{tr}(\overline{W}_n-\widehat{\overline{W}}_n)^2>\epsilon\right]<4\epsilon\,.$$

On the complementary event, $\frac{1}{n}\mathrm{tr}(\overline{W}_n-\widehat{\overline{W}}_n)^2 \leqslant \epsilon$, picking f Lipschitz with Lipschitz constant C, we obtain that

$$\begin{aligned}
\left|\langle f_{\overline{W}_n},g\rangle-\langle f_{\widehat{\overline{W}}_n},g\rangle\right| &\leqslant C\frac{1}{n}\sum_{i=1}^{n}|\lambda_i(\overline{W}_n)-\lambda_i(\widehat{\overline{W}}_n)|\\
&\leqslant C\sqrt{\frac{1}{n}\mathrm{tr}(\overline{W}_n-\widehat{\overline{W}}_n)^2}\leqslant C\sqrt{\epsilon}\,.
\end{aligned}$$

Note that the first of the two inequalities above is due to the triangle inequality, while the second is a result of the Cauchy-Schwartz inequality.

The conclusion is that

$$P\left(\left|\langle f_{\overline{W}_n},g\rangle-\langle f_{\widehat{\overline{W}}_n},g\rangle\right|>C\sqrt{\epsilon}\right)<4\epsilon\,,$$

and thus the ESD of $f_{\overline{W}_n}$ converges weakly to the same limit as that of $f_{\widehat{\overline{W}}_n}$: the semicircle law.

3.3. Additional notes and context. With a bit more work (as we will see in Section 4), the variance of the moments of the ESDs can be shown to be of order $O(1/n^2)$. This fact, as hinted at in [2], is sufficient to prove almost sure convergence to the semicircle (e.g., [4]). The Hoffman-Wielandt theorem is a celebrated perturbation theory result in linear algebra whose proof uses the Birkhoff theorem on extremal points of the convex set of doubly stochastic matrices (for a detailed explanation see [23]).

3.4. Problems.

Problem 3.4.1. In this problem, you might use the following theorem.

Theorem 3.4.2 (Cauchy Interlacing Theorem). *Let* A *be a Hermitian matrix of order* n *and* B *be a principal submatrix of* A *of order* $n-1$. *If* $\lambda_n(A) \leqslant \lambda_{n-1}(A) \leqslant \cdots \leqslant \lambda_1(A)$ *are the eigenvalues of* A *and* $\lambda_{n-1}(B) \leqslant \lambda_{n-2}(B) \leqslant \cdots \leqslant \lambda_1(B)$ *are those of* B, *then*

$$\lambda_{i+1}(A) \leqslant \lambda_i(B) \leqslant \lambda_i(A),$$

for all $1 \leqslant i \leqslant n-1$.

(1) If $A = B + uu^*$, where u is a complex column vector of dimension n, B is a positive definite Hermitian matrix of order n, show that

$$\lambda_{i+1}(A) \leqslant \lambda_i(B) \leqslant \lambda_i(A),$$

for all $1 \leqslant i \leqslant n-1$. Hint: you might use the following matrix identity:

$$\begin{bmatrix} B^{\frac{1}{2}} \\ u^* \end{bmatrix} \cdot \begin{bmatrix} B^{\frac{1}{2}} & u \end{bmatrix} = \begin{bmatrix} B & B^{\frac{1}{2}}u \\ u^*B^{\frac{1}{2}} & u^*u \end{bmatrix}.$$

(2) Let A, B be two Hermitian matrices. Let $N_I(A)$ be the number of eigenvalues of A in the interval I. Show that

$$|N_I(A+B) - N_I(A)| \leqslant 1$$

for any interval $I \subset \mathbb{R}$ if $\operatorname{rank}(B) = 1$.

(3) Prove the following rank inequality: Let A, B be two $n \times n$ Hermitian matrices. Then

$$\|F^A - F^B\| \leqslant \frac{1}{n}\operatorname{rank}(A-B),$$

where F^A is the cumulative function of the empirical spectral distribution of A and $\|f\| = \sup_x |f(x)|$.

Problem 3.4.3. Given a real number $p = p(n), 0 \leqslant p \leqslant 1$, the Erdös-Rényi random graph $G(n,p)$ of n vertices is obtained by connecting each pair of vertices randomly and indepedently with probability p. In this problem, we will prove the semicircle law these graphs.

Let G be a graph on n vertices, the adjacency matrix A of G is a $n \times n$ matrix with $a_{ij} = 1$ if there is an edge between vertices i and j and $a_{ij} = 0$ otherwise.

We assume $np \to \infty$ and $\sup_n p(n) < 1$. Let A_n be the adjacency matrix of $G(n,p)$, J_n be the $n \times n$ matrix all of whose entries are 1, $\sigma^2 = p(1-p)$ and $M_n = \frac{A_n - pJ_n}{\sigma}$.

(1) Show that

$$\lim_{n\to\infty} \frac{1}{n}\mathbb{E}\left[\operatorname{tr}\left(\frac{M_n}{\sqrt{n}}\right)^{2k}\right] = C_k,$$

$$\lim_{n\to\infty} \frac{1}{n}\mathbb{E}\left[\operatorname{tr}\left(\frac{M_n}{\sqrt{n}}\right)^{2k+1}\right] = 0,$$

where C_k is the k-th Catalan number.

(2) Prove that the empirical spectral distribution of $\frac{A_n}{\sqrt{n}\sigma}$ converges in distribution to the semicircle law.

4. Beyond the Semicircle Law: A Central Limit Theorem

So far, we have seen two ways in which the semicircle law is true, for a large class of ensembles (satisfying Assumption 3.2.1). There is yet another way to look at the Wigner matrix ESDs convergence to the semicircle law, once that may bring a more familiar perspective: the interpretation of this convergence as a *Law of Large Numbers*.

Assume that we have a sequence of random Wigner matrices satisfying Assumption 3.2.1, and let's revisit the convergence in probability issue: we can state it as "for any $g \in C_b(\mathbb{R})$, asymptotically almost surely (i.e., with probability tending to 1 as $n \to \infty$),

$$\frac{1}{n}\sum_{i=1}^{n} g(\lambda_i(\overline{W}_n)) - \int_{-2}^{2} g(x)d\sigma(x) \to 0 .$$

Recall the weak Law of Large Numbers: given $X_1, X_2, \ldots, X_n, \ldots$, an infinite sequence of independent samples from a distribution with mean μ, asymptotically almost surely,

$$\frac{1}{n}\sum_{i=1}^{n} X_i - \mu \to 0 \text{ , as } n \to \infty .$$

The parallels between the matrix case and the classical case should be apparent, even though there are some obvious differences: the $\lambda_i(\overline{W}_n)$ are not independent and are not identically distributed. Given this parallel though, the next question may be how large the fluctuations from this law are, i.e., to continue the trail of thought, is there a *Central Limit Theorem* equivalent? Again, recall that the "classical" version of the Central Limit Theorem says the following: given an infinite sequence $X_1, X_2, \ldots, X_n, \ldots$ of independent samples from a distribution with mean μ and variance σ^2, then

$$\frac{\sum_{i=1}^{n} X_i}{\sqrt{n}\sigma} \xrightarrow{d} N(0,1) ,$$

where the convergence above is in distribution. (Recall that the proof of the Central Limit Theorem, or CLT, is one of the first instances of the moment method use.)

We will show that something akin to the CLT is true for the spectra of Wigner matrices, more specifically, for the linear statistics of eigenvalues of Wigner matrices. Given a smooth function f, we define the (centered) linear statistic

$$X_{n,f} = \operatorname{tr}(f(\overline{W}_n)) - \mathbb{E}(\operatorname{tr}(f(\overline{W}_n))) . \tag{4.0.1}$$

Under various assumptions for the distributions of the variables and for the smoothness of the function, one may show that

$$X_{n,f} \xrightarrow{d} N(0, \sigma_f^2) ,$$

where the definition of σ_f^2 can be made precise in terms of Fourier coefficients of f; moreover, the covariances can be computed explicitly. Thus, the linear statistics in the limit define a Gaussian process on the line (see, e.g., [24] in the case of β-Hermitian ensembles). Of note is the fact that the basis of functions which diagonalizes the covariance matrix is the Chebyshev polynomials T_n; as it turns out, the variables X_{n,T_k} become independent in the limit as $n \to \infty$.

Although this phenomenon can be shown for various classes of smooth functions f (e.g., analytic, in the context of Wishart matrices [6]; differentiable or Lipschitz [3]; belonging to a Sobolev fractional space [24, 29]), we will focus here on proving it for f being a polynomial.

Definition 4.0.2. The following definitions serve to provide a compact form for expressions that will appear frequently in formulae for asymptotic covariances.

(1) For positive r and non-negative s, denote by $\mathcal{P}_r(s)$ the set of all weak compositions or partitions κ of s with r parts: $\kappa = (\kappa_1, \ldots, \kappa_r)$ with $\sum_{i=1}^{r} \kappa_i = s$ and each $\kappa_i \geqslant 0$. Then set

$$\mathbf{P}_r(s) = \sum_{\kappa \in \mathcal{P}_r(s)} \prod_{i=1}^{r} C_i .$$

(2) For any positive integer m, define $\mathcal{I}(m) = \{r \equiv m \mod (2) \mid 3 \leqslant r \leqslant m\}$ and then, whenever $k + \ell$ is even, set

$$\mathbf{S}(k, \ell) = \sum_{r \in \mathcal{I}(\frac{k+\ell}{2})} \frac{2k\ell}{r} \mathbf{P}_r\left(\frac{k-r}{2}\right) \mathbf{P}_r\left(\frac{l-r}{2}\right) .$$

Note, in particular, that

$$\mathbf{S}(k, k) = \sum_{r \in \mathcal{I}(k)} \frac{2k^2}{r} \mathbf{P}_r\left(\frac{k-r}{2}\right)^2 .$$

Theorem 4.0.3. *Let $\{W_n\}_n$ be a sequence of Wigner matrices with Assumption 2.1.2. Let $f(x) = x^k$ and define $X_{n,k} = X_{n,f}$ as given above. Then*

$$X_{n,k} \xrightarrow{d} N(0, \sigma_k^2) ,$$

where σ_k is independent of n. Specifically

$$\sigma_k^2 = \begin{cases} \frac{k^2}{2} C_{\frac{k}{2}}^2 \left(\mathbb{E}(w_{12}^4) - 1\right) + \mathbf{S}(k, k), & \text{if } k \text{ is even;} \\ k^2 C_{\frac{k-1}{2}}^2 \mathbb{E}(w_{11}^2) + \mathbf{S}(k, k), & \text{if } k \text{ is odd.} \end{cases}$$

Remark 4.0.4. Note that, unlike the "classical" CLT, there is no scaling of the centered variable by $\sqrt{n}$. In the classical case, the variables that are averaged are independent samples, each with $O(1)$ variance. Hence it is to be expected that the total variance will be $O(\sqrt{n})$. In the case of linear statistics of Wigner matrices, besides the fact that the eigenvalues are not independent, there is a lot less variance in the individual eigenvalues (they stay very close to their "classical locations", which are deterministic [17]).

Proof. To prove Theorem 4.0.3, we will once again use the method of moments. $X_{n,k}$ is a centered variable, and thus what we need is to show that its moments converge to the moments of a centered normal variable of variance σ_k^2. We will first show how to calculate the variance, and then extend this to all moments.

We have already examined the variance when we proved that the moments of the ESDs converge, although then we were examining the variance not of $X_{n,k}$, but that of $\text{tr}(\overline{W}_n^k)$.

Note that the variance is

$$\mathbb{E}\left[\left(\text{tr}(\overline{W}_n^k) - \mathbb{E}[\text{tr}((\overline{W}_n^k)]\right)^2\right] = \frac{1}{n^k}\sum_{I,J} \text{Cov}[w_I w_J] \ ,$$

where I and J are as in Section 2.

We are thus examining

$$\sum_{I,J \in \mathcal{I}} (\mathbb{E}(w_I w_J) - \mathbb{E}(w_I)\mathbb{E}(w_J)) \ ,$$

where $\mathcal{I}$ is the set of all k-tuples. This will involve looking at the graphs G_I, G_J, their disjoint union $G_{I \cup J}$, as well as the walks on G_I and G_J given by the words w_I and w_J.

Recall that, as the variables w_{ij} are centered, in order for a pair of words (w_I, w_J) to have highest-order contribution, each edge in the graph $G_{I \cup J}$ must be present at least twice in the join of walks. Thus, it follows that if v is the number of vertices and e is the number of edges in $G_{I \cup J}$, $v - 1 \leqslant e \leqslant k$.

For the same reasons as in Section 2 (given v labeled vertices, the number of pairs of graphs/words/walks one may construct on them is independent of n, while the choices for labels is $(n_v) := n(n-1)\dots(n-v+1) = n^v(1+o(1))$), we can write

$$\begin{aligned}\frac{1}{n^k}\sum_{I,J\in\mathcal{I}} (\mathbb{E}(w_I w_J) - \mathbb{E}(w_I)\mathbb{E}(w_J)) &= \frac{1}{n^k}\sum \prod_{\ell=n-v+1}^{n} \ell \cdot Q(k,v) \ , \\ &= \frac{1}{n^k}\sum_{v=1}^{k+1} n^v(1+o(1))\cdot Q(k,v) \ ,\end{aligned}$$

where $Q(k,v)$ is independent of n (and thus bounded).

Hence it makes sense to find the largest contributor to this sum, and at first glance that would be $Q(k, k+1)$. However, that turns out to be 0, as we see below.

Consider a term in which $v = k+1$. Each edge must be walked on exactly once. This implies that $G_{I \cup J}$ is a tree, and hence G_I and G_J must also be trees. But since w_I and w_J represent closed walks on trees, this implies that each edge in $G_{I \cup J}$ is walked on either zero or two times in the walks represented by w_I, respectively w_J. Hence there can be no overlap between G_I and G_J, or else the overlapping edge (walked on twice in the join of walks) would have to be walked on once in G_I and once in G_J.

But if there is no overlap between G_I and G_J, it follows that the words w_I and w_J are independent, hence $\text{Cov}[w_I w_J] = 0$. Thus, terms with $v = k+1$ have zero contribution to the calculation of the variance, and hence $Q(k, k+1) = 0$.

We have to now turn our attention to the next-highest term, which is $Q(k,k)$ (that is, $v = k$).

Case 1: k even. We have seen that there must be some overlap between G_I and G_J. As $v-1 \leqslant e \leqslant k$, and $v = k$, the possibilities for e (the number of edges in $G_{I \cup J}$) yield the following cases.

Case 1a): $e = k = v$. Then the graph $G_{I \cup J}$ has only one cycle or a loop, and each edge (including the loop, if it exists) is walked on exactly twice.

Also, for such a pair of words to produce a non-zero contribution, there must be overlap. Suppose there is a loop (corresponding to the presence of a w_{ii} term in w_I or in w_J). Since both w_I and w_J describe closed walks on a tree, or on a tree with a loop, and every edge is walked on twice, the only possibility for overlap is that the overlap is the loop (which would be walked on once in each walk). But this would imply that k is odd; contradiction. Thus the contribution in the case when the graph $G_{I \cup J}$ is a tree with a loop is zero.

Hence the only case to analyze here is that of a "bracelet" graph with dangling tree "charms" (see Figure 4.0.5). In this case, since every edge must be walked on exactly twice, while the two walks are closed walks on a tree or a bracelet graph, the overlap must contain the entire cycle (and nothing else). Note also that in this case the cycle will be walked on once in each graph, and hence for reasons of parity its length must be even.

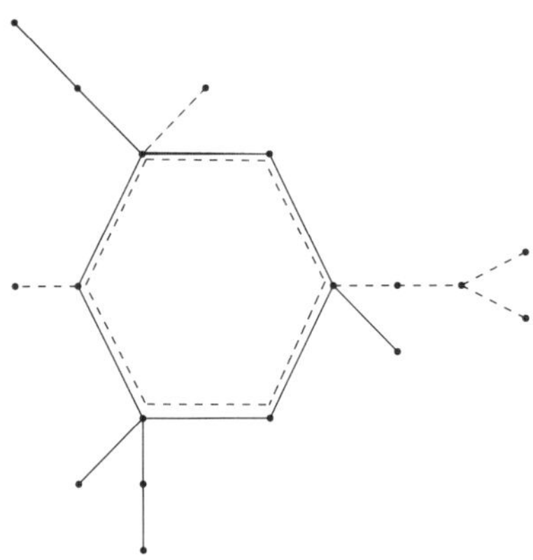

FIGURE 4.0.5. Two "bracelet" graphs whose cycles overlap and the dangling tree "charms" are different; here the length of the cycle is 6 and the total length of either of the two closed walks is 18.

Let $4 \leqslant r \leqslant k$ be the (even) length of the cycle in $G_{I \cup J}$.

The dangling trees divide in two subsets, each with the same number of edges; half, i.e., $(k-r)/2$, belong to G_I, the others to G_J. We can think of each vertex of the cycle as having its own pair $(T_{i,I}, T_{i,J})$ of trees, possibly empty, one from I, the other from J. Call their edge sizes $k_{i,I}$ and $k_{i,J}$.

One must have

$$\sum_{i=1}^{r} k_{i,I} = \sum_{i=1}^{r} k_{i,J} = \frac{k-r}{2} .$$

For each pair of partitions of $(k-r)/2$, one has a total number of possible combinations of trees which is

$$\prod_{i=1}^{r} C_{k_{i,I}} \prod_{i=1}^{r} C_{k_{i,J}} .$$

Recall that the correspondence between the walk and the tree is a bijection given by depth-first search.

Given a pair of labeled bracelet graphs which share the cycle and have the same number of vertices and edges, we can make the following choices for "gluing together" closed walks w_I and w_J:

- there are k^2 of ways of choosing starting vertices in both walks;
- there are 2 choices of direction along the cycle (same for both walks, or opposite); and
- a rotational invariance overcount. Up to this point, each pair of walks would be counted exactly once in each simultaneous cyclic permutation of $(k_{1,I}, k_{2,I}, \ldots, k_{r,I})$ and $(k_{1,J}, k_{2,J}, \ldots, k_{r,J})$ (and nowhere else), so we must divide by r.

Summing this over all possible choices of labels as well as the even cycle length r, we obtain $\mathbf{S}(k,k)$. Since in this case each edge is walked on exactly twice, the multiplicative contribution of each edge to the covariance is $\mathbb{E}(w_{12}^2) = 1$, so the count above is exact.

Case 1b): $e = k-1 = v-1$. In this case the graph must be a tree in which either one edge has multiplicity 4, or two edges have multiplicity 3. The latter case is impossible, because a multiplicity of 3 for an edge has to mean that in one of the two walks, the edge is walked on once or three times; this is impossible in a closed walk on a tree.

Thus, the only possibility remaining in this case is that $G_{I\cup J}$ is a tree with an edge walked on four times, and since both walks given by w_I and w_J are closed walks on a tree, and since they overlap, this means that the overlap has to be on that specific edge, and that both walks are simply closed walks on a tree with each edge walked on exactly twice.

Thus, the graph $G_{I\cup J}$ represents a "gluing together" over one edge of two trees, each with $k/2$ edges. The choices then are

- $C_{k/2}^2$ for a pair of trees with $k/2$ edges;
- $(k/2)^2$ over which edge of each tree one chooses to glue together, and
- 2, for the orientation of the gluing.

Hence in this case the contribution is

$$2\left(\frac{k}{2}\right)^2 C_{k/2}^2 \left(\mathbb{E}(w_{12}^4) - 1\right) ,$$

the last term representing the presence of the term $\mathbb{E}(w_e^4) - (\mathbb{E}(w_e^2))^2$ in the covariance, where e stands in for the labels of the vertices defining the edge along which one glues, and taking into account that $\mathbb{E}(w_e^2) = 1$. (Recall that all other edges have multiplicative contribution 1 to the expectation.)

Putting this all together, it follows that in Case 1 (k even), the variance converges to σ_k^2, as given in Theorem 4.0.3.

Figure 4.0.6 shows examples of pairs of walks with non-negligible contribution.

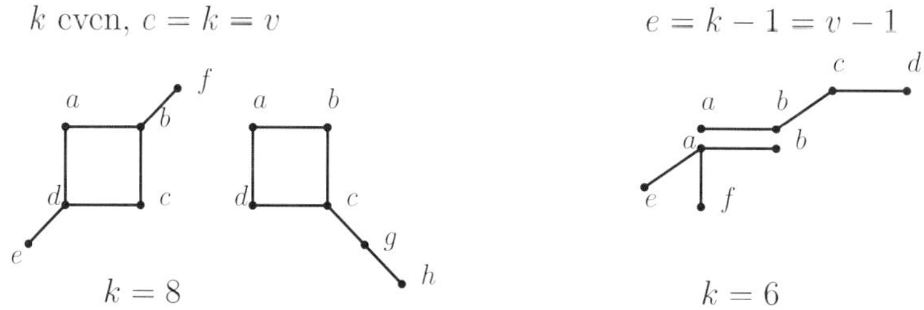

FIGURE 4.0.6. The above are possible examples of walks with non-negligible contribution for k even, for both cases 1a) and 1b). For 1a), on the left, one may take the pairs $I_1 = (e, d, a, b, f, b, c, d, e)$ and $I_2 = (a, b, c, g, h, g, c, d, a)$. For 1b), on the right, one may take $I_1 = (e, a, f, a, b, a, e)$ and $I_2 = (d, c, b, a, b, c, d)$.

Case 2: k odd. Recall that the graphs G_I and G_J must overlap, else the contribution from such terms to the covariance is 0. As $v = k$, the possibilities for e, the number of edges in $G_{I \cup J}$, are as follows.

Case 2a): $e = k = v$. The graph has a single cycle or a loop, and every edge (or loop) is walked on exactly twice.

If the graph has one cycle, the reasoning and the count are just as in the case when k is even, except that now the cycle must be of odd length to match the parity of k and we again obtain $\mathbf{S}(k.k)$.

If the graph has a loop, for the same reasons as before, the overlap must happen exactly on the loop, and now we have two trees with one loop each, glued on the loop. The number of edges in either tree must thus be $(k-1)/2$.

The choices for this case are:

- a choice of 2 trees with $(k-1)/2$ edges; and
- a choice of k spots where to insert the loop in each tree.

Once these choices are made, the two vertices where the loop was placed are then identified (the loops are glued). This provides a $k^2 C_{(k-1)/2}^2$ contribution to the covariance. This contribution comes with the special weight $\mathbb{E}(w_{11}^2)$, since we have not made assumptions over what that is, and also because in the covariance expression, the existence of the multiplicity-one loop means that in this case $\mathbb{E}(w_I) = \mathbb{E}(w_J) = 0$.

In total, this means that such walks contribute $k^2 C_{(k-1)/2}^2 + \mathbf{S}(k, k)$ to the covariance. The count above is exact due to the fact that the multiplicative contribution to the covariance, for each edge, is $\mathbb{E}(w_{12}^2) = 1$ (since each edge is walked on exactly twice).

Case 2b): $e = v - 1 = k - 1$. In this case the graph $G_{I \cup J}$ must be a tree; either one of the edges has multiplicity 4 (is walked on 4 times), or there are exactly two edges of multiplicity 3. In either case all the other edges have multiplicity 2.

Neither of these two cases are in fact possible; since the graph $G_{I \cup J}$ is a tree, both graphs G_I and G_J must also be trees, and since the words w_I and w_J induce closed walks on these trees, their lengths k must be even; but k is odd.

To conclude, it follows that in Case 2 (k odd), the asymptotical value of the variance is once again given by Theorem 4.0.3.

Figure 4.0.7 shows two examples of walks with non-negligible contribution.

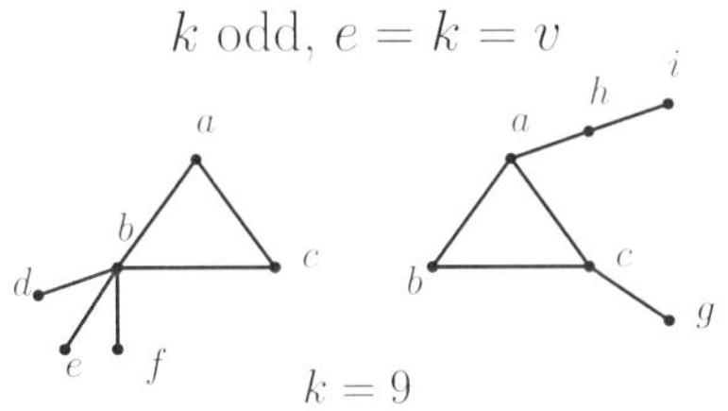

FIGURE 4.0.7. The above are possible examples of walks with non-negligible contribution for k odd, case 2a) (as noted above, case 2b) is not possible). One may take $I_1 = (d, b, a, c, b, f, b, e, b, d)$ and $I_2 = (b, a, h, i, h, a, c, g, c, b)$.

To prove Theorem 4.0.3, one does not only need the variance, but also the higher moments. However, higher moments reduce to the variance calculation, as we show below.

For higher moments, we are interested in the quantity

$$\mathbb{E}\left[\left(\overline{W}_n^k - \mathbb{E}(\overline{W}_n^k)\right)^l\right] = \mathbb{E}\left[\frac{1}{n^{kl/2}} \sum_{I_1, I_2, \dots I_l} \prod_{j=1}^{l} (w_{I_j} - \mathbb{E}(w_{I_j}))\right] \tag{4.0.8}$$

Theorem 4.0.9. *With the notation above, for any positive integer* l,

$$\mathbb{E}\left[\left(\overline{W}_n^k - \mathbb{E}(\overline{W}_n^k)\right)^l\right] \to \begin{cases} 0\,, & \text{if } l \text{ is odd,} \\ (\sigma_k^2)^{l/2}(l-1)!!\,, & \text{if } l \text{ is even.} \end{cases}$$

The crux of the proof reduces to Lemma 2.1.34 from [2] which we give here without proof using the notations given there, i.e., I_j is the k-tuplet of indices corresponding to the word w_j, G_{I_j} is the graph that corresponds to both, etc.

Lemma 4.0.10 (Lemma 2.1.34, [2]). *For an l-tuplet of words $(w_1, w_2, \dots, w_l)$, consider the union graph $G_{I_1 \cup I_2 \cup \dots \cup I_l}$. Let c be the number of components in this graph, and v be the number of vertices. Then*

$$c \leqslant \lfloor l/2 \rfloor\,, \text{ and } v \leqslant \lfloor (k+1)l/2 \rfloor + c - l\,.$$

Although we do not prove this lemma here, we will offer some intuition for it. In addition to the graph $G_{I_1 \cup I_2 \cup \dots \cup I_l}$, consider also the dependency graph H among these l words, i.e., the graph with vertex set $[l]$, and where an edge is drawn between vertices a and b if the word w_a and w_b are not independent, i.e., if G_{I_a} and G_{I_b} share edges. Note that c is also the number of connected components in H. Moreover, note that if there are any isolated vertices in H, then the corresponding word is independent from all the others, and hence, by the same argument as when $l = 2$, the contribution to the covariance from such an l-tuplet of words is zero.

As there are no isolated vertices, the number of components is therefore at most $l/2$. That takes care of the first part of the Lemma above. The second part is more subtle. Consider a spanning forest for H; the Lemma says that any edge in the spanning forest causes at least another vertex choice to be lost. If there were no edges (all vertices isolated), the requirement that each edge in $G_{I_1 \cup I_2 \cup \ldots \cup I_l}$ is repeated would yield a total of $\lfloor (k+1)l/2 \rfloor$ vertices; as the number of connected components is actually c, the number of edges in the spanning forest is $l - c$ and hence that is a lower bound on the number of vertex choices lost. Overall, this says that the total number of vertices is at most $v \leqslant \lfloor (k+1)l/2 \rfloor + c - l$.

Just as before, in order to find the highest-term order, we need to maximize the number of vertex choices. The Lemma above tell us that the maximal number of vertices happens when $c = \lfloor l/2 \rfloor$, which can only happen for l even, and which indicates that the dependency graph H is perfect matching. The number of vertices (again, maximal!) in this case is precisely $kl/2$, for a total order of vertex choices of $n^{kl/2}$, and this cancels precisely the power of n present in front of the sum of (4.0.8). It follows that any other l-tuples $(I_1, \ldots, I_l)$ will have negligible contribution, as $n \to \infty$.

Now that we know the only case that is not asymptotically negligible is that when H is a matching, we can see that there are $(l-1)!!$ ways of realizing that matching using the indices of the words $w_1, \ldots, w_l$, that the contributions from each pair of words matched are all independent, and that each contribution is given by the corresponding calculation we have shown for $l = 2$ (for a pair of words). Hence the total contribution to the variance will be asymptotically given by $(\sigma_k^2)^{l/2}(l-1)!!$, and Theorem 4.0.9 is proven. This shows that all the moments of $X_{n,k}$ converge asymptotically to the moments of $N(0, \sigma_k^2)$, and the proof of Theorem 4.0.3 is complete. □

Remark 4.0.11. Theorem 4.0.3, together with the covariance calculation that the reader is guided through in Problem 4.2.1, are sufficient to conclude that, for any collection of numbers $\{k_1, k_2, \ldots, k_m\}$, the variables $(X_{n,k_1}, X_{n,k_2}, \ldots, X_{n,k_m})$ converge to a multivariate Gaussian with covariance given by the formulae of Problem 4.2.1.

4.1. Additional notes and context. One of the first results in the study of central limit theorem for traces of powers of random matrices is [25], and this remains the "classical" introduction to the subject; we have relied on it in part here as well.

Extensions of the CLT from monomials (corresponding to traces of powers) to analytic functions can be obtained via [32, 33], or, through Stieltjes functions methods rather than moments, can be found in [7] for Wigner matrices and in [6, 31] for Wishart matrices. In [3] the distribution of the entries is generalized to softer moment conditions (distributions satisfy log-Sobolev inequalities) and linear statistics of Lipschitz, differentiable functions. For other classes of matrices (i.e., general β-Hermitian ensembles), the paper [24] uses analytical methods to obtain an explicit, diagonalized version of the CLT which works for linear statistics of functions in a fractional Sobolev space. For specific kinds of tridiagonal β-Laguerre and β-Jacobi ensembles, the method of moments has been worked out in [14] and [15].

Finally, one of the most extensive generalizations to fractional Sobolev spaces can be found in [29].

4.2. Problems.

Problem 4.2.1. In this problem we will compute the asymptotic covariance for $X_{n,k}$ and $X_{n,l}$ for $k, l \geqslant 1$.

Let $\{W_n\}_n$ be a sequence of Wigner matrices $W_n = (w_{ij})_{1\leqslant i,j\leqslant n}$. Following the notations of this section, let

$$\sigma_{kl} = \lim_{n\to\infty} \mathrm{Cov}\,[X_{n,k}, X_{n,l}] = \lim_{n\to\infty} \frac{1}{n^{(k+l)/2}} \sum_{I,J} \mathrm{Cov}[w_I w_J]\,,$$

where $I = (i_1, i_2, \ldots, i_k)$, $J = (j_1, j_2, \ldots, j_l)$, $w_I = w_{i_1 i_2} \cdots w_{i_{k-1} i_k} w_{i_k i_1}$. and $w_J = w_{j_1 j_2} \cdots w_{j_{l-1} j_l} w_{j_l j_1}$.

Let $G_I, G_J, G_{I\cup J}$ be the corresponding graphs of the closed walks I, J and their disjoint union, and let v, e be the number of vertices and edges of $G_{I\cup J}$.

(1) Show that if $k+l$ is odd, the number of vertices in any $G_{I\cup J}$ with nonzero contribution is strictly smaller than $(k+l)/2$; thus the number of vertex choices for is not high enough to overcome the scaling by $n^{(k+l)/2}$, and hence the limiting covariance in this case is 0.

(2) Now assume $k+l$ is even. To compute the nonzero terms in the limit, show that it is enough to count the contribution of graphs $G_{I\cup J}$ with $v = \frac{k+l}{2}$, and $v-1 \leqslant e \leqslant v$. (Hint: for nonzero terms in the limit, show that $e \leqslant k$. Since G must be connected, $e \geqslant v-1$.)

(3) Assume $v = \frac{k+l}{2}$, and that k, l are even.
   (a) If $e = v = \frac{k+l}{2}$, show that $G_{I\cup J}$ has only one cycle and no loops.
   (b) Let r be the length of the cycle, $4 \leqslant r \leqslant \frac{k+l}{2}$. Show that the contribution of all possible graphs $G_{I\cup J}$ with $e = v = \frac{k+l}{2}$ to $\sum_{I,J} \mathrm{Cov}[w_I, w_J]$ is $n^{(k+l)/2}(1+o(1))\,\mathbf{S}(k,l)$.
   (c) If $e = v-1 = \frac{k+l}{2} - 1$, show that the contribution of all possible graphs $G_{I\cup J}$ to the covariance, in the limit, is

$$\frac{kl}{2} C_{k/2} C_{l/2} (\mathbb{E} w_{12}^4 - 1).$$

(d) Conclude that for even k, l,

$$\sigma_{kl} = \frac{kl}{2} C_{k/2} C_{l/2} (\mathbb{E} w_{12}^4 - 1) + \mathbf{S}(k, l) .$$

(4) Assume $v = \frac{k+l}{2}$, and that k, l are odd.

(a) Show that for all graphs $G_{I \cup J}$ with non-negligible asymptotic contribution, $e = v = \frac{k+l}{2}$, therefore $G_{I \cup J}$ must have exactly a cycle or a loop.

(b) If $G_{I \cup J}$ has a cycle, show that the contribution of all possible such $G_{I \cup J}$ in the limit is $\mathbf{S}(k, l)$.

(c) For all graphs $G_{I \cup J}$ with a loop, show that the asymptotic contribution is $klC_{(k-1)/2} C_{(l-1)/2}$

(d) Conclude that for odd k, l,

$$\sigma_{kl} = klC_{(k-1)/2} C_{(l-1)/2} \, \mathbb{E} w_{11}^2 + \mathbf{S}(k, l) .$$

5. One more dimension: minor processes and the Gaussian Free Field

In this last section we show how to add one more layer of complexity on understanding the shape of Wigner spectra.

One may think of Theorem 2.2.2 as a "Law of Large Numbers", of sorts, for the ESDs (as opposed to studying the fluctuations, which leads to the "Central Limit Theorem"). This convergence of a sequence of random distributions (the ESDs) to a deterministic one (the semicircle law) can be thought of as defining a zero-dimensional process. By contrast, convergence of the centered linear statistics to a Gaussian process on the line can be seen as a one-dimensional phenomenon. To add one more dimension, then, one has to consider more than just one matrix; one must study a collection of overlapping, large Wigner matrices. Investigating joint eigenvalues distributions (or statistics) of overlapping matrices is not a new phenomenon and has gone by the names of "corner process" or "minor process", (see [12], [18]).

The simplest way to think of these overlapping Wigner matrices is as follows. Suppose we have an infinite, symmetric double array of real variables satisfying the conditions of Assumption 2.1.2 (moments of all kinds). Suppose further that all the off-diagonal variables are centered, independent, and identically distributed, and that all the diagonal variables are centered, independent, and identically distributed. Suppose further that

$$\mathbb{E}(w_{11}^2) = 2 \, , \mathbb{E}(w_{12}^2) = 1 \, , \mathbb{E}(w_{12}^4) = 3 \, .$$

Then any finite principal submatrix of this array is a Wigner matrix satisfying Assumption 2.1.2 and in addition having fixed variances on the diagonal and fixed 4th moment on the off-diagonal. These last two moment assumptions will be necessary, as they ensure that these moments match those of the Gaussian Orthogonal Ensemble and are crucial in obtaining the connection to the Gaussian Free Field.

Nota Bene: if instead of real variables one considers complex or quaternion ones, the 4th moment must be fixed to 2, respectively, 3/2—to match the Gaussian Unitary, respectively, the Gaussian Symplectic Ensembles.

Let L be a (large) integer. We will be interested in examining the $L \times L$ upper left corner of this infinite array; more specifically, we will consider its principal submatrices.

Given some finite integer k, we extract, for $i = 1, 2, \ldots, k$, principal submatrices W_i that have sizes $n_i(L)$, and that overlap in pairs on $n_{ij}(L)$ rows and columns. Assume that $n_i(L)/L \to b_i$, $n_{ij}(L)/L \to c_{ij}$ as $L \to \infty$ ($b_i, c_{ij} \in [0,1]$ for all $1 \leqslant i, j \leqslant n$).

Unlike in the previous sections (see 4.0.1 and Theorem 4.0.3) we will now define $\overline{W}_i = \frac{1}{\sqrt{L}} W_i$, for $i = 1, 2, \ldots, k$ (the scaling, instead of being with respect to the size of W_i, is with respect to L). In the following, we study the limiting behavior of $(X_{L,p_1}^{\overline{W}_1}, X_{L,p_2}^{\overline{W}_2}, \ldots, X_{L,p_k}^{\overline{W}_k})$ for any set of k natural numbers $p_1, \ldots, p_k$. Here we define for suitable functions f

$$X_{L,f}^{\overline{W}_i} = \mathrm{tr}(f(\overline{W}_i)) - \mathbb{E}(\mathrm{tr}(f(\overline{W}_i))), \tag{5.0.1}$$

and for integers p_i we use the notation

$$X_{L,p_i}^{\overline{W}_i} = \mathrm{tr}((\overline{W}_i)^{p_i}) - \mathbb{E}(\mathrm{tr}((\overline{W}_i)^{p_i})),$$

similar with the definition 4.0.1.

The theorem below is due to Alexei Borodin and it has appeared, in a more general form, in [10]. We note that the definition of $x(z)$ (and $x(w)$) appearing below is given explicitly in the change of coordinates Ω^{-1} explained in Figure 5.1.7.

Theorem 5.0.2. *The set of random variables* $(X_{L,p_1}^{\overline{W}_1}, X_{L,p_2}^{\overline{W}_2}, \ldots, X_{L,p_k}^{\overline{W}_k})$ *converges in distribution and with all the moments to the zero-mean* k*-dimensional random variable* $(\xi_1, \xi_2, \ldots, \xi_k)$ *with covariance* $\mathbb{E}[\xi_i \xi_j]$ *given by*

$$\frac{2p_ip_j}{\pi} \oint\limits_{\substack{|z|^2=b_i\\ \Im z>0}} \oint\limits_{\substack{|w|^2=b_i\\ \Im w>0}} (x(z))^{p_i-1}((x(w))^{p_j-1} \frac{1}{2\pi} \ln \left| \frac{c_{ij} - zw}{c_{ij} - z\overline{w}} \right| \frac{dx(z)}{dz} \frac{dx(w)}{dw} \, dz \, dw.$$

The proof of Theorem 5.0.2 follows closely in the footsteps of that of Theorem 4.0.3, generalizing the one-matrix calculation we have done there; since in the limit the moments are Gaussian, this is sufficient to show convergence in distribution. A smart use of basic complex analysis yields the double contour integral over semicircles.

Remark 5.0.3. Two things of note in the formula above: first, the integration involves the *derivatives* of the monomials, and second, the logarithmic term suggest a strong connection to the Gaussian Free Field (see Section 5.1). The first of these is an indication that by looking at the process defined on $X_{L,p_i}^{\overline{W}_i}$ we are not quite considering the right quantity; this will be introduced in the next section as the height function.

5.1. The Gaussian Free Field, the Height Function, and a Pullback. The following is a very brief introduction to the Gaussian Free Field.

Definition 5.1.1. The Gaussian Free Field (GFF) on $U \subset \mathbb{C}$ is a random distribution (or random generalized function) F_U which associates to every smooth function g a centered Gaussian variable $\langle F_U, g\rangle$, with covariances given by

$$\mathbb{E}(\langle F_U, g_1\rangle, \langle F_U, g_2\rangle) = \int_U \int_U g_1(z) G(z,w) g_2(w) dz dw \,,$$

where G is the Green's kernel for the Laplacian operator on U with Dirichlet boundary conditions (zero boundary).

When we talk about the Gaussian Free Field on $\mathbb{H}$, we will denote it by F.

Remark 5.1.2. An alternate way to understand F_U is to consider the following formal expansion:

$$F_U = \sum_k \xi_k \frac{\varphi_k}{\sqrt{\lambda_k}} \,,$$

where φ_k are the eigenfunctions of $-\Delta$ on U with 0 boundary conditions, λ_k are the corresponding eigenvalues, and ξ_k are independent, identically distributed standard Gaussians.

Note that F_U is not a random *function*, because its variance is infinite; however $\langle F_U, g\rangle$ is a centered Gaussian variable with finite variance.

Notably, if one picks $U = \mathbb{H}$, the upper half plane, then the Green's function is

$$G(z,w) = -\frac{1}{2\pi} \log \left| \frac{z-w}{z-\overline{w}} \right| \,. \tag{5.1.3}$$

Another way to look at the variance of the GFF is through the lense of the Dirichlet inner product. It's an easy calculation following from Remark 5.1.2 that, when one considers the Dirichlet inner product on smooth functions in U

$$\langle f, g\rangle_{\nabla, U} := \frac{1}{2\pi} \int_U \nabla f \cdot \nabla g \,, \tag{5.1.4}$$

the variance of $\langle F_U, g\rangle$ can be shown to be $2\pi \langle f, g\rangle_{\nabla, U}$.

All of the above are useful facts to know about the GFF on a domain, but to make sense of the formula of Theorem 5.0.2, we need to also understand how it works under composition with maps changing the domain.

Given Ω a bijection from U to $\mathbb{H}$, the composition $F \circ \Omega$ is the generalized Gaussian field on U with covariance given by the kernel

$$G_U(z,w) = -\frac{1}{2\pi} \log \left| \frac{\Omega(z) - \Omega(w)}{\Omega(z) - \Omega(\overline{w})} \right| \,,$$

and integrals of $F \circ \Omega$ with respect to measures $d\mu$ are obtained as

$$\int_U F \circ \Omega d\mu = \int_{\mathbb{H}} F d\Omega(\mu) \,,$$

where $F \circ \Omega$ is the *pullback of* F *by* Ω.

As we see, there is already a strong suggestion in the formula of Theorem 5.0.2 of a pullback of F; to understand the nature of this pullback, we need to define the height function and see what domain is being mapped onto the upper half plane.

We start with defining the generic height function.

Definition 5.1.5. For any matrix W and $I \subset \mathbb{R}$, let N_I^W be the number of eigenvalues of W in I. Consider now the $L \times L$ array of variables we described above, and let W_y be the upper $\lfloor y \rfloor \times \lfloor y \rfloor$ left corner of the array.

Define the (generic) height function on $\mathbb{R} \times \mathbb{R}_{\geqslant 1}$:

$$H(x,y) = \sqrt{\frac{\pi}{2}}\, N_{[x,\infty)}(W_y)\,. \tag{5.1.6}$$

(In general, the scaling by $\sqrt{\frac{\pi}{2}}$ does not appear as part of the definition, but it will allow for "cleaner" definitions of domains.)

The height function is a step function, with jumps at the eigenvalues of W_y for every y. Since W_y is a Wigner matrix, we expect that for y large (say $y = O(L)$) the jumps (which occur at the eigenvalues of W_y) will take place at $x = O(\sqrt{L})$. We will therefore scale the domain to reflect these facts: $(x,y) \mapsto \left(\frac{1}{\sqrt{L}}x, \frac{1}{L}y\right)$, and so the domain changes from $\mathbb{R} \times \mathbb{R}_{\geqslant 1}$ to $\mathbb{R} \times \mathbb{R}_{>0}$, the upper half plane $\mathbb{H}$.

More precisely, we can say that for L large, $x \sim \sqrt{L}$ and $y \sim L$, after rescaling, the jumps (or places of growth) in the height function will, with ovewhelming probability, be concentrated inside the parabola $y = x^2/4$, or–if one prefers–the domain

$$U := \{(x,y) \in \mathbb{R} \times \mathbb{R}_{>0} \mid -2\sqrt{y} \leqslant x \leqslant 2\sqrt{y}\}\,.$$

One may construct the bijection

$$\Omega \,:\, (x,y) \mapsto \frac{x}{2} + i\sqrt{y - \left(\frac{x}{2}\right)^2}\,,$$

which maps the domain U onto $\mathbb{H}$. Its inverse is

$$\Omega^{-1}(z) = (x(z), y(z)) = (2\Re z, |z|^2)\,.$$

See Figure 5.1.7 for a depiction of this bijection and its inverse.

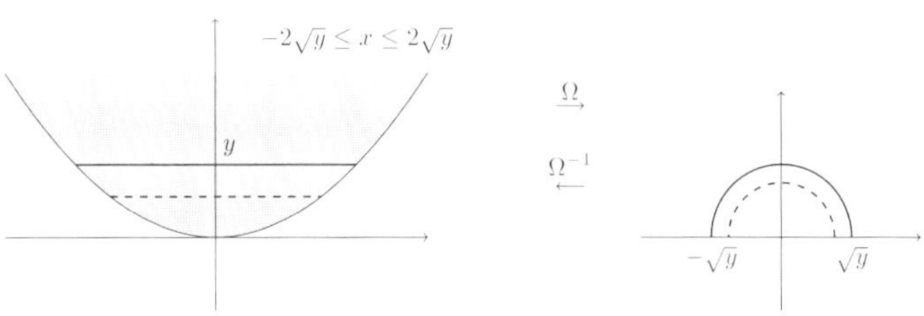

FIGURE 5.1.7. The bijection $\Omega \,:\, (x,y) \mapsto \frac{x}{2} + i\sqrt{y - (\frac{x}{2})^2}$ mapping $U = \{(x,y) \in \mathbb{R} \times \mathbb{R}_{>0} \mid -2\sqrt{y} \leqslant x \leqslant 2\sqrt{y}\}$ onto $\mathbb{H}$, as well as its inverse $\Omega^{-1} = (x(z), y(z)) = (2\Re z, |z|^2)$. This map allows us to define as in Theorem 5.1.9 convergence of the moments of the height function to moments of the pullback of the GFF by Ω^{-1}.

This map allows us to define a rescaled version of the height function which is defined on $\mathbb{H}$.

Definition 5.1.8. For z in the upper half-plane, define the rescaled and centered height function as

$$H^{\Omega}(z) = H(\sqrt{L}x(z), Ly(z)) - \mathbb{E}\left(H(\sqrt{L}x(z), Ly(z))\right),$$

using the inverse map Ω^{-1}.

As a consequence of Theorem 5.0.2, one has then the following.

Theorem 5.1.9. *All finite mixed moments of $H^{\Omega}(z)$ converge in the sense of finite dimensional distributions to the mixed moments of a corresponding set of pullbacks by Ω^{-1} of pairwise correlated Gaussian Free Fields, with covariances given by the kernels*

$$C_{ij}(z,w) = \frac{1}{2\pi}\ln\left|\frac{c_{ij} - zw}{c_{ij} - z\bar{w}}\right|,$$

following the notations above.

Remark 5.1.10. The existence of a family of correlated Gaussian Free Fields each defined on $\mathbb{H}$ with Dirichlet boundary conditions and with covariances given as above is not hard to see and can also be found in [10].

Finally, one more interesting way to look at it involves a slight tweak to the height function. Consider the centered version of the $H(x,y)$ given by (5.1.6):

$$\overline{H}(x,y) = \sqrt{\frac{2}{\pi}}H(x,y) - \mathbb{E}\left(\sqrt{\frac{2}{\pi}}H(x,y)\right),$$

With the definition above, we can understand now why, in the calculation of the covariances from Theorem 5.0.2, the functions under the integral are the derivatives of the monomials whose covariances are being computed.

Lemma 5.1.11. *Let $y > 0$ and for any $x \in \mathbb{R}$. For suitable f (e.g., f compactly supported or f polynomial),*

$$\int_{\mathbb{R}} f'(x)\overline{H}(x,y)dx = X_{L,f}^{\overline{W}_y},$$

where $X_{L,f}^{\overline{W}_y}$ follows (5.0.1).

The proof for the above follows immediately by doing integration by parts. Once one takes into acount that $\overline{H}(x,y) \to 0$ as $|x| \to \infty$, together with concentration properties of the Wigner matrices, one may extend this to $C_b(\mathbb{R})$ functions.

5.2. Additional Notes and Context. In [11], nested GOE/GUE submatrices executing a Dyson Brownian motion are connected to a 3-dimensional generalized Gaussian process; the proof uses the method of moments and extends to Wigner matrices and their stochastic evolution. Following [10], the GFF has been found in other contexts and for other types of random matrices, e.g. [16] (which relaxes moment conditions of entry distributions to existence of $4+\epsilon$ moments and extends convergence to a class of functions in a fractional Sobolev space, following [29]), [20] (which deals with adjacency matrices of random regular graphs), [12] (which examines β-Jacobi matrices), etc.

5.3. Problems.

Problem 5.3.1. Let μ be a positive, finite measure on the real line. The Stieltjes transform of μ is the function

$$S_\mu(z) = \int_{\mathbb{R}} \frac{\mu(dx)}{x-z}, z \in \mathbb{C} \setminus \mathbb{R}.$$

Let $\hat{C}(z) := 1 + \sum_{k=1}^{\infty} z^k C_k$, where C_k is the k-th Catalan number, and define

$$\sigma(x) = \frac{1}{2\pi}\sqrt{4-x^2}\mathbf{1}_{|x|\leqslant 2}.$$

Show that for $|z| < \frac{1}{4}$,

$$\hat{C}(z) = \frac{-1}{\sqrt{z}} S_\sigma\left(\frac{1}{\sqrt{z}}\right).$$

Problem 5.3.2. Let $U \subset \mathbb{R}^2$ be a bounded domain, and let $\mathcal{D}(U)$ be the set of compactly supported, C^∞ functions in D. For $f, g \in \mathcal{D}(U)$, define the Dirichlet inner product by (5.1.4), and show that it is conformally invariant: if $\psi : U \to U'$ is a conformal map, then for all $f, g \in \mathcal{S}(U')$,

$$\int_{U'} \nabla f \cdot \nabla g = \int_U \nabla(f \circ \psi) \cdot \nabla(g \circ \psi).$$

6. Acknowledgements

I would like to thank PCMI and the organizers of the 2017 PCMI Summer School on Random Matrices for the opportunity to lecture on this material and have it included in what is likely to be a key reference book for students of Random Matrix Theory.

I am very grateful to Ivan Corwin and the publishers for showing me tremendous patience, and to an anonymous referee for the careful reading of these notes and the insightful comments offered.

Finally, special thanks go to my Teaching Assistant Yizhe Zhu for all his help, including closely proofreading these notes, drawing figures and working with me on choosing problems.

References

[1] M. Abramowitz and I. Stegun, *Handbook of mathematical functions with formulas, graphs, and mathematical tables*, National Bureau of Standards Applied Mathematics Series, vol. 55, For sale by the Superintendent of Documents, U.S. Government Printing Office, Washington, D.C., 1964. MR0167642 ↑44

[2] G. Anderson, A. Guionnet, and O. Zeitouni, *An introduction to random matrices*, Cambridge Studies in Advanced Mathematics, vol. 118, Cambridge University Press, Cambridge, 2010. MR2760897 ↑42, 44, 55, 63

[3] G. Anderson and O. Zeitouni, *A CLT for a band matrix model*, Probab. Theory Related Fields **134** (2006), no. 2, 283–338. MR2222385 ↑41, 48, 58, 65

[4] L. Arnold, *On Wigner's semicircle law for the eigenvalues of random matrices*, Z. Wahrscheinlichkeitstheorie und Verw. Gebiete **19** (1971), 191–198. MR0348820 ↑41, 55

[5] Z. D. Bai, *Methodologies in spectral analysis of large-dimensional random matrices, a review*, Statist. Sinica **9** (1999), no. 3, 611–677. With comments by G. J. Rodgers and Jack W. Silverstein; and a rejoinder by the author. MR1711663 ↑41

[6] Z. D. Bai and J. W. Silverstein, *CLT for linear spectral statistics of large-dimensional sample covariance matrices*, Ann. Probab. **32** (2004), no. 1A, 553–605. MR2040792 ↑41, 58, 65

[7] Z. D. Bai and J. Yao, *On the convergence of the spectral empirical process of Wigner matrices*, Bernoulli **11** (2005), no. 6, 1059–1092. MR2189081 ↑65

[8] Z. Bai and J. Silverstein, *Spectral analysis of large dimensional random matrices*, Second, Springer Series in Statistics, Springer, New York, 2010. MR2567175 ↑44

[9] C. Bordenave, *A new proof of friedman's second eigenvalue theorem and its extension to random lifts*, 2015, pp. 1–45. ↑49

[10] A. Borodin, *CLT for spectra of submatrices of Wigner random matrices*, Mosc. Math. J. **14** (2014), no. 1, 29–38, 170. MR3221945 ↑67, 70

[11] A. Borodin, *CLT for spectra of submatrices of Wigner random matrices, II: Stochastic evolution*, Random matrix theory, interacting particle systems, and integrable systems, 2014, pp. 57–69. MR3380682 ↑70

[12] A. Borodin and V. Gorin, *General β-Jacobi corners process and the Gaussian free field*, Comm. Pure Appl. Math. **68** (2015), no. 10, 1774–1844. MR3385342 ↑66, 71

[13] F. Chung, L. Lu, and V. Vu, *The spectra of random graphs with given expected degrees*, Internet Math. **1** (2004), no. 3, 257–275. MR2111009 ↑41, 48

[14] I. Dumitriu and A. Edelman, *Global spectrum fluctuations for the β-Hermite and β-Laguerre ensembles via matrix models*, J. Math. Phys. **47** (2006), no. 6, 063302, 36. MR2239975 ↑41, 48, 65

[15] I. Dumitriu and E. Paquette, *Global fluctuations for linear statistics of β-Jacobi ensembles*, Random Matrices Theory Appl. **1** (2012), no. 4, 1250013, 60. MR3039374 ↑41, 48, 65

[16] I. Dumitriu and E. Paquette, *Spectra of overlapping Wishart matrices and the Gaussian free field*, Random Matrices Theory Appl. **7** (2018), no. 2, 1850003, 21. MR3786884 ↑70

[17] L. Erdős and H.-T. Yau, *Universality of local spectral statistics of random matrices*, Bull. Amer. Math. Soc. (N.S.) **49** (2012), no. 3, 377–414. MR2917064 ↑58

[18] P. Forrester and E. Nordenstam, *The anti-symmetric GUE minor process*, Mosc. Math. J. **9** (2009), no. 4, 749–774, 934. MR2663989 ↑66

[19] Z. Füredi and J. Komlós, *The eigenvalues of random symmetric matrices*, Combinatorica **1** (1981), no. 3, 233–241. MR637828 ↑48

[20] S. Ganguly and S. Pal, *The random transposition dynamics on random regular graphs and the gaussian free field* (2015), 1–43 pp., available at 1409.7766v2. ↑71

[21] U. Grenander, *Probabilities on algebraic structures*, John Wiley & Sons, Inc., New York-London; Almqvist & Wiksell, Stockholm-Göteborg-Uppsala, 1963. MR0206994 ↑41

[22] A. Guionnet, *Large deviations upper bounds and central limit theorems for non-commutative functionals of Gaussian large random matrices*, Ann. Inst. H. Poincaré Probab. Statist. **38** (2002), no. 3, 341–384. MR1899457 ↑41

[23] R. Horn and C. Johnson, *Matrix analysis*, Second, Cambridge University Press, Cambridge, 2013. MR2978290 ↑55

[24] K. Johansson, *On fluctuations of eigenvalues of random Hermitian matrices*, Duke Math. J. **91** (1998), no. 1, 151–204. MR1487983 ↑41, 48, 57, 58, 65

[25] D. Jonsson, *Some limit theorems for the eigenvalues of a sample covariance matrix*, J. Multivariate Anal. **12** (1982), no. 1, 1–38. MR650926 ↑41, 64

[26] V. A. Marčenko and L. A. Pastur, *Distribution of eigenvalues in certain sets of random matrices*, Mat. Sb. (N.S.) **72 (114)** (1967), 507–536. MR0208649 ↑41, 48

[27] B. McKay, *The expected eigenvalue distribution of a large regular graph*, Linear Algebra Appl. **40** (1981), 203–216. MR629617 ↑41, 48

[28] M. L. Mehta, *Random matrices*, Third, Pure and Applied Mathematics (Amsterdam), vol. 142, Elsevier/Academic Press, Amsterdam, 2004. MR2129906 ↑42

[29] M. Shcherbina, *Central limit theorem for linear eigenvalue statistics of orthogonally invariant matrix models*, Zh. Mat. Fiz. Anal. Geom. **4** (2008), no. 1, 171–195, 204. MR2404179 ↑58, 65, 71

[30] D. Shlyakhtenko, *Random Gaussian band matrices and freeness with amalgamation*, Internat. Math. Res. Notices **20** (1996), 1013–1025. MR1422374 ↑41

[31] J. Silverstein and Z. D. Bai, *On the empirical distribution of eigenvalues of a class of large-dimensional random matrices*, J. Multivariate Anal. **54** (1995), no. 2, 175–192. MR1345534 ↑41, 65

[32] Ya. Sinaĭ and A. Soshnikov, *Central limit theorem for traces of large random symmetric matrices with independent matrix elements*, Bol. Soc. Brasil. Mat. (N.S.) **29** (1998), no. 1, 1–24. MR1620151 ↑49, 65

[33] Ya. Sinaĭ and A. Soshnikov, *A refinement of Wigner's semicircle law in a neighborhood of the spectrum edge for random symmetric matrices*, Funktsional. Anal. i Prilozhen. **32** (1998), no. 2, 56–79, 96. MR1647832 ↑49, 65

[34] S. Sodin, *Random matrices, nonbacktracking walks, and orthogonal polynomials*, J. Math. Phys. **48** (2007), no. 12, 123503, 21. MR2377835 ↑49

[35] A. Soshnikov, *Universality at the edge of the spectrum in Wigner random matrices*, Comm. Math. Phys. **207** (1999), no. 3, 697–733. MR1727234 ↑49

[36] R. Stanley, *Enumerative combinatorics. Vol. 2*, Cambridge Studies in Advanced Mathematics, vol. 62, Cambridge University Press, Cambridge, 1999. With a foreword by Gian-Carlo Rota and appendix 1 by Sergey Fomin. MR1676282 ↑44, 47

[37] L. Tran, V. Vu, and K. Wang, *Sparse random graphs: eigenvalues and eigenvectors*, Random Structures Algorithms **42** (2013), no. 1, 110–134. MR2999215 ↑41

[38] H. F. Trotter, *Eigenvalue distributions of large Hermitian matrices; Wigner's semicircle law and a theorem of Kac, Murdock, and Szegö*, Adv. in Math. **54** (1984), no. 1, 67–82. MR761763 ↑41

[39] E. P. Wigner, *Characteristic vectors of bordered matrices with infinite dimension*, Ann. of Math. (3) **62** (1955), no. 3, 548–564. MR0083848 ↑41, 42

[40] E. P. Wigner, *On the distribution of the roots of certain symmetric matrices*, Ann. of Math. (2) **67** (1958), 325–327. MR0095527 ↑41, 42

[41] J. Wishart, *The generalized product moment distribution in samples from a normal multivariate population*, Biometrika A **20** (1928), 32–43. ↑42

Department of Mathematics, University of Washington, Seattle, WA 98195
Email address: dumitriu@uw.edu

IAS/Park City Mathematics Series
Volume 26, Pages 75–158
https://doi.org/10.1090/pcms/026/00844

The Matrix Dyson Equation and its Applications for Random Matrices

László Erdős

Abstract. These lecture notes are a concise introduction of recent techniques to prove local spectral universality for a large class of random matrices. The general strategy is presented following the recent book with H.T. Yau (Erdős and Yau, 2017). We extend the scope of this book by focusing on new techniques developed to deal with generalizations of Wigner matrices that allow for non-identically distributed entries and even for correlated entries. This requires to analyze a system of nonlinear equations, or more generally a nonlinear matrix equation called the *Matrix Dyson Equation (MDE)*. We demonstrate that stability properties of the MDE play a central role in random matrix theory. The analysis of MDE is based upon joint works with J. Alt, O. Ajanki, D. Schröder and T. Krüger that are supported by the ERC Advanced Grant, RANMAT 338804 of the European Research Council.

Contents

1 Introduction 76
1.1 Random matrix ensembles 78
1.2 Eigenvalue statistics on different scales 81
2 Tools 90
2.1 Stieltjes transform 90
2.2 Resolvent 93
2.3 The semicircle law for Wigner matrices via the moment method 94
3 The resolvent method 96
3.1 Probabilistic step 97
3.2 Deterministic stability step 104
4 Models of increasing complexity 106
4.1 Basic setup 106
4.2 Wigner matrix 107
4.3 Generalized Wigner matrix 108
4.4 Wigner type matrix 109

2010 *Mathematics Subject Classification.* Primary 15B52; Secondary 82B44.
Key words and phrases. Park City Mathematics Institute, Random matrix, Matrix Dyson Equation, local semicircle law, Dyson sine kernel, Wigner-Dyson-Mehta conjecture, Tracy-Widom distribution, Dyson Brownian motion.
Partially supported by ERC Advanced Grant, RANMAT 338804.

4.5 Correlated random matrix 111
4.6 The precise meaning of the approximations 113
5 Physical motivations 115
5.1 Basics of quantum mechanics 115
5.2 The "grand" universality conjecture for disordered quantum systems 116
5.3 Anderson model 118
5.4 Random band matrices 120
5.5 Mean field quantum Hamiltonian with correlation 121
6 Results 122
6.1 Properties of the solution to the Dyson equations 122
6.2 Local laws for Wigner-type and correlated random matrices 129
6.3 Bulk universality and other consequences of the local law 131
7 Analysis of the vector Dyson equation 139
7.1 Existence and uniqueness 139
7.2 Bounds on the solution 140
7.3 Regularity of the solution and the stability operator 144
7.4 Bound on the stability operator 146
8 Analysis of the matrix Dyson equation 148
8.1 Properties of the solution to the MDE 148
8.2 The saturated self-energy matrix 149
8.3 Bound on the stability operator 150
9 Ideas of the proof of the local laws 153
9.1 Structure of the proof 153
9.2 Probabilistic part of the proof 154
9.3 Deterministic part of the proof 154

1. Introduction

> *"Perhaps I am now too courageous when I try to guess the distribution of the distances between successive levels (of energies of heavy nuclei). Theoretically, the situation is quite simple if one attacks the problem in a simpleminded fashion. The question is simply what are the distances of the characteristic values of a symmetric matrix with random coefficients."*
>
> Eugene Wigner on the Wigner surmise, 1956

The cornerstone of probability theory is the fact that the collective behavior of many independent random variables exhibits universal patterns; the obvious examples are the *law of large numbers (LLN)* and the *central limit theorem (CLT)*. They assert that the normalized sum of N independent, identically distributed (i.i.d.) random variables $X_1, X_2, \ldots, X_N \in \mathbb{R}$ converge to their common expectation value:

$$\frac{1}{N}(X_1 + X_2 + \ldots + X_N) \to \mathbb{E}X_1 \tag{1.0.1}$$

as $N \to \infty$, and their centered average with a $\sqrt{N}$ normalization converges to the centered Gaussian distribution with variance $\sigma^2 = \text{Var}(X)$:

$$S_N := \frac{1}{\sqrt{N}} \sum_{i=1}^{N} (X_i - \mathbb{E}X_i) \Longrightarrow \mathcal{N}(0, \sigma^2).$$

The convergence in the latter case is understood in distribution, i.e. tested against any bounded continuous function Φ:

$$\mathbb{E}\Phi(S_N) \to \mathbb{E}\Phi(\xi),$$

where ξ is an $\mathcal{N}(0, \sigma^2)$ distributed normal random variable.

These basic results directly extend to random vectors instead of scalar valued random variables. The main question is: what are their analogues in the non-commutative setting, e.g. for matrices? Focusing on their spectrum, what do eigenvalues of typical large random matrices look like? Is there a deterministic limit of some relevant random quantity, like the average in case of the LLN (1.0.1). Is there some stochastic *universality pattern* arising, similarly to the ubiquity of the Gaussian distribution in Nature owing to the central limit theorem?

These natural questions could have been raised from pure curiosity by mathematicians, but historically random matrices first appeared in statistics (Wishart in 1928 [79]), where empirical covariance matrices of measured data (samples) naturally form a random matrix ensemble and the eigenvalues play a crucial role in principal component analysis. The question regarding the universality of eigenvalue statistics, however, appeared only in the 1950's in the pioneering work [78] of Eugene Wigner. He was motivated by a simple observation looking at data from nuclear physics, but he immediately realized a very general phenomenon in the background. He noticed from experimental data that gaps in energy levels of large nuclei tend to follow the same statistics irrespective of the material. Quantum mechanics predicts that energy levels are eigenvalues of a self-adjoint operator, but the correct Hamiltonian operator describing nuclear forces was not known at that time. Instead of pursuing a direct solution of this problem, Wigner appealed to a phenomenological model to explain his observation. His pioneering idea was to model the complex Hamiltonian by a random matrix with independent entries. All physical details of the system were ignored except one, the *symmetry type*: systems with time reversal symmetry were modeled by real symmetric random matrices, while complex Hermitian random matrices were used for systems without time reversal symmetry (e.g. with magnetic forces). This simple-minded model amazingly reproduced the correct gap statistics. Eigenvalue gaps carry basic information about possible excitations of the quantum systems. In fact, beyond nuclear physics, random matrices enjoyed a renaissance in the theory of disordered quantum systems, where the spectrum of a non-interacting electron in a random impure environment was studied. It turned out that eigenvalue statistics is one of the basic signatures of the celebrated *metal-insulator*, or *Anderson* transition in condensed matter physics [13].

1.1. Random matrix ensembles Throughout these notes we will consider $N \times N$ square matrices of the form

$$H = H^{(N)} = \begin{pmatrix} h_{11} & h_{12} & \dots & h_{1N} \\ h_{21} & h_{22} & \dots & h_{2N} \\ \vdots & \vdots & & \vdots \\ h_{N1} & h_{N2} & \dots & h_{NN} \end{pmatrix}. \tag{1.1.1}$$

The entries are real or complex random variables constrained by the symmetry

$$h_{ij} = \bar{h}_{ji}, \qquad i, j = 1, \dots, N,$$

so that $H = H^*$ is either Hermitian (complex) or symmetric (real). In particular, the eigenvalues of H, $\lambda_1 \leqslant \lambda_2 \leqslant \dots \leqslant \lambda_N$ are real and we will be interested in their statistical behavior induced by the randomness of H as the size of the matrix N goes to infinity. Hermitian symmetry is very natural from the point of view of physics applications and it makes the problem much more tractable mathematically. Nevertheless, there has recently been an increasing interest in non-hermitian random matrices as well motivated by systems of ordinary differential equations with random coefficients arising in biological networks (see, e.g. [11,38] and references therein).

There are essentially two customary ways to define a probability measure on the space of $N \times N$ random matrices that we now briefly introduce. The main point is that either one specifies the distribution of the matrix elements directly or one aims at a basis-independent measure. The prototype of the first case is the Wigner ensembles and we will be focusing on its natural generalizations in these notes. The typical example of the second case are the invariant ensembles. We will briefly introduce them now.

1.1.2. Wigner ensemble The most prominent example of the first class is the traditional **Wigner matrix**, where the matrix elements h_{ij} are i.i.d. random variables subject to the symmetry constraint $h_{ij} = \overline{h_{ji}}$. More precisely, Wigner matrices are defined by assuming that

$$\mathbb{E} h_{ij} = 0, \qquad \mathbb{E}|h_{ij}|^2 = \frac{1}{N}. \tag{1.1.3}$$

In the *real symmetric case*, the collection of random variables $\{h_{ij} : i \leqslant j\}$ are independent, identically distributed, while in the *complex hermitian case* the distributions of $\{\mathrm{Re}\, h_{ij}, \mathrm{Im}\, h_{ij} : 1 \leqslant i < j \leqslant N\}$ and $\{\sqrt{2} h_{ii} : i = 1, 2, \dots, N\}$ are independent and identical.

The common variance of the matrix elements is the single parameter of the model; by a trivial rescaling we may fix it conveniently. The normalization $1/N$ chosen in (1.1.3) guarantees that the typical size of the eigenvalues remain of order 1 even as N tends to infinity. To see this, we may compute the expectation of the trace of H^2 in two different ways:

$$\mathbb{E}\sum_i \lambda_i^2 = \mathbb{E}\,\mathrm{Tr}\,H^2 = \mathbb{E}\sum_{ij} |h_{ij}|^2 = N \tag{1.1.4}$$

indicating that $\lambda_i^2 \sim 1$ on average. In fact, much stronger bounds hold and one can prove that

$$\|H\| = \max_i |\lambda_i| \to 2, \qquad N \to \infty,$$

in probability.

In these notes we will focus on Wigner ensembles and their extensions, where we will drop the condition of identical distribution and we will weaken the independence condition. We will call them *Wigner type* and *correlated ensembles*. Nevertheless, for completeness we also present the other class of random matrices.

1.1.5. Invariant ensembles The ensembles in the second class are defined by the measure

$$\mathbb{P}(H)\,dH := Z^{-1}\exp\big(-\frac{\beta}{2}N\,\mathrm{Tr}\,V(H)\big)\,dH. \tag{1.1.6}$$

Here $dH = \prod_{i\leqslant j} dh_{ij}$ is the flat Lebesgue measure on $\mathbb{R}^{N(N+1)/2}$ (in case of complex Hermitian matrices and $i < j$, dh_{ij} is the Lebesgue measure on the complex plane $\mathbb{C}$ instead of $\mathbb{R}$). The (potential) function $V : \mathbb{R} \to \mathbb{R}$ is assumed to grow mildly at infinity (some logarithmic growth would suffice) to ensure that the measure defined in (1.1.6) is finite. The parameter β distinguishes between the two symmetry classes: $\beta = 1$ for the real symmetric case, while $\beta = 2$ for the complex hermitian case – for traditional reason we factor this parameter out of the potential.

Finally, Z is the normalization factor to make $P(H)\,dH$ a probability measure. Similarly to the normalization of the variance in (1.1.3), the factor N in the exponent in (1.1.6) guarantees that the eigenvalues remain order one even as $N \to \infty$. This scaling also guarantees that empirical density of the eigenvalues will have a deterministic limit without further rescaling.

Probability distributions of the form (1.1.6) are called **invariant ensembles** since they are invariant under the orthogonal or unitary conjugation (in case of symmetric or Hermitian matrices, respectively). For example, in the Hermitian case, for any fixed unitary matrix U, the transformation

$$H \to U^*HU$$

leaves the distribution (1.1.6) invariant thanks to $\mathrm{Tr}\,V(U^*HU) = \mathrm{Tr}\,V(H)$ and that $d(U^*HU) = dH$.

An important special case is when V is a quadratic polynomial, after shift and rescaling we may assume that $V(x) = \frac{1}{2}x^2$. In this case

$$\begin{aligned}\mathbb{P}(H)\,dH &= Z^{-1}\exp\Big(-\frac{\beta}{4}N\sum_{ij}|h_{ij}|^2\Big)\,dH \\ &= Z^{-1}\prod_{i<j}\exp\Big(-\frac{\beta}{2}N|h_{ij}|^2\Big)\,dh_{ij}\prod_{i}\exp\Big(-\frac{\beta}{4}Nh_{ii}^2\Big)\,dh_{ii},\end{aligned}$$

i.e. the measure factorizes and it is equivalent to independent Gaussians for the matrix elements. The factor N in the definition (1.1.6) and the choice of β ensure that we recover the normalization (1.1.3). (A pedantic reader may notice that the normalization of the diagonal element for the real symmetric case is off by a factor of 2, but this small discrepancy plays no role.) The invariant Gaussian ensembles, i.e. (1.1.6) with $V(x) = \frac{1}{2}x^2$, are called **Gaussian orthogonal ensemble (GOE)** for the real symmetric case ($\beta = 1$) and **Gaussian unitary ensemble (GUE)** for the complex hermitian case ($\beta = 2$).

Wigner matrices and invariant ensembles form two different universes with quite different mathematical tools available for their studies. In fact, these two classes are almost disjoint because the Gaussian ensembles are the only invariant Wigner matrices. This is the content of the following lemma:

Lemma 1.1.7 ([28] or Theorem 2.6.3 [65]). *Suppose that the real symmetric or complex Hermitian matrix ensembles given in* (1.1.6) *have independent entries* h_{ij}, $i \leqslant j$. *Then* $V(x)$ *is a quadratic polynomial,* $V(x) = ax^2 + bx + c$ *with* $a > 0$. *This means that apart from a trivial shift and normalization, the ensemble is GOE or GUE.*

The significance of the Gaussian ensembles is that they allow for explicit calculations that are not available for Wigner matrices with general non-Gaussian single entry distribution. In particular the celebrated **Wigner-Dyson-Mehta correlation functions** can be explicitly obtained for the GOE and GUE ensembles. Thus the typical proof of identifying the eigenvalue correlation function for a general matrix ensemble goes through universality: one first proves that the correlation function is independent of the distribution, hence it is the same as GUE/GOE, and then, in the second step, one computes the GUE/GOE correlation functions. This second step has been completed by Gaudin, Mehta and Dyson in the 60's by an ingenious calculation, see e.g. the classical treatise by Mehta [65].

One of the key ingredients of the explicit calculations is the surprising fact that the joint (symmetrized) density function of the eigenvalues, $p(\lambda_1, \lambda_2, \dots, \lambda_N)$ can be computed explicitly for any invariant ensemble. It is given by

$$p_N(\lambda_1, \lambda_2, \dots, \lambda_N) = \text{const.}\prod_{i<j}(\lambda_i - \lambda_j)^\beta e^{-\frac{\beta}{2}N\sum_{j=1}^N V(\lambda_j)}. \tag{1.1.8}$$

where the constant ensures the normalization, but its exact value is typically unimportant.

Remark 1.1.9. In other sections of these notes we usually label the eigenvalues in increasing order so that their probability density, denoted by $\widetilde{p}_N(\boldsymbol{\lambda})$, is defined on the set

$$\Xi^{(N)} := \{\lambda_1 \leqslant \lambda_2 \leqslant \ldots \leqslant \lambda_N\} \subset \mathbb{R}^N.$$

For the purpose of (1.1.8), however, we dropped this restriction and we consider $p_N(\lambda_1, \lambda_2, \ldots, \lambda_N)$ to be a symmetric function of N variables, $\boldsymbol{\lambda} = (\lambda_1, \ldots, \lambda_N)$ on $\mathbb{R}^N$. The relation between the ordered and unordered densities is clearly $\widetilde{p}_N(\boldsymbol{\lambda}) = N!\, p_N(\boldsymbol{\lambda}) \cdot \mathbf{1}(\boldsymbol{\lambda} \in \Xi^{(N)})$.

The emergence of the *Vandermonde determinant* in (1.1.8) is a result of integrating out the "angle" variables in (1.1.6), i.e., the unitary matrix in the diagonalization of $H = U\Lambda U^*$. This is a remarkable formula since it gives a direct access to the eigenvalue distribution. In particular, it shows that the eigenvalues are strongly correlated. For example, no two eigenvalues can be too close to each other since the corresponding probability is suppressed by the factor $\lambda_j - \lambda_i$ for any $i \neq j$; this phenomenon is called the **level repulsion**. We remark that level repulsion also holds for Wigner matrices with smooth distribution [39] but its proof is much more involved.

In fact, one may view the ensemble (1.1.8) as a statistical physics question by rewriting p_N as a classical Gibbs measure of a N point particles on the line with a logarithmic mean field interaction:

$$p_N(\boldsymbol{\lambda}) = (\text{const.})e^{-\beta N \mathcal{H}(\boldsymbol{\lambda})} \tag{1.1.10}$$

with a Hamiltonian

$$\mathcal{H}(\boldsymbol{\lambda}) = \frac{1}{2}\sum_i V(\lambda_i) - \frac{1}{N}\sum_{i<j} \log|\lambda_j - \lambda_i|.$$

This ensemble of point particles with logarithmic interactions is also called **log-gas**. We remark that viewing the Gibbs measure (1.1.10) as the starting point and forgetting about the matrix ensemble behind, the parameter β does not have to be 1 or 2; it can be any positive number, $\beta > 0$, and it has the interpretation of the inverse temperature. We will not pursue general invariant ensembles in these notes.

1.2. Eigenvalue statistics on different scales The normalization both in (1.1.3) and (1.1.6) is chosen in such a way that the typical eigenvalues remain of order 1 even in the large N limit. In particular, the typical distance between neighboring eigenvalues is of order $1/N$. We distinguish two different scales for studying eigenvalues: *macroscopic* and *microscopic* scales. With our scaling, the macroscopic scale is order one and on this scale we detect the cumulative effect of cN eigenvalues with some positive constant c. In contrast, on the microscopic scales individual eigenvalues are detected; this scale is typically of order $1/N$. However, near the spectral edges, where the density of eigenvalues goes to zero, the typical eigenvalue spacing hence the microscopic scale may be larger. Some phenomena (e.g. fluctuations of linear statistics of eigenvalues) occur on various *mesoscopic* scales that lie between the macroscopic and the microscopic scales.

1.2.1. Eigenvalue density on macroscopic scales: global laws The first and simplest question is to determine the eigenvalue density, i.e. the behavior of the *empirical eigenvalue density* or **empirical density of states**

$$\mu_N(dx) := \frac{1}{N}\sum_i \delta(x-\lambda_i)\,dx \tag{1.2.2}$$

in the large N limit. This is a random measure, but under very general conditions it converges to a deterministic measure, similarly to self-averaging property encoded in the law of large numbers (1.0.1).

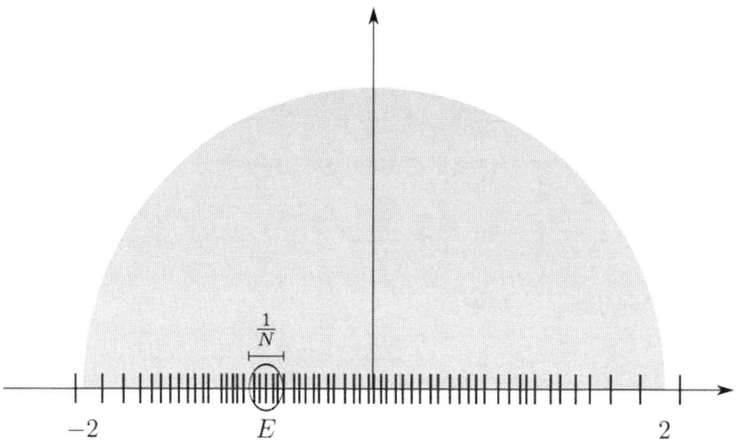

FIGURE 1.2.3. Semicircle law and eigenvalues of a GUE random matrix of size $N = 60$.

For Wigner ensemble, the empirical distribution of eigenvalues converges to the **Wigner semicircle law**. To formulate it more precisely, note that the typical spacing between neighboring eigenvalues is of order $1/N$, so in a fixed interval $[a, b] \subset \mathbb{R}$, one expects macroscopically many (of order N) eigenvalues. More precisely, it can be shown (first proof was given by Wigner [78]) that for any fixed $a \leqslant b$ real numbers,

$$\lim_{N\to\infty} \frac{1}{N}\#\{i : \lambda_i \in [a,b]\} = \int_a^b \rho_{sc}(x)\,dx, \qquad \rho_{sc}(x) := \frac{1}{2\pi}\sqrt{(4-x^2)_+}, \tag{1.2.4}$$

where $(a)_+ := \max\{a, 0\}$ denotes the positive part of the number a. Alternatively, one may formulate the Wigner semicircle law as the weak convergence in probability of the empirical distribution μ_N to the semicircle distribution, $\rho_{sc}(x)\,dx$. This means that the limit

$$\int_{\mathbb{R}} f(x)\mu_N(dx) = \frac{1}{N}\sum_i f(\lambda_i) \to \int_{\mathbb{R}} f(x)\rho_{sc}(x)\,dx, \qquad N\to\infty$$

holds in probability for any bounded continuous function f, i.e.,

$$\mathbb{P}\Big(\Big|\int_{\mathbb{R}} f(x)\mu_N(dx) - \int_{\mathbb{R}} f(x)\rho_{sc}(x)\,dx\Big| \geqslant \varepsilon\Big) \to 0$$

for any $\varepsilon > 0$ as $N \to \infty$.

Note that the emergence of the semicircle density is already a certain form of universality: the common distribution of the individual matrix elements is "forgotten"; the density of eigenvalues is asymptotically always the same, independently of the details of the distribution of the matrix elements.

We will see that for a more general class of Wigner type matrices with zero expectation but not identical distribution a similar limit statement holds for the empirical density of eigenvalues, i.e. there is a deterministic density function $\rho(x)$ such that

$$\int_{\mathbb{R}} f(x)\mu_N(dx) = \frac{1}{N}\sum_i f(\lambda_i) \to \int f(x)\rho(x)\,dx, \qquad N \to \infty \tag{1.2.5}$$

holds. The density function ρ thus approximates the empirical density, so we will call it **asymptotic density (of states).** In general it is not the semicircle density, but is determined by the second moments of the matrix elements and it is independent of other details of the distribution. For independent entries, the variance matrix

$$S = (s_{ij})_{i,j=1}^N, \qquad s_{ij} := \mathbb{E}|h_{ij}|^2 \tag{1.2.6}$$

contains all necessary information. For matrices with correlated entries, all relevant second moments are encoded in the linear operator

$$\mathcal{S}[R] := \mathbb{E}HRH, \qquad R \in \mathbb{C}^{N\times N}$$

acting on $N \times N$ matrices. It is one of the key questions in random matrix theory to compute the asymptotic density ρ from the second moments; we will see that the answer requires solving a system of nonlinear equations, that will be commonly called the **Dyson equation**. The explicit solution leading to the semicircle law is available only for Wigner matrices, or a little bit more generally, for ensembles with the property

$$\sum_j s_{ij} = 1 \qquad \text{for any } i. \tag{1.2.7}$$

These are called **generalized Wigner ensembles** and have been introduced in [46].

For invariant ensembles, the self-consistent density $\rho = \rho_V$ depends on the potential function V. It can be computed by solving a convex minimization problem, namely it is the unique minimizer of the functional

$$I(\nu) = \int_{\mathbb{R}} V(t)\nu(t)\,dt - \int_{\mathbb{R}}\int_{\mathbb{R}} \log|t-s|\nu(s)\nu(t)\,dt\,ds.$$

In both cases, under some mild conditions on the variances S or on the potential V, respectively, the asymptotic density ρ is compactly supported.

1.2.8. Eigenvalues on mesoscopic scales: local laws The Wigner semicircle law in the form (1.2.4) asymptotically determines the number of eigenvalues in a fixed interval $[a, b]$. The number of eigenvalues in such intervals is comparable with N. However, keeping in mind the analogy with the law of large numbers, it is natural to raise the question whether the same asymptotic relation holds if the length of the interval $[a, b]$ shrinks to zero as $N \to \infty$. To expect a deterministic answer, the interval should still contain many eigenvalues, but this would be guaranteed by $|b - a| \gg 1/N$. This turns out to be correct and the local semicircle law asserts that

$$\lim_{N\to\infty} \frac{1}{2N\eta} \#\{i : \lambda_i \in [E - \eta, E + \eta]\} = \rho_{sc}(E) \tag{1.2.9}$$

uniformly in $\eta = \eta_N$ as long as $N^{-1+\varepsilon} \leqslant \eta_N \leqslant N^{-\varepsilon}$ for any $\varepsilon > 0$ and E is not at the edge, $|E| \neq 2$. Here we considered the interval $[a, b] = [E - \eta, E + \eta]$, i.e. we fixed its center and viewed its length as an N-dependent parameter. (The N^{ε} factors can be improved to some $(\log N)$-power.)

1.2.10. Eigenvalues on microscopic scales: universality of local eigenvalue statistics Wigner's original observation concerned the distribution of the distances between consecutive (ordered) eigenvalues, or **gaps**. In the bulk of the spectrum, i.e. in the vicinity of a fixed energy level E with $|E| < 2$ in case of the semicircle law, the gaps have a typical size of order $1/N$ (at the spectral edge, $|E| = 2$, the relevant microscopic scale is of order $N^{-2/3}$, but we will not pursue edge behavior in these notes). Thus the corresponding rescaled gaps have the form

$$g_i := N\rho(\lambda_i)(\lambda_{i+1} - \lambda_i), \tag{1.2.11}$$

where ρ is the asymptotic density, e.g. $\rho = \rho_{sc}$ for Wigner matrices. Wigner predicted that the fluctuations of the gaps are universal and their distribution is given by a new law, the *Wigner surmise*. Thus there exists a random variable ξ, depending only on the symmetry class $\beta = 1, 2$, such that

$$g_i \Longrightarrow \xi$$

in distribution, for any gap away from the edges, i.e., if $\varepsilon N \leqslant i \leqslant (1 - \varepsilon)N$ with some fixed $\varepsilon > 0$. This might be viewed as the random matrix analogue of the central limit theorem. Note that universality is twofold. First, the distribution of g_i is independent of the index i (as long as λ_i is away from the edges). Second, more importantly, the limiting gap distribution is independent of the distribution of the matrix elements, similarly to the universal character of the central limit theorem.

However, the gap universality holds much more generally than the semicircle law: the rescaled gaps (1.2.11) follow the same distribution as the gaps of the GUE or GOE (depending on the symmetry class) essentially for any random matrix ensemble with "sufficient" amount of randomness. In particular, it holds for invariant ensembles, as well as for Wigner type and correlated random matrices, i.e. for very broad extensions of the original Wigner ensemble. In fact, it holds much beyond the traditional realm of random matrices; it is conjectured to hold for any random matrix describing a disordered quantum system in the *delocalized regime*, see Section 5.2 later.

The universality on microscopic scales can also be expressed in terms of the appropriately rescaled **correlation functions**. In fact, in this way the formulas are more explicit. First we define the correlation functions.

Definition 1.2.12. Let $p_N(\lambda_1, \lambda_2, \ldots, \lambda_N)$ be the joint symmetrized probability distribution of the eigenvalues. For any $n \geqslant 1$, the n-point correlation function is defined by

$$p_N^{(n)}(\lambda_1, \lambda_2, \ldots, \lambda_n) := \int_{\mathbb{R}^{N-n}} p_N(\lambda_1, \ldots, \lambda_n, \lambda_{n+1}, \ldots \lambda_N)\, d\lambda_{n+1} \ldots d\lambda_N. \tag{1.2.13}$$

The significance of the correlation functions is that with their help one can compute the expectation value of any symmetrized observable. For example, for any bounded continuous test function O of two variables we have, directly from the definition of the correlation functions, that

$$\frac{1}{N(N-1)} \mathbb{E} \sum_{i \neq j} O(\lambda_i, \lambda_j) = \int_{\mathbb{R} \times \mathbb{R}} O(\lambda_1, \lambda_2) p_N^{(2)}(\lambda_1, \lambda_2)\, d\lambda_1\, d\lambda_2, \tag{1.2.14}$$

where the expectation is w.r.t. the probability density p_N or in this case w.r.t. the original random matrix ensemble. Similar formula holds for observables of any number of variables. In particular, the global law (1.2.5) implies that the one point correlation function converges to the asymptotic density

$$p_N^{(1)}(x)\, dx \to \rho(x)\, dx$$

weakly, since

$$\int_{\mathbb{R}} O(x) p_N^{(1)}(x)\, dx = \frac{1}{N} \mathbb{E} \sum_i O(\lambda_i) \to \int O(x) \rho(x)\, dx.$$

Correlation functions are difficult to compute in general, even if the joint density function p_N is explicitly given as in the case of the invariant ensembles (1.1.8). Naively one may think that computing the correlation functions in this latter case boils down to an elementary calculus exercise by integrating out all but a few variables. However, that task is complicated.

As mentioned, one may view the joint density of eigenvalues of invariant ensembles (1.1.8) as a Gibbs measure of a log-gas and here β can be any positive number (inverse temperature). The universality of correlation functions is a valid question for all β-log-gases that has been positively answered in [17,21–23,69] by showing that for a sufficiently smooth potential V (in fact $V \in C^4$ suffices) the correlation functions depend only on β and are independent of V. We will not pursue general invariant ensembles in these notes.

The logarithmic interaction is of long range, so the system (1.1.10) is strongly correlated and standard methods of statistical mechanics to compute correlation functions cannot be applied. The computation is quite involved even for the simplest Gaussian case, and it relies on sophisticated identities involving Hermite orthogonal polynomials. These calculations have been developed by Gaudin, Mehta and Dyson in the 60's and can be found, e.g. in Mehta's book [65]. Here we just present the result for the most relevant $\beta = 1, 2$ cases.

We fix an energy E in the bulk, i.e., $|E| < 2$, and we rescale the correlation functions by a factor $N\rho$ around E to make the typical distance between neighboring eigenvalues 1. These rescaled correlation functions then have a universal limit:

Theorem 1.2.15. *For GUE ensembles, the rescaled correlation functions converge to the determinantal formula with the sine kernel,* $S(x) := \frac{\sin \pi x}{\pi x}$*, i.e.*

$$\frac{1}{[\rho_{sc}(E)]^n} p_N^{(n)}\Big(E + \frac{\alpha_1}{N\rho_{sc}(E)}, E + \frac{\alpha_2}{N\rho_{sc}(E)}, \dots, E + \frac{\alpha_n}{N\rho_{sc}(E)}\Big) \rightharpoonup q_{GUE}^{(n)}(\boldsymbol{\alpha}) := \det\big(S(\alpha_i - \alpha_j)\big)_{i,j=1}^n \tag{1.2.16}$$

as weak convergence of functions in the variables $\boldsymbol{\alpha} = (\alpha_1, \dots, \alpha_n)$.

Formula (1.2.16) holds for the GUE case. The corresponding expression for GOE is more involved [12, 65]

$$q_{GOE}^{(n)}(\boldsymbol{\alpha}) := \det\big(K(\alpha_i - \alpha_j)\big)_{i,j=1}^n, \qquad K(x) := \begin{pmatrix} S(x) & S'(x) \\ -\frac{1}{2}\operatorname{sgn}(x) + \int_0^x S(t)\,dt & S(x) \end{pmatrix}. \tag{1.2.17}$$

Here the determinant is understood as the trace of the *quaternion determinant* after the canonical correspondence between quaternions $a \cdot \mathbf{1} + b \cdot \mathbf{i} + c \cdot \mathbf{j} + d \cdot \mathbf{k}$, $a, b, c, d \in \mathbb{C}$, and 2×2 complex matrices given by

$$\mathbf{1} \leftrightarrow \begin{pmatrix} 1 & 0 \\ 0 & 1 \end{pmatrix} \qquad \mathbf{i} \leftrightarrow \begin{pmatrix} i & 0 \\ 0 & -i \end{pmatrix} \qquad \mathbf{j} \leftrightarrow \begin{pmatrix} 0 & 1 \\ -1 & 0 \end{pmatrix} \qquad \mathbf{k} \leftrightarrow \begin{pmatrix} 0 & i \\ i & 0 \end{pmatrix}.$$

Note that the limit in (1.2.16) is universal in the sense that it is independent of the energy E. However, universality also holds in a much stronger sense, namely that the local statistics (limits of rescaled correlation functions) depend only on the symmetry class, i.e. on β, and are independent of any other details. In particular, they are always given by the sine kernel (1.2.16) or (1.2.17) not only for the Gaussian case but for any Wigner matrices with arbitrary distribution of the matrix elements, as well as for any invariant ensembles with arbitrary potential V. This is the **Wigner-Dyson-Mehta (WDM) universality conjecture**, formulated precisely in Mehta's book [65] in the late 60's.

The WDM conjecture for invariant ensembles has been in the focus of very intensive research on orthogonal polynomials with general weight function (the Hermite polynomials arising in the Gaussian setup have Gaussian weight function). It motivated the development of the Riemann-Hilbert method [47], that was originally brought into this subject by Fokas, Its and Kitaev [47], and the universality of eigenvalue statistics was established for large classes of invariant ensembles by Bleher-Its [19] and by Deift and collaborators [28–30]. The key element of this success was that invariant ensembles, unlike Wigner matrices, have explicit formulas (1.1.8) for the joint densities of the eigenvalues. With the help of the Vandermonde structure of these formulas, one may express the eigenvalue correlation functions as determinants whose entries are given by functions of orthogonal polynomials.

For Wigner ensembles, there are no explicit formulas for the joint density of eigenvalues or for the correlation functions statistics and the WDM conjecture was open for almost fifty years with virtually no progress. The first significant advance in this direction was made by Johansson [58], who proved the universality for *complex* Hermitian matrices under the assumption that the common distribution of the matrix entries has a substantial Gaussian component, i.e., the random matrix H is of the form $H = H_0 + aH^G$ where H_0 is a general Wigner matrix, H^G is the GUE matrix, and a is a certain, not too small, positive constant independent of N. His proof relied on an explicit formula by Brézin and Hikami [25, 26] that uses a certain version of the Harish-Chandra-Itzykson-Zuber formula [57]. These formulas are available for the complex Hermitian case only, which restricted the method to this symmetry class.

Exercise 1.2.18. *Verify formula* (1.2.14).

1.2.19. The three step strategy The WDM conjecture in full generality has recently been resolved by a new approach called the **three step strategy** that has been developed in a series of papers by Erdős, Schlein, Yau and Yin between 2008 and 2013 with a parallel development by Tao and Vu. A detailed presentation of this method can be found in [45], while a shorter summary was presented in [43].

This approach consists of the following three steps:

Step 1. Local semicircle law: It provides an a priori estimate showing that the density of eigenvalues of generalized Wigner matrices is given by the semicircle law at very small microscopic scales, i.e., down to spectral intervals that contain N^{ε} eigenvalues.

Step 2. Universality for Gaussian divisible ensembles: It proves that the local statistics of *Gaussian divisible ensembles* $H_0 + aH^G$ are the same as those of the Gaussian ensembles H^G as long as $a \geqslant N^{-1/2+\varepsilon}$, i.e., already for very small a.

Step 3. Approximation by a Gaussian divisible ensemble: It is a type of "density argument" that extends the local spectral universality from Gaussian divisible ensembles to all Wigner ensembles.

The conceptually novel point is Step 2. The eigenvalue distributions of the Gaussian divisible ensembles, written in the form $e^{-t/2}H_0 + \sqrt{1-e^{-t}}H^G$, are the same as that of the solution of a *matrix valued Ornstein-Uhlenbeck (OU) process* H_t

$$dH_t = \frac{d\mathbf{B}_t}{\sqrt{N}} - \frac{1}{2}H_t\, dt, \qquad H_{t=0} = H_0, \tag{1.2.20}$$

for any time $t \geqslant 0$, where $\mathbf{B}_t$ is a matrix valued standard Brownian motion of the corresponding symmetry class (The OU process is preferable over its rescaled version $H_0 + aH^G$ since it keeps the variance constant). Dyson [31] observed half a century ago that the dynamics of the eigenvalues $\lambda_i = \lambda_i(t)$ of H_t is given by an interacting stochastic particle system, called the *Dyson Brownian motion (DBM)*, where the eigenvalues are the particles:

$$(1.2.21)\quad d\lambda_i = \sqrt{\frac{\beta}{2}}\frac{1}{\sqrt{N}}dB_i + \Big(-\frac{\lambda_i}{2} + \frac{1}{N}\sum_{j\neq i}\frac{1}{\lambda_i - \lambda_j}\Big)dt, \qquad i = 1, 2, \dots, N.$$

Here dB_i are independent white noises.

In addition, the invariant measure of this dynamics is exactly the eigenvalue distribution of GOE or GUE, i.e. (1.1.8) with $V(x) = \frac{1}{2}x^2$. This invariant measure is thus a Gibbs measure of point particles in one dimension interacting via a long range logarithmic potential. In fact, β can be any positive parameter, the corresponding DBM (1.2.21) may be studied even if there is no invariant matrix ensemble behind. Using a heuristic physical argument, Dyson remarked [31] that the DBM reaches its "local equilibrium" on a short time scale $t \gtrsim N^{-1}$. We call this *Dyson's conjecture*, although it was rather an intuitive physical picture than an exact mathematical statement. Step 2 gives a precise mathematical meaning of this vague idea. The key point is that by applying local relaxation to *all* initial states (within a reasonable class) simultaneously, Step 2 generates a large set of random matrix ensembles for which universality holds. For the purpose of universality, this set is sufficiently dense so that any Wigner matrix H is sufficiently close to a Gaussian divisible ensemble of the form $e^{-t/2}H_0 + \sqrt{1-e^{-t}}H^G$ with a suitably chosen H_0.

We note that in the Hermitian case, Step 2 can be circumvented by using the Harish-Chandra-Itzykson-Zuber formula. This approach was followed by Tao and Vu [75] who gave an alternative proof of universality for Wigner matrices in the Hermitian symmetry class as well as for the real symmetric class but only under a certain moment matching condition.

The three step strategy has been refined and streamlined in the last years. By now it has reached a stage when the content of Step 2 and Step 3 can be presented as a very general "black-box" result that is model independent assuming that Step 1, the local law, holds. The only model dependent ingredient is the local law. Hence to prove local spectral universality for a new ensemble, one needs to verify the local law. Thus in these lecture notes we will focus on the recent developments in the direction of the local laws.

We will discuss generalizations of the original Wigner ensemble to relax the basic conditions *"independent, identically distributed"*. First we drop the identical distribution and allow the variances $s_{ij} = \mathbb{E}|h_{ij}|^2$ to vary. The simplest class is the **generalized Wigner matrices**, defined in (1.2.7), which still leads to the Wigner semicircle law. The next level of generality is to allow arbitrary matrix of variances S. The density of states is not the semicircle any more and we need to solve a genuine **vector Dyson equation** to find the answer. The most general case discussed in these notes are correlated matrices, where different matrix elements have nontrivial correlation that leads to a **matrix Dyson equation**. In all cases we keep the mean field assumption, i.e. the typical size of the matrix elements is $|h_{ij}| \sim N^{-1/2}$. Since Wigner's vision on the universality of local eigenvalue statistics predicts the same universal behavior for a much larger class of hermitian random matrices (or operators), it is fundamentally important to extend the validity of the mathematical proofs as much as possible beyond the Wigner case.

We remark that there are several other directions to extend the Wigner ensemble that we will not discuss here in details, we just mention some of them with a few references, but we do not aim at completeness; apologies for any omissions. First, in these notes we will assume very high moment conditions on the matrix elements. These make the proofs easier and the tail probabilities of the estimates stronger. Several works have focused on *lowering the moment assumption* [2,53,59] and even considering *heavy tailed distributions* [18,20]. An important special case is the class of *sparse matrices* such as adjacency matrix of Erdős-Rényi random graphs and d-regular graphs [15,16,34,36,56]. Another direction is to remove the condition that the matrix elements are centered; this ensemble often goes under the name of *deformed Wigner matrices*. One typically separates the expectation and writes $H = A + W$, where A is a deterministic matrix and W is a Wigner matrix with centered entries. Diagonal deformations (A is diagonal) are easier to handle, this class was considered even for a large diagonal in [60,61,64,66]. The general A was considered in [54]. Finally, a very challenging direction is to depart from the mean field condition, i.e. allow some matrix elements to be much bigger than $N^{-1/2}$. The ultimate example is the *random band matrices* that goes towards the random Schrödinger operators [14,32,33,35,68,70–73].

1.2.22. User's guide These lecture notes were intended to Ph.D students and postdocs with general interest in analysis and probability; we assume knowledge of these areas on a beginning Ph.D. level. The overall style is informal, the proof of many statements are only sketched or indicated. Several technicalities are swept under the rug – for the precise theorems the reader should consult with the original papers. We emphasise conveying the main ideas in a colloquial way.

In Section 2 we collected basic tools from analysis such as Stieltjes transform and resolvent. We also introduce the semicircle law. We outline the moment method that was traditionally important in random matrices, but we will not rely on it in these notes, so this part can be skipped. In Section 3 we outline the main method to obtain local laws, the resolvent approach and we explain in an informal way its two constituents; the probabilistic and deterministic parts. In Section 4 we introduce four models of Wigner-like ensembles with increasing complexity and we informally explain the novelty and the additional complications for each model. Section 5 on the physical motivations to study these models is a detour. Readers interested only in the mathematical aspects may skip this section. Section 6 contains our main results on the local law formulated in a mathematically precise form. We did not aim at presenting the strongest results and the weakest possible conditions; the selection was guided to highlight some key phenomena. Some consequences of these local laws are also presented with sketchy proofs. Section 7 and 8 contain the main mathematical part of these notes, here we give a more detailed analysis of the vector and the matrix Dyson equation and their stability properties. In these sections we aim at rigorous presentation although not every proof contains all details. Finally, in Section 9 we present the main ideas of the proof of the local laws based on stability results on the Dyson equation.

These lecture notes are far from being a comprehensive text on random matrices. Many key issues are left out and even those we discuss will be presented in their simplest form. For more interested readers, we refer to the recent book [45] that focuses on the three step strategy and discusses all steps in details. For readers interested in other aspects of random matrix theory, in addition to the classical book of Mehta [65], several excellent works are available that present random matrices in a broader scope. The books by Anderson, Guionnet and Zeitouni [12] and Pastur and Shcherbina [67] contain extensive material starting from the basics. Tao's book [74] provides a different aspect to this subject and is self-contained as a graduate textbook. Forrester's monograph [48] is a handbook for any explicit formulas related to random matrices. Finally, [8] is an excellent comprehensive overview of diverse applications of random matrix theory in mathematics, physics, neural networks and engineering.

Notational conventions. In order to focus on the essentials, we will not follow the dependence of various constants on different parameters. In particular, we will use the generic letters C and c to denote positive constants, whose values may change from line to line and which may depend on some fixed basic parameters of the model. For two positive quantities A and B, we will write $A \lesssim B$ to indicate that there exists a constant C such that $A \leqslant CB$. If A and B are comparable in the sense that $A \lesssim B$ and $B \lesssim A$, then we write $A \sim B$. In informal explanations, we will often use $A \approx B$ which indicates closeness in a not precisely specified sense. We introduce the notation $[\![A, B]\!] := \mathbb{Z} \cap [A, B]$ for the set of integers between any two real numbers $A < B$. We will usually denote vectors in $\mathbb{C}^N$ by boldface letters; $\mathbf{x} = (x_1, x_2, \ldots, x_N)$.

Acknowledgement. A special thank goes to Torben Krüger for many discussions and suggestions on the presentation of this material as well as for his careful proofreading and invaluable comments. I am also very grateful to both referees for many constructive suggestions.

2. Tools

2.1. Stieltjes transform In this section we introduce our basic tool, the **Stieltjes transform** of a measure. We denote the open upper half of the complex plane by

$$\mathbb{H} := \{z \in \mathbb{C} \ : \ \operatorname{Im} z > 0.\}$$

Definition 2.1.1. Let μ be a Borel probability measure on $\mathbb{R}$. Its Stiltjes transform at a **spectral parameter** $z \in \mathbb{H}$ is defined by

$$m_\mu(z) := \int_{\mathbb{R}} \frac{d\mu(x)}{x - z}. \tag{2.1.2}$$

Exercise 2.1.3. *The following three properties are straightforward to check:*

i) *The Stieltjes transform $m_\mu(z)$ is analytic on $\mathbb{H}$ and it maps $\mathbb{H}$ to $\mathbb{H}$, i.e. $\operatorname{Im} m_\mu(z) > 0$.*
ii) *We have $-i\eta m_\mu(i\eta) \to 1$ as $\eta \to \infty$.*
iii) *We have the bound*

$$|m_\mu(z)| \leqslant \frac{1}{\operatorname{Im} z}.$$

In fact, properties i)-ii) characterize the Stieltjes transform in a sense that if a function $m : \mathbb{H} \to \mathbb{H}$ satisfies i)–ii), then there exists a probability measure μ such that $m = m_\mu$ (for the proof, see e.g. Appendix B of [77]; it is also called the Nevanlinna's representation theorem).

From the Stieltjes transform one may recover the measure:

Lemma 2.1.4 (Inverse Stieltjes transform). *Suppose that μ is a probability measure on $\mathbb{R}$ and let m_μ be its Stieltjes transform. Then for any $a < b$ we have*

$$\lim_{\eta \to 0} \frac{1}{\pi} \int_a^b \operatorname{Im} m_\mu(E + i\eta)\, dE = \mu(a, b) + \frac{1}{2}\big[\mu(\{a\}) + \mu(\{b\})\big]$$

Furthermore, if μ is absolutely continuous with respect to the Lebesgue measure, i.e. $\mu(dE) = \mu(E)\, dE$ with some density function $\mu(E) \in L^1$, then

$$\frac{1}{\pi} \lim_{\eta \to 0+} \operatorname{Im} m_\mu(E + i\eta) \to \mu(E)$$

pointwise for almost every E.

In particular, Lemma 2.1.4 guarantees that $m_\mu = m_\nu$ if and only of $\mu = \nu$, i.e. the Stieltjes transform uniquely characterizes the measure. Furthermore, pointwise convergence of a sequence of Stieltjes transforms is equivalent to weak convergence of the measures. More precisely, we have

Lemma 2.1.5. *Let μ_N be a sequence of probability measures and let $m_N(z) = m_{\mu_N}(z)$ be their Stieltjes transforms. Suppose that*

$$\lim_{N \to \infty} m_N(z) =: m(z)$$

exists for any $z \in \mathbb{H}$ and $m(z)$ satisfies property ii), i.e. $-i\eta m(i\eta) \to 1$ as $\eta \to \infty$. Then there exists a probability measure μ such that $m = m_\mu$ and μ_N converges to μ in distribution.

The proof can be found e.g. in [51] and it relies on Lemma 2.1.4 and Montel's theorem. The converse of Lemma 2.1.5 is trivial: if the sequence μ_N converges in distribution to a probability measure μ, then clearly $m_N(z) \to m_\mu(z)$ pointwise, since the Stieltjes transform for any fixed $z \in \mathbb{H}$ is just the integral of the continuous bounded function $x \to (x - z)^{-1}$. Note that the additional condition ii) is a compactness (tightness) condition, it prevents that part of the measures μ_N escape to infinity in the limit.

All these results are very similar to the Fourier transform (characteristic function)

$$\phi_\mu(t) := \int_{\mathbb{R}} e^{-itx}\mu(dx)$$

of a probability measure. In fact, there is a direct connection between them;

$$\int_0^\infty e^{-\eta t} e^{itE} \phi_\mu(t)\,dt = i\int_{\mathbb{R}} \frac{d\mu(x)}{x - E - i\eta} = i m_\mu(E + i\eta)$$

for any $\eta > 0$ and $E \in \mathbb{R}$. In particular, due to the regularizing factor $e^{-t\eta}$, the large t behavior of the Fourier transform $\phi(t)$ is closely related to the small $\eta \sim 1/t$ behavior of the Stieltjes transform.

Especially important is the imaginary part of the Stieltjes transform since

$$\operatorname{Im} m_\mu(z) = \int_{\mathbb{R}} \frac{\eta}{|x - E|^2 + \eta^2}\mu(dx), \qquad z = E + i\eta,$$

which can also be viewed as the convolution of μ with the Cauchy kernel on scale η:

$$P_\eta(E) = \frac{\eta}{E^2 + \eta^2},$$

indeed

$$\operatorname{Im} m_\mu(E + i\eta) = (P_\eta \star \mu)(E).$$

Up to a normalization $1/\pi$, the Cauchy kernel is an approximate delta function on scale η. Clearly

$$\int_{\mathbb{R}} \frac{1}{\pi} P_\eta(E)\,dE = 1$$

and the overwhelming majority of its mass is supported on scale η:

$$\int_{|E| \geqslant K\eta} \frac{1}{\pi} P_\eta(E) \leqslant \frac{2}{K}$$

for any K. Due to standard properties of the convolution, the *moral of the story* is that $\operatorname{Im} m_\mu(E + i\eta)$ **resolves the measure** μ **on a scale** η **around an energy** E.

Notice that the small η regime is critical; it is the regime where the integral in the definition of the Stieltjes transform (2.1.2) becomes more singular, and properties of the integral more and more depend on the local smoothness properties of the measure. In general, the regularity of the measure μ on some scales $\eta > 0$ is directly related to the Stieltjes transform $m(z)$ with $\operatorname{Im} z \approx \eta$.

The Fourier transform $\phi_\mu(t)$ of μ for large t also characterizes the local behavior of the measure μ on scales $1/t$, We will nevertheless work with the Stieltjes transform since for hermitian matrices (or self-adjoint operators in general) it is directly related to the resolvent, it is relatively easy to handle and it has many convenient properties.

Exercise 2.1.6. *Prove Lemma* 2.1.4 *by using Fubini's theorem and Lebesgue density theorem.*

2.2. Resolvent Let $H = H^*$ be a hermitian matrix, then its resolvent at spectral parameter $z \in \mathbb{H}$ is defined as

$$G = G(z) = \frac{1}{H - z}, \qquad z \in \mathbb{H}.$$

In these notes, the spectral parameter z will always be in the upper half plane, $z \in \mathbb{H}$. We usually follow the convention that $z = E + i\eta$, where $E = \operatorname{Re} z$ will often be referred as "energy" alluding to the quantum mechanical interpretation of E.

Let μ_N be the normalized *empirical measure of the eigenvalues* of H:

$$\mu_N(dx) = \frac{1}{N} \sum_{i=1}^{N} \delta(\lambda_i - x)\, dx.$$

Then clearly the normalized trace of the resolvent is

$$\frac{1}{N} \operatorname{Tr} G(z) = \frac{1}{N} \sum_{i=1}^{N} \frac{1}{\lambda_i - z} = \int_{\mathbb{R}} \frac{\mu_N(dx)}{x - z} = m_{\mu_N}(z) =: m_N(z)$$

exactly the Stieltjes transform of the empirical measure. This relation justifies why we focus on the Stieltjes transform; based upon Lemma 2.1.5, if we could identify the (pointwise) limit of $m_N(z)$, then the asymptotic eigenvalue density ρ would be given by the inverse Stieltjes transform of the limit.

Since μ_N is a discrete (atomic) measure on small $(1/N)$ scales, it may behave very badly (i.e. it is strongly fluctuating and may blow up) for η smaller than $1/N$, depending on whether there happens to be an eigenvalue in an η-vicinity of $E = \operatorname{Re} z$. Since the eigenvalue spacing is (typically) of order $1/N$, for $\eta \ll 1/N$ there is no approximately deterministic ("self-averaging") behavior of m_N. However, as long as $\eta \gg 1/N$, we may hope a *law of large number phenomenon*; this would be equivalent to the fact that the eigenvalue density does not have much fluctuation above its inter-particle scale $1/N$. The local law on m_N down to the smallest possible (optimal) scale $\eta \gg 1/N$ will confirm this hope.

In fact, the resolvent carries much more information than merely its trace. In general the resolvent of a hermitian matrix is a very rich object: it gives information on the eigenvalues and eigenvectors for energies near the real part of the spectral parameter. For example, by spectral decomposition we have

$$G(z) = \sum_i \frac{|\mathbf{u}_i\rangle\langle\mathbf{u}_i|}{\lambda_i - z}$$

where $\mathbf{u}_i$ are the (ℓ^2-normalized) eigenvectors associated with λ_i. (Here we used the Dirac notation $|\mathbf{u}_i\rangle\langle\mathbf{u}_i|$ for the orthogonal projection to the one-dimensional space spanned by $\mathbf{u}_i$.) For example, the diagonal matrix elements of the resolvent at z are closely related to the eigenvectors with eigenvalues near $E = \operatorname{Re} z$:

$$G_{xx} = \sum_i \frac{|\mathbf{u}_i(x)|^2}{\lambda_i - z}, \qquad \operatorname{Im} G_{xx} = \sum_i \frac{\eta}{|\lambda_i - E|^2 + \eta^2} |\mathbf{u}_i(x)|^2.$$

Notice that for very small η, the factor $\eta/(|\lambda_i - E|^2 + \eta^2)$ effectively reduces the sum from all $i = 1, 2, \ldots, N$ to those indices where λ_i is η-close to E; indeed this factor changes from the very large value $1/\eta$ to a very small value η as i moves away. Roughly speaking

$$\operatorname{Im} G_{xx} = \sum_i \frac{\eta}{|\lambda_i - E|^2 + \eta^2} |\mathbf{u}_i(x)|^2 \approx \sum_{i : |\lambda_i - E| \lesssim \eta} \frac{\eta}{|\lambda_i - E|^2 + \eta^2} |\mathbf{u}_i(x)|^2.$$

This idea can be made rigorous at least as an upper bound on each summand. A physically important consequence will be that one may directly obtain ℓ^∞ bounds on the eigenvectors: for any fixed $\eta > 0$ we have

$$\|\mathbf{u}_i\|_\infty^2 := \max_x |\mathbf{u}_i(x)|^2 \leqslant \eta \cdot \max_x \max_{E \in \mathbb{R}} \operatorname{Im} G_{xx}(E + i\eta). \tag{2.2.1}$$

In other words, if we can control diagonal elements of the resolvent on some scale $\eta = \operatorname{Im} z$, then we can prove an $\sqrt{\eta}$-sized bound on the max norm of the eigenvector. The strongest result is always the smallest possible scale. Since the local law will hold down to scales $\eta \gg 1/N$, in particular we will be able to establish that $\operatorname{Im} G_{xx}(E + i\eta)$ remains bounded as long as $\eta \gg 1/N$, thus we will prove the **complete delocalization** of the eigenvectors:

$$\|\mathbf{u}_i\|_\infty \leqslant \frac{N^\varepsilon}{\sqrt{N}} \tag{2.2.2}$$

for any $\varepsilon > 0$ fixed, independent of N, and with very high probability. Note that the bound (2.2.2) is optimal (apart from the N^ε factor) since clearly

$$\|\mathbf{u}\|_\infty \geqslant \frac{\|\mathbf{u}\|_2}{\sqrt{N}}$$

for any $\mathbf{u} \in \mathbb{C}^N$.

We also note that if $\operatorname{Im} G_{xx}(E + i\eta)$ can be controlled only for energies in a fixed subinterval $I \subset \mathbb{R}$, e.g. the local law holds only for all $E \in I$, the we can conclude complete delocalization for those eigenvectors whose eigenvalues lie in I.

2.3. The semicircle law for Wigner matrices via the moment method This section introduces the traditional moment method to identify the semicircle law. We included this material for historical relevance, but it will not be needed later hence it can be skipped at first reading.

For large z one can expand m_N as follows

$$m_N(z) = \frac{1}{N} \operatorname{Tr} \frac{1}{H - z} = -\frac{1}{Nz} \sum_{m=0}^{\infty} \operatorname{Tr} \left(\frac{H}{z} \right)^m, \tag{2.3.1}$$

so after taking the expectation, we need to compute traces of high moments of H:

$$\mathbb{E}\, m_N(z) = \sum_{k=0}^{\infty} z^{-(2k+1)} \frac{1}{N} \mathbb{E} \operatorname{Tr} H^{2k}. \tag{2.3.2}$$

Here we tacitly used that the contributions of odd powers are algebraically zero, which clearly holds at least if we assume that h_{ij} have symmetric distribution for simplicity. Indeed, in this case H^{2k+1} and $(-H)^{2k+1}$ have the same distribution, thus

$$\mathbb{E}\,\mathrm{Tr}\,H^{2k+1} = \mathbb{E}\,\mathrm{Tr}(-H)^{2k+1} = -\mathbb{E}\,\mathrm{Tr}\,H^{2k+1}.$$

The computation of even powers, $\mathbb{E}\,\mathrm{Tr}\,H^{2k}$, reduces to a combinatorial problem. Writing out

$$\mathbb{E}\,\mathrm{Tr}\,H^{2k} = \sum_{i_1,i_2,\dots i_{2k}} \mathbb{E} h_{i_1 i_2} h_{i_2 i_3} \dots h_{i_{2k} i_1},$$

one notices that, by $\mathbb{E} h_{ij} = 0$, all those terms are zero where at least one $h_{i_j i_{j+1}}$ stands alone, i.e. is not paired with itself or its conjugate. This restriction poses a severe constraint on the relevant index sequences $i_1, i_2, \dots, i_{2k}$. For the terms where an exact pairing of all the $2k$ factors is available, we can use $\mathbb{E}|h_{ij}|^2 = N^{-1}$ to see that all these terms contribute by N^{-k}. There are terms where three or more h's coincide, giving rise to higher moments of h, but their combinatorics is of lower order. Following Wigner's classical calculation (called the **moment method**, see e.g. [12]), one needs to compute the number of relevant index sequences that give rise to a perfect pairing and one finds that the leading term is given by the Catalan numbers, i.e.

$$\frac{1}{N}\mathbb{E}\,\mathrm{Tr}\,H^{2k} = \frac{1}{k+1}\binom{2k}{k} + O_k\Big(\frac{1}{N}\Big). \tag{2.3.3}$$

Notice that the N-factors cancelled out in the leading term.

Thus, continuing (2.3.2) and neglecting the error terms, we get

$$\mathbb{E}\, m_N(z) \approx -\sum_{k=0}^{\infty} \frac{1}{k+1}\binom{2k}{k} z^{-(2k+1)}, \tag{2.3.4}$$

which, after some calculus, can be identified as the Laurent series of the function $\frac{1}{2}(-z+\sqrt{z^2-4})$. The approximation becomes exact in the $N \to \infty$ limit. Although the expansion (2.3.1) is valid only for large z, given that the limit is an analytic function of z, one can extend the relation

$$\lim_{N\to\infty} \mathbb{E} m_N(z) = \frac{1}{2}(-z+\sqrt{z^2-4}) \tag{2.3.5}$$

by analytic continuation to the whole upper half plane $z = E + i\eta$, $\eta > 0$. It is an easy exercise to see that this is exactly the Stieltjes transform of the semicircle density, i.e.,

$$m_{sc}(z) := \frac{1}{2}(-z+\sqrt{z^2-4}) = \int_{\mathbb{R}} \frac{\rho_{sc}(x)\,dx}{x-z}, \qquad \rho_{sc}(x) = \frac{1}{2\pi}\sqrt{(4-x^2)_+}. \tag{2.3.6}$$

The square root function is chosen with a branch cut in the segment $[-2,2]$ so that $\sqrt{z^2-4} \sim z$ at infinity. This guarantees that $\mathrm{Im}\, m_{sc}(z) > 0$ for $\mathrm{Im}\, z > 0$.

Exercise 2.3.7. *As a simple calculus exercise, verify* (2.3.6). *Either use integration by parts, or compute the moments of the semicircle law and verify that they are given by the Catalan numbers, i.e.*

$$\int_{\mathbb{R}} x^{2k} \rho_{sc}(x)\,dx = \frac{1}{k+1}\binom{2k}{k}. \tag{2.3.8}$$

Since the Stieltjes transform identifies the measure uniquely, and pointwise convergence of Stieltjes transforms implies weak convergence of measures, we obtain

$$\mathbb{E}\,\rho_N(dx) \rightharpoonup \rho_{sc}(x)\,dx. \tag{2.3.9}$$

The relation (2.3.5) actually holds with high probability, that is, for any z with $\operatorname{Im} z > 0$,

$$\lim_{N\to\infty} m_N(z) = \frac{1}{2}(-z + \sqrt{z^2-4}), \tag{2.3.10}$$

in probability, implying a similar strengthening of the convergence in (2.3.9). In the next sections we will prove this limit with an effective error term via the resolvent method.

The semicircle law can be identified in many different ways. The *moment method* sketched above utilized the fact that the moments of the semicircle density are given by the Catalan numbers (2.3.8), which also emerged as the normalized traces of powers of H, see (2.3.3). The *resolvent method* relies on the fact that m_N approximately satisfies a self-consistent equation,

$$m_N(z) \approx -\frac{1}{z + m_N(z)}, \tag{2.3.11}$$

that is very close to the quadratic equation that m_{sc} from (2.3.6) exactly satisfies:

$$m_{sc}(z) = -\frac{1}{z + m_{sc}(z)}. \tag{2.3.12}$$

Comparing these two equations, one finds that $m_N(z) \approx m_{sc}(z)$. Taking inverse Stieltjes transform, one concludes the semicircle law. In the next section we give more details on (2.3.11).

In other words, in the resolvent method the semicircle density emerges via a specific relation for its Stieltjes transform. The key relation (2.3.12) is the simplest form of the **Dyson equation**, or a **self-consistent equation** for the trace of the resolvent: later we will see a Dyson equation for the entire resolvent. It turns out that the resolvent approach allows us to perform a much more precise analysis than the moment method, especially in the short scale regime, where $\operatorname{Im} z$ approaches to 0 as a function of N. Since the Stieltjes transform of a measure at spectral parameter $z = E + i\eta$ essentially identifies the measure around E on scale $\eta > 0$, a precise understanding of $m_N(z)$ for small $\operatorname{Im} z$ will yield a local version of the semicircle law.

3. The resolvent method

In this section we sketch the two basic steps of the resolvent method for the simplest Wigner case but we will already make remarks preparing for the more complicated setup. The first step concerns the derivation of the approximate equation (2.3.11). This is a probabilistic step since $m_N(z)$ is a random object and even in the best case (2.3.11) can hold only with high probability. In the second step we compare the approximate equation (2.3.11) with the exact equation (2.3.12) to conclude that m_N and m_{sc} are close. We will view (2.3.11) as a perturbation of (2.3.12), so this step is about a stability property of the exact equation and it is a deterministic problem.

3.1. Probabilistic step There are essentially two ways to obtain (2.3.11); either by Schur complement formula or by cumulant expansion. Typically the Schur method gives more precise results since it can be easier turned into a full asymptotic expansion, but it heavily relies on the independence of the matrix elements and that the resolvent of H is essentially diagonal. We now discuss these methods separately.

3.1.1. Schur complement method The basic input is the following well-known formula from linear algebra:

Lemma 3.1.2 (Schur formula). *Let* A, B, C *be* $n \times n$, $m \times n$ *and* $m \times m$ *matrices. We define* $(m+n) \times (m+n)$ *matrix* D *as*

$$D := \begin{pmatrix} A & B^* \\ B & C \end{pmatrix} \tag{3.1.3}$$

and $n \times n$ *matrix* $\widehat{D}$ *as*

$$\widehat{D} := A - B^* C^{-1} B. \tag{3.1.4}$$

Then $\widehat{D}$ *is invertible if* D *is invertible and for any* $1 \leqslant i, j \leqslant n$, *we have*

$$(D^{-1})_{ij} = (\widehat{D}^{-1})_{ij} \tag{3.1.5}$$

for the corresponding matrix elements. □

We will use this formula for the resolvent of H. Recall that $G_{ij} = G_{ij}(z)$ denotes the matrix element of the resolvent

$$G_{ij} = \left(\frac{1}{H - z} \right)_{ij}.$$

Let $H^{[i]}$ denote the i-th minor of H, i.e. the $(N-1) \times (N-1)$ matrix obtained from H by removing the i-th row and column:

$$H^{[i]}_{ab} := h_{ab}, \qquad a, b \neq i.$$

Similarly, we set

$$G^{[i]}(z) := \frac{1}{H^{[i]} - z}$$

to be the resolvent of the minor. For $i = 1$, H has the block-decomposition

$$H = \begin{pmatrix} h_{11} & [\mathbf{a}^1]^* \\ \mathbf{a}^1 & H^{[1]} \end{pmatrix},$$

where $\mathbf{a}^i \in \mathbb{C}^{N-1}$ is the i-th column of H without the i-th element.

Using Lemma 3.1.2 for $n = 1$, $m = N - 1$ we have

$$G_{ii} = \frac{1}{h_{ii} - z - [\mathbf{a}^i]^* G^{[i]} \mathbf{a}^i}, \tag{3.1.6}$$

where

$$[\mathbf{a}^i]^* G^{[i]} \mathbf{a}^i = \sum_{k,l \neq i} h_{ik} G^{[i]}_{kl} h_{li}. \tag{3.1.7}$$

Here and below, we use the convention that unspecified summations always run from 1 to N.

Now we use the fact that for Wigner matrices $\mathbf{a}^i$ and $H^{[i]}$ are independent. So in the quadratic form (3.1.7) we can condition on the i-th minor and momentarily consider only the randomness of the i-th column. Set $i = 1$ for notational simplicity. Then we have a quadratic form of the type

$$\mathbf{a}^* B \mathbf{a} = \sum_{k,l=2}^{N} \bar{a}_k B_{kl} a_l$$

where $B = G^{[1]}$ is considered as a fixed deterministic matrix and $\mathbf{a}$ is a random vector with centered i.i.d. components and $\mathbb{E}|a_k|^2 = 1/N$. We decompose it into its expectation w.r.t. $\mathbf{a}$, denoted by $\mathbb{E}_{\mathbf{a}}$, and the fluctuation:

$$\mathbf{a}^* B \mathbf{a} = \mathbb{E}_{\mathbf{a}} \mathbf{a}^* B \mathbf{a} + Z, \qquad Z := \mathbf{a}^* B \mathbf{a} - \mathbb{E}_{\mathbf{a}} \mathbf{a}^* B \mathbf{a}. \tag{3.1.8}$$

The expectation gives

$$\mathbb{E}_{\mathbf{a}} \mathbf{a}^* B \mathbf{a} = \mathbb{E}_{\mathbf{a}} \sum_{k,l=2}^{N} \bar{a}_k B_{kl} a_l = \frac{1}{N} \sum_{k=2}^{N} B_{kk} = \frac{1}{N} \operatorname{Tr} B,$$

where we used that a_k and a_l are independent, $\mathbb{E}_{\mathbf{a}} \bar{a}_k a_l = \delta_{kl} \cdot \frac{1}{N}$, so the double sum collapses to a single sum. Neglecting the fluctuation Z for a moment (see an argument later), we have from (3.1.6) that

$$G_{11} = -\frac{1}{z + \frac{1}{N} \operatorname{Tr} G^{[1]} + \text{error}}, \tag{3.1.9}$$

where we also included the small $h_{11} \sim N^{-1/2}$ into the error term. Furthermore, it is easy to see that $\frac{1}{N} \operatorname{Tr} G^{[1]}$ and $\frac{1}{N} \operatorname{Tr} G$ are close to each other, this follows from a basic fact from linear algebra that the eigenvalues of H and its minor $H^{[1]}$ **interlace** (see Exercise 3.1.17).

Similar formula holds for each i, not only for $i = 1$. Summing them up, we have

$$\frac{1}{N} \operatorname{Tr} G \approx -\frac{1}{z + \frac{1}{N} \operatorname{Tr} G},$$

which is exactly (2.3.11), modulo the argument that the fluctuation Z is small. Notice that we were aiming only at $\frac{1}{N} \operatorname{Tr} G$, but in fact the procedure gave us more. After approximately identifying $\frac{1}{N} \operatorname{Tr} G \approx \frac{1}{N} \operatorname{Tr} G^{[1]}$ with m_{sc}, we can feed this information back to (3.1.9) to obtain information for each diagonal matrix element of the resolvent:

$$G_{11} \approx -\frac{1}{z+m_{sc}} = m_{sc},$$

i.e. not only the trace of G are close to m_{sc}, but each diagonal matrix element.

What about the off-diagonals? It turns out that they are small. The simplest argument to indicate this is using the **Ward identity** that is valid for resolvents of any self-adjoint operator T:

$$\sum_j \Big|\big(\frac{1}{T-z}\big)_{ij}\Big|^2 = \frac{1}{\operatorname{Im} z} \operatorname{Im}\big(\frac{1}{T-z}\big)_{ii}. \tag{3.1.10}$$

We recall that the imaginary part of a matrix M is given by $\operatorname{Im} M = \frac{1}{2i}(M - M^*)$ and notice that $(\operatorname{Im} M)_{aa} = \operatorname{Im} M_{aa}$ so there is no ambiguity in the notation of its diagonal elements. Notice that the summation in (3.1.10) is removed at the expense of a factor $1/\operatorname{Im} z$. So if $\eta = \operatorname{Im} z \gg 1/N$ and diagonal elements are controlled, the Ward identity is a substantial improvement over the naive bound of estimating each of the N terms separately. In particular, applying (3.1.10) for G, we get

$$\sum_j |G_{ij}|^2 = \frac{1}{\operatorname{Im} z} \operatorname{Im} G_{ii}.$$

Since the diagonal elements have already been shown to be close to m_{sc}, this implies that

$$\frac{1}{N}\sum_j |G_{ij}|^2 \approx \frac{\operatorname{Im} m_{sc}}{N \operatorname{Im} z},$$

i.e. on average we have

$$|G_{ij}| \lesssim \frac{1}{\sqrt{N\eta}}, \qquad i \neq j.$$

With a bit more argument, one can show that this relation holds for every $j \neq i$ and not just on average up to a factor N^{ε} with very high probability. We thus showed that the resolvent G of a Wigner matrix is close to the m_{sc} times the identity matrix I, very roughly

$$G(z) \approx m_{sc}(z) I. \tag{3.1.11}$$

Such relation must be treated with a certain care, since G is a large matrix and the sloppy formulation in (3.1.11) does not indicate in which sense the closeness $\approx$ is meant. It turns our that it holds in **normalized trace sense**:

$$\frac{1}{N} \operatorname{Tr} G \approx m_{sc},$$

in **entrywise sense**:

$$G_{ij} \approx m_{sc}\delta_{ij} \tag{3.1.12}$$

for every fixed i, j; and more generally in **isotropic sense**:

$$\langle \mathbf{x}, G\mathbf{y}\rangle \approx m_{sc}\langle \mathbf{x}, \mathbf{y}\rangle$$

for every fixed (deterministic) vectors $\mathbf{x}, \mathbf{y} \in \mathbb{C}^N$. In all cases, these relations are meant with very high probability. But (3.1.11) does **not** hold in operator norm sense since

$$\|G(z)\| = \frac{1}{\eta}, \qquad \text{while} \quad \|m_{sc} I\| = |m_{sc}| \sim O(1)$$

even if $\eta \to 0$. One may not invert (3.1.11) either, since the relation

$$H - z \approx \frac{1}{m_{sc}} I \tag{3.1.13}$$

is very wrong, in fact

$$H - z \approx -z$$

if we disregard small off-diagonal elements as we did in (3.1.12). The point is that the cumulative effects of many small off diagonal matrix elements substantially changes the matrix. In fact, using (2.3.12), the relation (3.1.12) in the form

$$\Big(\frac{1}{H-z}\Big)_{ij} \approx \frac{1}{-z - m_{sc}(z)} \delta_{ij} \tag{3.1.14}$$

exactly shows how much the spectral parameter must be shifted compared to the naive (and wrong) approximation $(H - z)^{-1} \approx -1/z$. This amount is $m_{sc}(z)$ and it is often called **self-energy shift** in the physics literature. On the level of the resolvent (and in the senses described above), the effect of the random matrix H can be simply described by this shift.

Finally, we indicate the mechanism that makes the fluctuation term Z in (3.1.8) small. We compute only its variance, higher moment calculations are similar but more involved:

$$\mathbb{E}_\mathbf{a}|Z|^2 = \sum_{mn} \sum_{kl} \mathbb{E}_\mathbf{a} \Big[a_m \bar{B}_{mn} \bar{a}_n - \mathbb{E}_\mathbf{a} a_m \bar{B}_{mn} \bar{a}_n \Big] \Big[\bar{a}_k B_{kl} a_l - \mathbb{E}_\mathbf{a} \bar{a}_k B_{kl} a_l \Big].$$

The summations run for all indices from 2 to N. Since $\mathbb{E}_\mathbf{a} a_m = 0$, in the terms with nonzero contribution we need to pair every a_m to another $\bar{a}_m$. For simplicity, here we assume that we work with the complex symmetry class and $\mathbb{E} a_m^2 = 0$ (i.e. the real and imaginary parts of each matrix elements h_{ij} are independent and identically distributed). If a_m is paired with $\bar{a}_n$ in the above sum, i.e. $m = n$, then this pairing is cancelled by the $\mathbb{E}_\mathbf{a} a_m \bar{B}_{mn} \bar{a}_n$ term. So a_i must be paired with an a from the other bracket and since $\mathbb{E} a^2 = 0$, it has to be paired with $\bar{a}_k$, thus $m - k$. Similarly $n - l$ and we get

$$\mathbb{E}_\mathbf{a}|Z|^2 = \frac{1}{N^2} \sum_{m \neq n} |B_{mn}|^2 + \mathbb{E}_\mathbf{a}|a|^4 \sum_m |B_{mm}|^2, \tag{3.1.15}$$

where the last term comes from the case when $m = n = k = l$. Assuming that the matrix elements h_{ij} have fourth moments in a sense that $\mathbb{E}|\sqrt{N} h_{ij}|^4 \leqslant C$, we have $\mathbb{E}_\mathbf{a}|a|^4 = O(N^{-2})$ in this last term and it is negligible. The main term in (3.1.15) has a summation over N^2 elements, so *a priori* it looks order one, i.e. too large. But in our application, B will be the resolvent of the minor, $B = G^{[1]}$, and we can use the Ward identity (3.1.10).

In our concrete application with $B = G^{[1]}$ we get

$$\begin{aligned}\mathbb{E}_{\mathbf{a}}|Z|^2 =& \frac{1}{N\eta}\frac{1}{N}\sum_m \operatorname{Im} B_{mm} + \frac{C}{N^2}\sum_m |B_{mm}|^2 \\ \leqslant& \frac{C}{N\eta}\frac{1}{N}\operatorname{Im}\operatorname{Tr} G^{[1]} \leqslant \frac{C}{N\eta}\operatorname{Im} m^{[1]} = O\Big(\frac{1}{N\eta}\Big),\end{aligned}$$

which is small, assuming $N\eta \gg 1$. To estimate the second term here we used that for the resolvent of any hermitian matrix T we have

$$\sum_m \Big|\Big(\frac{1}{T-z}\Big)_{mm}\Big|^2 \leqslant \frac{1}{\eta}\operatorname{Im}\operatorname{Tr}\frac{1}{T-z} \tag{3.1.16}$$

by spectral calculus. We also used that the traces of G and $G^{[1]}$ are close:

Exercise 3.1.17. *Let* H *be any hermitian matrix and* $H^{[1]}$ *its minor. Prove that their eigenvalues interlace, i.e. they satisfy*

$$\lambda_1 \leqslant \mu_1 \leqslant \lambda_2 \leqslant \mu_2 \leqslant \ldots \leqslant \mu_{N-1} \leqslant \lambda_N,$$

where the λ*'s and* μ*'s are the eigenvalues of* H *and* $H^{[1]}$*, respectively. Conclude from this that*

$$\Big|\operatorname{Tr}\frac{1}{H-z} - \operatorname{Tr}\frac{1}{H^{[1]}-z}\Big| \leqslant \frac{1}{\operatorname{Im} z}$$

Exercise 3.1.18. *Prove the Ward identity* (3.1.10) *and the estimate* (3.1.16) *by using the spectral decomposition of* $T = T^*$.

3.1.19. Cumulant expansion Another way to prove (2.3.11) starts with the defining identity of the resolvent: $HG = I + zG$ and computes its expectation:

$$\mathbb{E}HG = I + z\mathbb{E}G. \tag{3.1.20}$$

Here H and G are not independent, but it has the structure that the basic random variable H multiplies a function of it viewing $G = G(H)$. In a single random variable h it looks like $\mathbb{E}hf(h)$. If h were a centered real Gaussian, then we could use the basic **integration by parts** identity of Gaussian variables:

$$\mathbb{E}hf(h) = \mathbb{E}h^2\mathbb{E}f'(h). \tag{3.1.21}$$

In our concrete application, when f is the resolvent whose derivative is its square, in the Gaussian case we have the formula

$$\mathbb{E}HG = -\mathbb{E}\widetilde{\mathbb{E}}\big[\widetilde{H}G\widetilde{H}\big]G, \tag{3.1.22}$$

where tilde denotes an independent copy of H. We may define a linear map $\mathcal{S}$ on the space of $N \times N$ matrices by

$$\mathcal{S}[R] := \widetilde{\mathbb{E}}\big[\widetilde{H}R\widetilde{H}\big], \tag{3.1.23}$$

then we can write (3.1.22) as

$$\mathbb{E}HG = -\mathbb{E}\mathcal{S}[G]G.$$

This indicates to smuggle the $\mathbb{E}HG$ term into $HG = I + zG$ and write it as

$$D = I + \big(z + \mathcal{S}[G]\big)G, \qquad D := HG + \mathcal{S}[G]G. \tag{3.1.24}$$

With these notations, (3.1.22) means that $\mathbb{E}D = 0$. Notice that the term $\mathcal{S}[G]G$ acts as a counter-term to balance HG.

Suppose we can prove that D is small with high probability, i.e. not only $\mathbb{E}D = 0$ but also $\mathbb{E}|D_{ij}|^2$ is small for any i, j, then

$$I + \big(z + \mathcal{S}[G]\big)G \approx 0. \tag{3.1.25}$$

So it is not unreasonable to hope that the solution G will be, in some sense, close to the solution M of the deterministic equation

$$I + \big(z + \mathcal{S}[M]\big)M = 0 \tag{3.1.26}$$

with the side condition that $\operatorname{Im} M := \frac{1}{2i}(M - M^*) \geqslant 0$ (positivity in the sense of hermitian matrices). It turns out that this equation in its full generality will play a central role in our analysis for much larger class of random matrices, see Section 4.5 later. The operator $\mathcal{S}$ is called the **self-energy operator** following the analogy explained around (3.1.14).

To see how $\mathcal{S}$ looks like, in the real Gaussian Wigner case (GOE) we have

$$\mathcal{S}[R]_{ij} = \widetilde{\mathbb{E}}\big[\widetilde{H}R\widetilde{H}\big]_{ij} = \widetilde{\mathbb{E}}\sum_{ab} \widetilde{h}_{ia}R_{ab}\widetilde{h}_{bj} = \delta_{ij}\frac{1}{N}\operatorname{Tr} R + \frac{1}{N}R_{ji}\mathbf{1}(i \neq j).$$

Plugging this relation back into (3.1.25) with $R = G$ and neglecting the second term $\frac{1}{N}G_{ji}$ we have

$$0 \approx I + \big(z + \frac{1}{N}\operatorname{Tr} G\big)G.$$

Taking the normalized trace, we end up with

$$1 + (z + m_N)m_N \approx 0, \tag{3.1.27}$$

i.e. we proved (2.3.11).

Exercise 3.1.28. *Prove* (3.1.21) *by a simple integration by parts and then use* (3.1.21) *to prove* (3.1.22)*. Formulate and prove the complex versions of these formulas (assume that* $\operatorname{Re} h$ *and* $\operatorname{Im} h$ *are independent).*

Exercise 3.1.29. *Compute the variance* $\mathbb{E}|D|^2$ *for a GOE/GUE matrix and conclude that it is small in the regime where* $N\eta \gg 1$ *(essentially as* $(N\eta)^{-1/2}$*). Compute* $\mathbb{E}\big|\frac{1}{N}\operatorname{Tr} D\big|^2$ *as well and show that it is essentially of order* $(N\eta)^{-1}$*.*

This argument so far heavily used that H is Gaussian. However, the basic integration by parts formula (3.1.21) can be extended to non-Gaussian situation. For this, we recall the **cumulants** of random variables. We start with a single random variable h. As usual, its moments are defined by

$$m_k := \mathbb{E}h^k,$$

and they are generated by the *moment generating function*

$$\mathbb{E}e^{th} = \sum_{k=0}^{\infty} \frac{t^k}{k!} m_k$$

(here we assume that all moments exist and even the exponential moment exists at least for small t). The **cumulants** κ_k of h are the Taylor coefficients of the *logarithm* of the moment generating function, i.e. they are defined by the identity

$$\log \mathbb{E} e^{th} = \sum_{k=0}^{\infty} \frac{t^k}{k!} \kappa_k.$$

The sequences of $\{m_k : k = 0, 1, 2 \ldots\}$ and $\{\kappa_k : k = 0, 1, 2 \ldots\}$ mutually determine each other; these relations can be obtained from formal power series manipulations. For example

$$\kappa_0 = m_0 = 1, \qquad \kappa_1 = m_1, \qquad \kappa_2 = m_2 - m_1^2, \qquad \kappa_3 = m_3 - 3m_2 m_1 + 2m_1^3, \ldots$$

and

$$m_1 = \kappa_1, \qquad m_2 = \kappa_2 + \kappa_1^2, \qquad m_3 = \kappa_3 + 3\kappa_2\kappa_1 + 2\kappa_1^3, \ldots$$

The general relations are given by

$$m_k = \sum_{\pi \in \Pi_k} \prod_{B \in \pi} \kappa_{|B|}, \qquad \kappa_k = \sum_{\pi \in \Pi_k} (-1)^{|\pi|-1} (|\pi| - 1)! \prod_{B \in \pi} m_{|B|}, \tag{3.1.30}$$

where Π_k is the set of all partitions of a k-element base set, say $\{1, 2, \ldots, k\}$. Such a π consists of a collection of nonempty, mutually disjoint sets $\pi = \{B_1, B_2, \ldots B_{|\pi|}\}$ such that $\cup B_i = \{1, 2, \ldots, k\}$ and $B_i \cap B_j = \emptyset$, $i \neq j$.

For Gaussian variables, all but the first and second cumulants vanish, that is, $\kappa_3 = \kappa_4 = \ldots = 0$, and this is the reason for the very simple form of the relation (3.1.21). For general non-Gaussian h we have

$$\mathbb{E} h f(h) = \sum_{k=0}^{\infty} \frac{\kappa_{k+1}}{k!} \mathbb{E} f^{(k)}(h). \tag{3.1.31}$$

Similarly to the Taylor expansion, one does not have to expand it up to infinity, there are versions of this formula containing only a finite number of cumulants plus a remainder term.

To see the formula (3.1.31), we use Fourier transform:

$$\hat{f}(t) = \int_{\mathbb{R}} e^{ith} f(h)\, dh, \qquad \hat{\mu}(t) = \int_{\mathbb{R}} e^{ith} \mu(dh) = \mathbb{E} e^{ith},$$

where μ is the distribution of h, then

$$\log \hat{\mu}(t) = \sum_{k=0}^{\infty} \frac{(it)^k}{k!} \kappa_k.$$

By Parseval identity (neglecting 2π's and assuming f is real)

$$\mathbb{E} h f(h) = \int_{\mathbb{R}} h f(h) \mu(dh) = i \int_{\mathbb{R}} \overline{\hat{f}'(t)} \hat{\mu}(t)\, dt.$$

Integration by parts gives

$$\begin{aligned} i \int_{\mathbb{R}} \overline{\hat{f}'(t)} \hat{\mu}(t)\, dt &= -i \int_{\mathbb{R}} \overline{\hat{f}(t)} \hat{\mu}'(t)\, dt = -i \int_{\mathbb{R}} \overline{\hat{f}(t)} \hat{\mu}(t) \big(\log \hat{\mu}(t) \big)'\, dt \\ &= \sum_{k=0}^{\infty} \frac{\kappa_{k+1}}{k!} \int_{\mathbb{R}} (it)^k \overline{\hat{f}(t)} \hat{\mu}(t)\, dt = \sum_{k=0}^{\infty} \frac{\kappa_{k+1}}{k!} \mathbb{E} f^{(k)}(h) \end{aligned}$$

by Parseval again.

So far we considered one random variable only, but joint cumulants can also be defined for any number of random variables. This becomes especially relevant beyond the independent case, e.g. when the entries of the random matrix have correlations. For the Wigner case, many of these formulas simplify, but it is useful to introduce joint cumulants in full generality.

If $\mathbf{h} = (h_1, h_2, \dots h_m)$ is a collection of random variables (with possible repetition), then

$$\kappa(\mathbf{h}) = \kappa(h_1, h_2, \dots h_m)$$

are the coefficients of the logarithm of the moment generating function:

$$\log \mathbb{E}e^{\mathbf{t}\cdot\mathbf{h}} = \sum_{\mathbf{k}=0}^{\infty} \frac{\mathbf{t}^{\mathbf{k}}}{\mathbf{k}!}\kappa_{\mathbf{k}}.$$

Here $\mathbf{t} = (t_1, t_2, \dots, t_n) \in \mathbb{R}^n$, and $\mathbf{k} = (k_1, k_2, \dots, k_n) \in \mathbb{N}^n$ is a multi index with n components and

$$\mathbf{t}^{\mathbf{k}} := \prod_{i=1}^{n} t_i^{k_i}, \qquad \mathbf{k}! = \prod_i k_i!, \qquad \kappa_{\mathbf{k}} = \kappa(h_1, h_1, \dots h_2, h_2, \dots),$$

where h_j appears k_j-times (order is irrelevant, the cumulants are fully symmetric functions in all their variables). The formulas (3.1.30) naturally generalize, see e.g. Appendix A of [37] for a good summary. The analogue of (3.1.31) is

$$\mathbb{E}h_1 f(\mathbf{h}) = \sum_{\mathbf{k}} \frac{\kappa_{\mathbf{k}+\mathbf{e}_1}}{\mathbf{k}!}\mathbb{E}f^{(\mathbf{k})}(\mathbf{h}), \qquad \mathbf{h} = (h_1, h_2, \dots, h_n), \tag{3.1.32}$$

where the summation is for all n-multi-indices and

$$\mathbf{k} + \mathbf{e}_1 = (k_1 + 1, k_2, k_3, \dots, k_n)$$

and the proof is the same.

We use these cumulant expansion formulas to prove that D defined in (3.1.24) is small with high probability by computing $\mathbb{E}|D_{ij}|^{2p}$ with large p. Written as

$$\mathbb{E}|D_{ij}|^{2p} = \mathbb{E}(HG + \mathcal{S}[G]G)_{ij} D_{ij}^{p-1}\bar{D}_{ij}^{p},$$

we may use (3.1.32) to do an integration by parts in the first H factor, considering everything else as a function f. It turns out that the $\mathcal{S}[G]G$ term cancels the second order cumulant and naively the effect of higher order cumulants are negligible since a cumulant of order k is $N^{-k/2}$. However, the derivatives of f can act on the $D^{p-1}\bar{D}^p$ part of f, resulting in a complicated combinatorics and in fact many cumulants need to be tracked, see [37] for an extensive analysis.

3.2. Deterministic stability step In this step we compare the approximate equation (2.3.11) satisfied by the empirical Stieltjes transform and the exact equation (2.3.12) for the self-consistent Stieltjes transform

$$m_N(z) \approx -\frac{1}{z + m_N(z)}, \qquad m_{sc}(z) = -\frac{1}{z + m_{sc}(z)}.$$

In fact, considering the format (3.1.25) and (3.1.27), sometimes it is better to relate the following two equations

$$1 + (z + m_N)m_N \approx 0, \qquad 1 + (z + m_{sc})m_{sc} = 0.$$

This distinction is irrelevant for Wigner matrices, where the basic object to investigate is m_N, a scalar quantity – multiplying an equation with it is a trivial operation. But already (3.1.24) indicates that there is an approximate equation for the entire resolvent G as well and not only for its trace and in general we are interested in resolvent matrix elements as well. Since inverting G is a nontrivial operation (see the discussion after (3.1.11)), the three possible versions of (3.1.24) are very different:

$$I + (z + \mathcal{S}[G])G \approx 0, \qquad G \approx -\frac{1}{z + \mathcal{S}[G]}, \qquad -\frac{1}{G} \approx z + \mathcal{S}[G]$$

In fact the last version is blatantly wrong, see (3.1.13). The first version is closer to the spirit of the cumulant expansion method, the second is closer to Schur formula method.

In both cases, we need to understand the stability of the equation

$$m_{sc}(z) = -\frac{1}{z + m_{sc}(z)} \quad \text{or} \quad 1 + (z + m_{sc})m_{sc} = 0$$

against a small additive perturbation. For definiteness, we look at the second equation and compare m_{sc} with m_ε, where m_ε solves

$$1 + (z + m_\varepsilon)m_\varepsilon = \varepsilon$$

for some small ε. Since these are quadratic equations, one may write up the solutions explicitly and compare them, but this approach will not work in the more complicated situations. Instead, we subtract these two equations and find that

$$(z + 2m_{sc})(m_\varepsilon - m_{sc}) + (m_\varepsilon - m_{sc})^2 = \varepsilon$$

We may also eliminate z using the equation $1 + (z + m_{sc})m_{sc} = 0$ and get

$$\frac{m_{sc}^2 - 1}{m_{sc}}(m_\varepsilon - m_{sc}) + (m_\varepsilon - m_{sc})^2 = \varepsilon. \tag{3.2.1}$$

This is a quadratic equation for the difference $m_\varepsilon - m_{sc}$ and its stability thus depends on the invertibility of the linear coefficient $(m_{sc}^2 - 1)/m_{sc}$, which is determined by the limiting equation only. If we knew that

$$|m_{sc}| \leqslant C, \qquad |m_{sc}^2 - 1| \geqslant c \tag{3.2.2}$$

with some positive constants c, C, then the linear coefficient would be invertible

$$\left| \left[\frac{m_{sc}^2 - 1}{m_{sc}} \right]^{-1} \right| \leqslant C/c \tag{3.2.3}$$

and (3.2.1) would imply that

$$|m_\varepsilon - m_{sc}| \leqslant C'\varepsilon$$

at least if we had an a priori information that $|m_\varepsilon - m_{sc}| \leqslant c/2C$. This a priori information can be obtained for large $\eta = \operatorname{Im} z$ easily since in this regime both m_{sc} and m_ε are of order η (we still remember that m_ε represents a Stieltjes transform). Then we can use a fairly standard continuity argument to reduce $\eta = \operatorname{Im} z$ and keeping $E = \operatorname{Re} z$ fixed to see that the bound $|m_\varepsilon - m_{sc}| \leqslant c/2C$ holds for small η as well, as long as the perturbation $\varepsilon = \varepsilon(\eta)$ is small.

Thus the key point of the stability analysis is to show that the inverse of the **stability constant** (later: operator/matrix) given in (3.2.3) is bounded. As indicated in (3.2.2), the control of the stability constant typically will have two ingredients: we need

(i) an upper bound on m_{sc}, the solution of the deterministic Dyson equation (2.3.12);

(ii) an upper bound on the inverse of $1 - m_{sc}^2$.

In the Wigner case, when m_{sc} is explicitly given (2.3.6), both bounds are easy to obtain. In fact, m_{sc} remains bounded for any z, while $1 - m_{sc}^2$ remains separated away from zero except near two special values of the spectral parameter: $z = \pm 2$. These are exactly the edges of the semicircle law, where an instability arises since here $m_{sc} \approx \pm 1$ (the same instability can be seen from the explicit solution of the quadratic equation).

We will see that it is not a coincidence: the edges of the asymptotic density ρ are always the critical points where the inverse of the stability constant blows up. These regimes require more careful treatment which typically consists in exploiting the fact that the error term D is proportional with the local density, hence it is also smaller near the edge. This additional smallness of D competes with the deteriorating upper bound on the inverse of the stability constant near the edge.

In these notes we will focus on the behavior in the bulk, i.e. we consider spectral parameters $z = E + i\eta$ where $\rho(E) \geqslant c > 0$ for fixed positive constants. This will simplify many estimates. The regimes where E is separated away from the support of ρ are even easier and we will not consider them here. The edge analysis is more complicated and we refer the reader to the original papers.

4. Models of increasing complexity

4.1. Basic setup In this section we introduce subsequent generalizations of the original Wigner ensemble. We also mention the key features of their resolvent that will be proven later along the local laws. The $N \times N$ matrix

$$H = \begin{pmatrix} h_{11} & h_{12} & \dots & h_{1N} \\ h_{21} & h_{22} & \dots & h_{2N} \\ \vdots & \vdots & & \vdots \\ h_{N1} & h_{N2} & \dots & h_{NN} \end{pmatrix} \tag{4.1.1}$$

will always be hermitian, $H = H^*$ and centered, $\mathbb{E}H = 0$. The distinction between real symmetric and complex hermitian cases play no role here; both symmetry classes are allowed. Many quantities, such as the distribution of H, the matrix of variances S, naturally depend on N, but for notational simplicity we will often omit this dependence from the notation.

We will always assume that we are in the mean field regime, i.e. the typical size of the matrix elements is of order $N^{-1/2}$ in a high moment sense:

$$\max_{ij} \mathbb{E}\big|\sqrt{N}h_{ij}\big|^p \leqslant \mu_p \tag{4.1.2}$$

for any p with some sequence of constants μ_p. This strong moment condition can be substantially relaxed but we will not focus on this direction.

4.2. Wigner matrix We assume that the matrix elements of H are independent (up to the hermitian symmetry) and identically distributed. We choose the normalization such that

$$\mathbb{E}|h_{ij}|^2 = \frac{1}{N},$$

see (1.1.4) for explanation. The asymptotic density of eigenvalues is the semicircle law, $\rho_{sc}(x)$ (1.2.4) and its Stieltjes transform $m_{sc}(z)$ is given explicitly in (2.3.6). The corresponding self-consistent (deterministic) equation (Dyson equation) is a **scalar equation**

$$1 + (z + m)m = 0, \qquad \operatorname{Im} m > 0,$$

that is solved by $m = m_{sc}$. The inverse of the stability "operator" is just the constant

$$\frac{1}{1 - m^2}, \qquad m = m_{sc}.$$

The resolvent $G(z) = (H - z)^{-1}$ is approximately constant diagonal in the **entrywise sense**, i.e.

$$G_{ij}(z) \approx \delta_{ij} m_{sc}(z). \tag{4.2.1}$$

In particular, the diagonal elements are approximately the same

$$G_{ii} \approx G_{jj} \approx m_{sc}(z).$$

This also implies that the normalized trace (Stieltjes transform of the empirical eigenvalue density) is close to m_{sc}

$$m_N(z) = \frac{1}{N} \operatorname{Tr} G(z) \approx m_{sc}(z), \tag{4.2.2}$$

which we often call an approximation in **average (or tracial) sense**.

Moreover, G is also diagonal in **isotropic sense**, i.e. for any vectors $\mathbf{x}, \mathbf{y}$ (more precisely, any sequence of vectors $\mathbf{x}^{(N)}, \mathbf{y}^{(N)} \in \mathbb{C}^N$) we have

$$G_{\mathbf{xy}} := \langle \mathbf{x}, G\mathbf{y} \rangle \approx m_{sc}(z) \langle \mathbf{x}, \mathbf{y} \rangle. \tag{4.2.3}$$

In Section 4.6 we will comment on the precise meaning of $\approx$ in this context, incorporating the fact that G is random.

If these relations hold for any fixed $\eta = \operatorname{Im} z$, independent of N, then we talk about **global law**. If they hold down to $\eta \geqslant N^{-1+\gamma}$ with some $\gamma \in (0,1)$, then we talk about **local law**. If $\gamma > 0$ can be chosen arbitrarily small (independent of N), than we talk about **local law on the optimal scale**.

4.3. Generalized Wigner matrix We assume that the matrix elements of H are independent (up to the hermitian symmetry), but not necessarily identically distributed. We define the matrix of variances as

$$S := \begin{pmatrix} s_{11} & s_{12} & \cdots & s_{1N} \\ s_{21} & s_{22} & \cdots & s_{2N} \\ \vdots & \vdots & & \vdots \\ s_{N1} & s_{N2} & \cdots & s_{NN} \end{pmatrix}, \qquad s_{ij} := \mathbb{E}|h_{ij}|^2. \tag{4.3.1}$$

We assume that

$$\sum_{j=1}^{N} s_{ij} = 1, \qquad \text{for every} \quad i = 1, 2, \ldots, N, \tag{4.3.2}$$

i.e., the deterministic $N \times N$ matrix of variances, $S = (s_{ij})$, is symmetric and doubly stochastic. The key point is that the row sums are all the same. The fact that the sum in (4.3.2) is exactly one is a chosen normalization. The original Wigner ensemble is a special case, $s_{ij} = \frac{1}{N}$.

Although generalized Wigner matrices form a bigger class than the Wigner matrices, the key results are exactly the same. The asymptotic density of states is still the semicircle law, G is constant diagonal in both the entrywise and isotropic senses:

$$G_{ij} \approx \delta_{ij} m_{sc} \qquad \text{and} \qquad G_{\mathbf{xy}} = \langle \mathbf{x}, G\mathbf{y} \rangle \approx m_{sc} \langle \mathbf{x}, \mathbf{y} \rangle.$$

In particular, the diagonal elements are approximately the same

$$G_{ii} \approx G_{jj}$$

and we have the same averaged law

$$m_N(z) = \frac{1}{N} \operatorname{Tr} G(z) \approx m_{sc}(z).$$

However, within the proof some complications arise. Although eventually G_{ii} turns out to be essentially independent of i, there is no a-priori complete permutation symmetry among the indices. We will need to consider the equations for each G_{ii} as a coupled system of N equations. The corresponding Dyson equation is a genuine **vector equation** of the form

$$1 + (z + (S\mathbf{m})_i) m_i = 0, \qquad i = 1, 2, \ldots N \tag{4.3.3}$$

for the unknown N-vector $\mathbf{m} = (m_1, m_2, \ldots, m_N)$ with $m_j \in \mathbb{H}$ and we will see that $G_{jj} \approx m_j$. The matrix S may also be called **self-energy matrix** according to the analogy explained around (3.1.14). Owing to (4.3.2), the solution to (4.3.3) is still the constant vector $m_i = m_{sc}$, but the stability operator depends on S and it is given by the matrix

$$1 - m_{sc}^2 S.$$

4.4. Wigner type matrix We still assume that the matrix elements are independent, but we impose no special algebraic condition on the variances S. For normalization purposes, we will assume that $\|S\|$ is bounded, independently of N, this guarantees that the spectrum of H also remains bounded. We only require an upper bound of the form

$$\max_{ij} s_{ij} \leqslant \frac{C}{N} \tag{4.4.1}$$

for some constant C. This is a typical mean field condition, it guarantees that no matrix element is too big. Notice that at this stage there is no requirement for a lower bound, i.e. some s_{ij} may vanish. However, the analysis becomes considerably harder if large blocks of S can become zero, so for pedagogical convenience later in these notes we will assume that $s_{ij} \geqslant c/N$ for some $c > 0$.

The corresponding Dyson equation is just the **vector Dyson equation** (4.3.3):

$$1 + (z + (S\mathbf{m})_i)m_i = 0, \qquad i = 1, 2, \dots N \tag{4.4.2}$$

but the solution is not the constant vector any more. We will see that the system of equations (4.4.2) still has a unique solution $\mathbf{m} = (m_1, m_2, \dots, m_N)$ under the side condition $m_j \in \mathbb{H}$, but the components of $\mathbf{m}$ may differ and they are not given by m_{sc} any more.

The components m_i approximate the diagonal elements of the resolvent G_{ii}. Correspondingly, their average

$$\langle \mathbf{m} \rangle := \frac{1}{N} \sum_i m_i, \tag{4.4.3}$$

is the Stieltjes transform of a measure ρ that approximates the empirical density of states. We will call this measure the **self-consistent density of states** since it is obtained from the self-consistent Dyson equation. It is well-defined for any finite N and if it has a limit as $N \to \infty$, then the limit coincides with the asymptotic density introduced earlier (e.g. the semicircle law for Wigner and generalized Wigner matrices). However, our analysis is more general and it does not need to assume the existence of this limit (see Remark 4.4.6 later).

In general there is no explicit formula for ρ, it has to be computed by taking the inverse Stieltjes transform of $\langle \mathbf{m}(z) \rangle$:

$$\rho(d\tau) = \lim_{\eta \to 0+} \frac{1}{\pi} \operatorname{Im} \langle \mathbf{m}(\tau + i\eta) \rangle \, d\tau. \tag{4.4.4}$$

No simple closed equation is known for the scalar quantity $\langle \mathbf{m}(z) \rangle$, even if one is interested only in the self-consistent density of states or its Stieltjes transform, the only known way to compute it is to solve (4.4.2) first and then take the average of the solution vector. Under some further conditions on S, the density of states is supported on finitely many intervals, it is real analytic away from the edges of these intervals and it has a specific singularity structure at the edges, namely it can have either square root singularity or cubic root cusp, see Section 6.1 later.

The resolvent is still approximately diagonal and it is given by the i-th component of $\mathbf{m}$:

$$G_{ij}(z) \approx \delta_{ij} m_i(z),$$

but in general

$$G_{ii} \not\approx G_{jj}, \qquad i \neq j.$$

Accordingly, the isotropic law takes the form

$$G_{\mathbf{xy}} = \langle \mathbf{x}, G\mathbf{y} \rangle \approx \langle \bar{\mathbf{x}}\mathbf{m}\mathbf{y} \rangle$$

and the averaged law

$$\frac{1}{N} \operatorname{Tr} G \approx \langle \mathbf{m} \rangle.$$

Here $\bar{\mathbf{x}}\mathbf{m}\mathbf{y}$ stands for the entrywise product of vectors, i.e., $\langle \bar{\mathbf{x}}\mathbf{m}\mathbf{y} \rangle = \frac{1}{N}\sum_i \bar{x}_i m_i y_i$.

The stability operator is

$$1 - \mathbf{m}^2 S, \tag{4.4.5}$$

where $\mathbf{m}^2$ is understood as an entrywise multiplication, so the linear operator $\mathbf{m}^2 S$ acts on any vector $\mathbf{x} \in \mathbb{C}^N$ as

$$[(\mathbf{m}^2 S)\mathbf{x}]_i := m_i^2 \sum_j s_{ij} x_j.$$

Notational convention. Sometimes we write the equation (4.4.2) in the concise vector form as

$$-\frac{1}{\mathbf{m}} = z + S\mathbf{m}.$$

Here we introduce the convention that for any vector $\mathbf{m} \in \mathbb{C}^N$ and for any function $f : \mathbb{C} \to \mathbb{C}$, the symbol $f(\mathbf{m})$ denotes the N-vector with components $f(m_j)$, that is,

$$f(\mathbf{m}) := \big(f(m_1), f(m_2), \dots, f(m_N)\big), \qquad \text{for any } \mathbf{m} = (m_1, m_2, \dots, m_N).$$

In particular, $1/\mathbf{m}$ is the vector of the reciprocals $1/m_i$. Similarly, the entrywise product of two N-vectors $\mathbf{x}, \mathbf{y}$ is denoted by $\mathbf{xy}$; this is the N-vector with components

$$(\mathbf{xy})_i := x_i y_i$$

and similarly for products of more than two factors. Finally $\mathbf{x} \leqslant \mathbf{y}$ for real vectors means $x_i \leqslant y_i$ for all i.

4.4.6. A remark on the density of states The Wigner type matrix is the first ensemble where the various concepts of density of states truly differ. The wording "density of states" has been used slightly differently by various authors in random matrix theory; here we use the opportunity to clarify this point. Typically, in the physics literature the **density of states** means the statistical average of the

empirical density of states μ_N defined in (1.2.2), i.e.

$$\mathbb{E}\mu_N(d\tau) = \mathbb{E}\frac{1}{N}\sum_{i=1}^{N}\delta(\lambda_i - \tau).$$

This object depends on N, but very often it has a limit (in a weak sense) as N, the system size, goes to infinity. The limit, if exists, is often called the **limiting (or asymptotic) density of states**.

In general it is not easy to find μ_N or its expectation; the vector Dyson equation is essentially the only way to proceed. However, the quantity computed in (4.4.4), called the **self-consistent density of states**, is not exactly the density of states, it is only a good approximation. The local law states that the empirical (random) eigenvalue density μ_N can be very well approximated by the self-consistent density of states, computed from the Dyson equation and (4.4.4). Here "very well" means in high probability and with an explicit error bound of size $1/N\eta$, i.e. on larger scales we have more precise bound, but we still have closeness even down to scales $\eta \geqslant N^{-1+\gamma}$. High probability bounds imply that also the density of states $\mathbb{E}\mu_N$ is close to the self-consistent density of states ρ, but in general they are not the same. Note that the significance of the local law is to approximate a random quantity with a deterministic one if N is large; there is no direct statement about any $N \to \infty$ limit. The variance matrix S depends on N and a-priori there is no relation between S-matrices for different N's.

In some cases a limiting version of these objects also exists. For example, if the variances s_{ij} arise from a deterministic nonnegative profile function $S(x, y)$ on $[0, 1]^2$ with some regularity, i.e.

$$s_{ij} = \frac{1}{N}S\Big(\frac{i}{N}, \frac{j}{N}\Big),$$

then the sequence of the self-consistent density of states $\rho^{(N)}$ have a limit. If the global law holds, then this limit must be the limiting density of states, defined as the limit of $\mathbb{E}\mu_N$. This is the case for Wigner matrices in a trivial way: the self-consistent density of states is always the semicircle for any N. However, the density of states for finite N is not the semicircle law; it depends on the actual distribution of the matrix elements, but decreasingly as N increases.

In these notes we will focus on computing the self-consistent density of states and proving local laws for fixed N; we will not consider the possible large N limits of these objects.

4.5. Correlated random matrix For this class we drop the independence condition, so the matrix elements of H may have nontrivial correlations in addition to the one required by the hermitian symmetry $h_{ij} = \bar{h}_{ji}$. The Dyson equation is still determined by the second moments of H, but the covariance structure of all matrix elements is not described by a matrix; but by a four-tensor. We already introduced in (3.1.23) the necessary "super operator"

$$\mathcal{S}[R] := \mathbb{E}HRH$$

acting linearly on the space of $N \times N$ matrices R. Explicitly

$$\mathcal{S}[R]_{ij} = \mathbb{E} \sum_{ab} h_{ia} R_{ab} h_{bj} = \sum_{ab} \left[\mathbb{E} h_{ia} h_{bj}\right] R_{ab}.$$

The analogue of the upper bound (4.4.1) is

$$\mathcal{S}[R] \leqslant C\langle R \rangle$$

for any positive definite matrix $R \geqslant 0$, where we introduced the notation

$$\langle R \rangle := \frac{1}{N} \operatorname{Tr} R.$$

In the actual proofs we will need a lower bound of the form $\mathcal{S}[R] \geqslant c\langle R \rangle$ and further conditions on the decay of correlations among the matrix elements of H.

The corresponding Dyson equation becomes a **matrix equation**

$$I + (z + \mathcal{S}[M])M = 0 \tag{4.5.1}$$

for the unknown matrix $M = M(z) \in \mathbb{C}^{N \times N}$ under the constraint that $\operatorname{Im} M \geqslant 0$. Recall that the imaginary part of any matrix is a hermitian matrix defined by

$$\operatorname{Im} M = \frac{1}{2i}(M - M^*).$$

In fact, one may add a hermitian **external source matrix** $A = A^*$ and consider the more general equation

$$I + (z - A + \mathcal{S}[M])M = 0. \tag{4.5.2}$$

In random matrix applications, A plays the role of the matrix of expectations, $A = \mathbb{E}H$. We will call (4.5.2) and (4.5.1) the **matrix Dyson equation** with or without external source. The equation (4.5.2) has a unique solution and in general it is a non-diagonal matrix even if A is diagonal. Notice that the Dyson equation contains only the second moments of the elements of H via the operator $\mathcal{S}$; no higher order correlations appear, although in the proofs of the local laws further conditions on the correlation decay are necessary.

The Stieltjes transform of the density of states is given by

$$\langle M(z) \rangle = \frac{1}{N} \operatorname{Tr} M(z).$$

The matrix $M = M(z)$ approximates the resolvent in the usual senses, i.e. we have

$$G_{ij}(z) \approx M_{ij}(z),$$

$$\langle \mathbf{x}, G\mathbf{y} \rangle \approx \langle \mathbf{x}, M\mathbf{y} \rangle,$$

and

$$\frac{1}{N} \operatorname{Tr} G \approx \langle M \rangle.$$

Since in general M is not diagonal, the resolvent G is not approximately diagonal any more. We will call M, the solution to the matrix Dyson equation (4.5.2), the **self-consistent Green function** or **self-consistent resolvent**.

The stability operator is of the form

$$I - \mathcal{C}_M \mathcal{S},$$

where $\mathcal{C}_M$ is the linear map acting on the space of matrices as

$$\mathcal{C}_M[R] := MRM.$$

In other words, the stability operator is the linear map $R \to R - M\mathcal{S}[R]M$ on the space of matrices.

The independent case (Wigner type matrix) is a special case of the correlated ensemble and it is interesting to exhibit their relation. In this case the super-operator $\mathcal{S}$ maps diagonal matrices to diagonal matrix. For any vector $\mathbf{v} \in \mathbb{C}^N$ we denote by $\mathrm{diag}(\mathbf{v})$ the $N \times N$ diagonal matrix with $(\mathrm{diag}(\mathbf{v}))_{ii} = v_i$ in the diagonal. Then we have, for the independent case with $s_{ab} := \mathbb{E}|h_{ab}|^2$ as before,

$$\big(\mathcal{S}[\mathrm{diag}(\mathbf{v})]\big)_{ij} = \sum_a \big[\mathbb{E}\bar{h}_{ai}h_{aj}\big]v_a = \delta_{ij}(S\mathbf{v})_i,$$

thus

$$\mathcal{S}[\mathrm{diag}(\mathbf{v})] = \mathrm{diag}(S\mathbf{v}).$$

Exercise 4.5.3. *Check that in the independent case, the solution* M *to* (4.5.1) *is diagonal,* $M = \mathrm{diag}(\mathbf{m})$*, where* $\mathbf{m}$ *solves the vector Dyson equation* (4.4.2). *Verify that the statements of the local laws formulated in the general correlated language reduce to those for the Wigner type problem. Check that the stability operator* $I - \mathcal{C}_M\mathcal{S}$ *restricted to diagonal matrices is equivalent to the stability operator* (4.4.5).

The following table summarizes the four classes of ensembles we discussed.

Name	Dyson Equation	For	Stability op	Feature
Wigner $\mathbb{E}\lvert h_{ij}\rvert^2 = s_{ij} = \frac{1}{N}$	$1 + (z+m)m = 0$	$m \approx \frac{1}{N}\operatorname{Tr} G$	$1 - m^2$	Scalar Dyson equation, $m = m_{sc}$ is explicit
Generalized Wigner $\sum_j s_{ij} = 1$	$1 + (z + S\mathbf{m})\mathbf{m} = 0$	$m_i \approx \frac{1}{N}\operatorname{Tr} G$	$1 - m^2 S$	Vector Dyson equation, Split S as $S^\perp + \lvert\mathbf{e}\rangle\langle\mathbf{e}\rvert$
Wigner-type s_{ij} arbitrary	$1 + (z + S\mathbf{m})\mathbf{m} = 0$	$m_i \approx G_{ii}$	$1 - \mathbf{m}^2 S$	Vector Dyson equation, $\mathbf{m}$ to be determined
Correlated matrix $\mathbb{E}h_{xy}h_{uw} \not\approx \delta_{xw}\delta_{yu}$	$I + (z + \mathcal{S}[M])M = 0$	$M_{ij} \approx G_{ij}$	$1 - M\mathcal{S}[\cdot]M$	Matrix Dyson equation Super-operator

We remark that in principle the averaged law (density of states) for generalized Wigner ensemble could be studied via a scalar equation only since the answer is given by the scalar Dyson equation, but in practice a vector equation is studied in order to obtain entrywise and isotropic information. However, Wigner-type matrices need a vector Dyson equation even to identify the density of states. Correlated matrices need a full scale matrix equation since the answer M is typically a non-diagonal matrix.

4.6. The precise meaning of the approximations In the previous sections we used the sloppy notation $\approx$ to indicate that the (random) resolvent G in various senses is close to a deterministic object. We now explain what we mean by that. Consider first (4.2.1), the entrywise statement for the Wigner case:

$$G_{ij}(z) \approx \delta_{ij} m_{sc}(z).$$

More precisely, we will see that

$$|G_{ij}(z) - \delta_{ij} m_{sc}(z)| \lesssim \frac{1}{\sqrt{N\eta}}, \qquad \eta = \operatorname{Im} z \tag{4.6.1}$$

holds. Here the somewhat sloppy notation $\lesssim$ indicates that the statement holds with very high probability and with an additional factor N^{ε}. The very precise form of (4.6.1) is the following: for any $\varepsilon, D > 0$ we have

$$\max_{ij} \mathbb{P}\Big(|G_{ij}(z) - \delta_{ij} m_{sc}(z)| \geqslant \frac{N^{\varepsilon}}{\sqrt{N\eta}}\Big) \leqslant \frac{C_{D,\varepsilon}}{N^{D}} \tag{4.6.2}$$

with some constant $C_{D,\varepsilon}$ independent of N, but depending on D, ε and the sequence μ_p bounding the moments in (4.1.2). We typically consider only spectral parameters with

$$|z| \leqslant C, \qquad \eta \geqslant N^{-1+\gamma} \tag{4.6.3}$$

for any fixed positive constants C and γ, and we encourage the reader to think of z satisfying these constraints, although our results are eventually valid for a larger set as well (the restriction $|z| \leqslant C$ can be replaced with $|z| \leqslant N^{C}$ and the lower bound on η is not necessary if $E = \operatorname{Re} z$ is away from the support of the density of states).

Notice that (4.6.2) is formulated for any fixed z, but the probability control is very strong, so one can extend the same bound to hold simultaneously for any z satisfying (4.6.3), i.e.

(4.6.4)

$$\mathbb{P}\Big(\exists z \in \mathbb{C} \,:\, |z| \leqslant C, \operatorname{Im} z \geqslant N^{-1+\gamma}, \ \max_{ij} |G_{ij}(z) - \delta_{ij} m_{sc}(z)| \geqslant \frac{N^{\varepsilon}}{\sqrt{N\eta}}\Big) \leqslant \frac{C_{D,\varepsilon}}{N^{D}}.$$

Bringing the maximum over i, j inside the probability follows from a simple union bound. The same trick does not work directly for bringing the maximum over all z inside since there are uncountable many of them. But notice that the function

$$z \to G_{ij}(z) - \delta_{ij} m_{sc}(z)$$

is Lipschitz continuous with a Lipschitz constant C/η^2 which is bounded by CN^2 in the domain (4.6.3). Therefore, we can first choose a very dense, say N^{-3}-grid of z values, apply the union bound to them and then argue with Lipschitz continuity for all other z values.

Exercise 4.6.5. *Make this argument precise, i.e. show that* (4.6.4) *follows from* (4.6.2).

Similar argument does not quite work for the isotropic formulation. While (4.2.3) holds for any fixed (sequences of) ℓ^2-normalized vectors $\mathbf{x}$ and $\mathbf{y}$, i.e. in its precise formulation we have

$$\mathbb{P}\Big(|\langle \mathbf{x}, G(z)\mathbf{y}\rangle - m_{sc}(z)\langle \mathbf{x}, \mathbf{y}\rangle| \geqslant \frac{N^{\varepsilon}}{\sqrt{N\eta}}\Big) \leqslant \frac{C_{D,\varepsilon}}{N^{D}} \tag{4.6.6}$$

for any fixed $\mathbf{x}, \mathbf{y}$ with $\|\mathbf{x}\|_2 = \|\mathbf{y}\|_2 = 1$, we cannot bring the supremum over all $\mathbf{x}, \mathbf{y}$ inside the probability. Clearly $\max_{\mathbf{x},\mathbf{y}} \langle \mathbf{x}, G(z)\mathbf{y} \rangle$ would give the norm of G which is $1/\eta$.

Furthermore, a common feature of all our estimates is that the local law in averaged sense is one order more precise than the entrywise or isotropic laws, e.g. for the precise form of (4.2.2) we have

$$\mathbb{P}\Big(\Big| \frac{1}{N} \operatorname{Tr} G(z) - m_{sc}(z) \Big| \geqslant \frac{N^{\varepsilon}}{N\eta} \Big) \leqslant \frac{C_{D,\varepsilon}}{N^D}. \tag{4.6.7}$$

5. Physical motivations

The primary motivation to study local spectral statistics of large random matrices comes from nuclear and condensed matter physics where the matrix models a quantum Hamiltonian and its eigenvalues correspond to energy levels. Other applications concern statistics (especially largest eigenvalues of sample covariance matrices of the form XX^* where X has independent entries), wireless communication and neural networks. Here we focus only on physical motivations.

5.1. Basics of quantum mechanics We start with summarizing the basic setup of quantum mechanics. A quantum system is described by a *configuration space* Σ, e.g. $\Sigma = \{\uparrow, \downarrow\}$ for a single spin, or $\Sigma = \mathbb{Z}^3$ for an electron hopping on an ionic lattice or $\Sigma = \mathbb{R}^3$ for an electron in vacuum. Its elements $x \subset \Sigma$ are called configurations and it is equipped with a natural measure (e.g. the counting measure for discrete Σ or the Lebesgue measure for $\Sigma = \mathbb{R}^3$). The *state space* is a complex Hilbert space, typically the natural L^2-space of Σ, i.e. $\ell^2(\Sigma) = \mathbb{C}^2$ in case of a single spin or $\ell^2(\mathbb{Z}^3)$ for an electron in a lattice. Its elements are called wave functions, these are normalized functions $\psi \in \ell^2(\Sigma)$, with $\|\psi\|_2 = 1$. The quantum wave function entirely describes the quantum state. In fact its overall phase does not carry measurable physical information; wave functions ψ and $e^{ic}\psi$ are indistinguishable for any real constant c. This is because only quadratic forms of ψ are measurable, i.e. only quantities of the form $\langle \psi, O\psi \rangle$ where O is a self-adjoint operator. The probability density $|\psi(x)|^2$ on the configuration space describes the probability to find the quantum particle at configuration x.

The dynamics of the quantum system, i.e. the process how ψ changes in time, is described by the *Hamilton operator*, which is a self-adjoint operator acting on the state space $\ell^2(\Sigma)$. If Σ is finite, then it is an $\Sigma \times \Sigma$ hermitian matrix. The matrix elements $H_{xx'}$ describe the quantum transition rates from configuration x to x'. The dynamics of ψ is described by the Schrödinger equation

$$i\partial_t \psi_t = H\psi_t$$

with a given initial condition $\psi_{t=0} := \psi_0$. The solution is given by $\psi_t = e^{-itH}\psi_0$. This simple formula is however, quite hard to compute or analyze, especially for large times. Typically one writes up the spectral decomposition of H in the form $H = \sum_n \lambda_n |\mathbf{v}_n\rangle\langle \mathbf{v}_n|$, where λ_n and $\mathbf{v}_n$ are the eigenvalues and eigenvectors of H, i.e. $H\mathbf{v}_n = \lambda_n \mathbf{v}_n$. Then

$$e^{-itH}\psi_0 = \sum_n e^{-it\lambda_n}\langle \mathbf{v}_n, \psi_0\rangle \mathbf{v}_n =: \sum_n e^{-it\lambda_n} c_n \mathbf{v}_n.$$

If ψ_0 coincides with one of the eigenvectors, $\psi_0 = \mathbf{v}_n$, then the sum above collapses and

$$\psi_t = e^{-itH}\psi_0 = e^{-it\lambda_n}\mathbf{v}_n.$$

Since the physics encoded in the wave function is insensitive to an overall phase, we see that eigenvectors remain unchanged along the quantum evolution.

Once ψ_0 is a genuine linear combination of several eigenvectors, quadratic forms of ψ_t become complicated:

$$\langle \psi_t, O\psi_t\rangle = \sum_{nm} e^{it(\lambda_m - \lambda_n)} \bar{c}_m c_n \langle \mathbf{v}_m, O\mathbf{v}_n, \rangle.$$

This double sum is highly oscillatory and subject to possible periodic and quasi-periodic behavior depending on the commensurability of the eigenvalue differences $\lambda_m - \lambda_n$. Thus the statistics of the eigenvalues carry important physical information on the quantum evolution.

The Hamiltonian H itself can be considered as an observable, and the quadratic form $\langle \psi, H\psi\rangle$ describes the *energy* of the system in the state ψ. Clearly the energy is a conserved quantity

$$\langle \psi_t, H\psi_t\rangle = \langle e^{-itH}\psi_0, He^{-itH}\psi_t\rangle = \langle \psi_0, H\psi_t\rangle.$$

The eigenvalues of H are called **energy levels** of the system.

Disordered quantum systems are described by random Hamiltonians, here the randomness comes from an external source and is often described phenomenologically. For example, it can represent impurities in the state space (e.g. the ionic lattice is not perfect) that we do not wish to (or cannot) describe with a deterministic precision, only their statistical properties are known.

5.2. The "grand" universality conjecture for disordered quantum systems The general belief is that disordered quantum systems with "sufficient" complexity are subject to a strong dichotomy. They exhibit one of the following two behaviors: they are either in the *insulating* or in the *conducting* phase. These two phases are also called **localization** and **delocalization** regime. The behavior may depend on the energy range: the same quantum system can be simultaneously in both phases but at different energies.

The **insulator (or localized regime)** is characterized by the following properties:

1) Eigenvectors are **spatially localized**, i.e. the overwhelming mass of the probability density $|\psi(x)|^2\,dx$ is supported in a small subset of Σ. More precisely, there exists an $\Sigma' \subset \Sigma$, with $|\Sigma'| \ll |\Sigma|$ such that

$$\int_{\Sigma\setminus\Sigma'} |\psi(x)|^2\,dx \ll 1$$

2) **Lack of transport**: if the state ψ_0 is initially localized, then it remains so (maybe on a larger domain) for all times. Transport is usually measured with the mean square displacement if Σ has a metric. For example, for $\Sigma = \mathbb{Z}^d$ we consider

$$\langle x^2 \rangle_t := \sum_{x \in \mathbb{Z}^3} x^2 |\psi_t(x)|^2, \tag{5.2.1}$$

then localization means that

$$\sup_{t \geqslant 0} \langle x^2 \rangle_t \leqslant C$$

assuming that at time $t = 0$ we had $\langle x^2 \rangle_{t=0} < \infty$. Strictly speaking this concept makes sense only if Σ is infinite, but one can require that the constant C does not depend on some relevant size parameter of the model.

3) Green functions have a **finite localization length** ℓ, i.e. the off diagonal matrix elements of the resolvent decays exponentially (again for $\Sigma = \mathbb{Z}^d$ for simplicity)

$$|G_{xx'}| \leqslant Ce^{-|x-x'|/\ell}.$$

4) **Poisson local eigenvalue statistics**: Nearby eigenvalues are statistically independent, i.e. they approximately form a Poisson point process after appropriate rescaling.

The **conducting (or delocalized) regime** is characterized by the opposite features:

1) Eigenvectors are **spatially delocalized**, i.e. the mass of the probability density $|\psi(x)|^2$ is not concentrated on a much smaller subset of Σ.

2) **Transport via diffusion**: The mean square displacement (5.2.1) grows diffusively, e.g. for $\Sigma = \mathbb{Z}^d$

$$\langle x^2 \rangle_t \approx Dt$$

with some nonzero constant D (diffusion constant) for large times. If Σ is a finite part of $\mathbb{Z}^d$, e.g. $\Sigma = [1, L]^d \cap \mathbb{Z}^d$, then this relation should be modified so that the growth of $\langle x^2 \rangle_t$ with time can last only until the whole Σ is exhausted.

3) The Green function does not decay exponentially, the localization length $\ell = \infty$.

4) **Random matrix local eigenvalue statistics**: Nearby eigenvalues are statistically strongly dependent, in particular there is a level repulsion. They approximately form a GUE or GOE eigenvalue point process after appropriate rescaling. The symmetry type of the approximation is the same as the symmetry type of the original model (time reversal symmetry gives GOE).

The most prominent simple example for the conducting regime is the Wigner matrices or more generally Wigner-type matrices. They represent a quantum system where hopping from any site $x \in \Sigma$ to any other site $x' \in \Sigma$ is statistically equally likely (Wigner ensemble) or at least comparably likely (Wigner type ensemble).

Thus, a convenient way to represent the conducting regime is via a complete graph as illustrated below in Figure 5.2.2. This graph has one vertex for each of the $N = |\Sigma|$ states and an edge joins each pair of states. The edges correspond to the matrix elements $h_{xx'}$ in (4.1.1) and they are independent. For Wigner matrices there is no specific spatial structure present, the system is completely homogeneous. Wigner type ensembles model a system with an inhomogeneous spatial structure, but it is still a **mean field** model since most transition rates are comparable. However, some results on Wigner type matrices allow zeros in the matrix of variances S defined in (4.3.1), i.e. certain jumps are explicitly forbidden.

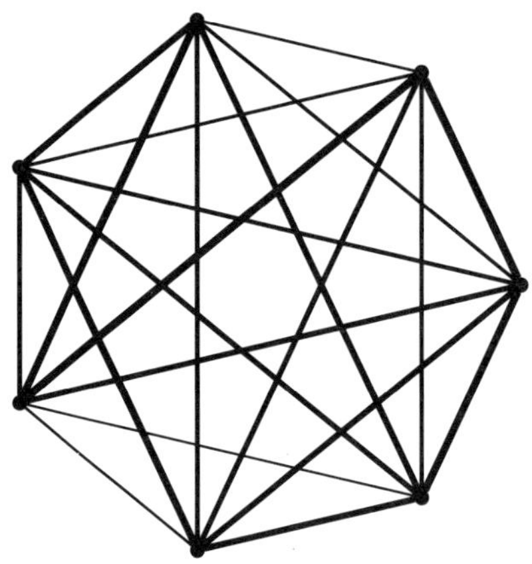

FIGURE 5.2.2. Graph schematically indicating the configuration space of $N = |\Sigma| = 7$ states with random quantum transition rates

The delocalization of the eigenvectors (item 1) was presented in (2.2.1), while item 4) is the WDM universality. The diffusive feature (item 2) is trivial since due to the mean field character, the maximal displacement is already achieved after $t \sim O(1)$. Thus the **Wigner matrix is in the delocalized regime.**

It is not so easy to present a non-trivial example for the insulator regime. A trivial example is if H is a diagonal matrix in the basis given by Σ, with i.i.d. entries in the diagonal, then items 1)–4) of the insulator regime clearly hold. Beyond the diagonal, even a short range hopping can become delocalized, for example the lattice Laplacian on $\mathbb{Z}^d$ has delocalized eigenvectors (plane waves). However, if the Laplacian is perturbed by a random diagonal, then localization may occur—this is the celebrated *Anderson metal-insulator transition* [13], which we now discuss.

5.3. Anderson model The prototype of the random Schrödinger operators is the Anderson model on the d-dimensional square lattice $\mathbb{Z}^d$. It consists of a Laplacian (hopping term to the neighbors) and a random potential:

$$H = \Delta + \lambda V \tag{5.3.1}$$

acting on $\ell^2(\mathbb{Z}^d)$. The matrix elements of the Laplacian are given by

$$\Delta_{xy} = \mathbf{1}(|x-y| = 1)$$

and the potential is diagonal, i.e.

$$V_{xy} = \delta_{xy} v_x,$$

where $\{v_x \ : \ i \in \mathbb{Z}^d\}$ is a collection of real i.i.d. random variables sitting on the lattice sites. For definiteness we assume that

$$\mathbb{E} v_x = 0, \qquad \mathbb{E} v_x^2 = 1$$

and λ is a coupling parameter. Notice that Δ is self-adjoint and bounded, while the potential at every site is bounded almost surely. For simplicity we may assume that the common distribution of v has bounded support, i.e. V, hence H are bounded operators. This eliminates some technical complications related to the proper definition of the self-adjoint extensions.

5.3.2. The free Laplacian For $\lambda = 0$, the spectrum is well known, the eigenvector equation $\Delta f - \mu f$, i.e.

$$\sum_{|y-x|=1} f_y = \mu f_x, \qquad \forall x \in \mathbb{Z}^d,$$

has plane waves parametrized by the d-torus, $k = (k_1, k_2, \dots, k_d) \in [-\pi, \pi]^d$ as eigenfunctions:

$$f_x = e^{ik \cdot x}, \qquad \mu = 2 \sum_{i=1}^{d} \cos k_i.$$

Although these plane waves are not ℓ^2-normalizable, they still form a complete system of generalized eigenvectors for the bounded self-adjoint operator Δ. The spectrum is the interval $[-2d, 2d]$ and it is a purely absolutely continuous spectrum (we will not need its precise definition if you are unfamiliar with it). Readers uncomfortable with unbounded domains can take a large torus $[-L, L]^d$, $L \in \mathbb{N}$, instead of $\mathbb{Z}^d$ as the configuration space. Then everything is finite dimensional, and the wave-numbers k are restricted to a finite lattice within the torus $[-\pi, \pi]^d$. Notice that the eigenvectors are still plane waves, in particular they are completely delocalized.

One may also study the time evolution $e^{it\Delta}$ (basically by Fourier transform) and one finds **ballistic behavior**, i.e. for the mean square displacement (5.2.1) one finds

$$\langle x^2 \rangle_t = \sum_{x \in \mathbb{Z}^d} x^2 |\psi_t(x)|^2 \sim Ct^2, \qquad \psi_t = e^{it\Delta} \psi_0$$

for large t. Thus for $\lambda = 0$ the system in many aspects is in the delocalized regime. Since randomness is completely lacking, it is not expected that other features of the delocalized regime hold, e.g. the local spectral statistics is not the one from random matrices – it is rather related to a lattice point counting problem. Furthermore, the eigenvalues have degeneracies, i.e. level repulsion, a main characteristics for random matrices, does not hold.

5.3.3. Turning on the randomness Now we turn on the randomness by taking some $\lambda \neq 0$. This changes the behavior of the system drastically in certain regimes. More precisely:

- In $d = 1$ **dimension** the system is in the localized regime as soon as $\lambda \neq 0$, see [52]
- In $d = 2$ **dimensions** On physical grounds it is conjectured that the system is localized for any $\lambda \neq 0$ [76]. No mathematical proof exists.
- In the most important physical $d = 3$ **dimensions** we expect a phase transition: The system is localized for large disorder, $|\lambda| \geqslant \lambda_0(d)$ or at the spectral edges [3, 49]. For small disorder and away from the spectral edges delocalization is expected but there is no rigorous proof. This is the celebrated **extended states or delocalization conjecture**, one of the few central holy grails of mathematical physics.

Comparing random Schrödinger with random matrices, we may write up the matrix of the $d = 1$ dimensional operator H (5.3.1) in the basis given by $\Sigma = [\![1, L]\!]$:

$$H = \Delta + \sum_{x=1}^{L} v_x = \begin{pmatrix} v_1 & 1 & & & & \\ 1 & v_2 & 1 & & & \\ & 1 & \ddots & & & \\ & & & \ddots & 1 & \\ & & & 1 & v_{L-1} & 1 \\ & & & & 1 & v_L \end{pmatrix}.$$

It is tridiagonal matrix with i.i.d. random variables in the diagonal and all ones in the minor diagonal. It is a **short range model** as immediate quantum transitions (jumps) are allowed only to the nearest neighbors. Structurally this H is very different from the typical Wigner matrix (5.2.2) where all matrix elements are roughly comparable (**mean field model**).

5.4. Random band matrices Random band matrices naturally interpolate between the mean field Wigner ensemble and the short range random Schrödinger operators. Let the state space be

$$\Sigma := [1, L]^d \cap \mathbb{Z}^d$$

a lattice box of linear size L in d dimensions. The total dimension of the state space is $N = |\Sigma| = L^d$. The entries of $H = H^*$ are centered, independent but not identically distributed – it is like the Wigner type ensemble, but without the mean field condition $s_{xy} = \mathbb{E}|h_{xy}|^2 \leqslant C/N$. Instead, we introduce a new parameter, $1 \leqslant W \leqslant L$ the **bandwidth** or the interaction range. We assume that the variances behave as

$$\mathbb{E}|h_{xy}|^2 = \frac{1}{W^d} f\Big(\frac{|x-y|}{W}\Big).$$

In $d = 1$ physical dimension the corresponding matrix is an $L \times L$ matrix with a nonzero band of width $2W$ around the diagonal. From any site a direct hopping of size W is possible, see the figure below with $L = 7$, $W = 2$:

$$H = \begin{pmatrix} * & * & * & 0 & 0 & 0 & 0 \\ * & * & * & * & 0 & 0 & 0 \\ * & * & * & * & * & 0 & 0 \\ 0 & * & * & * & * & * & 0 \\ 0 & 0 & * & * & * & * & * \\ 0 & 0 & 0 & * & * & * & * \\ 0 & 0 & 0 & 0 & * & * & * \end{pmatrix}$$

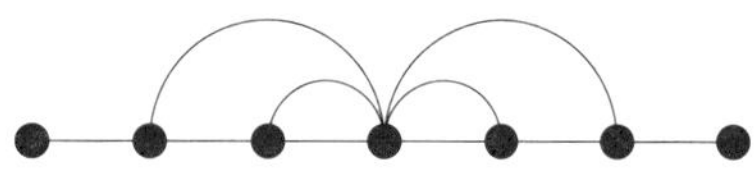

Clearly $W = L$ corresponds to the Wigner ensemble, while $W = 1$ is very similar to the random Schrödinger with its short range hopping. The former is delocalized, the latter is localized, hence there is a transition with can be probed by changing W from 1 to L. The following table summarizes "facts" from physics literature on the transition threshold:

Anderson metal-insulator transition occurs at the following thresholds:

$W \sim L^{1/2}$	$(d = 1)$	Supersymmetry [50]
$W \sim \sqrt{\log L}$	$(d = 2)$	Renormalization group scaling [1]
$W \sim W_0(d)$	$(d \geqslant 3)$	extended states conjecture [13]

All these conjectures are mathematically open, the most progress has been done in $d = 1$. It is known that we have localization in the regime $W \ll L^{1/8}$ [68] and delocalization for $W \gg L^{4/5}$ [35]. The two point correlation function of the characteristic polynomial was shown to be given by the Dyson sine kernel up to the threshold $W \gg L^{1/2}$ in [70].

In these lectures we restrict our attention to mean field models, i.e. band matrices will not be discussed. We nevertheless mentioned them because they are expected to be easier than the short range random Schrödinger operators and they still exhibit the Anderson transition in a highly nontrivial way.

5.5. Mean field quantum Hamiltonian with correlation Finally we explain how correlated random matrices with a certain correlation decay are motivated. We again equip the state space Σ with a metric to be able to talk about "nearby" states. It is then reasonable to assume that h_{xy} and $h_{xy'}$ are correlated if y and y' are close with a decaying correlation as $\text{dist}(y, y')$ increases.

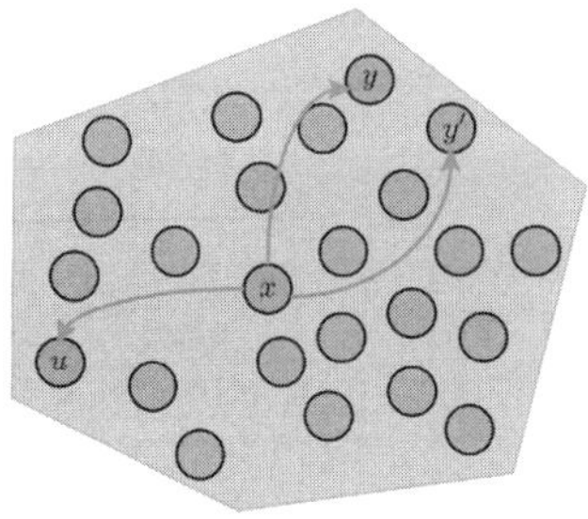

For example, in the figure h_{xy} and $h_{xy'}$ are strongly correlated but h_{xy} and h_{xu} are not (or only very weakly) correlated. We can combine this feature with an inhomogeneous spatial structure as in the Wigner-type ensembles.

6. Results

Here we list a few representative results with precise conditions. The results can be divided roughly into three categories:

- Properties of the solution of the Dyson equation, especially the singularity structure of the density of states and the boundedness of the inverse of the stability operator. This part of the analysis is deterministic.
- Local laws, i.e. approximation of the (random) resolvent G by the solution of the corresponding Dyson equation with very high probability down to the optimal scale $\eta \gg 1/N$.
- Bulk universality of the local eigenvalue statistics on scale $1/N$.

6.1. Properties of the solution to the Dyson equations

6.1.1. Vector Dyson equation First we focus on the vector Dyson equation (4.4.2) with a general symmetric variance matrix S motivated by Wigner type matrices:

$$-\frac{1}{\mathbf{m}} = z + S\mathbf{m}, \qquad \mathbf{m} \in \mathbb{H}^N, \quad z \in \mathbb{H} \tag{6.1.2}$$

(recall that the inverse of a vector $1/\mathbf{m}$ is understood component wise, i.e. $1/\mathbf{m}$ is an N vector with components $(1/\mathbf{m})_i = 1/m_i$). We may add an external source which is real vector $\mathbf{a} \in \mathbb{R}^N$ and the equation is modified to

$$-\frac{1}{\mathbf{m}} = z - \mathbf{a} + S\mathbf{m}, \qquad \mathbf{m} \in \mathbb{H}^N, \quad z \in \mathbb{H}, \tag{6.1.3}$$

but we will consider the $\mathbf{a} = 0$ case for simplicity. We equip the space $\mathbb{C}^N$ with the maximum norm,

$$\|\mathbf{m}\|_\infty := \max_i |m_i|,$$

and we let $\|S\|_\infty$ be the matrix norm induced by the maximum norm of vectors.

We start with the existence and uniqueness result for (6.1.2), see e.g. Proposition 2.1 in [4]:

Theorem 6.1.4. *The equation* (6.1.2) *has a unique solution* $\mathbf{m} = \mathbf{m}(z)$ *for any* $z \in \mathbb{H}$. *For each* $i \in [\![1, N]\!]$ *there is a probability measure* $\nu_i(dx)$ *on* $\mathbb{R}$ *(called* **generating measure***) such that* m_i *is the Stieltjes transform of* ν_i*:*

$$m_i(z) = \int_{\mathbb{R}} \frac{\nu_i(d\tau)}{\tau - z}, \tag{6.1.5}$$

and the support of all ν_i *lie in the interval* $[-2\|S\|_\infty^{1/2}, 2\|S\|_\infty^{1/2}]$. *In particular we have the trivial upper bound*

$$\|\mathbf{m}(z)\|_\infty \leqslant \frac{1}{\eta}, \qquad \eta = \operatorname{Im} z. \tag{6.1.6}$$

Recalling that the **self-consistent density of states** was defined in (4.4.3) via the inverse Stieltjes transform of $\langle \mathbf{m} \rangle = \frac{1}{N} \sum m_i$, we see that

$$\rho = \langle \boldsymbol{\nu} \rangle = \frac{1}{N} \sum_i \nu_i.$$

We now list two assumptions on S, although for some results we will need only one of them:

- **Boundedness:** We assume that there exists two positive constants c, C such that

$$\frac{c}{N} \leqslant s_{ij} \leqslant \frac{C}{N} \tag{6.1.7}$$

- **Hölder regularity:**

$$|s_{ij} - s_{i'j'}| \leqslant \frac{C}{N} \left[\frac{|i - i'| + |j - j'|}{N} \right]^{1/2} \tag{6.1.8}$$

We remark that the lower bound in (6.1.7) can be substantially weakened, in particular large zero blocks are allowed. For example, we may assume only that S has a substantial diagonal, i.e. $s_{ij} \geqslant \frac{c}{N} \cdot \mathbf{1}(|i - j| \leqslant \varepsilon N)$ with some fixed positive c, ε, but for simplicity of the presentation we follow (6.1.7).

The Hölder regularity (6.1.8) expresses a regularity on the order N scale in the matrix. It can be understood in the easiest way if we imagine that the matrix elements s_{ij} come from a macroscopic profile function $S(x, y)$ on $[0, 1] \times [0, 1]$ by the formula

$$s_{ij} = \frac{1}{N} S\left(\frac{i}{N}, \frac{j}{N}\right). \tag{6.1.9}$$

It is easy to check that if $S : [0, 1]^2 \to \mathbb{R}_+$ is Hölder continuous with a Hölder exponent $1/2$, then (6.1.8) holds. In fact, the Hölder regularity condition can also be weakened to **piecewise 1/2-Hölder regularity** (with finitely many pieces), in that case we assume that s_{ij} is of the form (6.1.9) with a profile function $S(x, y)$ that is piecewise Hölder continuous with exponent $1/2$, i.e. there exists a fixed (N-independent) partition $I_1 \cup I_2 \cup \ldots \cup I_n = [0, 1]$ of the unit interval into smaller intervals such that

$$\max_{ab} \sup_{x, x' \in I_a} \sup_{y, y' \in I_b} \frac{|S(x, y) - S(x', y')|}{|x - x'|^{1/2} + |y - y'|^{1/2}} \leqslant C. \tag{6.1.10}$$

The main theorems summarizing the properties of the solution to (6.1.2) are the following. The first theorem assumes only (6.1.7) and it is relevant in the bulk. We will prove it later in Section 6.1.11.

Theorem 6.1.11. *Suppose that* S *satisfies* (6.1.7). *Then we have the following bounds:*
(6.1.12)

$$\|\mathbf{m}(z)\|_\infty \lesssim \frac{1}{\rho(z) + dist(z, supp\rho)}, \qquad \rho(z) \lesssim \operatorname{Im} \mathbf{m}(z) \lesssim (1+|z|^2)\|\mathbf{m}(z)\|_\infty^2 \rho(z).$$

The second theorem additionally assumes (6.1.8), but the result is much more precise, in particular a complete analysis of singularities is possible.

Theorem 6.1.13. *[Theorem 2.6 in* [6]*] Suppose that* S *satisfies* (6.1.7) *and it is Hölder continuous* (6.1.8) *[or piecewise Hölder continuous* (6.1.10)*]. Then we have the following:*

(i) *The generating measures have Lebesgue density,* $\nu_i(d\tau) = \nu_i(\tau)\,d\tau$ *and the* **generating densities** ν_i *are uniformly 1/3-Hölder continuous, i.e.*

$$\max_i \sup_{\tau \neq \tau'} \frac{|\nu_i(\tau) - \nu_i(\tau')|}{|\tau - \tau'|^{1/3}} \leqslant C'. \tag{6.1.14}$$

(ii) *The set on which* ν_i *is positive is independent of* i*:*

$$\mathfrak{S} := \{\tau \in \mathbb{R} \,:\, \nu_i(\tau) > 0\}$$

and it is a union of finitely many open intervals. If S *is Hölder continuous in the sense of* (6.1.8), *then* $\mathfrak{S}$ *consist of a single interval.*

(iii) *The restriction of* $\mathbf{v}(\tau)$ *to* $\mathbb{R} \setminus \partial\mathfrak{S}$ *is analytic in* τ *(as a vector-valued function).*

(iv) *At the (finitely many) points* $\tau_0 \in \partial\mathfrak{S}$ *the generating density has one of the following two behaviors:*

CUSP: *If* τ_0 *is at the intersection of the closure of two connected components of* $\mathfrak{S}$, *then* $\mathbf{v}$ *has a cubic root singularity, i.e.*

$$\nu_i(\tau_0 + \omega) = c_i|\omega|^{1/3} + O(|\omega|^{2/3}) \tag{6.1.15}$$

with some positive constants c_i.

EDGE: *If* τ_0 *is not a cusp, then it is the right or left endpoint of a connected component of* $\mathfrak{S}$ *and* $\mathbf{v}$ *has a square root singularity at* τ_0*:*

$$\nu_i(\tau_0 \pm \omega) = c_i \omega^{1/2} + O(\omega), \qquad \omega \geqslant 0, \tag{6.1.16}$$

with some positive constants c_i.

The positive constant C' *in* (6.1.14) *depends only on the constants* c *and* C *in the conditions* (6.1.7) *and* (6.1.8) *[or* (6.1.10)*], in particular it is independent of* N. *The constants* c_i *in* (6.1.15) *and* (6.1.16) *are also uniformly bounded from above and below, i.e.,* $c'' \leqslant c_i \leqslant C''$, *with some positive constants* c'' *and* C'' *that, in addition to* c *and* C, *may also depend on the distance between the connected components of the generating density.*

Some of these statements will be proved in Section 7. We now illustrate this theorem by a few pictures. The first picture indicates a nontrivial S-profile (different shades indicate different values in the matrix) and the corresponding self-consistent density of states.

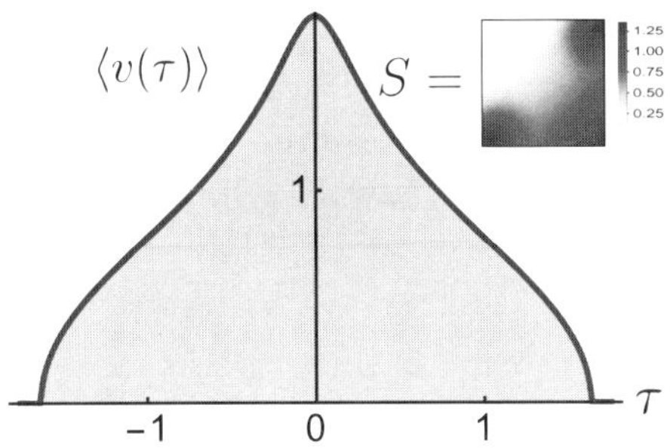

In particular, we see that in general the density of states is not the semicircle if $\sum_j s_{ij} \neq \text{const}$.

The next pictures show how the support of the self-consistent density of states splits via cusps as the value of s_{ij} slowly changes. Each matrix below the pictures is the corresponding variance matrix S represented as a 4×4 block matrix with $(N/4) \times (N/4)$ blocks with constant entries. Notice that the corresponding continuous profile function $S(x, y)$ is only piecewise Hölder (in fact, piecewise constant). As the parameter in the diagonal blocks increases, a small gap closes at a cusp, then it develops a small local minimum.

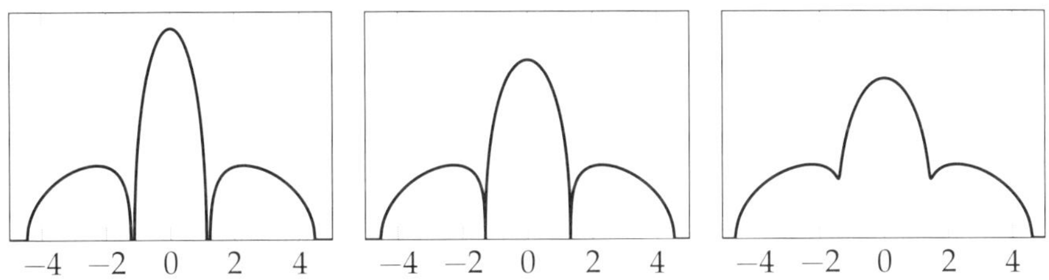

Small gap Exact cusp Small minimum

$$\frac{1}{N}\begin{pmatrix} .07 & 1 & 1 & 1 \\ 1 & .07 & .07 & .07 \\ 1 & .07 & .07 & .07 \\ 1 & .07 & .07 & .07 \end{pmatrix} \quad \frac{1}{N}\begin{pmatrix} .1 & 1 & 1 & 1 \\ 1 & .1 & .1 & .1 \\ 1 & .1 & .1 & .1 \\ 1 & .1 & .1 & .1 \end{pmatrix} \quad \frac{1}{N}\begin{pmatrix} .13 & 1 & 1 & 1 \\ 1 & .13 & .13 & .13 \\ 1 & .13 & .13 & .13 \\ 1 & .13 & .13 & .13 \end{pmatrix}.$$

Cusps and splitting of the support are possible only if there is a discontinuity in the profile of S. If the above profile is smoothed out (indicated by the narrow shaded region in the schematic picture of the matrix below), then the support becomes a single interval with a specific smoothed out "almost cusp".

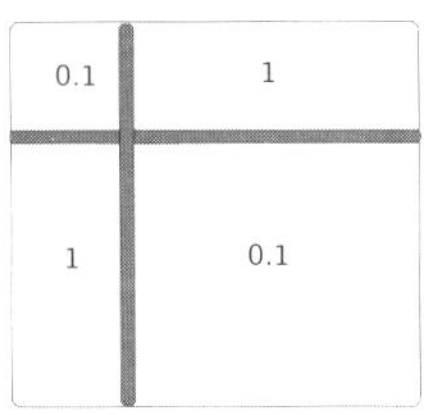

Finally we show the universal shape of the singularities and near singularities in the self-consistent density of states. The first two pictures are the edges and cusps, below them the approximate form of the density near the singularity in terms of the parameter $\omega = \tau - \tau_0$, compare with (6.1.16) and (6.1.15):

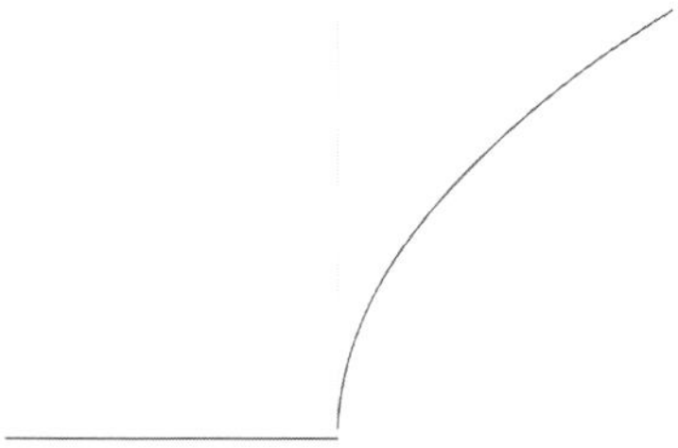

Edge, $\sqrt{\omega}$ singularity

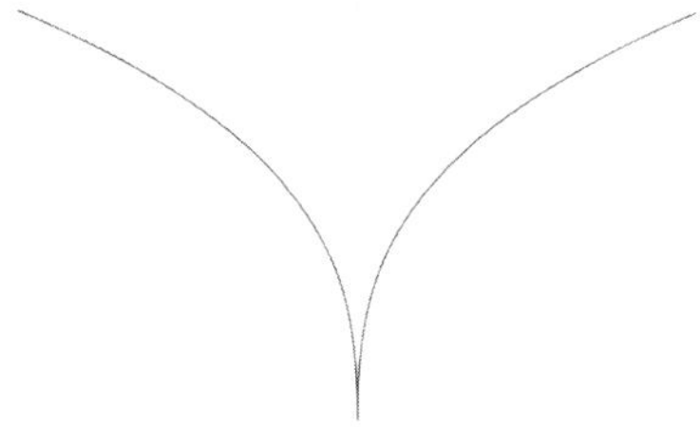

Cusp, $|\omega|^{1/3}$ singularity

The next two pictures show the asymptotic form of the density right before and after the cusp formation. The relevant parameter t is an appropriate rescaling of ω; the size of the gap (after the cusp formation) and the minimum value of the density (before the cusp formation) set the relevant length scales on which the universal shape emerges:

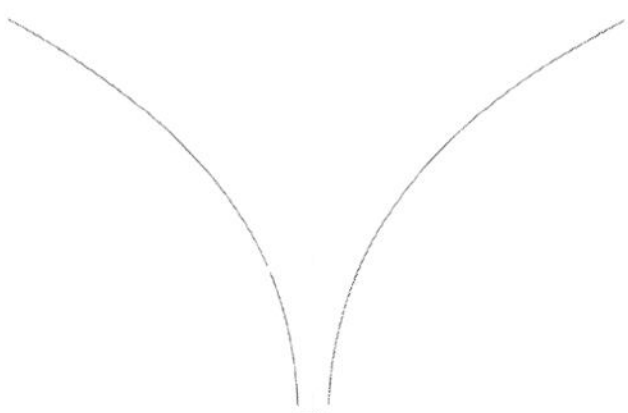

Small-gap

$$\frac{(2+t)t}{1+(1+t+\sqrt{(2+t)t})^{2/3}+(1+t-\sqrt{(2+t)t})^{2/3}}$$

$$t := \frac{|\omega|}{\text{gap}},$$

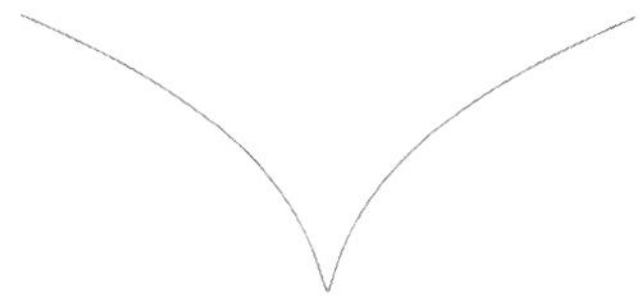

Smoothed cusp

$$\frac{\sqrt{1+t^2}}{(\sqrt{1+t^2}+t)^{2/3}+(\sqrt{1+t^2}-t)^{2/3}-1} - 1$$

$$t := \frac{|\omega|}{(\text{minimum of } \rho)^3}$$

We formulated the vector Dyson equation in a discrete setup for N unknowns but it can be considered in a more abstract setup as follows. For a measurable space $\mathfrak{A}$ and a subset $\mathbb{D} \subseteq \mathbb{C}$ of the complex numbers, we denote by $\mathcal{B}(\mathfrak{A}, \mathbb{D})$ the space of bounded measurable functions on $\mathfrak{A}$ with values in $\mathbb{D}$. Let $(\mathfrak{X}, \pi(dx))$ be a measure space with bounded positive (non-zero) measure π. Suppose we are given a real valued $a \in \mathcal{B}(\mathfrak{X}, \mathbb{R})$ and a non-negative, symmetric, $s_{xy} = s_{yx}$, function $s \in \mathcal{B}(\mathfrak{X}^2, \mathbb{R}_0^+)$. Then we consider the *quadratic vector equation (QVE)*,

$$-\frac{1}{m(z)} = z - a + Sm(z), \qquad z \in \mathbb{H}, \tag{6.1.17}$$

for a function $m : \mathbb{H} \to \mathcal{B}(\mathfrak{X}, \mathbb{H})$, $z \mapsto m(z)$, where $S : \mathcal{B}(\mathfrak{X}, \mathbb{C}) \to \mathcal{B}(\mathfrak{X}, \mathbb{C})$ is the integral operator with kernel s,

$$(Sw)_x := \int s_{xy} w_y \pi(dy), \qquad x \in \mathfrak{X},\ w \in \mathcal{B}(\mathfrak{X}, \mathbb{C}).$$

We equip the space $\mathcal{B}(\mathfrak{X},\mathbb{C})$ with its natural supremum norm,

$$\|w\| := \sup_{x\in\mathfrak{X}} |w_x|, \qquad w \in \mathcal{B}(\mathfrak{X},\mathbb{C}).$$

With this norm $\mathcal{B}(\mathfrak{X},\mathbb{C})$ is a Banach space. All results stated in Theorem 6.1.13 are valid in this more general setup, for details, see [4]. The special case we discussed above corresponds to

$$\mathfrak{X} := \Big\{\frac{1}{N},\frac{2}{N},\dots,\frac{N}{N}\Big\}, \qquad \pi(dx) = \frac{1}{N}\sum_{i=1}^{N}\delta\Big(x-\frac{i}{N}\Big).$$

The scaling here differs from (6.1.9) by a factor of N, since now $s_{xy} = S(x,y)$, $x,y \in \mathfrak{X}$ in which case there is an infinite dimensional limiting equation with $\mathfrak{X} = [0,1]$ and $\pi(dx)$ being the Lebesgue measure. If s_{ij} comes from a continuous profile, (6.1.9), then in the $N \to \infty$ limit, the vector Dyson equation becomes

$$-\frac{1}{m_x(z)} = z + \int_0^1 S(x,y)m_y(z)\,dy, \qquad x \in [0,1], \quad z \in \mathbb{H}.$$

6.1.18. Matrix Dyson equation The matrix version of the Dyson equation naturally arises in the study of correlated random matrices, see Section 3.1.19 and Section 4.5. It takes the form

(6.1.19) $$I + (z + \mathcal{S}[M])M = 0, \qquad \operatorname{Im} M > 0, \quad \operatorname{Im} z > 0, \qquad \text{(MDE)}$$

where we assume that $\mathcal{S} : \mathbb{C}^{N\times N} \to \mathbb{C}^{N\times N}$ is a linear operator that is

1) symmetric with respect to the Hilbert-Schmidt scalar product. In other words, $\operatorname{Tr} R^*\mathcal{S}[T] = \operatorname{Tr} \mathcal{S}[R]^*T$ for any matrices $R, T \in \mathbb{C}^{N\times N}$;
2) positivity preserving, i.e. $\mathcal{S}[R] \geqslant 0$ for any $R \geqslant 0$.

Somewhat informally we will refer to linear maps on the space of matrices as **superoperators** to distinguish them from usual matrices.

Originally, $\mathcal{S}$ is defined in (3.1.23) as a covariance operator of a hermitian random matrix H, but it turns out that (6.1.19) can be fully analyzed solely under these two conditions 1)–2). It is straightforward to check that $\mathcal{S}$ defined in (3.1.23) satisfies the conditions 1) and 2).

Similarly to the vector Dyson equation (6.1.3) one may add an external source $A = A^* \in \mathbb{C}^{N\times N}$ and consider

(6.1.20) $$I + (z - A + \mathcal{S}[M])M = 0, \qquad \operatorname{Im} M > 0$$

but these notes will be restricted to $A = 0$. We remark that instead of finite dimensional matrices, a natural extension of (6.1.20) can be considered on a general von Neumann algebra, see [9] for an extensive study.

The matrix Dyson equation (6.1.20) is a generalization of the vector Dyson equation (6.1.3). Indeed, if $\operatorname{diag}(\mathbf{v})$ denotes the diagonal matrix with the components of the vector $\mathbf{v}$ in the diagonal, then (6.1.20) reduces to (6.1.3) with the identification $A = \operatorname{diag}(\mathbf{a})$, $M = \operatorname{diag}(\mathbf{m})$ and $\mathcal{S}(\operatorname{diag}(\mathbf{m})) = \operatorname{diag}(S\mathbf{m})$. The solution $\mathbf{m}$ to the vector Dyson equation was controlled in the maximum norm $\|\mathbf{m}\|_\infty$;

for the matrix Dyson equation the analogous natural norm is the Euclidean matrix (or operator) norm, $\|M\|_2$, given by

$$\|M\|_2 = \sup\{\|M\mathbf{x}\|_2 \,:\, \mathbf{x} \in \mathbb{C}^N,\ \|\mathbf{x}\|_2 = 1\}.$$

Clearly, for diagonal matrices we have $\|\mathrm{diag}(\mathbf{m})\|_2 = \|\mathbf{m}\|_\infty$. Correspondingly, the natural norm on the superoperator $\mathcal{S}$ is the norm $\|\mathcal{S}\|_2 := \|\mathcal{S}\|_{2\to 2}$ induced by the Euclidean norm on matrices, i.e.

$$\|\mathcal{S}\|_2 := \sup\{\|\mathcal{S}[R]\|_2 \,:\, R \in \mathbb{C}^{N\times N},\ \|R\|_2 = 1\}.$$

Similarly to Theorem 6.1.4, we have an existence and uniqueness result for the solution (see [55]) moreover, we have a Stieltjes transform representation (Proposition 2.1 of [5]):

Theorem 6.1.21. *For any $z \in \mathbb{H}$, the MDE (6.1.19) with the side condition* $\mathrm{Im}\, M > 0$ *has a unique solution $M = M(z)$ that is analytic in the upper half plane. The solution admits a Stieltjes transform representation*

$$M(z) = \int_{\mathbb{R}} \frac{V(d\tau)}{\tau - z} \tag{6.1.22}$$

where $V(d\tau)$ is a positive semidefinite matrix valued measure on $\mathbb{R}$ with normalization $V(\mathbb{R}) = I$. In particular

$$\|M(z)\|_2 \leqslant \frac{1}{\mathrm{Im}\, z}. \tag{6.1.23}$$

The support of this measure lies in $[-2\|\mathcal{S}\|_2^{1/2}, 2\|\mathcal{S}\|_2^{1/2}]$.

The solution M is called the **self-consistent Green function** or **self-consistent resolvent** since it will be used as a computable deterministic approximation to the random Green function G.

From now on we assume the following *flatness* condition on $\mathcal{S}$ that is the matrix analogue of the boundedness condition (6.1.7):

Flatness condition: The operator $\mathcal{S}$ is called **flat** if there exists two positive constants, c, C, independent of N, such that

$$c\langle R\rangle \leqslant \mathcal{S}[R] \leqslant C\langle R\rangle, \qquad \text{where} \quad \langle R\rangle := \frac{1}{N}\,\mathrm{Tr}\, R \tag{6.1.24}$$

holds for any positive definite matrix $R \geqslant 0$.

Under this condition we have the following quantitative results on the solution M (Proposition 2.2 and Proposition 4.2 of [5]):

Theorem 6.1.25. *Assume that $\mathcal{S}$ is flat, then the holomorphic function $\langle M\rangle : \mathbb{H} \to \mathbb{H}$ is the Stieltjes transform of a Hölder continuous probability density ρ w.r.t. the Lebesgue measure:*

$$\langle V(d\tau)\rangle = \rho(\tau)\, d\tau$$

i.e.

$$|\rho(\tau_1) - \rho(\tau_2)| \leqslant C|\tau_1 - \tau_2|^{\varepsilon} \tag{6.1.26}$$

with some Hölder regularity exponent ε*, independent of* N *(*$\varepsilon = 1/100$ *would do). The density* ρ *is called the* **self-consistent density of states.** *Furthermore,* ρ *is real analytic on the open set* $\mathfrak{S} := \{\tau \in \mathbb{R}\ ;\ \rho(\tau) > 0\}$ *which is called the* **self-consistent bulk spectrum**. *For the solution itself we also have*

$$\|M(z)\|_2 \leqslant \frac{C}{\rho(z) + dist(z, \mathfrak{S})} \tag{6.1.27}$$

and

$$c\rho(z) \leqslant \operatorname{Im} M(z) \leqslant C\|M(z)\|_2^2 \rho(z),$$

where $\rho(z)$ *is the harmonic extension of* $\rho(\tau)$ *to the upper half plane. In particular, in the bulk regime of spectral parameters, where* $\rho(\operatorname{Re} z) \geqslant \delta$ *for some fixed* $\delta > 0$*, we see that* $M(z)$ *is bounded and* $\operatorname{Im} M(z)$ *is comparable (as a positive definite matrix) with* $\rho(z)$*.*

Notice that unlike in the analogous Theorem 6.1.13 for the vector Dyson equation, here we do not assume any regularity on $\mathcal{S}$, but the conclusion is weaker. We do not get Hölder exponent 1/3 for the self-consistent density of states ρ. Furthermore, cusp and edge analysis would also require further conditions on $\mathcal{S}$. Since in the correlated case we focus on the bulk spectrum, i.e. on spectral parameters z with $\operatorname{Re} z \in \mathfrak{S}$, we will not need detailed information about the density near the spectral edges. A detailed analysis of the singularity structure of the solution to (6.1.20), in particular a theorem analogous to Theorem 6.1.13, has been given in [9]. The corresponding edge universality for correlated random matrices was proven in [10].

6.2. Local laws for Wigner-type and correlated random matrices We now state the precise form of the local laws.

Theorem 6.2.1 (Bulk local law for Wigner type matrices, Corollary 1.8 from [7]). *Let* H *be a centered Wigner type matrix with bounded variances* $s_{ij} = \mathbb{E}|h_{ij}|^2$ *i.e.* (6.1.7) *holds. Let* $\mathbf{m}(z)$ *be the solution to the vector Dyson equation* (6.1.3). *If the uniform moment condition* (4.1.2) *for the matrix elements, then the local law in the bulk holds. If we fix positive constants* $\delta, \gamma, \varepsilon$ *and* D*, then for any spectral parameter* $z = \tau + i\eta$ *with*

$$\rho(\tau) \geqslant \delta, \qquad \eta \geqslant N^{-1+\gamma} \tag{6.2.2}$$

we have the entrywise local law

$$\max_{ij} \mathbb{P}\Big(\big|G_{ij}(z) - \delta_{ij} m_i(z)\big| \geqslant \frac{N^{\varepsilon}}{\sqrt{N\eta}}\Big) \leqslant \frac{C}{N^D}, \tag{6.2.3}$$

and, more generally, the isotropic law that for non-random normalized vectors $\mathbf{x}, \mathbf{y} \in \mathbb{C}^N$,

$$\max_{ij} \mathbb{P}\Big(\big|\langle \mathbf{x}, G(z)\mathbf{y}\rangle - \langle \mathbf{x}, \mathbf{m}(z)\mathbf{y}\rangle\big| \geqslant \frac{N^{\varepsilon}}{\sqrt{N\eta}}\Big) \leqslant \frac{C}{N^D}. \tag{6.2.4}$$

Moreover for any non-random vector $\mathbf{w} = (w_1, w_2, \dots) \in \mathbb{C}^N$ *with* $\max_i |w_i| \leqslant 1$ *we have the averaged local law*

$$\mathbb{P}\Big(\Big|\frac{1}{N}\sum_i w_i\big[G_{ii}(z) - m_i(z)\big]\Big| \geqslant \frac{N^{\varepsilon}}{N\eta}\Big) \leqslant \frac{C}{N^D}, \tag{6.2.5}$$

in particular (with $w_i = 1$*) we have*

$$\mathbb{P}\Big(\big|\frac{1}{N}\operatorname{Tr} G(z) - \langle \mathbf{m}(z)\rangle\big| \geqslant \frac{N^{\varepsilon}}{N\eta}\Big) \leqslant \frac{C}{N^D}. \tag{6.2.6}$$

The constant C *in* (6.2.3)–(6.2.6) *is independent of* N *and the choice of* w_i*, but it depends on* $\delta, \gamma, \varepsilon, D$*, the constants in* (6.1.7) *and the sequence* μ_p *bounding the moments in* (4.1.2).

As we explained around (4.6.4), in the entrywise local law (6.2.3) one may bring both superma on i, j and on the spectral parameter z inside the probability, i.e. one can guarantee that $G_{ij}(z)$ is close to $m_i(z)\delta_{ij}$ simultaneously for all indices and spectral parameters in the regime (6.2.2). Similarly, z can be brought inside the probability in (6.2.4) and (6.2.5), but the isotropic law (6.2.4) cannot hold simultaneously for all $\mathbf{x}, \mathbf{y}$ and similarly, the averaged law (6.2.5) cannot simultaneously hold for all $\mathbf{w}$.

We formulated the local law only under the boundedness condition (6.1.7) but only in the bulk of the spectrum for simplicity. Local laws near the edges and cusps require much more delicate analysis and some type of regularity on s_{ij}, e.g. the $1/2$-Hölder regularity introduced in (6.1.8) would suffice. Much easier is the regime outside of the spectrum. The precise statement is found in Theorem 1.6 of [7].

For the correlated matrix we have the following local law from [5]:

Theorem 6.2.7 (Bulk local law for correlated matrices)**.** *Consider a random hermitian matrix* $H \in \mathbb{C}^{N\times N}$ *with correlated entries. Define the self-energy super operator* $\mathcal{S}$ *as*

$$\mathcal{S}[R] = \mathbb{E}\big[HRH\big] \tag{6.2.8}$$

acting on any matrix $R \in \mathbb{C}^{N\times N}$*. Assume that the flatness condition* (6.1.24) *and the moment condition* (4.1.2) *hold. We also assume an exponential decay of correlations in the form*

$$Cov\Big(\phi(W_A); \psi(W_B)\Big) \leqslant C(\phi, \psi)\ e^{-d(A,B)}. \tag{6.2.9}$$

Here $W = \sqrt{N}H$ *is the rescaled random matrix,* A, B *are two subsets of the index set* $[\![1, N]\!] \times [\![1, N]\!]$*, the distance* d *is the usual Euclidean distance between the sets* $A \cup A^t$ *and* $B \cup B^t$ *and* $W_A = (w_{ij})_{(i,j)\in A}$*, see figure below. Let* M *be the self-consistent Green function, i.e. the solution of the matrix Dyson equation* (6.1.19) *with* $\mathcal{S}$ *given in* (6.2.8)*, and consider a spectral parameter in the bulk, i.e.* $z = \tau + i\eta$ *with*

$$\rho(\tau) \geqslant \delta, \qquad \eta \geqslant N^{-1+\gamma}. \tag{6.2.10}$$

Then for any non-random normalized vectors $\mathbf{x}, \mathbf{y} \in \mathbb{C}^N$ *we have the isotropic local law*

$$\mathbb{P}\Big(\big|\langle \mathbf{x}, G(z)\mathbf{y}\rangle - \langle \mathbf{x}, M(z)\mathbf{y}\rangle\big| \geqslant \frac{N^{\varepsilon}}{\sqrt{N\eta}}\Big) \leqslant \frac{C}{N^D}, \tag{6.2.11}$$

in particular we have the entrywise law

$$\mathbb{P}\Big(\big|G_{ij}(z) - M_{ij}(z)\big| \geqslant \frac{N^{\varepsilon}}{\sqrt{N\eta}}\Big) \leqslant \frac{C}{N^D}, \tag{6.2.12}$$

for any i, j. *Moreover for any fixed (deterministic) matrix* T *with* $\|T\| \leqslant 1$, *we have the averaged local law*

$$\mathbb{P}\Big(\Big|\frac{1}{N}\operatorname{Tr} T\big[G(z) - M(z)\big]\Big| \geqslant \frac{N^{\varepsilon}}{N\eta}\Big) \leqslant \frac{C}{N^D}. \tag{6.2.13}$$

The constant C *is independent of* N *and the choice of* $\mathbf{x}, \mathbf{y}$, *but it depends on* $\delta, \gamma, \varepsilon, D$, *the constants in* (6.1.24) *and the sequence* μ_p *bounding the moments in* (4.1.2).

In our recent paper [37], we substantially relaxed the condition on the correlation decay (6.2.9) to the form

$$\operatorname{Cov}\Big(\phi(W_A); \psi(W_B)\Big) \leqslant \frac{C(\phi, \psi)}{1 + d^2} e^{-d/N^{1/4}}, \qquad d = d(A, B),$$

and a similar condition on higher order cumulants, see [37] for the precise forms.

In Theorem 6.2.7, we again formulated the result only in the bulk, but similar (even stronger) local law is available for energies τ that are separated away from the support of ρ.

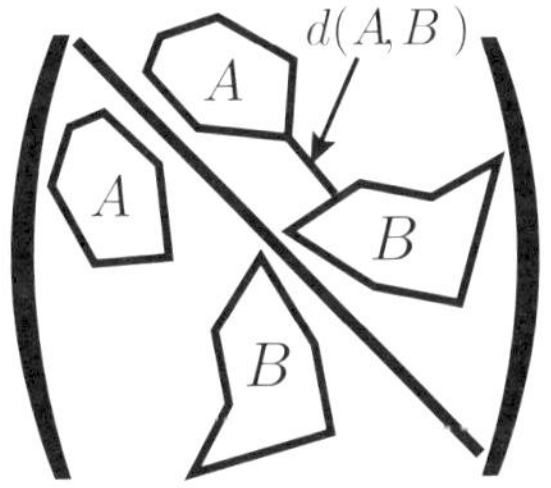

In these notes we will always assume that H is centered, $\mathbb{E}H = 0$ for simplicity, but our result holds in the general case as well. In that case $\mathcal{S}$ is given by

$$\mathcal{S}[R] = \mathbb{E}\big[(H - \mathbb{E}H)R(H - \mathbb{E}H)\big]$$

and M solves the MDE with external source $A := \mathbb{E}H$, see (6.1.20).

6.3. Bulk universality and other consequences of the local law In this section we give precise theorems of three important consequences of the local law. We will formulate the results in the simplest case, in the bulk. We give some sketches of the proofs. Complete arguments for these results can be found in the papers [7] and [5,37].

6.3.1. Delocalization The simplest consequence of the entrywise local law is the delocalization of the eigenvectors as explained in Section 2.2. The precise formulation goes as follows:

Theorem 6.3.2 (Delocalization of bulk eigenvectors). *Let* H *be a Wigner type or, more generally, a correlated random matrix, satisfying the conditions of Theorem* 6.2.1 *or Theorem* 6.2.7, *respectively. Let* ρ *be the self-consistent density of states obtained from solving the corresponding Dyson equation. Then for any* $\delta, \gamma > 0$ *and* $D > 0$ *we have*

$$\mathbb{P}\Big(\exists \mathbf{u}, \lambda, \ H\mathbf{u} = \lambda\mathbf{u}, \ \ \|\mathbf{u}\|_2 = 1 \ \ \rho(\lambda) \geqslant \delta, \ \ \|\mathbf{u}\|_\infty \geqslant N^{-\frac{1}{2}+\gamma}\Big) \leqslant \frac{C}{N^D}.$$

Sketch of the proof. The proof was basically given in (2.2.1). The local laws guarantee that $\operatorname{Im} G_{jj}(z)$ is close to its deterministic approximant, $m_i(z)\delta_{ij}$ or $M_{ij}(z)$, these statements hold for any $E = \operatorname{Re} z$ in the bulk and for $\eta \geqslant N^{-1+\gamma}$. Moreover, (6.1.12) and (6.1.27) show that in the bulk regime both $|\mathbf{m}|$ and $\|M\|$ are bounded. From these two information we conclude that $\operatorname{Im} G_{jj}(z)$ is bounded with very high probability. □

6.3.3. Rigidity The next standard consequence of the local law is the **rigidity of eigenvalues**. It states that with very high probability the eigenvalues in the bulk are at most $N^{-1+\gamma}$-distance away from their **classical locations** predicted by the corresponding quantiles of the self-consistent density of states, for any $\gamma > 0$. This error bar $N^{-1+\gamma}$ reflects that typically the eigenvalues are almost as close to their deterministically prescribed locations as the typical level spacing N^{-1}. This is actually an indication of a very strong correlation; e.g. if the eigenvalues were completely uncorrelated, i.e. given by a Poisson point process with intensity N, then the typical fluctuation of the location of the points would be $N^{-1/2}$.

Since local laws at spectral parameter $z = E + i\eta$ determine the local eigenvalue density on scale η, it is very natural that a local law on scale $\eta = \operatorname{Im} z$ locates individual eigenvalues with η-precision. Near the edges and cusps the local spacing is different ($N^{-2/3}$ and $N^{-3/4}$, respectively), and the corresponding rigidity result must respect this. For simplicity, here we state only the bulk result, as we did for the local law as well; for results at the edge and cusp, see [7].

Given the self-consistent density ρ, for any energy E, define

$$i(E) := \left\lceil N \int_{-\infty}^{E} \rho(\omega)\, d\omega \right\rceil \tag{6.3.4}$$

to be the index of the N-quantile closest to E. Alternatively, for any $i \in [\![1, N]\!]$ one could define $\gamma_i = \gamma_i^{(N)}$ to be the i-th N-quantile of ρ by the relation

$$\int_{-\infty}^{\gamma_i} \rho(\omega)\, d\omega = \frac{i}{N},$$

then clearly $\gamma_{i(E)}$ is (one of) the closest N-quantile to E as long as E is in the bulk, $\rho(E) > 0$.

Theorem 6.3.5 (Rigidity of bulk eigenvalues). *Let H be a Wigner type or, more generally, a correlated random matrix, satisfying the conditions of Theorem 6.2.1 or Theorem 6.2.7, respectively. Let ρ be the self-consistent density of states obtained from solving the corresponding Dyson equation. Fix any $\delta, \varepsilon, D > 0$. For any energy E in the bulk, $\rho(E) \geqslant \delta$, we have*

$$\mathbb{P}\left(|\lambda_{i(E)} - E| \geqslant \frac{N^{\varepsilon}}{N}\right) \leqslant \frac{C}{N^D}. \tag{6.3.6}$$

Sketch of the proof. The proof of rigidity from the local law is a fairly standard procedure by now, see Chapter 11 of [45], or Lemma 5.1 [7] especially tailored to our situation. The key step is the following *Helffer-Sjöstrand formula* that expresses integrals of a compactly supported function f on the real line against a (signed) measure ν with bounded variation in terms of the Stieltjes transform of ν. (Strictly speaking we defined Stieltjes transform only for probability measures, but the concept can be easily extended since any signed measure with bounded variation can be written as a difference of two non-negative measures, and thus Stieltjes transform extends by linearity).

Let χ be a compactly supported smooth cutoff function on $\mathbb{R}$ such that $\chi \equiv 1$ on $[-1,1]$. Then the Cauchy integral formula implies

$$f(\tau) = \frac{1}{2\pi}\int_{\mathbb{R}^2} \frac{i\eta f''(\sigma)\chi(\eta) + i(f(\sigma) + i\eta f'(\sigma))\chi'(\eta)}{\tau - \sigma - i\eta}\, d\sigma\, d\eta. \tag{6.3.7}$$

Thus for any real valued smooth f the **Helffer-Sjöstrand formula** states that

$$\int_{\mathbb{R}} f(\tau)\nu(d\tau) = -\frac{1}{2\pi}(L_1 + L_2 + L_3) \tag{6.3.8}$$

with

$$L_1 = \int_{\mathbb{R}^2} \eta f''(\sigma)\chi(\eta)\,\mathrm{Im}\, m(\sigma + i\eta)\, d\sigma\, d\eta$$

$$L_2 = \int_{\mathbb{R}^2} f'(\sigma)\chi'(\eta)\,\mathrm{Im}\, m(\sigma + i\eta)\, d\sigma\, d\eta$$

$$L_3 = \int_{\mathbb{R}^2} \eta f'(\sigma)\chi'(\eta)\,\mathrm{Re}\, m(\sigma + i\eta)\, d\sigma\, d\eta$$

where $m(z) = m_\nu(z)$ is the Stieltjes transform of ν. Although this formula is a simple identity, it plays an essential role in various problems of spectral analysis. One may apply it to develop functional calculation (functions of a given self-adjoint operator) in terms of the its resolvents [27].

For the proof of the eigenvalue rigidity, the formula (6.3.8) is used for $\nu := \mu_N - \rho$, i.e. for the difference of the empirical and the self-consistent density of states. Since the normalized trace of the resolvent is the Stieltjes transform of the empirical density of states, the averaged local law (6.2.6) (or (6.2.13) with $T = 1$) states that

$$|m_\nu(\tau + i\eta)| \leqslant \frac{N^\varepsilon}{N\eta} \qquad \eta \geqslant N^{-1+\gamma} \tag{6.3.9}$$

with very high probability for any τ with $\rho(\tau) \geqslant \delta$. Now we fix two energies, τ_1 and τ_2 in the bulk and define f to be the characteristic function of the interval $[\tau_1, \tau_2]$ smoothed out on some scale η_0 at the edges, i.e.

$$f|_{[\tau_1,\tau_2]} = 1, \qquad f|_{\mathbb{R}\setminus[\tau_1-\eta_0,\tau_2+\eta_0]} = 0$$

with derivative bounds $|f'| \leqslant C/\eta_0$, $|f''| \leqslant C/\eta_0^2$ in the transition regimes

$$J := [\tau_1 - \eta_0, \tau_1] \cup [\tau_2, \tau_2 + \eta_0].$$

We will choose $\eta_0 = N^{-1+\gamma}$. Then it is easy to see that L_2 and L_3 are bounded by $N^{-1+\varepsilon+C\gamma}$ since $\chi'(\eta)$ is supported far away from 0, say on $[1,2] \cup [-2,-1]$, hence, for example

$$|L_2| \lesssim \int_1^2 d\eta \int_J d\sigma\, \frac{1}{\eta_0}\, |\chi'(\eta)| \frac{N^\varepsilon}{N\eta} \leqslant N^{-1+\varepsilon+2\gamma}$$

using that $|J| \leqslant 2\eta_0 \lesssim N^{-1+\gamma}$. A similar direct estimate does not work for L_1 since it would give

$$|L_1| \lesssim \int_0^\infty d\eta \int_J d\sigma\, \eta\, \frac{1}{\eta_0^2}\, \chi(\eta) \frac{N^\varepsilon}{N\eta} \leqslant N^{\varepsilon+3\gamma}. \tag{6.3.10}$$

Even this estimate would need a bit more care since the local law (6.3.9) does not hold for η smaller than $N^{-1+\gamma}$, but here one uses the fact that for any positive measure μ, the (positive) function $\eta \to \eta \operatorname{Im} m_\mu(\sigma + i\eta)$ is monotonously increasing, so the imaginary part of the Stieltjes transform at smaller η-values can be controlled by those at larger η values. Here it is crucial that L_1 contains only the imaginary part of the Stieltjes transforms and not the entire Stieltjes transform. The argument (6.3.10), while does not cover the entire L_1, it gives a sufficient bound on the small η regime:

$$\begin{aligned}
&\int_0^{\eta_0} d\eta \int_J d\sigma \eta f''(\sigma)\chi(\eta) \operatorname{Im} m(\sigma + i\eta) \\
&\quad \leqslant \int_0^{\eta_0} d\eta \int_J d\sigma\, |f''(\sigma)| \eta_0 \operatorname{Im} m(\sigma + i\eta_0) \\
&\quad \leqslant N^{-1+\varepsilon+3\gamma}.
\end{aligned}$$

To improve (6.3.10) by a factor $1/N$ for $\eta \geqslant \eta_0$, we integrate by parts before estimating. First we put one σ-derivative from f'' to $m_\gamma(\sigma + i\eta)$, then the ∂_σ derivate is switched to ∂_η derivative, then another integration by parts, this time in η removes the derivative from m_γ. The boundary terms, we obtain formulas similar to L_2 and L_3 that have already been estimated.

The outcome is that

$$\int_{\mathbb{R}} f(\tau) \left[\mu_N(d\tau) - \rho(\tau)\, d\tau\right] \leqslant N^{-1+\varepsilon'} \tag{6.3.11}$$

for any $\varepsilon' > 0$ with very high probability, since ε and γ can be chosen arbitrarily small positive numbers in the above argument. If f were exactly the characteristic function, then (6.3.11) would imply that

$$\frac{1}{N} \#\{\lambda_j \in [\tau_1, \tau_2]\} = \int_{\tau_1}^{\tau_2} \rho(\omega)\, d\omega + O(N^{-1+\varepsilon'}) \tag{6.3.12}$$

i.e. it would identify the eigenvalue counting function down to the optimal scale. Estimating the effects of the smooth cutoffs is an easy technicality. Finally, (6.3.12) can be easily turned into (6.3.6), up to one more catch. So far we assumed that τ_1, τ_2 are both in the bulk since the local law was formulated in the bulk and (6.3.12) gave the number of eigenvalues in any interval with endpoints in the bulk.

The quantiles appearing in (6.3.6), however, involve semi-infinite intervals, so one also needs a local law well outside of the bulk. Although in Theorems 6.2.1 and 6.2.7 we formulated local laws in the bulk, similar, and typically even easier estimates are available for energies far away from the support of ρ. In fact, in the regime where $\operatorname{dist}(\tau, \operatorname{supp}\rho) \geqslant \delta$ for some fixed $\delta > 0$, the analogue (6.3.9) is improved to

$$(6.3.13)\qquad |m_\nu(\tau + i\eta)| \leqslant \frac{N^\varepsilon}{N} \qquad \eta > 0,$$

makingo the estimates on L_j's even easier when τ_1 or τ_2 is far from the bulk. □

6.3.14. Universality of local eigenvalue statistics The universality of the local distribution of the eigenvalues is the main coveted goal of random matrix theory. While local laws and rigidity are statements where random quantities are compared with deterministic ones, i.e. they are, in essence, law of large number type results (even if not always formulated in that way), the universality is about the emergence and ubiquity of a new distribution.

We will formulate universality in two forms: on the level of correlation functions and on the level of individual gaps. While these formulations are "morally" equivalent, technically they require quite different proofs.

We need to strengthen a bit the assumption on the lower bound on the variances in (6.1.7) for complex hermitian Wigner type matrices H. In this case we define the real symmetric 2×2 matrix

$$\sigma_{ij} := \begin{pmatrix} \mathbb{E}(\operatorname{Re} h_{ij})^2 & \mathbb{E}(\operatorname{Re} h_{ij})(\operatorname{Im} h_{ij}) \\ \mathbb{E}(\operatorname{Re} h_{ij})(\operatorname{Im} h_{ij}) & \mathbb{E}(\operatorname{Im} h_{ij})^2 \end{pmatrix}$$

for every i, j and we will demand that

$$(6.3.15)\qquad \sigma_{ij} \geqslant \frac{c}{N}$$

with some $c > 0$ uniformly for all i, j in the sense of quadratic forms on $\mathbb{R}^2$. Similarly, for correlated matrices the flatness condition (6.1.24) is strengthened to the requirement that there is a constant $c > 0$ such that

$$(6.3.16)\qquad \mathbb{E}|\operatorname{Tr} BH|^2 \geqslant c \operatorname{Tr} B^2$$

for any real symmetric (or complex hermitian, depending on the symmetry class of H) deterministic matrix B.

Theorem 6.3.17 (Bulk universality). *Let H be a Wigner type or, more generally, a correlated random matrix, satisfying the conditions of Theorem 6.2.1 or Theorem 6.2.7, respectively. For Wigner type matrices in the complex hermitian symmetry class we additionally assume (6.3.15). For correlated random matrices, we additionally assume (6.3.16).*

Let ρ be the self-consistent density of states obtained from solving the corresponding Dyson equation. Let $k \in \mathbb{N}$, $\delta > 0, E \in \mathbb{R}$ with $\rho(E) \geqslant \delta$ and let $\Phi \colon \mathbb{R}^k \to \mathbb{R}$ be a compactly supported smooth test function. Then for some positive constants c and C, depending on Φ, δ, k, we have the following:

(i) [Universality of correlation functions] Denote the k-point correlation function of the eigenvalues of H by $p_N^{(k)}$ (see (1.2.13)) and denote the corresponding k-point correlation function of the GOE/GUE-point process by $\Upsilon^{(k)}$. Then

$$(6.3.18)\qquad \left| \int_{\mathbb{R}^k} \Phi(\mathbf{t}) \left[\frac{1}{\rho(E)^k} p_N^{(k)}\Big(E + \frac{\mathbf{t}}{N\rho(E)}\Big) - \Upsilon_k(\mathbf{t}) \right] d\mathbf{t} \right| \leqslant CN^{-c}.$$

(ii) [Universality of gap distributions] Recall that $i(E)$ *is the index of the* N*-th quantile in the density* ρ *that is closest to the energy* E (6.3.4). *Then*

$$
\begin{aligned}
\Bigg| &\mathbb{E}\Phi\Big(\big(N\rho(\lambda_{k(E)})[\lambda_{k(E)+j}-\lambda_{k(E)}]\big)_{j=1}^{k}\Big) \\
&- \mathbb{E}_{GOE/GUE}\Phi\Big(\big(N\rho_{sc}(0)[\lambda_{\lceil N/2\rceil+j}-\lambda_{\lceil N/2\rceil}]\big)_{j=1}^{k}\Big)\Bigg| \leqslant CN^{-c},
\end{aligned}
\tag{6.3.19}
$$

where the expectation $\mathbb{E}_{GOE/GUE}$ *is taken with respect to the Gaussian matrix ensemble in the same symmetry class as* H.

Short sketch of the proof. The main method to prove universality is the *three-step strategy* outlined in Section 1.2.19. The first step is to obtain a local law which serves as an *a priori* input for the other two steps and it is the only model dependent step. The second step is to show that a small Gaussian component in the distribution already produces the desired universality. The third step is a perturbative argument to show that removal of the Gaussian component does not change the local statistics. There have been many theorems of increasing generality to complete the second and third steps and by now very general "black-box" theorems exist that are model-independent.

The **second step** relies on the local equilibration properties of the Dyson Brownian motion introduced in (1.2.21). The latest and most general formulation of this idea concerns universality of deformed Wigner matrices of the form

$$H_t = V + \sqrt{t}W,$$

where V is a deterministic matrix and W is a GOE/GUE matrix. In applications V itself is a random matrix and in H_t an additional independent Gaussian component is added. But for the purpose of local equilibration of the DBM, hence for the emergence of the universal local statistics, only the randomness of W is used, hence one may condition on V. The main input of the following result is that the local eigenvalue density of V must be controlled in a sense of lower and upper bounds on the imaginary part of the Stieltjes transform m_V of the empirical eigenvalue density of V. In practice this is obtained from the local law with very high probability in the probability space of V.

Theorem 6.3.20 ([62, 63]). *Choose two* N*-dependent parameters,* L, ℓ *for which we have* $1 \gg L^2 \gg \ell \gg N^{-1}$ *(here the notation* $\gg$ *indicates separation by an* N^{ε} *factor for an arbitrarily small* $\varepsilon > 0$*). Suppose that around a fixed energy* E_0 *in a window of size* L *the local eigenvalue density of* V *on scale* ℓ *is controlled, i.e.*

$$c \leqslant \operatorname{Im} m_V(E + i\eta) \leqslant C, \qquad E \in (E_0 - L, E_0 + L), \qquad \eta \in [\ell, 10]$$

(in particular, E_0 *is in the bulk of* V*). Assume also that* $\|V\| \leqslant N^C$*. Then for any* t *with* $N^{\varepsilon}\ell \leqslant t \leqslant N^{-\varepsilon}L^2$ *the bulk universality of* H_t *around* E_0 *holds both in the sense of correlation functions at fixed energy* (6.3.18) *and in sense of gaps* (6.3.19).

Theorem 6.3.20 in this general form appeared in [63] (gap universality) and in [62] (correlation functions universality at fixed energy). These ideas have been developed in several papers. Earlier results concerned Wigner or generalized Wigner matrices and proved correlation function universality with a small energy averaging [40, 41], fixed energy universality [24] and gap universality [44]. Averaged energy and gap universality for random matrices with general density profile were also proven in [42] assuming more precise information on m_V that are available from the optimal local laws.

Finally, the **third step** is to remove the small Gaussian component by realizing that the family of matrices of the form $H_t = V + \sqrt{t}W$ to which Theorem 6.3.20 applies is sufficiently rich so that for any given random matrix H there exists a matrix V and a small t so that the local statistics of H and $H_t = V + \sqrt{t}W$ coincide. We will use this result for some t with $t = N^{-1+\gamma}$ with a small γ. The time t has to be much larger than ℓ and ℓ has to be much larger than N^{-1} since below that scale the local density of V (given by $\operatorname{Im} m_V(E + i\eta)$) is not bounded. But t cannot be too large either otherwise the comparison result cannot hold.

Note that the local statistics is not compared directly with that of V; this would not work even for Wigner matrices V and even if we used the Ornstein Uhlenbeck process, i.e. $H_t = e^{-t/2}V + \sqrt{1-e^{-t}}W$ (for Wigner matrices V the OU process has the advantage that it preserves not only the first but also the second moments of H_t). But for any given Wigner-type ensemble H one can find a random V and an independent Gaussian W so that the first three moments of H and $H_t = e^{-t/2}V + \sqrt{1-e^{-t}}W$ coincide and the fourth moments are very close; this freedom is guaranteed by the lower bound on s_{ij} and σ_{ij} (6.3.15).

The main perturbative result is the following *Green function comparison theorem* that allows us to compare *expectations* of reasonable functions of the Green functions of two different ensembles whose first four moments (almost) match (the idea of matching four moments in random matrices was introduced in [75]). The key point is that $\eta = \operatorname{Im} z$ can be slightly below the critical threshold $1/N$: the expectation regularizes the possible singularity. Here is the prototype of such a theorem:

Theorem 6.3.21 (Green function comparison). [46] *Consider two Wigner type ensembles H and $\widetilde{H}$ such that their first two moments are the same, i.e. the matrices of variances coincide, $S = \widetilde{S}$ and the third and fourth moments almost match in a sense that*

$$\big|\mathbb{E}h_{ij}^s - \mathbb{E}\widehat{h}_{ij}^s\big| \leqslant N^{-2-\delta}, \qquad s = 3, 4 \tag{6.3.22}$$

(for the complex hermitian case all mixed moments of order 3 and 4 should match). Define a sequence of interpolating Wigner-type matrices $H_0, H_1, H_2, \dots$ such that $H_0 = H$, then in H_1 the h_{11} matrix element is replaced with $\widetilde{h}_{11}$, in H_2 the h_{11} and h_{12} elements are replaced with $\widetilde{h}_{11}$ and $\widetilde{h}_{12}$, etc., i.e. we replace one by one the distribution of the matrix elements. Suppose that the Stieltjes transform on scale $\eta = N^{-1+\gamma}$ is bounded for all these interpolating matrices and for any $\gamma > 0$. Set now $\eta' := N^{-1-\gamma}$ and let Φ a

smooth function with moderate growth. Then

$$\left|\mathbb{E}\Phi\big(G(E+i\eta')\big) - \widetilde{\mathbb{E}}\Phi\big(G(E+i\eta')\big)\right| \leqslant N^{-\delta+C\gamma} \tag{6.3.23}$$

and similar multivariable versions also hold.

In the applications, choosing γ sufficiently small, we could conclude that the distribution of the Green functions of H and $\widetilde{H}$ on scale even **below the eigenvalue spacing** are close. On this scale local correlation functions can be identified, so we conclude that the local eigenvalue statistics of H and $\widetilde{H}$ are the same. This will conclude step 3 of the three step strategy and finish the proof of bulk universality, Theorem 6.3.17. □

Idea of the proof of Theorem 6.3.21. The proof of (6.3.23) is a "brute force" resolvent and Taylor expansion. For simplicity, we first replace Φ by its finite Taylor polynomial. Moreover, we consider only the linear term for illustration in this proof. We estimate the change of $\mathbb{E}G(E+i\eta')$ after each replacement; we need to bound each of them by $o(N^{-2})$ since there are of order N^2 replacements. Fix an index pair i,j. Suppose we are at the step when we change the (ij)-th matrix element h_{ij} to $\widetilde{h}_{ij}$. Let R denote the resolvent of the matrix with (ij)-th and (ji)-th elements being zero, in particular R is independent of h_{ij}. It is easy to see from the local law that $\max_{ab}|R_{ab}(E+i\eta)| \lesssim 1$ for any $\eta \geqslant N^{-1+\gamma}$ and therefore, by the monotonicity of $\eta \to \eta\,\mathrm{Im}\, m(E+i\eta)$ we find that $|R_{ab}(E+i\eta') \lesssim N^{2\gamma}$. Then simple resolvent expansion gives, schematically, that

$$G = R + Rh_{ij}R + Rh_{ij}Rh_{ij}R + Rh_{ij}Rh_{ij}Sh_{ij}R + Rh_{ij}Rh_{ij}Rh_{ij}Rh_{ij}R + \dots \tag{6.3.24}$$

and a similar expansion for $\widetilde{G} = G(h_{ij} \leftrightarrow \widetilde{h}_{ij})$ where all h_{ij} is replaced with $\widetilde{h}_{ij}$ (strictly speaking we need to replace h_{ij} and $h_{ji} = \bar{h}_{ij}$ simultaneously due to hermitian symmetry, but we neglect this). We do the expansion up to the fourth order terms (counting the number of h's). The naive size of a third order term, say, $Rh_{ij}Rh_{ij}Rh_{ij}R$ is of order $N^{-3/2+8\gamma}$ since every h_{ij} is of order $N^{-1/2}$. However, the difference in $\mathbb{E}$ and $\widetilde{\mathbb{E}}$-expectations of these terms are of order $N^{-2-\delta}$ by (6.3.22). Thus for the first four terms (fully expanded ones) in (6.3.24) it holds that

$$\mathbb{E}G - \widetilde{\mathbb{E}}\widetilde{G} = O(N^{-2-\delta+C\gamma}) + \text{fifth and higher order terms}$$

But all fifth and higher order terms have at least five h factors so their size is essentially $N^{-5/2}$, i.e. negligible, even without any cancellation between G and $\widetilde{G}$. Finally, we need to repeat this one by one replacement N^2 times, so we arrive at a bound of order $N^{-\delta+C\gamma}$. This proves (6.3.23). □

Exercise 6.3.25. *For a given real symmetric matrix V let H_t solve the SDE*

$$dH_t = \frac{dB_t}{\sqrt{N}}, \qquad H_{t=0} = V$$

where $B_t = B(t)$ is a standard real symmetric matrix valued Brownian motion, i.e. the matrix elements $b_{ij}(t)$ for $i<j$ as well as $b_{ii}(t)/\sqrt{2}$ are independent standard Brownian

motions and $b_{ij}(t) = b_{ji}(t)$. *Prove that the eigenvalues of* H_t *satisfy the following coupled system of stochastic differential equations (Dyson Brownian motion):*

$$d\lambda_a = \sqrt{\frac{2}{N}}\, dB_a + \frac{1}{N}\sum_{b\neq a} \frac{1}{\lambda_a - \lambda_b}\, dt, \qquad a \in [\![1, N]\!]$$

where $\{B_a : a \in [\![1, N]\!]\}$ *is a collection of independent standard Brownian motions with initial condition* $\lambda_a(t=0)$ *given by the eigenvalues of* V. *Hint: Use first and second order perturbation theory to differentiate the eigenvalue equation* $H\mathbf{u}_a = \lambda_a \mathbf{u}_a$ *with the side condition* $\langle \mathbf{u}_a, \mathbf{u}_b\rangle = \delta_{ab}$, *then use Ito formula (see Section 12.2 of* [45]*). Ignore the complication that Ito formula cannot be directly used due to the singularity; for a fully rigorous proof, see Section 4.3.1 of* [12].

7. Analysis of the vector Dyson equation

In this section we outline the proof of a few results concerning the vector Dyson equation (6.1.2)

$$-\frac{1}{\mathbf{m}} = z + S\mathbf{m}, \qquad z \in \mathbb{H}, \quad \mathbf{m} \in \mathbb{H}^N, \tag{7.0.1}$$

where $S = S^t$ is symmetric, bounded, $\|S\|_\infty \leqslant C$ and has nonnegative entries.

We recall the convention that $1/\mathbf{m}$ denotes a vector in $\mathbb{C}^N$ with components $1/m_j$. Similarly, the relation $\mathbf{u} \leqslant \mathbf{v}$ and the product $\mathbf{uv}$ of two vectors are understood in coordinate-wise sense.

7.1. Existence and uniqueness We sketch the existence and uniqueness result, i.e. Theorem 6.1.4, a detailed proof can be found in Chapter 4 [4]. To orient the reader here we only mention that it is a fix-point argument for the map

$$\Phi(\mathbf{u}) := -\frac{1}{z + S\mathbf{u}} \tag{7.1.1}$$

that maps $\mathbb{H}^N$ to $\mathbb{H}^N$ for any fixed $z \in \mathbb{H}$. Denoting by

$$D(\zeta, \omega) := \frac{|\zeta - \omega|^2}{(\operatorname{Im}\zeta)(\operatorname{Im}\omega)}, \qquad \zeta, \omega \in \mathbb{H}$$

the standard hyperbolic metric on the upper half plane, one may check that Φ is a contraction in this metric. More precisely, for any fixed constant η_0, we have the bound

$$\max_j D\Big(\Phi(\mathbf{u})_j, \Phi(\mathbf{w})_j\Big) \leqslant \Big(1 + \frac{\eta_0^2}{\|S\|}\Big)^{-2} \max_j D(u_j, w_j) \tag{7.1.2}$$

assuming that $\operatorname{Im} z \geqslant \eta_0$ and both $\mathbf{u}$ and $\mathbf{w}$ lie in a large compact set

$$B_{\eta_0} := \Big\{\mathbf{u} \in \mathbb{H}^N \,:\, \|\mathbf{u}\|_\infty \leqslant \frac{1}{\eta_0}, \quad \inf_j \operatorname{Im} u_j \geqslant \frac{\eta_0^2}{(2 + \|S\|)^2}\Big\}, \tag{7.1.3}$$

that is mapped by Φ into itself. Here $\|\mathbf{u}\| = \max_j |u_j|$. Once setting up the contraction properly, the rest is a straightforward fixed point theorem. The representation (6.1.5) follows from the Nevanlinna's theorem as mentioned after Definition 2.1.1.

Given (6.1.5), we recall that $\rho = \langle \mathbf{v} \rangle = \frac{1}{N}\sum_j v_j$ is the self-consistent density of states. We consider its harmonic extension to the upper half plane and continue to denote it by ρ:

$$\rho = \rho(z) := \frac{\eta}{\pi}\int_{\mathbb{R}} \frac{\rho(d\tau)}{|x-E|^2+\eta} = \frac{1}{\pi}\langle \operatorname{Im} \mathbf{m}(z)\rangle, \qquad z = E + i\eta. \tag{7.1.4}$$

Exercise 7.1.5. *Check directly from* (7.0.1) *that the solution satisfies the additional condition of the Nevanlinna's theorem, i.e. that for every* j *we have* $i\eta m_j(i\eta) \to -1$ *as* $\eta \to \infty$. *Moreover, check that* $|m_j(z)| \leqslant 1/\operatorname{Im} z$.

Exercise 7.1.6. *Prove that the support of all measures* v_i *lie in* $[-2\sqrt{\|S\|_\infty}, 2\sqrt{\|S\|_\infty}]$.
Hint: suppose $|z| > 2\sqrt{\|S\|_\infty}$*, then check the following implication:*

$$\textit{If}\quad \|\mathbf{m}(z)\|_\infty < \frac{|z|}{2\|S\|_\infty}, \quad \textit{then}\quad \|\mathbf{m}(z)\|_\infty < \frac{2}{|z|}$$

and apply a continuity argument to conclude that $\|\mathbf{m}(z)\|_\infty < \frac{2}{|z|}$ *holds unconditionally. Taking the imaginary part of* (7.0.1) *conclude that* $\operatorname{Im} \mathbf{m}(E+i\eta) \to 0$ *as* $\eta \to 0$ *for any* $|E| > 2\sqrt{\|S\|_\infty}$.

Exercise 7.1.7. *Prove the inequality* (7.1.2)*, i.e. that* Φ *is indeed a contraction on* B_{η_0}.
Hint: Prove and then use the following properties of the metric D:

1) *The metric* D *is invariant under linear fractional transformations of* $\mathbb{H}$ *of the form*

$$f(z) = \frac{az+b}{cz+d}, \qquad z \in \mathbb{H}, \qquad \begin{pmatrix} a & b \\ c & d \end{pmatrix} \in SL_2(\mathbb{R}).$$

2) *Contraction: for any* $z, w \in \mathbb{H}$ *and* $\lambda > 0$ *we have*

$$D(z+i\lambda, w+i\lambda) = \Big(1+\frac{\lambda}{\operatorname{Im} z}\Big)^{-1}\Big(1+\frac{\lambda}{\operatorname{Im} w}\Big)^{-1} D(z,w);$$

3) *Convexity: Let* $\mathbf{a} = (a_1, \ldots, a_N) \in \mathbb{R}_+^N$*, then*

$$D\Big(\sum_i a_i u_i, \sum_i a_i w_i\Big) \leqslant \max_i D(u_i, w_i), \qquad \mathbf{u}, \mathbf{w} \in \mathbb{H}^N.$$

7.2. Bounds on the solution Now we start the quantitative analysis of the solution and we start with a result on the boundedness in the bulk. We introduce the maximum norm and the ℓ^p norms on $\mathbb{C}^N$ as follows:

$$\|\mathbf{u}\|_\infty = \max_j |u_j|, \qquad \|\mathbf{u}\|_p^p := \frac{1}{N}\sum_j |u_j|^2 = \langle |\mathbf{u}|^p \rangle.$$

The procedure to bound $\mathbf{m}$ is that we first obtain an ℓ^2-bound which usually requires less conditions. Then we enhance it to an ℓ^∞ bound. First we obtain a bound that is useful in the bulk but deteriorates as the self-consistent density vanishes, e.g. at the edges and cusps. Second, we improve this bound to one that is also useful near the edges/cusps but this requires some additional regularity condition on s_{ij}. In these notes we will not aim at the most optimal conditions, see [6] and [4] for the detailed analysis.

7.2.1. Bounds useful in the bulk

Theorem 7.2.2. *[Bounds on the solution] Given lower and upper bounds of the form*

$$\frac{c}{N} \leqslant s_{ij} \leqslant \frac{C}{N}, \tag{7.2.3}$$

as in (6.1.7)), *we have*

$$\|\mathbf{m}\|_2 \lesssim 1, \qquad |\mathbf{m}(z)| \lesssim \frac{1}{\rho(z) + dist(z, supp\rho)}, \qquad \frac{1}{|\mathbf{m}(z)|} \lesssim 1 + |z|,$$

and

$$\rho(z) \lesssim \operatorname{Im} \mathbf{m}(z) \lesssim (1+|z|)^2 \|\mathbf{m}(z)\|_\infty^2 \rho(z),$$

where we recall that $\lesssim$ *indicates a bound up to an unspecified multiplicative constant that is independent of* N *(also, recall that the last three inequalities are understood in coordinate-wise sense).*

Proof. For simplicity, in the proof we assume that $|z| \lesssim 1$; the large z regime is much easier and follows directly from the Stieltjes transform representation of $\mathbf{m}$. Taking the imaginary part of the Dyson equation (7.0.1), we have

$$\frac{\operatorname{Im} \mathbf{m}}{|\mathbf{m}|^2} = \eta + S \operatorname{Im} \mathbf{m}. \tag{7.2.4}$$

Using the lower bound from (7.2.3), we get

$$S \operatorname{Im} \mathbf{m} \geqslant c\langle \operatorname{Im} \mathbf{m} \rangle \geqslant c\rho$$

thus

$$\operatorname{Im} \mathbf{m} \geqslant c|\mathbf{m}|^2 \rho. \tag{7.2.5}$$

Taking the average of both sides and dividing by $\rho > 0$, we get $\|\mathbf{m}\|_2 \lesssim 1$. Using $\operatorname{Im} \mathbf{m} \leqslant |\mathbf{m}|$, we immediately get an upper bound on $|\mathbf{m}| \lesssim 1/\rho$. The alternative bound

$$|\mathbf{m}(z)| \lesssim \frac{1}{\operatorname{dist}(z, \operatorname{supp} \rho)}$$

follows from the Stieltjes transform representation (6.1.5).

Next, we estimate the rhs. of (7.0.1) trivially, we have

$$\frac{1}{|m_i|} \leqslant |z| + \sum_j s_{ij} |m_j| \lesssim |z| + \|\mathbf{m}\|_1 \leqslant |z| + \|\mathbf{m}\|_2 \lesssim 1$$

using Hölder inequality in the last but one step. This gives the upper bound on $1/|\mathbf{m}|$.

Using this bound, we can conclude from (7.2.5) that $\rho \lesssim \operatorname{Im} \mathbf{m}$. The upper bound on $\operatorname{Im} \mathbf{m}$ also follows from (7.2.4) and (7.2.3):

$$\frac{\operatorname{Im} \mathbf{m}}{|\mathbf{m}|^2} \leqslant \eta + S \operatorname{Im} \mathbf{m} \leqslant \eta + \langle \operatorname{Im} \mathbf{m} \rangle \lesssim \eta + C\rho.$$

Using that

$$\rho(z) \gtrsim \frac{\eta}{(1+|z|)^2}$$

which can be easily checked from (7.1.4) and the boundedness of the support of ρ, we conclude the two-sided bounds on $\operatorname{Im}\mathbf{m}$. □

Notice two weak points when using this relatively simple argument. First, the lower bound in (7.2.3) was heavily used, although much less assumption is sufficient. We will not discuss these generalizations in these notes, but see Theorem 2.11 of [4] and remarks thereafter addressing this issue. Second, the upper bound on $|\mathbf{m}|$ for small η is useful only inside the self-consistent bulk spectrum or away from the support of ρ, it deteriorates near the edges of the spectrum. In the next sections we remedy this situation.

7.2.6. Unconditional ℓ^2-bound away from zero Next, we present a somewhat surprising result that shows that an ℓ^2-bound on the solution, $\|\mathbf{m}(z)\|_2$, away from the only critical point $z = 0$ is possible *without any condition on* S. The spectral parameter $z = 0$ is clearly critical, e.g. if $S = 0$, the solution $\mathbf{m}(z) = -1/z$ blows up. Thus to control the behavior of $\mathbf{m}$ around $z \approx 0$ one needs some non degeneracy condition on S. We will not address the issue of $z \approx 0$ in these notes, but we remark that a fairly complete picture was obtained in Chapter 6 of [4] using the concept of *fully indecomposability*.

Before presenting the ℓ^2-bound away from zero, we introduce an important object, the **saturated self-energy operator**, that will also play a key role later in the stability analysis:

Definition 7.2.7. Let S be a symmetric matrix with nonnegative entries and let $\mathbf{m} = \mathbf{m}(z) \in \mathbb{H}^N$ solve the vector Dyson equation (7.0.1) for some fixed spectral parameter $z \in \mathbb{H}$. The matrix $F = (F_{ij})$ with

$$F_{ij} := |m_i| s_{ij} |m_j|$$

acting as

$$\text{(7.2.8)} \qquad F\mathbf{u} = |\mathbf{m}| S\big(|\mathbf{m}|\mathbf{u}\big), \quad \text{i.e.} \quad \big(F\mathbf{u}\big)_i = |m_i| \sum_j s_{ij} |m_j| u_j$$

on any vector $\mathbf{u} \in \mathbb{H}^N$, is called the **saturated self-energy operator**.

Suppose that S has strictly positive entries. Since $m_i \neq 0$ from (7.0.1), clearly F has also positive entries, and $F = F^*$. Thus the Perron-Frobenius theorem applies to F, and it guarantees that F has a single largest eigenvalue r (so that for any other eigenvalue λ we have $|\lambda| < r$) and the corresponding eigenvector $\mathbf{f}$ has positive entries: $F\mathbf{f} = r\mathbf{f}, \quad \mathbf{f} > 0$. Moreover, since F is symmetric, we have $\|F\|_2 = r$ for the usual Euclidean matrix norm of F.

Proposition 7.2.9. *Suppose that S has strictly positive entries and let* $\mathbf{m}$ *solve* (7.0.1) *for some* $z = E + i\eta \in \mathbb{H}$. *Then the norm of the saturated self-energy operator is given by*

$$\text{(7.2.10)} \qquad \|F\|_2 = 1 - \eta \frac{\langle \mathbf{f}|\mathbf{m}|\rangle}{\big\langle \mathbf{f}\frac{\operatorname{Im}\mathbf{m}}{|\mathbf{m}|}\big\rangle},$$

in particular $\|F\|_2 < 1$. *Moreover,*

$$\|\mathbf{m}(z)\|_2 \leqslant \frac{2}{|z|}. \tag{7.2.11}$$

We remark that for the bounds $\|F\|_2 < 1$ and (7.2.11) it is sufficient if S has nonnegative entries instead of positive entries; the proof requires a bit more care, see Lemma 4.5 [4].

Proof. Taking the imaginary part of (7.0.1) and multiplying it by $|\mathbf{m}|$, we have

$$\frac{\operatorname{Im}\mathbf{m}}{|\mathbf{m}|} = \eta|\mathbf{m}| + |\mathbf{m}|S\Big(|\mathbf{m}|\frac{\operatorname{Im}\mathbf{m}}{|\mathbf{m}|}\Big) = \eta|\mathbf{m}| + F\frac{\operatorname{Im}\mathbf{m}}{|\mathbf{m}|}. \tag{7.2.12}$$

Scalar multiply this equation by $\mathbf{f}$, use the symmetry of F and $F\mathbf{f} = \|F\|_2\mathbf{f}$ to get

$$\Big\langle \mathbf{f}, \frac{\operatorname{Im}\mathbf{m}}{|\mathbf{m}|}\Big\rangle = \eta\langle \mathbf{f}|\mathbf{m}|\rangle + \Big\langle \mathbf{f}, F\frac{\operatorname{Im}\mathbf{m}}{|\mathbf{m}|}\Big\rangle = \eta\langle \mathbf{f}|\mathbf{m}|\rangle + \|F\|_2\Big\langle \mathbf{f}, \frac{\operatorname{Im}\mathbf{m}}{|\mathbf{m}|}\Big\rangle,$$

which is equivalent to (7.2.10) (note that $\langle , \rangle$ as a binary operation is the scalar product while $\langle \cdot \rangle$ is the averaging).

For the bound on $\mathbf{m}$, we write (7.0.1) as $-z\mathbf{m} = 1 + \mathbf{m}S\mathbf{m}$, so taking the ℓ^2-norm, we have

$$\|\mathbf{m}\|_2 \leqslant \frac{1}{|z|}(1 + \|\mathbf{m}S\mathbf{m}\|_2) \leqslant \frac{1}{|z|}\big(1 + \big\||\mathbf{m}|S|\mathbf{m}|\big\|_2\big) = \frac{1}{|z|}(1 + \|F\mathbf{1}\|_2) \leqslant \frac{2}{|z|},$$

where $\mathbf{1} = (1, 1, 1, \dots)$, note that $\|\mathbf{1}\|_2 = 1$ and we used (7.2.10) in the last step. □

7.2.13. Bounds valid uniformly in the spectrum In this section we introduce an extra regularity assumption that enables us to control $\mathbf{m}$ uniformly throughout the spectrum, including edges and cusps. For simplicity, we restrict our attention to the special case when s_{ij} originates from a piecewise continuous nonnegative profile function $S(x, y)$ defined on $[0, 1] \times [0, 1]$, i.e. we assume

$$s_{ij} = \frac{1}{N}S\Big(\frac{i}{N}, \frac{j}{N}\Big). \tag{7.2.14}$$

We will actually need that S is piecewise $1/2$-Hölder continuous (6.1.10).

Theorem 7.2.15. *Assume that* s_{ij} *is given by* (7.2.14) *with a piecewise Hölder-1/2 continuous function* S *with uniform lower and upper bounds* $c \leqslant S(x, y) \leqslant C$. *Then for any* $R > 0$ *and for any* $|z| \leqslant R$ *we have*

$$|\mathbf{m}(z)| \sim 1, \qquad \operatorname{Im} m_i(z) \sim \operatorname{Im} m_j(z),$$

where the implicit constants in the $\sim$ *relation depend only on* c, C *and* R. *In particular, all components of* $\operatorname{Im}\mathbf{m}$ *are comparable, hence*

$$\operatorname{Im} m_i \sim \langle \operatorname{Im}\mathbf{m}\rangle = \rho. \tag{7.2.16}$$

We mention that this theorem also holds under weaker conditions. Piecewise $1/2$-Hölder continuity can be replaced by a weaker condition called *component regularity*, see Assumption (C) in [6]. Furthermore, the uniform lower bound of $S(x, y)$ can be replaced with a condition called *diagonal positivity* see Assumption (A) in [6] but we omit these generalizations here.

Proof. We have already obtained an ℓ^2-bound $\|\mathbf{m}\|_2 \lesssim 1$ in Theorem 7.2.2. Now we consider any two indices i, j, evaluate (7.0.1) at these points and subtract them. From

$$-\frac{1}{m_i} = z + (S\mathbf{m})_i, \qquad -\frac{1}{m_j} = z + (S\mathbf{m})_j$$

we thus obtain

$$\left|\frac{1}{m_i}\right| \leqslant \left|\frac{1}{m_j}\right| + \sum_k |s_{ik} - s_{jk}||m_k| \leqslant \left|\frac{1}{m_j}\right| + \|\mathbf{m}\|_2 \Big(N \sum_k |s_{ik} - s_{jk}|^2\Big)^{1/2}.$$

Using (7.2.14) and the Hölder continuity (for simplicity assume $n = 1$), we have

$$N \sum_k |s_{ik} - s_{jk}|^2 \leqslant \frac{1}{N} \sum_k \Big|S\Big(\frac{i}{N}, \frac{k}{N}\Big) - S\Big(\frac{j}{N}, \frac{k}{N}\Big)\Big|^2 \leqslant C\frac{|i-j|}{N},$$

thus

$$\left|\frac{1}{m_i}\right| \leqslant \left|\frac{1}{m_j}\right| + C'\sqrt{\frac{|i-j|}{N}}.$$

Taking the reciprocal and squaring it we have for every fixed j that

$$\frac{1}{N}\sum_i \left[\frac{1}{\left|\frac{1}{m_j}\right| + C'\sqrt{\frac{|i-j|}{N}}}\right]^2 \leqslant \frac{1}{N}\sum_i |m_i|^2 = \|\mathbf{m}\|_2^2 \lesssim 1.$$

The left hand side is can be estimated from below by

$$\frac{1}{N}\sum_i \left[\frac{1}{\left|\frac{1}{m_j}\right| + C'\sqrt{\frac{|i-j|}{N}}}\right]^2 \gtrsim \frac{1}{N}\sum_i \frac{1}{\frac{1}{|m_j|^2} + \frac{|i-j|}{N}} \gtrsim \log |m_j|.$$

Combining the last two inequalities, this shows the uniform upper bound

$$|\mathbf{m}| \lesssim 1.$$

The lower bound is obtained from

$$\left|\frac{1}{m_i}\right| = |z + \sum_j s_{ij} m_j| \leqslant |z| + \frac{C}{N}\sum |m_j| \lesssim 1$$

using the upper bound $|\mathbf{m}| \lesssim 1$ and $s_{ij} \lesssim 1/N$. This proves $|\mathbf{m}| \sim 1$.

To complete the proof, note that comparability of the components of $\mathrm{Im}\,\mathbf{m}$ now follows from the imaginary part of (7.0.1), $|\mathbf{m}| \sim 1$ and from $S(\mathrm{Im}\,\mathbf{m}) \sim \langle \mathrm{Im}\,\mathbf{m}\rangle$:

$$\frac{\mathrm{Im}\,\mathbf{m}}{|\mathbf{m}|^2} = \eta + S(\mathrm{Im}\,\mathbf{m}) \quad \Longrightarrow \quad \mathrm{Im}\,\mathbf{m} \sim \eta + \langle \mathrm{Im}\,\mathbf{m}\rangle. \qquad \square$$

7.3. Regularity of the solution and the stability operator In this section we prove some parts of the regularity Theorem 6.1.13. We will not go into the details of the edge and cusp analysis here, see [6] for a shorter qualitative analysis and [4] for the full quantitative analysis of all possible singularities. Here we will only show the 1/3-Hölder regularity (6.1.14). We will use this opportunity to introduce and analyze the key stability operator of the problem which then will also be used in the random matrix part of our analysis.

It is to keep in mind that the small $\eta = \operatorname{Im} z$ regime is critical; typically bounds of order $1/\eta$ or $1/\eta^2$ are easy to obtain but these are useless for local analysis (recall that η indicates the scale of the problem). For the fine regularity properties of the solution, one needs to take $\eta \to 0$ with uniform controls. For the random matrix part, we will take η down to $N^{-1+\gamma}$ for any small $\gamma > 0$, so any $1/\eta$ bound would not be affordable.

Proof of (i) and (iii) from Theorem 6.1.13. We differentiate (7.0.1) with respect to z (note that $\mathbf{m}(z)$ is real analytic by (6.1.5) for any $z \in \mathbb{H}$).

$$-\frac{1}{\mathbf{m}} = z + S\mathbf{m} \quad \Longrightarrow \quad \frac{\partial_z \mathbf{m}}{\mathbf{m}^2} = 1 + S\partial_z \mathbf{m} \quad \Longrightarrow \quad \partial_z \mathbf{m} = \frac{1}{1 - \mathbf{m}^2 S}\mathbf{m}^2. \tag{7.3.1}$$

The (z-dependent) linear operator $1 - \mathbf{m}^2 S$ is called the **stability operator**. We will later prove the following main bound on this operator:

Lemma 7.3.2 (Bound on the stability operator)**.** *Suppose that for any* $z \in \mathbb{H}$ *with* $|z| \leqslant C$ *we have* $|\mathbf{m}(z)| \sim 1$. *Then*

$$\left\| \frac{1}{1 - \mathbf{m}^2 S} \right\|_2 \lesssim \frac{1}{\rho(z)^2} = \frac{1}{\langle \operatorname{Im} \mathbf{m} \rangle^2}. \tag{7.3.3}$$

In fact, the same bound also holds in the $\ell^\infty \to \ell^\infty$ *norm, i.e.*

$$\left\| \frac{1}{1 - \mathbf{m}^2 S} \right\|_\infty \lesssim \frac{1}{\rho(z)^2} = \frac{1}{\langle \operatorname{Im} \mathbf{m} \rangle^2}. \tag{7.3.4}$$

By Theorem 7.2.15 we know that under conditions of Theorem 6.1.13, we have $\|\mathbf{m}\| \sim 1$, so the lemma is applicable.

Assuming this lemma for the moment, and using that $\mathbf{m}$ is analytic on $\mathbb{H}^N$, we conclude from (7.3.4) that

$$|\partial_z \operatorname{Im} \mathbf{m}| = \frac{1}{2}|\partial_z \mathbf{m}| \lesssim \frac{1}{\langle \operatorname{Im} \mathbf{m} \rangle^2} \sim \frac{1}{(\operatorname{Im} \mathbf{m})^2},$$

i.e. the derivative of $(\operatorname{Im} \mathbf{m}(z))^3$ is bounded. Thus $z \to \operatorname{Im} \mathbf{m}(z)$ is a $1/3$-Hölder regular function on the open upper half plane with a uniform Hölder constant. Therefore $\operatorname{Im} \mathbf{m}(z)$ extends to the real axis as a $1/3$-Hölder continuous function. This proves (6.1.14). Moreover, it is real analytic away from the edges of the self-consistent spectrum $\mathfrak{S} = \{\tau \in \mathbb{R} : \rho(\tau) > 0\}$; indeed on $\mathfrak{S}$ it satisfies an analytic ODE (7.3.1) with bounded coefficients by (7.3.4) while outside of the closure of $\mathfrak{S}$ the density is zero. □

Exercise 7.3.5. *Assume the conditions of Theorem* 6.1.13, *i.e.* (6.1.7) *and that* S *is piecewise Hölder continuous* (6.1.10). *Prove that the saturated self-energy operator has norm 1 on the imaginary axis exactly on the support of the self-consistent density of states. In other words,*

$$\lim_{\eta \to 0+} \|F(E + i\eta)\|_2 = 1 \quad \textit{if and only if} \quad E \in \operatorname{supp} \rho.$$

Hint: First prove that the Stieltjes transform of a 1/3-Hölder continuous function with compact support is itself 1/3-Hölder continuous up to the real line.

7.4. Bound on the stability operator

Proof of Lemma 7.3.2. The main mechanism for the stability bound (7.3.3) goes through the operator $F = |\mathbf{m}|S(|\mathbf{m}|\cdot)$ defined in (7.2.8). We know that F has a single largest eigenvalue, but in fact under the condition (7.2.3) this matrix has a substantial gap in its spectrum below the largest eigenvalue. To make this precise, we start with a definition:

Definition 7.4.1. For a hermitian matrix T the **spectral gap** $\mathrm{Gap}(T)$ is the difference between the two largest eigenvalues of $|T| = \sqrt{TT^*}$. If $\|T\|_2$ is a degenerate eigenvalue of $|T|$, then the gap is zero by definition.

The following simple lemma shows that matrices with nonnegative entries tend to have a positive gap:

Lemma 7.4.2. *Let $T = T^*$ have nonnegative entries, $t_{ij} = t_{ji} \geqslant 0$ and let $\mathbf{h}$ be the Perron-Frobenius eigenvector, $T\mathbf{h} = \|T\|_2\mathbf{h}$ with $\mathbf{h} \geqslant 0$. Then*

$$\mathit{Gap}(T) \geqslant \left(\frac{\|\mathbf{h}\|_2}{\|\mathbf{h}\|_\infty}\right) \cdot \min_{ij} t_{ij}.$$

Exercise 7.4.3. *Prove this lemma. Hint: Set $\|T\|_2 = 1$ and take a vector $\mathbf{u} \perp \mathbf{h}$, $\|\mathbf{u}\|_2 = 1$. Verify that*

$$\langle \mathbf{u}, (1 \pm T)\mathbf{u}\rangle = \frac{1}{2}\sum_{ij} t_{ij}\left[u_i\left(\frac{h_j}{h_i}\right)^{1/2} \pm u_i\left(\frac{h_i}{h_j}\right)^{1/2}\right]^2$$

and estimate it from below.

Applying this lemma to F, we have the following:

Lemma 7.4.4. *Assume* (7.2.3) *and let $|z| \leqslant C$. Then F has norm of order one, it has uniform spectral gap;*

$$\|F\|_2 \sim 1, \qquad \mathit{Gap}(F) \sim 1;$$

and its ℓ^2-normalized Perron-Frobenius eigenvector, $\mathbf{f}$ with $F\mathbf{f} = \|F\|_2\mathbf{f}$, has comparable components

$$\mathbf{f} \sim 1.$$

Proof. We have already seen that $\|F\|_2 \leqslant 1$. The lower bound $\|F\|_2 \gtrsim 1$ follows from $F_{ij} = |m_i|s_{ij}|m_j| \gtrsim 1/N$, in fact $F_{ij} \sim N^{-1}$, thus $\|F\mathbf{1}\|_2 \gtrsim 1$. For the last statement, we write $\mathbf{f} = \|F\|_2^{-1}F\mathbf{f} \sim F\mathbf{f} \sim \langle\mathbf{f}\rangle$ and then by normalization obtain $1 = \|\mathbf{f}\|_2 \sim \langle\mathbf{f}\rangle \sim \mathbf{f}$. Finally the statement on the gap follows from Lemma 7.4.2 and that $\|\mathbf{f}\|_\infty \sim \|\mathbf{f}\|_2$. □

Armed with this information on F, we explain how F helps to establish a bound on the stability operator. Using the polar decomposition $\mathbf{m} = e^{i\varphi}|\mathbf{m}|$, we can write for any vector $\mathbf{w}$

$$(1 - \mathbf{m}^2 S)\mathbf{w} = |\mathbf{m}|(1 - e^{2i\varphi}F)|\mathbf{m}|^{-1}\mathbf{w}. \tag{7.4.5}$$

Since $|\mathbf{m}| \sim 1$, it is sufficient to invert $1 - e^{2i\boldsymbol{\varphi}}F$ or $e^{-2i\boldsymbol{\varphi}} - F$. Since F has a real spectrum, this latter matrix should intuitively be invertible unless $\sin 2\varphi \approx 0$. This intuition is indeed correct if $\mathbf{m}$ and thus $e^{2i\boldsymbol{\varphi}}$ were constant; the general case is more complicated.

Assume first that we are in the generalized Wigner case, when $\mathbf{m} = m_{sc} \cdot \mathbf{1}$, i.e. the solution is a constant vector with components $m := m_{sc}$. Writing $m = |m|e^{i\varphi}$ with some phase φ, we see that

$$1 - m^2 S = 1 - e^{2i\varphi}F.$$

Since F is hermitian and has norm bounded by 1, it has spectrum in $[-1, 1]$. So without the phase the inverse of $1 - F$ would be quite singular (basically, we would have $\|F\|_2 \approx 1 - c\eta$, see (7.2.10) at least in the bulk spectrum). The phase $e^{2i\varphi}$ however rotates F out of the real axis, see the picture.

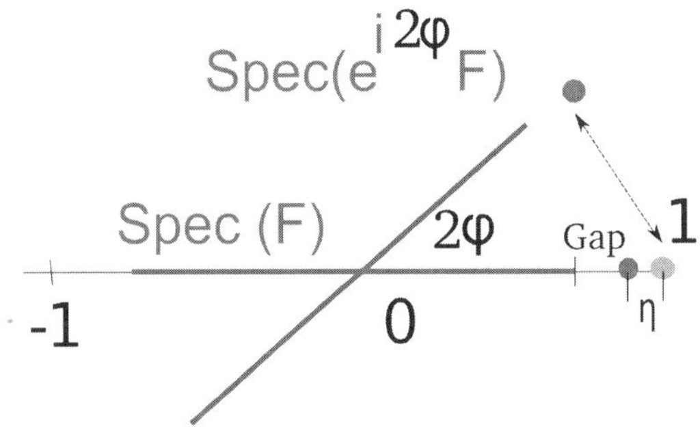

The distance of 1 from the spectrum of F is tiny, but from the spectrum of $e^{2i\varphi}F$ is comparable with $\varphi \sim \operatorname{Im} m = \rho$:

$$\left\| \frac{1}{1 - m^2 S} \right\|_2 = \left\| \frac{1}{1 - e^{2i\varphi}F} \right\|_2 \sim \frac{C}{|\varphi|} \sim \frac{C}{\rho}$$

in the regime where $|\varphi| \leqslant \pi/2$ thanks to the gap in the spectrum of F both below 1 and above -1. In fact this argument indicates a better bound of order $1/\varphi \sim 1/\rho$ and not only its square in (7.3.3).

For the general case, when $\mathbf{m}$ is not constant, such a simple argument does not work, since the rotation angles φ_j from $m_j = e^{i\varphi_j}|m_j|$ now depend on the coordinate j, so there is no simple geometric relation between the spectrum of F and that of $\mathbf{m}^2 S$. In fact the optimal bound in general is $1/\rho^2$ and not $1/\rho$.

To obtain it, we still use the identity

$$(1 - \mathbf{m}^2 S)\mathbf{w} = e^{2i\boldsymbol{\varphi}}|\mathbf{m}|(e^{-2i\boldsymbol{\varphi}} - F)|\mathbf{m}|^{-1}\mathbf{w}, \tag{7.4.6}$$

and focus on inverting $e^{-2i\boldsymbol{\varphi}} - F$. We have the following general lemma:

Lemma 7.4.7. *Let T be hermitian with $\|T\|_2 \leqslant 1$ and with top normalized eigenvector $\mathbf{f}$, i.e. $T\mathbf{f} = \|T\|_2\mathbf{f}$. For any unitary operator U we have*

$$\left\| \frac{1}{U - T} \right\|_2 \leqslant \frac{C}{\mathit{Gap}(T) \cdot |1 - \|T\|_2 \langle \mathbf{f}, U\mathbf{f} \rangle|}. \tag{7.4.8}$$

A simple calculation shows that this lemma applied to $T = F$ and $U = \left(|\mathbf{m}|/\mathbf{m}\right)^2$ yields the bound C/ρ^2 for the inverse of $e^{-2i\varphi} - F$ since

$$\left|1 - \|T\|_2\langle \mathbf{f}, U\mathbf{f}\rangle\right| \geqslant \operatorname{Re}\left[1 - \left\langle \frac{\mathbf{m}^2\mathbf{f}^2}{|\mathbf{m}|^2}\right\rangle\right] = 2\left\langle \frac{(\operatorname{Im}\mathbf{m})^2\mathbf{f}^2}{|\mathbf{m}|^2}\right\rangle \sim \langle \operatorname{Im}\mathbf{m}\rangle^2.$$

This proves the ℓ^2-stability bound (7.3.3) in Lemma 7.3.2. Improving this bound to the stability bound (7.3.4) in ℓ^∞ is left as the following exercise. □

Exercise 7.4.9. *By using* $|\mathbf{m}(z)| \sim 1$ *and* (6.1.7), *prove* (7.3.4) *from* (7.3.3). *Hint: show that for any matrix* R *such that* $1 - R$ *is invertible, we have*

$$\frac{1}{1-R} = 1 + R + R\frac{1}{1-R}R, \tag{7.4.10}$$

and apply this with $R = \mathbf{m}^2 S$.

Sketch of proof of Lemma 7.4.7. For details, see Appendix B of [6]. The idea is that one needs a lower bound on $\|(U - T)\mathbf{w}\|_2$ for any ℓ^2-normalized $\mathbf{w}$. Split $\mathbf{w}$ as $\mathbf{w} = \langle \mathbf{f}, \mathbf{w}\rangle\mathbf{f} + P\mathbf{w}$, where P is the orthogonal projection to the complement of $\mathbf{f}$. We will frequently use that

$$\|TP\mathbf{w}\|_2 \leqslant \left[\|T\| - \operatorname{Gap}(T)\right]\|P\mathbf{w}\|_2, \tag{7.4.11}$$

following from the definition of the gap. Setting $\alpha := \left|1 - \|T\|_2\langle \mathbf{f}, U\mathbf{f}\rangle\right|$, we distinguish three cases

(i) $16\|P\mathbf{w}\|_2^2 \geqslant \alpha$;
(ii) $16\|P\mathbf{w}\|_2^2 < \alpha$ and $\alpha \geqslant \|PU\mathbf{f}\|_2^2$;
(iii) $16\|P\mathbf{w}\|_2^2 < \alpha$ and $\alpha < \|PU\mathbf{f}\|_2^2$.

In regime (i) we use a crude triangle inequality $\|(U - T)\mathbf{w}\|_2 \geqslant \|\mathbf{w}\|_2 - \|T\mathbf{w}\|_2$, the splitting of w and (7.4.11). In regime (ii) we first project $(U - T)\mathbf{w}$ onto the $\mathbf{f}$ direction: $\|(U - T)\mathbf{w}\|_2 \geqslant |\langle \mathbf{f}, (1 - U^*T)\mathbf{w}\rangle|$ and estimate. Finally in regime (iii) we first project $(U - T)\mathbf{w}$ onto the P direction $\|(U - T)\mathbf{w}\|_2 \geqslant \|P(U - T)\mathbf{w}\|_2$ and estimate.

Exercise 7.4.12. *Complete the analysis of all these three regimes and finish the proof of Lemma 7.4.7.* □

8. Analysis of the matrix Dyson equation

8.1. Properties of the solution to the MDE In this section we analyze the matrix Dyson equation introduced in (6.1.19)

$$I + (z + \mathcal{S}[M])M = 0, \qquad \operatorname{Im} M > 0, \quad \operatorname{Im} z > 0, \qquad (\mathrm{MDE}) \tag{8.1.1}$$

where we assume that $\mathcal{S} : \mathbb{C}^{N\times N} \to \mathbb{C}^{N\times N}$ is a symmetric and positivity preserving linear map. In many aspects the analysis goes parallel to that of the vector Dyson equation and we will highlight only the main complications due to the matrix character of this problem.

The proof of the existence and uniqueness result, Theorem 6.1.21, is analogous to the vector case using the Caratheodory metric, so we omit it, see [55]. The

Stieltjes transform representation (6.1.22) can also be proved by reducing it to the scalar case (Exercise 8.1.5). The self-consistent density of states is defined as before:

$$\rho(d\tau) = \frac{1}{\pi}\langle V(d\tau)\rangle = \frac{1}{\pi N}\operatorname{Tr} V(d\tau),$$

and its harmonic extension is again denoted by $\rho(z) = \frac{1}{\pi}\langle \operatorname{Im} M(z)\rangle$.

From now on we assume the flatness condition (6.1.24) on $\mathcal{S}$. We have the analogue of Theorem 7.2.2 on various bounds on M that can be proven in a similar manner. The role of the ℓ^2-norm, $\|m\|_2$ in the vector case will be played by the (normalized) Hilbert-Schmidt norm, i.e. $\|M\|_{hs} := \big(\frac{1}{N}\operatorname{Tr} MM^*\big)^{1/2}$ as it comes from the natural scalar product structure on matrices. The role of the supremum norm of $|\mathbf{m}|$ in the vector case will be played by the operator norm $\|M\|_2$ in the matrix case and similarly the supremum norm of $1/|\mathbf{m}|$ is replaced with $\|M^{-1}\|_2$.

Theorem 8.1.2. *[Bounds on M] Assuming the flatness condition* (6.1.24), *we have*

$$(8.1.3)\qquad \|M\|_{hs} \lesssim 1, \quad \|M(z)\|_2 \lesssim \frac{1}{\rho(z) + dist(z, supp(\rho))}, \qquad \|M^{-1}(z)\|_2 \lesssim 1 + |z|$$

and

$$(8.1.4)\qquad \rho(z) \lesssim \operatorname{Im} M(z) \lesssim (1+|z|)^2 \|M(z)\|_2^2 \rho(z)$$

where $\|T\|_{hs} := \big(\frac{1}{N}\operatorname{Tr} TT^*\big)^{1/2}$ *is the normalized Hilbert-Schmidt norm.*

Exercise 8.1.5. *Prove that if* $M(z)$ *is an analytic matrix-valued function on the upper half plane,* $z \in \mathbb{H}$, *such that* $\operatorname{Im} M(z) > 0$, *and* $i\eta M(i\eta) \to -I$ *as* $\eta \to \infty$, *then* $M(z)$ *has a Stieltjes transform representation of the form* (6.1.22). *Hint: Reduce the problem to the scalar case by considering the quadratic form* $\langle \mathbf{w}, M(z)\mathbf{w}\rangle$ *for* $\mathbf{w} \in \mathbb{C}$.

Exercise 8.1.6. *Prove Theorem* 8.1.2 *by mimicking the corresponding proof for the vector case but watching out for the non commutativity of the matrices.*

8.2. The saturated self-energy matrix We have seen in the vector Dyson equation that the stability operator $1 - \mathbf{m}^2 S$ played a central role both in establishing regularity of the self-consistent density of states and also in establishing the local law. What is the matrix analogue of this operator? Is there any analogue for the saturated self-energy operator F defined in Definition 7.2.7 ?

The matrix responsible for the stability can be easily found, mimicking the calculation (7.3.1) by differentiating (8.1.1) wrt. z

$$(8.2.1)\qquad I + (z + \mathcal{S}[M])M = 0 \implies (I + \mathcal{S}[\partial_z M])M + (z + \mathcal{S}[M])\partial_z M = 0$$
$$\implies \partial_z M = (1 - M\mathcal{S}[\cdot]M)^{-1} M^2.$$

where we took the inverse of the "super operator" $1 - M\mathcal{S}[\cdot]M$. We introduce the notation $\mathcal{C}_T$ for the operator **"sandwiching by a matrix** T", that acts on any matrix R as

$$\mathcal{C}_T[R] := TRT.$$

With this notation we have $1 - M\mathcal{S}[\cdot]M = 1 - \mathcal{C}_M\mathcal{S}$ that acts on $N \times N$ matrices as $(1 - \mathcal{C}_M\mathcal{S})[R] = R - M\mathcal{S}(R)M$.

The boundedness of the inverse of the stability operator, $1 - \mathbf{m}^2 S$ in the vector case, relied crucially on finding a symmetrized version of the operator $\mathbf{m}^2 S$, the saturated self-energy operator (Definition 7.2.8), for which spectral theory can be applied, see the identity (7.4.6). This will be the heart of the proof in the following section where we control the spectral norm of the inverse of the stability operator. Note that spectral theory in the matrix setup means to work with the Hilbert space of matrices, equipped with the Hilbert-Schmidt scalar product. We denote by $\|\cdot\|_{sp} := \|\cdot\|_{hs \to hs}$ the corresponding norm of superoperators viewed as linear maps on this Hilbert space.

8.3. Bound on the stability operator The key technical result of the analysis of the MDE is the following lemma:

Lemma 8.3.1. *Assuming the flatness condition* (6.1.24), *we have, for* $|z| \leqslant C$,

$$\left\| (1 - \mathcal{C}_{M(z)}\mathcal{S})^{-1} \right\|_{sp} \lesssim \frac{1}{\left[\rho(z) + dist(z, supp(\rho))\right]^C} \tag{8.3.2}$$

with some universal constant ($C = 100$ *would do).*

Similarly to the argument in Section 7.3 for the vector case, the bound (8.3.2) directly implies Hölder regularity of the solution and it implies (6.1.26). It is also the key estimate in the random matrix part of the proof of the local law.

Proof of Lemma 8.3.1. In the vector case, the saturated self-energy matrix F naturally emerged from taking the imaginary part of the Dyson equation and recognizing a Perron-Frobenius type eigenvector of the form $\frac{\operatorname{Im} \mathbf{m}}{|\mathbf{m}|}$, see (7.2.12). This structure was essential to establish the bound $\|F\|_2 \leqslant 1$. We proceed similarly for the matrix case to find the analogous super operator $\mathcal{F}$ that has to be symmetric and positivity preserving in addition to having a "useful" Perron-Frobenius eigenequation. The imaginary part of the MDE in the form

$$-\frac{1}{M} = z + \mathcal{S}[M]$$

is given by

$$\frac{1}{M^*} \operatorname{Im} M \frac{1}{M} = \eta + \mathcal{S}[\operatorname{Im} M], \quad \Longrightarrow \quad \operatorname{Im} M = \eta M^* M + M^* \mathcal{S}[\operatorname{Im} M] M. \tag{8.3.3}$$

What is the analogue of $\frac{\operatorname{Im} \mathbf{m}}{|\mathbf{m}|}$ in this equation that is positive, but this time as a matrix? "Dividing by $|M|$" is a quite ambiguous operation, not just because the matrix multiplication is not commutative, but also for the fact that for non-normal matrices, the absolute value of a general matrix R is not defined in a canonical way. The standard definition is $|R| = \sqrt{R^* R}$, which leads to the polar decomposition of the usual form $R = U|R|$ with some unitary U, but the alternative definition $\sqrt{RR^*}$ would also be equally justified. But they are not the same, and this ambiguity would destroy the symmetry of the attempted super operator $\mathcal{F}$ if done naively.

Instead of guessing the right form, we just look for the matrix version of $\frac{\operatorname{Im}\mathbf{m}}{|\mathbf{m}|}$ in the form $\frac{1}{Q}\operatorname{Im} M\frac{1}{Q^*}$ with some matrix Q yet to be found. Then we can rewrite (8.3.3) (for $\eta = 0$ for simplicity) as

$$\frac{1}{Q}\operatorname{Im} M\frac{1}{Q^*} = \frac{1}{Q}M^*\frac{1}{Q^*}Q^*\mathcal{S}\Big[Q\,\frac{1}{Q}(\operatorname{Im} M)\frac{1}{Q^*}\,Q^*\Big]Q\frac{1}{Q}M\frac{1}{Q^*}.$$

We write it in the form

$$X = Y^*Q^*\mathcal{S}[QXQ^*]QY, \quad \text{with} \quad X := \frac{1}{Q}\,(\operatorname{Im} M)\frac{1}{Q^*}, \quad Y := \frac{1}{Q}M\frac{1}{Q^*}$$

i.e.

$$X = Y^*\mathcal{F}[X]Y, \qquad \text{with} \quad \mathcal{F}[\cdot] := Q^*\mathcal{S}[Q\cdot Q^*]Q.$$

With an appropriate Q, this operator will be the correct saturated self-energy operator. Notice that $\mathcal{F}$ is positivity preserving.

To get the Perron-Frobenius structure, we need to "get rid" of the Y and Y^* above; we have a good chance if we require that Y be unitary, $YY^* = Y^*Y = I$. The good news is that $X = \operatorname{Im} Y$ and if Y is unitary, then X and Y commute (check this fact). We thus arrive at

$$X = \mathcal{F}[X].$$

Thus the Perron-Frobenius argument applies and we get that $\mathcal{F}$ is bounded in spectral norm:

$$\|\mathcal{F}\|_{sp} \leqslant 1$$

Actually, if $\eta > 0$, then we get a strict inequality.

Using the definition of $\mathcal{F}$ and that $M = QYQ^*$ with some unitary Y, we can also write the operator $\mathcal{C}_M\mathcal{S}$ appearing in the stability operator in terms of $\mathcal{F}$. Indeed, for any matrix R

$$M\mathcal{S}[R]M = QYQ^*\mathcal{S}\Big[Q\frac{1}{Q}R\frac{1}{Q^*}Q^*\Big]QYQ^* = QY\mathcal{F}\Big[\frac{1}{Q}R\frac{1}{Q^*}\Big]YQ^*$$

so

$$R - M\mathcal{S}[R]M = Q\Big(1 - Y\mathcal{F}[\cdot]Y\Big)\Big[\frac{1}{Q}R\frac{1}{Q^*}\Big]Q^*.$$

Thus

$$I - \mathcal{C}_M\mathcal{S} = \mathcal{K}_Q(I - \mathcal{C}_Y\mathcal{F})\mathcal{K}_Q^{-1}, \tag{8.3.4}$$

where for any matrix T we defined the super operator $\mathcal{K}_T$ acting on any matrix R as $\mathcal{K}_T[R] := TRT^*$ to be the symmetrized analogue of the sandwiching operator $\mathcal{C}_T$. The formula (8.3.4) is the matrix analogue of (7.4.5).

Thus, assuming that $Q \sim 1$ in a sense that $\|Q\|_2 \lesssim 1$ and $\|Q^{-1}\|_2 \lesssim 1$, we have

$$I - \mathcal{C}_M\mathcal{S} \text{ is stable} \quad \Longleftrightarrow \quad I - \mathcal{C}_Y\mathcal{F} \text{ is stable} \quad \Longleftrightarrow \quad \mathcal{C}_{Y^*} - \mathcal{F} \text{ is stable}$$

bringing our stability operator into the form of a "unitary minus bounded self-adjoint" to which Lemma 7.4.7 (in the Hilbert space of matrices) will apply.

To complete this argument, all we need is a "symmetric polar decomposition" of M in the form $M = QYQ^*$, where Y is unitary and $Q \sim 1$ knowing that $M \sim 1$.

We will give this decomposition explicitly. Write $M = A + iB$ with $A := \operatorname{Re} M$ and $B := \operatorname{Im} M > 0$. Then we can write

$$M = \sqrt{B}\Big(\frac{1}{\sqrt{B}}A\frac{1}{\sqrt{B}} + i\Big)\sqrt{B}$$

and now we make the middle factor unitary by dividing its absolute value:

$$M = \sqrt{B}WYW\sqrt{B} =: QYQ^*$$

$$W := \Big[1 + \Big(\frac{1}{\sqrt{B}}A\frac{1}{\sqrt{B}}\Big)^2\Big]^{\frac{1}{4}}, \qquad Y := \frac{\frac{1}{\sqrt{B}}A\frac{1}{\sqrt{B}} + i}{W^2}.$$

In the regime, where $c \leqslant B \leqslant C$ and $\|A\|_2 \leqslant C$, we have

$$Q = \sqrt{B}W \sim 1$$

in the sense that $\|Q\|_2 \lesssim 1$ and $\|Q^{-1}\|_2 \lesssim 1$. In our application, we use the upper bound (8.1.3) for $\|M\|_2$ and the lower bound on $B = \operatorname{Im} M$ from (8.1.4). This gives a control on both $\|Q\|_2$ and $\|Q^{-1}\|_2$ as a certain power of $\rho(z)$ and this will be responsible for parts of the powers collected in the right hand side of (8.3.2). In this proof here we focus only on the bulk, so we do not intend to gain the additional term $\operatorname{dist}(z, \operatorname{supp}\rho)$ that requires a slightly different argument. The result is

$$\Big\|\frac{1}{1 - \mathcal{C}_M\mathcal{S}}\Big\|_{sp} \leqslant \frac{1}{\rho(z)^C}\Big\|\frac{1}{\mathcal{U} - \mathcal{F}}\Big\|_{sp} \tag{8.3.5}$$

with $\mathcal{U} := \mathcal{C}_{Y^*}$.

We remark that $\mathcal{F}$ can also be written as follows:

$$\mathcal{F} = \mathcal{K}_Q^*\mathcal{S}\mathcal{K}_Q = \mathcal{C}_W\mathcal{C}_{\sqrt{\operatorname{Im} M}}\mathcal{S}\mathcal{C}_{\sqrt{\operatorname{Im} M}}\mathcal{C}_W. \tag{8.3.6}$$

Finally, we need to invert $\mathcal{C}_{Y^*} - \mathcal{F}$ effectively with the help of Lemma 7.4.7. Since $\mathcal{F}$ is positivity preserving, a Perron-Frobenius type theorem (called the Krein-Rutman theorem in more general Banach spaces) applied to $\mathcal{F}$ yields that it has a normalized *eigenmatrix* F with eigenvalue $\|\mathcal{F}\|_{sp} \leqslant 1$. The following lemma collects information on $\mathcal{F}$ and F, similarly to Lemma 7.4.4:

Lemma 8.3.7. *Assume the flatness condition* (6.1.24) *and let $\mathcal{F}$ be defined by* (8.3.6). *Then $\mathcal{F}$ has a unique normalized eigenmatrix corresponding to its largest eigenvalue*

$$\mathcal{F}[F] = \|\mathcal{F}\|_{sp}F, \qquad \|F\|_{hs} = 1, \qquad \|\mathcal{F}\|_{sp} \leqslant 1.$$

Furthermore

$$\|\mathcal{F}\|_{sp} = 1 - \frac{\langle F, \mathcal{C}_W[\operatorname{Im} M]\rangle}{\langle F, W^{-2}\rangle}\operatorname{Im} z,$$

the eigenmatrix F has bounds

$$\frac{1}{\|M\|_2^7} \leqslant F \leqslant \|M\|_2^6$$

and $\mathcal{F}$ has a spectral gap:

$$\operatorname{Spec}(\mathcal{F}/\|\mathcal{F}\|_{sp}) \subset [-1+\theta, 1-\theta] \cup \{1\}, \qquad \theta \geqslant \|M\|_2^{-42} \tag{8.3.8}$$

(the explicit powers do not play any significant role).

We omit the proof of this lemma (see Lemma 4.6 of [5]), its proof is similar but more involved than that of Lemma 7.4.4, especially the noncommutative analogue of Lemma 7.4.2 needs substantial changes (this is given in Lemma A.3 in [5]).

Armed with the bounds on $\mathcal{F}$ and F, we can use Lemma 7.4.7 with T playing the role of $\mathcal{F}$ and $\mathcal{U} := \mathcal{C}_{Y^*}$ playing the role of U:

$$\left\| \frac{1}{\mathcal{U} - \mathcal{F}} \right\|_{sp} \lesssim \frac{1}{\operatorname{Gap}(\mathcal{F}) \; \left|1 - \|\mathcal{F}\| \langle F, \mathcal{U}(F)\rangle\right|}.$$

We already had a bound on the gap of $\mathcal{F}$ in (8.3.8). As a last step, we prove the estimate

$$\left|1 - \langle F, \mathcal{U}(F)\rangle\right| = \left|1 - \langle F, Y^* F Y^*\rangle\right| \geqslant \frac{\rho^2(z)}{\|M\|_2^4} \geqslant \rho(z)^6.$$

Exercise 8.3.9. *Prove these last two bounds by using* $\left|1 - \langle F, Y^* F Y^*\rangle\right| \geqslant \langle F, \mathcal{C}_{\operatorname{Im} Y^*} F\rangle$, *using the definition of* Y *and various bounds on* M *from Theorem* 8.1.2.

Combining (8.3.5) with these last bounds and with the bound on the gap of $\mathcal{F}$ (8.3.8) we complete the proof of Lemma 8.3.1 (without the $\operatorname{dist}(z, \operatorname{supp}\rho)$ part). □

Exercise 8.3.10. *Prove the matrix analogue of the unconditional bound* (7.2.11), *i.e. if* M *solves the MDE* (8.1.1), *where we only assume that* $\mathcal{S}$ *is symmetric and positivity preserving, then* $\|M\|_{hs} \leqslant \frac{2}{|z|}$. *(Hint: use the representation* $M = QYQ^*$ *to express* $YQ^*Q = -\frac{1}{z}(1 + Y\mathcal{F}(Y))$ *and take Hilbert-Schmidt norm on both sides.)*

9. Ideas of the proof of the local laws

In this section we sketch the proof of the local laws. We will present the more general correlated case, i.e. Theorem 6.2.7 and we will focus on the entrywise local law (6.2.12).

9.1. Structure of the proof In Section 3 around (3.1.24) we already outlined the main idea. Starting from $HG = I + zG$, we have the identity

$$I + (z + \mathcal{S}[G])G = D, \qquad D := HG + \mathcal{S}[G]G, \tag{9.1.1}$$

and we compare it with the matrix Dyson equation

$$I + (z + \mathcal{S}[M])M = 0. \tag{9.1.2}$$

The first (probabilistic) part of the proof is a good bound on D, the second (deterministic) part is to use the stability of the MDE to conclude from these two equations that $G - M$ is small. The first question is in which norm should one estimate these quantities?

Since D is still random, it is not realistic to estimate it in operator norm, in fact $\|D\|_2 \gtrsim 1/\eta$ with high probability. To see this, consider the simplest Wigner case,

$$D = I + \Big(z + \frac{1}{N}\operatorname{Tr} G\Big)G.$$

Let λ be the closest eigenvalue to $\operatorname{Re} z$ with normalized eigenvector $\mathbf{u}$. Note that typically $|\operatorname{Re} z - \lambda| \lesssim 1/N$ and $\eta \gg 1/N$, thus $\|G\mathbf{u}\|_2 = 1/|\lambda - z| \sim 1/\eta$ (suppose that $\operatorname{Re} z$ is away from zero). From the local law we know that $\frac{1}{N}\operatorname{Tr} G \sim m_{sc} \sim 1$ and $z + m_{sc} \sim 1$. Thus

$$\|D\mathbf{u}\|_2 = \Big\|\mathbf{u} + \Big(z + \frac{1}{N}\operatorname{Tr} G\Big) G\mathbf{u}\Big\|_2 \sim \|G\mathbf{u}\|_2 \sim 1/\eta.$$

The appropriate weaker norm is the entrywise maximum norm defined by

$$\|T\|_{\max} := \max_{ij} |T_{ij}|.$$

9.2. Probabilistic part of the proof In the maximum norm we have the following

Theorem 9.2.1. *Under the conditions of Theorem 6.2.7, for any $\gamma, \varepsilon, D > 0$ we have the following high probability statement for some $z = E + i\eta$ with $|z| \leqslant 1000$, $\eta \geqslant N^{-1+\gamma}$:*

$$\mathbb{P}\Big(\|D(z)\|_{\max} \geqslant \frac{N^{\varepsilon}}{\sqrt{N\eta}} \Big) \leqslant \frac{C}{N^D}, \tag{9.2.2}$$

i.e. all matrix elements D_{ij} are small simultaneously for all spectral parameters.

We will omit the proof, which is a tedious calculation and whose basic ingredients were sketched in Section 3. For the Wigner type matrices or for correlated matrices with fast (exponential) correlation decay as in [5] one may use the Schur complement method together with concentration estimates on quadratic functionals of independent or essentially independent random vectors (Section 3.1.1). For more general correlations or if nonzero expectation of H is allowed, then we may use the cumulant method (Section 3.1.19). In both cases, one establishes a high moment bound on $\mathbb{E}|D_{ij}|^p$ via a detailed expansion and then one concludes a high probability bound via Markov inequality.

9.3. Deterministic part of the proof In the second (deterministic) part of the proof we compare (9.1.1) and (9.1.2). From these two equations we have

$$(I - M\mathcal{S}[\cdot]M)[G - M] = MD + M\mathcal{S}[G - M](G - M), \tag{9.3.1}$$

so by inverting the super operator $I - M\mathcal{S}[\cdot]M = I - \mathcal{C}_M\mathcal{S}$, we get

$$G - M = \frac{1}{I - \mathcal{C}_M\mathcal{S}}\big[MD\big] + \frac{1}{I - \mathcal{C}_M\mathcal{S}}\Big[M\mathcal{S}[G - M](G - M)\Big]. \tag{9.3.2}$$

Not only is $\|M\|$ is bounded, see (8.1.3), but also both

$$\|M\|_\infty := \max_i \sum_j |M_{ij}| \qquad \text{and} \qquad \|M\|_1 := \max_j \sum_i |M_{ij}| \tag{9.3.3}$$

are bounded. This information is obvious for Wigner type matrices, when M is diagonal. For correlated matrices with fast correlation decay it requires a somewhat involved additional proof that we do not repeat here, see Theorem 2.5 of [5]. Slow decay needs another argument [37].

Furthermore, we know that in the bulk spectrum the inverse of the stability operator is bounded in spectral norm (8.3.2), i.e. when the stability operator is considered mapping matrices with Hilbert Schmidt norm. We may also consider its norm in the other two natural norms, i.e. when the space of matrices is equipped with the maximum norm (9.3.3) and the Euclidean matrix norm $\|\cdot\|$.

We find the boundedness of the inverse of the stability operator in these two other norms as well since we can prove (see Exercise 9.3.7)

$$\Big\|\frac{1}{I-\mathcal{C}_M\mathcal{S}}\Big\|_\infty + \Big\|\frac{1}{I-\mathcal{C}_M\mathcal{S}}\Big\|_2 \lesssim \Big\|\frac{1}{I-\mathcal{C}_M\mathcal{S}}\Big\|_{sp}. \tag{9.3.4}$$

Note that the bound on the first term in the left hand side is the analogue of the estimate from Exercise 7.4.9. Using all this , we obtain from (9.3.2) that

$$\|G-M\|_{max} \lesssim \|D\|_{max} + \|G-M\|_{max}^2,$$

where $\lesssim$ includes factors of $\rho(z)^{-C}$, which are harmless in the bulk. From this quadratic inequality we easily obtain that

$$\|G-M\|_{max} \lesssim \|D\|_{max}, \tag{9.3.5}$$

assuming a weak bound $\|G-M\|_{max} \ll 1$. This latter information is obtained by a continuity argument in the imaginary part of the spectral parameter. We fix an E in the bulk, $\rho(E) > 0$ and consider $(G-M)(E+i\eta)$ as a function of η. For large η we know that both G and M are bounded by $1/\eta$, hence they are small, so the weak bound $\|G-M\|_{max} \ll 1$ holds. Then we conclude that (9.3.5) holds for large η. Since $\|D\|_{max}$ is small, at least with very high probability, see (9.2.2), we obtain that the strong bound

$$\|G-M\|_{max} \lesssim \frac{N^\varepsilon}{\sqrt{N\eta}} \tag{9.3.6}$$

also holds. Now we may reduce the value of η a bit using the fact that the function $\eta \to (G-M)(E+i\eta)$ is Lipschitz continuous with Lipschitz constant C/η^2. So we know that $\|G-M\|_{max} \ll 1$ for this smaller η value as well. Thus (9.3.5) can again be applied and together with (9.2.2) we get the strong bound (9.3.6) for this reduced η as well. We continue this "small-step" reduction as long as the strong bound implies the weak bound, i.e. as long as $N\eta \gg N^{2\varepsilon}$, i.e. $\eta \gg N^{-1+2\varepsilon}$. Since $\varepsilon > 0$ is arbitrary we can go down to the scales $\eta \geqslant N^{-1+\gamma}$ for any $\gamma > 0$. Some care is needed in this argument, since the smallness of $\|D\|_{max}$ holds only with high probability, so in every step we lose a set of small probability. This is, however, affordable by the union bound since the probability of the events where D is not controlled is very small, see (9.2.2).

The proof of the averaged law (6.2.13) is similar. Instead of the maximum norm, we use averaged quantities of the form $\langle TD\rangle = \frac{1}{N}\operatorname{Tr} TD$. In the first, probabilistic step instead of (9.2.2) we prove that for any fixed deterministic matrix T we have

$$|\langle TD\rangle| \leqslant \frac{N^\varepsilon}{N\eta}\|T\|$$

with very high probability. Notice that averaged quantities can be estimated with an additional $(N\eta)^{-1/2}$ power better; this is the main reason why averaged law (6.2.13) has a stronger control than the entrywise or the isotropic laws.

Exercise 9.3.7. *Prove* (9.3.4). *Hint: consider the identity* (7.4.10) *with* $R = \mathcal{C}_M\mathcal{S}$ *and use the smoothing properties of the self-energy operation* $\mathcal{S}$ *following from* (6.1.24) *and the boundedness of* M *in all three relevant norms.*

References

[1] E. Abrahams, P.W. Anderson, D.C. Licciardello, and T.V. Ramakrishnan, *Scaling theory of localization: Absence of quantum diffusion in two dimensions*, Phys. Rev. Lett. **42** (1979), 673. ↑121

[2] A. Aggarwal, *Bulk universality for generalized Wigner matrices with a few moments*, Preprint Arxiv 1612.00421 (2016). ↑89

[3] M. Aizenman and S. Molchanov, *Localization at large disorder and at extreme energies: an elementary derivation*, Comm. Math. Phys. **157** (1993), no. 2, 245–278. MR1244867 ↑120

[4] O. Ajanki, L. Erdős, and T. Krüger, *Quadratic vector equations on complex upper half-plane* (2015), available at 1506.05095. ↑122, 127, 139, 140, 142, 143, 144

[5] Oskari H. Ajanki, László Erdős, and Torben Krüger, *Stability of the matrix Dyson equation and random matrices with correlations*, Probab. Theory Related Fields **173** (2019), no. 1-2, 293–373, DOI 10.1007/s00440-018-0835-z, available at 1604.08188. MR3916109 ↑128, 130, 131, 153, 154

[6] O. Ajanki, L. Erdős, and T. Krüger, *Singularities of solutions to quadratic vector equations on complex upper half-plane*, Comm. Pure Appl. Math. **70** (2017), no. 9, 1672–1705. ↑124, 140, 143, 144, 148

[7] O. Ajanki, L. Erdős, and T. Krüger, *Universality for general Wigner-type matrices*, Prob. Theor. Rel. Fields **169** (2017), no. 3-4, 667–727. ↑129, 130, 131, 132

[8] G. Akemann, J. Baik, and P. Di Francesco (eds.), *The Oxford handbook of random matrix theory*, Oxford University Press, Oxford, 2011. MR2920518 ↑90

[9] J. Alt, L. Erdős, and T. Krüger, *The Dyson equation with linear self-energy: spectral bands, edges and cusps* (2018), available at 1804.07752. ↑127, 129

[10] J. Alt, L. Erdős, T. Krüger, and D. Schröder, *Correlated random matrices: band rigidity and edge universality* (2018), available at 1804.07744. ↑129

[11] A. Amir, N. Hatano, and D. R. Nelson, *Non-Hermitian localization in biological networks*, Phys. Rev. E **93** (2016), 042310. ↑78

[12] G. W. Anderson, A. Guionnet, and O. Zeitouni, *An introduction to random matrices*, Cambridge Studies in Advanced Mathematics, vol. 118, Cambridge University Press, Cambridge, 2010. MR2760897 ↑86, 90, 95, 139

[13] P. W. Anderson, *Absence of diffusion in certain random lattices*, Phys. Rev. **109** (1958Mar), 1492–1505. ↑78, 118, 121

[14] Z. Bao and L. Erdős, *Delocalization for a class of random block band matrices*, Probab. Theory Related Fields (2016), 1–104. ↑89

[15] R. Bauerschmidt, J. Huang, A. Knowles, and H.-T. Yau, *Bulk eigenvalue statistics for random regular graphs* (May 2015), available at 1505.06700. ↑89

[16] R. Bauerschmidt, A. Knowles, and H.-T. Yau, *Local semicircle law for random regular graphs* (March 2015), available at 1503.08702. ↑89

[17] F. Bekerman, A. Figalli, and A. Guionnet, *Transport maps for β-matrix models and universality*, Comm. Math. Phys. **338** (2015), no. 2, 589–619. MR3351052 ↑85

[18] F. Benaych-Georges and S. Péche, *Localization and delocalization for heavy tailed band matrices*, Annales de l'Institut Henri Poincaré (B) Probabilités et Statistiques, Institute Henri Poincaré **50(4)** (2014), 1385–1403. ↑89

[19] P. Bleher and A. Its, *Semiclassical asymptotics of orthogonal polynomials, Riemann-Hilbert problem, and universality in the matrix model*, Ann. of Math. (2) **150** (1999), no. 1, 185–266. MR1715324 ↑87

[20] C. Bordenave and A. Guionnet, *Delocalization at small energy for heavy-tailed random matrices* (2016), available at 1603.08845. ↑89

[21] P. Bourgade, L. Erdős, and H.-T. Yau, *Bulk universality of general β-ensembles with non-convex potential*, J. Math. Phys. **53** (2012), no. 9, 095221, 19. MR2905803 ↑85

[22] P. Bourgade, L. Erdős, and H.-T. Yau, *Edge universality of beta ensembles*, Comm. Math. Phys. **332** (2014), no. 1, 261–353. MR3253704 ↑85

[23] P. Bourgade, L. Erdős, and H.-T. Yau, *Universality of general β-ensembles*, Duke Math. J. **163** (2014), no. 6, 1127–1190. MR3192527 ↑85

[24] P. Bourgade, L. Erdős, H.-T. Yau, and J. Yin, *Fixed energy universality for generalized Wigner matrices*, Comm. Pure Appl. Math. (2015), 1–67. ↑137

[25] E. Brézin and S. Hikami, *Correlations of nearby levels induced by a random potential*, Nuclear Phys. B **479** (1996), no. 3, 697–706. MR1418841 ↑87

[26] E. Brézin and S. Hikami, *Spectral form factor in a random matrix theory*, Phys. Rev. E **55** (1997), no. 4, 4067–4083. MR1449379 ↑87

[27] E. B. Davies, *The functional calculus*, J. London Math. Soc. **52** (1995), no. 1, 166–176. MR1345723 ↑133

[28] P. Deift, *Orthogonal polynomials and random matrices: a Riemann-Hilbert approach*, Courant Lecture Notes in Mathematics, vol. 3, New York University, Courant Institute of Mathematical Sciences, New York; American Mathematical Society, Providence, RI, 1999. MR1677884 ↑80, 87

[29] P. Deift and D. Gioev, *Universality at the edge of the spectrum for unitary, orthogonal, and symplectic ensembles of random matrices*, Comm. Pure Appl. Math. **60** (2007), no. 6, 867–910. MR2306224 ↑87

[30] P. Deift, T. Kriecherbauer, K. T-R McLaughlin, S. Venakides, and X. Zhou, *Strong asymptotics of orthogonal polynomials with respect to exponential weights*, Comm. Pure Appl. Math. **52** (1999), no. 12, 1491–1552. MR1711036 ↑87

[31] F. J. Dyson, *A Brownian-motion model for the eigenvalues of a random matrix*, J. Math. Phys. **3** (1962), 1191–1198. MR0148397 ↑88

[32] L. Erdős and A. Knowles, *Quantum diffusion and delocalization for band matrices with general distribution*, Ann. Henri Poincaré **12** (2011), no. 7, 1227–1319. MR2846669 ↑89

[33] L. Erdős, A. Knowles, and H.-T. Yau, *Averaging fluctuations in resolvents of random band matrices*, Ann. Henri Poincaré **14** (2013), no. 8, 1837–1926. MR3119922 ↑89

[34] L. Erdős, A. Knowles, H.-T. Yau, and J. Yin, *Spectral statistics of Erdős-Rényi Graphs II: Eigenvalue spacing and the extreme eigenvalues*, Comm. Math. Phys. **314** (2012), no. 3, 587–640. MR2964770 ↑89

[35] L. Erdős, A. Knowles, H.-T. Yau, and J. Yin, *Delocalization and diffusion profile for random band matrices*, Comm. Math. Phys. **323** (2013), no. 1, 367–416. MR3085669 ↑89, 121

[36] L. Erdős, A. Knowles, H.-T. Yau, and J. Yin, *Spectral statistics of Erdős-Rényi graphs I: Local semicircle law*, Ann. Probab. **41** (2013), no. 3B, 2279–2375. MR3098073 ↑89

[37] L. Erdős, T. Krüger, and D. Schröder, *Random matrices with slow correlation decay* (2017), available at `1705.10661`. ↑104, 131, 154

[38] László Erdős, Torben Krüger, and David Renfrew, *Power law decay for systems of randomly coupled differential equations*, SIAM J. Math. Anal. **50** (2018), no. 3, 3271–3290, DOI 10.1137/17M1143125, available at `1708.01546`. MR3816180 ↑78

[39] L. Erdős, B. Schlein, and H.-T. Yau, *Wegner estimate and level repulsion for Wigner random matrices*, Int. Math. Res. Not. **2010** (2010), no. 3, 436–479. MR2587574 ↑81

[40] L. Erdős, B. Schlein, and H.-T. Yau, *Universality of random matrices and local relaxation flow*, Invent. Math. **185** (2011), no. 1, 75–119. MR2810797 ↑137

[41] L. Erdős, B. Schlein, H.-T. Yau, and J. Yin, *The local relaxation flow approach to universality of the local statistics for random matrices*, Ann. Inst. H. Poincaré Probab. Statist. **48** (2012), no. 1, 1–46. MR2919197 ↑137

[42] L. Erdős and K. Schnelli, *Universality for random matrix flows with time-dependent density*, Ann. Inst. H. Poincaré Probab. Statist. **53** (2015), no. 4, 1606–1656. ↑137

[43] L. Erdős and H.-T. Yau, *Universality of local spectral statistics of random matrices*, Bull. Amer. Math. Soc. (N.S.) **49** (2012), no. 3, 377–414. MR2917064 ↑87

[44] L. Erdős and H.-T. Yau, *Gap universality of generalized Wigner and β-ensembles*, J. Eur. Math. Soc. (JEMS) **17** (2015), no. 8, 1927–2036. MR3372074 ↑137

[45] L. Erdős and H.-T. Yau, *Dynamical approach to random matrix theory*, Vol. 28, Courant Lecture Notes in Mathematics, 2017. ↑87, 90, 132, 139

[46] László Erdős, Horng-Tzer Yau, and Jun Yin, *Bulk universality for generalized Wigner matrices*, Probab. Theory Related Fields **154** (2011), no. 1-2, 341–407. ↑83, 137

[47] A. S. Fokas, A. R. Its, and A. V. Kitaev, *The isomonodromy approach to matrix models in 2D quantum gravity*, Comm. Math. Phys. **147** (1992), no. 2, 395–430. MR1174420 ↑86

[48] P. J. Forrester, *Log-gases and random matrices*, London Mathematical Society Monographs Series, vol. 34, Princeton University Press, Princeton, NJ, 2010. MR2641363 ↑90

[49] J. Fröhlich and T. Spencer, *Absence of diffusion in the Anderson tight binding model for large disorder or low energy*, Comm. Math. Phys. **88** (1983), no. 2, 151–184. MR696803 ↑120

[50] Y. V. Fyodorov and A. D. Mirlin, *Scaling properties of localization in random band matrices: a σ-model approach*, Phys. Rev. Lett. **67** (1991), no. 18, 2405–2409. MR1130103 ↑121

[51] J. Geronimo and T. Hill, *Necessary and sufficient condition that the limit of Stieltjes transforms is a Stieltjes transform*, J. Approx. Theory **121** (2003), 54–60. ↑91

[52] I. Ya. Goldsheid, S. A. Molchanov, and L. A. Pastur, *A pure point spectrum of the stochastic one-dimensional Schrödinger equation*, Funkt. Anal. Appl. **11** (1977), 1–10. ↑120

[53] F. Götze, A. Naumov, and A. Tikhomirov, *Local semicircle law under moment conditions. Part I: The Stieltjes transform* (October 2015), available at 1510.07350. ↑89

[54] Y. He, A. Knowles, and R. Rosenthal, *Isotropic self-consistent equations for mean-field random matrices*, Probab. Th. Rel. Fields **171** (2018), no. 1–2, 203–249. ↑89

[55] J. W. Helton, R. R. Far, and R. Speicher, *Operator-valued Semicircular Elements: Solving A Quadratic Matrix Equation with Positivity Constraints*, Internat. Math. Res. Notices **2007** (2007). ↑128, 148

[56] J. Huang, B. Landon, and H.-T. Yau, *Bulk universality of sparse random matrices*, J. Math. Phys. **56** (2015), no. 12, 123301, 19. MR3429490 ↑89

[57] C. Itzykson and J. B. Zuber, *The planar approximation. II*, J. Math. Phys. **21** (1980), no. 3, 411–421. MR562985 ↑87

[58] K. Johansson, *Universality of the local spacing distribution in certain ensembles of Hermitian Wigner matrices*, Comm. Math. Phys. **215** (2001), no. 3, 683–705. MR1810949 ↑87

[59] K. Johansson, *Universality for certain Hermitian Wigner matrices under weak moment conditions*, Ann. Inst. H. Poincaré Probab. Statist. **48** (2012), no. 1, 47–79. MR2919198 ↑89

[60] A. Knowles and J. Yin, *The isotropic semicircle law and deformation of Wigner matrices*, Comm. Pure Appl. Math. **66** (2013), no. 11, 1663–1750. MR3103909 ↑89

[61] A. Knowles and J. Yin, *Anisotropic local laws for random matrices* (October 2014), available at 1410.3516. ↑89

[62] B. Landon, P. Sosoe, and H.-T. Yau, *Fixed energy universality of Dyson Brownian motion* (2016), available at 1609.09011. ↑136, 137

[63] B. Landon and H.-T. Yau, *Convergence of local statistics of Dyson Brownian motion*, Comm. Math. Phys **355** (2017), no. 3, 949–1000. ↑136, 137

[64] J. O. Lee, K. Schnelli, B. Stetler, and H.-T. Yau, *Bulk universality for deformed Wigner matrices*, Ann. Probab. **44** (2016), no. 3, 2349–2425. MR3502606 ↑89

[65] M. L. Mehta, *Random matrices*, Second, Academic Press, Inc., Boston, MA, 1991. MR1083764 ↑80, 86, 90

[66] S. O'Rourke and V. Vu, *Universality of local eigenvalue statistics in random matrices with external source*, Random Matrices: Theory and Applications **03** (2014), no. 02. ↑89

[67] L. Pastur and M. Shcherbina, *Eigenvalue distribution of large random matrices*, Mathematical Surveys and Monographs, vol. 171, American Mathematical Society, Providence, RI, 2011. MR2808038 ↑90

[68] J. Schenker, *Eigenvector localization for random band matrices with power law band width*, Comm. Math. Phys. **290** (2009), no. 3, 1065–1097. MR2525652 ↑89, 121

[69] M. Shcherbina, *Change of variables as a method to study general β-models: bulk universality*, J. Math. Phys. **55** (2014), no. 4, 043504, 23. MR3390602 ↑85

[70] T. Shcherbina, *On the second mixed moment of the characteristic polynomials of 1D band matrices*, Comm. Math. Phys. **328** (2014), no. 1, 45–82. MR3196980 ↑89, 121

[71] T. Shcherbina, *Universality of the local regime for the block band matrices with a finite number of blocks*, J. Stat. Phys. **155** (2014), no. 3, 466–499. MR3192170 ↑89

[72] T. Shcherbina, *Universality of the second mixed moment of the characteristic polynomials of the 1D band matrices: real symmetric case*, J. Math. Phys. **56** (2015), no. 6, 063303, 23. MR3369897 ↑89

[73] S. Sodin, *The spectral edge of some random band matrices*, Ann. of Math. (2) **172** (2010), no. 3, 2223–2251. MR2726110 ↑89

[74] T. Tao, *Topics in random matrix theory*, Graduate Studies in Mathematics, vol. 132, American Mathematical Society, Providence, RI, 2012. MR2906465 ↑90

[75] T. Tao and V. Vu, *Random matrices: universality of local eigenvalue statistics*, Acta Math. **206** (2011), no. 1, 127–204. MR2784665 ↑88, 137

[76] D. Vollhardt and P. Wölfle, *Diagrammatic, self-consistent treatment of the Anderson localization problem in* $d \leqslant 2$ *dimensions*, Phys. Rev. B **22** (1980), 4666–4679. ↑120

[77] J. Weidmann, *Linear Operators in Hilbert Spaces*, Springer Verlag, New York, 1980. ↑91

[78] E. P. Wigner, *Characteristic vectors of bordered matrices with infinite dimensions*, Ann. of Math. (2) **62** (1955), 548–564. MR0077805 ↑77, 82

[79] J. Wishart, *The generalised product moment distribution in samples from a normal multivariate population*, Biometrika **20A** (1928), no. 1/2, 32–52. ↑77

Institute of Science and Technology (IST) Austria, Am Campus 1, A-3400, Klosterneuburg, Austria
Email address: lerdos@ist.ac.at

IAS/Park City Mathematics Series
Volume 26, Pages 159–212
https://doi.org/10.1090/pcms/026/00845

Counting equilibria in complex systems via random matrices

Yan V. Fyodorov

Abstract. How many equilibria will a large complex system, modeled by N randomly coupled autonomous nonlinear differential equations typically have? How many of those equilibria are stable being local attractors of nearby trajectories? These questions arise in many applications and can be partly answered within the framework of a model introduced in [32] by employing the methods of Random Matrix Theory (RMT) applied to real asymmetric matrices from the Gaussian Elliptic Ensemble. An efficient approach to the problem developed by Gerard Ben Arous, Boris Khoruzhenko and the author in an unpublished manuscript [9] exploits the ideas of the Large Deviation Theory in the RMT context. The lectures aim to outline these recent developments in an informal style typical for Theoretical Physics.

Contents

1 May model of a complex system: an introduction 159
2 Large−N asymptotics and large deviations for the Ginibre ensemble 169
3 Counting multiple equilibria via Kac-Rice formulas 177
4 Mean number of equilibria: asymptotic analysis for large deviations 183
5 Appendix: Supersymmetry and characteristic polynomials of real Ginibre matrices. 192
6 Exercises with hints 195

1. May model of a complex system: an introduction

Will diversity make a food chain more or less stable? The prevailing view in the mid-twentieth century was that diverse ecosystems have greater resilience to recover from events displacing the system from equilibrium and hence are more stable. This ‘ecological intuition’ was challenged by Robert May in 1972 [46]. At that time, computer simulations suggested that large complex systems assembled at random might become unstable as the system complexity increases [35]. May’s 1972 paper complemented that work with an analytic investigation of the neighbourhood stability of a model ecosystem whereby N species at equilibrium are subject to random interactions.

The time evolution of large complex systems, of which model ecosystems give one example, is often described within the general mathematical framework of coupled first-order nonlinear ordinary differential equations (ODEs).

$$\frac{d\mathbf{x}}{dt} = \mathbf{F}(\mathbf{x}), \quad \mathbf{x} = (x_1, \ldots, x_N) \in \mathbb{R}_N \tag{1.1}$$

The choice of the vector field $\mathbf{F}(\mathbf{x}) = (F_1(\mathbf{x}), \ldots, F_N(\mathbf{x}))$ and hence detailed properties of the phase space trajectories strongly depend on specific area of applications and vary considerably from model to model. In populational ecology a basic model is the multispecies Lotka-Volterra system [47] for population densities $x_i \geqslant 0$, with the choice $F_i = x_i(r_i + \sum_{j=1}^{N} J_{ij} x_j)$ where r_i is the intrinsic growth/death rate of the population (for $r_i > 0$ and $r_i < 0$, respectively), and the parameters J_{ij} model the influence of growth in population j on growth rate in population i. In that setting J_{ii} are usually negative to ensure a self-regulation mechanism, whereas $J_{i \neq j}$ can be of any sign, reflecting either competitive (with $J_{i \neq j} < 0$), mutualistic (with $J_{i \neq j} > 0$) or predator-prey (with $J_{ij} > 0$ and $J_{ji} < 0$) relations between the living organisms. In general, even when the entries J_{ij} and J_{ji} of interaction matrix are of the same sign, there seems to be no clear reason to assume interactions to be symmetric.

Another area where equation (1.1) had been most popular and well-studied numerically was the research of neural networks consisting of randomly interconnected neural units [51], with a particular choice of $F_i = -x_i + \sum_{j \neq i}^{n} J_{ij} S(x_j)$ where $S(x)$ is an odd sigmoid function (i.e. $S(-x) = -S(x)$, $S'(x) \geqslant 0$, $S(\infty) < \infty$) representing the synaptic nonlinearity. In that context J_{ij} are representing the synaptic connectivity between neuron i and j and can be of any sign, either excitatory ($J_{ij} > 0$) or inhibitory ($J_{ij} < 0$). Other examples of this sort include machine learning [33], complex gene regulatory networks [16, 44], and catalytic reaction networks [52].

Study of any autonomous dynamical system (1.1) traditionally starts with a "local stability analysis". The latter amounts to determining all possible points of equilibria defined as zeroes of the vector field: $\mathbf{F}(\mathbf{x}) = 0$ and analysing the dynamical behaviour in the vicinity of each those points by Taylor-expanding and subsequently replacing the non-linear interaction functions near the equilibrium with their linear approximations. In the context of a generic system, the Hartman-Grobner theorem [39] then asserts that the neighbourhood stability of a typical equilibrium is governed by its linear approximation. It is along these lines that May suggested to look at the model linear system, thinking of y_i as the deviations from a local equilibrium:

$$\frac{dy_j}{dt} = -\mu y_j + \sum_{k=1}^{N} J_{jk}\, y_k\,, \quad j = 1, \ldots, N\,, \tag{1.2}$$

to shed some light on stability of large complex nonlinear systems. Here $J = (J_{jk})$ is the coupling matrix and $\mu > 0$ ensures that in the absence of interactions, i.e., when all $J_{jk} = 0$, the system (1.2) is self-regulating: if disturbed from the equilibrium $y_1 = y_2 = \ldots = y_N = 0$ it returns back with some characteristic relaxation time set by μ. In ecological context $y_j(t)$ is interpreted as the variation about the equilibrium value, $y_j = 0$, in the population density of species j at time t. The coupling matrix J, called the *community* matrix in ecology, has elements J_{jk} which measure the per capita effect of species k on species j at the presumed equilibrium. Generically, the community matrix is asymmetric, $J_{jk} \neq J_{kj}$. The stability properties of linear model $\mathbf{y}(t) = e^{(-\mu \mathbf{I} + \mathbf{J})t}\mathbf{y}(0)$ is obviously controlled by spectrum of the matrix $\mathbf{J}$. In particular, according to Hartman-Grobner theorem the equilibrium at $\mathbf{y} = 0$ becomes unstable if, and only if the matrix $-\mu\mathbf{I} + \mathbf{J}$ has eigenvalues with positive real part. Note however that even for a stable equilibrium of linear systems a transient behaviour with a temporal growth of initial perturbation may be observed due to non-orthogonality of eigenvectors of asymmetric community matrices, see [37].

In reality, the detailed description of the vector fields $\mathbf{F}(\mathbf{x})$ is rarely available for large enough systems of considerable complexity in problems of practical interest. At the same time it is natural to assume that in order to understand *generic* qualitative properties of the global dynamics of large systems of ODE's shared by many models of similar type it may be enough to retain only a few characteristic structural features of the vector field, treating the rest as random. In part such approach is methodologically inspired by the paradigm of *universality* of complex systems behaviour well-accepted in modern physics, in particular, by undisputed successes of the Random Matrix Theory (RMT). Such theory manages to describe in a single conceptual framework [54] many properties of systems of very diverse nature, such as energy levels of heavy nuclei, zeroes of the Riemann zeta-function and distances between tightly parked cars.

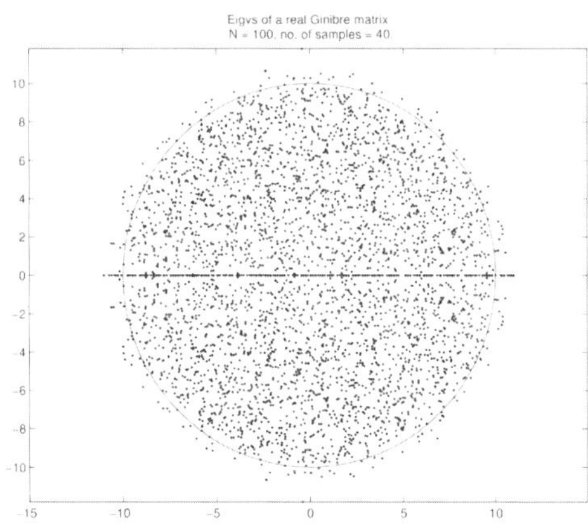

FIGURE 1.3. Eigenvalues of Real Ginibre matrix of size $N = 100$ and variance $\alpha = 1$. As typical eigenvalue of $\mathbf{J}$ with the largest real part grows as $\alpha\sqrt{N}$ instability inevitably occurs with growing N as long as $\mu < \mu_c = \alpha\sqrt{N}$.

Along these lines May considered an ensemble of community matrices J assembled at random, whereby the matrix elements J_{jk} are sampled from a probability distribution, which for simplicity may be assumed Gaussian, with zero mean and a prescribed variance α^2. The corresponding matrix **J** is said to belong to the real Ginibre ensemble which we will denote in this set of lectures as GinOE (see Figure 1.3). Invoking early studies on eigenvalues of GinOE by Ginibre [36] May claimed that for large N the largest real part of the eigenvalues of J is *typically* equal to $\alpha\sqrt{N}$. Obviously, the model's stability is then controlled by the ratio $m = \mu/(\alpha\sqrt{N})$. For N large, the system (1.2) will almost certainly be stable if $m > 1$ and unstable if $m < 1$, with a sharp transition between the two types of behaviour with changing either μ, α or N. In particular, for fixed μ, α the system (1.2) will almost certainly become unstable for N sufficiently large. May himself frequently referred to this instability mechanism as the 'May-Wigner theorem', in an obvious tribute to one of the RMT founding fathers Eugene Wigner.

Despite the simplistic character of May's model, his pioneering work gave rise to a long standing 'stability versus diversity' debate, which is not fully settled yet, see e.g. [5], and played a fundamental role in theoretical ecology by prompting ecologists to think about special features of real multi-species ecosystems that help such systems to remain stable. Variations of May's model are still being discussed nowadays in the context of neighbourhood stability, see [4,5] and references therein. Moreover, May's stability analysis in fact became a paradigm in complex systems of diverse nature; for example, it was used in attempts to understand behaviour of financial ecosystems [20,38]. Following the May's ideas in a wider context, attempts to develop understanding of structures of randomly assembled communities of highly diverse species is an active field of research, see [24] and references therein.

One obvious limitation of the neighbourhood stability analysis is that it gives no insight into the behaviour of the model beyond the instability threshold. Hence May's model has only limited bearing on the dynamics of populations operating out-of-equilibrium. An instability does not necessarily imply lack of persistence: populations could coexist thanks to limit cycles or chaotic attractors, which typically originate from unstable equilibrium points. Important questions posing serious challenges in this area then relate to the classification of equilibria by stability, to the study of their basins of attraction, and to other features of their global dynamics [5].

In a recent paper [32] a simple *nonlinear* extension of the May model was proposed by retaining only the bare essentials - nonlinearity and stochasticity. Much in the spirit of the May's original approach, the suggested model turned out to be simple enough to allow for an analytic treatment yet at the same time rich enough to exhibit a non-trivial behaviour. In particular, it captures an instability transition of the May-Wigner type, but now on the global scale. It also sheds additional light on the nature of this transition by relating it to an exponential explosion in the number of unstable equilibria. Interestingly, despite the nonlinear setting of the problem the properties of Ginibre-like random matrices again play a central role in its analysis.

The goal of these lectures is to give a quite detailed, though heuristic (i.e. informal 'theoretical physics' style) account of the ideas and methods introduced by Boris Khoruzhenko and the present author in [32], and further essentially developed by Gerard Ben Arous, Boris Khoruzhenko and the author in an unpublished manuscript [9]. To achieve this one needs first to develop the necessary background knowledge on eigenvalues of Ginibre Ensemble (for the standard case of unit variance $\alpha = 1$ the corresponding matrices will be denoted by G). The next chapter will aim exactly at this, using, in particular, the review paper [42] for some background material. We will eventually see however that in the case of a generic non-gradient nonlinear autonomous dynamics the relevant random matrix ensemble turns out to be a generalization of that of Ginibre, known as the real Gaussian Elliptic Ensemble. Some technical details helping to better understand mathematical facts mentioned in the lectures are formulated as exercises, and collected (frequently with hints, and sometimes with solutions) in the exercises in § 6.

Acknowledgements: the author is most grateful to Gerard Ben Arous and Boris Khoruzhenko for a collaboration whose results [32] and [9] formed the basis for these lectures. Jean-Philippe Bouchaud is acknowledged for his early insightful comments hinting to possible existence of the line (4.15) and overall encouraging interest in the project. The author also would like to thank Mihail Poplavskyi for discussing some aspects of the material presented, and especially for his indispensable help with preparing exercises for this lecture course, and delivering their solutions to interested students. Giulio Biroli and Chiara Cammarota are acknowledged for a possibility to test this set of lectures on students with background in Theoretical Physics in the most pleasant environment of Beg Rohu Summer School 2017. Finally, the author is very grateful to the organizers and participants of the PCMI Summer School 2017 for creating a stimulating atmosphere, and for the financial support of his participation in the event, in particular from the NSF grant DMS:1441467. The research at King's College London was supported by EPSRC grant EP/N009436/1 "The many faces of random characteristic polynomials".

Real Ginibre ensemble: a concise review We consider $N \times N$ square matrices $G \in \mathcal{M}_N(\mathbb{R})$ with independent identically distributed matrix elements

$$G_{j,k} \sim N(0,1).$$

For this ensemble we will use a notation GinOE, underlying the orthogonal symmetry of the distribution. We will also denote by the angular brackets $\langle \cdots \rangle_{\text{GinOE}}$ the expectation of any function $F : \mathbb{R}^{N\times N} \to \mathbb{C}$ with respect the associated probability distribution, and will frequently omit the corresponding subscript.

Remark 1.4. One can also introduce in a similar way the so called complex Ginibre ensemble (GinUE) defined by

$$G_{j,k} = g_{j,k}^{(1)} + i \cdot g_{j,k}^{(2)}, \quad \text{with i.i.d. } g_{j,k}^{(\cdot)} \sim N(0, 1/2),$$

as well as the so-called quaternion (GinSE) ensemble. Both GinUE and GinSE are largely immaterial for the purposes of these lectures.

Assigning the Dyson's index $\beta = 1, 2, 4$ to GinOE, GinUE and GinSE, respectively one can write for all three ensembles the Joint Probability Density (JPD) with respect to flat Lebesgue measure in the form

$$P_\beta(G) = \left(\frac{\beta}{2\pi}\right)^{\frac{\beta}{2}N^2} \exp\left\{-\frac{\beta}{2}\operatorname{Tr} GG^*\right\}. \tag{1.5}$$

Let $z_1, \ldots, z_N$ be eigenvalues of $G \in$ GinOE. We always assume that eigenvalues are ordered in the following way: real eigenvalues in descending order are followed by pairs of conjugate complex eigenvalues, where pairs are ordered in lexicographical order. We may assume that there is no multiple eigenvalues because of

Lemma 1.6. *For every* N*, the set of elements of* $\mathcal{M}_N(\mathbb{R})$ *with multiple eigenvalues has zero Lebesgue measure in* $\mathbb{R}^{N\times N}$ *(see the exercises in § 6).*

Density of real and complex eigenvalues via partial Schur decompositions Let x be a real eigenvalue of a matrix G with real entries. Then, as is well-known, see e.g. [18, 19], it is always possible to represent the matrix G as

$$G = P\tilde{G}P^T, \quad \tilde{G} = \begin{pmatrix} x & \mathbf{w}^T \\ 0 & G_1 \end{pmatrix} \tag{1.7}$$

for some real $(N-1)$- component column vector $\mathbf{w}$ and a real matrix G_1 of size $(N-1)\times(N-1)$, whereas the matrix P is symmetric and orthogonal: $P^T = P$ and $P^2 = I_N$ (in fact P is parametrized by a vector of $(N-1)$ real components via the so-called *Housholder reflection*). Such representation of G is known as the *incomplete Schur decomposition* corresponding to the real eigenvalue x obviously shared by the matrices G and $\tilde{G}$. Next we can exploit the following

Proposition 1.8 (see [19]). *Assume that the JPD of* G *is given by (1.5) with* $\beta = 1$*, and apply the transformation (1.7). Then the joint probability density of elements of the matrix* $\tilde{G}$ *is given by*

$$\mathcal{P}(\tilde{G})d\tilde{G} = \mathcal{C}_{1,N}|\det(xI_{N-1} - G_1)|\, e^{-\frac{1}{2}(x^2+\mathbf{w}^T\mathbf{w}+\operatorname{Tr} G_1G_1^T)}\, dx\, d\mathbf{w}\, dG_1\,. \tag{1.9}$$

with some normalization constant $\mathcal{C}_{1,N}$.

It is now elementary to integrate out the variables $\mathbf{w}$ and finally arrive to the marginal probability density $R_1^{(r)}(x)$ of the real eigenvalue x (equivalent to the first correlation function/mean density of purely real eigenvalues for the Ginibre ensemble discussed in the end of this chapter):

Proposition 1.10. *The probability density of a real eigenvalue* x *for real Ginibre matrices of size* $N \times N$ *has the following representation:*

$$R_1^{(r)}(x) = \mathcal{C}_{2,N} e^{-\frac{1}{2}x^2} \langle |\det(xI_{N-1} - G_1)| \rangle_{GinOE,N-1} \tag{1.11}$$

where the averaging goes over the Ginibre $\beta = 1$ *ensemble of matrices* G_1 *of the reduced size* $(N-1)\times(N-1)$*, and* $\mathcal{C}_{2,N}$ *is the appropriate normalization constant (which can be actually shown to be equal to* $\frac{1}{2^{N/2}\Gamma(N/2)}$*).*

To derive a closed-form expression for $R_1^{(r)}(x)$ one obviously needs to evaluate the mean modulus of the characteristic polynomial featuring in the right-hand side of (1.11). One way of doing this was suggested by Edelman in [19], which eventually lead to the following expression:

$$R_1^{(r)}(x) = \frac{\Gamma(N-1, x^2)}{\sqrt{2\pi}(N-2)!} + \frac{x^{N-1}}{\sqrt{2\pi}(N-2)!} e^{-x^2/2} \int_0^x e^{-\frac{u^2}{2}} u^{N-2}\, du \tag{1.12}$$

where we used the incomplete Γ–function defined as

$$\Gamma(N, a) = (N-1)!\, e^{-a} \sum_{n=0}^{N-1} \frac{a^n}{n!} = \int_a^\infty e^{-t}\, t^{N-1}\, dt\,. \tag{1.13}$$

In fact there exists an alternative route to arrive to (1.12) in the framework of the so-called 'supersymmetry approach' which uses Berezin integration over anti-commuting variables. Such a method has its own merits. In particular, recently it was demonstrated to provide an access to statistics of the 'left' and 'right' eigenvectors associated with the real eigenvalue x of the Ginibre ensemble [26]. For completeness we give an account of that approach in the Appendix.

Another possibility for the real matrix G is to have a pair of complex conjugate eigenvalues $z = x + iy$, $\bar{z} = x - iy$, with $y > 0$. Then the corresponding incomplete Schur decomposition is known to be described by the following

Proposition 1.14. *(see [18]) Let* $G \in \mathcal{M}_N(\mathbb{R})$ *be a real valued matrix with a pair of complex conjugate* $x \pm iy$ *eigenvalues. Then it can be represented as*

$$G = Q\tilde{G}Q^T, \quad \tilde{G} = \left(\begin{array}{c|c} \begin{pmatrix} x & b \\ -c & x \end{pmatrix} & W \\ \hline 0 & G_2 \end{array}\right), \quad bc > 0,\ b \geqslant c, \quad y = \sqrt{bc}.$$

Here the matrix Q *is symmetric and orthogonal:* $Q^T = Q$ *and* $Q^2 = I_N$ *(given by a product of two Householder reflections, and parametrized by* $2N-3$ *real parameters), whereas* W *is a real* $(N-2)\times 2$ *matrix with* $2(N-2)$ *real entries and* G_2 *is* $(N-2)\times(N-2)$*. The joint probability density of elements of the matrix* $\tilde{G}$ *is given in this case by*

$$\mathcal{P}(\tilde{G})d\tilde{G} = \mathcal{C}_{3,N} \det\left[(xI_{N-2} - G_2)^2 + y^2 I_{N-2}\right] e^{-\frac{1}{2}\operatorname{Tr} G_2 G_2^T}\, dG_2$$
$$\times\, (b-c)\, e^{-\frac{1}{2}(2x^2 + b^2 + c^2 + \operatorname{Tr} WW^T)}\, dx\, db\, dc\, dW\,.$$

with some normalization constant $\mathcal{C}_{3,N}$.

Now it is straightforward to integrate out the variables in W, and then after introducing $\delta = b - c > 0$ and changing variables $(b, c) \to (y, \delta)$ (so that $dbdc = \frac{2y}{\sqrt{\delta^2+4y^2}} dy d\delta$) to integrate out δ as well. As the result one arrives to the following representation for the joint probability density $R_1^{(c)}(x, y)$ of variables x and y (or, equivalently, to the mean density of complex eigenvalues of the Ginibre ensemble):

Proposition 1.15. *The probability density of a complex eigenvalue* $z = x + iy$ *for real Ginibre matrices of size* $N \times N$ *has the following representation:*

$$\begin{aligned} R_1^{(c)}(x, y) &= \mathcal{C}_{4,N}\, |y| e^{-\frac{1}{2}(x^2-y^2)} \operatorname{erfc}(\sqrt{2}|y|) \\ &\times \left\langle \det\left[(xI_{N-1} - G_2)^2 + y^2 I_{N-1} \right] \right\rangle_{GinOE,N-2} \end{aligned} \tag{1.16}$$

where $\operatorname{erfc}(x) = \frac{2}{\sqrt{\pi}} \int_x^\infty e^{-t^2} dt$, *the averaging goes over the Ginibre* $\beta = 1$ *ensemble of matrices* G_2 *of the reduced size* $(N-2) \times (N-2)$, *and* $\mathcal{C}_{4,N}$ *is the appropriate normalization constant.*

Note that in (1.16) we assumed that sign of y can be arbitrary, so that the density should be obviously symmetric with respect to the complex conjugation. Again, the ensemble average of the determinant entering (1.16) can be relatively simply found using anticommuting variables (see the Appendix), with the final result for the mean density of complex eigenvalues given by

$$R_1^{(c)}(x, y) = \frac{2|y|}{\sqrt{2\pi}} e^{2y^2} \operatorname{erfc}(\sqrt{2}|y|) \frac{\Gamma(N-1, x^2+y^2)}{(N-2)!} \tag{1.17}$$

The method of partial Schur decomposition reveals properties of a single real eigenvalue of the real Ginibre ensemble, or of a complex conjugate pair. As frequently happens, one is interested in understanding more general properties reflecting correlations between different eigenvalues. Those can be revealed by continuing the Schur procedure by exposing further eigenvalues/pairs.

To that end let us introduce sets

$$\mathcal{X}_{L,M} = \left\{ G \in \mathbb{R}^{N \times N} : |\text{spec}(G) \cap \mathbb{R}| = L, |\text{spec}(G) \cap \mathbb{C}_+| = M \right\}.$$

i.e. G has exactly L real eigenvalues and M pairs of complex conjugate eigenvalues. Then

$$\mathcal{M}_N(\mathbb{R}) = \bigsqcup_{L+2M=N} \mathcal{X}_{L,M}.$$

Below we study eigenvalues distribution for every set $\mathcal{X}_{L,M}$ independently. Let us fix L and M. For every $G \in \mathcal{X}_{L,M}$ we write its eigenvalues as

$$(\lambda_1, \dots, \lambda_L, x_1 + iy_1, x_1 - iy_1, \dots, x_M + iy_M, x_M - iy_M)$$

with $\lambda_1 > \lambda_2 > \ldots > \lambda_L$ and $x_1 \geqslant x_2 \geqslant \ldots \geqslant x_M$ and $y_j > 0, j = 1, \ldots, M$.

Theorem 1.18 (see, [18], [43], and the exercises in § 6)**.** *Let* $(z_1, z_2, \dots, z_N)$ *be the ordered set of either real or pairs of complex conjugate eigenvalues of matrix* G *taken randomly from the Real Ginibre Ensemble. Then the eigenvalues Joint Probability Density* *(conditional) function is given by*

(1.19) $$\mathcal{P}_1^{(L,M)}(z_1,\ldots,z_N) = 2^M \left(2^{N(N+1)/4}\prod_{j=1}^{N}\Gamma(j/2)\right)^{-1}$$

$$\times \prod_{1\leqslant j<k\leqslant N}|z_j - z_k|\exp\Big\{-\frac{1}{2}\sum_{1\leqslant j\leqslant N} z_j^2\Big\}\prod_{1\leqslant j\leqslant N}\sqrt{\operatorname{erfc}\left(\sqrt{2}\,|\mathrm{Im} z_j|\right)}\mathbf{1}_{G\in\mathcal{X}_{L,M}}.$$

Remark 1.20 (see, [36]). Similarly to the above one can derive eigenvalues JPDF in the case of GinUE (and eventually GinSE) ensembles. Corresponding density for GinUE is given below

(1.21) $$\mathcal{P}_2(z_1,\ldots,z_N) = \frac{1}{(2\pi)^N\prod_{k=1}^{N-1}k!}\prod_{j<k}|z_j-z_k|^2\exp\Big\{-\sum_j |z_j|^2\Big\},$$

Hence, those matrices with real eigenvalues in this case have zero Lebesgue measure.

Integrable structure of real Ginibre ensemble. The most important property of GinOE was revealed in the last decade by several groups of authors in [3,11,21,22, 49,50] and amounts to an *integrable structure* of the ensemble. Namely, ensemble averages for a class of multiplicative test functions turn out to be given by *Pfaffians*. The method of reducing multiple integrals of particular types to Pfaffians has its roots in the paper [13]. In the present context one may quote

Theorem 1.22 (see, e.g. [49], [50], and the exercises in § 6). *Let* N *be even and* $f:\mathbb{C}\to\mathbb{C}$ *be an arbitrary (up to some conditions ensuring the existence of integrals involved) function of complex argument. Define* $F:\mathbb{R}^{N\times N}\to\mathbb{C}$ *by an identity*

$$F(G) = \prod_{1\leqslant j\leqslant N} f(z_j).$$

where z_j *are eigenvalues of* G. *Let* $\{P_{j-1}(z)\}_{j=1}^N$ *be a set of monic polynomials in* z *of degree* $j-1$. *Then the expectation value or 'ensemble averaged value"of* F *is given by*

$$\langle F\rangle_{GinOE} = \left(2^{N(N+1)/4}\prod_{j=1}^{N}\Gamma(j/2)\right)^{-1}\mathrm{Pf}\,(U),$$

where U *is an* $N\times N$ *skew-symmetric matrix with*

(1.23) $$U_{j,k} = \int_{\mathbb{C}^2} d^2z_1 d^2z_2 f(z_1)f(z_2)\mathcal{F}(z_1,z_2)\,P_{k-1}(z_1)\,P_{j-1}(z_2)$$

and the skew-symmetric measure

(1.24) $$\mathcal{F}(z_1 = x_1+iy_1, z_2 = x_2+iy_2) =$$
$$e^{-(z_1^2+z_2^2)/2}\left[2i\delta^2(z_1-\bar{z}_2)\,\mathrm{sgn}(y_1)\operatorname{erfc}(|y_1|\sqrt{2}) + \delta(y_1)\delta(y_2)\,\mathrm{sgn}\,(x_2-x_1)\right].$$

where we understand $\delta^2(z-u)$ *as* $\delta(\mathrm{Re}z-\mathrm{Re}u)\,\delta(\mathrm{Im}z-\mathrm{Im}u)$ *for some fixed* $u\in\mathbb{C}$.

Remark 1.25. In the case of odd N one should take U to be skew symmetric matrix of size $N+1 \times N+1$ with the last column equal to

$$U_{j,N+1} = \int_{\mathbb{R}} f(t)\, P_{j-1}(t)\, e^{-t^2/2}, j = \overline{1,N}.$$

The last row is defined due to skew-symmetric structure of U. In what follows we tacitly assume N to be an even integer. See [13] for more detail.

Definition 1.26. For any skew symmetric matrix A of size $2n \times 2n$ we write Pf A to denote its Pfaffian given by

$$\mathrm{Pf}\, A = \frac{1}{2^n n!} \sum_{\sigma \in S_{2n}} \mathrm{sgn}(\sigma) \prod_{j=1}^{n} A_{\sigma(2j-1),\sigma(2j)},$$

where the summation is taken over all permutations of $(1, 2, \dots, 2n)$.

Proposition 1.27. *One has* $\mathrm{Pf}^2 A = \det A$ *for any skew symmetric matrix* A *of size* $2n \times 2n$ *(see the exercises in § 6).*

Let $\mathcal{P}(z_1, \dots, z_N)$ stand for the *unconditional* JPDF given by the sum of all conditional JPDFs (1.19). Introduce the k−point correlation functions

$$R_k(z_1, z_2, \dots, z_k) = \frac{N!}{(N-k)!} \int_{\mathbb{C}^{N-k}} \mathcal{P}(z_1, \dots, z_N) \prod_{j=k+1}^{N} d^2 z_j. \tag{1.28}$$

In particular, $R_1(z) = N \int_{\mathbb{C}^{N-1}} \mathcal{P}(z_1, \dots, z_N)\, d^2 z_2, \dots, d^2 z_N$ is proportional to the mean eigenvalue density $\langle \rho(z) \rangle = \left\langle \frac{1}{N} \sum_{i=1}^{N} \delta^2 (z - z_i) \right\rangle$.

It turns out that the Theorem (1.22) eventually provides an efficient tool of calculating all possible correlation functions. For example, consider for some fixed $u \in \mathbb{C}$, $f(z) = 1 + \delta^2(z-u)$. Then the corresponding average

$$\langle F \rangle_{\mathrm{GinOE}} = \left\langle \prod_{j=1}^{N} \left(1 + \delta^2 (z_j - u)\right) \right\rangle_{\mathrm{GinOE}} = 1 + N \langle \rho(u) \rangle,$$

since all averages containing products of δ -functions will be zero, because they correspond to matrices with an eigenvalue of multiplicity greater than one.

In fact, by using Theorem 1.22 with judiciously chosen functions f one is able to calculate similarly all the correlation functions explicitly, see below and the exercises in § 6.

Theorem 1.29 (see [11], [50]). *Define the kernel*

$$\mathcal{K}_N(z_1, z_2) = \sum_{k=1}^{N} \sum_{l=1}^{N} \left[U^{-1}\right]_{kl} P_{k-1}(z_1)\, P_{j-1}(z_2) \tag{1.30}$$

where the antisymmetric matrix U *is defined in (1.23) taking* $f(z) = 1 + \sum_{j=1}^{n} \delta^2(z - z_j)$. *Then*

$$R_n(z_1, \dots, z_n) = \mathrm{Pf}\,(Q_{kl}), \quad Q_{kl} = \begin{pmatrix} K_{kl} & G_{kl} \\ -G_{lk} & W_{kl} \end{pmatrix}, \quad k, l = 1, 2, \dots, n. \tag{1.31}$$

That is the Pfaffian of $2n \times 2n$ matrix built of the n^2 matrices Q_{kl} whose entries are expressed in terms of the kernel (1.34) and the skew-symmetric measure (1.24) as

$$K_{kl} = \mathcal{K}_N(z_k, z_l), \quad G_{k,l} = \int_C \mathcal{K}_N(z_k, z)\mathcal{F}(z, z_l)\, d^2z,$$

$$W_{kl} = -\mathcal{F}(z_k, z_l) + \int_{C^2} \mathcal{F}(z_k, z)\mathcal{K}_N(z, z')\mathcal{F}(z', z_l)\, d^2z\, d^2z'$$

and symmetries $K_{kl} = -K_{lk}$, $W_{kl} = -W_{lk}$. See the exercises in § 6.

Consider the simplest example, $n = 1$ where by skew symmetry $K_{11} = W_{11} = 0$. This yields the one-point correlation function (proportional to the mean density) in the form

$$R_1(z_1) = \text{Pf}\begin{pmatrix} 0 & G_{11} \\ -G_{11} & 0 \end{pmatrix} = G_{11}$$

implying

$$R_1(z_1) = \int_C \mathcal{K}_N(z_1, z)\mathcal{F}(z, z_1)\, d^2z = R_1^{(c)}(z_1) + \delta(y)R_1^{(r)}(x_1) \tag{1.32}$$

in terms of the densities of complex and purely real eigenvalues, respectively. In particular, in view of (1.24)

$$R_1^{(c)}(z) = \mathcal{K}_N(z, \bar{z}) \cdot 2e^{-\frac{z^2+\bar{z}^2}{2}} \operatorname{erfc}\left(|y|\sqrt{2}\right) \operatorname{sgn}(y_1) \tag{1.33}$$

The kernel $\mathcal{K}_N(z_1, z_2)$ can be found explicitly with a judiciously chosen set of monic polynomials in (1.30), but the calculation is technically quite complicated, see the exercises in § 6. To bypass that calculation one may rely on comparison with the earlier derived formula (1.17) and deduce that actually the result must be rather simple:

$$\mathcal{K}_N(z_1, z_2) = \frac{z_1 - z_2}{2\sqrt{2\pi}} e^{z_1 z_2} \frac{\Gamma(N-1, z_1 z_2)}{(N-2)!}, \tag{1.34}$$

This ends our review of the real Ginibre ensemble for finite N, and now we proceed to analysing the regime of actual interest to our problem, namely $N \gg 1$.

2. Large$-$N asymptotics and large deviations for the Ginibre ensemble

For the goals of the present Lectures, that is for understanding the stability of a large complex nonlinear system, we are mostly interested in understanding the behaviour of the eigenvalues of G in the limit of large $N \gg 1$, especially concentrating on characteristics of the right-most eigenvalues. In every realization of the Ginibre Ensemble the positions of its eigenvalues are traditionally characterized by the empirical spectral counting measure $\mu_N(z)$ such that $d\mu_N = \rho_N(z)d^2z$ where $d^2z = d\,(\text{Re}z)\; d\,(\text{Im}z) = dx\, dy$ and the density associated with the empirical measure is given by $\rho_N(z) = \frac{1}{N}\sum_{i=1}^{N} \delta^2(z - z_i)$. In particular, it is easy to check that the mean value of such density in the complex plane given for finite N by (1.17), with $R_1^{(c)}(z) = N\langle\rho_N(z)\rangle$, will be described asymptotically for $N \gg 1$

and $|z| \sim \sqrt{N}$ by the so-called 'circular law'

$$R_1^{c(eq)}(z) = \begin{cases} \frac{1}{\pi}, & |z| < \sqrt{N} \\ 0, & \text{otherwise} \end{cases} \tag{2.1}$$

where the notation *'eq'* used above refers to an interpretation of such density as one related to the so-called 'equilibrium measure', see below.

To see the validity of (2.1) one can use the easily verifiable identity (see the exercises in § 6):

$$\lim_{N\to\infty} \frac{\Gamma(N-1, Na)}{(N-2)!} = 1, \text{ if } 0 \leqslant a < 1 \text{ and } 0 \text{ if } a > 1, \tag{2.2}$$

and the form (2.1) indeed emerges from (1.17) by the rescaling $x = \tilde{x}\sqrt{N}$ and $y = \tilde{y}\sqrt{N}$ with finite $\tilde{x} < \infty, \tilde{y} < \infty$ and using the large-N asymptotics

$$\operatorname{erfc}\left(\sqrt{2}|\tilde{y}|\sqrt{N}\right) \approx \sqrt{\frac{1}{2\pi N}\frac{1}{|\tilde{y}|}}e^{-2N\tilde{y}^2}. \tag{2.3}$$

One also finds asymptotically a constant mean density of real eigenvalues inside the same circle: $R_1^{r(eq)}(x) = \frac{1}{\sqrt{2\pi}}$ for $|x| < \sqrt{N}$ and zero outside. This implies that for large $N \gg 1$ a typical Ginibre matrix has of the order of $O(\sqrt{N})$ real eigenvalues uniformly filling in the interval $|x| < \sqrt{N}$, whereas of the order of $O(N)$ eigenvalues uniformly fill in the interior of the circle $|z| \leqslant \sqrt{N}$.[1] Coming back to the issue of the right-most eigenvalue, we see that the mean value of $x_m = \max_i[\text{Re} z_i]$ should be given by the end of support for the mean density, so that $\langle x_m \rangle \approx \sqrt{N}$. In fact, it is well known that not only the mean density, but the whole measure $\mu_N(z)$ weakly converges to the non-random 'equilibrium measure' $\mu_{eq}(z)$ with the density (2.1), so in a certain sense deviations of x_m from $\sqrt{N}$ should be small, which was precisely the basis for May's conclusions on the threshold of linear stability.

It turns out that it is possible to characterize quite precisely how much the empirical measure $\mu_N(z)$ for large $N \gg 1$ deviates from $\mu_{eq}(z)$ (and x_m from its expectation $\sqrt{N}$) by using the framework of the so-called Large Deviation Theory (LDT) developed for GinOE in [10]. Before discussing the LDT results we need to make the following remark. Eventually, it is more convenient to work with the rescaled version of the Ginibre matrices $Z = G/\sqrt{N}$ for which eigenvalues in large $N \gg 1$ limit fill in the interior of the unit circle, with the equilibrium mean density given by

$$\rho_{eq}(z) = \begin{cases} \frac{1}{\pi}, & |z| < 1, \\ 0, & \text{otherwise.} \end{cases} \tag{2.4}$$

For such an arrangement of eigenvalues an average separation between the neighbouring eigenvalues in the complex plane is of the order of $N^{-1/2}$. From now on

[1] The latter statement is correct if we are not interested in what is going on close to the real line, for $|y| \sim 1 \ll \sqrt{N}$. At such distances one finds inside the circle a certain depletion of complex eigenvalues: $R_1^{c(eq)}(z) \approx \sqrt{2/\pi}|y|e^{2y^2}\operatorname{erfc}(\sqrt{2}|y|)$.

we will assume such a rescaling done. It is natural to expect that nontrivial correlations between different eigenvalues $z_1, \ldots, z_N$ of the rescaled matrices Z exist only on the scale $|z_i - z_j| \sim N^{-1/2}$. In particular, one may expect that if the values of the variables $z_1 \neq z_2 \ldots \neq z_k$ are distinct and fixed, in the limit $N \to \infty$ one should have asymptotic factorization of the corresponding higher correlation functions defined in (1.28) (see the exercises in § 6):

$$R_k(z_1, z_2, \ldots, z_k) \sim \prod_{l=1}^{k} R_1(z_l). \tag{2.5}$$

This fact can be verified by employing the Theorem. **1.29**, and later on will be used when developing the LDT for the right-most eigenvalue of Z.

Now we recall, still very informally, a few very basic facts behind the Large Deviation Theory. Let N be an integer (eventually considered to be large $N \gg 1$) and $\alpha_N \to \infty$ be any sequence growing with N to infinity. Consider a random variable x_N indexed by N, and taking values in some space E, so that x_N is characterized by the probability $P_N(x_N \in A)$ for x_N to be in any measurable subset $A \in E$. Then, in a nutshell, the distribution P_N satisfies the Large Deviation Principle (LDP) with the speed α_N and the rate function $0 \leqslant I(x) < \infty$ if as long as $N \to \infty$

$$P_N(x_N \in A) \approx \exp\Big\{-\alpha_N \inf_A I(x)\Big\}, \quad \forall A \subset E.$$

Here the symbol $\approx$ describes the leading asymptotic behaviour, i.e. disregards both the pre-exponential multiplicative factors and subleading exponential terms which grow with N slower than α_N. As the simplest example, let the random variable $x_N \in \mathbb{R}$ be real and characterized by the probability density $p_N(x)$ with respect to the Lebesgue measure. Then LDP is essentially equivalent to writing

$$P_N\left(x_N \in [x, x + dx)\right) = p_N(x)\, dx \approx \exp\Big\{-\alpha_N I(x)\Big\}\, dx.$$

Having this picture in mind, let us denote E_μ the space of all normalized probability measures $\mu(z)$ symmetric (with respect to complex conjugation) equipped with metric (distance) compatible with weak convergence of measures. Consider any (open) subset $B \subset E_\mu$ and let μ_N be the normalized eigenvalue counting measure for the rescaled Ginibre ensemble Z on the complex plane $\mathbb{C}$, with the density $\rho_N(z) = \frac{1}{N}\sum_{i=1}^{N} \delta^2(z - z_i)$. Then for large $N \gg 1$ the probability $\mathcal{P}_N(B) = \text{Prob}[\mu_N \in B]$ can be shown to behave as required by LDP with the speed $\alpha_N \sim N^2$:

$$\mathcal{P}_N(B) \approx \exp\Big\{-N^2 \inf_B \mathcal{J}[\mu]\Big\}, \tag{2.6}$$

with the LDT rate functional $\mathcal{J}[\mu_N]$ given by

$$\mathcal{J}[\mu] = \frac{1}{2}\int_{\mathbb{C}} |z|^2 d\mu(z) - \frac{1}{2}\int_{\mathbb{C}^2} \log|z - w| d\mu(z) d\mu(w) - \frac{3}{8}. \tag{2.7}$$

One can further show that the above functional is nonnegative: $\mathcal{J}[\mu] \geqslant 0$, and has a unique minimizer. Moreover, it takes its minimal value zero precisely at the equilibrium measure $\mu(z) = \mu_{eq}(z)$ with the density from (2.4), that is $\mathcal{J}[\mu_{eq}] = 0$. For the precise meaning and proof of these statements see [10].

The above formulae show that deviations from the equilibrium density are exponentially penalized, and the penalty rate is proportional to N^2. One can informally illustrate the origin of (2.7) by noting that for the rescaled Ginibre ensemble the Eq.(1.19) can be rewritten as

$$
\begin{aligned}
&\mathcal{P}_1^{(L,M)}(z_1,\dots,z_N)\\
&= \mathcal{C}_{N,M} \prod_{1\leqslant j<k\leqslant N} |z_j - z_k| \exp\Big\{-\frac{N}{2}\sum_{1\leqslant j\leqslant N} |z_j|^2\Big\}\\
&\quad\times \prod_{1\leqslant j\leqslant M} \sqrt{N}|\mathrm{Im} z_j|\,\mathrm{erfc}\left(\sqrt{2N}\,|\mathrm{Im} z_j|\right)\exp\Big\{2N(\mathrm{Im} z_j)^2\Big\}\mathbf{1}_{G\in\mathcal{X}_{L,M}}.
\end{aligned}
\tag{2.8}
$$

where the first product (over $j < k$) is assumed to go over distinct eigenvalues and pairs, whereas the second product goes only over distinct pairs of complex conjugate eigenvalues. The full JPD of all eigenvalues is obtained by summing up all contributions for $M = 0, 1, \dots, N/2$. Now we note that by using empirical measure $d\mu_N$ the first factor can be informally rewritten as

$$
\begin{aligned}
&\prod_{1\leqslant j<k\leqslant N} |z_j - z_k| \exp\Big\{-\frac{N}{2}\sum_{1\leqslant j\leqslant N} |z_j|^2\Big\}\\
&\approx \exp\Big\{-N^2\left[\frac{1}{2}\int_{\mathbb{C}} |z|^2\, d\mu_N(z) - \frac{1}{2}\int_{\mathbb{C}^2} \ln|z-w|\,\mu_N(z)d\mu_N(w)\right]\Big\},
\end{aligned}
$$

disregarding for the sake of the illustration the obvious problem on the diagonal $w = z$. After replacing $\mu_N \to \mu_{eq}$ the above expression shows the same structure as the rate functional (2.7). Taking into account that 'typically' the number of complex conjugate pairs is $M \sim N$, and using the bound $|y|e^{2y^2}\mathrm{erfc}(\sqrt{2}|y|) \leqslant C_1$ for some constant C_1, one also can verify that the product over $j = 1, \dots, M$ in Eq.(2.8) induces factors of the subleading order $O(\exp\{N\})$, whereas an accurate account of all multiplicative constants gives the factor $\exp\left\{\frac{3}{8}N^2\right\}$ leading to the LDP rate functional (2.7), see [10] for a controlled derivation.

Now it should be evident that an event displacing the eigenvalue with the largest real part $x_m = \mathrm{Max}_i\{x_i = \mathrm{Re} z_i\}$ to any value *to the left* of the rightmost point $\mathrm{Re} z = 1$ of the support of the equilibrium measure (2.4) necessarily implies shoving all the remaining $N-1$ eigenvalues to the left as well, so that the corresponding measure μ_N is *macroscopically* distinct from μ_{eq}. Then the Eq.(2.6) implies that such displacement may only happen with an exponentially small probability of the order $\mathrm{Prob}(x_m < x) \sim \exp\{-C(x)N^2\}$, with some $C(x) > 0$ as long as $x < 1$. In contrast, the event of displacing the eigenvalue with the largest real part x_m to any value *to the right* of $\mathrm{Re} z = 1$ can be done by simply splitting off a *single* eigenvalue from the rest, without effecting rest of the eigenvalues, so for large enough N we will still have $\mu_N \approx \mu_{eq}$, to the leading order. We conclude that the knowledge of LDP for measures Eq.(2.6) is not enough to control the latter situation.

Now we will outline below how to establish the LDT for the right-most eigenvalue to be found beyond the right-edge of the 'equilibrium circle'. It can be shown to have the form

$$\text{Prob}(x_m > x) \approx e^{-NL(x)}, \quad x > 1 \tag{2.9}$$

with a rate function $L(x) > 0$ for $x > 1$ which is minimized at $L(x = 1) = 0$. We thus see that such event is also exponentially penalized, however with the penalty rate linear in N.

To verify (2.9) we will consider the probability density

$$\text{Prob}\,(x_m \in (x, x + dx)) = p_N(x)\,dx$$

for any $x > 1$ and follow the ideas of [23] where a similar question was addressed for Hermitian matrices with real eigenvalues (see [8] for an earlier and different way of treating that case). There exists an obvious relation between the probability density $p_N(x)$ of the right-most eigenvalue position and the probability $E_N(x)$ to have no eigenvalues with $x_i = \text{Re}\, z_i > x$:

$$p_N(x) = \frac{dE_N}{dx}, \quad E_N(x) = \left\langle \prod_{i=1}^{N} \chi(x - x_i) \right\rangle$$

where $\chi(u) = 1$ for $u > 0$ and $\chi(u) = 0$ otherwise and we omit the subscipt 'GinOE' by angular brackets here and henceforth. As $\chi(u) = 1 - \chi(-u)$ one can further expand

$$\begin{aligned} E_N(x) &= \left\langle \prod_{i=1}^{N} [1 - \chi(x_i - x)] \right\rangle \\ &= 1 - \left\langle \sum_{i=1}^{N} \chi(x_i - x) \right\rangle + \left\langle \sum_{i \neq j}^{N} \chi(x_i - x)\chi(x_j - x) \right\rangle - \dots \\ &= 1 - \int_C R_1(z)\chi\,(\text{Re}\, z - x)\; d^2z \\ &\quad + \frac{1}{2!}\int_{C^2} R_2(z_1, z_2)\chi\,(\text{Re}\, z_1 - x)\,\chi\,(\text{Re}\, z_2 - x)\; d^2z_1\, d^2z_2 + \dots \end{aligned}$$

where we have used the definition of the correlation functions (1.28). Clearly, (2.1) or, equivalently, (2.4) imply that $R_1(z)$ decays to zero as $N \to \infty$ outside the support of the equilibrium density (we will characterize the rate of this decay quantitatively below) so that $R_1(z)\chi\,(\text{Re}\, z - x) \ll 1$ for $x > 1$. Using the 'asymptotic independence' (2.5) we therefore may heuristically argue[2] that all the terms which involve higher order correlation functions $R_k(z_1, \dots, z_k)$ with $k > 1$ are

[2]To make the above heuristic argument rigorous one needs to prove some proper bounds for the $k-$th correlation function with arguments being at a finite distance from each other. The author is not aware of such a proof in the literature, though its existence seems highly plausible.

negligible for $x > 1$, and to the leading order approximation using $z = x + iy$ we have

$$p_N(x > 1) \approx \int_{\mathbb{C}} R_1(z)\delta\left(\mathrm{Re}\, z - x\right)\, d^2z = N\int_{\mathbb{R}} \langle \rho_N(x,y)\rangle\, dy,$$

where $\langle \rho_N(x,y)\rangle = R_1^{(c)}(x,y) + \delta(y)R_1^{(r)}(x)$.

We see that the problem is reduced to finding the leading asymptotics of the mean density beyond the right edge of the equilibrium measure support (which for the rescaled GinOE is at $x = 1$, but for the original GinOE happens at $x = \sqrt{N}$). Towards this task we again consider (1.17) and (1.12) after rescaling $x = \tilde{x}\sqrt{N}$, but now employing a more refined large-N asymptotics than (2.2) (see the exercises in § 6):

$$\frac{\Gamma(N-1, Na)}{(N-2)!} \approx \frac{1}{\sqrt{2\pi N}}\frac{1}{a(a-1)}e^{-N[a-1-\ln a]}, \quad a > 1.$$

Using it one finds that keeping only the leading exponential terms

$$\int_{-\infty}^{\infty} R_1^{(c)}(\tilde{x}\sqrt{N}, y)dy \sim e^{-2N\mathcal{L}(\tilde{x})}, \quad R_1^{(r)}(\tilde{x}\sqrt{N}) \sim e^{-N\mathcal{L}(\tilde{x})}, \quad \tilde{x} > 1 \tag{2.10}$$

with

$$\mathcal{L}(\tilde{x}) = \frac{\tilde{x}^2 - 1}{2} - \ln\tilde{x} > 0$$

thus verifying (2.9) with $L(x) = \mathcal{L}(x)$.

When looking at a distinctly different limiting behaviour of the density of real eigenvalues $R_1^{(r)}(\tilde{x}\sqrt{N})$ for $\tilde{x} < 1$ and for $\tilde{x} > 1$ one may be interested in characterizing how these two formulas match when $\tilde{x} = 1$. To answer this question in full detail one can identify a crossover going on in the so-called 'edge scaling' regime: $\tilde{x} = 1 + \frac{\delta}{\sqrt{N}}$ with δ being of order of unity (note that such scaling ensures that the exponent $N\mathcal{L}(\tilde{x})$ becomes of order of unity, $N\mathcal{L}(\tilde{x}) \approx \delta^2$). After using the appropriate asymptotics of the incomplete Γ-function valid for a fixed δ:

$$\lim_{N\to\infty} \frac{\Gamma\left(N-1, N\left(1+\delta N^{-1/2}\right)\right)}{(N-2)!} = \frac{1}{\sqrt{2\pi}}\int_{\delta}^{\infty} e^{-\frac{v^2}{2}}\, dv = \mathrm{erfc}(\sqrt{2}\delta)$$

one then finds

$$R_1^{(r)}(x) \approx \frac{1}{2\sqrt{2\pi}}\left[1 - \mathrm{erf}(\delta\sqrt{2}) + \frac{1}{\sqrt{2}}e^{-\delta^2}\left(1 + \mathrm{erf}(\delta)\right)\right], \quad x = \sqrt{N} + \delta \tag{2.11}$$

In particular, for $\delta \to +\infty$ the asymptotic decay law

$$R_1^{(r)}(x = \sqrt{N} + \delta, \delta \to \infty) \approx \frac{1}{2\sqrt{\pi}}e^{-\delta^2}$$

precisely matches the behaviour $R_1^r(\tilde{x}\sqrt{N}) \sim e^{-N\mathcal{L}(\tilde{x})}$ outside the limiting 'Ginibre circle' $\tilde{x} > 1$, whereas for $\delta \to -\infty$ the expression (2.11) tends to the constant limiting value $\frac{1}{\sqrt{2\pi}}$ matching the expected 'bulk' behaviour for the mean density of real eigenvalues inside the circle, see a discussion after (2.3).

A nonlinear generalization of the May's model Our goal is to attempt to go beyond the linearized May model (1.2). To this end we suggest to consider a system of N coupled non-linear autonomous ODEs in the form

$$\frac{dx_i}{dt} = -\mu x_i + f_i(x_1, \dots, x_N), \quad i = 1, \dots, N, \tag{2.12}$$

where $\mu > 0$ and the components $f_i(\mathbf{x})$ of the vector field $\mathbf{f} = (f_1, \dots, f_N)$ are zero mean random functions (assumed to be smooth in every realization) of the state vector $\mathbf{x} = (x_1, \dots, x_N)$. We again will use the brackets $\langle \cdots \rangle$ to denote averaging over the associated distributions. To put this model in the context of our earlier discussion, it is easy to see that if $\mathbf{x}_*$ is an *equilibrium* of (2.12) (i.e., if $-\mu\mathbf{x}_* + \mathbf{f}(\mathbf{x}_*) = \mathbf{0}$) then in the *immediate* neighbourhood of that equilibrium the system (2.12) reduces to May's model (1.2) with $\mathbf{y} = \mathbf{x} - \mathbf{x}_*$ and $J_{jk} = (\partial f_j / \partial x_k)|_{\mathbf{x}_*}$.

Our next goal is to chose a generic nonlinear model which would be rich enough to allow a description of the May-Wigner instability as a feature of its global rather than local phase portrait, yet simple enough to allow analytical insights.

To visualise the global picture, it is helpful to consider first a special case of a gradient-descent flow, characterised by the existence of a potential function $V(\mathbf{x})$ such that $\mathbf{f} = -\nabla V$. In this case, the system (2.12) can be rewritten as $d\mathbf{x}/dt = -\nabla L$, with $L(\mathbf{x}) = \mu|\mathbf{x}|^2/2 + V(\mathbf{x})$ being the associated Lyapunov function describing the effective landscape. In the domain of L, the state vector $\mathbf{x}(t)$ moves in the direction of the steepest descent, i.e., perpendicular to the level surfaces $L(\mathbf{x}) = h$ towards ever smaller values of h. This provides a useful geometric intuition. The term $\mu|\mathbf{x}|^2/2$ represents the globally confining parabolic potential, i.e., a deep well on the surface of $L(\mathbf{x})$, which does not allow $\mathbf{x}$ to escape to infinity. At the same time the random potential $V(\mathbf{x})$ may generate many local minima of $L(\mathbf{x})$ (shallow wells) which will play the role of attractors for our dynamical system. Moreover, if the confining term is strong enough then the full landscape will only be a small perturbation of the parabolic well, typically with a single stable equilibrium located close to $\mathbf{x} = 0$. In the opposite case of relatively weak confining term, the disorder-dominated landscape will be characterised by a complicated random topology with many points of equilibria, both stable and unstable. Note that in physics, complicated energy landscapes is a generic feature of glassy systems with intriguingly slow long-time relaxation and non-equilibrium dynamics, see e.g. [6]. From that angle the properties of the random landscapes exemplified by the above Lyapunov function $L(\mathbf{x})$ and related models attracted considerable attention [7,27,30,31].

The above picture of a gradient-descent flow is however only a very special case since the generic systems of ODEs (2.12) are not gradient. The latter point can easily be understood in the context of model ecosystems. For, by linearising a gradient flow in a vicinity of any equilibrium, one always obtains a symmetric community matrix, whilst the community matrices of model ecosystems are in general asymmetric. Note also a discussion of an interplay between non-gradient dynamics in random environment and glassy behaviour in [15].

To allow for a suitable level of generality we therefore suggest to choose the N–dimensional vector field $\mathbf{f}(\mathbf{x})$ as a sum of 'gradient' and non-gradient (sometimes called 'solenoidal' in physical literature) contributions:

$$f_i(\mathbf{x}) = -\frac{\partial V(\mathbf{x})}{\partial x_i} + \frac{1}{\sqrt{N}} \sum_{j=1}^{N} \frac{\partial A_{ij}(\mathbf{x})}{\partial x_j}, \quad i = 1, \dots, N, \tag{2.13}$$

where we require the matrix $A(\mathbf{x})$ to be antisymmetric: $A_{ij} = -A_{ji}$. Such a representation for fields $\mathbf{f} : \mathbb{R}^3 \to \mathbb{R}^3$, smooth enough and vanishing at infinity sufficiently fast, into curl-free and divergence-free parts $\mathbf{f} = \nabla V + \nabla \times \mathbf{A}$, with $V : \mathbb{R}^3 \to \mathbb{R}$ and $\mathbf{A} : \mathbb{R}^3 \to \mathbb{R}^3$ is the well-known Helmholtz decomposition. The meaning of this decomposition is that vector fields can be generically divided into a conservative irrotational component, sometimes called 'longitudinal', whose gradient connects the attractors or repellers and a solenoidal 'incompressible' curl field, also called 'transversal'. Note that in the case of bounded domains a nonzero 'harmonic' (i.e. simultaneously irrotational and incompressible) component may be present; however this component is always zero if the flow on the boundary is zero.

A natural generalization of the Helmholtz decomposition to higher dimensions is the so-called Hodge decomposition of differential forms, see e.g. [1].

Theorem 2.14 (Hodge decomposition). *Let $\mathcal{M}$ be a compact, boundaryless, oriented Riemannian manifold, and $\Omega^k(\mathcal{M})$ be the space of smooth, weakly differentiable k–forms on $\mathcal{M}$, d stands for the exterior derivative operator, and δ for the codifferential operator. Then any form $\omega \in \Omega^k(\mathcal{M})$ can be uniquely decomposed as*

$$\omega = d\alpha + \delta\beta + \gamma.$$

where $\alpha \in \Omega^{k-1}(\mathcal{M})$, $\beta \in \Omega^{k+1}(\mathcal{M})$ and $\gamma \in \mathcal{H}^k(\mathcal{M})$ is harmonic, i.e. simultaneously $d\gamma = 0$ and $\delta\gamma = 0$.

We are here interested in $k = 1$ since with any vector field $\mathbf{f} = (f_1, \dots, f_N)$ one can associate a differential 1–form as $\omega = \sum_i f_i dx^i$. Then the role of α will be played by a scalar function (0–form) $V(\mathbf{x})$ with $d\alpha = \sum_i \partial_i V \, dx^i$. The role of β will be played by some 2–form which we can always write in the form $\beta = \sum_{lk}^N A_{lk} dx^l \bigwedge dx^k$, for some antisymmetric $A_{lk}(\mathbf{x}) = -A_{kl}(\mathbf{x})$. To define the action of the codifferential operator δ one needs to specify the Riemannian metric g_{ij} on $\mathcal{M}$. We will use the $\mathbb{R}^N$ with the Euclidean metric[3] $g_{ij} = \delta_{ij}$ for the role of $\mathcal{M}$ in which case the action of the codifferential operator is especially simple: $\delta\beta = \sum_{i,j} \partial_j A_{ji} dx^i$. We see that our choice (2.13) amounts to neglecting the harmonic component which can be justified either by appropriately chosen boundary conditions at infinity, or simply restricting our consideration to the class of fields with no harmonic component.

[3] Though $\mathbb{R}^N$ is not compact, with due effort the Theorem 2.14 can be extended to this case as well.

Correspondingly, we will call $V(\mathbf{x})$ the scalar potential and the matrix $A(\mathbf{x})$ the vector potential. The normalising factor $1/\sqrt{N}$ in front of the sum on the right-hand side in (2.13) ensures that the transversal and longitudinal parts of $\mathbf{f}(\mathbf{x})$ are of the same order of magnitude for large N.

Finally, to make the model as simple as possible and amenable to a rigorous and detailed mathematical analysis we choose the scalar potential $V(\mathbf{x})$ and its components $A_{ij}(\mathbf{x})$, $i < j$ to be statistically independent, zero mean Gaussian random fields, with smooth realisations and the additional assumptions of *homogeneity* (translational invariance) and *isotropy* reflected in the covariance structure:

$$\langle V(\mathbf{x})V(\mathbf{y})\rangle = \nu^2\Gamma_V\left(|\mathbf{x}-\mathbf{y}|^2\right); \tag{2.15}$$

$$\langle A_{ij}(\mathbf{x})A_{nm}(\mathbf{y})\rangle = a^2\Gamma_A\left(|\mathbf{x}-\mathbf{y}|^2\right)\left(\delta_{in}\delta_{jm}-\delta_{im}\delta_{jn}\right). \tag{2.16}$$

Here the angular brackets $\langle ...\rangle$ stand for the ensemble average over all realisations of $V(\mathbf{x})$ and $A(\mathbf{x})$, and δ_{in} is the Kronecker delta: $\delta_{in} = 1$ if $i = n$ and zero otherwise.

For simplicity, we also assume that the functions $\Gamma_V(r)$ and $\Gamma_A(r)$ do not depend on N. This implies [48]

$$\Gamma_\sigma(r) = \int_0^\infty \exp\left\{-s\,r\right\}\gamma_\sigma(s)ds, \quad \sigma = A, V,$$

where the 'radial spectral' densities $\gamma_\sigma(s) \geqslant 0$ have finite total mass:

$$\int_0^\infty \gamma_\sigma(s)ds < \infty.$$

We normalize these densities by requiring that $\Gamma_\sigma''(0) = \int_0^\infty s^2\gamma_\sigma(s)ds = 1$. We assume the finiteness of second derivatives everywhere, which should be enough to ensure that our fields are smooth enough in any realization. The ratio

$$\tau = \nu^2/(\nu^2+a^2), \quad 0 \leqslant \tau \leqslant 1, \tag{2.17}$$

is a dimensionless measure of the relative strengths of the longitudinal and transversal components of $\mathbf{f}(\mathbf{x})$: if $\tau = 0$ then $\mathbf{f}(\mathbf{x})$ is divergence free and if $\tau = 1$ it is curl free. We also define the 'May ratio' of the relaxation rate to a characteristic value set by the interactions as

$$m = \mu/\mu_c, \quad \mu_c = 2\sqrt{N(a^2+\nu^2)} \tag{2.18}$$

The phase portrait of our system is controlled precisely by these two parameters: m and τ.

3. Counting multiple equilibria via Kac-Rice formulas

The non-linear system (2.12) may have multiple equilibria whose number $\mathcal{N}_{eq}$ and locations $\mathbf{x}_*^1, \ldots, \mathbf{x}_*^{\mathcal{N}_{eq}}$ depend on the realization of the random field $\mathbf{f}(\mathbf{x})$. For the above model with smooth realizations of random vector fields one may safely assume that above locations are isolated, that is one can find $\mathcal{N}_{eq}$ nonoverlapping balls such that each ball contains inside only a single equilibrium.

In this situation it seems natural to define the function

$$\mathcal{P}_{eq}(\mathcal{J}) = \left\langle \frac{1}{\mathcal{N}_{eq}} \sum_{k=1}^{\mathcal{N}_{eq}} \prod_{i,j} \delta\left(\mathcal{J}_{ij} + \mu\delta_{ij} - \frac{\partial f_i}{\partial x_j}|_{\mathbf{x}_*^k}\right)\right\rangle \tag{3.1}$$

which may be interpreted as the probability density for the entries $\mathcal{J}_{ij}$ of the Jacobian $\mathcal{J}$ sampled over all critical points. One of the central points of our approach is a possibility to rewrite the right-hand side of (3.1) using the Kac-Rice type identity (see the exercises in § 6 for the simplest version of this type of identities) valid for smooth enough functions $G(\mathbf{x})$:

$$\left\langle \sum_{k=1}^{\mathcal{N}_{eq}} G\left(\mathbf{x}_*^k\right) \right\rangle = \left\langle \int_{\mathbb{R}^N} d\mathbf{x}\, G(\mathbf{x})\, \delta\left(-\mu\mathbf{x} + \mathbf{f}(\mathbf{x})\right) |\det \mathcal{J}(\mathbf{x})| \right\rangle$$

where $\mathcal{J}(\mathbf{x}) = -\mu\mathbf{1} + \frac{\partial f}{\partial \mathbf{x}}$, with $\delta(\mathbf{x})$ being the multivariate Dirac δ-function and $d\,\mathbf{x}$ is the volume element in $\mathbb{R}^N$. In this way we see that

$$\mathcal{P}_{eq}(\mathcal{J}) = \left\langle \frac{1}{\mathcal{N}_{eq}} \int_{\mathbb{R}^N} d\mathbf{x}\delta\left(\mathcal{J} - \mathcal{J}(\mathbf{x})\right) \delta\left(-\mu\mathbf{x} + \mathbf{f}(\mathbf{x})\right) |\det \mathcal{J}(\mathbf{x})| \right\rangle$$

where the (random) number of equilibria is formally given by the integral

$$\mathcal{N}_{eq} = \int_{\mathbb{R}^N} \delta\left(-\mu\mathbf{x} + \mathbf{f}(\mathbf{x})\right) |\det \mathcal{J}(\mathbf{x})| d\mathbf{x}. \tag{3.2}$$

The number of stable equilibria $\mathcal{N}_{st}$ is given by a similar integral:

$$\mathcal{N}_{st} = \int_{\mathbb{R}^N} \delta\left(-\mu\mathbf{x} + \mathbf{f}(\mathbf{x})\right) |\det \mathcal{J}(\mathbf{x})| \chi\left(\operatorname{Re}\mathcal{J} < 0\right) d\mathbf{x} \tag{3.3}$$

where the factor $\chi\left(\operatorname{Re}\mathcal{J} < 0\right)$ ensures that all eigenvalues of the Jacobian matrix have negative real parts.

This, in particular, implies that the probability p_{st} for a given equilibrium to be stable is simply given by

$$p_{st} = \int_{\mathbb{R}^{N\times N}} \mathcal{P}_{eq}(\mathcal{J}) \chi\left(\operatorname{Re}\mathcal{J} < 0\right) d\mathcal{J} = \left\langle \frac{\mathcal{N}_{st}}{\mathcal{N}_{eq}} \right\rangle.$$

It seems rather challenging to evaluate p_{st} due to correlations between the random denominator $\mathcal{N}_{eq}$ and the numerator $\mathcal{N}_{st}$. To get some insights in the problem we will be using as a proxy the annealed version $p_{st}^{(a)} = \langle \mathcal{N}_{st} \rangle / \langle \mathcal{N}_{eq} \rangle$. Our main goal will be therefore to develop methods to evaluate $\langle \mathcal{N}_{st} \rangle$ and $\langle \mathcal{N}_{eq} \rangle$ for our model, with emphasis on the asymptotic analysis for $N \gg 1$. An important issue of how well $p_{st}^{(a)}$ approximates p_{st} remains at present open, though informal arguments indicate that the approximation is benign.

The Kac-Rice formulas (3.2)-(3.3) yield the ensemble average of $\mathcal{N}_{eq}$ and $\mathcal{N}_{st}$ in terms of that of the modulus of the spectral determinant of the matrix $(J_{ij})_{i,j=1}^N$, $J_{ij} = \partial f_i / \partial x_j$. We begin with

$$\langle \mathcal{N}_{eq} \rangle = \int_{\mathbb{R}^N} \langle \delta\left(-\mu\mathbf{x} + \mathbf{f}(\mathbf{x})\right) \left| \det\left(-\mu\delta_{ij} + J_{ij}(\mathbf{x})\right)\right| \rangle\, d\mathbf{x}. \tag{3.4}$$

By our assumptions (2.15) – (2.16) the random field $\mathbf{f}(\mathbf{x})$ is homogeneous and isotropic. For such fields samples of $\mathbf{f}$ and J taken in one and the same spatial point $\mathbf{x}$ are uncorrelated, $\langle f_l \cdot \partial f_i/\partial x_j\rangle = 0$ for all i, j, l (see the exercises in § 6). This is well known and can be checked by straightforward differentiation. In addition, the field $\mathbf{f}$ is Gaussian, hence the $\mathbf{f}(\mathbf{x})$ and $J_{ij}(\mathbf{x})$ being uncorrelated are actually statistically independent. This simplifies the evaluation of the integral in (3.4) considerably. Indeed, the statistical average in (3.4) factorizes and, moreover, due to stationarity the factor $\langle|\det\left(-\mu\delta_{ij} + J_{ij}(\mathbf{x})\right)|\rangle$ does not vary with the position $\mathbf{x}$. Introducing the Fourier-representation of the Dirac δ–function it remains only to evaluate

$$\langle\delta(-\mu\mathbf{x} + \mathbf{f}(\mathbf{x}))\rangle = \int_{\mathbb{R}^N} \frac{d\mathbf{k}}{(2\pi)^N}\, e^{-i\mu\mathbf{k}\cdot\mathbf{x}} \langle e^{i\mathbf{k}\cdot\mathbf{f}(\mathbf{x})}\rangle. \tag{3.5}$$

At every spatial point $\mathbf{x}$ the vector $\mathbf{f}(\mathbf{x})$ is Gaussian with uncorrelated and identically distributed components, implying according to (2.13)-(2.15) (see the exercises in § 6)

$$\langle f_i(\mathbf{x}) f_j(\mathbf{x})\rangle = \delta_{ij}\sigma^2, \quad \sigma^2 = 2v^2|\Gamma_V'(0)| + 2a^2|\Gamma_A'(0)|\frac{N-1}{N}.$$

Therefore $\langle e^{i\mathbf{k}\cdot\mathbf{f}(\mathbf{x})}\rangle = e^{-\sigma^2|\mathbf{k}|^2/2}$, and evaluating the integral on the right-hand side in (3.5) one arrives at

$$\langle\mathcal{N}_{eq}\rangle = \frac{1}{\mu^N}\left\langle\left|\det\left(-\mu\delta_{ij} + J_{ij}\right)\right|\right\rangle, \tag{3.6}$$

thus bringing the original non-linear problem into the realms of the random matrix theory.

The probability distribution defining the ensemble of random matrices J can easily be determined in a closed form. Indeed, the matrix entries $J_{ij} = \partial f_i/\partial x_j$ are zero mean Gaussian variables and their covariance structure at a spatial point $\mathbf{x}$ can be obtained from (2.15)–(2.16) by repeated differentiations:

$$\left\langle J_{ij} J_{nm}\right\rangle = \alpha^2[(1+\epsilon_N)\delta_{in}\delta_{jm} + (\tau - \epsilon_N)(\delta_{jn}\delta_{im} + \delta_{ij}\delta_{mn})],$$

where $\epsilon_N = (1-\tau)/N$ and $\alpha = 2\sqrt{v^2 + a^2}$. As we are interested only in the leading behaviour of the mean number of equilibria in the limit $N \gg 1$, it is enough to keep only the leading order in the covariance structure, which is equivalent to writing

$$J_{ij} = \alpha(X_{ij} + \sqrt{\tau}\delta_{ij}\xi), \tag{3.7}$$

where X_{ij}, $i, j = 1, \ldots, N$ are zero mean Gaussian random variables with the covariance structure

$$\left\langle X_{ij} X_{nm}\right\rangle = \delta_{in}\delta_{jm} + \tau\delta_{jn}\delta_{im}, \tag{3.8}$$

and ξ is a standard Gaussian real variable, $\xi \sim N(0,1)$, which is statistically independent of all $X = (X_{ij})$. Note that for the divergence free fields $\mathbf{f}(\mathbf{x})$ (i.e., if $\tau = 0$) the matrix entries J_{ij} are statistically independent, exactly as in the May's model. On the other side, if $\mathbf{f}(\mathbf{x})$ has a longitudinal component so that $\tau > 0$ this induces positive correlations between the pairs of matrix entries J_{ij} and J_{ji} symmetric about the main diagonal: $\left\langle X_{ij} X_{ji}\right\rangle = \tau$ if $i \neq j$. Such distributions of the community matrix has also been used in the neighbourhood stability analysis of model ecosystems [4]. Finally, in the limiting case of curl free fields ($\tau = 1$), the matrix J is real symmetric: $J_{ij} = J_{ji}$.

The representation (3.7) comes in handy as it allows one, after a straightforward change of variables, to express (3.6) as an integral:

$$\langle \mathcal{N}_{eq} \rangle = \frac{N^{\frac{N}{2}}}{m^N} \int_{\mathbb{R}} \langle | \det (x\, \delta_{ij} - X_{ij}) | \rangle_{X_N} \frac{e^{-\frac{Nt^2}{2}} dt}{\sqrt{2\pi/N}}, \tag{3.9}$$

where we used (2.18) and introduced the parameter $x = \sqrt{N}(m + t\sqrt{\tau})$, whereas the angle brackets $\langle \ldots \rangle_{X_N}$ denote the averaging over the real random $N \times N$ matrices X defined in (3.8), see also (3.11) below. The same procedure yields also the ensemble average of the number of stable equilibria $\langle \mathcal{N}_{st} \rangle$ in the form

$$\langle \mathcal{N}_{st} \rangle = \frac{N^{\frac{N}{2}}}{m^N} \int_{\mathbb{R}} \langle | \det (x\, \delta_{ij} - X_{ij}) | \chi_x(X) \rangle_{X_N} \frac{e^{-\frac{Nt^2}{2}} dt}{\sqrt{2\pi/N}}, \tag{3.10}$$

where $\chi_x(X) = 1$, if all N eigenvalues of matrix X have real parts less than the spectral parameter x, and $\chi_x(X) = 0$ otherwise.

It is obvious that we need to concentrate on the matrices X_{ij}. This one-parameter family of random asymmetric matrices known as the real 'Gaussian Elliptic Ensemble' interpolates between the Gaussian Orthogonal Ensemble of real symmetric matrices (GOE, $\tau = 1$) and the real Ginibre ensemble of fully asymmetric matrices (GinOE, $\tau = 0$) and as such has enjoyed considerable interest in the literature in recent years [2,21]. In the next section we give a very brief overview of this ensemble, emphasizing similarities with the Ginibre case studied by us in detail earlier.

Real Gaussian elliptic ensemble: summary of the main features The joint probability density function $\mathcal{P}_N(X)$ of the matrix entries in the elliptic ensemble of real Gaussian random matrices X of size $N \times N$ is given by

$$\mathcal{P}_N(X) = \mathcal{Z}_N(\tau)^{-1} \exp\left\{ -\frac{1}{2(1-\tau^2)} \operatorname{Tr} \left(XX^T - \tau X^2 \right) \right\}, \tag{3.11}$$

where $\mathcal{Z}_N(\tau) = (2\pi)^{\frac{N^2}{2}} (1-\tau)^{N(N-1)/4} (1+\tau)^{N(N+1)/4}$ is the normalisation constant and $\tau \in [0, 1)$. It is straightforward to verify that the covariance of matrix entries X_{ij} is given by the expression specified in (3.8).

The JPD of complex eigenvalues z_i, $i = 1, \ldots, N$ for every set $\mathcal{X}_{L,M}$ can be immediately written down by simple modifications of the Theorem (**1.18**) by noticing that $\operatorname{Tr}\left(X^2\right) = \sum_i z_i^2$ so that

$$\mathcal{P}_\tau^{(L,M)}(z_1, \ldots, z_N) = \mathcal{C}_{L,M}(\tau)^{-1} \prod_{1 \leqslant j < k \leqslant N} |z_j - z_k|$$

$$\times \exp\left\{ -\frac{1}{2(1+\tau)} \sum_{1 \leqslant j \leqslant N} z_j^2 \right\} \prod_{1 \leqslant j \leqslant N} \sqrt{\operatorname{erfc}\left(\sqrt{\frac{2}{1-\tau^2}} |\operatorname{Im} z_j| \right)} \mathbf{1}_{G \in \mathcal{X}_{L,M}}.$$

where $\mathcal{C}_{L,M}(\tau)$ are the appropriate normalisation constants. Correspondingly, as shown in [21] all the essential features of the Pfaffian structure described in the Theorems (**1.22**) and (**1.29**) retain their validity up to replacing the skew-symmetric measure $\mathcal{F}(z_1, z_2)$ with the following modified expression:

$$\mathcal{F}_\tau(z_1 = x_1 + iy_1, z_2 = x_2 + iy_2) = e^{-\frac{1}{1+\tau}(z_1^2+z_2^2)/2}$$
$$\times \left[2i\delta^2(z_1 - \bar{z}_2)\,\mathrm{sgn}(y_1)\,\mathrm{erfc}\left(|y_1| \sqrt{\frac{2}{1-\tau^2}} \right) + \delta(y_1)\delta(y_2)\,\mathrm{sgn}\,(x_2 - x_1) \right].$$

The kernel $\mathcal{K}_N(z_1, z_2)$ again can be found explicitly, in terms of which all the correlation functions can be readily represented. In this way, after some further work, one recovers the analogues of (1.17) and (1.12). Namely, denote

$$\psi_k^{(\tau)}(z) = e^{-\frac{z^2}{2(1+\tau)}} h_k^{(\tau)}(z)$$

where $h_k^{(\tau)}(z)$ are rescaled complex Hermite polynomials, see the exercises in § 6:

$$\text{(3.12)} \qquad \begin{aligned} h_k^{(\tau)}(z) &= \frac{1}{\sqrt{\pi}} \int_{\mathbb{R}} e^{-t^2} \left(z \pm it\sqrt{2\tau} \right)^k dt \\ &= \frac{(\pm i\sqrt{N})^k}{\sqrt{\pi}} \sqrt{\frac{N}{2\tau}} \int_{\mathbb{R}} e^{-\frac{N}{2\tau}(u \pm iz/\sqrt{N})^2} u^k\, du. \end{aligned}$$

Then, for $z = x + iy$ one finds

$$\text{(3.13)} \qquad R_1^{(c)}(x, y) = 2i\,\mathrm{sign}\,(y)\,\mathrm{erfc}\left(\sqrt{\frac{2}{1-\tau^2}} |y| \right) K_N^{(c)}(z, \bar{z}),$$

where

$$\text{(3.14)} \qquad \mathcal{K}_N^{(c)}(z, \bar{z}) = \frac{1}{2(1+\tau)\sqrt{2\pi}} \sum_{j=0}^{N-2} \frac{\psi_{j+1}^{(\tau)}(\bar{z})\psi_j^{(\tau)}(z) - \psi_j^{(\tau)}(\bar{z})\psi_{j+1}^{(\tau)}(z)}{j!}.$$

whereas the density of real eigenvalues takes the form

$$\text{(3.15)} \qquad \begin{aligned} R_1^{(r)}(x) &= \frac{1}{\sqrt{2\pi}} \sum_{k=0}^{N-1} \frac{|\psi_k^{(\tau)}(x)|^2}{k!} \\ &+ \frac{1}{\sqrt{2\pi}(1+\tau)(N-1)!} \psi_N^{(\tau)}(x) \int_0^x \psi_{N-1}^{(\tau)}(u)\, du. \end{aligned}$$

It is easy to check that for $\tau = 0$ the formulas (3.13) and (3.15) are reduced to (1.17) and (1.12), respectively.

Having in mind that we are mostly interested in extracting asymptotics for large $N \gg 1$, we rescale $x \to x\sqrt{N}, y \to y\sqrt{N}$ and by manipulating the integral representation for Hermite polynomials (3.12), rewrite (3.14) for $z = x + iy$ as

$$\text{(3.16)} \quad \mathcal{K}_N^{(c)}(z, \bar{z}) = i \frac{\left(\frac{N}{2\pi}\right)^{3/2}}{\sqrt{2\tau}(1+\tau)} \iint_{\mathbb{R}^2} dp\,dq\, \frac{\Gamma\left(N-2, \frac{N}{2}(p^2 - q^2)\right)}{(N-2)!} e^{-N[A_y(p) + B_x(q)]},$$

where

$$A_y(p) = \frac{1-\tau}{2\tau} p^2 + \frac{p\sqrt{2}}{\tau} y, \quad B_x(q) = \frac{1+\tau}{2\tau} q^2 + \frac{iq\sqrt{2}}{\tau} x,$$

and we used the incomplete Γ−function defined in (1.13). This allows one to extract the leading $N \gg 1$ behaviour of the integrals in (3.16) by the saddle-point method, taking (2.2) into account, and using the large-N asymptotics (2.3). In this way one quickly arrives to the limiting 'equilibrium' density of complex eigenvalues $\lim_{N\to\infty}\left\langle \rho_N^{(c)}(x,y)\right\rangle$ known as the elliptic law, which we present below for the 'rescaled' ensemble $X/\sqrt{N}$:

$$\rho_N(z) \to \rho_{eq}(z) = \begin{cases} \frac{1}{\pi(1-\tau^2)}, & \frac{(\mathrm{Re}\, z)^2}{(1+\tau)^2} + \frac{(\mathrm{Im}\, z)^2}{(1-\tau)^2} \leqslant 1 \\ 0, & \text{otherwise} \end{cases} \tag{3.17}$$

which replaces the 'circular law' (2.4) to the case of general $0 \leqslant \tau < 1$. This perfectly agrees with the numerical simulations, see the Figure 3.18.

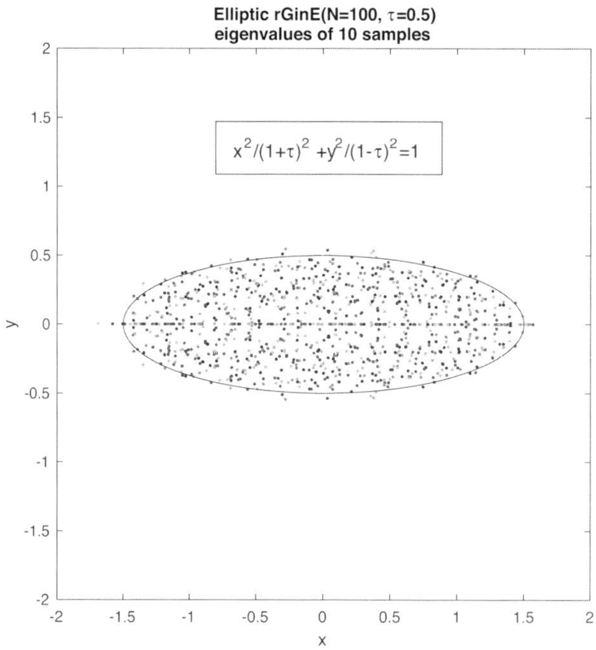

FIGURE 3.18. Eigenvalues of Real Elliptic matrix of size $N = 100$ and $\tau = 0.5$. The notation rGinE is the same as GinOE used elsewhere in the lectures.

One further needs to extend the LDT developed in [10] in order to describe how much the empirical measure $\mu_N(z)$ of the Elliptic Ensemble deviates from $\mu_{eq}(z)$ for large $N \gg 1$. By combining the knowledge of the LDT rate functional (2.7) for pure Ginibre case with the so-called Laplace-Varadhan Theorem (discussed later on in these lectures, see (4.5) below) one can show, see [9], that the probability $\mathcal{P}_N(B) = \mathrm{Prob}[\mu_N \in B]$ for a general elliptic case behaves again as predicted by (2.6) with the LDT rate functional now given by

$$\mathcal{J}_\tau(\mu) = \frac{1}{2}\left[\int_{\mathbb{C}}\left(\frac{x^2}{1+\tau} + \frac{y^2}{1-\tau}\right) d\mu(z) - \int_{\mathbb{C}^2} \log|z-w| d\mu(z) d\mu(w)\right] - 3/8. \tag{3.19}$$

One can further show that the above functional is non-negative: $\mathcal{J}[\mu] \geqslant 0$, and has a unique minimizer given by the equilibrium measure with the density uniform in the ellipse with half-axes $1 \pm \tau$, in full agreement with Eq.(3.17). Finally, analysing the asymptotic decay of the density of real eigenvalues inside the ellipse, and at its crossover 'edge scaling' regime one finds:

$$R_1^{(r)}(x) \approx \begin{cases} \frac{1}{\sqrt{1-\tau^2}\sqrt{2\pi}}, & \frac{|x|}{1+\tau} < \sqrt{N} \\ \frac{1}{2\sqrt{1-\tau^2}\sqrt{2\pi}}\left[1 - \operatorname{erf}(\delta_\tau\sqrt{2}) + \frac{e^{-\delta_\tau^2}}{\sqrt{2}}\left(1 + \operatorname{erf}(\delta_\tau)\right)\right], & \frac{x}{1+\tau} = \sqrt{N} + \delta \end{cases}$$

for finite δ, and we used the notation $\delta_\tau = \delta\sqrt{\frac{1+\tau}{1-\tau}}$ in the above. In particular, in the second line we have for $\delta \to -\infty$ the constant limiting value given by $1/\sqrt{2\pi(1-\tau^2)}$ matching the first line. On the other hand, for $\delta \to +\infty$ the asymptotic decay law

$$R_1^{(r)}\left(x = (1+\tau)\left(\sqrt{N} + \delta\right), \delta \to \infty\right) \approx \frac{1}{2\sqrt{1-\tau^2}\sqrt{\pi}} e^{-\delta^2 \frac{1+\tau}{1-\tau}} \tag{3.20}$$

matches the behaviour $R_1^{(r)}(x > (1+\tau)) \sim e^{-NL_\tau^{(r)}(x)}$ where one can show that now ((cf. Eq.2.10) for $\tau = 0$)

$$L^{(r)}(x) = -\frac{1}{2} + \frac{x^2}{2(1+\tau)} - \frac{1}{8\tau}(x - \sqrt{x^2 - 4\tau})^2 - \ln\frac{x + \sqrt{x^2 - 4\tau}}{2}.$$

Similarly, for the density of real parts of complex eigenvalues to the right of the ellipse, that is for $x > 1 + \tau$ in scaled variables, one finds the exponential decay of the density with twice larger rate: $L^{(c)}(x) = 2L^{(r)}(x)$. From this information one finds the analogue of the LDT for the right-most eigenvalue of the Real Elliptic Ensemble to be found beyond the right-edge of the 'equilibrium ellipse', which is again of the form analogous to (2.9):

$$\operatorname{Prob}(x_m > x) \approx e^{-NL(x)}, \quad x > 1 + \tau \tag{3.21}$$

with the rate

$$L(x) = \min\left\{L^{(c)}(x), L^{(r)}(x)\right\},$$

and one immediately sees that fluctuations of real eigenvalues far beyond the equilibrium support are more probable than that of the complex-conjugate pairs, as real eigenvalues have twice as smaller rate of decay. Hence the large deviation tail will be controlled by these events.

4. Mean number of equilibria: asymptotic analysis for large deviations

By shifting and rescaling the integration variables we can rewrite (3.9) for $\mathcal{N}_{eq}$ and (3.10) for $\mathcal{N}_{st}$ in the form most suitable for the asymptotic analysis which we for brevity present as

$$\langle \mathcal{N}_\sigma \rangle = \frac{1}{m^N} \int_{\mathbb{R}} \mathcal{D}_N^{(\sigma)}(x) \, \frac{e^{-\frac{N(x-m)^2}{2\tau}} dx}{\sqrt{2\pi\tau/N}}. \tag{4.1}$$

Here we introduced the label σ taking two values, $\sigma = \mathrm{st}$ or $\sigma = \mathrm{eq}$, and define correspondingly

(4.2) $$\mathcal{D}_N^{(eq)}(x) = \langle |\det(x\,\delta_{ij} - Z_{ij})| \rangle_Z,$$

(4.3) $$\mathcal{D}_N^{(st)}(x) = \langle |\det(x\,\delta_{ij} - Z_{ij})|\, \chi_x(Z) \rangle_Z$$

where $\langle \cdots \rangle_Z$ indicates expectation over the distribution of matrices $Z = X/\sqrt{N}$ from the rescaled Real Elliptic Ensemble with JPD (cf. 3.11)

$$\mathcal{P}(Z) = \mathcal{C}_N(\tau) e^{-\frac{N}{2(1-\tau^2)}[\mathrm{Tr} Z Z^T - \tau \mathrm{Tr} Z^2]}, \quad \tau \in [0,1].$$

Here $\mathcal{C}_N(\tau)$ is the appropriate normalization constant and the indicator function in (4.3) $\chi_x(Z) = 1$, if all N complex eigenvalues z_i of the matrix Z have real parts $\mathrm{Re} z_i$ less than the real parameter x, and $\chi_x(Z) = 0$ otherwise. We will also use below the notation $\chi(\mu_N \in B)$ for an indicator function equal to zero if the measure μ_N is not in a set B and $\chi(\mu_N \in B) = 1$ otherwise.

To be able to use the LDT in this setting we further rewrite Eqs.(4.2) for $\mathcal{D}_N^{(\sigma)}(x)$ in terms of the spectral measure μ_N whose density is the empirical counting function $\rho_N(z) = \sum_i \delta^2(z - z_i)$. If B_x stands for the set of the measures supported in the complex z–plane to the left of $\mathrm{Re}\, z = x$ we then can write

$$\mathcal{D}_N^{eq}(x) = \left\langle e^{N\Phi^{(x)}(\mu_N)} \right\rangle_Z, \quad \mathcal{D}_N^{(st)}(x) = \left\langle e^{N\Phi^{(x)}(\mu_N)} \chi(\mu_N \in B_x) \right\rangle_Z,$$

where we introduced the 'logarithmic potential' functional associated with any measure μ as

(4.4) $$\Phi^{(x)}(\mu) = \int_{\mathbb{C}} \ln |x - z|\, d\mu(z).$$

One of the most important results of the Large Deviation Theory is known as the Laplace-Varadhan Theorem which we now illustrate in its most basic manifestation. Namely, assume that the distribution P_N of a real random variable x_N satisfies the LDP with the speed α_N and the rate function $I(x)$, take any bounded real function $F(x_N)$ and consider the functional

(4.5) $$\mathcal{Z}_N(F) := \left\langle e^{-\tilde{\alpha}_N F(x_N)} \right\rangle_P$$

where $\tilde{\alpha}_N \to \infty$ as $N \to \infty$, though in general $\tilde{\alpha}_N \neq \alpha_N$. The most simple version of the Laplace-Varadhan Theorem asserts that for $\tilde{\alpha}_N = \alpha_N$

$$\lim_{N \to \infty} \alpha_N^{-1} \log \mathcal{Z}_N = -\inf_E (F + I)$$

or, equivalently, that $\mathcal{Z}_N \approx e^{-\alpha_N \inf_E (F+I)}$ for $N \gg 1$. For the simplest case of a random real x with the probability measure P_N so that

$$\mathcal{Z}_N(F) := \int_{\mathbb{R}} e^{-\alpha_N F(x)} dP_N(x)$$

(in particular, for P_N having a continuous density: $dP_N(x) = p_N(x)dx$) this statement immediately follows from applying the Laplace steepest descent method to evaluation of the above integral. One should appreciate, however, that the actual statement of the theorem is much stronger as it also holds true for random variables of much more general nature, like random measures. Note that the condition of boundedness of $F(x)$ can be further relaxed. Note also that if $\frac{\tilde{\alpha}_N}{\alpha_N} \to 0$ as $N \to \infty$ asymptotically $\mathcal{Z}_N \approx e^{-\tilde{\alpha}(N)F(x_*)}$ where x_* is the global minimizer of the rate $I(x)$.

Using the last fact in our setting and recalling the LDP (2.6) and (3.19) for the Elliptic Ensemble with the speed $\alpha_N = N^2 \gg N$), immediately implies that

$$\begin{aligned} \mathcal{D}_N^{(eq)}(x) &= \left\langle e^{N\Phi^{(x)}(\mu)} \right\rangle_Z \approx e^{N\Phi(x)}, \\ \Phi(x) &:= \Phi^{(x)}(\mu_{eq}) = \int_C \ln|x-z|\, d\mu_{eq}(z), \end{aligned} \tag{4.6}$$

since the elliptic law measure $\mu_{eq}(z)$, see Eq.(3.17), is the global minimizer for the LDP rate function over all possible random measures. Here we ignored for simplicity the fact that the 'logarithmic potential' functional defined in (4.4) is not trivially bounded due to logarithmic singularities (the step can be justified with due effort). The integral in Eq.(4.6) can be performed explicitly, and yields (see the exercises in § 6),

$$\Phi(x) = \begin{cases} \frac{1}{8\tau}\left(|x| - \sqrt{x^2-4\tau}\right)^2 + \ln\frac{|x|+\sqrt{x^2-4\tau}}{2}, & |x| > 1+\tau \\ \frac{x^2}{2(1+\tau)} - \frac{1}{2}, & |x| < 1+\tau \end{cases} \tag{4.7}$$

This allows us to approximate the expression (4.1) for $\langle \mathcal{N}_{eq} \rangle$ with the required accuracy (that is, disregarding the subleading pre-exponential factors in the integrand) as

$$\langle \mathcal{N}_{eq} \rangle \approx \frac{1}{m^N} \int_{\mathbb{R}} e^{-N\mathcal{L}_{eq}(x)} \frac{dx}{\sqrt{2\pi\tau/N}}, \quad \mathcal{L}_{eq}(x) = \frac{(x-m)^2}{2\tau} - \Phi(x)$$

and representing $\mathbb{R} = B_- \cup A \cup B_+$ with

$$A = \{|x| < 1+\tau\}, \quad B_- = \{x < -(1+\tau)\}, \quad B_+ = \{x > 1+\tau\}$$

write it further as a sum of three contributions:

$$\langle \mathcal{N}_{eq} \rangle \approx \left\langle \mathcal{N}_{eq}^{A} \right\rangle + \left\langle \mathcal{N}_{eq}^{B_-} \right\rangle + \left\langle \mathcal{N}_{eq}^{B_+} \right\rangle$$

where

$$\left\langle \mathcal{N}_{eq}^{A} \right\rangle = \frac{1}{m^N} \int_A e^{-N\mathcal{L}_{eq}(x)} \frac{dx}{\sqrt{2\pi\tau/N}}, \quad \left\langle \mathcal{N}_{eq}^{B_\pm} \right\rangle = \frac{1}{m^N} \int_{B_\pm} e^{-N\mathcal{L}_{eq}(x)} \frac{dx}{\sqrt{2\pi\tau/N}}.$$

Now the integrals can be treated by the standard Laplace steepest descent method. For $x \in A$ we have due to (4.7)

$$\mathcal{L}_{eq}(x) = \frac{(x-m)^2}{2\tau} - \frac{x^2}{2(1+\tau)} + \frac{1}{2}, \quad \frac{d}{dx}\mathcal{L}_{eq}(x) = \frac{x}{\tau(1+\tau)} - \frac{m}{\tau},$$

so that $\frac{d}{dx}\mathcal{L}_{eq}(x) = 0$ at $x = m(1+\tau) := x_*$. Since $\frac{d^2\mathcal{L}_{eq}(x)}{dx^2} = \frac{1}{\tau(1+\tau)} > 0$ we see that for $0 \leqslant m < 1$, the function $\mathcal{L}_{eq}(x)$ is minimized in the interior the interval of integration as $x^* \in (0, 1+\tau)$, and $\mathcal{L}_{eq}(x_*) = \frac{1}{2}(m^2-1)$. At the same time, for $m > 1$, the function is minimized at the right end of the interval where

$\mathcal{L}_{eq}(x = 1+\tau) = \frac{(1-m)^2}{2\tau} - (1-m)$. This gives, asymptotically for large $N \gg 1$,

$$\left\langle \mathcal{N}_{eq}^{A} \right\rangle \approx \begin{cases} e^{-N\left(\frac{1}{2}(m^2-1)+\ln m\right)}, & 0 \leqslant m < 1 \\ e^{-N\left(\frac{(1-m)^2}{2\tau}-(1-m)+\ln m\right)}, & m > 1 \end{cases} \tag{4.8}$$

Turning our attention to the integral for $\left\langle \mathcal{N}_{eq}^{B_+} \right\rangle$ we have by applying (4.7) for $x > 1+\tau$, the equations

$$\frac{d}{dx}\mathcal{L}_{eq}(x) = \frac{1}{\tau}\left(\frac{x+\sqrt{x^2-4\tau}}{2} - m\right) \quad \text{and}$$

$$\frac{d^2}{dx^2}\mathcal{L}_{eq}(x) = \frac{1}{2\tau}\frac{1}{\sqrt{x^2-4\tau}}(x+\sqrt{x^2-4\tau}),$$

and look for local minimum satisfying $\frac{d}{dx}\mathcal{L}_{eq}(x) = 0$. In the integration domain $x > 1+\tau$ we have $x^2 > 4\tau$ and it is convenient to parametrize $x = 2\sqrt{\tau}\cosh\theta$ with $\theta \geqslant 0$ so that the equation for the position θ_{**} of minimum in $\theta-$variable takes the form $m/\sqrt{\tau} = e^{\theta_{**}}$. The solution $\theta_{**} = \ln\left(m/\sqrt{\tau}\right) > 0$ exists as long as $m > \sqrt{\tau}$, implying finally that

$$x_{**} = m + \frac{\tau}{m}$$

for which one finds

$$\frac{d^2}{dx^2}\mathcal{L}_{eq}(x_{**}) = \frac{m}{\tau}\frac{1}{m-\frac{\tau}{m}} > 0$$

Therefore the function $\mathcal{L}_{eq}(x)$ has its global minimum at $x = x_{**}$ as long as $m > \sqrt{\tau}$. The condition $x_{**} > 1+\tau$ is equivalent to $(m-1) > (m-1)\cdot(\tau/m)$. If $m > 1$ the last condition implies $m > \tau$ which is always satisfied as $\tau < 1$. Therefore for $m > 1$ the integral is dominated by the minimum, and one finds by direct calculations $\mathcal{L}_{eq}(x_{**}) = -\ln m$. In the case $0 \leqslant m < 1$ the condition $x_{**} > 1+\tau$ implies $m < \tau$ which is incompatible with $m > \sqrt{\tau}$, so that the minimum is not operative any longer. In that case one finds $\frac{d}{dx}\mathcal{L}_{eq}(x) > 0$ for any real x so the function increases monotonically and we conclude that the integral over the domain $x \in (1+\tau, \infty)$ for any $0 \leqslant m < 1$ is dominated by the boundary of integration $x = 1+\tau$ where $\mathcal{L}_{eq}(x = 1+\tau) = \frac{(1-m)^2}{2\tau} - (1-m)$. Combining everything, we arrive to the large-N asymptotics in the form

$$\left\langle \mathcal{N}_{eq}^{B_+} \right\rangle \approx \begin{cases} e^{-N\left(\frac{(1-m)^2}{2\tau}-(1-m)+\ln m\right)} & 0 \leqslant m < 1, \\ O(1) & m > 1. \end{cases} \tag{4.9}$$

Comparing (4.9) to (4.8) we see that $\langle \mathcal{N}_{eq}^{B_+} \rangle \ll \langle \mathcal{N}_{eq}^{A} \rangle$ for $0 < m < 1$, whereas for $m > 1$ the order is opposite: $\langle \mathcal{N}_{eq}^{B_+} \rangle \gg \langle \mathcal{N}_{eq}^{A} \rangle$. It is easy also to check that for $m > 0$ the contribution of $\langle \mathcal{N}_{eq}^{B_-} \rangle$ is negligible, so that one finally sees that the mean number of equilibria satisfies

$$\lim_{N\to\infty} \frac{1}{N} \ln \langle \mathcal{N}_{eq} \rangle = \Sigma_{eq}(m)$$

where $\Sigma_{eq}(m) = \frac{1}{2}(m^2-1) - \ln m > 0$ for all $0 < m < 1$ and $\Sigma_{eq}(m) = 0$ for $m > 1$. Therefore, if $m < 1$ then $\langle \mathcal{N}_{eq} \rangle$ grows exponentially with N with the rate independent of the value of τ. A more detailed analysis [32] not relying upon the Large Deviation Principle allows one to find the leading sub-exponential factor in such growth which turned out to depend on τ . Moreover, the same analysis reveals that if $m > 1$ then

$$\lim_{N\to\infty} \langle \mathcal{N}_{eq} \rangle = 1.$$

showing that on average—indeed with probability tending to one—the system possesses a single equilibrium which can be shown to be stable. Moreover, for large but finite values of N one can find a smooth interpolating formula between exponential in N behaviour for the number of equilibria for $m < 1$, and a single equilibrium at $m > 1$. Such crossover regime occurs in a small vicinity of $m = 1$ where $|1-m| \sim N^{-1/2}$ and $\langle \mathcal{N}_{eq} \rangle$ depends on the scaling variable $\kappa = (m-1)\sqrt{N}$ via a crossover function whose explicit form is determined in [32].

One can also use the same LDT methods to find the leading exponential asymptotics for $\mathcal{D}_N^{(st)}(x)$ which turns out to be given by

$$\mathcal{D}_N^{(st)}(x) \approx \begin{cases} e^{N\Phi_1(x)}, & x > 1+\tau \\ e^{-N^2 K_\tau(x) + N\Phi_2(x)}, & x < 1+\tau \end{cases} \tag{4.10}$$

where the function $\Phi_1(x)$ defined for $x > 1+\tau$ coincides with $\Phi(x)$ from the upper line of (4.7). The explicit forms of the functions $K_\tau(x)$ and $\Phi_2(x)$ are not actually needed for our purposes apart from the following properties:

(1) The function $K_\tau(x)$ defined for all $x \leqslant 1+\tau$ has its *minimum* at $x = 1+\tau$ where $K_\tau(x = 1+\tau) = 0$ and for any fixed $0 \leqslant \tau < 1$ the behaviour close to the point of minimum is at least quadratic. More precisely, we have $K_\tau(x) = \mathcal{K}'' \cdot (x-1-\tau)^q + O\big((x-1-\tau)^{q+1}\big)$ for some $q \geqslant 2$ and some positive, τ–dependent constant $\mathcal{K}''$.

(2) The function $\Phi_1(x)$ defined for $x > 1+\tau$ and $\Phi_2(x)$ defined for $x < 1+\tau$ satisfy a *continuity* property at the point $x = 1+\tau$:

$$\lim_{x\to(1+\tau)+0} \Phi_2(x) = \lim_{x\to(1+\tau)-0} \Phi_1(x) = \frac{\tau}{2} \tag{4.11}$$

where we used Eq.(4.7) and employed ± 0 to indicate the limit from the right/left, respectively.

The line of reasoning leading to (4.10) is somewhat long, and we refer to [9] for the details of the procedure. To give an informal flavor of the ideas, we first notice that in a typical realization all eigenvalues will be located to the left of the line $\operatorname{Re} z = x$, provided $x > 1+\tau$. Hence, for such values of x we will have $\chi(\mu_N \in B_x) = 1$ with probability close to one, as in view of (3.21)

$$\operatorname{Prob}\{\mu_N \in B_x\} = \langle \chi(\mu_N \in B_x) \rangle = 1 - \operatorname{Prob}(x_m > x) \approx 1 - e^{-NL(x)}, \quad x > 1+\tau$$

This immediately shows why for $x > 1+\tau$ we should expect the approximate equality $\mathcal{D}_N^{(st)}(x) \approx \mathcal{D}_N^{(eq)}(x)$ implying the first line of (4.10). At the same time, for $x < 1+\tau$ we have $\chi(\mu_N \in B_x) = 1$ with a small probability given by the Large Deviation Theory, see (2.6):

$$\text{Prob}\{\mu_N \in B_x\} \approx e^{-N^2 K_\tau(x)}, \quad K_\tau(x) = \inf_{\mu_N \in B_x} [\mathcal{J}_\tau(\mu)] = \mathcal{J}_\tau(\mu_x)$$

where we denoted by μ_x a measure minimizing the rate functional $\mathcal{J}_\tau(\mu)$ in (3.19) under the above constraints, for a given values of x and τ. Finding the corresponding density $\rho_x(z)$ explicitly is a highly nontrivial exercise in potential theory, which for the present case is only solved for the special case of purely gradient flow $\tau = 1$ [17]. Still, one can write

$$\mathcal{D}_N^{(st)}(x) \approx e^{N\Phi_2(x)} e^{-N^2 K_\tau(x)}$$

giving the bottom line of (4.10). Continuity is expected as for $x \to 1+\tau$ we necessarily must have $\mu_x \to \mu_{eq}$ and $\Phi_2(x) \to \Phi^{(x)}(\mu_{eq}) = \Phi_1(x)$.

Having Eqs. (4.10) at our disposal, it is a straightforward task to analyse exponential asymptotics of $\langle \mathcal{N}_{st} \rangle$ in (4.1). First, we again subdivide the integration domain $\mathbb{R}$ into two parts: $C_- = (-\infty, 1+\tau)$ and $C_+ = (1+\tau, \infty)$ aiming to extract the large-N asymptotics of $\langle \mathcal{N}_{st} \rangle$ from

$$\langle \mathcal{N}_{st} \rangle = \left\langle \mathcal{N}_{st}^{(-)} \right\rangle + \left\langle \mathcal{N}_{st}^{(+)} \right\rangle'$$

Ignoring the sub-exponential terms in the bottom and top lines of (4.10), we have

$$\left\langle \mathcal{N}_{st}^{(-)} \right\rangle \approx \frac{1}{m^N} \int_{C_-} e^{-N^2 K_\tau(x) - N\mathcal{L}_-(x)} \frac{dx}{\sqrt{2\pi\tau/N}}, \tag{4.12}$$

with $\mathcal{L}_-(x) = \frac{(x-m)^2}{2\tau} - \Phi_2(x)$ and

$$\left\langle \mathcal{N}_{st}^{(+)} \right\rangle \approx \frac{1}{m^N} \int_{C_+} e^{-N\mathcal{L}_+(x)} \frac{dx}{\sqrt{2\pi\tau/N}},$$

with $\mathcal{L}_+(x) = \frac{(x-m)^2}{2\tau} - \Phi_1(x)$. Note that $\left\langle \mathcal{N}_{st}^{(+)} \right\rangle$ for $N \gg 1$ is obviously the same as $\left\langle \mathcal{N}_{eq}^{B_+} \right\rangle$ whose asymptotics is given by (4.9).

Turning our attention now to the integral featuring in Eq.(4.12) we immediately conclude that for large $N \gg 1$ the integral, when evaluated by the Laplace method, will be dominated by the vicinity of the boundary $x = 1+\tau$ where the function $\mathcal{K}_\tau(x)$ has its minimum (and equal to zero). The second term $N\mathcal{L}_-(x = 1+\tau)$ will however define the value of the integrand at this point. Using the *continuity* property Eq.(4.11) it is easy to verify that the leading exponential asymptotics of $\left\langle \mathcal{N}_{st}^{(-)} \right\rangle$ are exactly the same as for $\left\langle \mathcal{N}_{st}^{(+)} \right\rangle$. This immediately implies that the large-N asymptotics for the mean number of stable equilibria is given by:

$$\lim_{N\to\infty} \frac{1}{N} \ln \langle \mathcal{N}_{st} \rangle = \Sigma_{st}(m;\tau), \tag{4.13}$$

where the 'complexity function' $\Sigma_{st}(m;\tau)$ for any values of the control parameters $0 < m < 1$ and $0 < \tau \leqslant 1$ is given explicitly by

$$\Sigma_{st}(m;\tau) = -\left[1 - m + \ln m + \frac{(1-m)^2}{2\tau}\right]. \tag{4.14}$$

Finally, as was already mentioned, the boundary case of purely gradient descent dynamics $\tau = 1$ is equivalent to counting minima of a certain random potential, see discussion in [32]. That counting can be done by several methods, with or without the use of the large deviations techniques, see [7,12,27,31], and the resulting complexity is exactly the same as given by Eq.(4.13) with $\tau = 1$.

We can immediately infer that in the 'topologically non-trivial' regime of the (m, τ) parameter plane (see Figure 4.16) there exists a curve $\tau_B(m)$ given explicitly by

$$\tau_B(m) = -\frac{1}{2}\frac{(1-m)^2}{1-m+\ln m}, \quad 0 < m \leqslant 1,\, 0 \leqslant \tau \leqslant 1 \tag{4.15}$$

such that for the parameter values below that line the 'complexity function' associated with the stable equilibria is negative: $\Sigma_{st}(m;\tau) < 0$, implying that for such values of m and τ the mean number of stable solutions is exponentially small. As the random variable $\mathcal{N}_{st}$ can take only non-negative integer values $\mathcal{N}_{st} = 0, 1, 2, \ldots$ this in turn implies that in a typical realization of random couplings $f_i(\mathbf{x})$ in Eq.(2.12) there are simply no stable equilibria at all, i.e. $\mathcal{N}_{st} = 0$. We conclude that the variable $\mathcal{N}_{st}$ takes positive integer values only in rare realizations, with exponentially small probability. It is natural to name the parameter range corresponding to such a type of the phase portrait as an 'absolute instability' regime. In contrast, for the parameter values above the curve $\tau_B(m)$ the complexity function Eq.(4.14) is positive so that stable equilibria are exponentially abundant. Still, since $\Sigma_{st}(m;\tau) < \Sigma_{eq}$ for any $m < 1$, the stable equilibria are exponentially rare among all possible equilibria. One may call the associated type of the phase portrait as the 'relative instability' regime.

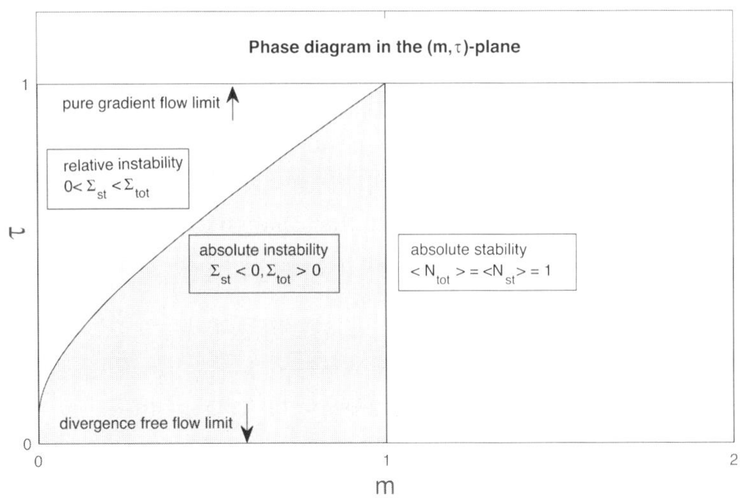

FIGURE 4.16. Regimes for the complexity of stable equilibria in (m, τ) plane. The 'absolute instabilty' regime below the line $\tau_B(m)$ where no stable equilibria typically exist is shaded.

Discussion, further developments and open questions In these lectures we attempted to explain the picture developed recently in [32] and [9] which suggests the following essential refinement of the May-Wigner instability transition phenomenon in a nonlinear setting. Define the parameter τ for the dynamical system Eq.(2.12) according to Eq.(2.17), and set it to a value smaller than unity so that the corresponding dynamics is not of the pure gradient descent type. Further define the value of the scaled relaxation rate m according to Eq.(2.18) and set it to a value $m > 1$ to place the system initially in the topologically trivial regime with a single stable equilibrium. Then with keeping τ fixed and decreasing the value of m our system will first experience an abrupt transition at $m = m_C = 1$ from the topologically trivial stable regime to the 'absolute instability' regime extending for the interval $m_B < m < 1$, and then finally transits to the 'relative instability' regime for $0 < m < m_B$, where the value m_B for a given τ is given by the unique solution of the equation $\tau_B(m) = \tau$, with the function $\tau_B(m)$ defined in Eq.(4.15). In particular it is evident that $m_B \to 1$ as $\tau \to 1$ so that the 'absolute instability' regime is expected to be absent for the purely gradient descent flow. Actually, the case $\tau = 1$ is equivalent to counting minima of certain random potentials, see [32], and was done earlier by several methods, with or without the use of the large deviations techniques, see [7,12,27,31]. The resulting complexity is exactly that given by Eq.(4.13) and Eq.(4.14) with $\tau = 1$. The case of purely gradient flow is clearly quite special. Indeed, not only the regime of 'absolute instability' is absent for $\tau = 1$ (i.e. $m_B = m_C$), but also the complexity $\Sigma_{st}(m;1)$ vanishes at the boundary of the corresponding 'relative instability' regime $m = m_B = m_C = 1$ cubically, $\Sigma_{st}(m;1) \propto (1-m)^3$ as $m \to 1$, whereas for any non-gradient dynamics it vanishes quadratically when approaching the value $m = m_B$. Such a peculiarity is related to the 'third order' nature [45] of the transition into the glassy phase for the associated random potential problem, see [27].

For any $\tau < 1$ the 'absolute instability' regime does exist, and for weakly non-gradient system $1 - \tau \ll 1$ its widths changes linearly: $1 - m_B \approx \frac{3}{2}(1-\tau)$. In the opposite limit $\tau \to 0$ the value m_B tends to zero exponentially: $m_B \sim \exp\left\{-\frac{1}{2\tau}\right\}$, so that for a purely solenoidal dynamics the 'relative instability' regime does not exist at all: there are no stable equilibria for any value of $0 < m < 1$. All these features are evident from the Figure 4.16.

Let us mention that a simple adaptation of the Large Deviations technique used for treating the mean number of stable equilibria allows more detailed classification of unstable equilibria by the value of the 'relative index' $\alpha = K/N$ where K is number of unstable directions (note that such definition does not differentiate between purely stable equilibria studied by us earlier and equilibria with zero *fraction* of unstable directions in the limit $N \to \infty$). Namely, let us fix some $\alpha \in [0,1]$ and denote $\langle \mathcal{N}_\alpha \rangle$ the mean number of equilibria with *no more* than αN unstable directions, and Σ_α the associated complexity: $\langle \mathcal{N}_\alpha \rangle \approx e^{N\Sigma_\alpha}$. The calculation presented in [9] reveals the existence of a characteristic value $m = m_\alpha < 1$

(defined as the unique root of the equation $\alpha = \frac{1}{\pi}[\arccos m - m\sqrt{1-m^2}]$) such that for $m \in [m_\alpha, 1]$ and *any* $0 \leqslant \tau \leqslant 1$ holds $\Sigma_\alpha = \Sigma_{eq}$. In other words, all equilibria in that (rectangular) domain of the (m, τ) parameter plane have asymptotically *no more* than αN unstable directions. On the other hand, for $0 < m < m_\alpha$ there exists a curve described by the expression

$$\tau_B^\alpha(m) = \frac{(m_\alpha - m)^2}{-1 - m_\alpha^2 + 2mm_\alpha - 2\ln m} \tag{4.17}$$

such that for all points (m, τ) *below* such a curve one finds $\Sigma_\alpha = 0$ (i.e. all equilibria have *more* than αN unstable directions), whereas above such a curve holds $0 < \Sigma_\alpha < \Sigma_{eq}$. Note that for any $\alpha > 0$ such curve can be shown to have a unique maximum inside the topologically non-trivial phase. This is in sharp contrast to $\alpha = 0$ case as $\tau_B^{\alpha\to 0}(m)$ can be shown to tend to $\tau_B(m)$ given by Eq.(4.15) and does not have a maximum, see Figure 4.16.

We expect the picture of instability transitions outlined above to be shared by other systems of randomly coupled autonomous ODE's with large number of degrees of freedom, such as, e.g. a model of neural network consisting of randomly interconnected neural units [51], or non-relaxational version of the spherical spin-glass model [15]. As was already mentioned, the model considered in [51] is essentially of the form Eq.(2.12) but with the particular choice of $f_i = \sum_j J_{ij} S(x_j)$ where S is an odd sigmoid function representing the synaptic nonlinearity and J_{ij} are independent centred Gaussian variables representing the synaptic connectivity between neuron i and j. Although being Gaussian, the corresponding (non-gradient) vector field is not homogeneous (in particular, the origin $\mathbf{x} = 0$ is *always* the point of an equilibrium) and thus seems rather different from our choice and not easily amenable to as detailed and controlled analysis as our model. Nevertheless, the paper [53] provided a compelling evidence about existence of a threshold in the coupling strength such that beyond such a threshold the single equilibrium at $\mathbf{x} = 0$ becomes unstable and exponentially many equilibria emerge instead for $\mathbf{x} \neq 0$. In fact, the total complexity rate Σ_{eq} estimated in [53] in the vicinity of the threshold turned out to be given by exactly the same expression as in our model. Moreover, by combining the methods of the present paper with ideas of [53] one can infer that the complexity of stable equilibria Σ_{st} in the vicinity of the same threshold should be shared by the two models as well. This observation points towards considerable universality of the conclusions based on our model Eq.(2.12).

Our results should be certainly considered as only first steps towards deeper quantitative analysis of dynamical behaviour of generic complex systems of many randomly coupled degrees of freedom. Earlier studies, starting from the classical paper [51] suggested that for systems of such type dynamics in the 'topologically nontrivial' regime should be predominantly chaotic, see [41, 53] and references therein. The absence of stable, attracting equilibria certainly corroborates this conclusion, though presence of stable periodic orbits in the phase space can not be excluded on those grounds either. In particular, one may hope to be able to shed some light on the dynamical features by incorporating insights from recent research on statistics of the Lyapunov exponents in random flows [40]. The influence of the non-gradient (also termed 'non-relaxational') component of the vector field on systems dynamics needs further clarification as well. On one hand, as we discovered above any admixture of such components very efficiently eliminates all stable equilibria when entering the 'topologically non-trivial' regime. On the other hand, the results of the paper [15] suggest that the influence of non-relaxational components on long-time 'aging' effects in dynamics of glassy-type models is relatively benign. This may imply that the dynamical dominance of exponentially abundant, though unstable equilibria with yet extensively many both stable and unstable directions may be enough for 'trapping' the system dynamics for a long time in the vicinity of such equilibria thus inducing aging phenomena, similar to the gradient descent case addressed in [14]. Clarification of these intriguing questions remains a serious challenge. Addressing statistical characteristics of $\mathcal{N}_{eq}$ and $\mathcal{N}_{st}$ beyond their mean values, and finally investigating similar questions in other models with non-relaxational dynamics (see e.g. [25, 34]) are within reach of the presently available methods and we hope to be able to address them in future publications.

5. Appendix: Supersymmetry and characteristic polynomials of real Ginibre matrices.

Our goal is to sketch the method of evaluating the following ensemble averages over Ginibre matrices, cf. (1.11) and (1.16),

$$D_1^{(N)} = \langle |\det(xI_{N-1} - G)| \rangle_{GinOE,N} \tag{5.1}$$

and

$$D_2^{(N)} = \left\langle \det\left[(xI_N - G)^2 + y^2 I_N\right]\right\rangle_{GinOE,N} \tag{5.2}$$

As the procedure is simpler for the equation (5.2), we start with the latter by rewriting

$$\begin{aligned} D_2^{(N)} &= \left\langle \det\left[G - (x+iy)I_N\right] \det\left[G^T - (x-iy)I_N\right]\right\rangle_{GinOE,N} \\ &= \left\langle \det\begin{pmatrix} 0 & i(G - zI_N) \\ i(G^T - \bar{z}I_N) & 0 \end{pmatrix}\right\rangle_{GinOE,N}. \end{aligned} \tag{5.3}$$

Let now $\Psi_1, \Psi_2, \Phi_1, \Phi_2$ be four column vectors with N anticommuting components each. Using the standard rules of Berezin integration one represents the determinant featuring in (5.3) as a Gaussian integral

$$D_2^{(N)} = \int d\Psi_1\, d\Psi_2\, d\Phi_1\, d\Phi_2 \left\langle e^{-i(\Psi_1^T, \Phi_1^T)\begin{pmatrix} 0 & G - zI_N \\ G^T - \bar{z}I_N & 0 \end{pmatrix}\begin{pmatrix} \Psi_2 \\ \Phi_2 \end{pmatrix}}\right\rangle_{GinOE,N} \tag{5.4}$$

which simplifies to

$$\int d\Psi_1\, d\Psi_2\, d\Phi_1\, d\Phi_2\; e^{iz\Psi_1^T\Phi_2+i\bar{z}\Phi_1^T\Psi_2}\left\langle e^{i\mathrm{Tr}G(\Phi_2\otimes\Psi_1^T)+i\mathrm{Tr}G^T(\Phi_2\otimes\Psi_1^T)}\right\rangle_{\mathrm{GinOE},N}$$

where $M = \mathbf{a}\otimes\mathbf{b}^T$ stands for the matrix with entries $M_{ij} = a_i b_j$. Ensemble averaging over the real Ginibre matrices can be easily performed by using the identity

$$\left\langle e^{-\mathrm{Tr}\,(GA+G^TB)}\right\rangle_{\mathrm{GinOE},N} = e^{\frac{1}{2}\mathrm{Tr}\,(A^TA+B^TB+2AB)}. \tag{5.5}$$

In our case $A = i\Phi_2\otimes\Psi_1^T$ and $B = \Phi_2\otimes\Psi_1^T$, which (remembering the anticommuting nature of entries of the vectors) implies $A^TA = B^TB = 0$ so that

$$D_2^{(N)} = \int d\Psi_1\, d\Psi_2\, d\Phi_1\, d\Phi_2\; e^{iz\Psi_1^T\Phi_2+i\bar{z}\Phi_1^T\Psi_2+(\Phi_1^T\Phi_2)(\Psi_1^T\Psi_2)}$$

After the ensemble average is performed, there exists only one term in the exponential in the integrand which is quartic in anticommuting variables, and it is of the form $(\Phi_1^T\Phi_2)\,(\Psi_1^T\Psi_2)$. The corresponding exponential factor is then represented as:

$$e^{(\Phi_1^T\Phi_2)(\Psi_1^T\Psi_2)} = \frac{1}{\pi}\int dq\, d\bar{q}\, e^{-|q|^2-q(\Psi_1^T\Psi_2)-\bar{q}(\Phi_1^T\Phi_2)},$$

where the formula above represents the simplest instance of what is generally known as the Hubbard-Stratonovich transformation. After such a representation is employed, it allows to perform the (by now, Gaussian) integration over the anticommuting vectors explicitly, yielding:

$$\begin{aligned} D_2^{(N)} &= \frac{1}{\pi}\int dq\, d\bar{q}\, e^{-|q|^2}\det\begin{pmatrix} q\,I_N & -izI_N \\ -i\bar{z}I_N & \bar{q}I_N \end{pmatrix} \\ &= \frac{1}{\pi}\int dq\, d\bar{q}\, e^{-|q|^2}(|q|^2+|z|^2)^N = 2\int_0^\infty dr\, r e^{-r^2}(r^2+|z|^2)^N \\ &= e^{|z|^2}\Gamma\left(N+1,|z|^2\right), \end{aligned} \tag{5.6}$$

where we used the definition (1.13) of the incomplete Γ-function. This shows that replacing in (1.16) the factor $D_2^{(N-2)}$ with its value according to (5.6) indeed reproduces (1.17) as expected.

Turning now our attention to the evaluation of the mean modulus of the determinant featuring in (5.1) we first represent as a limit of a parameter-dependent regularized expression:

$$\mathcal{D}_1^{(N)} = \lim_{\epsilon\to 0}\langle|\det(xI_{N-1}-G)|_\epsilon\rangle_{\mathrm{GinOE},N},$$

where we introduced

$$|\det(xI_{N-1}-G)|_\epsilon := \frac{\det\begin{pmatrix} 0 & i(x\,I_N-G) \\ i(xI_N-G^T) & 0 \end{pmatrix}}{\det^{1/2}\begin{pmatrix} \epsilon\, I_N & i(xI_N-G) \\ i(xI_N-G^T) & \epsilon\, I_N \end{pmatrix}}. \tag{5.7}$$

with a regularization parameter $\epsilon > 0$. Now one can further use a form of the standard Gaussian integral well-defined for any positive $\epsilon > 0$ to represent the denominator in (5.7) as a Gaussian integral over two N−vectors $\mathbf{S}_{1,2}$ with commuting components :

$$
\begin{aligned}
&\det{}^{-1/2}\begin{pmatrix} \epsilon\, I_N & i(xI_N - G) \\ i(xI_N - G^T) & \epsilon\, I_N \end{pmatrix} \\
&\propto \int_{\mathbb{R}^N} d\mathbf{S}_1 \int_{\mathbb{R}^N} d\mathbf{S}_2\, e^{-\frac{1}{2}\begin{pmatrix}\mathbf{S}_1^T & \mathbf{S}_2^T\end{pmatrix}\begin{pmatrix} \epsilon\, I_N & i(x\, I_N - G) \\ i(xI_N - G^T) & \epsilon\, I_N \end{pmatrix}\begin{pmatrix}\mathbf{S}_1 \\ \mathbf{S}_2\end{pmatrix}}.
\end{aligned} \tag{5.8}
$$

where $\propto$ here and below stands for (temporaly) ignored N−dependent multiplicative constants product of which will be restored in the very end of the procedure. Combining (5.8) with the representation of the numerator in (5.7) via a gaussian integral over anticommuting N− vectors, cf. (5.4), we get the so-called 'supersymmetric' integral representation for the ratio. The ensemble average over Ginibre matrices can be easily done by exploiting (5.5), and after straightforward manipulations one arrives at

$$
\begin{aligned}
|\det(xI_{N-1} - G)|_\epsilon \propto & \int_{\mathbb{R}^N} d\mathbf{S}_1 \int_{\mathbb{R}^N} d\mathbf{S}_2\, e^{-\frac{\epsilon}{2}(\mathbf{S}_1^T\mathbf{S}_1+\mathbf{S}_2^T\mathbf{S}_2)-i\frac{x}{2}(\mathbf{S}_1^T\mathbf{S}_2+\mathbf{S}_2^T\mathbf{S}_1)} \\
\times\, e^{-\frac{1}{2}(\mathbf{S}_1^T\mathbf{S}_1)(\mathbf{S}_2^T\mathbf{S}_2)} & \int \frac{dq\, d\overline{q}}{\pi} e^{-|q|^2} \det\begin{pmatrix} \overline{q}\, I_N & ixI_N + \mathbf{S}_1 \otimes \mathbf{S}_2^T \\ ixI_N + \mathbf{S}_2 \otimes \mathbf{S}_1^T & qI_N \end{pmatrix}
\end{aligned}
$$

A straightforward calculation shows that the determinant in the above expression is equal to

$$
\begin{aligned}
&(|q|^2 + x^2)^{N-2}\Big[(|q|^2 + x^2)^2 \\
&+(|q|^2 + x^2)\left(-2ix(\mathbf{S}_1^T\mathbf{S}_2) - (\mathbf{S}_1^T\mathbf{S}_1)(\mathbf{S}_2^T\mathbf{S}_2)\right) + x^2\left((\mathbf{S}_1^T\mathbf{S}_1)(\mathbf{S}_2^T\mathbf{S}_2) - (\mathbf{S}_1^T\mathbf{S}_2)^2\right)\Big]
\end{aligned}
$$

and the integration over $q, \overline{q}$ is performed via using the polar coordinates, as in our first example.In the remaining integrations, the integrand depends only on the entries of a positive semidefinite real symmetric matrix

$$
\hat{Q} = \begin{pmatrix} Q_1 & Q \\ Q & Q_2 \end{pmatrix}, \quad Q_1 = \mathbf{S}_1^T\mathbf{S}_1,\ Q_2 = \mathbf{S}_2^T\mathbf{S}_2,\ Q = \mathbf{S}_1^T\mathbf{S}_2
$$

A useful trick suggested in [29] in such cases is to pass from the pair of vectors $(\mathbf{S}_1, \mathbf{S}_2)$ to the matrix $\hat{Q}$ as a new integration variable. This change is non-singular for $N \geqslant 2$, incurs a Jacobian factor proportional to $\det \hat{Q}^{(N-3)/2}$ (see the Appendix D of [28]) and brings us to a representation of the following form:

$$
\begin{aligned}
|\det(xI_{N-1} - G)|_\epsilon \propto e^{x^2} & \int_{\hat{Q}\geqslant 0} d\hat{Q}\, \det \hat{Q}^{(N-3)/2} e^{-\frac{1}{2}\epsilon(Q_1+Q_2) - ixQ - \frac{1}{2}Q_1Q_2} \\
\times & \left[\Gamma(N+1, x^2) + \Gamma(N, x^2)(-2ixQ - Q_1Q_2) + \Gamma(N-1, x^2)x^2 \det Q\right].
\end{aligned} \tag{5.9}
$$

Employing a convenient parametrization of the positive semidefinite matrix $\hat{Q} \geqslant 0$ given by

$$\hat{Q} = \begin{pmatrix} Q_1 & Q \\ Q & \frac{r^2+Q^2}{Q_1} \end{pmatrix}, \quad d\hat{Q} := dQ_1 dQ_2 dQ = 2\frac{dQ_1}{Q_1} r\,dr\,dQ,$$

with $-\infty \leqslant Q \leqslant \infty$, $0 \leqslant Q_1, r < \infty$ and further changing $Q_1 \to \epsilon Q_1$ brings (5.9) to the form

$$\begin{aligned} &|\det(xI_{N-1} - G)|_\epsilon \\ &\propto e^{x^2}\int_{\mathbb{R}_+} \frac{dQ_1}{Q_1} e^{-\frac{\epsilon^2}{2}Q_1} \int_{\mathbb{R}_+} dr\, r^{N-2} e^{-\frac{r^2}{2}\left(1+\frac{1}{Q_1}\right)} \int_{\mathbb{R}} \frac{dQ}{\sqrt{2\pi}} e^{-ixQ-\frac{1}{2}Q^2\left(1+\frac{1}{Q_1}\right)} \\ &\qquad \times \left[\Gamma(N+1,x^2) + \Gamma(N,x^2)(-2ixQ - Q^2 - r^2) + \Gamma(N-1,x^2)x^2r^2\right] \end{aligned}$$

which allows us to perform the integrals over Q and r explicitly, see Appendix A of [26] for more detail, finally arriving at the following expression (and restoring the normalization constant)

$$\begin{aligned} |\det(xI_{N-1} - G)|_\epsilon &= \frac{e^{x^2}}{2^{N/2}\Gamma\left(\frac{N}{2}\right)} \int_{\mathbb{R}_+} \frac{dQ_1}{Q_1^2} e^{-\frac{\epsilon^2}{2}Q_1} \\ &\times e^{-\frac{x^2}{2\left(1+\frac{1}{Q_1}\right)}} \left(\frac{Q_1}{1+Q_1}\right)^{\frac{N+2}{2}} \left[\Gamma(N+1,x^2) - x^2\frac{Q_1}{1+Q_1}\Gamma(N,x^2)\right]. \end{aligned} \tag{5.10}$$

Such a representation now allows one to take the limit $\epsilon \to 0$ safely, and the integration over Q_1 in (5.10) can be easily performed introducing $u = |x|\sqrt{\frac{Q_1}{1+Q_1}}$ as the new integration variable. One gets in this way the final result

$$\langle|\det(xI_N - G)|\rangle_{Gin_1,N} = \frac{2}{2^{N/2}\Gamma\left(\frac{N}{2}\right)}\left[e^{\frac{x^2}{2}}\Gamma\left(N,x^2\right) + |x|^N \int_0^{|x|} e^{-\frac{u^2}{2}} u^{N-1}\,du\right]$$

which after replacing $N \to (N-1)$ can be substituted in (1.11) to produce (1.12).

6. Exercises with hints

Lecture 1

Exercise 6.1 (Matrices with multiple eigenvalues). Let $\mathcal{A}$ be a subset of $\mathcal{M}_N(\mathbb{R})$ consisting of matrices with multiple eigenvalues. Show that

$$\mu_L(\mathcal{A}) = 0,$$

where μ_L is a Lebesgue measure in $\mathbb{R}^{N^2}$.

Remark. The statement can be easily generalized to complex matrices, real symmetric, Hermitian, etc.

Remark. Probability distribution introduced for the Real Ginibre Ensemble is absolutely continuous with respect to flat Lebesgue measure and this yield that $\mathcal{A}$ has zero measure with respect to GinOE distribution.

Hints

- Show that matrix $M \in \mathcal{M}_N(\mathbb{R})$ has a multiple eigenvalue iff its characteristic polynomial has a multiple root.
- For an arbitrary polynomial $p(\lambda) = \sum_{k=0}^{N} a_k \lambda^k = a_N \prod_{j=1}^{N} (\lambda - \lambda_j)$ let us define its discriminant by
$$D(p) := a_N^{2N-2} \prod_{i<j}^{N} (\lambda_i - \lambda_j)^2.$$
Use Vieta's formulas to prove that discriminant of a polynomial can be written as a multivariate polynomial in terms of coefficients $\{a_k\}_{k=0}^{N}$.
- Use the above to conclude, that the set of matrices $M \in \mathcal{M}_N(\mathbb{R})$ with multiple eigenvalues corresponds to a surface of positive co-dimension in space $\mathbb{R}^{N^2}$.

Exercise 6.2 (Joint Probability Distributions of eigenvalues (conditional)). Let $(z_1, z_2, \ldots, z_N)$ be the ordered set of either real or pairs of complex conjugate eigenvalues of matrix G taken randomly from the Real Ginibre Ensemble. We aim to show that Joint Probability Density function (conditional) of eigenvalues is given by
$$p_1^{(L,M)}(z_1, \ldots, z_N) = C_{L,M}^{(N)} \prod_{1 \leqslant j < k \leqslant N} |z_j - z_k| \exp\Big\{-\frac{1}{2} \sum_{1 \leqslant j \leqslant N} z_j^2\Big\}$$
$$\times \prod_{1 \leqslant j \leqslant N} \sqrt{\operatorname{erfc}\left(\sqrt{2}\,|\mathrm{Im} z_j|\right)} \mathbf{1}_{G \in \mathcal{X}_{L,M}},$$
with
$$C_{L,M}^{(N)} = 2^M \left(2^{N(N+1)/4} \prod_{j=1}^{N} \Gamma(j/2)\right)^{-1}.$$

(1) Start with 2×2 real matrix and show that under an assumption of simple spectrum it can be uniquely decomposed as (Real Schur decomposition)
$$G = O(\Lambda + R) O^T,$$
where $O \in O(2)$ is an orthogonal matrix with non-negative entries in the first row and either
$$\Lambda = \begin{pmatrix} \lambda_1 & 0 \\ 0 & \lambda_2 \end{pmatrix} \text{ and } R = \begin{pmatrix} 0 & r \\ 0 & 0 \end{pmatrix} \quad \text{or} \quad \Lambda = \begin{pmatrix} \alpha & \beta \\ -\gamma & \alpha \end{pmatrix} \text{ and } R = 0_{2\times 2},$$
depending on whether G has real or complex eigenvalues. Here all parameters are real with $\lambda_1 > \lambda_2$, $\beta \geqslant \gamma$ and $\beta\gamma > 0$.

(2) Let us define the set $X_{L,M}$ to be the set of all $G \in \mathbb{R}^{N \times N}$ having exactly L real eigenvalues and M conjugate pairs of complex eigenvalues. Then
$$\mathcal{M}_2(\mathbb{R}) = X_{2,0} \bigcup X_{0,1}.$$

Take $F : \mathcal{M}_2(\mathbb{R}) \to \mathbb{C}$ to be arbitrary function depending only on eigenvalues. Perform change of variables $G \to (O, \Lambda, R)$ (and in the case of complex eigenvalues further change $\Lambda \to$ eigenvalues $z, \bar{z}$) in both integrals

$$\int_{X_{2,0}} F(G)\, dP(G) \quad \text{and} \quad \int_{X_{0,1}} F(G)\, dP(G).$$

Compute corresponding Jacobians and integrate out all variables except eigenvalues to obtain expressions for $p^{(2,0)}(\lambda_1, \lambda_2)$ and $p^{(0,1)}(z, \bar{z})$.

(3) Derive JPDF expression for a general N by first decomposing the matrix into quasi upper triangular form. Then perform the change of variables and calculate the Jacobian. Finally integrate out all terms independent of eigenvalues.

Hints

(1)
- In the case of two real eigenvalues rewrite the decomposition in a form $GO = O(\Lambda + R)$ and analyse the first column of matrix O.
- For another case start with writing matrix G in the basis

$$I = \begin{pmatrix} 1 & 0 \\ 0 & 1 \end{pmatrix}, \quad J = \begin{pmatrix} 0 & 1 \\ -1 & 0 \end{pmatrix}, \quad K = \begin{pmatrix} 1 & 0 \\ 0 & -1 \end{pmatrix}, \quad L = \begin{pmatrix} 0 & 1 \\ 1 & 0 \end{pmatrix}.$$

  Find $O^T \cdot \ldots \cdot O$ for basis elements and conclude that there exists a unique O such that $O^T G O$ has a shape of Λ.
  Jacobians of size 4×4 in both cases can be significantly simplified by first using row/column operations.

(2)
- Prove that any matrix $G \in X_{L,M}$ can be uniquely decomposed as $G = O(\Lambda + R) O^T$, where Λ is a block-diagonal matrix with L blocks of size 1×1, corresponding real eigenvalues, and M blocks of size 2×2, corresponding pairs of complex conjugate eigenvalues.
- Rewrite DG in terms of new variables in a matrix form. Using properties of matrix O find expressions for $dG_{j,k}$. Finally, compute the corresponding Jacobian

$$\frac{DG}{DO\, D\Lambda\, DR}.$$

- Consider the integral

$$\int_{X_{L,M}} F(G)\, dP(G),$$

  for any function F depending only on eigenvalues. Perform change of variables and integrate out variables R and O by using explicit expression for GinOE probability distribution.
- For complex eigenvalues perform one more change of variables $(\alpha, \beta, \gamma) \to z = x + iy$ and obtain the result.

Exercise 6.3 (Basic properties of Pfaffians). For any skew symmetric matrix A of size $2n \times 2n$ we write $\operatorname{Pf} A$ to denote its Pfaffian given by

$$\operatorname{Pf} A = \frac{1}{2^n n!} \sum_{\sigma \in S_{2n}} \operatorname{sgn}(\sigma) \prod_{j=1}^{n} a_{\sigma(2j-1),\sigma(2j)},$$

where the summation is taken over all permutations of $(1, 2, \ldots, 2n)$.

Prove the following statements

(1) $\operatorname{Pf} A = -\operatorname{Pf} A^{(j,k)}$, where $A^{(j,k)}$ is a matrix obtained from A by exchanging j- th and k -th columns and rows;

(2) Let $\psi_1, \psi_2, \ldots, \psi_{2n}$ be a sequence of anticommuting variables. Show that

$$\int d\psi_1 \ldots d\psi_{2n} \exp\Big\{-\frac{1}{2} \sum_{j,k=1}^{2n} A_{j,k} \psi_j \psi_k\Big\} = \operatorname{Pf} A,$$

where integral is a standard Berezin integral;

(3) Let $\psi_1, \psi_2, \ldots, \psi_{2n}$ and $\phi_1, \phi_2, \ldots, \phi_{2n}$ be two sequences of anticommuting variables. Show that

$$\int d\psi_1 d\phi_1 d\psi_2 d\phi_2 \ldots d\psi_{2n} d\phi_{2n} \exp\Big\{-\sum_{j,k=1}^{2n} \psi_j A_{j,k} \phi_k\Big\} = \det A.$$

(4) $\operatorname{Pf}^2 A = \det A$;

(5) For arbitrary matrix B and skew-symmetric A of size $2n \times 2n$

$$\operatorname{Pf} BAB^T = \operatorname{Pf} A \det B;$$

(6) Let $a_{i,j} = 0$ for $2|i-j$ (chessboard pattern). Show that $\operatorname{Pf} A = \det A'$, where A' is a matrix of size $n \times n$ with entries $a'_{i,j} = a_{2i-1,2j}$.

(7) Let $a_{i,j} = \operatorname{sgn}(j-i) \dfrac{b_i}{b_j}$. Then

$$\operatorname{Pf} A = \prod_{j=1}^{N/2} \frac{b_{2j-1}}{b_{2j}}.$$

Hints

(2)-(3) Use a Taylor expansion for the exponential and show that non-vanishing terms in the integrand exactly match the definitions of $\operatorname{Pf} A$ (or $\det A$).

(4)-(6) Apply results of (1) and (2) and use properties of Berezin integral.

(7) Use row/column operations to reduce the Pfaffian to a simpler form and then use recursive definition

$$\operatorname{Pf}(A) = \sum_{\substack{j=1\\ j\neq i}}^{2n} (-1)^{i+j+1+\theta(i-j)} a_{ij} \operatorname{Pf}\left(A^{(i,j)}\right),$$

where index i can be selected arbitrarily, $\theta(x)$ is the Heaviside step function, and $A^{(i,j)}$ denotes the matrix A with both the i-th and j-th rows and columns removed.

Exercise 6.4 (Pfaffian structure of GinOE averages). Let N be even and $f : \mathbb{C} \to \mathbb{C}$ be an arbitrary (up to some integrability conditions) function of complex argument. Define $F : \mathbb{R}^{N \times N} \to \mathbb{C}$ by an identity

$$F(G) = \prod_{1 \leqslant j \leqslant N} f(z_j).$$

Let $\{P_{j-1}(z)\}_{j=1}^{N}$ be a set of monic polynomials of degree $j-1$. Show that

$$\langle F \rangle_{\text{GinOE}} = \left(2^{N(N+1)/4} \prod_{j=1}^{N} \Gamma(j/2) \right)^{-1} \text{Pf}\,(U),$$

where U is an $N \times N$ skew-symmetric matrix with

$$U_{j,k} = \int d^2 z_1 d^2 z_2 f(z_1) f(z_2) \mathcal{F}(z_1, z_2) P_{k-1}(z_1) P_{j-1}(z_2)$$

and the skew-symmetric measure

$$\mathcal{F}(z_1 = x_1 + iy_1, z_2 = x_2 + iy_2) = \exp\left\{-(z_1^2 + z_2^2)/2\right\}$$
$$\times \left[2i\delta^2(z_1 - \bar{z}_2)\,\text{sgn}\,(y_1)\, \text{erfc}\left(|y_1|\sqrt{2}\right) + \delta(y_1)\delta(y_2)\,\text{sgn}\,(x_2 - x_1) \right].$$

Hints

- In the formula for JPDF of eigenvalues rewrite
$$\prod_{1 \leqslant j < k \leqslant N} |z_j - z_k|$$
as a Vandermonde determinant with some prefactor.
- Using elementary row&column operations show that powers of z_j's in the Vandermonde determinant can be exchanged by any family of monic polynomials.
- Rewrite the determinant in terms of a Berezin integral and integrate out half of Grassmann variables.
- Write $\langle F \rangle_{\text{GinOE}}$ as an integral and show that the integrand is given by N-th power of some linear combination of pairwise products of anticommuting Grassmann variables.
- Use a definition of the Pfaffian in terms of a Berezin integral to obtain the result.

Exercise 6.5 (Kernel of a Pfaffian Point Process corresponding to GinOE). Define the kernel

$$\mathcal{K}_N(z_1, z_2) = \sum_{j,k=1}^{N} \left(\left(U^{(0)} \right)^{-1} \right)_{j,k} P_{j-1}(z_1) P_{k-1}(z_2),$$

where a skew symmetric matrix $U^{(0)} = U|_{f \equiv 1}$, i.e.

$$U_{j,k}^{(0)} = \int_{\mathbb{C}^2} \mathcal{F}(z_1, z_2) P_{k-1}(z_1) P_{j-1}(z_2)\, d^2 z_1 d^2 z_2.$$

Show that

$$R_n(z_1,\ldots,z_n) = \operatorname{Pf} Q, \quad Q_{k,l} = \begin{pmatrix} K_{k,l} & G_{k,l} \\ -G_{l,k} & W_{k,l} \end{pmatrix},$$

where

$$K_{k,l} = \mathcal{K}(z_k, z_l)$$

$$G_{k,l} = \int_{\mathbb{C}} \mathcal{K}(z_k, z)\, \mathcal{F}(z, z_l)\, d z,$$

$$W_{k,l} = -\mathcal{F}(z_k, z_l) + \iint_{\mathbb{C}} \mathcal{F}(z_k, z)\, \mathcal{K}(z, z')\, \mathcal{F}(z', z_l)\, d z d z'.$$

Hints: Take $f : \mathbb{C} \to \mathbb{C}$ defined by

$$f(\zeta) = 1 + \sum_{\ell=1}^{n} c_\ell \delta(\zeta - z_\ell).$$

Calculate corresponding average $\langle F \rangle_{\text{GinOE}}$ and consider it as a polynomial in terms of $\{c_\ell\}_{\ell=1}^{n}$. Show that the coefficient in front of

$$\prod_{j=1}^{k} c_{\ell_j}$$

coincides with the correlation function $R_k\left(z_{\ell_1}, z_{\ell_2}, \ldots, z_{\ell_k}\right)$. Then use the above result to represent the average as a Pfaffian and expand it to get a polynomial in terms of $\{c_\ell\}_{\ell=1}^{n}$. Compare the corresponding coefficients to obtain the result. You are advised to use the Cauchy-Binet theorem for manipulations with Pfaffians.
Solution: Let us take $F : \mathcal{M}\left(\mathbb{R}^N\right) \to \mathbb{C}$ defined by

$$F(G) = \prod_{j=1}^{N} f(\lambda_j),$$

where $\lambda_1, \ldots, \lambda_N$ are the eigenvalues (not necessarily real) of G and

$$f(u) = 1 + \sum_{l=1}^{n} c_l \delta^2(u - z_l),$$

for some n -tuple of distinct points $(z_1, z_2, \ldots, z_n) \subset \mathbb{C}$. Our aim is to calculate $\langle F \rangle_{\text{GinOE}}$. We will focus only on the simplest case $n = 2$. An interested reader may wish to consult the review [50] and the paper [11] for a general case (which requires some additional combinatorial ideas to succeed).

Let us first write

$$\langle F \rangle_{\text{GinOE}} = \left\langle \prod_{j=1}^{N} \left(1 + \sum_{\ell=1}^{2} \delta(\lambda_j - z_\ell)\right)\right\rangle_{\text{GinOE}}.$$

Expanding the product inside the brackets one can see a sum of several δ - functions products. Obviously, terms with at least two identical z_ℓ's would not contribute to the sum as they correspond to matrices with multiple eigenvalues (see Exercise 6.1) and yield vanishing contribution. Therefore

$$\begin{aligned}\langle F\rangle_{\text{GinOE}} &= 1+\sum_{\ell=1}^{2} c_\ell \left\langle \sum_{j=1}^{N} \delta\left(\lambda_j - z_\ell\right)\right\rangle_{\text{GinOE}} + c_1 c_2 \left\langle \sum_{j\neq k}^{N} \delta\left(\lambda_j - z_1\right)\right\rangle_{\text{GinOE}} \\ &= 1 + c_1 R_1\left(z_1\right) + c_2 R_1\left(z_2\right) + c_1 c_2 R_2\left(z_1, z_2\right).\end{aligned}$$

Using the earlier established Pfaffian structure of GinOE averages we can also get

$$\begin{aligned}\langle F\rangle_{\text{GinOE}} &= \operatorname{Pf}\left(U_{j,k}\right) \\ &= C_N^{-1} \operatorname{Pf}\left(\int d^2\zeta_1 d^2\zeta_2 f\left(\zeta_1\right) f\left(\zeta_2\right) \mathcal{F}\left(\zeta_1, \zeta_2\right) P_{k-1}\left(\zeta_1\right) P_{j-1}\left(\zeta_2\right)\right) \\ &= \operatorname{Pf}^{-1} U^{(0)} \operatorname{Pf}\left(U\right),\end{aligned}$$

where $U^{(0)}$ is a matrix defined above. Matrix elements $U_{j,k}$ can be written as

$$U_{j,k} = U_{j,k}^{(0)} + c_1 U_{j,k}^{(1)}\left(z_1\right) + c_2 U_{j,k}^{(1)}\left(z_2\right) + c_1 c_2 U_{j,k}^{(2)}\left(z_1, z_2\right)$$

with

$$\begin{aligned}U_{j,k}^{(1)}\left(z\right) &= \int d^2\zeta \mathcal{F}\left(\zeta, z\right)\left(P_{j-1}\left(z\right) P_{k-1}\left(\zeta\right) - P_{j-1}\left(\zeta\right) P_{k-1}\left(z\right)\right) \\ U_{j,k}^{(2)}\left(z, w\right) &= \mathcal{F}\left(z, w\right)\left(P_{k-1}\left(z\right) P_{j-1}\left(w\right) - P_{j-1}\left(z\right) P_{k-1}\left(w\right)\right).\end{aligned}$$

Now one needs to rewrite $\operatorname{Pf} U$ as a polynomial in terms of c_1 and c_2 and compare with the above. Note that the Pfaffian is of size $N \times N$. Therefore, first we will use

Lemma (Cauchy-Binet formula). Let B and C be a $2n \times 2n$ and $2N \times 2N$ invertible skew symmetric matrices. Then for any matrix A of size $2n \times 2N$

$$\frac{\operatorname{Pf}\left(C^{-T} - A^T B A\right)}{\operatorname{Pf}\left(C^{-T}\right)} = \frac{\operatorname{Pf}\left(B^{-T} - A C A^T\right)}{\operatorname{Pf}\left(B^{-T}\right)} = \operatorname{Pf}\begin{pmatrix} B^{-T} & -A \\ A^T & C^{-T}\end{pmatrix}$$

Let us consider the following matrices

$$C = U^{-T}$$

$$A^T = \begin{pmatrix} \sqrt{c_1} P_0\left(z_1\right) & \sqrt{c_1}\int d^2\zeta \mathcal{F}\left(\zeta, z_1\right) P_0\left(\zeta\right) & \sqrt{c_2} P_0\left(z_2\right) & \sqrt{c_2}\int d^2\zeta \mathcal{F}\left(\zeta, z_2\right) P_0\left(\zeta\right) \\ \sqrt{c_1} P_1\left(z_1\right) & \sqrt{c_1}\int d^2\zeta \mathcal{F}\left(\zeta, z_1\right) P_1\left(\zeta\right) & \sqrt{c_2} P_1\left(z_2\right) & \sqrt{c_2}\int d^2\zeta \mathcal{F}\left(\zeta, z_2\right) P_1\left(\zeta\right) \\ \vdots & \vdots & \vdots & \vdots \\ \sqrt{c_1} P_{N-1}\left(z_1\right) & \sqrt{c_1}\int d^2\zeta \mathcal{F}\left(\zeta, z_1\right) P_{N-1}\left(\zeta\right) & \sqrt{c_2} P_{N-1}\left(z_2\right) & \sqrt{c_2}\int d^2\zeta \mathcal{F}\left(\zeta, z_2\right) P_{N-1}\left(\zeta\right)\end{pmatrix}$$

$$B = \begin{pmatrix} 0 & -1 & \sqrt{c_1 c_2}\mathcal{F}\left(z_1, z_2\right) & 0 \\ 1 & 0 & 0 & 0 \\ -\sqrt{c_1 c_2}\mathcal{F}\left(z_1, z_2\right) & 0 & 0 & -1 \\ 0 & 0 & 1 & 0\end{pmatrix}$$

Then it is easy to check that

$$U = C^{-T} - A^T B A,$$

and after applying the Cauchy-Binet formula we arrive to

$$\langle F \rangle_{\text{GinOE}} = \text{Pf}\left(B^{-T} - ACA^{T}\right),$$

where we have used the fact that $\text{Pf}\,B = 1$. Inverse of the matrix B has the form

$$B^{-T} = \begin{pmatrix} 0 & -1 & 0 & 0 \\ 1 & 0 & 0 & -\sqrt{c_1c_2}\mathcal{F}(z_1, z_2) \\ 0 & 0 & 0 & -1 \\ 0 & \sqrt{c_1c_2}\mathcal{F}(z_1, z_2) & 1 & 0 \end{pmatrix}.$$

Finally, calculating the product ACA^{T} we obtain

$$\langle F \rangle_{\text{GinOE}} = \text{Pf}\begin{pmatrix} J + c_1Q_{1,1} & \sqrt{c_1c_2}Q_{1,2} \\ \sqrt{c_1c_2}Q_{2,1} & J + c_2Q_{2,2} \end{pmatrix},$$

where

$$J = \begin{pmatrix} 0 & 1 \\ -1 & 0 \end{pmatrix}$$

is the standard symplectic matrix and $Q_{i,j}$ were defined above. Comparing last identity with

$$\langle F \rangle_{\text{GinOE}} = 1 + c_1R_1(z_1) + c_2R_1(z_2) + c_1c_2R_2(z_1, z_2)$$

one arrives to the required result.

Exercise 6.6 (Skew orthogonal polynomials)**.** Consider the family of monic polynomials $P_{2k}(z) = z^{2k}$ and $P_{2k+1}(z) = z^{2k+1} - 2kz^{2k-1}$. Show that corresponding matrix $U^{(0)}$ has a block-diagonal structure, more precisely

$$U^{(0)}_{j,k} = \int_{\mathbb{C}^2} \mathcal{F}(z_1, z_2)\, P_{k-1}(z_1)\, P_{j-1}(z_2)\, d^2z_1 d^2z_2$$

$$= \begin{cases} 2\sqrt{2\pi}\ 2p!, & k = 2p+1, \quad j = 2p+2, \\ -2\sqrt{2\pi}\ 2p!, & k = 2p+2, \quad j = 2p+1, \\ 0, & \text{otherwise.} \end{cases}$$

Hints

- Split the integral into two parts: an integration along the real line (corresponding to $\delta(y_1)\,\delta(y_2)$ part of $\mathcal{F}$) and in the complex plane (corresponding to $\delta(y_1 + y_2)\,\delta(x_1 - x_2)$ part of $\mathcal{F}$).
- Show that both integrals vanish if j and k are of the same parity.
- For $j = 2q$ and $k = 2p - 1$ calculate the "real line" integral.
- For $j = 2q$ and $k = 2p - 1$ perform integration by parts in the "complex plane" integral to find a connection between

$$\left(U^{(0)}_{2q,2p-1}\right)_{\text{complex}} \text{ and } \left(U^{(0)}_{2p,2q-1}\right)_{\text{complex}}.$$

- For $j = 2q$ and $k = 2p - 1$ perform a <u>different</u> integration by parts in the "complex plane" integral to find <u>another</u> connection between values of
$$\left(u^{(0)}_{2q,2p-1}\right)_{\text{complex}} \text{ and } \left(u^{(0)}_{2p,2q-1}\right)_{\text{complex}}.$$
Combine the last three results to obtain the statement.

Lecture 2

Exercise 6.7 (Incomplete Γ-function). The incomplete Γ-function $\Gamma(M, a)$ is defined by

$$\Gamma(M, a) = \int_a^\infty e^{-t} t^{M-1} dt.$$

Using the Laplace method for asymptotic analysis of integrals depending on a (big) parameter show that

(1)
$$\lim_{M\to\infty} \frac{\Gamma(M-1, Ma)}{\Gamma(M-1)} = \begin{cases} 1, & 0 \leqslant a < 1, \\ \frac{1}{2}, & a = 1 \\ 0, & \text{otherwise.} \end{cases}$$

(2) For large M and $a > 1$ one has asymptotically
$$\frac{\Gamma(M-1, Ma)}{\Gamma(M-1)} \approx \frac{1}{\sqrt{2\pi M}} \frac{1}{a(a-1)} e^{-M(a-1-\log a)}.$$

Hints: Rewrite scaled and normalized incomplete Γ -function as

$$\frac{\Gamma(M-1, Ma)}{\Gamma(M-1)} = \frac{M^{M-1}}{\Gamma(M-1)} \int_a^\infty e^{-2s} e^{-(M-2)(s-\log s)} ds,$$

and find such a point $s_0 \in [a, \infty)$ that a small neighbourhood of the point produces the dominant contribution to the integral.

Exercise 6.8 (Factorization of correlation functions). Let $R_{N,n}(z_1, z_2, \ldots, z_n)$ be the nth correlation function for GinOE(N) ensemble rescaled by $\sqrt{N}$ (meaning that eigenvalues are supported inside the unit disk). Show that if in the limit $N \to \infty$ one has $z_i \neq z_j$, for $i \neq j$ then

$$R_{N,n}(z_1, z_2, \ldots, z_n) \approx \prod_{j=1}^n R_{N,1}(z_j)$$

Hint: Show that the corresponding matrices $Q_{k,l}$ for $k \neq l$ decay fast as $N \to \infty$.

Lecture 3

Exercise 6.9 (The simplest case of the Kac-Rice formula). Let $f : \mathbb{R} \to \mathbb{R}$ be a deterministic, smooth enough function. Assume that it has finitely many distinct simple zeros $(x_1, x_2, \ldots, x_n) \subset \mathbb{R}$.

(1) Let $\delta(x)$ be a standard Dirac δ- function. Show that

$$\delta(f(x)) = \sum_{k=1}^{n} \frac{\delta(x - x_k)}{|f'(x_k)|}.$$

(2) Prove that (Kac-Rice)

$$\#\{\text{Zeros of } f(x) \text{ in } (a,b)\} = \int_a^b \delta(f(x)) \left|f'(x)\right| dx.$$

Hints:

(1) Consider δ -function as a linear functional on a space of smooth functions given by

$$\delta[h] = \int_{\mathbb{R}} \delta(s) h(s) ds$$

and perform change of variables.

(2) Use the result of (1).

Exercise 6.10 (Properties of random Gaussian vector fields **f**). Let a vector field $\mathbf{f} : \mathbb{R}^N \to \mathbb{R}^N$ be given by

$$\mathbf{f}(\mathbf{x}) = -\nabla V(\mathbf{x}) + \frac{1}{\sqrt{N}} \nabla \times \mathbf{A}(\mathbf{x}),$$

for a smooth function $V : \mathbb{R}^N \to \mathbb{R}$ and a skew-symmetric matrix $\mathbf{A} : \mathbb{R}^N \to \mathbb{R}^{N^2}$. Assume that V and $\mathbf{A}$ are statistically independent, zero mean Gaussian random fields with covariance structure given by

$$\langle V(\mathbf{x}) V(\mathbf{y}) \rangle = v^2 \Gamma_V \left(|\mathbf{x} - \mathbf{y}|^2 \right),$$

$$\left\langle A_{i,j}(\mathbf{x}) A_{n,m}(\mathbf{y}) \right\rangle = a^2 \Gamma_A \left(|\mathbf{x} - \mathbf{y}|^2 \right) \left(\delta_{i,n}\delta_{j,m} - \delta_{i,m}\delta_{j,n} \right).$$

Show that

(1) $\left\langle f_k(\mathbf{x}) \partial f_i / \partial x_j(\mathbf{x}) \right\rangle = 0$, for any $i, j, k = \overline{1, N}$;

(2) $\left\langle f_j(\mathbf{x}) f_k(\mathbf{x}) \right\rangle = \delta_{j,k}\sigma^2$ with $\sigma^2 = 2v^2 \left|\Gamma_V'(0)\right| + 2a^2 \left|\Gamma_A'(0)\right| \frac{N-1}{N}$;

Remark. Use without the proof the following fact on Gaussian random fields: Let $\mathbf{G}(\mathbf{x}) : \mathbb{R}^N \to \mathbb{R}^d$ be an isotropic, mean zero Gaussian random field with covariance function ρ, meaning that

$$\langle \mathbf{G}(\mathbf{x}) \mathbf{G}(\mathbf{y}) \rangle = \rho \left(|\mathbf{x} - \mathbf{y}|^2 \right).$$

Then, $\rho'(0) < 0$.

Solution. Let us first write the definition of the vector field f component-wise as

$$f_j(\mathbf{x}) = -\frac{\partial V(\mathbf{x})}{\partial x_j} + \frac{1}{\sqrt{N}} \sum_{k=1}^{N} \frac{\partial A_{j,k}(\mathbf{x})}{\partial x_k}.$$

(1) Let

$$B_{j,k,l} = \left\langle f_j(\mathbf{x}) \frac{df_k(\mathbf{x})}{dx_l} \right\rangle.$$

Using explicit expressions for f_k and its derivatives we obtain for $B_{j,k,l}$

$$\left\langle \left(-\frac{\partial V(\mathbf{x})}{\partial x_j} + \frac{1}{\sqrt{N}} \sum_{p=1}^{N} \frac{\partial A_{j,p}(\mathbf{x})}{\partial x_p} \right) \cdot \left(-\frac{\partial^2 V(\mathbf{x})}{\partial x_k \partial x_l} + \frac{1}{\sqrt{N}} \sum_{q=1}^{N} \frac{\partial^2 A_{k,q}(\mathbf{x})}{\partial x_q \partial x_l} \right) \right\rangle$$

$$= \left\langle \frac{\partial V(\mathbf{x})}{\partial x_j} \frac{\partial^2 V(\mathbf{x})}{\partial x_k \partial x_l} \right\rangle + \frac{1}{N} \left\langle \sum_{p,q=1}^{N} \frac{\partial A_{j,p}(\mathbf{x})}{\partial x_p} \frac{\partial^2 A_{k,q}(\mathbf{x})}{\partial x_q \partial x_l} \right\rangle,$$

where we used statistical independence of V and A and the fact they are centred Gaussian variables. Each of the averages can be obtained by differentiating corresponding covariance functions of Gaussian fields, for example:

$$\left\langle \frac{\partial V(\mathbf{x})}{\partial x_j} \frac{\partial^2 V(\mathbf{x})}{\partial x_k \partial x_l} \right\rangle = \frac{\partial}{\partial x_j} \frac{\partial}{\partial y_k} \frac{\partial}{\partial y_l} \left(\nu^2 \Gamma_V \left(|\mathbf{x} - \mathbf{y}|^2 \right) \right) \Big|_{\mathbf{y}=\mathbf{x}}.$$

After one differentiates the function $\Gamma_V\left(|\mathbf{x}-\mathbf{y}|^2\right)$ with respect to three variables (possibly not distinct) one gets a sum of several terms, each of them containing at least one factor of the form $(x_p - y_p)$ for some p. As the result, after substituting $\mathbf{y} = \mathbf{x}$ all of them must vanish, yielding

$$B_{j,k,l} = 0, \text{for} \quad \forall j,k,l = \overline{1,N}.$$

(2) Analogously to the above.

Exercise 6.11 (Hermite and rescaled Hermite polynomials)**.** Let $\{H_k(\cdot)\}_{k=0}^{\infty}$ be the sequence of Hermite polynomials defined by

$$H_k(x) = (-1)^k e^{x^2} \frac{d^k}{dx^k} e^{-x^2}.$$

Show that

(1) $\{H_k\}_{k=0}^{\infty}$ form an orthogonal basis in the space of polynomials with inner product

$$(f,g) = \int_{\mathbb{R}} f(x)\, g(x)\, e^{-x^2} dx,$$

and calculate the norms $\|H_k\|^2$.

(2)

$$H_k(x) = \frac{2^k}{\sqrt{\pi}} \int_{\mathbb{R}} e^{-t^2} (x+it)^k \, dt.$$

(3) Let $\left\{h_k^{(\tau)}\right\}_{k=0}^{\infty}$ be a sequence of rescaled Hermite polynomials defined by

$$h_k^{(\tau)}(z) = \frac{1}{\sqrt{\pi}} \int_{\mathbb{R}} e^{-t^2} \left(z + it\sqrt{2\tau} \right)^k dt.$$

Show that polynomials $C_k(z)$ given by

$$C_k(z) = \begin{cases} h_k, & k \text{ is even} \\ h_k - (k-1)\, h_{k-2}, & k \text{ is odd}. \end{cases}$$

are skew-orthogonal with respect to the skew-product

$$(f, g) = \int_{\mathbb{C}^2} \mathcal{F}_{(\tau)}(z_1, z_2)\, f(z_1)\, g(z_2)\, d^2 z_1 d^2 z_2.$$

Hints:

(1) Write corresponding scalar product for polynomials H_k and H_m as

$$(H_k, H_m) = (-1)^m \int_{\mathbb{R}} H_k(x) \frac{d^m}{dx^m} e^{-x^2} dx, \quad m \geqslant k$$

and use integration by parts m times to move all derivatives to H_k.

(2) Show that the r.h.s satisfies the recursive relation

$$H_{k+1}(x) = 2x H_k(x) - H_k'(x),$$

and that the same is true for originally defined Hermite polynomials.

(3) This part can be done similarly to the Exercise 6.6. Calculate the inner product $I_{j,k} = \left(h_k^{(\tau)}, h_k^{(\tau)}\right)$ in several steps:

- Show that both integrals vanish if j and k are of the same parity using symmetry properties of Hermite polynomials.
- Take $j = 2p+1$ and $k = 2q$ and split the integral into two parts: an integration along the real line $R_{p,q}$ and in the complex plane $C_{p,q}$. Use integration by parts with respect to x and y with the identities

$$\frac{d\, h_k^{(\tau)}(z)}{dz} = k h_{k-1}^{(\tau)}(z), \quad z h_k^{(\tau)}(z) = h_{k+1}^{(\tau)}(z) + k h_{k-1}^{(\tau)}(z),$$

to show

$$R_{p,q} = \frac{1}{(1+\tau)(2p+2)} R_{p+1,q} + \frac{\tau}{1+\tau} R_{p,q} + \frac{1}{p+1} \int_{\mathbb{R}} e^{-\frac{x^2}{1+\tau}} h_{2p+2}^{(\tau)}(x)\, h_{2q}^{(\tau)}(x)\, dx,$$

$$R_{p,q} = \frac{1}{(1+\tau)(2q+1)} R_{p,q+1} + \frac{\tau}{1+\tau} R_{p,q} - \frac{2}{2q+1} \int_{\mathbb{R}} e^{-\frac{x^2}{1+\tau}} h_{2p+1}^{(\tau)}(x)\, h_{2q+1}^{(\tau)}(x)\, dx.$$

- Consider "modified complex" integral given by

$$\widehat{C}_{p,q} = 2i \int_{\mathbb{R}} dx \int_0^\infty dy\, e^{\frac{y^2 - x^2}{1+\tau}} \operatorname{erfc}\left(y \sqrt{\frac{2}{1-\tau^2}}\right)$$
$$\times \left[h_{2p+2}^{(\tau)}(x+iy)\, h_{2q-1}^{(\tau)}(x-iy) - h_{2q-1}^{(\tau)}(x+iy)\, h_{2p+2}^{(\tau)}(x-iy)\right],$$

that differs from $C_{p,q}$ by changing $2p+1 \to 2p+2$ and $2q \to 2q-1$. Use integration by parts with respect to x and y to show that

$$C_{p+1,q} = (2p+2)\,C_{p,q} + 2\,(1+\tau)\int_{\mathbb{R}} e^{-\frac{x^2}{1+\tau}} h^{(\tau)}_{2p+1}(x)\, h^{(\tau)}_{2q+1}(x)\,dx - 2\sqrt{2\pi}\,(2p+2)!\,(1+\tau)\,\delta_{p+1,q},$$

$$C_{p,q+1} = (2q+1)\,C_{p,q} - 2\,(1+\tau)\int_{\mathbb{R}} e^{-\frac{x^2}{1+\tau}} h^{(\tau)}_{2p+1}(x)\, h^{(\tau)}_{2q+1}(x)\,dx + 2\sqrt{2\pi}\,(2q+1)!\,(1+\tau)\,\delta_{p,q}.$$

- Conclude that the total integral $I_{p,q} = R_{p,q} + C_{p,q}$ solves the system

$$I_{p+1,q} = (2p+2)\,I_{p,q} - 2\sqrt{2\pi}\,(2p+2)!\,(1+\tau)\,\delta_{p+1,q},$$
$$I_{p,q+1} = (2q+1)\,I_{p,q} + 2\sqrt{2\pi}\,(2q+1)!\,(1+\tau)\,\delta_{p,q},$$

with the initial condition $I_{0,0} = -2\sqrt{2\pi}\,(1+\tau)$. Show that the unique solution of the system is given by

$$I_{p,q} = \begin{cases} -2^{p+q+3/2} q!\,\Gamma\left(p+\frac{1}{2}\right)(1+\tau), & p \geqslant q \\ 0, & \text{otherwise.} \end{cases}$$

Exercise 6.12 (Equilibrium density of eigenvalues for real Gaussian Elliptic Ensemble). Consider real Gaussian Elliptic Ensemble of even size N and asymmetry parameter τ.

(1) Show that the eigenvalues form a Pfaffian Point Process in complex plane;

(2) Show that the corresponding kernel of the process is expressed in terms of

$$\mathcal{K}^{(\tau)}_N(z,\zeta) = \frac{1}{2\,(1+\tau)\,\sqrt{2\pi}} \sum_{k=0}^{N-2} \frac{\psi^{(\tau)}_{k+1}(\zeta)\,\psi^{(\tau)}_k(z) - \psi^{(\tau)}_{k+1}(z)\,\psi^{(\tau)}_k(\zeta)}{k!},$$

where $\psi^{(\tau)}_k(z) = e^{-\frac{z^2}{2(1+\tau)}} h^{(\tau)}_k(z)$ are rescaled Hermite functions;

(3) Using the explicit expression for the kernel $\mathcal{K}^{(\tau)}_N(z,\zeta)$ show that the density of eigenvalues distribution converges when $N \to \infty$ to 'equilibrium' measure given by

$$d\mu_{eq}(z) = \begin{cases} \dfrac{d^2 z}{\pi\,(1-\tau^2)}, & \dfrac{\mathrm{Re}^2 z}{(1+\tau)^2} + \dfrac{\mathrm{Im}^2 z}{(1-\tau)^2} \leqslant 1, \\ 0, & \text{otherwise,} \end{cases}$$

Hints:

(1) Repeat the steps given in the Exercise 6.4. Note that the difference between $\mathcal{P}^{(L,M)}$ and $\mathcal{P}^{(L,M)}_\tau$ comes only from the weight function and therefore the Pfaffian structure stemming from the Vandermonde determinant is preserved.

(2) Use the result analogous to the Exercise 6.5, but with the choice of skew orthogonal polynomials from the Exercise 6.11.

(3) Use an integral representation of the rescaled Hermite polynomials (cf. Exercise 6.11) to show

$$\rho_{N,\tau}(z = x + iy) = \int_{\mathbb{C}} \mathcal{K}_N^{(\tau)}(\bar{z}, \zeta)\, \mathcal{F}(\zeta, z)\, d\zeta$$

$$= \frac{i}{\pi\tau(1+\tau)} e^{\frac{y^2-x^2}{1+\tau}} \operatorname{erfc}\left(\sqrt{\frac{2}{1-\tau^2}}\, y\right) \int_{\mathbb{R}^2} dt_1 dt_2 e^{-\frac{t_1^2}{2\tau}} F_N(\bar{z} + it_1, z + it_2),$$

where

$$F_N(w, z) = \frac{z - w}{2\sqrt{2\pi}} \sum_{j=0}^{N-2} \frac{(wz)^j}{\Gamma(j+1)}.$$

Rescale by $z = \sqrt{N}\zeta$ to analyse asymptotically the corresponding integral.

Lecture 4

Exercise 6.13 (Logarithmic potential for the Real Ginibre and Elliptic Ensembles). Let x be real. Show that the 'logarithmic potential'

$$\Phi^{(\mu)}(x) = \int_{\mathbb{C}} \log|x - z|\, d\mu(z)$$

(1) for the equilibrium measure of GinOE ensemble

$$d\mu_{\mathrm{GinOE}}(z) = \begin{cases} \dfrac{d^2 z}{\pi}, & \mathrm{Re}^2 z + \mathrm{Im}^2 z \leqslant 1, \\ 0, & \text{otherwise}, \end{cases}$$

is given by

$$\Phi^{(\mu_{\mathrm{GinOE}})}(x) = \begin{cases} \log|x|, & |x| > 1 \\ \dfrac{|x|^2 - 1}{2}, & |x| \leqslant 1 \end{cases}$$

(2) for the equilibrium measure of real Elliptic ensemble

$$d\mu_{\mathrm{rEll}}(z) = \begin{cases} \dfrac{d^2 z}{\pi(1-\tau^2)}, & \dfrac{\mathrm{Re}^2 z}{(1+\tau)^2} + \dfrac{\mathrm{Im}^2 z}{(1-\tau)^2} \leqslant 1, \\ 0, & \text{otherwise}, \end{cases}$$

is given by

$$\Phi^{(\mu_{\mathrm{rEll}})}(x) = \begin{cases} \dfrac{1}{8\tau}\left(|x| - \sqrt{x^2 - 4\tau}\right)^2 + \log\dfrac{|x| + \sqrt{x^2 - 4\tau}}{2}, & |x| > 1 + \tau \\ \dfrac{|x|^2}{2(1+\tau)} - \dfrac{1}{2}, & |x| \leqslant 1 + \tau \end{cases}$$

Hints:

(1) Use mean value theorem for harmonic functions.

(2) First use the residue theorem to verify that

$$I_{A>B>0}(x) = \frac{1}{2\pi}\int_0^{2\pi} \frac{d\theta}{x - A\cos\theta - iB\sin\theta} = \frac{1}{\sqrt{x^2 - A^2 + B^2}}, \quad x > A$$

and zero otherwise. Then using the parametrization $\operatorname{Re} z = (1+\tau)r\cos\theta$, $\operatorname{Im} z = (1-\tau)r\sin\theta$ relate $\frac{d}{dx}\Phi^{(\mu)}$ to the above integral. Finally fix the logarithmic potential by evaluating separately its value at $x = 0$ using the fact that for $A > B$.

$$\frac{1}{\pi}\int_0^{\pi} \ln\left[A + B\cos\phi\right] d\phi = \ln\frac{A + \sqrt{A^2 - B^2}}{2}.\text{the}$$

Solution for (1):

We will use the following form of mean value theorem: Let u be a harmonic function in the open disk $|z - z_0| < r$, then

$$u(z_0) = \frac{1}{2\pi}\int_0^{2\pi} u\left(z_0 + re^{i\theta}\right) d\theta.$$

One can easily check, that $\log|z|$ is harmonic in $\mathbb{C}/\{0\}$. We rewrite the logarithmic potential using polar coordinates.

$$\Phi^{(\mu_{\text{GinOE}})}(x) = \frac{1}{\pi}\int_0^1 r\,dr \int_0^{2\pi} \log\left|x - re^{i\theta}\right| d\theta.$$

If $|x| > 1$, then $\log|z|$ is harmonic in $B_r(\xi)$ for all $r \in [0,1]$. And therefore we can apply mean value theorem to get

$$\Phi^{(\mu_{\text{GinOE}})}(x) = \frac{2\pi\log|x|}{\pi}\int_0^1 r\,dr = \log|x|.$$

If $|x| \leqslant 1$, then we can still use mean value theorem until r reaches $|x|$. Therefore we split the integral into two parts and obtain

$$\Phi^{(\mu_{\text{GinOE}})}(x) = \frac{1}{\pi}\left(\int_0^{|x|} + \int_{|x|}^1\right) r\,dr \int_0^{2\pi} \log\left|x - re^{i\theta}\right| d\theta.$$

We deal with the first integral in a similar as above.

$$\Phi_1^{(\mu_{\text{GinOE}})}(x) = \frac{2\pi\log|x|}{\pi}\int_0^{|x|} r\,dr = x^2\log|x|.$$

In the second one, while integrating with respect to θ, we interchange the roles of ξ and r to be able to apply mean value theorem.

$$\Phi_2^{(\mu_{\text{GinOE}})}(x) = \frac{1}{\pi}\int_{|x|}^1 r\,dr \int_0^{2\pi} \log\left|xe^{-i\theta} - r\right| d\theta = 2\int_{|x|}^1 r\log r\,dr = \frac{|x|^2 - 1}{2} - x^2\log|x|.$$

Adding last two expressions we obtain the result.

References

[1] R. Abraham, J. E. Marsden, and T. Ratiu, *Manifolds, tensor analysis, and applications*, 2nd ed., Applied Mathematical Sciences, vol. 75, Springer-Verlag, New York, 1988. MR960687 ↑176

[2] G. Akemann and M. J. Phillips, *The interpolating Airy kernels for the* $\beta = 1$ *and* $\beta = 4$ *elliptic Ginibre ensembles*, J. Stat. Phys. **155** (2014), no. 3, 421–465. MR3192169 ↑180

[3] Gernot Akemann and Eugene Kanzieper, *Integrable structure of Ginibre's ensemble of real random matrices and a Pfaffian integration theorem*, J. Stat. Phys. **129** (2007), no. 5-6, 1159–1231. MR2363393 ↑167

[4] S. Allesina and Si Tang, *Stability criteria for complex ecosystems*, Nature **483** (2012), 205 –208. ↑162, 180

[5] S. Allesina and Si Tang, *The stability complexity relationship at age 40: a random matrix perspective*, Popul. Ecol. **57** (2015), 63–75. ↑162

[6] L. Angelani, G. Parisi, G. Ruocco, and G. Viliani, *Potential Energy Landscape and long-time dynamics in a simple model glass*, Phys. Rev. E **61** (2000), 1681–1691. ↑175

[7] Antonio Auffinger, Gérard Ben Arous, and Jiří Černý, *Random matrices and complexity of spin glasses*, Comm. Pure Appl. Math. **66** (2013), no. 2, 165–201. MR2999295 ↑175, 189, 190

[8] G. Ben Arous, A. Dembo, and A. Guionnet, *Aging of spherical spin glasses*, Probab. Theory Related Fields **120** (2001), no. 1, 1–67. MR1856194 ↑173

[9] G. Ben Arous, Y. V. Fyodorov, and B. A. Khoruzhenko, *Counting stable equilibria of large complex systems* (201p). Unpublished. ↑159, 163, 182, 187, 190

[10] Gérard Ben Arous and Ofer Zeitouni, *Large deviations from the circular law*, ESAIM Probab. Statist. **2** (1998), 123–134. MR1660943 ↑170, 172, 182

[11] A. Borodin and C. D. Sinclair, *The Ginibre Ensemble of Real Random Matrices and its Scaling Limits*, Commun. Math. Phys. **291** ((2009)), no. 1, 177–224. ↑167, 168, 200

[12] A. J. Bray and D. Dean, *The statistics of critical points of Gaussian fields on large-dimensional spaces*, Phys. Rev. Lett. **98** (2007), 150201. ↑189, 190

[13] de Bruijn N. G., *On some multiple integrals involving determinants.*, J. Indian Math.Soc. New Series **19** (1955), 133–151. ↑167, 168

[14] L. F. Cugliandolo and J. Kurchan, *Analytical Solution of the off-equilibrium dynamics of a long-range spin-glass model*, Phys. Rev. Lett. **71** (1993), 173-176. ↑192

[15] L. F. Cugliandolo, J. Kurchan, P. Le Doussal, and L. Peliti, *Glassy behaviour in disordered systems with nonrelaxational dynamics*, Phys. Rev. Lett. **78** (1997), 350–353. ↑175, 191, 192

[16] H. de Jong, *Modeling and Simulation of Genetic Regulatory Systems: A Literature Review*, J Comp. Biol. **9** (2002), 67–107. ↑160

[17] David S. Dean and Satya N. Majumdar, *Large deviations of extreme eigenvalues of random matrices*, Phys. Rev. Lett. **97** (2006), no. 16, 160201, 4. MR2274338 ↑188

[18] Alan Edelman, *The probability that a random real Gaussian matrix has* k *real eigenvalues, related distributions, and the circular law*, J. Multivariate Anal. **60** (1997), no. 2, 203–232. MR1437734 ↑164, 165, 166

[19] Alan Edelman, Eric Kostlan, and Michael Shub, *How many eigenvalues of a random matrix are real?*, J. Amer. Math. Soc. **7** (1994), no. 1, 247–267. MR1231689 ↑164, 165

[20] J. Doyne Farmer and Spyros Skouras, *An ecological perspective on the future of computer trading*, Quant. Finance **13** (2013), no. 3, 325–346. MR3038193 ↑162

[21] Peter J. Forrester and Taro Nagao, *Skew orthogonal polynomials and the partly symmetric real Ginibre ensemble*, J. Phys. A **41** (2008), no. 37, 375003, 19. MR2430570 ↑167, 180, 181

[22] P. J. Forrester and T. Nagao, *Eigenvalue statistics of the real Ginibre ensemble*, Phys. Rev. Lett. **99** (2007), no. 5, 050603. ↑167

[23] Peter J. Forrester, *Spectral density asymptotics for Gaussian and Laguerre* β*-ensembles in the exponentially small region*, J. Phys. A **45** (2012), no. 7, 075206, 17. MR2881075 ↑173

[24] Y. Fried, D. A. Kessler, and N. M. Shnerb, *Communities as cliques*, Scientific Reports **6** (2016), 35648. ↑162

[25] Y. V. Fyodorov, *Topology trivialization transition in random non-gradient autonomous ODEs on a sphere*, J. Stat. Mech. Theory Exp. **12** (2016), 124003, 21. MR3598251 ↑192

[26] Yan V. Fyodorov, *On statistics of bi-orthogonal eigenvectors in real and complex Ginibre ensembles: combining partial Schur decomposition with supersymmetry*, Comm. Math. Phys. **363** (2018), no. 2, 579–603, available at `1710.04699`. MR3851824 ↑165, 195

[27] Y. V. Fyodorov and C. Nadal, *Critical Behavior of the Number of Minima of a Random Landscape at the Glass Transition Point and the Tracy-Widom Distribution*, Phys. Rev. Lett. **109** (2012), 167203. ↑175, 189, 190

[28] Yan V. Fyodorov and Eugene Strahov, *Characteristic polynomials of random Hermitian matrices and Duistermaat-Heckman localisation on non-compact Kähler manifolds*, Nuclear Phys. B **630** (2002), no. 3, 453–491. MR1902873 ↑194

[29] Yan V. Fyodorov, *Negative moments of characteristic polynomials of random matrices: Ingham-Siegel integral as an alternative to Hubbard-Stratonovich transformation*, Nuclear Phys. B **621** (2002), no. 3, 643–674. MR1877952 ↑194

[30] Yan V. Fyodorov, *Complexity of random energy landscapes, glass transition, and absolute value of the spectral determinant of random matrices*, Phys. Rev. Lett. **92** (2004), no. 24, 240601, 4. MR2115095 ↑175

[31] Yan V. Fyodorov and Ian Williams, *Replica symmetry breaking condition exposed by random matrix calculation of landscape complexity*, J. Stat. Phys. **129** (2007), no. 5-6, 1081–1116. MR2363390 ↑175, 189, 190

[32] Yan V. Fyodorov and Boris A. Khoruzhenko, *Nonlinear analogue of the May-Wigner instability transition*, Proc. Natl. Acad. Sci. USA **113** (2016), no. 25, 6827–6832. MR3521630 ↑159, 162, 163, 187, 189, 190

[33] Tobias Galla and J. Doyne Farmer, *Complex dynamics in learning complicated games*, Proc. Natl. Acad. Sci. USA **110** (2013), no. 4, 1232–1236. MR3037098 ↑160

[34] X. Garcia, *On the number of equilibria with a given number of unstable directions* (2017), available at `1709.04021`. ↑192

[35] M. R. Gardner and W. R. Ashby, *Connectance of Large Dynamic (Cybernetic) Systems: Critical Values for Stability*, Nature **228** (1970), 784. ↑159

[36] Jean Ginibre, *Statistical ensembles of complex, quaternion, and real matrices*, J. Mathematical Phys. **6** (1965), 440–449. MR0173726 ↑162, 167

[37] Jacek Grela, *What drives transient behaviour in complex systems?*, Phys. Rev. E **96** (2017), no. 2, 022316, available at `1705.08758`. ↑161

[38] A.G. Haldane and R. M. May, *Systemic risk in banking ecosystems*, Nature **469** (2011), 351–355. ↑162

[39] Philip Hartman, *Ordinary differential equations*, Classics in Applied Mathematics, vol. 38, Society for Industrial and Applied Mathematics (SIAM), Philadelphia, PA, 2002. Corrected reprint of the second (1982) edition [Birkhäuser, Boston, MA; MR0658490 (83e:34002)]; With a foreword by Peter Bates. MR1929104 ↑160

[40] J. R. Ipsen and H. Schomerus, *Isotropic Brownian motions over complex fields as a solvable model for May-Wigner stability analysis*, J. Phys. A **49** (2016), no. 38, 385201, 14. MR3546769 ↑192

[41] J. Kadmon and H. Sompolinsky, *Transition to Chaos in Random Neuronal Networks*, Phys. Rev. X **5** (2015), no. 4, 041030. ↑191

[42] Boris A. Khoruzhenko and Hans-Jürgen Sommers, *Non-Hermitian ensembles*, The Oxford handbook of random matrix theory, Oxford Univ. Press, Oxford, 2011, pp. 376–397. MR2932638 ↑163

[43] Nils Lehmann and Hans-Jürgen Sommers, *Eigenvalue statistics of random real matrices*, Phys. Rev. Lett. **67** (1991), no. 8, 941–944. MR1121461 ↑166

[44] Z. Li, S. Bianco, Z. Zhang, and C. Tang, *Generic Properties of Random Gene Regulatory Networks*, Quant. Biol. **1** (2013), no. 4, 253–260. ↑160

[45] Satya N. Majumdar and Grégory Schehr, *Top eigenvalue of a random matrix: large deviations and third order phase transition*, J. Stat. Mech. Theory Exp. **1** (2014), P01012, 31. MR3172178 ↑190

[46] R. M. May, *Will a Large Complex System be Stable?*, Nature **238** (1972), 413–414. ↑159

[47] R. M. May, *Stability and complexity in model ecosystems*, Princeton University Press, Princeton, NJ, 1972. ↑160

[48] I. J. Schoenberg, *Metric spaces and completely monotone functions*, Ann. of Math. (2) **39** (1938), no. 4, 811–841. MR1503439 ↑177

[49] Christopher D. Sinclair, *Averages over Ginibre's ensemble of random real matrices*, Int. Math. Res. Not. IMRN **5** (2007), Art. ID rnm015, 15. MR2341601 ↑167

[50] Hans-Jürgen Sommers and Waldemar Wieczorek, *General eigenvalue correlations for the real Ginibre ensemble*, J. Phys. A **41** (2008), no. 40, 405003, 24. MR2439268 ↑167, 168, 200

[51] H. Sompolinsky, A. Crisanti, and H.-J. Sommers, *Chaos in random neural networks*, Phys. Rev. Lett. **61** (1988), no. 3, 259–262. MR949871 ↑160, 191

[52] Peter F. Stadler, Walter Fontana, and John H. Miller, *Random catalytic reaction networks*, Phys. D **63** (1993), no. 3-4, 378–392. MR1210013 ↑160

[53] G. Wainrib and J. Touboul, *Topological and Dynamical Complexity of Random Neural Networks*, Phys. Rev. Lett. **110** (2013), 118101. ↑191

[54] Gernot Akemann, Jinho Baik, and Philippe Di Francesco (eds.), *The Oxford handbook of random matrix theory*, Oxford University Press, Oxford, 2011. MR2920518 ↑161

Department of Mathematics, King's College London, London WC2R 2LS, United Kingdom
Email address: yan.fyodorov@kcl.ac.uk

IAS/Park City Mathematics Series
Volume 26, Pages 213–250
https://doi.org/10.1090/pcms/026/00846

A Short Introduction to Operator Limits of Random Matrices

Diane Holcomb and Bálint Virág

Abstract. These are notes to a four-lecture minicourse given at the 2017 PCMI Summer Session on Random Matrices. It is a quick introduction to the theory of large random matrices through limits that preserve their operator structure, rather than just their eigenvalues. This structure takes the role of exact formulas, and allows for results in the context of general β-ensembles. Along the way, we cover a non-computational proof of the Wiegner semicircle law, quick proofs for the Füredi-Komlós result on the top eigenvalue, the BBP phase transition, as well as local convergence of the soft-edge process and tail asymptotics for the TW_β distribution.

1. The Gaussian Ensembles

1.1. The Gaussian Orthogonal and Unitary Ensembles. One of the earliest appearances of random matrices in mathematics was due to Eugene Wigner in the 1950's. Let G be an $n \times n$ matrix with independent standard normal entries. Consider the matrix

$$M_n = \frac{G + G^t}{\sqrt{2}}.$$

This distribution on symmetric matrices is called the Gaussian Orthogonal Ensemble, because it is invariant under orthogonal conjugation. For any orthogonal matrix OM_nO^{-1} has the same distribution as M_n. To check this, note that OG has the same distribution as G by the rotation invariance of the Gaussian column vectors, and the same is true for OGO^{-1} by the rotation invariance of the row vectors. To finish note that orthogonal conjugation commutes with symmetrization.

Starting instead with a matrix with independent standard complex Gaussian entries we would get the Gaussian Unitary ensemble. To see how the eigenvalues behave, we recall the following classical theorem.

Theorem 1.1.1 (see e.g. [2]). *Suppose M_n has GOE or GUE distribution then M_n has eigenvalue density*

$$f(\lambda_1, \ldots, \lambda_n) = \frac{1}{Z_{n,\beta}} \prod_{k=1}^{n} e^{-\frac{\beta}{4}\lambda_k^2} \prod_{i<j} |\lambda_i - \lambda_j|^\beta \tag{1.1.2}$$

2010 *Mathematics Subject Classification.* Primary 60B20; Secondary15B52.
Key words and phrases. Park City Mathematics Institute.

with $\beta = 1$ *for the GOE and* $\beta = 2$ *for the GUE.*

For convenience set $\Lambda = \Lambda_n = \{\lambda_i\}_{i=1}^n$ the set of eigenvalues of the GOE or GUE. This notation will be used later to denote the eigenvalues or points in whatever random matrix model is being discussed at the time.

From the density in Theorem 1.1.1 we can see that this is a model for n particles that would like to be Gaussian, but the Vandermonde term $\prod_{i<j} |\lambda_i - \lambda_j|^\beta$ pushes them apart. Note that $\mathrm{Tr}\ M_n^2/n^2 \to 1$ in probability (the sum of squares of Gaussians), so the empirical quadratic mean of the eigenvalues is asymptotically $\sqrt{n}$, rather than order 1. The interaction term has a very strong effect.

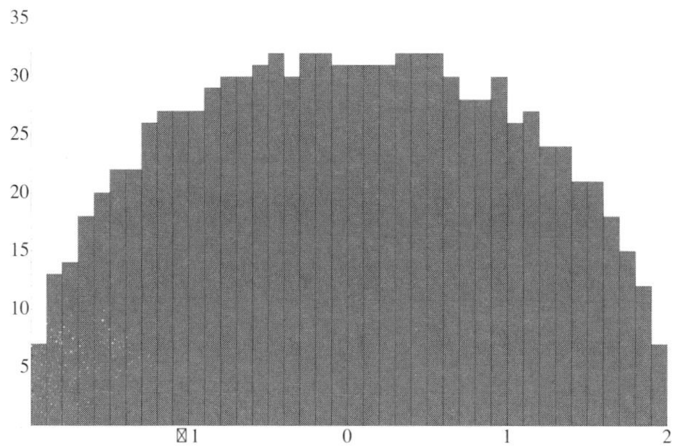

FIGURE 1.1.3. Rescaled eigenvalues of a 1000×1000 GOE matrix

We begin by introducing a tridiagonal matrix model that has the same joint density as the Gaussian ensembles. This model and Jacobi matrices more generally share many characteristics with differential operator theory including Sturm-Liouville theory. In this section we derive the tridiagonal model for the Gaussian unitary ensemble and then give tridiagonal models for a wider class of models called β-ensembles. In Section 2 we begin by introducing two different notions of graph convergence which are then used to prove the Wigner semicircle law. In Section 3 we study the behavior of the largest eigenvalue of the GOE under rank one perturbations. The behavior depends on the strength of the perturbation and is called the Baik-Ben Arous-Péché transition. In Section 4 we introduce the notion of local convergence and give a proof of local convergence at the soft edge. We also study the tail behavior of the smallest eigenvalue of the Stochastic Airy Operator which is the $\beta > 0$ generalization of the Tracy-Widom 1,2, and 4 laws. Section 5 gives a partial overview of results that are proved using operator methods including other local limit theorems.

1.2. Tridiagonalization and spectral measure. The spectral measure of a matrix at a vector (which we will take to be the first coordinate vector) is a measure supported on the eigenvalues that reflects the local structure of the matrix there. This is more easily seen in the case where the matrix is the adjacency matrix of a (possibly weighted) graph. In this case the spectral measure at coordinate vector j gives information about the graph in the neighborhood of vertex j.

Definition 1.2.1. The spectral measure σ_A of a symmetric matrix A at the first coordinate vector is the measure for which

$$\int x^k d\sigma_A = A^k_{11}.$$

From this definition it is not clear that the spectral measure exists. The advantage is that this definition can be extended to the operator setting in many cases. For bounded operators (e.g. matrices) uniqueness is clear since probability measures with compact support are determined by their moments (see e.g. Kallenberg [21]). For finite matrices, the following alternative explicit definition proves existence.

Exercise 1.2.2. Check that if $\lambda_1, \ldots, \lambda_n$ are the eigenvalues of A then

$$\sigma_A = \sum_i \delta_{\lambda_i} \varphi_i(1)^2,$$

where φ_i is the ith normalized eigenvector of A.

The spectral measure is a complete invariant for a certain set of symmetries. For this, first recall something more familiar.

Two symmetric matrices are equivalent if they have the same eigenvalues with multiplicity. This equivalence is well understood: two matrices are equivalent if and only if they are conjugates by an orthogonal matrix. In group theory language, the equivalence classes are the orbits of the conjugation action of the orthogonal group. There is a canonical representative in each class, a diagonal matrix with non-increasing diagonals, and the set of eigenvalues is a complete invariant.

Is there a similar characterization for matrices with the same spectral measure? The answer is yes, for a generic class of matrices.

Definition 1.2.3. A vector v is cyclic for an $n \times n$ matrix A if $v, Av, \ldots, A^{n-1}v$ is a basis for the vector space $\mathbb{R}^n$.

Theorem 1.2.4. *Let A and B be two Hermitian matrices for which the first coordinate vector is cyclic. Then $\sigma_A = \sigma_B$ if an only if $O^{-1}AO = B$ where O is orthogonal matrix fixing the first coordinate vector.*

Let's find a nice set of representatives for the class of matrices for which the first coordinate vector is cyclic.

Definition 1.2.5. A Jacobi matrix is a real symmetric tridiagonal matrix with positive off-diagonals.

Theorem 1.2.6. *For every Hermitian matrix A there exists a unique Jacobi matrix J such that $\sigma_J = \sigma_A$.*

Proof of Existence. It is possible to conjugate a symmetric matrix to a Jacobi matrix by hand. Write our matrix in a block form,

$$A = \left[\begin{array}{c|c} a & b^t \\ \hline b & C \end{array}\right].$$

Now let O be an $(n-1) \times (n-1)$ orthogonal matrix, and let

$$Q = \left[\begin{array}{c|c} 1 & 0 \\ \hline 0 & O \end{array}\right].$$

Then Q is orthogonal and

$$QAQ^t = \left[\begin{array}{c|c} a & (Ob)^t \\ \hline Ob & OCO^t \end{array}\right].$$

Now we can choose the orthogonal matrix O so that Ob is in the direction of the first coordinate vector, namely $Ob = |b|\mathbf{e}_1$.

An explicit option for O is the following Householder reflection:

$$Ov = v - 2\frac{\langle v, w\rangle}{\langle w, w\rangle} w \qquad \text{where} \qquad w = b - |b|\mathbf{e}_1$$

Check that $OO^t = I$, $Ob = |b|\mathbf{e}_1$.

Therefore

$$QAQ^t = \left[\begin{array}{c|cccc} a & |b| & 0 & \dots & 0 \\ \hline |b| & & & & \\ 0 & & & & \\ \vdots & & & OCO^t & \\ 0 & & & & \end{array}\right].$$

Repeat the previous step, but this time choosing the first two rows and columns to be 0 except having 1's in the diagonal entries, and than again until the matrix becomes tridiagonal. □

There are a lot of choices that can be made for the orthogonal matrices during the tridiagonalization. However, these choices do not affect the final result. J is unique, as shown in the following exercise.

Exercise 1.2.7. Show that two Jacobi matrices with the same spectral measure are equal. (Hint: express the moments $J^k_{1,1}$ of the spectral measure as sums over products of matrix entries.)

The procedure presented above may have a familiar feeling. It turns out that Gram-Schmidt is lurking in the background:

Exercise 1.2.8. Suppose that the first coordinate vector $\mathbf{e}_1$ is cyclic. Apply Gram-Schmidt to the vectors $(\mathbf{e}_1, A\mathbf{e}_1, ..., A^{n-1}\mathbf{e}_1)$ to get a new orthonormal basis. Show that A written in this basis will be a Jacobi matrix.

1.2.1. Tridiagonalization and the GOE. The tridiagonalization procedure can be applied to the GOE matrix, as pioneered by Trotter [33]. The invariance of the distribution under independent orthogonal transformations yields a tractable Jacobi matrix.

Proposition 1.2.9 (Trotter [33]). *Let* A *be* GOE_n. *There exists a random orthogonal matrix fixing the first coordinate vector* e *so that*

$$OAO^t = \begin{bmatrix} a_1 & b_1 & 0 & \dots & & 0 \\ b_1 & a_2 & \ddots & & & \\ 0 & \ddots & \ddots & \ddots & & \\ \vdots & & \ddots & a_{n-1} & b_{n-1} \\ 0 & & & b_{n-1} & a_n \end{bmatrix}$$

with $a_i \sim \mathcal{N}(0,2)$ *and* $b_i \sim \chi_{n-i}$ *and independent. In particular,* OAO^t *has the same spectral measure as* A.

Let v be a vector of independent $\mathcal{N}(0,1)$ random variables of length k, then the length of the vector has χ distribution with parameter k, $\chi_k \stackrel{d}{=} \|v\|$. The density of a χ random variable for $k > 0$ is given by

$$f_{\chi_k}(x) = \frac{1}{2^{\frac{k}{2}-1}\Gamma(k/2)} x^{k-1} e^{-x^2/2},$$

where $\Gamma(x)$ is the Gamma function.

Proof. The tridiagonalization algorithm above can be applied to the random matrix. After the first step, OCO^t will be independent of a, b and have a GOE distribution. This is because GOE is invariant by conjugation with a fixed O, and O is only a function of b. The independence propagates throughout the algorithm meaning each rotation defined produces the relevant tridiagonal terms and an independent submatrix. □

Exercise 1.2.10. Let X be an $n \times m$ matrix with $X_{i,j} \sim \mathcal{N}(0,1)$ (not symmetric nor Hermitian). The distribution of this matrix is invariant under left and right multiplication by independent orthogonal matrices. Show that such a matrix X may be lower bidiagonalized such that the distribution of the singular values is the same for both matrices. Note that the singular values of a matrix are unchanged by multiplication by a orthogonal matrix.

(1) Start by coming up with a matrix that right multiplied with A gives you a matrix where the first row is 0 except the 11 entry.
(2) What can you say about the distribution of the rest of the matrix after this transformation to the first row?
(3) Next apply a left multiplication. Continue using right and left multiplication to finish the bidiagonalization.

Let's consider the spectral measure as a map $J \mapsto \sigma_J$ from Jacobi matrices of dimension n to probability measures on at most n points. We have seen that this map is one-to-one. First we see that in fact spectral measures in the image are supported on exactly n points.

Exercise 1.2.11. Show that a Jacobi matrix cannot have an eigenvector whose first coordinate is zero. Conclude that all eigenspaces are 1-dimensional.

Second, for the set of such probability measures, the map $J \mapsto \sigma_J$ is onto. This is left as an exercise.

Exercise 1.2.12. For every probability measure σ supported on exactly n points there exists an $n \times n$ symmetric matrix with spectral measure σ. The existence part of Theorem 1.2.6 then implies that there exists a Jacobi matrix with spectral measure σ.

1.3. β-ensembles. Let

$$A_n = \frac{1}{\sqrt{\beta}} \begin{bmatrix} a_1 & b_1 & 0 & \dots & 0 \\ b_1 & a_2 & \ddots & & \\ 0 & \ddots & \ddots & \ddots & \\ \vdots & & \ddots & a_{n-1} & b_{n-1} \\ 0 & & & b_{n-1} & a_n \end{bmatrix}, \tag{1.3.1}$$

that is a tridiagonal matrix with $a_1, a_2, ..., a_n \sim N(0,2)$ on the diagonal and $b_1, ..., b_{n-1}$ with $b_k \sim \chi_{\beta(n-k)}$ and everything independent. We will frequently use the notation $a_i = \mathcal{N}_i$ in order to refer more directly to the distribution of the random variable. Recall that if $z_1, z_2, ...$ are independent standard normal random variables, then $z_1^2 + \dots + z_k^2 \sim \chi_k^2$.

If $\beta = 1$ then A_n is similar to a GOE matrix (the joint density of the eigenvalues is the same). If $\beta = 2$ then A_n is similar to a GUE matrix.

Theorem 1.3.2 (Dumitriu, Edelman, [12]). *If $\beta > 0$ then the joint density of the eigenvalue of A_n is given by*

$$f(\lambda_1, ..., \lambda_n) = \frac{1}{Z_{n,\beta}} e^{-\frac{\beta}{4}\sum_{i=1}^n \lambda_i^2} \prod_{1 \leqslant i < j \leqslant n} |\lambda_i - \lambda_j|^\beta.$$

The spectral measure of a Jacobi matrix may be written as $\sigma_J = \sum_{j=1}^n \delta_{\lambda_j} q_j^2$, where $\lambda_1, ..., \lambda_n$ are the eigenvalues of the matrix and the $q_1, ..., q_n$ are the associated the spectral weights with $\sum q_i^2 = 1$. Recall that the map $J \mapsto \sigma_J$ is a bijection, so one possible strategy is to use this map to compute the distribution of the eigenvalues and spectral weights from the Jacobi matrix entries by the change-of-variable formula. This is possible since the Jacobian determinant can be computed

Dumitriu and Edelman used direct computation of the Jacobian of the map $(\overline{\lambda}, \overline{q}) \to (\overline{a}, \overline{b})$ to prove Theorem 1.3.2. This computation may be simplified by working through the moments, which have a simple connection to both representations:

$$m_k = \int x^k d\sigma = \sum \lambda_i^k q_i^2$$

One can look at maps from both sets to $(m_1, \dots, m_{2n-1})$. Notice that $2n-1$ moments are required as there are $2n-1$ variables in both the spectral and tridiagonal basis. These are simple transformations and one can write down the appropriate matrices and then find their determinants. These computations can be found in [25], and yield the following.

Theorem 1.3.3 (Dumitriu, Edelman, Krishnapur, Rider, Virág, [12,25]). *Let V a real-valued function, and $\overline{a}, \overline{b}$ are chosen from then density proportional to*

$$\exp\left(-TrV(J)\right) \prod_{k=1}^{n-1} b_{n-k}^{k\beta-1}$$

assuming such a denisty exists (i.e. the total integral is finite). Then the eigenvalues have distribution

$$f(\lambda_1, .., \lambda_n) = \frac{1}{Z_{n,\beta}} \exp\left(-\sum_i V(\lambda_i)\right) \prod_{i<j} |\lambda_i - \lambda_j|^\beta \tag{1.3.4}$$

and the q_i are independent of the λ with $(q_1, \dots, q_n) = (\varphi_1(1)^2, \dots, \varphi_n(1)^2)$ have Dirichlet$(\frac{\beta}{2}, \dots, \frac{\beta}{2})$ distribution.

Exercise 1.3.5. Show that when $V(x) = x^4$, the sequence $\{(a_i, b_i), i \geqslant 1\}$ with the distribution from the theorem forms a time-inhomogeneous Markov Chain.

A result like this holds for general polynomial V, though one needs to take bigger blocks of (a_i, b_i). This is exploited in [25] to get universality for the top eigenvalue.

2. Graph Convergence and the Wigner semicircle law

The approach of operator limits is a case of the "objective method" pioneered by Aldous. In limit theories, it is best to understand the limit of a high-level structured object, such as a matrix or a graph, rather than just its statistics (such as eigenvalues, or graph-related quantities such as triangle density).

In this section we will use Jacobi matrices together with graph convergence in order to give proofs of the Wigner semicircle law. The graphs themselves are operators through their adjacency matrices. While it is not helpful here to formalize this correspondence, we still think of graph convergence as an example of operator limits.

We begin by defining the spectral measure of a graph and give an introduction to different notions of graph convergence. We finish by proving the Wigner semicircle law in two different ways using different notions of graph convergence.

2.1. Graph convergence. The proof of the Wigner semicircle law given here will rest on a graph convergence argument. First we introduce the notions of convergence needed for the proof. Examples will make the convergence easier to understand.

We will be considering rooted graphs (G, ρ) where the root ρ is a marked vertex. The spectral measure of (G, ρ) is the spectral measure of the adjacency matrix A of G at the coordinate corresponding to ρ (which we will often just take to be the first entry).

Note that the kth moment of the spectral measure is just the number of paths of length k starting and ending at the root. In particular, moments up to $2k$ of the spectral measure are determined by the k-neighborhood of ρ in G.

The definition of the spectral measure (1.2.1) works even for infinite graphs, but it is again not a priori clear that it exists or it is unique.

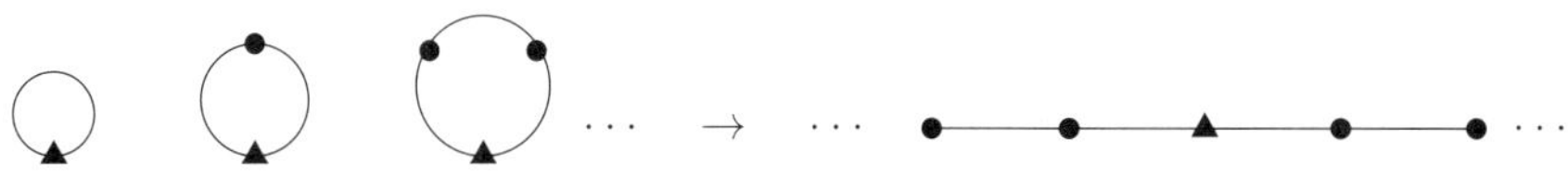

FIGURE 2.1.1. Rooted convergence: n-cycles to $\mathbb{Z}$

Definition 2.1.2. Rooted convergence. A sequence of rooted graphs (G_n, ρ) converges to a limit (G, ρ) if for any radius r, the ball of radius r in G_n about ρ equals that in G for all large enough n.

Examples 2.1.3. Two examples:

(1) The n cycle with any vertex chosen as the root converges to $\mathbb{Z}$.

(2) The k by k grid with vertices at the intersection. With a vertex at the center of the grid as the root, we get convergence to $\mathbb{Z}^2$.

For bounded degree graphs G_n, if (G_n, ρ_n) converges to (G, ρ) in the sense of rooted convergence, then by definitions, the moments of the spectral measures σ_n converge. This implies two things. First, since the spectral measures graphs with degrees bounded by b are supported on $[-b, b]$, σ_n have subsequential weak limits on $[-b, b]$. But such measures are determined by their moments, so the limit of σ_n exists, and is the spectral measure of (G, ρ).

Since any bounded degree rooted infinite graph is a rooted limit of balls around the root, we get

Proposition 2.1.4. *Bounded degree rooted infinite graphs have unique spectral measure defined by* (1.2.1).

Exercise 2.1.5. Consider a straight line with n vertices spaced evenly rooted at the left end point which converges to $\mathbb{Z}_+$ in the limit. The limit of the spectrum is the Wigner semicircle law, since the moments are Dyck paths. But one can prove this directly, since the path of length n is easy to diagonalize. This is an example where the spectral measure and the eigenvalue distribution have different limits.

Exercise 2.1.6. Consider the random d-regular graph (G_n, ρ) on n vertices in the configuration model (for a given degree sequence, choose a uniform random matching on the half edges attached to each vertex). Show that (G_n, ρ) converges to the d-regular infinite tree in probability.

In the case where there is no designated root we will need a different notion of convergence.

Definition 2.1.7 (Benjamini-Schramm, [4]). For a sequence of graphs $\{G_n\}$ choose a root vertex uniformly at random. We say that the graphs converge in the **Benjamini-Schramm** sense if this random sequence of rooted graphs converges in distribution with respect to rooted convergence to a random rooted graph.

Benjamini-Schramm convergence is equivalent to convergence of local statistics. This is the following statement. For every finite rooted graph (K, ρ) and every r, the proportion of vertices in G_n whose ball radius r is rooted-isomorphic to (K, ρ) converges to the probability that the ball of radius r in the limit is rooted-isomorphic to (K, ρ).

Benjamini-Schramm limits of finite graphs are *unimodular*. For the special case of randomly rooted regular graphs (G, o) unimodularity means that if for a uniformly chosen random neighbor ν of ρ in G, the triple (G, ρ, ν) has the same distribution as (G, ν, ρ). For the general case, in order to define unimodularity the distributions have to first be biased by the degree of the root, see e.g. [27].

Exercise 2.1.8. Show that if G is a fixed connected finite regular graph with a random vertex ρ, then (G, ρ) is unimodular if and only if ρ has uniform distribution.

The most intriguing open problem in this area is whether all infinite unimodular random graphs are Benjamini-Schramm limits. Those that are called sofic. For more on this see [1].

Proposition 2.1.9. *Let G be a fixed finite graph and choose a root ρ uniformly at random from the vertex set $V(G)$. This defines a random rooted graph and its associated random spectral measure σ. Then $E\sigma = \mu$ is the eigenvalue distribution.*

Proof. Recall that for the spectral measure of a matrix (and so a graph) we have

$$\sigma_{(G,\rho)} = \sum_{i=1}^{n} \delta_{\lambda_i} \varphi_i^2(\rho)$$

Since φ_i is of length one, we have

$$\sum_{\rho \in V(G)} \varphi_i(\rho)^2 = 1$$

hence

$$E\sigma_{G,\rho} = \frac{1}{n} \sum_{\rho \in V(G)} \sigma_{G,\rho} = \mu_G.$$

□

Example 2.1.10. The following are examples of Banjamini-Schramm convergence:

(1) A cycle graph converges to the graph of $\mathbb{Z}$.
(2) A path of length n converges to the graph of $\mathbb{Z}$.
(3) Large box of $\mathbb{Z}^d$ converges to the full $\mathbb{Z}^d$ lattice.

Notice that for the last two examples the probability of being in a neighborhood of the edge goes to 0 and so the limiting graph doesn't see the edge effects.

Exercise 2.1.11. A sequence of d-regular graphs G_n with n vertices is of essentially large girth if for every k the number of k-cycles in G_n is $o(n)$. Show that G_n is essentially large girth if and only if it Benjamini-Schramm converges to the d-regular tree.

Exercise 2.1.12. Show that for $d \geqslant 3$ the d-regular tree is not the Benjamini-Schramm limit of finite trees. (Hint: consider the expected degree).

How is Benjamini-Schramm convergence related to the eigenvalue distribution? First, an exercise abound random probability measures, useful here and in the sequel.

Exercise 2.1.13. Let ν_n be a sequence of random probability measures, and assume that $\nu_n \to \nu$ in distribution with respect to the weak topology.

(1) Show that $E\nu_n \to E\nu$ weakly.

(2) Assume that ν is deterministic. Let $\mathcal{F}_n$ be an arbitrary sequence of σ-fields. Define the random probability measure $E(\nu_n|F_n)$ by $E(\nu_n|F_n)(A) = E(\nu_n(A)|F_n)$ for measurable sets A. Show that $E(\nu_n|F_n) \to \nu$ in probability with respect to the weak topology.

Assume that $G_n \to (G, \rho)$ in the Benjamini-Schramm sense. By the continuity theorem (applied to the rooted convergence topology), the distribution of the random spectral measures σ_n converges to the distribution of $\sigma_{G,\rho}$.

By Exercise 2.1.13 (1) this implies that the eigenvalue distributions converge as well:

Theorem 2.1.14. *Let G_n be a sequence of graphs with eigenvalue distributions μ_n. If $G_n \to (G, \rho)$ in the Benjamini-Schram sense, then $\mu_n = E\sigma_n \to E\sigma_{G,\rho}$ weakly.*

One can consider a more general setting of weighted graphs. This corresponds to general symmetric matrices A. In this case we require that the neighborhoods stabilize and the weights also converge. Everything above goes through.

Example 2.1.15 (The spectral measure of $\mathbb{Z}$). We use Benjamini-Schramm convergence of the cycle graph to $\mathbb{Z}$. First we compute the spectral measure of the n-cycle G_n. We get that $A = T + T^t$ where

$$T = \begin{bmatrix} 0 & 1 & & & \\ & 0 & 1 & & \\ & & \ddots & \ddots & \\ & & & \ddots & 1 \\ 1 & & & & 0 \end{bmatrix}.$$

The eigenvalues of T are the nth roots of unity η_i. Since $A = T + T^{-1}$ the eigenvalues of A are $\eta_i + \eta_i^{-1} = 2\Re\eta_i$. Geometrically, these are projections of the $2\eta_i$, that is points uniformly spaced on the circle of radius 2, to the real line.

In the limit, the measure converges to the projection of the uniform measure on that circle, also called the arcsine distribution

$$\sigma_{\mathbb{Z}} = \frac{1}{2\pi\sqrt{4-x^2}}\mathbf{1}_{x\in[-2,2]}\,dx.$$

Exercise 2.1.16. Let B_n be the unweighted finite binary tree with n levels. Suppose a vertex is chosen uniformly at random from the set of vertices. Give the distribution of the limiting graph.

2.2. Wigner's semicircle law.

Theorem 2.2.1 (Wigner's semicircle Law). *Let Λ_n have β-Hermite distribution and let $\mu_n = \frac{1}{n}\sum_{j=1}^{n}\delta_{\lambda_j/\sqrt{n}}$ be the empirical distribution of $\Lambda_n/\sqrt{n}$. Then*

$$\mu_n \Rightarrow \mu_{sc}, \qquad \textit{where} \qquad \frac{d\mu_{sc}}{dx} = \frac{1}{2\pi}\sqrt{4-x^2}\mathbf{1}_{x\in[-2,2]}\,dx.$$

The following exercise provides the necessary tools to give several different proofs of Wigner's semicircle law. You can attempt the exercise first or read on in order to see more details of the proof.

Exercise 2.2.2. Let A be a rescaled $n\times n$ Dumitriu-Edelman tridiagonal matrix

$$A = \frac{1}{\sqrt{\beta n}}\begin{bmatrix} a_1 & b_1 & & & \\ b_1 & a_2 & b_2 & & \\ & b_2 & \ddots & \ddots & \\ & & \ddots & a_{n-1} & b_{n-1} \\ & & & b_{n-1} & a_n \end{bmatrix}, \qquad b_i \sim \chi_{\beta(n-i)}, \quad a_i \sim \mathcal{N}(0,2)$$

all independent, and suppose that A is the adjacency matrix of a weighted graph.

(1) Draw the graph with adjacency matrix A. (There can be loops)
(2) Suppose a root for your graph is chosen uniformly at random, what is the limiting distribution of your graph?
(3) What is the limiting spectral measure of the graph rooted at the vertex corresponding to the first row and column?
(4) What is the limiting spectral measure of the unweighted graph?

Note that a Jacobi matrix can be thought of as the adjacency matrix of a weighted path with loops. For all of the proofs of the Wigner semi-circle law we will use the graph with the adjacency matrix given by the rescaled Dumitriu-Edelman model given in Exercise 2.2.2, see Figure 2.2.3.

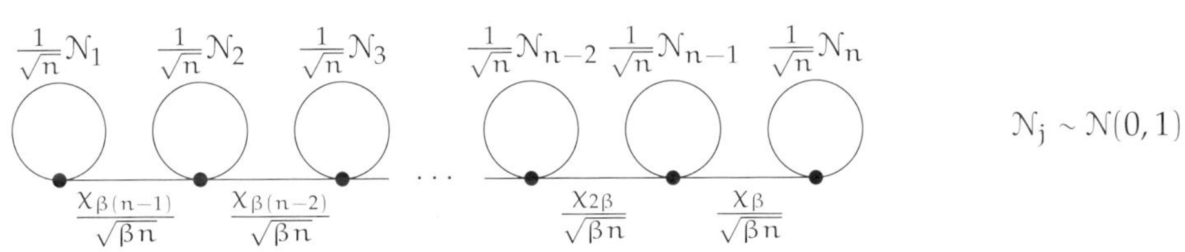

FIGURE 2.2.3. Unrooted rescaled Dumitriu-Edelman graph

Exercise 2.2.4. Check that $\chi_n - \sqrt{n} \xrightarrow[n\to\infty]{(d)} \mathcal{N}(0,1/2)$.

Proof 1. [37]

Take the graph associated to the rescaled Dumitriu-Edelman tridiagonal matrix shown in Figure 2.2.3, and then take a Benjamini-Schramm limit.

The convergence is in probability to a random rooted graph. Note that there are two levels of randomness, one coming from the fact that we take a limit of random weighted graphs (not just weighted graphs) and the second from the fact that even deterministic graphs have Benjamini-Schramm limits given by a random rooted graph. The first kind of randomness is lost in the limit, hence the convergence in probability.

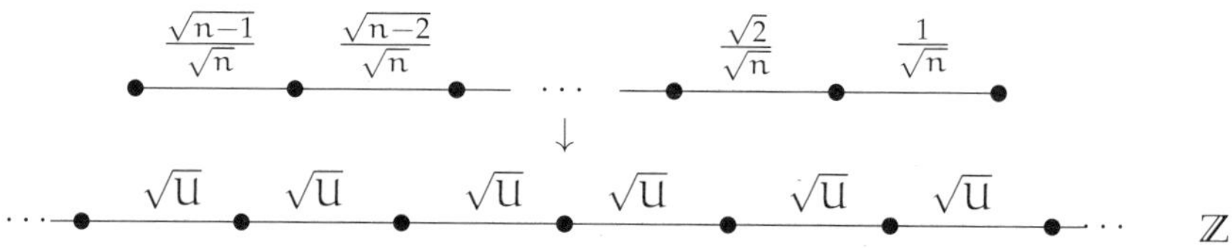

FIGURE 2.2.5. Terms from the graph visible in Benjamini-Schramm convergence.

Notice that an application of Exercise 2.2.4 shows that it is enough to consider the graph with no loops and deterministic edge labels $\sqrt{\frac{n-k}{n}}$. See Figure 2.2.5.

What is the limit? The structure clearly converges to $\mathbb{Z}$. The edge weights in a randomly rooted neighborhood converge in distribution to $\sqrt{U}$ for a single random variable U uniform in $[0,1]$. Let σ denote the spectral measure $\mathbb{Z}$ with these edge weights. Theorem 2.1.14 and continuity theorem implies that $\mu_n \to E\sigma$ in probability.

Now σ is a rescaling of the spectral measure of $\mathbb{Z}$ by $\sqrt{U}$. By Exercise 2.1.15, σ is the scaled arcsine measure given by the projection of a circle of radius $\sqrt{U}$.

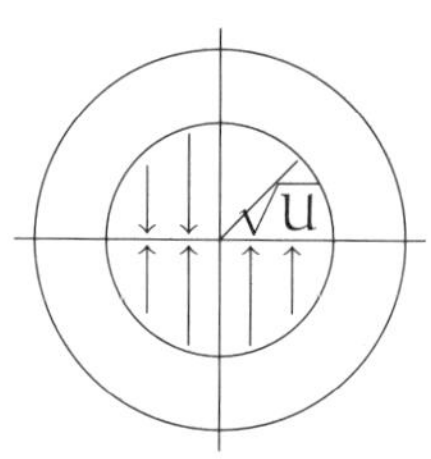

FIGURE 2.2.6. The role of U in the limiting distribution and projection

The distribution of a point chosen with radius $\sqrt{U}$ and uniform angle is in fact a uniform random point in the disk. Thus $E\sigma$ is the projection of the uniform measure on the disk to the real line. See figure 2.2.6. This is the semicircle law.

$$\frac{d\mu_{sc}}{dx} = \frac{1}{2\pi}\sqrt{4-x^2}\mathbf{1}_{x\in[-2,2]}\,dx. \qquad \square$$

Proof 2. Going back to the full matrix model of GOE, let A_n be $1/\sqrt{n}$ times the tridiagonalization of GOE from a uniformly chosen random vertex v. This vertex will correspond to the first coordinate in the tridiagonalization. Let $\mathcal{F}_n$ be a sigma-field generated by the randomness in the nth GOE, but not the choice of the vertex. Then with $\sigma_n = \sigma_{A_n,v}$ the empirical eigenvalue distribution of GOE satisfies

$$\mu_n = E(\sigma_n|\mathcal{F}_n).$$

It suffices to show that σ_n converges to the semicircle law in probability, and conclude by Exercise 2.1.13 (2).

A_n has $1/\sqrt{n}$ times the distribution (1.3.1). Consider the rooted limit of (A_n, v). In Figure 2.2.3 v is the leftmost point on the graph. This random weighted graph converges in probability with respect to the rooted convergence topology to $\mathbb{Z}_+$.

Therefore by the continuity theorem and Theorem 2.1.14

$$\sigma_n \to \sigma_{\mathbb{Z}_+} = \rho_{sc}$$

in probability, as required. □

The limit of the spectral measure at the first vertex should have nothing to do with the limit of the eigenvalue distribution in the general case. The Jacobi matrices that we get in the case of the GOE are special.

3. The top eigenvalue and the Baik-Ben Arous-Péché transition

3.1. The top eigenvalue. The eigenvalue distribution of the GOE converges after scaling by $\sqrt{n}$ to the Wigner semicircle law. From this, it follows that the top eigenvalue, $\lambda_1(n)$ satisfies for every $\varepsilon > 0$

$$P\left(\frac{\lambda_1(n)}{\sqrt{n}} > 2 - \varepsilon\right) \to 1,$$

the 2 here is the top of the support of the semicircle law. However, the matching upper bound does not follow and needs more work. This is the content of the following theorem.

Theorem 3.1.1 (Füredi-Komlós [14]).

$$\frac{\lambda_1(n)}{\sqrt{n}} \to 2 \textit{ in probability.}$$

This holds for more general entry distributions in the original matrix model; we have a simple proof for the GOE case.

Lemma 3.1.2. *If J is a Jacobi matrix (a's diagonal, b's off-diagonal) then*

$$\lambda_1(J) \leqslant \max_i(a_i + b_i + b_{i-1}).$$

Here we take the convention $b_0 = b_n = 0$.

Proof. Observe that J may be written as

$$J = -AA^t + \text{diag}(a_i + b_i + b_{i-1}),$$

where

$$A = \begin{bmatrix} 0 & \sqrt{b_1} & & & \\ & -\sqrt{b_1} & \sqrt{b_2} & & \\ & & -\sqrt{b_2} & \sqrt{b_3} & \\ & & & \ddots & \ddots \end{bmatrix}$$

and AA^t is nonnegative definite. So for the top eigenvalues we have

$$\lambda_1(J) \leqslant -\lambda_1(AA^t) + \lambda_1(\mathrm{diag}(a_i + b_i + b_{i-1})) \leqslant \max_i (a_i + b_i + b_{i-1}).$$

We used subadditivity of λ_1, which follows from the Rayleigh quotient representation. □

Applying this to our setting we get that

$$\lambda_1(\mathrm{GOE}) \leqslant \max_i (N_i, \chi_{n-i} + \chi_{n-i+1}) \leqslant 2\sqrt{n} + c\sqrt{\log n} \tag{3.1.3}$$

the right inequality is an exercise (using the Gaussian tails in χ) and holds with probability tending to 1 if c is large enough. This completes the proof of Theorem 3.1.1.

This shows that the top eigenvalue cannot go further than an extra $\sqrt{\log n}$ outside of the spectrum. Indeed we will see that

$$\lambda_1(\mathrm{GOE}) = 2\sqrt{n} + TW_1 n^{-1/6} + o(n^{-1/6})$$

for some distribution TW_1, so the bound above is not optimal.

3.2. Baik-Ben Arous-Péché transition. The approach taken here is a version of a section in Bloemendal's PhD thesis [7].

Historically random matrices have been used to study correlations in data sets. To see whether correlations are significant, one compares to a case in which data is sampled randomly without correlations.

Wishart in the 20s considered matrices $X_{n\times m}$ with independent normal entries and studied the eigenvalues of XX^t. The rank-1 perturbations below model the case where there is one significant trend in the data, but the rest is just noise. We consider the case $n = m$. A classical result is the following.

Theorem 3.2.1 (BBP transition). *Let X_n be an $n \times n$ matrix with independent $\mathcal{N}(0,1)$ entries, then*

$$\frac{1}{n}\lambda_1\left(X\,\mathrm{diag}(1+a^2, 1, 1, \dots, 1)X^t\right) \xrightarrow[n\to\infty]{P} \varphi(a)^2$$

where

$$\varphi(a) = \begin{cases} 2 & a \leqslant 1 \\ a + \frac{1}{a} & a \geqslant 1. \end{cases}$$

Heuristically, correlation in the population appears in the asymptotics in the top eigenvalue of the sample only if it is sufficiently large, $a > 1$. Otherwise, it gets washed out by the fake correlations coming from noise. We will prove the GOE analogue of this theorem, and leave the Wishart case as an exercise.

What about the distributional limit of the top eigenvalue? When $a < 1$ the distribution is unchanged from the unperturbed case, the limit being Tracy-Widom. When $a > 1$ the top eigenvalue separates and has limiting Gaussian fluctuations. Near the point $a = 1$ a deformed Tracy-Widom distribution appears, see [3], [5].

The GOE analogue answers the following question. Suppose that we add a common nontrivial mean to the entries of a GOE matrix. When does this influence the top eigenvalue on the semicircle scaling?

Theorem 3.2.2 (Top eigenvalue of GOE with nontrivial mean).

$$\frac{1}{\sqrt{n}}\lambda_1(\mathrm{GOE}_n + \frac{a}{\sqrt{n}}11^t) \xrightarrow[n\to\infty]{P} \varphi(a)$$

where 1 *is the all-1 vector, and* 11^t *is the all-1 matrix.*

It may be surprising how little change in the mean in fact changes the top eigenvalue!

We will not use the next exercise in the proof of 3.2.2, but it does show where the function φ comes from and motivates the proof for the GOE case.

Exercise 3.2.3 (BBP for $\mathbb{Z}^+$)**.** For an infinite graph, we can define λ_1 by Rayleigh quotients using the adjacency matrix A

$$\lambda_1(G) = \sup_v \frac{\langle v, Av\rangle}{\|v\|_2^2}.$$

(1) Show that λ_1 is at most the maximal degree in G.
(2) Prove that for $a \leqslant 1$

$$\lambda_1(\mathbb{Z}^+ + \text{loop of weight } a \text{ on } 0) = \varphi(a).$$

Hint To prove the lower bound, use specific test functions. When $a > 1$, note that there is an eigenvector $(1, a^{-1}, a^{-2}, ...)$ with eigenvalue $a + \frac{1}{a}$. When $a \leqslant 1$ use the indicator of a large interval. The upper bound for $a > 1$ is more difficult; use rooted convergence and interlacing.

We will need the following result.

Exercise 3.2.4. Let A be a symmetric matrix, let v be a vector of ℓ^2-norm at least 1, and let $x \in \mathbb{R}$ so that $\|Av - xv\| \leqslant \varepsilon$. Then there is an eigenvalue λ of A with $|\lambda - x| \leqslant \varepsilon$. Hint: consider the inverse of $A - Ix$.

Proof of Theorem 3.2.2. The first observation is that because the GOE is an invariant ensemble, we can replace 11^t by vv^t for any vector v having the same length as the vector 1. We can replace the perturbation with $\sqrt{n}a\mathbf{e}_1\mathbf{e}_1^t$. Such perturbations commute with tridiagonalization.

Therefore we can consider Jacobi matrices of the form

$$J(a) = \frac{1}{\sqrt{n}}\begin{bmatrix} a\sqrt{n} + N_1 & \chi_{n-1} & & \\ \chi_{n-1} & N_2 & \chi_{n-2} & \\ & \ddots & \ddots & \ddots \end{bmatrix}$$

Case 1: $a \leqslant 1$. Since the perturbation is positive, we only need an upper bound. We use the maximum bound from before. For $i = 1$, the first entry, there was space of size $\sqrt{n}$ below $2\sqrt{n}$. For $i = 1$ the max bound still holds.

Case 2: $a > 1$ Now fix k and let $v = (1, 1/a, 1/a^2, ..., 1/a^k, 0, ..., 0)$. Thus the error from the noise will be of order $1/\sqrt{n}$ so that

$$\left\| J(a)v - v(a + \frac{1}{a}) \right\| \leqslant ca^{-k}$$

with probability tending to 1.

By Exercise 3.2.4, $J(a)$ has an eigenvalue λ^* that is ca^{-k}-close to $a + 1/a$.

We now need to check that this eigenvalue will actually be the maximum.

Exercise 3.2.5. If A, P are asymmetric matrices, with $P \geqslant 0$ of rank 1, then the eigenvalues of A and $A + P$ interlace and the shift under perturbation is to the right. Hint: use the Courant-Fisher characterization.

By interlacing,

$$\lambda_2(J(a)) \leqslant \lambda_1(J(0)) = 2 + o(1) < a + 1/a - ca^k$$

if we chose k large enough. Thus the eigenvalue λ^* we identified must be λ_1. $\square$

4. The Stochastic Airy Operator

4.1. Global and local scaling. In the Wigner semicircle law the rescaled eigenvalues $\{\lambda_i/\sqrt{n}\}_{i=1}^n$ accumulate on a compact interval and so in the limit become indistinguishable from each other. For the local interactions between eigenvalues, the behavior of individual points has to prevail in the limit.

To make a guess at the correct spacing required to see individual points in the limit we begin by pretending that they are quantiles of the Wigner semicircle law. When n is large we get that for $a < b \in [-2, 2]$

$$\#\Lambda_n \cap [a\sqrt{n}, b\sqrt{n}] \approx n \int_a^b d\mu_{sc}(x) = n \int_a^b \frac{1}{2\pi}\sqrt{4 - x^2}dx.$$

So we expect that for $a \in (-2, 2)$ the process $\sqrt{n(4 - a^2)}(\Lambda_n - a\sqrt{n})$ should have average spacing $\frac{1}{2\pi}$.

Exercise 4.1.1. Show that typical spacing at the edge $2\sqrt{n}$ of Λ_n has order $n^{-1/6}$.

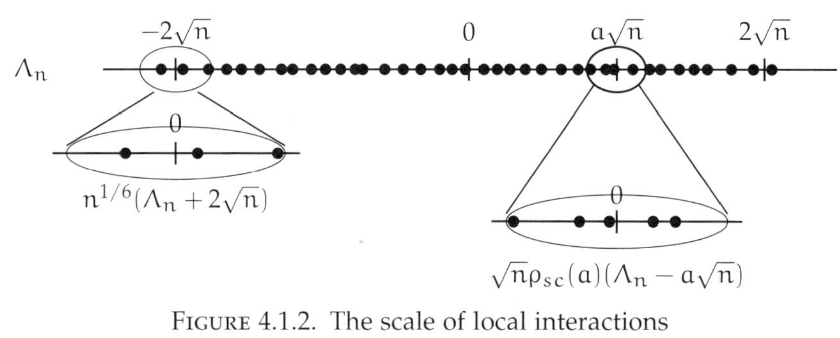

FIGURE 4.1.2. The scale of local interactions

The correct scales needed to obtain a local limit are give in Figure 4.1.2. These notes will focus on the convergence result for the edge of the spectrum. For the statement for the bulk and more on the operator viewpoint, see Section 5.

4.2. The heuristic convergence argument at the edge. The goal here is to understand the limiting top eigenvalue of the Hermite β ensembles in terms of a random operator. To do this we look at the geometric structure of the tridiagonal matrix. Jacobi matrices are frequently associated with differential equations and sometimes studied under the name of discrete Schrödinger operators. To see the connection with Schrödinger operators consider the following example:

$$A = \begin{bmatrix} 0 & 1 & & & & \\ 1 & 0 & 1 & & & \\ & 1 & \ddots & \ddots & & \\ & & \ddots & \ddots & \ddots & \\ & & & \ddots & \ddots & 1 \\ & & & & 1 & 0 & 1 \\ & & & & & 1 & 0 \end{bmatrix}$$

The semi-infinite version of this is frequently called the **discrete Laplacian**. To understand this name let m be large for $f : \mathbb{R}^+ \to \mathbb{R}$ define the discretization $v_f = (f(0), f(1/m), f(2/m), ...)^t$. Then $B = m^2(A-2I)$ acts as a discrete second derivative on f, in the sense that $Bv_f \approx v_{f''}$ as $m \to \infty$. For this to hold in the first entry we need to further assume that f satisfies a Dirichlet boundary condition $f(0) = 0$. This convergence argument may be easily extended to matrices of the form $(A-2I)+D$ where D is a semi-infinite diagonal matrix with entries $D_k = V(\frac{k}{m})$ for some function $V : \mathbb{R} \to \mathbb{R}$. In this case the matrices converge to the Schrödinger operator $\Delta + V$.

Now apply this type of convergence argument to the tridiagonal model for the β-Hermite ensemble. To start we first need to determine which portions of the matrix contribute to the behavior of the largest eigenvalue. Recall the Dumitriu-Edelman matrix model A_n for the β-Hermite ensemble defined in equation (1.3.1). Take $u = c_1\mathbf{e}_1 + c_2\mathbf{e}_2 + \cdots c_n\mathbf{e}_n$ where $\mathbf{e}_k$ is the kth coordinate vector, and observe that if we assume the c_k vary smoothly we have

$$A_n u = \frac{1}{\sqrt{\beta}} \sum_{k=1}^{n} (c_{k-1}a_k + c_{k+1}a_k + c_k b_k)\mathbf{e}_k \approx \sum_{k=1}^{n} 2c_k\sqrt{n-k}\,\mathbf{e}_k.$$

We are interested in which eigenvectors u give us $A_n u = (2\sqrt{n} + o(1))u$. This calculation suggests thar these eigenvectors should be concentrated on the first $k = o(n)$ coordinates. This suggests that the top corner of the matrix determines the behavior of the top eigenvalue.

Returning to the β-Hermite case, by Exercise 2.2.4, for $k \ll n$ we have

$$\chi_{n-k} \approx \sqrt{\beta(n-k)} + \mathcal{N}(0,1/2) \approx \sqrt{\beta}(\sqrt{n} - \frac{k}{2\sqrt{n}}) + \mathcal{N}(0,1/2)$$

We can use this expansion to break the matrix $m^\gamma(2\sqrt{n}I - J)$ approximately into the sum of terms

$$(4.2.1)\quad m^\gamma\sqrt{n}\begin{bmatrix} 2 & -1 & & \\ -1 & 2 & -1 & \\ & -1 & 0 & \ddots \\ & & \ddots & \ddots \\ & & & \ddots \end{bmatrix} + \frac{m^\gamma}{2\sqrt{n}}\begin{bmatrix} 0 & 1 & & & \\ 1 & 0 & 2 & & \\ & 2 & 0 & 3 & \\ & & 3 & 0 & \ddots \\ & & & \ddots & \ddots \end{bmatrix} + \frac{m^\gamma}{\sqrt{\beta}}\begin{bmatrix} N_1 & \tilde{N}_1 & & \\ \tilde{N}_1 & N_2 & \tilde{N}_2 & \\ & \tilde{N}_2 & N_3 & \ddots \\ & & \ddots & \ddots \\ & & & \ddots \end{bmatrix}$$

and assume that we have $m = n^\alpha$ for some α. What choice of α should we make? For the first term we want

$$m^\gamma\sqrt{n}\begin{bmatrix} 2 & -1 & & \\ -1 & 2 & -1 & \\ & -1 & 0 & \ddots \\ & & \ddots & \ddots \end{bmatrix}$$

to behave like a second derivative. This means that $m^\gamma\sqrt{n} = m^2$ which gives $2\alpha = \alpha\gamma + 1/2$. A similar analysis can be done on the second term. This term should behave like multiplication by t. For this we want $\frac{m^\gamma}{\sqrt{n}} = \frac{1}{m}$ which gives $\alpha\gamma - 1/2 = -\alpha$. Solving this system we get $\alpha = 1/3$ and $\gamma = 1/2$. For the noise term, multiplication by it should yield a distribution (in the Schwarz sense), which means that its integral over intervals should be of order 1. In other words, the average of m noise terms times m^γ should be of order 1. This gives $\gamma = 1/2$, consistent with the previous computations.

This means that we need to look at the section of the matrix that is $m = n^{1/3}$ and we rescale by $n^{1/6}$. That is we look at the matrix

$$H_n = n^{1/6}(2\sqrt{n}I - A_n)$$

acting on functions with mesh size $n^{-1/3}$.

Exercise 4.2.2. Show that in this scaling, the second matrix in the expansion above has the same limit as the diagonal matrix with $0, 2, 4, 6, 8....$ on the diagonal (scaled the same way).

Conclusion. H_n acting on functions with this mesh size behaves like a differential operator. That is

$$(4.2.3)\qquad H_n = n^{1/6}(2\sqrt{n} - A_n) \approx -\partial_x^2 + x + \frac{2}{\sqrt{\beta}}b'_x = \mathrm{SAO}_\beta,$$

here b'_x is white noise. This operator will be called the Stochastic Airy operator (SAO_β). We also set the boundary condition to be Dirichlet. This conclusion can be made precise. The heuristics are due to Edelman and Sutton [13], and the proof to Ramírez, Rider, and Virág [29].

There are two problems at this point that must be overcome in order to make this convergence rigorous. The first is that we need be able to make sense of that limiting operator. The second is that the matrix even embedded as an operator on step functions acts on a different space that the SAO_β so we need to make sense of what the convergence statement should be.

Remarks on operator convergence

(1) Embed $\mathbb{R}^n$ into $L^2(\mathbb{R})$ via

$$\mathbf{e}_i \mapsto \sqrt{m}\mathbf{1}_{[\frac{i-1}{m},\frac{i}{m})}.$$

This gives an embedding of the matrix A_n acting on a subspace of $L^2(\mathbb{R}^+)$.

(2) It is not clear what functions the Stochastic Airy Operator acts on at this point. Certainly nice functions multiplied by the derivative of Brownian motion will not be functions, but distributions. The only way we get nice functions as results if this is cancelled out by the second derivative. Nevertheless, the domain of SAO_β can be defined.

In any case, these operators act on two completely different sets of functions. The matrix acts on piecewise constant functions, while SAO_β acts on some exotic functions.

(3) The nice thing is that if there are no zero eigenvalues, both H_n^{-1} and J^{-1} can be defined in their own domains, and the resulting operators have compact extensions to the entire L^2.

We will not do this in these notes, but the sense of convergence that can be shown is

$$\|H_n^{-1} - SAO_\beta^{-1}\|_{2\to2} \to 0.$$

This is called norm resolvent convergence, and it implies convergence of eigenvalues and eigenvectors if the limit has discrete simple spectrum. See e.g. Chapter 7 [31].

(4) The simplest way to deal with the limiting operator and the issues of white noise is to think of it as a bilinear form. This is the approach we follow in the next section. The kth eigenvalue can be identified using the Courant-Fisher characterization.

Exercise 4.2.4. We will consider cases where a matrix $A_{n\times n}$ can be embedded as an operator acting on the space of step function, with mesh size $1/m_n$. In particular, we can encode these step functions by vectors $v_f = [f(\frac{1}{m_n}), f(\frac{2}{m_n}), ..., f(\frac{n}{m_n})]^t$.

Let A be the matrix

$$A = \begin{bmatrix} -1 & 1 & & \\ & -1 & 1 & \\ & & \ddots & \ddots \\ & & & -1 \end{bmatrix}.$$

For which k_n we get $k_n A v_f \to v_{f'}$?

Exercise 4.2.5. Let A be the diagonal matrix with diagonal entries $(1, 4..., n^2)$. Find a k_n such that $k_n A v_f$ converges to something nontrivial. What is k_n and what does the limit converge to?

Exercise 4.2.6. Let J be a Jacobi matrix (tridiagonal with positive off-diagonal entries) and v be an eigenvector with eigenvalue λ. The number of times that v changes sign is equal to the number of eigenvalues above λ. More generally the equation $Jv = \lambda v$ determines a recurrence for the entries of v. If we run this recurrence for an arbitrary λ (not necessarily an eigenvalue) and count the number of times that v changes sign this still gives the number of eigenvalues greater than λ.

(1) Based on this give a description of the number of eigenvalues in the interval $[a, b]$.
(2) Suppose that $v^t = (v_1, \ldots, v_n)$ solves the recurrence defined by $Jv = \lambda v$. What is the recurrence for $r_k = v_{k+1}/v_k$? What are the boundary conditions for r that would make v an eigenvector?

4.3. The bilinear form SAO_β. Recall the Airy operator

$$A = -\partial_x^2 + x$$

acting on $f \in L^2(\mathbb{R}^+)$ with boundary condition $f(0) = 0$. The equation $Af = 0$ has two solutions $\mathrm{Ai}(x)$ and $\mathrm{Bi}(x)$, called Airy functions. Note that the solution of $(A - \lambda)f = 0$ is just a shift of these functions by λ.

Since only Ai^2 is integrable, the eigenfunctions of A are the shifts of Ai with the eigenvalues the amount of the shift. The kth zero of the Ai function is at $z_k = -\left(\frac{3}{2}\pi k\right)^{2/3} + o(1)$, therefore to satisfy the boundary conditions the shift must place a 0 at 0, so the kth eigenvalue is given by

$$\lambda_k = -z_k = \left(\frac{3}{2}\pi k\right)^{2/3} + o(1). \tag{4.3.1}$$

The asymptotics are classical.

For the Airy operator A and a.e. differentiable, continuous functions f with $f(0) = 0$ we can define

$$\|f\|_*^2 := \langle Af, f\rangle = \int_0^\infty f^2(x)x + f'(x)^2\, dx. \tag{4.3.2}$$

Let L^* be the space of functions with $\|f\|_* < \infty$.

Exercise 4.3.3. Show that there is $c > 0$ so that

$$\|f\|_2 \leqslant c\|f\|_*$$

for every $f \in L^*$. In particular, $L^* \subset L^2$.

Recall the Rayleigh quotient characterization of the eigenvalues λ_1 of A.

$$\lambda_1 = \inf_{f \in L^*, \|f\|_2 = 1} \langle Af, f\rangle.$$

More generally, the Courant-Fisher characterization is

$$\lambda_k = \inf_{W \subset L^*, \dim W = k} \; \sup_{f \in W, \|f\|_2 = 1} \langle Af, f\rangle,$$

where the infimum is over subspaces B.

For two operators $A \leqslant B$ if for all $f \in L^*$

$$\langle f, Af \rangle \leqslant \langle f, Bf \rangle.$$

Exercise 4.3.4. If $A \leqslant B$, then $\lambda_k(A) \leqslant \lambda_k(B)$.

Our next goal is to define the bilinear form associated with the Stochastic Airy operator on functions in L^*. Clearly, the only missing part is to define

$$\int_0^\infty f^2(x) b'(x)\, dx.$$

At this point you could say that this is defined in terms of stochastic integration, but the standard L^2 theory is not strong enough – we need it to be defined in the almost sure sense for all functions in L^*. We could define it in the following way:

$$\langle f, b'f \rangle \text{``}=\text{''} \int_0^\infty f^2(x) b'(x) dx = -\int_0^\infty 2f'(x)f(x)b(x)dx.$$

This is now a perfectly fine integral, but it may not converge. The main idea will be to write b as its average together with an extra term.

$$b(x) = \int_x^{x+1} b(s)ds + \tilde{b}(x) = \bar{b}(x) + \tilde{b}(x).$$

In this decomposition we get that $\bar{b}$ is differentiable and $\tilde{b}$ is small. The average term decouples quickly (at time intervals of length 1), so this term is analogous to a sequence of i.i.d. random variables. We define the inner product in terms of this decomposition as follows.

$$\langle f, b'f \rangle := \langle f, \bar{b}'f \rangle - 2\langle f', \tilde{b}f \rangle$$

It follows from Lemma 4.3.7 below that the integrals on the right hand side are well defined.

Exercise 4.3.5. There exists a random constant C so that we have the following inequality of functions:

$$|\bar{b}'|, |\tilde{b}| \leqslant C\sqrt{\log(2+x)} \tag{4.3.6}$$

Now we return to the Stochastic Airy operator, the following lemma will give us that the operator is bounded from below.

Lemma 4.3.7. *For every $\varepsilon > 0$ there exists random C so that in the positive definite order,*

$$\pm b' \leqslant \varepsilon A + CI,$$

and therefore

$$-CI + (1-\varepsilon)A \leqslant SAO_\beta \leqslant (1+\varepsilon)A + CI.$$

The upper bound here implies that the bilinear form is defined for all functions $f \in L^*$.

Proof. For $f \in L^*$ by our definition,

$$\langle f, b'f \rangle = \langle f, \bar{b}'f \rangle - 2\langle f', \tilde{b}f \rangle.$$

Now using bounds of the form $-2yz \leqslant y^2/\varepsilon + z^2\varepsilon$ we get the upper bound

$$\langle f, (\bar{b}' + \tilde{b}^2/\varepsilon)f\rangle + \varepsilon\|f'\|^2.$$

By Exercise 4.3.5 there exists a random constant C so that

$$\bar{b}' + \tilde{b}^2/\varepsilon \leqslant \varepsilon x + C.$$

We get the desired bound for $+b'$, and the same arguments works for $-b'$. □

The above Lemma implies that the eigenvalues of Stochastic Airy should behave asymptotically the same as those of the Airy operator with the same boundary condition. From the discussion at the start of Section 4.3 we will get the following asymptotic result.

Corollary 4.3.8. *The eigenvalues of SAO_β satisfy*

$$\frac{\lambda_k^\beta}{k^{2/3}} \to \left(\frac{2\pi}{3}\right)^{2/3} \qquad \text{a.s.}$$

Proof. It suffices to show that a.s. for every rational $\varepsilon > 0$ there exists $C_\varepsilon > 0$ so that

$$(1-\varepsilon)\lambda_k - C_\varepsilon \leqslant \lambda_k^\beta \leqslant (1+\varepsilon)\lambda_k + C_\varepsilon,$$

where the λ_k are the Airy eigenvalues (4.3.1). But this follows from the operator inequality of Lemma 4.3.7 and Exercise 4.3.4. □

One way to view the above Corollary is through the empirical distribution of the eigenvalues as $k \to \infty$. In this view the "density" behaves like $\sqrt{\lambda}$. More precisely, the number of eigenvalues less than λ is of order $\lambda^{3/2}$. This is the Airy-β version of the Wigner semicircle law. Only the edge of the semicircle appears here.

4.4. Convergence to the Stochastic Airy Operator. The goal of this section is to give a rigorous convergence argument for the extreme eigenvalues to those of the limiting operator. To avoid technicalities in the exposition, we will use a simplified model, which has the features of the tridiagonal beta ensembles. Consider the $n \times n$ matrix

$$(4.4.1)\quad H_n = n^{2/3}\begin{bmatrix} 2 & -1 & & \\ -1 & 2 & -1 & \\ & -1 & 2 & \ddots \\ & & \ddots & \ddots \end{bmatrix} + n^{-1/3}\operatorname{diag}(1,2,3,\dots) + \operatorname{diag}(N_{n,1}, N_{n,2},\dots)$$

Here for each n the $N_{n,i}$ are independent centered normal random variables with variance $\frac{4}{\beta}n^{-1/3}$. This is a simplified version of (4.2.1).

We couple the randomness by setting

$$N_{n,i} = b(in^{-1/3}) - b((i-1)n^{-1/3})$$

for a fixed Brownian motion b which, here for notational simplicity, has variance $4/\beta$. From now on we fix b and our arguments will be deterministic, so we drop the a.s. notation.

We now embed the domains $\mathbb{R}^n$ of H_n into $L^2(\mathbb{R}^+)$ by the map

$$e_i \mapsto n^{1/6}\mathbf{1}_{[\frac{i-1}{n^{1/3}},\frac{i}{n^{1/3}})},$$

and denote by $\tilde{\mathbb{R}}^n$ the isometric image of $\mathbb{R}^n$ in this embedding. Let $-\Delta_n$, x_n and b_n be the images of the three matrix terms on the right of (4.4.1) under this map, respectively. For $f \in \tilde{\mathbb{R}}^n$, let

$$\|f\|_{*n}^2 = \langle f, (-\Delta_n + x_n)f\rangle$$

and recall the L^* norm $\|f\|_*$ from (4.3.2).

We will need some standard analysis Lemmas.

Exercise 4.4.2. Let $f \in L^*$ of compact support. Let f_n be its orthogonal projection to $\tilde{\mathbb{R}}^n$. Then $f_n \to f$ in L^2, and $\langle f_n, H_n f_n\rangle \to \langle f, Hf\rangle$ where $H = \mathrm{SAO}_\beta$.

Let $\lambda_{n,k}, \lambda_k$ denote the kth lowest eigenvalue of H_n and the Stochastic Airy Operator $H = \mathrm{SAO}_\beta = -\partial_x^2 + x + b'$, respectively.

Proposition 4.4.3. $\limsup \lambda_{n,1} \leqslant \lambda_1$.

Proof. For $\varepsilon > 0$ let f be of compact support and norm 1 so that $\langle f, Hf\rangle \leqslant \lambda_1 + \varepsilon$. Let f_n be the projection of f to $\tilde{\mathbb{R}}^n$. Then by Exercise 4.4.2 we have

$$\lambda_{n,1} \leqslant \frac{\langle f_n, H_n f_n\rangle}{\|f_n\|^2} \to \langle f, Hf\rangle \leqslant \lambda_1 + \varepsilon.$$

Since ε is arbitrary, the claim follows. □

For the upper bound, we need a tightness argument for eigenvectors and eigenvalues.

Exercise 4.4.4. Show that for every $\varepsilon > 0$ there is a random constant C so that

$$\pm b_n \leqslant \varepsilon(-\Delta_n + x_n) + CI$$

in the positive definite order for all n. Hint: use a version of the argument in Lemma 4.3.7.

Note that this exercise implies

$$H_n \geqslant (1-\varepsilon)(-\Delta_n + x_n) - CI$$

and since $-\Delta_n + x_n$ is positive definite, it follows that $\lambda_{n,1} \geqslant -C$, which is a Füredi-Komlós type bound, but now of the right order! (Compare to 3.1.3).

Exercise 4.4.5. Show that if $\tilde{f}_n \to f$ uniformly on compact subsets with f_n differentiable and $f' \in L^2$, then

$$\liminf_{n\to\infty} \|f_n'\| \geqslant \|f'\|.$$

Exercise 4.4.6. Recall that b_n is defined to be the image of b under the embedding defined above. Show that for $\tilde{b}_n$ and $\bar{b}_n$ defined as before we have that

$$\tilde{b}_n \to \tilde{b} \qquad \text{and} \qquad \bar{b}_n' \to \bar{b}'$$

converge uniformly on compact subsets.

Proposition 4.4.7. *Let* $f_n \in \tilde{\mathbb{R}}^n$ *with* $\|f_n\|_{*n} \leqslant c$ *for all* n. *Then* f_n *has a subsequential limit* f *in* L^2 *so that along that subsequence*

$$\liminf \langle f_n, H_n f_n \rangle \geqslant \langle f, Hf \rangle.$$

Proof. Let

$$\tilde{f}_n(x) = \int_0^x \Delta_n f_n(s) ds.$$

Exercise 4.4.8. Show that $\tilde{f}_n - f_n \to 0$ uniformly on compact subsets.

Note that by Cauchy-Schwarz

$$\left|\tilde{f}_n(t+s) - \tilde{f}_n(t)\right| = \left|\int_t^{t+s} \Delta f_n(x) dx\right| \leqslant \sqrt{s}\|\Delta f_n\|, \qquad \text{with} \quad f_n(0) = 0.$$

Therefore the $\tilde{f}_n$ form an equicontinuous family of functions and an application of the Arzela-Ascoli theorem gives us that there exists a continuous f and subsequence such that $\tilde{f}_n \to f$ uniformly on compacts. Moreover we have that $\tilde{f}_n'$ are in L^2 and so there exists $g \in L^2$ and a further subsequence along which $\tilde{f}_n' \to g$ weakly in L^2. This follows from the fact that the balls are weak*-compact. By testing against indicators of intervals we can show that we must have $f' = g$.

From the previous exercise we get the same convergence statements for the f_n and Δf_n.

Recalling the definition of the bilinear form we need to prove several different convergence statements. First observe that

$$\liminf_{n\to\infty} \langle f_n, x_n f_n \rangle \geqslant \langle f, xf \rangle.$$

This follows directly from the positivity of the integrand and Fatou's lemma. That second term $\liminf_{n\to\infty} \|\Delta f_n\|_n \geqslant \|f'\|_2$ follows from Exercise 4.4.5. For the final two terms involving $\bar{b}$ and $\bar{b}'$ we will need to make use of the L^* bounds to cut off the integral at some large number K.

We will first consider the term $\int f_n^2 \bar{b}_n' dx$. For K large enough we have that

$$\left|\int_0^\infty f_n^2 \bar{b}_n' dx - \int_0^K f_n^2 \bar{b}_n' dx\right| \leqslant \int_K^\infty f_n^2 (C + \sqrt{x}) dx \leqslant \int_K^\infty f_n^2 x dx \leqslant \frac{C + \sqrt{K}}{K} \|f_n\|_*.$$

This error may be made arbitrarily small, therefore it will be enough to show the necessary inequality on compact subsets of $\mathbb{R}_+$. This we do by observing that $f_n \to f$ and $\bar{b}_n' \to \bar{b}'$ uniformly on compacts. The dominated convergence theorem implies convergence of the integrals. The following exercise completes the proof. □

Exercise 4.4.9. Prove that $\langle f_n, \bar{b}_n \Delta f_n \rangle \to \langle f, \bar{b} f' \rangle$. Use the same method of cutting off the integral at large K and use convergence on compact subsets.
Hint: $2\int_K^\infty |fg| ds \leqslant \varepsilon \|f\|_2^2 + \frac{1}{\varepsilon}\|g\|_2^2$.

Proposition 4.4.10. $\liminf \lambda_{n,1} \geqslant \lambda_1$.

Proof. By Exercise 4.4.4, in the positive definite order,

$$H_n \leqslant (1+\varepsilon)(-\Delta_n + x_n) + CI$$

but since $\Delta_n + x$ is nonnegative definite, $\lambda_{1,n} \leqslant C$.

Now let $(f_n, \lambda_{n,1})$ be the eigenvector, lowest eigenvalue pair for H_n, so that $\|f_n\| = 1$. Then by Exercise 4.4.4

$$(1-\varepsilon)\|f_n\|_{*n} \leqslant \langle f_n, H_n f_n\rangle + C = \lambda_{n,1} + C \leqslant 2C.$$

Now consider a subsequence along which $\lambda_{n,1}$ converges to its lim inf. By Exercise 4.4.7 we can find a further subsequence of f_n so that $f_n \to f$ in L^2, and

$$\liminf \lambda_{n,1} = \liminf \langle f_n, H_n f_n\rangle \geqslant \langle f, Hf\rangle \geqslant \lambda_1,$$

as required. □

Exercise 4.4.11. Modify the proofs above using the Courant-Fisher characterization to show that for every k, we have $\lambda_{n,k} \to \lambda_k$.

4.5. Tails of the Tracy Widom$_\beta$ distribution.

Definition 4.5.1. We define the Tracy-Widom-β distribution

$$TW_\beta = -\lambda_1(SAO_\beta)$$

In the case $\beta = 1, 2$, and 4 this is consistent with the classical definition. In these cases the soft edge or Airy process may be characterized as a determinantal or Pfaffian process. Tracy and Widom express the law of the lowest eigenvalue in terms of a Painlevé transcendent [32].

The tails are asymmetric. Our methods can be used to show that as $a \to \infty$ the right tail satisfies

$$P(TW_\beta > a) = \exp\left(-\frac{2+o(1)}{3}\beta a^{3/2}\right),$$

see [29]. Here we show that the left tail satisfies the following.

Theorem 4.5.2 ([29]).

$$P(TW_\beta < -a) = \exp\left(-\frac{\beta + o(1)}{24}a^3\right) \qquad \textit{as} \quad a \to \infty.$$

Proof of the upper bound. Suppose we have $\lambda_1 > a$, then for all $f \in L^*$ we have

$$\langle f, A_\beta f\rangle \geqslant a\|f\|_2^2.$$

Therefore we are interested in the probability

$$P\left(\|f'\|_2^2 + \|\sqrt{x}f\|_2^2 + \frac{2}{\sqrt{\beta}}\int f^2 b' dx \geqslant a\|f\|_2^2\right)$$

The first two terms are deterministic, and for f fixed the third term is a Paley-Wiener integral. In particular, it has centered normal distribution with variance

$$\frac{4}{\beta}\int f^4 dx = \frac{4}{\beta}\|f\|_4^4.$$

This leads us to computing

$$P\left(\|f'\|_2^2 + \|\sqrt{x}f\|_2^2 + N\|f\|_4^2 \geqslant a\|f\|_2^2\right),$$

where N is a normal random variable with variance $4/\beta$. Using the standard tail bound for a normal random variable we get

$$\text{(4.5.3)}\quad P\big(\|f'\|_2^2+\|\sqrt{x}f\|_2^2+N\|f\|_4^2 \geqslant a\|f\|_2^2\big) \leqslant 2\exp\left(-\frac{\beta(a\|f\|_2^2-\|f'\|_2^2-\|f\sqrt{x}\|_2^2)^2}{8\|f\|_4^4}\right)$$

and we want to optimize over possible choices of f. It turns out the optimal f will have small derivative, so we will drop the derivative term and then optimize the remaining terms. That is we wish to maximize

$$\frac{(a\|f\|_2^2-\|f\sqrt{x}\|_2^2)^2}{\|f\|_4^4}.$$

With some work we can show that the optimal function will be approximately $f(x) \approx \sqrt{(a-x)^+}$. This needs to be modified a bit in order to keep the derivative small, so we replace the function at the ends of its support by linear pieces:

$$f(x) = \sqrt{(a-x)^+} \wedge (a-x)^+ \wedge x\sqrt{a}.$$

We can check that

$$a\|f\|_2^2 \sim \frac{a^3}{2} \qquad \|f\| = O(a) \qquad \|\sqrt{x}f\|^2 \sim \frac{a^3}{6} \qquad \|f\|_4^4 \sim \frac{a^3}{3}.$$

Using these values in equation (4.5.3) gives us the correct upper bound. □

Proof of the lower bound. We begin by introducing the Riccati transform: Suppose we have an operator

$$L = -\partial_{xx} + V(x),$$

then the eigenvalue equation is

$$\lambda f = (-\partial_{xx} + V(x))f.$$

We can pick a λ and attempt to solve this equation. The left boundary condition is given, so one can check if the solution satisfies $f \in L^2$, in which case we get an eigenfunction. Most of the time this won't be true, but we can still gain information by studying these solutions. To study this problem we first make the transformation

$$p = \frac{f'}{f}, \qquad \text{which gives} \qquad p' = V(x) - \lambda - p^2, \qquad p(0) = \infty.$$

The following is standard part of the theory for Schrödinger operators of the form SAO_β, although some technical work is needed because the potential is irregular.

Proposition 4.5.4. *Choose λ, we will have $\lambda \leqslant \lambda_1$ if an only if the solution to the Ricatti equation does not blow up.*

The slope field looks as follows. When $V(x) = x$ and $\lambda = 0$ there is a right facing parabola $p^2 = x$ where the upper branch is attracting and the lower branch is repelling. The drift will be negative outside the parabola and positive inside. Shifting the initial condition to the left is equivalent to shifting the λ to the right, so this picture may be used to consider the problem for all λ.

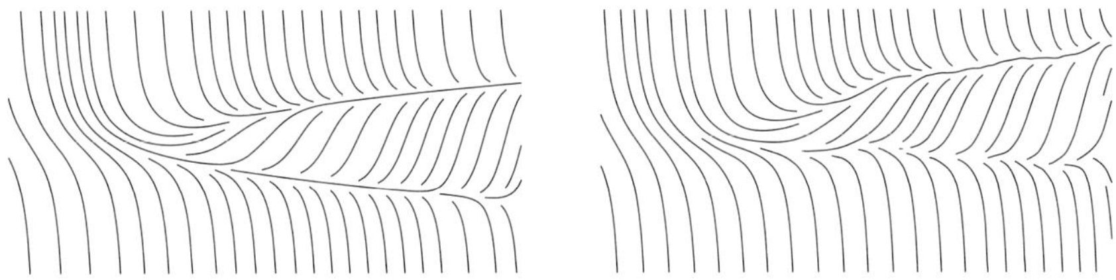

FIGURE 4.5.5. Drift trajectories for p and the random ODE p − B

Now replace $V(x) = x$ by $V(x) = x + \frac{2}{\sqrt{\beta}} b'$. The solution of the Ricatti equation is now an Itô diffusion given by

$$dp(x) = (x + \lambda - p(x)^2)dx + \frac{2}{\sqrt{\beta}} db_x, \qquad p(0) = \infty. \tag{4.5.6}$$

In this case there is some positive chance of the diffusion moving against the drift, including crossing the parabola. Drift trajectories for this slope field and an example of the random slope field for the ODE satisfied by p − B are given in Figure 4.5.5. If we use $P_{-\lambda,y}$ to denote the probability measure associated with starting our diffusion with initial condition $p(-\lambda) = y$, then we get

$$P(\lambda_1 > a) = P_{-a,+\infty}(\text{p does not blow up}).$$

Because diffusion solution paths do not cross, we can bound this below by starting our particle at 1.

$$P_{-a,+\infty}(\text{p does not blow up}) \geqslant P_{-a,1}(\text{p does not blow up}).$$

We now bound this below by requiring that our diffusion stays in $p(x) \in [0,2]$ on the interval $x \in [-a, 0)$ and then choosing convergence to the upper edge of the parabola after 0. This gives

$$\begin{aligned} &P_{-a,1}(\text{p does not blow up}) \\ \geqslant &P_{-a,1}(\text{p stays in } [0,2] \text{ for } x < 0) \cdot P_{0,0}(\text{p does not blow up}). \end{aligned}$$

The second probability is a positive constant not depending on a. We focus on the first event.

A Girsanov change of measure can be used to determine the probability. This change of measure moves us to working on the space where p is replaced by a standard Brownian motion (started at 1). The Radon-Nikodym derivative of this change of measure may be computed explicitly. We compute

$$\begin{aligned} &E_{-a,1}[\mathbf{1}(p_x \in [0,2], x \in (-a,0))] \\ &= E_{-a,1}\left[\exp\left(\frac{\beta}{4}\int_{-a}^0 (x - b^2)db - \frac{\beta}{8}\int_{-a}^0 (x-b^2)^2 dx\right)\mathbf{1}(b_x \in [0,2], x \in (-a,0))\right]. \end{aligned}$$

Notice that when b stays in $[0,2]$, the density term can be controlled

$$\frac{\beta}{4}\int_{-a}^0 (x - b^2)db \sim O(a), \qquad \text{and} \qquad \frac{\beta}{8}\int_{-a}^0 (x-b^2)^2 dx \approx -\frac{\beta}{24}a^3,$$

while the probability of staying in $[0,2]$ is only exponentially small in a. This gives us the desired lower bound. □

5. Related Results

This section will give a brief partial survey of other work that makes use of the tridiagonal matrix models and operator convergence techniques that were introduced in these notes. We will discuss two other local limits that appear in the bulk and the hard-edge of a random matrix model. We will also briefly review results that can be obtained about the limiting processes, connections to sum laws and large deviations, connections to Painelevé, and an alternate viewpoint for operator convergence.

5.1. The Bulk Limit In Section 4 we proved a limit result about the local behavior of the β-Hermite ensemble at the edge of the spectrum. A similar result can be obtained for the local behavior of the spectrum near a point $a\sqrt{n}$ where $|a| < 2$.

The limiting process is the spectrum of the self-adjoint random differential operator $\mathtt{Sine}_\beta$ given by

(5.1.1) $$f \mapsto 2R_t^{-1}\begin{pmatrix} 0 & -\frac{d}{dt} \\ \frac{d}{dt} & 0 \end{pmatrix} f, \qquad f : [0,1) \to \mathbb{R}^2,$$

where R_t is the positive definite matrix representation of hyperbolic Brownian motion with variance $4/\beta$ in logarithmic time. This operator is associated with a canonical system, see de Branges, [11]. It gives a link between the Montgomery-Dyson conjecture about the Sine_2 process and the non-trivial zeros of the Riemann zeta function, the Hilbert-Pólya conjecture and de Brange's attempt to prove the Riemann hypothesis, see [34].

To be more specific, we have

(5.1.2) $$R_t = \frac{1}{2y} X^t_{s(t)} X_{s(t)}, \qquad s(t) = -\log(1-t).$$

where X satisfies the SDE

$$dX = \begin{pmatrix} 0 & dB_1 \\ 0 & dB_2 \end{pmatrix} X, \qquad X_0 = I,$$

and B_1, B_2 are two indepenent copies of Brownian motion with variance $4/\beta$. The boundary conditions are $f(0) \| (1,0)$ and when $\beta > 2$ also $f(1_-) \| X_\infty^{-1}(0,1)^t$. The ratio of entries in $X_t^{-1}(0,1)^t$ performs a hyperbolic Brownian motion in the Poincare half plane representation, see [35].

Theorem 5.1.3 ([34],[35]). *Let Λ_n have β-Hermite distribution and $a \in (-2,2)$ then*

$$\sqrt{4-a^2}\sqrt{n}(\Lambda_n - a\sqrt{n}) \Rightarrow \mathrm{Sine}_\beta$$

where Sine_β *is the point process of eigenvalues of the* $\mathtt{Sine}_\beta$ *operator.*

Remark 5.1.4. Local limit theorems including the bulk limit given in Theorem 5.1.3 were originally proved for $\beta = 1$ and 2 and stated using the integrable structure of the GOE and GUE. The GUE eigenvalues form a determinantal point process, and the GOE eigenvalues form a Pfaffian point process with kernels constructed from Hermite polynomials. The limiting processes may be identified to looking at the limit of the kernel in the appropriate scale. A version for circular ensembles with $\beta > 0$ is proved in [24].

The original description of the bulk limit process for the β-Hermite ensemble was through a process called the Browning Carousel first introduced by Valkó and Virág in [34]. The limiting process introduced there could also be described in terms of a system of coupled stochastic differential equations which gave the counting function of the process. In particular let α_λ satisfy

$$d\alpha_\lambda = \lambda \frac{\beta}{4} e^{-\frac{\beta}{4}t} dt + \operatorname{Re}\left[(e^{-i\alpha_\lambda} - 1)dZ\right], \tag{5.1.5}$$

where $Z_t = X_t + iY_t$ with X and Y standard Brownian motions and $\alpha_\lambda(0) = 0$. The α_λ are coupled through the noise term. Define $N_\beta(\lambda) = \frac{1}{2\pi}\lim_{t\to\infty}\alpha_\lambda(t)$, then $N_\beta(\lambda)$ is the counting function for Sine_β. This characterization is the one used to prove all of the results about Sine_β presented in Section 5.4.

The circular unitary β-ensemble is a distribution on $\mathbb{C}^n$ with joint density proportional to

$$\prod_{i<j} |\lambda_i - \lambda_j|^\beta$$

with respect to length measure on the unit circle. The local convergence to the Sine_β process was first shown by Killip and Stoiciu [24]. Using the Killip-Nenciu [22] representation, Valkó-Virág [35] show that the opeator $\mathtt{Circ}_{n,\beta}$ given by (5.1.1) with hyperbolic Brownian motion replaced by a certain hyperbolic random walk, has eigenvalues that are liftings of these λ_i to the universal cover $\mathbb{R}$. So the convergence of random matrices reduces to convergence of random walks!

In fact, the inverses of the finite-n and limiting operators can be coupled so that they are close in Hilbert-Schmidt norm.

Theorem 5.1.6 ([36]). *There exists a coupling so that for all large n we have*

$$\|\mathtt{Circ}_{\beta,n}^{-1} - \mathtt{Sine}_\beta^{-1}\|_{HS} \leqslant \frac{\log^6 n}{n}.$$

Also, if $\cdots < \lambda_{n,-1} < \lambda_{n,0} < 0 < \lambda_{n,1} < \cdots$ *are the eigenvalues of* $\mathtt{Circ}_{\beta,n}$ *and* λ_k *are the points of* Sine_β *ordered in the same way then for all large n we have*

$$\sum_{k\in\mathbb{Z}} \left(\frac{1}{\lambda_k} - \frac{1}{\lambda_{n,k}}\right)^2 \leqslant \frac{\log^6 n}{n}.$$

Moreover as $n\to\infty$ *we have a.s.*

$$\max_{|k|\leqslant \frac{n^{1/4}}{\log^2 n}} |\lambda_k - \lambda_{n,k}| \to 0.$$

It is the strongest coupling known so far, and it's open whether one can do better than the exponent $1/4$.

5.2. The Hard–edge Limit There is another exciting local behavior that has a general $\beta > 0$ limit process description. This process does not appear as a limit of the β-Hermite ensemble, but does for the related β-Laguerre ensemble. Consider a rectangular matrix $n \times p$ matrix X_n with $p \geqslant n$ and $x_{i,j} \sim \mathcal{N}(0,1)$ all independent. The matrix

$$M_n = X_n X_n^t$$

is a symmetric matrix which may be thought of as a sample covariance matrix for a population with independent normally distributed traits. As in the case of the Gaussian ensembles we could have started with complex entries and looked instead at XX^* to form a Hermitian matrix. The eigenvalues of this matrix have distribution

$$f_{L,\beta}(\lambda_1, \dots, \lambda_n) = \frac{1}{Z_{\beta,n,p}} \prod_{i=1}^{n} \lambda_i^{\frac{\beta}{2}(p-n+1)-1} e^{-\frac{\beta}{2}\lambda_i} \prod_{j<k} |\lambda_j - \lambda_k|^{\beta}, \tag{5.2.1}$$

with $\beta = 1$. This generalizes to the β-Laguerre ensemble which is a set of points with density $f_{L,\beta}$ for any $\beta > 0$. The matrix model M_n is part of a wider class of random matrix models called Wishart matrices. This class of models was originally introduced by Wishart in the 1920's.

As in the case of the Gaussian ensembles there is a limiting spectral measure when the eigenvalues are in the correct scale.

Theorem 5.2.2 (Marchenko-Pastur law). *Let $\lambda_1, \dots, \lambda_n$ have β-Laguerre distribution,*

$$\nu_n = \frac{1}{n} \sum_{i=1}^{n} \delta_{\lambda_i/n},$$

and suppose that $\frac{n}{p} \to \gamma \in (0,1]$. Then as $n \to \infty$

(5.2.3)

$$\nu_n \Rightarrow \sigma_{mp}, \qquad \textit{where} \qquad \frac{d\sigma_{mp}}{dx} = \rho_{mp}(x) = \frac{\sqrt{(\gamma_+ - x)(x - \gamma_-)}}{2\pi\gamma x} \mathbf{1}_{[\gamma_-,\gamma_+]},$$

and $\gamma_\pm = (1 \pm \sqrt{\gamma})^2$.

Notice that this density can display different behavior at the lower end point depending on the value of γ. For any $\gamma < 1$ we get that the lower edge has the same $\sqrt{x}$ type behavior that we see at the edge of the semi-circle distribution. In this case the local limit is again the Airy$_\beta$ process discussed in Section 4. We get something different if $\gamma = 1$. This gives us $\gamma_- = 0$ and the density simplifies to

$$\rho_{mp}(x) = \frac{1}{2\pi} \sqrt{\frac{4-x}{x}} \mathbf{1}_{[0,4]}.$$

In this case the lower edge has an asymptote at 0. In the case where $p - n \to \infty$ this is conjectured to still produce soft-edge behavior and there are limited results in this direction. In the case where $p - n \to a$ as $n \to \infty$ we obtain a different edge process at the lower edge called a hard-edge process. The name derives from the fact that the process occurs when the spectrum of a random matrix is forced against some hard constraint. Recalling the full matrix models for $\beta = 1, 2$ we observe that the matrices are positive definite. This gives a hard lower constraint of 0 for the eigenvalues. If p is close to n than this hard constraint on the lower edge will be felt and so result in different local behavior.

We begin by defining the positive random differential operator

$$\mathfrak{G}_{\beta,a} = -e^{(a+1)x + \frac{2}{\sqrt{\beta}} b(x)} \frac{d}{dx}\left[e^{-ax - \frac{2}{\sqrt{\beta}} b(x)} \frac{d}{dx}\right], \tag{5.2.4}$$

where $b(x)$ is a standard Brownian motion.

Theorem 5.2.5 (Ramírez, Rider, [28]). *Let $0 < \lambda_1 < \lambda_2 < \cdots$ have β–Laguerre distribution with $p - n = a$ and let $\Lambda_1(a) < \Lambda_2(a) < \cdots$ be the eigenvalues of the Stochastic Bessel Operator $\mathfrak{G}_{\beta,a}$ on the positive half-line with Dirichlet boundary conditions, then*

$$\{n\lambda_1, n\lambda_2, ..., n\lambda_k\} \Rightarrow \{\Lambda_1(a), \Lambda_2(a), ..., \Lambda_k(a)\}$$

(jointly in law) for any fixed $k < \infty$ as $n \to \infty$.

Remark 5.2.6. This result was originally conjectured with a different formulation by Edelman and Sutton [13] using intuition similar to the method of proof used for the soft edge. The actual result is proved instead by working with the inverses and a natural embedding of matrices as integral operators with piece-wise constant kernels.

5.3. Universality of local processes Recall the definition of β-ensembles introduced in (1.3.4) with the general potential function $V(x)$. The three local processes that we have discussed capture the local behavior for a wide range of these models. In particular for β-ensembles where the limiting spectral density has a single measure of support and is non-vanishing in the interval as long as $V(x)$ grows fast enough it can be proved that these are the correct limit processes. This was showed first for the bulk process by Bourgade, Erdős, and Yau [8]. For the soft edge this was showed by two groups with slightly different conditions on V and β. Bourgade, Erdős, and Yau use analytical techniques involving Stieltjes transforms [9], while Krishnapur, Rider, and Virág give a proof that makes use of the operator convergence structure studied in these notes [25]. Finally a universality result for the hard edge was shown again using operator methods related to those introduced in these notes by Rider and Waters [30].

5.4. Properties of the limit processes For the Stochastic Airy Operator (4.2.3) we saw that it was useful to have a family of stochastic differential equations that characterizes the point process. The SDEs for SAO_β came from considering the Ricatti equation. We can build a similar family of diffusions for the Stochastic Bessel Operator introduced in (5.2.4). There is also a description for the counting function of the bulk process in terms of SDEs which was given in (5.1.5). These characterizations are be used to prove the results introduced in this section.

We will begin by discussing the result for the Sine_β process. The first two results are asymptotic results for the number of points in a large interval $[0, \lambda]$. Let

$N_\beta(\lambda)$ denote the number of points of Sine_β in $[0,\lambda]$. By looking at the integrated expression of α_λ we can check that $EN_\beta(\lambda) = \frac{\lambda}{2\pi}$ we consider fluctuation around the mean.

Theorem 5.4.1 (Kritchevski, Valkó, Virág [26]). *As $\lambda \to \infty$ we have*

$$\frac{1}{\sqrt{\log \lambda}}\left(N_\beta(\lambda) - \frac{\lambda}{2\pi}\right) \Rightarrow \mathcal{N}(0, \frac{2}{\beta\pi^2})$$

This result describes the distribution of the fluctuations on the scale of $\sqrt{\log \lambda}$ there are other regimes. In particular for fluctuation on the order of $c\lambda$ we have the following.

Theorem 5.4.2 (Holcomb, Valkó [19]). *The rescaled counting function $N_\beta(\lambda)/\lambda$ satisfies a large deviation principle with scale λ^2 and a good rate function $\beta I_{\mathrm{Sine}_\beta}(\rho)$ which can be written in terms of elliptic integrals.*

Roughly speaking, this means for large λ

$$P(N_\beta(\lambda) \sim \rho\lambda) \sim e^{-\lambda^2 I_{\mathrm{Sine}_\beta}(\rho)}.$$

Remark 5.4.3. Results similar to Theorems 5.4.1 and 5.4.2 may be shown for the hard edge process. The key observation is that there is an SDE description for the counting function that may be treated using mostly the same techniques as those used for the α_λ diffusion that characterizes the Sine_β process [17].

The next result give the asymptotic probability of having a large number of points in a small interval.

Theorem 5.4.4 (Holcomb, Valkó [20]). *Fix $\lambda_0 > 0$. Then there exists c depending only on β and λ_0 such that for any $n \geqslant 1$ and $0 < \lambda \leqslant \lambda_0$ we have*

$$P(N_\beta(\lambda) \geqslant n) \leqslant e^{-\frac{\beta}{2}n^2\log(\frac{n}{\lambda}) + cn\log(n+1)\log(\frac{n}{\lambda}) + cn^2}. \tag{5.4.5}$$

Moreover, there exists an $n_0 \geqslant 1$ so that for any $n \geqslant n_0$ and for any λ with $0 < \lambda \leqslant \lambda_0$ we also have

$$P(N_\beta(\lambda) = n) \geqslant e^{-\frac{\beta}{2}n^2\log(\frac{n}{\lambda}) - cn\log(n+1)\log(\frac{n}{\lambda}) - cn^2}. \tag{5.4.6}$$

The previous three results focused on the number of points in a single interval. In this situation we have the advantage that the α_λ diffusion satisfies a simplified SDE

$$d\alpha_\lambda = \lambda\frac{\beta}{4}e^{-\frac{\beta}{4}t}dt + 2\sin\left(\frac{\alpha_\lambda}{2}\right)dB_t^{(\lambda)},$$

where the Brownian motion that appears $B^{(\lambda)}$ depends on the choice of parameter. The next two results require information multiple values of λ and so this simplification cannot be used. The first is a result on the maximum deviation of the counting function. This is closely related to questions on the maximum of $\mathrm{Im}\log\Phi_n(x)$ where $\Phi_n(x)$ is the characteristic polynomial of the $n \times n$ tridiagonal model.

Theorem 5.4.7 (Holcomb, Paquette [18]).

$$\frac{\max_{0\leqslant\lambda\leqslant x}[N_\beta(\lambda)+N_\beta(-\lambda)-\frac{\lambda}{\pi}]}{\log x} \xrightarrow[x\to\infty]{P} \frac{2}{\sqrt{\beta\pi}}.$$

The next result is a type of rigidity for the Sine_β point process.

Definition 5.4.8. A point process X on a complete separable metric space E is rigid if and only if for all bounded Borel subsets B of E, the number of points $X(B)$ in B is measurable with respect to the σ-algebra $\Sigma_{E\setminus B}$. Here $\Sigma_{E\setminus B}$ is the σ-algebra generated by all of the random variables $X(A)$ with $A \subset E\setminus B$.

A way of thinking about this is that if we have complete information about a point process X outside of a set B, then this determines the number of points in B. Notice that in a finite point process with n points this notion of rigidity follows immediately since we must have $X(B) + X(E\setminus B) = n$.

Theorem 5.4.9 (Chhaibi, Najnudel [10]). *The* Sine_β *point process is rigid in the sense of Definition* 5.4.8.

5.5. Spiked matrix models and more on the BBP transition Recall that is Section 3 we studied the impact of a rank one perturbation on the top eigenvalue of the GOE. In that case we studied the case where the perturbation was strong enough to be seen at the scale of the empirical spectral density. That is that the location of the top eigenvalue when scaled down by $\sqrt{n}$ depends on the strength of the perturbation. These types of results may be refined further to consider the impact of such a perturbation at the level of the local interactions. As in Section 3, we may look at two types of perturbations.

(1) For additive ensembles, we study the perturbations

$$\mathrm{GOE}_n + \frac{a}{\sqrt{n}}11^t.$$

Here GOE_n is the $n\times n$ full matrix model, and 1 is the all-1 vector, while 11^t is the all-1 matrix.

(2) For multiplicative type ensembles, we take $X_{n\times m}$ be an $n\times m(n)$ matrix with $n < m(n)$ and independent $\mathcal{N}(0,1)$ entries, then we study the perturbations

$$X\,\mathrm{diag}(1+a^2,1,1,...,1)X^t.$$

Here we will focus on the additive case. In this case it can be shown that if $\mathbf{T}$ is the tridiagonal matrix obtained by tridiagonalizing a GOE, then the corresponding tridiagonal model will have the form $\mathbf{T} + (a\beta n + aY)\mathbf{e}_1\mathbf{e}_1^t$ for some random variable Y with $EY = 0$ and EY^2 bounded (here $\mathbf{e}_1\mathbf{e}_1^t$ is the $n\times n$ matrix with a 1 in the top left corner and 0 everywhere else). In the analogue to the soft edge limit if we have $n^{1/3}(1-a) \to w \in (-\infty,\infty]$ then the top eigenvalues will converge to the eigenvalues of the Stochastic Airy operator, but with a modified boundary condition.

Exercise 5.5.1. Let

$$T = m_n^2 \begin{bmatrix} 1+\frac{w}{m_n} & -1 & & & \\ -1 & 2 & -1 & & \\ & -1 & 2 & \ddots & \\ & & \ddots & \ddots & -1 \\ & & & -1 & \ddots \end{bmatrix}.$$

Show that $Tv_f \to -\partial_x^2$ for $v_f = [f(0), f(\frac{1}{m_n}), f(\frac{2}{m_n}), ..., f(\frac{n}{m_n})]$ with an appropriate boundary condition for f. Determine what the boundary condition should be.

We denote by $\mathcal{H}_{\beta,w}$ the Stochastic Airy Operator defined in equation (4.2.3), with boundary condition $f'(0) = wf(0)$, a Neumann or Robin condition with the $w = \infty$ case corresponding to the Dirichlet condition of the original soft edge process $f(0) = 0$.

Theorem 5.5.2 (Bloemendal, Virág, [5]). *Let $a_n \in \mathbb{R}$, and $G_n \sim \mathrm{GOE}_n + \frac{a_n}{\sqrt{n}} 11^t$, and suppose that $n^{1/3}(1 - a_n) \to w \in (-\infty, \infty]$ and $n \to \infty$. Let $\lambda_1 > \lambda_2 > \cdots > \lambda_n$ be the ordered eigenvalues of G_n. Then jointly for $k = 1, 2, ...$ in the sense of finite dimensional distributions we have*

$$n^{1/6}(2\sqrt{n} - \lambda_k) \Rightarrow \Lambda_k \qquad \text{as } n \to \infty,$$

where $\Lambda_1 < \Lambda_2 < \cdots$ are the eigenvalues of $\mathcal{H}_{1,w}$.

As in the case of the eigenvalue problem for the stochastic Airy operator that was originally studied (with boundary condition $f(0) = 0$) we may study the Ricatti process introduced in (4.5.6). The new boundary condition for $\mathcal{H}_{\beta,w}$ leads us to the same diffusion with a different boundary condition

$$dp_\lambda = (x + \lambda - p^2)dx + \frac{d}{\sqrt{\beta}} db_x, \qquad p(0) = w.$$

Relationships between the law of the perturbed eigenvalues and the original Tracy-Widom distributions may be studied using the space-time generator for this SDE.

The generator gives a boudnary value problem representation for the Tracy-Widom$_\beta$ distribution. This can be solved explicitly for $\beta = 2, 4$. This gives fast derivations of the famous Painlevé representation for the Tracy-Widom distributions without the use of determinantal formulas. See [5,6] for further results and details. It is not known how to deduce Painlevé formulas for the Sine$_\beta$ process directly, even for $\beta = 2$.

5.6. Sum rules via large deviations Sum rules are a family of relationships used and studied in the field of orthogonal polynomials that give a relationship between a functional on a subset of probability measures and the recurrence (or Jacobi) coefficients of the orthogonal polynomials. It was recently recognized by Gamboa, Nagel, and Rouault that these relationships can be obtained using large deviation theory for random matrices [15]. This is a beautiful example of the power of large deviation theory as well as a demonstration of the relationship between the Jacobi data and spectral data of an operator. Here we will only state the result for the semicircle distribution, but the methods have been used for a wider range of models including the Marchenko-Pastur law and matrix valued measures. The proof of the theorem for the Hermite/semicircle case is originally due to Killip and Simon [23] using different methods. The real advance here is the recognition that Large Deviations may be used to prove sum rules. These techniques may then be used on a wider range of models.

Before introducing the theorem statement we introduce the Kullback-Leibler divergence or relative entropy between two probability measures μ and ν

$$\mathcal{K}(\mu|\nu) = \begin{cases} \int_{\mathbb{R}} \log \frac{d\mu}{d\nu} d\mu & \text{if } \mu \ll \nu \\ \infty & \text{otherwise.} \end{cases} \tag{5.6.1}$$

Here $\mu \ll \nu$ means that μ is absolutely continuous with respect to ν.

Now returning to the semicircle distribution we note that the Jacobi coefficients for the semicircle measure are given by

$$a_k^{sc} = 0, \qquad b_k^{sc} = 1, \qquad \text{for } k \geqslant 1. \tag{5.6.2}$$

The corresponding orthogonal polynomials are the Chebyshev polynomials of the second kind. Now suppose that μ is a probability measure on $\mathbb{R}$ with Jacobi coefficients $\{a_k, b_k\}_{k \geqslant 1}$ and define

$$\mathcal{I}_H(\mu) = \sum_{k \geqslant 1} \frac{a_k^2}{2} + b_k - 1 - \log b_k. \tag{5.6.3}$$

Now suppose that $\operatorname{supp}(\mu) = I \cup \{\lambda_i^-\}_{i=1}^{N-} \cup \{\lambda_i^+\}_{i=1}^{N+}$ where we have $I \subset [-2, 2]$, $\lambda_1^- < \lambda_2^- < \cdots < -2$ and $\lambda_1^+ > \lambda_2^+ > \cdots > 2$. and define

$$\mathcal{F}_H^+(x) = \begin{cases} \int_2^x \sqrt{t^2 - 4}dt & \text{if } x \geqslant 2 \\ \infty \text{ otherwise,} \end{cases} \tag{5.6.4}$$

and $\mathcal{F}_H^-(x) = \mathcal{F}_H^+(-x)$.

Theorem 5.6.5 (Killip and Simon [23], Gamboa, Nagel, and Rouault [15]). *Let* J *be a Jacobi matrix with diagonal entries* $a_1, a_2, \ldots \in \mathbb{R}$ *and off-diagonal entries* $b_1, b_2, \ldots > 0$ *satisfying* $\sup_k b_k + \sup_k |a_k| < \infty$ *and let* μ *be the associated spectral measure. Then* $\mathcal{I}_H(\mu)$ *is infinite if*

$$\operatorname{supp}(\mu) \neq I \cup \bigcup_{i=1}^{N-} \{\lambda_i^-\} \cup \bigcup_{i=1}^{N+} \{\lambda_i^+\}$$

as given above. If μ *has the desired support structure, then*

$$\mathcal{I}_H(\mu) = \mathcal{K}(\mu_{sc}|\mu) + \sum_{i=1}^{N^+} \mathcal{F}_H^+(\lambda_i^+) + \sum_{i=1}^{N^-} \mathcal{F}_H^-(\lambda_i^-)$$

where both sides may be infinite simultaneously.

This is the same $\mathcal{J}_H$ given in (5.6.3). This gives us that for measures that are "close enough" to semicircular (where $\mathcal{J}_H(\mu)$ is finite) we get that

$$\sum_{k\geqslant 1} \frac{a_k^2}{2} + b_k - 1 - \log b_k = \mathcal{K}(\mu_{sc}|\mu) + \sum_{i=1}^{N^+} \mathcal{F}_H^+(\lambda_i^+) + \sum_{i=1}^{N^-} \mathcal{F}_H^-(\lambda_i^-).$$

The important observation is that $\mathcal{J}_H(\mu)$ is a large deviation rate function for the appropriate large deviation problem. Because the spectral data and the Jacobi coefficient data can both be used to describe the asymptotic likelihood of the same event the rate functions must coincide. See more details and more sum rules in (e.g. for Marchenko-Pastur) in [15].

5.7. The Stochastic Airy semigroup The idea here will be to show convergence of the moment generating function tridiagonal matrix model to the operator $e^{-\frac{T}{2}SAO_\beta}$. This work by Gorin and Shkolnikov makes use of the moment method to prove this alternate version of convergence [16]. We begin by considering the tridiagonal matrix model with the coefficients reversed

$$M_N = \frac{1}{\sqrt{\beta}} \begin{bmatrix} a_1 & b_1 & & & \\ b_1 & a_2 & b_2 & & \\ & b_2 & a_3 & \ddots & \\ & & \ddots & \ddots & b_{N-1} \\ & & & b_{N-1} & a_n \end{bmatrix}, \qquad b_k \sim \chi_{\beta k}, \quad a_k \sim \mathcal{N}(0,2). \tag{5.7.1}$$

Take $A \subset \mathbb{R}_{\geqslant 0}$, $T > 0$, and define $[M_{N,A}]_{i,j} = [M_N]_{i,j}\mathbf{1}(\frac{N-i+1/2}{N^{1/3}}, \frac{N-j+1/2}{N^{1/3}} \in A)$. Then we study the moments

$$\mathcal{M}(T,A,N) = \frac{1}{2}\left(\left(\frac{M_{N,A}}{2\sqrt{N}}\right)^{\lfloor TN^{2/3}\rfloor} + \left(\frac{M_{N,A}}{2\sqrt{N}}\right)^{\lfloor TN^{2/3}\rfloor - 1}\right).$$

Theorem 5.7.2 (Gorin and Shkolnikov, [16]). *There exist an almost surely symmetric non-negative trace class operator $\mathcal{U}_A(T)$ on $L^2(\mathbb{R}_{\geqslant 0})$ with $\mathcal{U}_{\mathbb{R}_{\geqslant 0}}(T) = e^{-\frac{T}{2}SAO_\beta}$ almost surely such that*

$$\lim_{N\to\infty} \mathcal{M}(T,A,N) = \mathcal{U}_A(T),$$

in the following senses for any $T \geqslant 0$:

(1) *Weak convergence: For any locally integrable f, g with subexponential growth at infinity, if $\pi_N f$ denotes the appropriate projection of f onto step functions, then*

$$\lim_{N\to\infty} (\pi_N f)^t \mathcal{M}(T,A,N)(\pi_N g) = \int_{\mathbb{R}_{\geqslant 0}} (\mathcal{U}_A(T)f)(x)g(x)dx$$

in distribution and in the sense of moments.

(2) *Convergence of traces: Moreover,*

$$\lim_{N\to\infty} \operatorname{Trace}(\mathcal{M}(T,A,N)) = \operatorname{Trace}(\mathcal{U}_A(T))$$

in distribution and in the sense of moments.

This is an alternate notion of operator convergence.

References

[1] D. Aldous and R. Lyons, *Processes on unimodular random networks*, Electron. J. Probab. **12** (2007), no. 54, 1454–1508. MR2354165 ↑221

[2] G. W. Anderson, A. Guionnet, and O. Zeitouni, *An introduction to random matrices*, Cambridge Studies in Advanced Mathematics, vol. 118, Cambridge University Press, Cambridge, 2010. MR2760897 ↑213

[3] J. Baik, G. Ben Arous, and S. Péché, *Phase transition of the largest eigenvalue for nonnull complex sample covariance matrices*, Ann. Probab. **33** (2005), no. 5, 1643–1697. MR2165575 ↑227

[4] I. Benjamini and O. Schramm, *Recurrence of distributional limits of finite planar graphs*, Electron. J. Probab. **6** (2001), no. 23, 13. MR1873300 ↑221

[5] A. Bloemendal and B. Virág, *Limits of spiked random matrices I*, Probab. Theory Related Fields **156** (2013), no. 3-4, 795–825. MR3078286 ↑227, 246

[6] A. Bloemendal and B. Virág, *Limits of spiked random matrices II*, Ann. Probab. **44** (2016), no. 4, 2726–2769. MR3531679 ↑246

[7] A. Bloemendal, *Finite Rank Perturbations of Random Matrices and Their Continuum Limits*, ProQuest LLC, Ann Arbor, MI, 2011. Thesis (Ph.D.)–University of Toronto (Canada). MR3004406 ↑226

[8] P. Bourgade, L. Erdős, and H.-T. Yau, *Bulk universality of general* β*-ensembles with non-convex potential*, J. Math. Phys. **53** (2012), no. 9, 095221, 19. MR2905803 ↑243

[9] P. Bourgade, L. Erdős, and H.-T. Yau, *Edge universality of beta ensembles*, Comm. Math. Phys. **332** (2014), no. 1, 261–353. MR3253704 ↑243

[10] R. Chhaibi and J. Najnudel, *Rigidity of the* Sine_β *process*, ArXiv e-prints (April 2018), available at 1804.01216. ↑245

[11] L. de Branges, *Hilbert spaces of entire functions*, Prentice-Hall, Inc., Englewood Cliffs, N.J., 1968. ↑240

[12] I. Dumitriu and A. Edelman, *Matrix models for beta ensembles*, J. Math. Phys. **43** (2002), no. 11, 5830–5847. MR1936554 ↑218, 219

[13] A. Edelman and B. D. Sutton, *From random matrices to stochastic operators*, J. Stat. Phys. **127** (2007), no. 6, 1121–1165. MR2331033 ↑230, 243

[14] Z. Füredi and J. Komlós, *The eigenvalues of random symmetric matrices*, Combinatorica **1** (1981), no. 3, 233–241. MR637828 ↑225

[15] F. Gamboa, J. Nagel, and A. Rouault, *Sum rules via large deviations*, J. Funct. Anal. **270** (2016), no. 2, 509–559. MR3425894 ↑247, 248

[16] V. Gorin and M. Shkolnikov, *Stochastic Airy semigroup through tridiagonal matrices*, Ann. Probab. **46** (2018), no. 4, 2287–2344. MR3813993 ↑248

[17] D. Holcomb, *The random matrix hard edge: rare events and a transition*, Electron. J. Probab. **23** (2018), 1–20. ↑244

[18] D. Holcomb and E. Paquette, *The maximum deviation of the* Sine_β *counting process*, Electron. Commun. Probab. **23** (2018), 1–13. ↑245

[19] D. Holcomb and B. Valkó, *Large deviations for the* Sine_β *and* Sch_τ *processes*, Probab. Theory Related Fields **163** (2015), no. 1-2, 339–378. MR3405620 ↑244

[20] D. Holcomb and B. Valkó, *Overcrowding asymptotics for the* sine_β *process*, Ann. Inst. Henri Poincaré Probab. Stat. **53** (2017), no. 3, 1181–1195. MR3689965 ↑244

[21] O. Kallenberg, *Random measures, theory and applications*, Probability Theory and Stochastic Modelling, vol. 77, Springer, Cham, 2017. MR3642325 ↑215

[22] R. a. N. Killip Irina, *Matrix models for circular ensembles*, International Mathematics Research Notices **2004** (2004), no. 50, 2665–2701. ↑241

[23] R. Killip and B. Simon, *Sum rules for Jacobi matrices and their applications to spectral theory*, Ann. of Math. (2) **158** (2003), no. 1, 253–321. MR1999923 ↑247

[24] R. Killip and M. Stoiciu, *Eigenvalue statistics for CMV matrices: from Poisson to clock via random matrix ensembles*, Duke Math. J. **146** (2009), no. 3, 361–399. MR2484278 ↑241

[25] M. Krishnapur, B. Rider, and B. Virág, *Universality of the stochastic Airy operator*, Comm. Pure Appl. Math. **69** (2016), no. 1, 145–199. MR3433632 ↑219, 243

[26] E. Kritchevski, B. Valkó, and B. Virág, *The scaling limit of the critical one-dimensional random Schrödinger operator*, Comm. Math. Phys. **314** (2012), no. 3, 775–806. MR2964774 ↑244

[27] R. Lyons and Y. Peres, *Probability on trees and networks*, Cambridge Series in Statistical and Probabilistic Mathematics, vol. 42, Cambridge University Press, New York, 2016. MR3616205 ↑221

[28] J. A. Ramí rez and B. Rider, *Diffusion at the random matrix hard edge*, Comm. Math. Phys. **288** (2009), no. 3, 887–906. MR2504858 ↑243

[29] J. A. Ramí rez, B. Rider, and B. Virág, *Beta ensembles, stochastic Airy spectrum, and a diffusion*, J. Amer. Math. Soc. **24** (2011), no. 4, 919–944. MR2813333 ↑230, 237

[30] B. Rider and P. Waters, *Universality of the Stochastic Bessel Operator*, ArXiv e-prints (October 2016), available at 1610.01637. ↑243

[31] B. Simon, *Operator theory*, A Comprehensive Course in Analysis, Part 4, American Mathematical Society, Providence, RI, 2015. MR3364494 ↑231

[32] C. A. Tracy and H. Widom, *Level-spacing distributions and the Airy kernel*, Comm. Math. Phys. **159** (1994), no. 1, 151–174. MR1257246 ↑237

[33] H. F. Trotter, *Eigenvalue distributions of large Hermitian matrices; Wigner's semicircle law and a theorem of Kac, Murdock, and Szegő*, Adv. in Math. **54** (1984), no. 1, 67–82. MR761763 ↑216, 217

[34] B. Valkó and B. Virág, *Continuum limits of random matrices and the Brownian carousel*, Invent. Math. **177** (2009), no. 3, 463–508. MR2534097 ↑240, 241

[35] B. Valkó and B. Virág, *The* Sine$_\beta$ *operator*, Invent. Math. **209** (2017), no. 1, 275–327. MR3660310 ↑240, 241

[36] B. Valkó and B. Virág, *Operator limit of the circular* β *ensemble*, ArXiv e-prints (2017), available at 1710.06988. ↑241

[37] B. Virág, *Operator limits of random matrices*, Proceedings of the International Congress of Mathematicians—Seoul 2014. Vol. IV, 2014, pp. 247–271. MR3727611 ↑224

KTH, Lindstedtsvägen 25, 100 44 Stockholm, Sweden
Email address: holcomb@kth.se

University of Toronto, 40 St George St., Toronto, ON, M5S 2E4, Canada
Email address: balint@math.toronto.edu

IAS/Park City Mathematics Series
Volume 26, Pages 251–301
https://doi.org/10.1090/pcms/026/00847

From the totally asymmetric simple exclusion process to the KPZ fixed point

Jeremy Quastel and Konstantin Matetski

Abstract. These notes are based on the article *The KPZ fixed point* of Matetski, Quastel and Remenik [22] and give a self contained exposition of construction of the KPZ fixed point which is a Markov process at the centre of the KPZ universality class. Starting from the Schütz's formula for transition probabilities of the totally asymmetric simple exclusion process, the method is by writing them as the biorthogonal ensemble/non-intersecting path representation found by Borodin, Ferrari, Prähofer and Sasamoto. We derive an explicit formula for the correlation kernel which involves transition probabilities of a random walk forced to hit a curve defined by the initial data. This in particular yields a Fredholm determinant formula for the multipoint distribution of the height function of the totally asymmetric simple exclusion process with arbitrary initial condition. In the 1:2:3 scaling limit the formula leads in a transparent way to a Fredholm determinant formula for the KPZ fixed point, in terms of an analogous kernel based on Brownian motion. The formula readily reproduces known special self-similar solutions such as the Airy_1 and Airy_2 processes.

1. The totally asymmetric simple exclusion process

The *totally asymmetric simple exclusion process* (TASEP) is a basic interacting particle system studied in non-equilibrium statistical mechanics. The system consists of particles performing totally asymmetric nearest neighbour random walks on the one-dimensional integer lattice with the exclusion rule. Each particle independently attempts to jump to the neighbouring site to the right at rate 1, the jump being allowed only if that site is unoccupied. More precisely, if we denote by $\eta \in \{0,1\}^{\mathbb{Z}}$ a particle configuration (where $\eta_x = 1$ if there is a particle at the site x, and $\eta_x = 0$ if the site is empty), then TASEP is a Markov process with infinitesimal generator acting on cylinder functions $f : \{0,1\}^{\mathbb{Z}} \to \mathbb{R}$ by

$$(Lf)(\eta) := \sum_{x \in \mathbb{Z}} \eta_x(1-\eta_{x+1})\big(f(\eta^{x,x+1}) - f(\eta)\big),$$

2010 *Mathematics Subject Classification.* Primary 60K35; Secondary 82C27.
Key words and phrases. TASEP, growth process, biorthogonal ensemble, determinantal point process, KPZ fixed point.

where $\eta^{x,x+1}$ denotes the configuration η with interchanged values at x and $x+1$:

$$\eta_y^{x,x+1} := \begin{cases} \eta_{x+1}, & \text{if } y = x, \\ \eta_x, & \text{if } y = x+1, \\ \eta_y, & \text{if } y \notin \{x, x+1\}. \end{cases}$$

See [20] for the proof of the non-trivial fact that this process is well-defined.

Exercise 1.0.1. *Prove that the following measures* μ *are invariant for TASEP, i.e. prove that* $\int (Lf)d\mu = 0$*:*

(1) *the Bernoulli product measures with any density* $\rho \in [0,1]$,

(2) *the Dirac measure on any configuration with* $\eta_x = 1$ *for* $x \geqslant x_0$, $\eta_x = 0$ *for* $x < x_0$.

It is known [20] *that these are the only invariant measures.*

The TASEP dynamics preserves the order of particles. Let us denote positions of particles at time $t \geqslant 0$ by

$$\cdots < X_t(2) < X_t(1) < X_t(0) < X_t(-1) < X_t(-2) < \cdots,$$

where $X_t(i) \in \mathbb{Z}$ is the position of the i-th particle. Adding $\pm\infty$ into the state space and placing a necessarily infinite number of particles at infinity allows for left- or right-finite data with no change of notation (the particles at $\pm\infty$ are playing no role in the dynamics). We follow the standard practice of ordering particles from the right; for right-finite data the rightmost particle is labelled 1.

TASEP is a particular case of the *asymmetric simple exclusion process* (ASEP) introduced by Spitzer in [30]. Particles in this model jump to the right with rate p and to the left with rate q such that $p+q=1$, following the exclusion rule. Obviously, in the case $p=1$ we get TASEP. In the case $p \in (0,1)$ the model becomes significantly more complicated comparing to TASEP, for example Schütz's formula described in Section 2 below cannot be written as a determinant which prevents the following analysis in the general case. ASEP is important because of the weakly asymmetric limit, which means to diffusively rescale the growth process introduced below as $\varepsilon^{1/2}h_{\varepsilon^{-2}t}(\varepsilon^{-1}z)$ while at the same time taking $q-p=\mathcal{O}(\varepsilon^{1/2})$, in order to obtain the KPZ equation [1].

1.1. The growth process. Of special interest in non-equilibrium physics is the *growth process* associated to TASEP. More precisely, let

$$X_t^{-1}(u) := \min\{k \in \mathbb{Z} : X_t(k) \leqslant u\}$$

denote the label of the rightmost particle which sits to the left of, or at, u at time t. The *TASEP height function* associated to X_t is given for $z \in \mathbb{Z}$ by

$$h_t(z) := -2\left(X_t^{-1}(z-1) - X_0^{-1}(-1)\right) - z, \tag{1.1.1}$$

which fixes $h_0(0) = 0$. The height function is a random walk path satisfying $h_t(z+1) = h_t(z) + \hat{\eta}_t(z)$ with $\hat{\eta}_t(z) = 1$ if there is a particle at z at time t and $\hat{\eta}_t(z) = -1$ if there is no particle at z at time t. We can also easily extend the height function to a continuous function of $x \in \mathbb{R}$ by linearly interpolating between the integer points.

Exercise 1.1.2. *Show that the dynamics of h_t is that local max's become local min's at rate 1; i.e. if $h_t(z) = h_t(z \pm 1) + 1$ then $h_t(z) \mapsto h_t(z) - 2$ at rate 1, the rest of the height function remaining unchanged (see the figure below). What happens if we consider ASEP?*

FIGURE 1.1.3. Evolution of TASEP and its height function.

Two standard examples of initial data for TASEP are the *step initial data* (when $X_0(k) = -k$ for $k \geqslant 1$) and d-*periodic initial data* (when $X_0(k) = -d(k-1)$ for $k \in \mathbb{Z}$) with $d \geqslant 2$. Analysis of TASEP with one of these initial data is much easier than in the general case. In particular the results presented in Sections 5 and 6 below were known from [6,7] and served as a starting point for our work.

2. Distribution function of TASEP

If the number of particles is finite, we can alternatively denote their positions

$$\vec{x} \in \Omega_N := \{x_N < \cdots < x_1\} \subset \mathbb{Z}^N,$$

where Ω_N is called the *Weyl chamber*. The transition probabilities for TASEP with a finite number of particles was first obtained in [29] using *(coordinate) Bethe ansatz*.

Proposition 2.0.1 (Schütz's formula). *The transition probability for $2 \leqslant N < \infty$ TASEP particles has a determinantal form*

$$\mathbb{P}(X_t = \vec{x} \,|\, X_0 = \vec{y}) = \det\left[F_{i-j}(x_{N+1-i} - y_{N+1-j}, t)\right]_{1 \leqslant i,j \leqslant N} \tag{2.0.2}$$

with $\vec{x}, \vec{y} \in \Omega_N$, and

$$F_n(x,t) := \frac{(-1)^n}{2\pi i} \oint_{\Gamma_{0,1}} dw \, \frac{(1-w)^{-n}}{w^{x-n+1}} e^{t(w-1)}, \tag{2.0.3}$$

where $\Gamma_{0,1}$ is any simple loop oriented anticlockwise which includes $w = 0$ and $w = 1$.

In the rest of this section we provide a proof of this result using Bethe ansatz and in Section 2.2 we show that Schütz's formula can alternatively be easily checked to satisfy the Kolmogorov forward equation.

2.1. Proof of Schütz's formula using Bethe ansatz. In this section we will prove Proposition 2.0.1 following the argument of [31].We will consider $N \geqslant 2$ particles in TASEP and derive the master (Kolmogorov forward) equation for the process

$X_t = (X_t(1), \cdots, X_t(N)) \in \Omega_N$, where Ω_N is the Weyl chamber defined above. For a function $F : \Omega_N \to \mathbb{R}$ we introduce the operator

$$(\mathcal{L}^{(N)}F)(\vec{x}) := -\sum_{k=1}^{N} \mathbf{1}_{\{x_k - x_{k+1} > 1\}} (\nabla_k^- F)(\vec{x}),$$

where $x_{N+1} = -\infty$ and ∇_k^- is the discrete derivative

$$\nabla^- f(z) := f(z) - f(z-1), \qquad f : \mathbb{Z} \to \mathbb{R}, \tag{2.1.1}$$

acting on the k-th argument of F. One can see that this is the infinitesimal generator of TASEP in the variables X_t. Thus, if

$$P_t^{(N)}(\vec{y}, \vec{x}) := \mathbb{P}(X_t = \vec{x} \,|\, X_0 = \vec{y})$$

is the transition probability of N particles of TASEP from $\vec{y} \in \Omega_N$ to $\vec{x} \in \Omega_N$, then *the master equation (=Kolmogorov forward equation)* is

$$\frac{d}{dt} P_t^{(N)}(\vec{y}, \cdot) = \mathcal{L}^{(N)} P_t^{(N)}(\vec{y}, \cdot), \qquad P_0^{(N)}(\vec{y}, \cdot) = \delta_{\vec{y}, \cdot}. \tag{2.1.2}$$

The idea of [2] was to rewrite (2.1.2) as a differential equation with constant coefficients and boundary conditions, i.e. if $u_t^{(N)} : \mathbb{Z}^N \to \mathbb{R}$ solves

$$\frac{d}{dt} u_t^{(N)} = -\sum_{k=1}^{N} \nabla_k^- u_t^{(N)}, \qquad u_0^{(N)}(\vec{x}) = \delta_{\vec{y}, \vec{x}}, \tag{2.1.3}$$

with the *boundary conditions*

$$\nabla_k^- u_t^{(N)}(\vec{x}) = 0, \quad \text{when} \quad x_k = x_{k+1} + 1, \tag{2.1.4}$$

then for $\vec{x}, \vec{y} \in \Omega_N$ one has

$$P_t^{(N)}(\vec{y}, \vec{x}) = u_t^{(N)}(\vec{x}). \tag{2.1.5}$$

Exercise 2.1.6. *Prove this by induction on* $N \geqslant 1$.

The strategy is now to find a general solution to the master equation (2.1.3) and then a particular one which satisfies the boundary and initial conditions. The method is known as *(coordinate) Bethe ansatz.*

Solution to the master equation. For a fixed $\vec{y} \in \mathbb{Z}^N$, we are going to find a solution to the equation (2.1.3). For this we will consider indistinguishable particles, so that for given positions of particles $\{x_1, \cdots, x_N\} \subset \mathbb{Z}$ we will look for a solution of the form

$$\sum_{\sigma \in S_N} u_t^{(N)}(\vec{x}_\sigma),$$

where S_N is the symmetric group and $\vec{x}_\sigma = (x_{\sigma(1)}, \cdots, x_{\sigma(N)})$. With this in mind we define the generating function

$$\phi_t^{(N)}(\vec{w}) := \frac{1}{|S_N|} \sum_{\vec{x} \in \mathbb{Z}^N} \sum_{\sigma \in S_N} \vec{w}^{\vec{x}_\sigma} u_t^{(N)}(\vec{x}_\sigma),$$

where $\vec{w} \in \mathbb{C}^N$, $\vec{w}^{\vec{x}} = z_1^{x_1} \cdots z_N^{x_N}$ and $|S_N| = N!$. Since we would like the identity (2.1.5) to hold, it is reasonable to assume that $\left|u_t^{(N)}(\vec{x})\right| \leqslant \min_i \frac{t^{x_i - y_i}}{(x_i - y_i)!}$ which guarantees locally absolute convergence of the sum above and all the following computations. Then (2.1.3) yields

$$
\begin{aligned}
\frac{d}{dt}\phi_t^{(N)}(\vec{w}) &= \frac{1}{|S_N|} \sum_{\vec{x} \in \mathbb{Z}^N} \sum_{\sigma \in S_N} \vec{w}^{\vec{x}_\sigma} \frac{d}{dt} u_t^{(N)}(\vec{x}_\sigma) \\
&= -\frac{1}{|S_N|} \sum_{\vec{x} \in \mathbb{Z}^N} \sum_{\sigma \in S_N} \vec{w}^{\vec{x}_\sigma} \sum_{k=1}^{N} \nabla_k^- u_t^{(N)}(\vec{x}_\sigma) \\
&= -\frac{1}{|S_N|} \sum_{k=1}^{N} \sum_{\vec{x} \in \mathbb{Z}^N} \sum_{\sigma \in S_N} \vec{w}^{\vec{x}_\sigma} \nabla_k^- u_t^{(N)}(\vec{x}_\sigma) \\
&= \frac{1}{|S_N|} \sum_{\vec{x} \in \mathbb{Z}^N} \sum_{\sigma \in S_N} \vec{w}^{\vec{x}_\sigma} u_t^{(N)}(\vec{x}_\sigma) \sum_{k=1}^{N} \left(w_{\sigma(k)} - 1\right) \\
&= \phi_t^{(N)}(\vec{w}) \sum_{k=1}^{N} \varepsilon(w_k),
\end{aligned}
$$

where $\varepsilon(w) := w - 1$ for $w \in \mathbb{C}$. From the last identity we conclude that

$$
\phi_t^{(N)}(\vec{w}) = C(\vec{w}) \prod_{k=1}^{N} e^{\varepsilon(w_k) t},
$$

for a function $C : \mathbb{C}^N \to \mathbb{C}$ which is independent of t, but can depend on $\vec{y}$. Then Cauchy's integral theorem gives a solution to the master equation

$$
\begin{aligned}
u_t^{(N)}(\vec{x}) &= \frac{1}{(2\pi i)^N} \sum_{\sigma \in S_N} \oint_{\Gamma_0} d\vec{w} \, \frac{\phi_t^{(N)}(\vec{w})}{\vec{w}_\sigma^{\vec{x}+1}} \\
&= \frac{1}{(2\pi i)^N} \sum_{\sigma \in S_N} \oint_{\Gamma_0} d\vec{w} \, C(\vec{w}) \prod_{k=1}^{N} \frac{e^{\varepsilon(w_k) t}}{w_{\sigma(k)}^{x_k + 1}},
\end{aligned}
\tag{2.1.7}
$$

where $\vec{x} + 1 = \left(x_1 + 1, \cdots, x_N + 1\right)$ and Γ_0 is a contour in $\mathbb{C}^N$ around the origin. Our next goal it to find C and Γ_0 such that this solution satisfies the initial and boundary conditions for (2.1.3).

Satisfying the boundary conditions. We are going to find functions C and a contour Γ_0 such that the solution (2.1.7) satisfies the boundary conditions (2.1.4). We will look for a solution in a more general form than (2.1.7). More precisely, we will consider functions $C_\sigma(z)$ depending on $\sigma \in S_N$, which gives us the *Bethe ansatz solution*

$$
u_t^{(N)}(\vec{x}) = \frac{1}{(2\pi i)^N} \sum_{\sigma \in S_N} \oint_{\Gamma_0} d\vec{w} \, C_\sigma(\vec{w}) \prod_{k=1}^{N} \frac{e^{\varepsilon(w_k) t}}{w_{\sigma(k)}^{x_k + 1}}.
\tag{2.1.8}
$$

In the case $x_k = x_{k+1}+1$ we denote $f(w) := 1-w^{-1}$, and the boundary condition (2.1.4) yields

$$\nabla_k^- u_t^{(N)}(\vec{x})$$
$$= -\frac{1}{(2\pi i)^N}\sum_{\sigma\in S_N}\oint_{\Gamma_0} d\vec{w}\prod_{i\neq k,k+1}\frac{C_\sigma(\vec{w})}{w_{\sigma(i)}^{x_i+1}}\frac{1-w_{\sigma(k)}^{-1}}{w_{\sigma(k)}^{x_k}w_{\sigma(k+1)}^{x_{k+1}+1}}\prod_{i=1}^N e^{\varepsilon(w_i)t}$$
$$= -\frac{1}{(2\pi i)^N}\sum_{\sigma\in S_N}\oint_{\Gamma_0} d\vec{w}\prod_{i\neq k,k+1}\frac{C_\sigma(\vec{w})}{w_{\sigma(i)}^{x_i+1}}\frac{f(w_{\sigma(k)})}{(w_{\sigma(k)}w_{\sigma(k+1)})^{x_k}}\prod_{i=1}^N e^{\varepsilon(w_i)t} = 0.$$

In particular, this identity holds if for all $\vec{w}\in\mathbb{C}^N$ we have

$$\sum_{\sigma\in S_N}\frac{C_\sigma(\vec{w})f(w_{\sigma(k)})}{(w_{\sigma(k)}w_{\sigma(k+1)})^{x_k}} = 0,$$

Let $T_k \in S_N$ be the transposition $(k, k+1)$, i.e. the permutation which interchanges the elements k and $k+1$. Then the last identity holds if we have

$$C_\sigma(\vec{w})f(w_{\sigma(k)}) + C_{T_k\sigma}(\vec{w})f(w_{\sigma(k+1)}) = 0.$$

In particular, one can see that the following functions satisfy this identity

$$C_\sigma(\vec{w}) = \operatorname{sgn}(\sigma)\prod_{i=1}^N f(w_{\sigma(i)})^i\,\psi(\vec{w}), \tag{2.1.9}$$

for any function $\psi : \mathbb{C}^N \to \mathbb{R}$. Thus we need to find a specific function ψ so that the initial condition in (2.1.3) is satisfied.

Satisfying the initial condition. We need to check the initial condition for $\vec{x}$ and $\vec{y}$ in Ω_N. Combining (2.1.8) with (2.1.9), the initial condition at $t = 0$ is given by

$$\frac{1}{(2\pi i)^N}\sum_{\sigma\in S_N}\oint_{\Gamma_0} d\vec{w}\frac{C_\sigma(\vec{w})}{w_\sigma^{\vec{x}+1}} = \delta_{\vec{y},\vec{x}}. \tag{2.1.10}$$

If $\mathrm{id}\in S_N$ is the identity permutation and $C_{\mathrm{id}}(\vec{w}) = \vec{w}^{\vec{y}}$ then obviously

$$\frac{1}{(2\pi i)^N}\oint_{\Gamma_0} d\vec{w}\frac{C_{\mathrm{id}}(\vec{w})}{w^{\vec{x}+1}} = \delta_{\vec{y},\vec{x}}.$$

For this to hold we need to choose the function ψ in (2.1.9) to be

$$\psi(\vec{w}) = \prod_{i=1}^N f(w_i)^{-i}w_i^{y_i}.$$

Thus, a candidate for the solution is given by

$$u_t^{(N)}(\vec{x}) = \frac{1}{(2\pi i)^N}\sum_{\sigma\in S_N}\operatorname{sgn}(\sigma)\oint_{\Gamma_{0,1}} d\vec{w}\prod_{k=1}^N\frac{f(w_k)^{k-\sigma(k)}e^{\varepsilon(w_k)t}}{w_{\sigma(k)}^{x_k-y_{\sigma(k)}+1}},$$

which can be written as Schütz's formula (2.0.2). It is obvious that the contour $\Gamma_{0,1}$ should go around 0 and 1, since otherwise the determinant in (2.0.2) will vanish when $\vec{x}$ and $\vec{y}$ are far enough.

In order to complete the proof we need to show that this solution satisfies the initial condition. To this end we notice that for $n \geqslant 0$ we have

$$F_{-n}(x,0) = \frac{(-1)^n}{2\pi i}\oint_{\Gamma_0} dw\, \frac{(1-w)^n}{w^{x+n+1}},$$

where Γ_0 is any contour around the origin, which in particular implies that $F_{-n}(x,0) = 0$ for $x < -n$ and $x > 0$, and $F_0(x,0) = \delta_{x,0}$. In the case $x_N < y_N$, we have $x_N < y_k$ for all $k = 1,\dots,N-1$, and $x_N - y_{N+1-j} < 1-j$ since $y \in \Omega_N$. This yields $F_{1-j}(x_N - y_{N+1-j}, 0) = 0$ and the determinant in (2.0.2) vanishes, because the matrix contains a row of zeros. If $x_N \geqslant y_N$, then we have $x_k > y_N$ for all $k = 1,\dots,N-1$, and all entries of the first column in the matrix from (2.0.2) vanish, except the first entry which equals δ_{x_N,y_N}. Repeating this argument for x_{N-1}, x_{N-2} and so on, we obtain that the matrix is upper-triangular with delta-functions at the diagonal, which shows that this solution satisfies the initial condition.

Remark 2.1.11. Similar computations yield the distribution function of ASEP [31]. Unfortunately, this distribution function doesn't have the determinantal form as (2.0.2) which makes its analysis significantly more complicated.

2.2. Direct check of Schütz's formula. We will show that the determinant in (2.0.2) satisfies the master equation (2.1.3) with the boundary conditions (2.1.4), providing an alternate proof to the one in Section 2.1. To this end we will use only the following properties of the functions F_n, which can be easily proved,

$$\text{(2.2.1)} \qquad \partial_t F_n(x,t) = -\nabla^- F_n(x,t), \qquad F_n(x,t) = -\nabla^+ F_{n+1}(x,t),$$

where $\nabla^- f(x) := f(x) - f(x-1)$ and $\nabla^+ f(x) := f(x+1) - f(x)$. Furthermore, it will be convenient to define the vectors

$$\text{(2.2.2)} \qquad H_i(x,t) := \big[F_{i-1}(x-y_N,t), \cdots, F_{i-N}(x-y_1,t)\big]'.$$

Then, denoting by $u_t^{(N)}(\vec{x})$ the right-hand side of (2.0.2), we can write

$$\begin{aligned}
\partial_t u_t^{(N)}(\vec{x}) &= \sum_{k=1}^N \det\big[\cdots, \partial_t H_k(x_{N+1-k},t), \cdots\big] \\
&= -\sum_{k=1}^N \det\big[\cdots, \nabla^- H_k(x_{N+1-k},t), \cdots\big] \\
&= -\sum_{k=1}^N \nabla_k^- \det\big[F_{i-j}(x_{N+1-i} - y_{N+1-j}, t)\big]_{1\leqslant i,j\leqslant N},
\end{aligned}$$

where the operators in the first and second sums are applied only to the k-th column, and where we made use of the first identity in (2.2.2) and multi-linearity of determinants. Here, ∇_k^- is as before the operator ∇^- acting on x_k.

Now, we will check the boundary conditions (2.1.4). If $x_k = x_{k+1} + 1$, then using again multi-linearity of determinants and the second identity in (2.2.2) we obtain

$$\begin{aligned}\nabla_k^- \det\big[F_{i-j}(x_{N+1-i} - y_{N+1-j}, t)\big] &= \det\big[\cdots, \nabla^- H_{N+1-k}(x_k, t), H_{N-k}(x_{k+1}), \cdots\big] \\ &= \det\big[\cdots, \nabla^+ H_{N+1-k}(x_k - 1, t), H_{N-k}(x_{k+1}), \cdots\big] \\ &= \det\big[\cdots, \nabla^+ H_{N+1-k}(x_{k+1}, t), H_{N-k}(x_{k+1}), \cdots\big] \\ &= \det\big[\cdots, -H_{N-k}(x_{k+1}, t), H_{N-k}(x_{k+1}), \cdots\big].\end{aligned}$$

The latter determinant vanishes, because the matrix has two equal columns. A proof of the initial condition was provided at the end of the previous section.

3. Determinantal point processes

In this section we provide some results on determinantal point processes, which can be found e.g. in [4,10,16]. These processes were studied first in [21] as 'fermion' processes and the name 'determinantal' was introduced in [9].

Definition 3.0.1. Let $\mathfrak{X}$ be a discrete space and let μ be a measure on $\mathfrak{X}$. A *determinantal point process* on the space $\mathfrak{X}$ with correlation kernel $\mathcal{K} : \mathfrak{X} \times \mathfrak{X} \to \mathbb{C}$ is a signed[1] measure $\mathcal{W}$ on $2^{\mathfrak{X}}$ (the power set of $\mathfrak{X}$), integrating to 1 and such that for any distinct points $x_1, \dots, x_n \in \mathfrak{X}$ one has the identity

$$\sum_{\substack{Y \subset \mathfrak{X}: \\ \{x_1, \dots, x_n\} \subset Y}} \mathcal{W}(Y) = \det\big[\mathcal{K}(x_i, x_j)\big]_{1 \leqslant i,j \leqslant n} \prod_{k=1}^{n} \mu(x_k), \tag{3.0.2}$$

where the sum runs over finite subsets of $\mathfrak{X}$.

The determinants on the right-hand side are called n-*point correlation functions* or *joint intensities* and denoted by

$$\rho^{(n)}(x_1, \dots, x_n) = \det\big[\mathcal{K}(x_i, x_j)\big]_{1 \leqslant i,j \leqslant n}. \tag{3.0.3}$$

One can easily see that these functions have the following properties: they are symmetric under permutations of arguments and vanish if $x_i = x_j$ for $i \neq j$.

Exercise 3.0.4. *In the case that $\mathcal{W}$ is a positive measure, show that if K is the kernel of the orthogonal projection onto a subset of dimension n, then the number of points in $\mathfrak{X}$ is almost surely equal to n.*

Usually, it is non-trivial to show that a process is determinantal. Below we provide several examples of determinantal point processes (these ones are not signed).

---

[1]In our analysis of TASEP we will be using only a counting measure μ assigning a unit mass to each element of $\mathfrak{X}$. However, a determinantal point process can be defined in full generality on a locally compact Polish space with a Radon measure (see [15]). Moreover, in contrast to the usual definition we define the measure $\mathcal{W}$ to be signed rather than a probability measure. This fact will be crucial in Section 4.1 below, and we should also note that all the properties of determinantal point processes which we will use don't require $\mathcal{W}$ to be positive.

Example 3.0.5 (Non-intersecting random walks). Let $X_i(t)$, $1 \leqslant i \leqslant n$, be independent time-homogeneous Markov chains on $\mathbb{Z}$ with one step transition probabilities $p_t(x, y)$ satisfying the identity $p_t(x, x-1) + p_t(x, x+1) = 1$ (i.e. every time each random walk makes one unit step either to the left or to the right). Let furthermore $X_i(t)$ be reversible with respect to a probability measure π on $\mathbb{Z}$, i.e. $\pi(x)p_t(x, y) = \pi(y)p_t(y, x)$ for all $x, y \in \mathbb{Z}$ and $t \in \mathbb{N}$. Then, conditioned on the events that the values of the random walks at times 0 and $2t$ are fixed, i.e. $X_i(0) = X_i(2t) = x_i$ for all $1 \leqslant i \leqslant n$ where each x_i is even, and no two of the random walks intersect on the time interval $[0, 2t]$, the configuration of mid-positions $\{X_i(t) : 1 \leqslant i \leqslant n\}$, with fixed t, is a determinantal point process on $\mathbb{Z}$ with respect to the measure π, i.e.

$$\mathbb{P}\big[X_i(t) = z_i, 1 \leqslant i \leqslant n\big] = \det\big[\mathcal{K}(z_i, z_j)\big]_{1 \leqslant i,j \leqslant n} \prod_{k=1}^{n} \pi(z_k), \tag{3.0.6}$$

where the probability is conditioned by the described event (assuming of course that its probability is non-zero). Here, the correlation kernel $\mathcal{K}$ is given by

$$\mathcal{K}(u, v) = \sum_{i=1}^{n} \psi_i(u)\phi_i(v), \tag{3.0.7}$$

where the functions ψ_i and ϕ_i are defined by

$$\psi_i(u) = \sum_{k=1}^{n} \left(A^{-\frac{1}{2}}\right)_{i,k} \frac{p_t(x_k, u)}{\pi(u)}, \qquad \phi_i(v) = \sum_{k=1}^{n} \left(A^{-\frac{1}{2}}\right)_{i,k} \frac{p_t(x_k, v)}{\pi(v)},$$

with the matrix A having the entries $A_{i,k} = \frac{p_{2t}(x_i, x_k)}{\pi(x_k)}$. Invertibility of the matrix A follows from the fact that the probability of the condition is non-zero and Karlin-McGregor formula (see Exercise 3.0.8 below). This result is a particular case of a more general result of [16] and it can be obtained from Karlin-McGregor formula similarly to [15, Cor. 4.3.3].

Exercise 3.0.8 (Karlin-McGregor formula [19]). *Let X_i, $1 \leqslant i \leqslant n$, be i.i.d. (time-inhomogeneous) Markov chains on $\mathbb{Z}$ with transition probabilities $p_{k,\ell}(s, t)$ satisfying $p_{k,k+1}(t, t+1) + p_{k,k-1}(t, t+1) = 1$ for all k and $t > 0$. Let us fix initial states $X_i(0) = k_i$ for $k_1 < k_2 < \cdots < k_n$ such that each k_i is even. Then the probability that at time t the Markov chains are at the states $\ell_1 < \ell_2 < \cdots < \ell_n$, and that no two of the chains intersect up to time t, equals $\det\big[p_{k_i,\ell_j}(0, t)\big]_{1 \leqslant i,j \leqslant n}$.*

Hint (this idea is due to S.R.S. Varadhan): for a permutation $\sigma \in S_n$ and $0 \leqslant s \leqslant t$, define the process

$$M_\sigma(s) = \prod_{i=1}^{n} \mathbb{P}\big(X_i(t) = \ell_{\sigma(i)} \big| X_i(s)\big), \tag{3.0.9}$$

which is a martingale with respect to the filtration generated by the Markov chains X_i. This implies that the process $M = \sum_{\sigma \in S_n} \text{sgn}(\sigma) M_\sigma$ is also a martingale. Obtain the Karlin-McGregor formula by applying the optional stopping theorem to M for a suitable stopping time.

Exercise 3.0.10. *Prove that the mid-positions* $\{X_i(t) : 1 \leqslant i \leqslant n\}$ *of the random walks defined in the previous example form a determinantal process with the correlation kernel* (3.0.7).

Example 3.0.11 (Gaussian unitary ensemble)**.** The most famous example of determinantal point processes is the *Gaussian unitary ensemble* (GUE) introduced by Wigner. Let us define the $n \times n$ matrix A to have i.i.d. standard complex Gaussian entries and let $H = \frac{1}{\sqrt{2}}(A + A^*)$. Then the eigenvalues $\lambda_1 > \lambda_2 > \cdots > \lambda_n$ of H form a determinantal point process on $\mathbb{R}$ with the correlation kernel

$$\mathcal{K}(x, y) = \sum_{k=0}^{n-1} H_k(x) H_k(y),$$

with respect to the Gaussian measure $d\mu(x) = \frac{1}{\sqrt{2\pi}} e^{-x^2/2} dx$, where H_k are Hermite polynomials which are orthonormal on $L^2(\mathbb{R}, \mu)$. A proof of this result can be found in [23, Ch. 3].

Example 3.0.12 (Aztec diamond tilings)**.** The Aztec diamond is a diamond-shaped union of lattice squares (see Figure 3.0.13). Let's now color some squares in gray following the pattern of a chess board and so that all the bottom left squares are colored. It is easy to see that the Aztec diamond can be perfectly covered by domino tilings, which are 2×1 or 1×2 rectangles, and the number of tilings grows exponentially in the width of the diamond. Let's draw a tiling uniformly from all possible tilings and let's mark gray left squares of horizontal dominos and gray bottom squares of vertical dominos. This random set is a determinantal point process on the lattice $\mathbb{Z}^2$ [17].

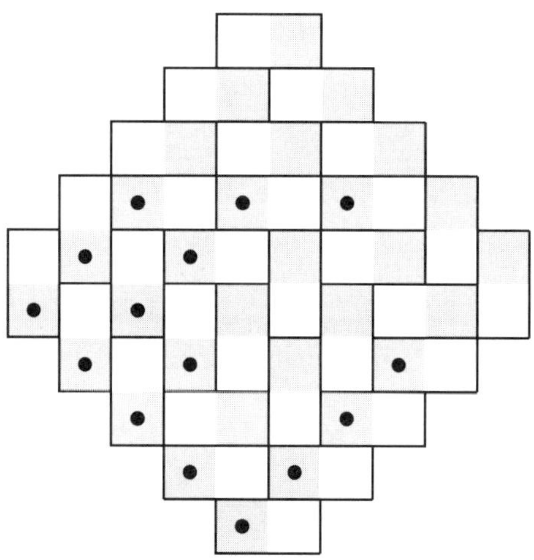

FIGURE 3.0.13. Aztec diamond tiling.

3.1. Probability of an empty region. A useful property of determinantal point processes is that the 'probability' (recall that the measure in Definition 3.0.1 is signed) of having an empty region is given by a Fredholm determinant.

Lemma 3.1.1. *Let $\mathcal{W}$ be a determinantal point process on a discrete set $\mathfrak{X}$ with a measure μ and with a correlation kernel $\mathcal{K}$. Then for a Borel set $B \subset \mathfrak{X}$ one has*

$$\sum_{X \subset \mathfrak{X} \setminus B} \mathcal{W}(X) = \det(I - \mathcal{K})_{\ell^2(B,\mu)},$$

where the latter is the Fredholm determinant defined by

$$\det(I - \mathcal{K})_{\ell^2(B,\mu)} := \sum_{n \geqslant 0} \frac{(-1)^n}{n!} \int_{B^n} \det\big[\mathcal{K}(y_i, y_j)\big]_{1 \leqslant i,j \leqslant n} d\mu(y_1) \cdots d\mu(y_n). \tag{3.1.2}$$

Proof. Using Definition 3.0.1 and the correlation functions (3.0.3) we can write

$$\begin{aligned}
\sum_{X \subset \mathfrak{X} \setminus B} \mathcal{W}(X) &= \sum_{X \subset \mathfrak{X}} \mathcal{W}(X) \prod_{x \in X} \Big((1 - \mathbf{1}_{\{B\}}(x)\Big) \\
&= \sum_{n \geqslant 0} \frac{(-1)^n}{n!} \sum_{X \subset \mathfrak{X}} \mathcal{W}(X) \sum_{\substack{x_1, \ldots, x_n \in X \\ x_i \neq x_j}} \prod_{k=1}^{n} \mathbf{1}_{\{B\}}(x_k) \\
&= \sum_{n \geqslant 0} \frac{(-1)^n}{n!} \sum_{\substack{y_1, \ldots, y_n \in B \\ y_i \neq y_j}} \sum_{X \subset \mathfrak{X}} \mathcal{W}(X) \sum_{x_1, \ldots, x_n \in X} \prod_{k=1}^{n} \mathbf{1}_{\{x_k = y_k\}} \\
&= \sum_{n \geqslant 0} \frac{(-1)^n}{n!} \sum_{\substack{y_1, \ldots, y_n \in B \\ y_i \neq y_j}} \sum_{\substack{X \subset \mathfrak{X} \\ \{y_1, \ldots, y_n\} \subset X}} \mathcal{W}(X) \\
&= \sum_{n \geqslant 0} \frac{(-1)^n}{n!} \sum_{\substack{y_1, \ldots, y_n \in B \\ y_i \neq y_j}} \rho^{(n)}(y_1, \ldots, y_n) \prod_{k=1}^{n} \mu(y_k) \\
&= \sum_{n \geqslant 0} \frac{(-1)^n}{n!} \int_{B^n} \det\big[\mathcal{K}(y_i, y_j)\big]_{1 \leqslant i,j \leqslant n} d\mu(y_1) \cdots d\mu(y_n) \\
&= \det(I - \mathcal{K})_{\ell^2(B,\mu)},
\end{aligned}$$

which is exactly our claim. Note, that the condition $y_i \neq y_j$ can be omitted, because $\rho^{(n)}$ vanishes on the diagonals. □

Exercise 3.1.3. *Prove that if $\mathfrak{X}$ is finite and μ is the counting measure, then the Fredholm determinant* (3.1.2) *coincides with the usual determinant.*

3.2. L-ensembles of signed measures. A more restrictive definition of a determinantal process was introduced in [10]. In order to simplify our notation, we take the measure μ in this section to be the counting measure and we will omit it in notation below.

With the notation of Definition 3.0.1, let us suppose we are given a function $L : \mathfrak{X} \times \mathfrak{X} \to \mathbb{C}$. For any finite subset $X = \{x_1, \ldots, x_n\} \subset \mathfrak{X}$ we define a symmetric minor $L_X := \big[L(x_i, x_j)\big]_{x_i, x_j \in X}$. Then one can define a (signed) measure on $\mathfrak{X}$, called the L-*ensemble*, by

$$W(X) = \frac{\det(L_X)}{\det(1 + L)_{\ell^2(\mathfrak{X})}}, \tag{3.2.1}$$

for $X \subset \mathfrak{X}$, if the Fredholm determinant $\det(1 + L)_{\ell^2(\mathfrak{X})}$ is non-zero (recall the definition (3.1.2)).

Exercise 3.2.2. *Check that the measure* $\mathcal{W}$ *defined in* (3.2.1) *integrates to* 1.

The requirement $\det(1 + L)_{\ell^2(\mathfrak{X})} \neq 0$ guarantees that there exists a unique function $(1 + L)^{-1} : \mathfrak{X} \times \mathfrak{X} \to \mathbb{C}$ such that $(1 + L)^{-1} * (1 + L) = 1$, where $*$ is the convolution on $\mathfrak{X}$ and $1 : \mathfrak{X} \times \mathfrak{X} \to \{0, 1\}$ is the identity function non-vanishing only on the diagonal. Furthermore, it was proved in [21] that the L-ensemble is a determinantal point process:

Proposition 3.2.3. *The measure* $\mathcal{W}$ *defined in* (3.2.1) *is a determinantal point process with correlation kernel* $\mathcal{K} = L(1 + L)^{-1} = 1 - (1 + L)^{-1}$.

Example 3.2.4 (Non-intersecting random walks). It is not difficult to see that the distribution of the mid-positions $\{X_i(t) : 1 \leqslant i \leqslant n\}$ of the random walks from Example 3.0.5 is the L-ensemble with the function

$$L(u, v) = \sum_{i=1}^{N} p_t(u, x_i) p_t(x_i, v).$$

The correlation kernel $\mathcal{K}$ can be computed from Proposition 3.2.3 and it coincides with (3.0.6).

Exercise 3.2.5. *Perform the computations from the previous example.*

3.3. Conditional L-ensembles. An L-ensemble can be conditioned by fixing certain values of the determinantal process. More precisely, consider a nonempty subset $\mathfrak{Z} \subset \mathfrak{X}$ and a given L-ensemble on $\mathfrak{X}$. We define a measure on $2^{\mathfrak{Z}}$, called *conditional* L-*ensemble*, in the following way:

$$\mathcal{W}(Y) = \frac{\det(L_{Y \cup \mathfrak{Z}^c})}{\det(1_{\mathfrak{Z}} + L)_{\ell^2(\mathfrak{Z})}}, \tag{3.3.1}$$

for any $Y \subset \mathfrak{Z}$, where $1_{\mathfrak{Z}}(x, y) = 1$ if and only if $x = y \in \mathfrak{Z}$, and $1_{\mathfrak{Z}}(x, y) = 0$ otherwise.

Exercise 3.3.2. *Prove that the measure* $\mathcal{W}$ *defined in* (3.3.1) *integrates to* 1.

Roughly speaking the definition (3.3.1) means that we restrict the L-ensemble by the condition that the values $\mathfrak{Z}^c$ are fixed. The following result is a generalisation of Proposition 3.2.3 and its proof can be found in [10, Prop. 1.2]:

Proposition 3.3.3. *The conditional* L-*ensemble is a determinantal point process on* $\mathfrak{Z}$ *with correlation kernel*

$$\mathcal{K} = 1_{\mathfrak{Z}} - (1_{\mathfrak{Z}} + L)^{-1}\big|_{\mathfrak{Z} \times \mathfrak{Z}}, \tag{3.3.4}$$

where $F\big|_{\mathfrak{Z} \times \mathfrak{Z}}$ *means restriction of the function* F *to the set* $\mathfrak{Z} \times \mathfrak{Z}$.

4. Biorthogonal representation of the correlation kernel

The formula (2.0.2) is not suitable for asymptotic analysis of TASEP, because the size of the matrix goes to ∞ as the number of particles N increases. To overcome this problem, the authors of [7] (and in its preliminary version [28]) wrote it as a Fredholm determinant, which can be then subject to asymptotic analysis.

In order to state this result, we need to make some definitions. For an integer $M \geqslant 1$, a fixed vector $\vec{a} \in \mathbb{R}^M$ and indices $n_1 < \ldots < n_M$ we introduce the projections

$$\chi_{\vec{a}}(n_j, x) := \mathbf{1}_{\{x > a_j\}}, \qquad \bar{\chi}_{\vec{a}}(n_j, x) := \mathbf{1}_{\{x \leqslant a_j\}}, \tag{4.0.1}$$

acting on $x \in \mathbb{Z}$, which we will also regard as multiplication operators acting on $\ell^2(\{n_1, \ldots, n_M\} \times \mathbb{Z})$. We will use the same notation if a is a scalar, writing

$$\chi_a(x) = 1 - \bar{\chi}_a(x) = \mathbf{1}_{\{x > a\}}. \tag{4.0.2}$$

Then from [7] we have the following result:

Theorem 4.0.3. *If TASEP starts with particles labeled $X_0(1) > X_0(2) > \ldots > X_0(N)$ and, for some $1 \leqslant M \leqslant N$, we let $1 \leqslant n_1 < n_2 < \cdots < n_M \leqslant N$ and $\vec{a} \in \mathbb{R}^M$. Then for $t > 0$ we have*

$$\mathbb{P}(X_t(n_j) > a_j,\ j = 1, \ldots, M) = \det(I - \bar{\chi}_{\vec{a}} K_t \bar{\chi}_{\vec{a}})_{\ell^2(\{n_1, \ldots, n_M\} \times \mathbb{Z})} \tag{4.0.4}$$

where the kernel K_t is given by

$$K_t(n_i, x_i; n_j, x_j) = -\phi^{n_j - n_i}(x_i, x_j)\mathbf{1}_{\{n_i < n_j\}} + \sum_{k=1}^{n_j} \Psi^{n_i}_{n_i - k}(x_i)\Phi^{n_j}_{n_j - k}(x_j), \tag{4.0.5}$$

and where $\phi(x, y) := \mathbf{1}_{\{x > y\}}$ and

$$\Psi^n_k(x) := \frac{1}{2\pi i} \oint_{\Gamma_0} dw \left(\frac{1 - w}{w}\right)^k \frac{e^{t(w-1)}}{w^{x - X_0(n-k) + 1}}. \tag{4.0.6}$$

Here, Γ_0 is any simple loop, oriented counterclockwise, which includes the pole at $w = 0$ but does not include $w = 1$. The functions Φ^n_k, $0 \leqslant k < n$, are defined implicitly by the following two properties:

(1) *the biorthogonality relation, for $0 \leqslant k, \ell < n$,*

$$\sum_{x \in \mathbb{Z}} \Psi^n_k(x)\Phi^n_\ell(x) = \mathbf{1}_{\{k = \ell\}}, \tag{4.0.7}$$

(2) *the spanning property*

$$\operatorname{span}\{\Phi^n_k : 0 \leqslant k < n\} = \operatorname{span}\{x^k : 0 \leqslant k < n\}, \tag{4.0.8}$$

which in particular implies that the function Φ^n_k is a polynomial of degree at most $n - 1$.

Remark 4.0.9. The problem with this result is that the functions Φ^n_k are not given explicitly. In special cases of periodic and step initial data exact integral expressions of these functions had been found in [7], [13] and [6] (the latter is for the discrete time TASEP, which can be easily adapted for continuous time). More precisely, for the step initial data $X_0(i) = -i$, $i \geqslant 1$, we have

$$\Phi_k^n(x) = \frac{1}{2\pi i} \oint_{\Gamma_0} dv \, \frac{(1-v)^{x+n}}{v^{k+1}} e^{tv},$$

and in the case of periodic initial data $X_0(i) = -di$, $i \geqslant 1$, with $d \geqslant 2$,

$$\Phi_k^n(x) = \frac{1}{2\pi i} \oint_{\Gamma_0} dv \, \frac{(1-dv)(2(1-v))^{x+dn-1}}{v(2^d(1-v)^{d-1}v)^k} e^{tv}.$$

The key new result in [18] is an expression for the functions Φ_k^n, and therefore the kernel K_t, for *arbitrary* initial data.

Remark 4.0.10. The functions F from (2.0.3) and Ψ from (4.0.6) are obviously related by the identity

$$\Psi_k^N(x) = (-1)^k F_{-k}(x - y_{N-k}, t), \tag{4.0.11}$$

for $0 \leqslant k \leqslant N$, so that all properties of F can be translated to Ψ. Moreover, one can see that if $n \leqslant 0$, then the function inside the integral in (2.0.3) has the only pole at $w = 0$, which yields

$$F_{n+1}(x,t) = -\sum_{y<x} F_n(y,t).$$

Writing this relation in terms of the functions Ψ_k^N, we get

$$\Psi_{N-k}^N(x) = \sum_{y<x} \Psi_{N+1-k}^{N+1}(x). \tag{4.0.12}$$

In the next section we provide a proof of this result following [7]. The main idea is to rewrite the problem in terms of non-intersecting random walks (or *vicious random walks*) whose configurations form a *Gelfand-Tsetlin pattern*[2]. The distribution of these random walks forms a determinantal point process whose correlation kernel is (4.0.5).

4.1. Non-intersecting random walks. Our aim in this section is to write Schütz's formula (2.0.2) in a form involving transition probabilities of non-intersecting random walks.

We start with rewriting the transition probabilities (2.0.2) in the following way:

Proposition 4.1.1. *For $\vec{x}, \vec{y} \in \Omega_N$, one has the following identity:*

$$\mathbb{P}(X_t = \vec{x} \,|\, X_0 = \vec{y}) = \sum_{\substack{\mathbf{z} \in GT_N: \\ z_1^n = x_n}} \det\left[\Psi_{N-j}^N(z_i^N)\right]_{1 \leqslant i,j \leqslant N}, \tag{4.1.2}$$

where the functions Ψ are defined in (4.0.6), where the sum runs over the domain GT_N of triangular arrays given by a Gelfand-Tsetlin pattern

$$GT_N = \left\{\mathbf{z} = (z_i^n)_{n,i} \colon z_i^n \in \mathbb{Z},\ 1 \leqslant i \leqslant n < N,\ z_i^{n+1} < z_i^n \leqslant z_{i+1}^{n+1}\right\},$$

[2]This property relates TASEP to random matrices, see [13].

with fixed values $z_1^n = x_n$ *for all* $n = 1, \cdots, N$ *(see Figure* 4.1.3 *for a graphical representation of* GT_4*).*

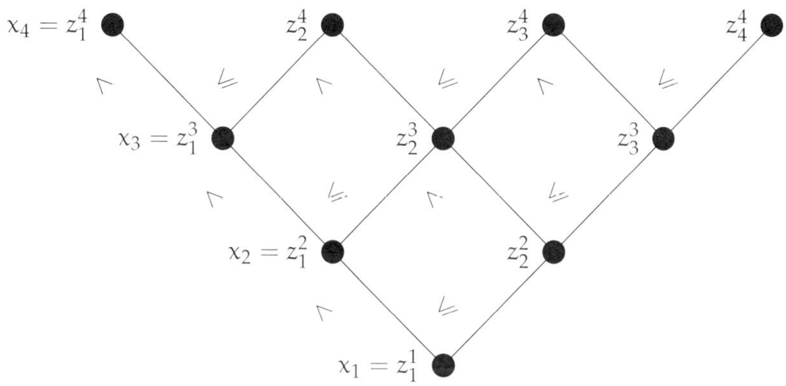

FIGURE 4.1.3. The Gelfand-Tsetlin pattern GT_4 with fixed values $z_1^n = x_n$.

Proof. This decomposition is obtained using only the identity

$$F_{n+1}(x,t) = \sum_{y \geqslant x} F_n(y,t), \tag{4.1.4}$$

which is the integrated form of the second equality in (2.2.1) combined with the fact that the convergence $\lim_{y \to +\infty} F_n(y,t) = 0$ holds fast enough. From Schütz's formula (2.0.2) we have

$$\mathbb{P}\big(X_t = \vec{x} \,|\, X_0 = \vec{y}\big) = \det \begin{bmatrix} F_0(z_1^N - y_N, t) & \cdots & F_{-N+1}(z_1^N - y_1, t) \\ \vdots & \ddots & \vdots \\ F_{N-1}(z_1^1 - y_N, t) & \cdots & F_0(z_1^1 - y_1, t) \end{bmatrix}, \tag{4.1.5}$$

where we renamed the variables $z_1^n := x_n$. Applying the property (4.1.4) twice to each entry of the last row we can rewrite it as

$$\begin{aligned} &\Big[\ F_{N-1}(z_1^1 - y_N, t) \quad \cdots \quad F_0(z_1^1 - y_1, t)\ \Big] \\ &= \sum_{z_2^2 \geqslant z_1^1} \Big[\ F_{N-2}(z_2^2 - y_N, t) \quad \cdots \quad F_{-1}(z_2^2 - y_1, t)\ \Big] \\ &= \sum_{z_2^2 \geqslant z_1^1} \sum_{z_3^3 \geqslant z_2^2} \Big[\ F_{N-3}(z_3^3 - y_N, t) \quad \cdots \quad F_{-2}(z_3^3 - y_1, t)\ \Big]. \end{aligned} \tag{4.1.6}$$

Applying furthermore the identity (4.1.4) to the penultimate row in (4.1.5) we obtain

$$\sum_{z_2^3 \geqslant z_1^2} \Big[\ F_{N-3}(z_2^3 - y_N, t) \quad \cdots \quad F_{-2}(z_2^3 - y_1, t)\ \Big].$$

Combining these two identities with multilinearity of determinant, the right-hand side of (4.1.5) equals

$$\sum_{z_2^2 \geqslant z_1^1} \sum_{z_3^3 \geqslant z_2^2} \sum_{z_2^3 \geqslant z_1^2} \det \begin{bmatrix} F_0(z_1^N - y_N, t) & \cdots & F_{-N+1}(z_1^N - y_1, t) \\ \vdots & \ddots & \vdots \\ F_{N\ 3}(z_1^3 - y_N, t) & \cdots & F_{-2}(z_1^3 - y_1, t) \\ F_{N-3}(z_2^3 - y_N, t) & \cdots & F_{-2}(z_2^3 - y_1, t) \\ F_{N-3}(z_3^3 - y_N, t) & \cdots & F_{-2}(z_3^3 - y_1, t) \end{bmatrix}.$$

The determinant is antisymmetric in the variables z_2^3 and z_3^3 (i.e. it changes sign if we swap z_2^3 and z_3^3), therefore the contribution of the symmetric part of the summation domain $\{z_3^3 \geqslant z_2^2\} \cap \{z_2^3 \geqslant z_1^2\}$ is zero. Since the symmetric part is $\{z_3^3 \geqslant z_2^2\} \cap \{z_2^3 \geqslant z_2^2\}$, we are left with the sum over $\{z_3^3 \geqslant z_2^2\} \cap \{z_2^3 \in [z_1^2, z_2^2)\}$. We iterate the same procedure for $k = 3, \ldots, N-1$, applying (4.1.4) to the last k rows and removing the sums over symmetric domains, and we get the formula

$$\mathbb{P}(X_t = \vec{x} \,|\, X_0 = \vec{y}) = \sum_{\substack{z \in GT_N: \\ z_1^n = x_n}} \det \left[F_{1-j}(z_i^N - y_{N+1-j}, t) \right]_{1 \leqslant i,j \leqslant N}.$$

Now, we can use the identity (4.0.11) to get

$$\begin{aligned} &\det \left[F_{1-j}(z_i^N - y_{N+1-j}, t) \right]_{1 \leqslant i,j \leqslant N} \\ &= \det \left[(-1)^{j-1} \Psi_{j-1}^N (z_i^N) \right]_{1 \leqslant i,j \leqslant N} \\ &= (-1)^{(1+2+\cdots+N)-N} \det \left[\Psi_{j-1}^N (z_i^N) \right]_{1 \leqslant i,j \leqslant N}, \end{aligned}$$

and we change the order of the columns of the matrix inside the determinant

$$\det \left[\Psi_{j-1}^N (z_i^N) \right]_{1 \leqslant i,j \leqslant N} = (-1)^{\lfloor N/2 \rfloor} \det \left[\Psi_{N-j}^N (z_i^N) \right]_{1 \leqslant i,j \leqslant N}.$$

It is not difficult to see that $(1+2+\cdots+N) - N + \lfloor N/2 \rfloor$ is an even integer so that the power of -1 equals 1. Hence, combining these identities we get

$$\det \left[F_{1-j}(z_i^N - y_{N+1-j}, t) \right]_{1 \leqslant i,j \leqslant N} = \det \left[\Psi_{N-j}^N (z_i^N) \right]_{1 \leqslant i,j \leqslant N},$$

which gives exactly our claim (4.1.2). □

We define the space of integer-valued triangular arrays

$$\Lambda_N := \{ \mathbf{z} = (z_i^n)_{n,i} : z_i^n \in \mathbb{Z},\ 1 \leqslant i \leqslant n \leqslant N \}, \tag{4.1.7}$$

then for a configuration $\mathbf{z} = (z_i^n)_{n,i} \in \Lambda_N$, for the function $\phi(x,y) = \mathbf{1}_{\{x>y\}}$ and for new values $z_n^{n-1} = +\infty$ (so that $\phi(z_n^{n-1}, y) = 1$ for all $y \in \mathbb{Z}$), the determinant $\det \left[\phi(z_i^{n-1}, z_j^n) \right]_{1 \leqslant i,j \leqslant n}$ is the indicator function for the inequalities in GT_N between the levels $n-1$ and n to hold. This yields

$$\prod_{n=1}^{N} \det \left[\phi(z_i^{n-1}, z_j^n) \right]_{1 \leqslant i,j \leqslant n} = \mathbf{1}_{\{z \in GT_N\}}. \tag{4.1.8}$$

Exercise 4.1.9. *Prove that the identity (4.1.8) holds.*

To every configuration $\mathbf{z} \in \Lambda_N$ we associate the weight

$$(4.1.10)\quad \mathcal{W}_N(\mathbf{z}) := \left(\prod_{n=1}^{N} \det\left[\phi(z_i^{n-1}, z_j^n)\right]_{1\leqslant i,j\leqslant n}\right) \det\left[\Psi_{N-j}^N(z_i^N)\right]_{1\leqslant i,j\leqslant N},$$

where we recall that the values $z_n^{n-1} = +\infty$ are fixed. Then the identities (4.1.2), (4.1.8) and (4.1.10) yield

$$(4.1.11)\quad \mathbb{P}(X_t = \vec{x} \,|\, X_0 = \vec{y}) = \sum_{\substack{\mathbf{z}\in\Lambda_N: \\ z_1^n = x_n}} \mathcal{W}_N(\mathbf{z}),$$

where the sum runs over the set Λ_N of integer-valued triangular arrays with fixed boundary values $z_1^n = x_n$ for all $n = 1, \ldots, N$.

The variables $\{z_i^n : i = 1, \ldots, n\}$ in (4.1.10) can be interpreted as the positions of particles labelled by $i = 1, \ldots, n$ at time n, so that $z_k^k, \ldots, z_k^n$ is the trajectory of particle k with the transition kernel ϕ (which can be made a probability kernel by multiplying by a power of 2). At time n there are n particles at positions $z_1^n, \ldots, z_n^n$, which make geometric jumps to the left at time $n+1$ conditioned by non-intersection (they are called *vicious random walks*). Furthermore, a new $(n+1)$-st particle is added at position $z_{n+1}^{n+1} \geqslant z_n^n$ at time $n+1$.

The measure $\mathcal{W}_N$ on Λ_N is not a probability measure, because the contribution coming from the functions Ψ can give a negative value. We will show that this is a determinantal measure, which in particular means that the probability that the sites $z_{k_i}^{n_i}$ for $i = 1, \ldots, M$ are occupied by the respective walks is proportional to

$$(4.1.12)\quad \det\left[\mathcal{K}_t\big((n_i, k_i, z_{k_i}^{n_i}), (n_j, k_j, z_{k_j}^{n_j})\big)\right]_{1\leqslant i,j\leqslant M},$$

for a correlation kernel $\mathcal{K}_t$ which will be obtained below.

4.2. The correlation kernel of the signed measure. We will prove in this section that the measure $\mathcal{W}_N$ defined in (4.1.10) on triangular arrays Λ_N is a determinantal point process and will find its correlation kernel, so that in particular the property (4.1.12) holds. Note that we will consider the measure $\mathcal{W}_N$ on the whole space Λ_N, and only after having found the correlation kernel we will fix the boundary values as in (4.1.11).

The measure $\mathcal{W}_N$ is a conditional L-ensemble. It will be more convenient to write the values of a triangular array as a one-dimensional array. More precisely, if we fix some point configuration $(z_i^n)_{n,i} \in \Lambda_N$, then every value $z_i^n = z$ can be identified with the triplet (n, i, z), where $n \in \{1, \cdots, N\}$ and $i \in \{1, \cdots, n\}$. In this way the values $(z_i^n)_{n,i}$ can be written as a one-dimensional array of size $N(N+1)/2$, parametrized by (n, i), e.g. in the case $N = 3$, we have that $(z_i^n)_{n,i}$ is equivalent to the array

(1,1)	(2,1)	(2,2)	(3,1)	(3,2)	(3,3)
z_1^1	z_1^2	z_2^2	z_1^3	z_2^3	z_3^3

With this idea in mind, we define a the point process which has the domain

$$\mathfrak{Z} := \{(n, i, z) : n \in \{1, \cdots, N\}, i \in \{1, \cdots, n\}, z \in \mathbb{Z}\}.$$

Since we would like to fix the initial values y_n of TASEP and the 'infinities' z_n^{n-1} (with $n = 1, \ldots, N$ in both cases), we need to add these points into the domain. In fact, it is sufficient to define the domain as

$$\mathfrak{X} = \{1, 2, \ldots, N\} \cup \mathfrak{Z}$$

where depending on the context a number $n \in \{1, 2, \ldots, N\}$ will refer to either z_n^{n-1} or the initial value y_n, and the determinantal point process will be conditioned by $\mathfrak{Z}^c = \mathfrak{X} \setminus \mathfrak{Z} = \{1, 2, \ldots, N\}$. Our aim is to define a function $L : \mathfrak{X} \times \mathfrak{X} \to \mathbb{R}$ such that for every set $Y \subset \mathfrak{Z}$ one has

$$\mathcal{W}_N(Y) = \det\left(L_{Y \cup \mathfrak{Z}^c}\right),$$

(where we use the notation from (3.2.1)) which means that $\mathcal{W}_N$ is a conditional L-ensemble. As we will see below, this conditioning on the set $\mathfrak{Z}^c$ corresponds to fixing the initial values y_i of TASEP and the 'infinities' z_n^{n-1}. Here, we don't divide the determinant by $\det(1_{\mathfrak{Z}} + L)_{\ell^2(\mathfrak{Z})}$ as in (3.3.1), because the measure $\mathcal{W}_N$ is already normalized, i.e. $\det(1_{\mathfrak{Z}} + L)_{\ell^2(\mathfrak{Z})} = 1$. Indeed, it follows from (4.1.11), that the integral of $\mathcal{W}_N$ over the whole state space is the probability that TASEP particles start from fixed positions and go to any possible configuration, which is obviously 1.

Using the equivalence between point configurations $(z_i^n)_{n,i} \in \Lambda_N$ and one-dimensional arrays described in the previous paragraph, every point configuration from $\mathfrak{X}$ can be obviously identified with an array as well (by adding new boxes indexed by 1, 2, ..., N). In the previous example we will have the array

1	2	3	$(1,1)$	$(2,1)$	$(2,2)$	$(3,1)$	$(3,2)$	$(3,3)$
$*$	$*$	$*$	z_1^1	z_1^2	z_2^2	z_1^3	z_2^3	z_3^3

Here, each entry z_i^n, labeled by a pair (n, i), corresponds to a point $(n, i, z_i^n) \in \mathfrak{Z}$, while by $*$ we mean the values indexed by 1, 2, $\cdots$, N. In this way, each entry of the array corresponds to a point in $\mathfrak{X}$. This means that $L_{\{\mathbf{z}\} \cup \mathfrak{Z}^c}$ (recall the notation from (3.2.1)) is in fact a function of two arguments each of which is either (n, i), such that $n \in \{1, \cdots, N\}$ and $i \in \{1, \cdots, n\}$, or $k \in \{1, \cdots, N\}$. So we can identify $L_{\{\mathbf{z}\} \cup \mathfrak{Z}^c}$ with a square matrix of size $N(N+1)/2 + N$. Now, we are going to define this matrix.

Writing the L-function in a matrix form. Let us denote by $\mathcal{M}_{n,m}$ the set of all $n \times m$ matrices with real entries. Then for $1 \leqslant n < m \leqslant N$ we define the function $W_{[n,m)} : \Lambda_N \to \mathcal{M}_{n,m}$ such that for a fixed particles configuration $\mathbf{z} = (z_i^n) \in \Lambda_N$ the matrix $W_{[n,m)}(\mathbf{z})$ is given by

$$\left[W_{[n,m)}(\mathbf{z})\right]_{i,j} = \phi^{m-n}(z_i^n, z_j^m) \mathbf{1}_{\{n<m\}}, \qquad 1 \leqslant i \leqslant n, \, 1 \leqslant j \leqslant m.$$

Similarly we define the function $\Psi^{(N)} : \Lambda_N \to \mathcal{M}_{N,N}$ to have the entries

$$\Big[\Psi^{(N)}(\mathbf{z})\Big]_{i,j} = \Psi^N_{N-j}(z^N_i), \qquad 1 \leqslant i, j \leqslant N,$$

where the functions ϕ and Ψ^N_{N-j} are as in the statement of Theorem 4.0.3. Finally, for $0 \leqslant m < N$ we define the function $E_m : \Lambda_N \to \mathcal{M}_{N,m+1}$ by the entries

$$\Big[E_m(\mathbf{z})\Big]_{i,j} = \begin{cases} \phi(z^m_{m+1}, z^{m+1}_j), & \text{if } i = m+1,\ 1 \leqslant j \leqslant m+1, \\ 0, & \text{otherwise.} \end{cases}$$

In fact, the function E_m should also have the values of z^m_{m+1} as arguments, but we prefer not to indicate this, since we will always fix these values to be infinity. With these objects at hand we identify the function $L_{\{\mathbf{z}\}\cup\mathfrak{Z}^c}$ (recall, that the configuration $\mathbf{z} \in \Lambda_N$ has been fixed) with the following square matrix in block form:

$$L_{\{\mathbf{z}\}\cup\mathfrak{Z}^c} = \begin{bmatrix} 0 & E_0 & E_1 & E_2 & \cdots & E_{N-1} \\ 0 & 0 & -W_{[1,2)} & 0 & \cdots & 0 \\ 0 & 0 & 0 & -W_{[2,3)} & \ddots & \vdots \\ \vdots & \vdots & \vdots & \ddots & \ddots & 0 \\ 0 & 0 & 0 & 0 & \cdots & -W_{[N-1,N)} \\ \Psi^{(N)} & 0 & 0 & 0 & \cdots & 0 \end{bmatrix}(\mathbf{z}), \tag{4.2.1}$$

where each block takes $\mathbf{z}$ as an argument and gives a usual matrix. The first N columns (resp. rows) of the matrix (4.2.1) are parametrized by the values $\{1, 2, \ldots, N\}$ and the succeeding columns (resp. rows) are parametrized by the pairs $\{(n, i) : 1 \leqslant n \leqslant N,\ 1 \leqslant i \leqslant n\}$ which have the lexicographic order.

Example 4.2.2. In the case $N = 2$, a configuration $\mathbf{z} \in \Lambda_2$ contains 3 values $\{z^1_1\} \cup \{z^1_2, z^2_2\}$, and the matrix $L_{\{\mathbf{z}\}\cup\mathfrak{Z}^c}$ is given by

$$L_{\{\mathbf{z}\}\cup\mathfrak{Z}^c} = \begin{array}{c} \\ 1 \\ 2 \\ (1,1) \\ (2,1) \\ (2,2) \end{array} \begin{array}{c} \begin{array}{ccccc} 1 & 2 & (1,1) & (2,1) & (2,2) \end{array} \\ \begin{bmatrix} 0 & 0 & \phi(z^0_1, z^1_1) & 0 & 0 \\ 0 & 0 & 0 & \phi(z^1_2, z^2_1) & \phi(z^1_2, z^2_2) \\ 0 & 0 & 0 & -\phi(z^1_1, z^2_1) & -\phi(z^1_1, z^2_2) \\ \Psi^2_1(z^2_1) & \Psi^2_0(z^2_1) & 0 & 0 & 0 \\ \Psi^2_1(z^2_2) & \Psi^2_0(z^2_2) & 0 & 0 & 0 \end{bmatrix} \end{array},$$

Ψ^N $\quad E_0 \quad -W_{[1,2)} \quad E_1$

where the 'infinities' z^0_1 and z^1_2 are fixed. In particular, it follows from the definition of the function ϕ in Theorem 4.0.3 that $\phi(z^0_1, z) = \phi(z^1_2, z) = 1$ for any $z \in \mathbb{Z}$.

The L-function defines the measure $\mathcal{W}_N$. We will show that one has the identity $\mathcal{W}(\mathbf{z}) = C\det(L_{\{\mathbf{z}\}\cup\mathfrak{Z}^c})$, where $C \in \{\pm 1\}$ (recall the definition (4.1.10)). To see this, we define the square matrices

$$\left[T_m\right]_{i,j} = \phi(z_i^m, z_j^{m+1}), \qquad 1 \leqslant i,j \leqslant m+1.$$

One can see that the first row of T_m coincides with the only non-zero row of $E_m(\mathbf{z})$ and the other rows of T_m, with $m \geqslant 1$, form a matrix coinciding with $-W_{[m,m+1)}(\mathbf{z})$. Then the matrix (4.2.1) is obtained by rows permutations from the following one:

$$\begin{bmatrix} 0 & T_0 & 0 & 0 & \cdots & 0 \\ 0 & 0 & T_1 & 0 & \cdots & 0 \\ 0 & 0 & 0 & T_2 & \cdots & 0 \\ \vdots & \vdots & \vdots & \vdots & \ddots & \vdots \\ 0 & 0 & 0 & 0 & \cdots & T_{N-1} \\ \Psi^{(N)}(\mathbf{z}) & 0 & 0 & 0 & \cdots & 0 \end{bmatrix},$$

(one does it by swapping the first row of T_2 with the previous two rows, then the first row of T_3 with the previous three rows, and so on). The determinant of the latter matrix, and hence of $L_{\{\mathbf{z}\}\cup\mathfrak{Z}^c}$, equals exactly $\mathcal{W}(\mathbf{z})$ defined in (4.1.10).

The correlation kernel. One can see that the minors (4.2.1) uniquely define the function $L : \mathfrak{X} \times \mathfrak{X} \to \mathbb{R}$. For example, in the case $N = 2$ above, one has

$$L\big((1),(1,1,z)\big) = \phi(z_1^0, z), \qquad L\big((1,1,y),(2,2,z)\big) = -\phi(y,z),$$
$$L\big((2,1,z),(2)\big) = \Psi_0^2(z), \qquad L\big((2,1,y),(1,1,z)\big) = 0.$$

Hence, the point measure $\mathcal{W}_N$ is a conditional L-ensemble with this function L. By Proposition 3.3.3, the point-measure $\mathcal{W}_N$ on $\mathfrak{Z}$ is determinantal with correlation kernel $\mathcal{K} : \mathfrak{Z} \times \mathfrak{Z} \to \mathbb{R}$ given by

$$\mathcal{K} = 1_{\mathfrak{Z}} - (1_{\mathfrak{Z}} + L)^{-1}\big|_{\mathfrak{Z}\times\mathfrak{Z}}. \tag{4.2.3}$$

In fact, we can compute the inverse of the operator above. To this end, we identify the function L with a function on Λ_N and with values in $\mathcal{M}_{N(N+3)/2,N(N+3)/2}$ so that the identities (4.2.1) hold. Then L can be written in the block form

$$L = \begin{bmatrix} 0 & B \\ C & D_0 \end{bmatrix}$$

with blocks B of size $N \times N(N+1)/2$, C of size $N(N+1)/2 \times N$, and D of size $N(N+1)/2 \times N(N+1)/2$ given by

$$B = \left[E_0, \ldots, E_{N-1}\right],$$
$$C = \left[0, \ldots, 0, (\Psi^{(N)})'\right]',$$

$$D_0 = \begin{bmatrix} 0 & -W_{[1,2)} & 0 & \cdots & 0 \\ 0 & 0 & -W_{[2,3)} & \ddots & \vdots \\ \vdots & \vdots & \ddots & \ddots & 0 \\ \vdots & \vdots & \ddots & \ddots & \vdots \\ 0 & 0 & 0 & \cdots & -W_{[N-1,N)} \\ 0 & 0 & 0 & \cdots & 0 \end{bmatrix}.$$

Defining furthermore the function $D = 1 + D_0$, the result of [10, Lem. 1.5] yields an expression for the correlation kernel $\mathcal{K}$. For the sake of completeness we provide a proof here. As for the function L, we will identify the functions B, C and D with the functions on $\mathfrak{X} \times \mathfrak{X}$. Moreover, we will denote for clarity the convolutions over the values in $\mathfrak{Z}$ by $\star$, and the convolution over $\{1, \cdots, N\}$ we will write as a usual product.

Lemma 4.2.4. *The operators* D *and* $M = B \star D^{-1} \star C$ *are invertible, and the correlation kernel* $\mathcal{K}$ *defined in* (4.2.3) *can be written as*

$$\mathcal{K} = 1_{\mathfrak{Z}} - D^{-1} + D^{-1} \star CM^{-1}B \star D^{-1}. \tag{4.2.5}$$

Proof. By the relationship (4.2.3), the claim will follow if we show that one has

$$(1_{\mathfrak{Z}} + L)^{-1} = \begin{bmatrix} -M^{-1} & M^{-1}B \star D^{-1} \\ D^{-1} \star CM^{-1} & D^{-1} - D^{-1} \star CM^{-1}B \star D^{-1} \end{bmatrix}, \tag{4.2.6}$$

where M is as in the lemma. The easiest way is to check that the matrix on the right-hand side multiplied by $1_{\mathfrak{Z}} + L$ is the identity matrix, i.e.

$$\begin{aligned}
&(1_{\mathfrak{Z}} + L) \begin{bmatrix} -M^{-1} & M^{-1}B \star D^{-1} \\ D^{-1} \star CM^{-1} & D^{-1} - D^{-1} \star CM^{-1}B \star D^{-1} \end{bmatrix} \\
&= \begin{bmatrix} 0 & B \\ C & D \end{bmatrix} \begin{bmatrix} -M^{-1} & M^{-1}B \star D^{-1} \\ D^{-1} \star CM^{-1} & D^{-1} - D^{-1} \star CM^{-1}B \star D^{-1} \end{bmatrix} \\
&= \begin{bmatrix} B \star D^{-1} \star CM^{-1} & B \star D^{-1} - B \star D^{-1}C \star M^{-1}B \star D^{-1} \\ -CM^{-1} + CM^{-1} & CM^{-1}B \star D^{-1} + 1 - CM^{-1}B \star D^{-1} \end{bmatrix} \\
&= \begin{bmatrix} MM^{-1} & B \star D^{-1} - MM^{-1}B \star D^{-1} \\ -CM^{-1} + CM^{-1} & CM^{-1}B \star D^{-1} + 1 - CM^{-1}B \star D^{-1} \end{bmatrix} \\
&= \begin{bmatrix} 1 & 0 \\ 0 & 1 \end{bmatrix},
\end{aligned}$$

which means that (4.2.6) indeed holds. Taking the restriction $(1_{\mathfrak{Z}} + L)^{-1}|_{\mathfrak{Z} \times \mathfrak{Z}}$ we get the right-bottom block of the matrix in (4.2.6) which combined with the definition (4.2.3) gives exactly the expression on the right-hand side of (4.2.5). □

The inverse of the matrix D can be computed very easily

$$D^{-1} = (1+D_0)^{-1} = \sum_{k\geqslant 0}(-1)^k D_0^{\star k} = \begin{bmatrix} 1 & W_{[1,2)} & \cdots & W_{[1,N)} \\ 0 & 1 & \ddots & \vdots \\ \vdots & \ddots & \ddots & W_{[N-1,N)} \\ 0 & 0 & 0 & 1 \end{bmatrix}$$

so $W_{[n,m)}\mathbf{1}_{\{n<m\}}$ is the submatrix of $1-D^{-1}$ with rows $(n,\cdot)$ and columns $(m,\cdot)$.

Exercise 4.2.7. *Prove that the inverse of D is indeed given by the matrix above.*

Moreover, we can easily compute

$$D^{-1}\star C = \begin{bmatrix} W_{[1,N)}\star\Psi^{(N)} \\ \vdots \\ W_{[N-1,N)}\star\Psi^{(N)} \\ \Psi^N \end{bmatrix},$$

as well as

$$B\star D^{-1} = \begin{bmatrix} E_0 & E_0\star W_{[1,2)}+E_1 & \cdots & \sum_{k=1}^{N-1} E_{k-1}\star W_{[k,N)} + E_{N-1} \end{bmatrix}.$$

Therefore the (n,m)-block of the correlation kernel $\mathcal{K}$ is given by

$$\begin{aligned} [\mathcal{K}]_{(n,\cdot),(m,\cdot)} &= -W_{[n,m)}\mathbf{1}_{\{n<m\}} \\ &\quad + W_{[n,N)}\star\Psi^{(N)}M^{-1}\left(\sum_{k=1}^{m-1} E_{k-1}\star W_{[k,m)} + E_{m-1}\right). \end{aligned} \tag{4.2.8}$$

It follows from the property (4.0.12) that for $\mathbf{z}\in\Lambda_N$ one has

$$\left[W_{[n,N)}\star\Psi^{(N)}(\mathbf{z})\right]_{i,j} = \left(\phi^{N-n} * \Psi^N_{N-j}\right)(z_i^n) = \Psi^n_{n-j}(z_i^n),$$

and it remains to evaluate the last part of (4.2.8). For the $N\times m$ matrix in the bracket in (4.2.8) we have

$$\left[\left(\sum_{k=1}^{m-1} E_{k-1}\star W_{[k,m)} + E_{m-1}\right)(\mathbf{z})\right]_{i,j} = \begin{cases} \phi^{m+1}(z_i^{i-1}, z_j^m), & 1\leqslant i\leqslant m, \\ 0, & m<i\leqslant N, \end{cases}$$

and we arrive at the expression

$$\begin{aligned} [\mathcal{K}(\mathbf{z})]_{(n,i),(m,j)} &= -\phi^{m-n}(z_i^n, z_j^m)\mathbf{1}_{\{n<m\}} \\ &\quad + \sum_{\ell,k=1}^{m} \Psi^n_{n-\ell}(z_i^n)\left[M^{-1}\right]_{\ell,k}\phi^{m+1}(z_k^{k-1}, z_j^m), \end{aligned}$$

where the matrix M is given by

$$[M]_{i,j} = \begin{cases} \left(\phi^{N-i} * \Psi^N_{N-j}\right)(z_i^{i-1}), & i<j, \\ 1, & i=j, \\ 0, & i>j. \end{cases} \tag{4.2.9}$$

Biorthogonalization of the correlation kernel. The functions $\phi^{m+1}(z_i^{i-1}, x)$, for $i = 1, \ldots, m$, form a basis of $\operatorname{span}\{1, x, \ldots, x^{m-1}\}$ (by considering z_i^{i-1} to be a fixed value). Furthermore, by the spanning property (4.0.8) in Theorem 4.0.3, the functions $\{\Phi^m_{m-1}(x), \ldots, \Phi^m_0(x)\}$ form a basis of this space as well. Hence, we can define a matrix $A_m \in \mathcal{M}_{m,m}$ which does a change of basis to the functions $\{\Phi^m_{m-1}(x), \ldots, \Phi^m_0(x)\}$, namely

$$\phi^{m+1}(z_i^{i-1}, x) = \sum_{\ell=1}^{m} [A_m]_{i,\ell}\, \Phi^m_{m-\ell}(x).$$

We convolve this equation with $\Psi^m_{m-j}(x)$ using the biorthogonality (4.0.7) to get

$$[A_m]_{i,j} = \left(\phi^{m-i+1} * \Psi^m_{m-j}\right)(z_i^{i-1}). \tag{4.2.10}$$

In particular, we have $A_N = M$, where the matrix M is defined in (4.2.9). Since M is invertible, the two properties from Theorem 4.0.3 indeed define the functions Φ^n_k uniquely. Thus we obtain

$$\sum_{k=1}^{m} [M^{-1}]_{\ell,k}\, \phi^{m+1}(z_k^{k-1}, z_j^m) = \sum_{k=1}^{m} [A_N^{-1}]_{\ell,k} \sum_{i=1}^{m} [A_m]_{k,i}\, \Phi^m_{m-i}(z_j^m). \tag{4.2.11}$$

We now claim that

$$\sum_{k=1}^{m} [A_N^{-1}]_{\ell,k} [A_m]_{k,i} = \delta_{\ell,i}, \tag{4.2.12}$$

for all $1 \leqslant \ell \leqslant N$ and $1 \leqslant i \leqslant m$. To prove this, we notice that

$$[A_m]_{k,i} = \left(\phi^{m-k+1} * \Psi^m_{m-i}\right)(z_k^{k-1}) = \left(\phi^{N-k+1} * \Psi^N_{N-i}\right)(z_k^{k-1}) = [A_N]_{k,i},$$

for $1 \leqslant k, i \leqslant m$, where the first equality is the definition (4.2.10). Thus, using the fact that A_N is upper-triangular (which follows from (4.2.9)), we obtain for $1 \leqslant i \leqslant m$:

$$\sum_{k=1}^{m} [A_N^{-1}]_{\ell,k} [A_m]_{k,i} = \sum_{k=1}^{m} [A_N^{-1}]_{\ell,k} [A_N]_{k,i} = \sum_{k=1}^{N} [A_N^{-1}]_{\ell,k} [A_N]_{k,i} = \delta_{\ell,i},$$

which is exactly our claim. This gives that the right-hand side of (4.2.11) is equal to $\Phi^m_{m-\ell}(z_j^m)$, and the (n, m)-th block of the correlation kernel $\mathcal{K}$ is given by

$$[\mathcal{K}(\mathbf{z})]_{(n,i),(m,j)} = -\phi^{m-n}(z_i^n, z_j^m)\mathbf{1}_{\{n<m\}} + \sum_{\ell=1}^{m} \Psi^n_{n-\ell}(z_i^n)\, \Phi^m_{m-\ell}(z_j^m).$$

We notice that the values i and j appear on the right-hand side only in the indices of $\mathbf{z}$. It means that if we go to the function $\mathcal{K} : \mathfrak{Z} \times \mathfrak{Z} \to \mathbb{R}$, then the values of i and j don't appear on the right-hand side at all. More precisely, we obtain

$$\mathcal{K}\big((n, i, x), (m, j, y)\big) = -\phi^{m-n}(x, y)\mathbf{1}_{\{m>n\}} + \sum_{\ell=1}^{m} \Psi^n_{n-\ell}(x)\, \Phi^m_{m-\ell}(y).$$

Since we are interested in the distribution of the particles z_1^n (see Figure 4.1.3), the kernel (4.0.5) is obtained by fixing the values $i = j = 1$ in the above expression.

Exercise 4.2.13. *Follow the proof of Lemma* 3.1.1 *to show that the identity* (4.0.4) *holds.*

5. Explicit formulas for the correlation kernel

Before the result of [22], the correlation kernels (4.0.5) were known only for a few special cases of initial data (step, see e.g. [14]; and periodic [6–8]), and hence only for those choices asymptotics could be performed in the TASEP and related cases, leading to the Tracy-Widom F_{GUE} and F_{GOE} one-point distributions, and then later to the Airy processes for multipoint distributions. Below we provide formulas for arbitrary initial data and their extensions as $N \to \infty$.

Throughout this and the following sections we use the same notation for integral operators and their kernels.

5.1. Finite initial data. We conjugate the kernels from Theorem 4.0.3 by powers of 2:

$$Q(x,y) := \frac{1}{2^{x-y}} \mathbf{1}_{\{x>y\}}, \tag{5.1.1}$$

as well as

$$\Psi^n_k(x) := \frac{1}{2\pi i} \oint_{\Gamma_0} dw \, \frac{(1-w)^k}{2^{x-X_0(n-k)} w^{x+k+1-X_0(n-k)}} e^{t(w-1)}. \tag{5.1.2}$$

Then the functions $\Phi^n_k(x)$, $k = 0, \ldots, n-1$, are defined implicitly by

(1) the biorthogonality relation $\sum_{x \in \mathbb{Z}} \Psi^n_k(x) \Phi^n_\ell(x) = \mathbf{1}_{\{k=\ell\}}$;

(2) $2^{-x}\Phi^n_k(x)$ with $0 \leqslant k < n$ form a basis of $\operatorname{span}\{x^k : 0 \leqslant k < n\}$.

We are interested in computing the kernel,

$$K_t(n_1, x_1; n_2, x_2) = -Q^{n_2-n_1}(x_1, x_2) \mathbf{1}_{\{n_1<n_2\}} + \sum_{k=1}^{n_2} \Psi^{n_1}_{n_1-k}(x_1) \Phi^{n_2}_{n_2-k}(x_2), \tag{5.1.3}$$

for $n_1, n_2 \in \mathbb{Z}_{\geqslant 1}$ and $x_1, x_2 \in \mathbb{Z}$. The initial data X_0 appears in a simple way in the functions Ψ^n_k, which can be computed explicitly. We note that Q is the transition matrix of a geometric random walk on $\mathbb{Z}$ and has a left inverse

$$Q^{-1} = I + 2\nabla^+, \tag{5.1.4}$$

where we recall that $\nabla^+ f(x) = f(x+1) - f(x)$. Moreover, for all $m, n \in \mathbb{Z}_{\geqslant 0}$ we have the identities

$$Q^{n-m} \Psi^n_{n-k} = \Psi^m_{m-k}, \qquad \Psi^n_k = R_t Q^{-k} \delta_{X_0(n-k)},$$

where $\delta_x(y) = \mathbf{1}_{x=y}$ is the Kronecker's delta, and

$$R_t := e^{-\frac{t}{2}(I+\nabla^-)}, \tag{5.1.5}$$

where $\nabla^- f(x) = f(x) - f(x-1)$. The operator R_t can also be defined through its integral kernel

$$R_t(x,y) = e^{-t} \frac{t^{x-y}}{2^{x-y}(x-y)!} \mathbf{1}_{x \geqslant y} = \frac{1}{2\pi i} \oint_{\Gamma_0} dw \, \frac{e^{t(w-1)}}{2^{x-y} w^{x-y+1}}, \tag{5.1.6}$$

valid for all $t \in \mathbb{R}$, with Γ_0 any counterclockwise contour, containing the origin.

Exercise 5.1.7. *Prove that these identities hold.*

In order to give exact formulas for the functions Φ^n_k from Theorem 4.0.3 we need to define the functions $h^n_k : \mathbb{Z}_{\geqslant 0} \times \mathbb{Z} \to \mathbb{R}$ as solutions to the initial-boundary value problem for the backwards heat equation

$$\begin{cases} (Q^*)^{-1} h^n_k(\ell, z) = h^n_k(\ell+1, z), & 0 \leqslant \ell < k,\ z \in \mathbb{Z}; \quad (5.1.8a) \\ h^n_k(k, z) = 2^{z - X_0(n-k)}, & z \in \mathbb{Z}; \quad (5.1.8b) \\ h^n_k(\ell, X_0(n-\ell)) = 0, & 0 \leqslant \ell < k; \quad (5.1.8c) \end{cases}$$

for $n \geqslant 1$ and $0 \leqslant k < n$ and where Q^* is the adjoint of Q.

Exercise 5.1.9. *Show that the dimension of* $\ker(Q^*)^{-1}$ *is* 1 *and use this to show existence and uniqueness of the solution to the equation* (5.1.8).

Remark 5.1.10. The expression $Q^* h^n_k(k,z)$ is divergent so you can not write $Q^* h^n_k(\ell+1, z) = h^n_k(\ell, z)$.

With these functions at hand we are ready to give exact formulas for Φ^n_k.

Theorem 5.1.11. *The functions* Φ^n_k *from Theorem* 4.0.3 *are given by*

$$\Phi^n_k(z) = \sum_{y \in \mathbb{Z}} h^n_k(0, y)\, R^{-1}_t(y, z), \tag{5.1.12}$$

where $R^{-1}_t = R_{-t}$ *and the latter is defined in* (5.1.5).

Proof. Before proving (5.1.12) we need to prove that $2^{-x} h^n_k(0, x)$ is a polynomial of degree k. We proceed by induction, starting with the base case $\ell = k$, and then decreasing $\ell = k-1, k-2, \ldots, 1$. Note first that, by (5.1.8b), $2^{-x} h^n_k(k, x)$ is a polynomial of degree 0. Assume now that $\tilde{h}^n_k(\ell, x) := 2^{-x} h^n_k(\ell, x)$ is a polynomial of degree $k - \ell$ for some $0 < \ell \leqslant k$. By (5.1.8a) and (5.1.4) we have

$$\tilde{h}^n_k(\ell, y) = 2^{-y}(Q^*)^{-1} h^n_k(\ell-1, y) = \tilde{h}^n_k(\ell-1, y-1) - \tilde{h}^n_k(\ell-1, y) \tag{5.1.13}$$

Taking $x \geqslant X_0(n-\ell+1)$ and summing gives

$$\tilde{h}^n_k(\ell-1, x) = -\sum_{y = X_0(n-\ell+1)+1}^{x} 2^{-y} h^n_k(\ell, y)$$

thanks to (5.1.8c), which by induction is a polynomial of degree $k - \ell + 1$. The function $\tilde{h}^n_k(\ell-1, x)\big|_{x \geqslant X_0(n-\ell+1)}$ has a unique polynomial extension to all $\mathbb{Z}$, which by uniqueness of solutions of (5.1.8) and (5.1.13) shows that $\tilde{h}^n_k(\ell-1, \cdot)$ is a polynomial of degree $k - \ell + 1$ as needed.

Now we check the biorthogonality relation of (5.1.12) with the functions Ψ^n_k. We have

$$\begin{aligned} \sum_{z \in \mathbb{Z}} \Psi^n_\ell(z) \Phi^n_k(z) &= \sum_{z_1, z_2 \in \mathbb{Z}} \sum_{z \in \mathbb{Z}} R_t(z, z_1) Q^{-\ell}(z_1, X_0(n-\ell)) h^n_k(0, z_2) R^{-1}_t(z_2, z) \\ &= \sum_{z \in \mathbb{Z}} Q^{-\ell}(z, X_0(n-\ell)) h^n_k(0, z) = (Q^*)^{-\ell} h^n_k(0, X_0(n-\ell)), \end{aligned}$$

where in the first equality we have used the fact that $2^{-x}h_k^n(0,x)$ is a polynomial together with the fact that the z_1 sum is finite to apply Fubini. For $0 \leqslant \ell \leqslant k$, we use all equations in (5.1.8) to get

$$(Q^*)^{-\ell} h_k^n(0, X_0(n-\ell)) = h_k^n(\ell, X_0(n-\ell)) = \mathbf{1}_{\{k=\ell\}}.$$

For $\ell > k$, we use (5.1.8a) and $2^z \in \ker(Q^*)^{-1}$ to obtain

$$(Q^*)^{-\ell} h_k^n(0, X_0(n-\ell)) = (Q^*)^{-(\ell-k-1)}(Q^*)^{-1} h_k^n(k, X_0(n-\ell)) = 0.$$

Thus, the biorthogonality property holds, and both conditions from Theorem 4.0.3 have been demonstrated. □

5.2. Correlation kernel as a transition probability. In this section we will perform summation in (5.1.3) and obtain a formula for the correlation kernel involving hitting probabilities of a random walk. We start with noting that it is sufficient to find the kernel (5.1.3) with $n_1 = n_2$.

Exercise 5.2.1. *Show that the kernel* $K_t^{(n)}(x_1, x_2) := K_t(n, x_1; n, x_2)$ *recovers the kernel* (5.1.3) *by*

$$K_t(n_i, \cdot; n_j, \cdot) = Q^{n_j - n_i}\left(-\mathbf{1}_{\{n_i < n_j\}} + K_t^{(n_j)}\right). \tag{5.2.2}$$

From the exercise, we can restrict our discussion to the kernel $K_t^{(n)}$. Using the functions h_k^n, introduced in (5.1.8), let us define

$$G_{0,n}(z_1, z_2) := \sum_{k=0}^{n-1} Q^{n-k}(z_1, X_0(n-k)) h_k^n(0, z_2), \tag{5.2.3}$$

so that (5.2.2) and (5.1.3) imply that the correlation kernel $K_t^{(n)}$ equals

$$K_t^{(n)} = R_t Q^{-n} G_{0,n} R_t^{-1}. \tag{5.2.4}$$

Next, we note that the functions h_k^n can be written as hitting probabilities of a random walk. More precisely, let Q^* (the adjoint of Q) be the transition kernel of the random walk B_m^* with $\mathrm{Geom}\left[\frac{1}{2}\right]$ jumps (strictly) to the right, i.e. the increments $B_{m+1}^* - B_m^* \sim \mathrm{Geom}\left[\frac{1}{2}\right]$ are iid, with the convention $\xi \sim \mathrm{Geom}\left[\frac{1}{2}\right] \iff \mathbb{P}[\xi = j] = (\frac{1}{2})^j \mathbf{1}_{\{j \geqslant 1\}}$. Then for $0 \leqslant \ell \leqslant k \leqslant n-1$ we define stopping times

$$\tau^{\ell,n} = \min\{m \in \{\ell, \dots, n-1\} \colon B_m^* > X_0(n-m)\},$$

with the convention that $\min \varnothing = \infty$. For $z \leqslant X_0(n-\ell)$ the function h_k^n can be written as

$$h_k^n(\ell, z) = \mathbb{P}_{B_{\ell-1}^* = z}(\tau^{\ell,n} = k). \tag{5.2.5}$$

Exercise 5.2.6. *Prove this identity.*

Exercise 5.2.7. *Suppose* X *is a random variable taking values in* $\mathbb{N}$ *with the* memoryless property,*i.e. for each pair of numbers* $m, n \in \mathbb{N}$ *one has*

$$\mathbb{P}(X \geqslant m + n \mid X > n) = \mathbb{P}(X \geqslant m).$$

Show that X *has a geometric distribution.*

From the memoryless property of the geometric distribution we get for all $y > X_0(n-k)$,

$$\mathbb{P}_{B^*_{-1}=z}\big(\tau^{0,n} = k,\, B^*_k = y\big) = 2^{X_0(n-k)-y}\,\mathbb{P}_{B^*_{-1}=z}\big(\tau^{0,n} = k\big), \tag{5.2.8}$$

and as a consequence, for $z_2 \leqslant X_0(n)$, we have

$$G_{0,n}(z_1, z_2) = \mathbb{P}_{B^*_{-1}=z_2}\big(\tau^{0,n} < n,\, B^*_{n-1} = z_1\big), \tag{5.2.9}$$

which is the probability for the walk starting at z_2 at time -1 to end up at z_1 after n steps, having hit the curve $\big(X_0(n-m)\big)_{m=0,\dots,n-1}$ in between.

Exercise 5.2.10. *Show that the identities* (5.2.8) *and* (5.2.9) *indeed hold.*

The next step is to obtain an expression along the lines of (5.2.9) which holds for all z_2, and not just for $z_2 \leqslant X_0(n)$. To this end, we need to define analytic extensions of the kernels Q^n. More precisely, for each fixed y_1, $2^{-y_2}Q^n(y_1, y_2)$ extends in y_2 to a polynomial $2^{-y_2}\bar{Q}^{(n)}(y_1, y_2)$ of degree $n-1$ with

$$\bar{Q}^{(n)}(y_1, y_2) := \frac{1}{2\pi i}\oint_{\Gamma_0} dw\, \frac{(1+w)^{y_1-y_2-1}}{2^{y_1-y_2}w^n}, \tag{5.2.11}$$

so that for $y_1 - y_2 \geqslant n$ we have $\bar{Q}^{(n)}(y_1, y_2) = Q^n(y_1, y_2)$.

Exercise 5.2.12. *Prove that the operators $\bar{Q}^{(n)}$ satisfy the following properties: for $n > 1$, $Q^{-1}\bar{Q}^{(n)} = \bar{Q}^{(n)}Q^{-1} = \bar{Q}^{(n-1)}$, but $Q^{-1}\bar{Q}^{(1)} = \bar{Q}^{(1)}Q^{-1} = 0$, and $\bar{Q}^{(n)}\bar{Q}^{(m)}$ is divergent (so the $\bar{Q}^{(n)}$ are no longer a group like Q^n).*

Let B_m be now a random walk with transition matrix Q (that is, B_m has $\mathrm{Geom}\big[\frac{1}{2}\big]$ jumps strictly to the left, i.e. $B_m - B_{m+1} \sim \mathrm{Geom}\big[\frac{1}{2}\big]$) for which we define the stopping time

$$\tau = \min\{m \geqslant 0 : B_m > X_0(m+1)\}. \tag{5.2.13}$$

Using this stopping time and the extension of Q^m we obtain:

Lemma 5.2.14. *For all $z_1, z_2 \in \mathbb{Z}$ we have the identity*

$$\begin{aligned} G_{0,n}(z_1, z_2) = \mathbf{1}_{\{z_1 > X_0(1)\}}\bar{Q}^{(n)}(z_1, z_2) \\ + \mathbf{1}_{\{z_1 \leqslant X_0(1)\}}\mathbb{E}_{B_0=z_1}\Big[\bar{Q}^{(n-\tau)}(B_\tau, z_2)\mathbf{1}_{\{\tau<n\}}\Big]. \end{aligned} \tag{5.2.15}$$

Proof. For $z_2 \leqslant X_0(n)$, the expression in (5.2.9) can be written as

$$\begin{aligned} &G_{0,n}(z_1, z_2) \\ &= \mathbb{P}_{B^*_{-1}=z_2}\big(\tau^{0,n} \leqslant n-1,\, B^*_{n-1} = z_1\big) \\ &= \mathbb{P}_{B_0=z_1}\big(\tau \leqslant n-1, B_n = z_2\big) \\ &= \sum_{k=0}^{n-1}\sum_{z > X_0(k+1)} \mathbb{P}_{B_0=z_1}\big(\tau = k,\, B_k = z\big)Q^{n-k}(z, z_2) \\ &= \mathbb{E}_{B_0=z_1}\Big[Q^{n-\tau}(B_\tau, z_2)\mathbf{1}_{\{\tau<n\}}\Big]. \end{aligned} \tag{5.2.16}$$

If $z_1 > X_0(1)$, one can verify that $\mathbb{E}_{B_0=z_1}\Big[Q^{n-\tau}(B_\tau, z_2)\mathbf{1}_{\{\tau<n\}}\Big] = Q^n(z_1,z_2)$, and so we get

$$G_{0,n}(z_1,z_2) = \mathbf{1}_{\{z_1>X_0(1)\}}Q^n(z_1,z_2) + \mathbf{1}_{\{z_1\leqslant X_0(1)\}}\mathbb{E}_{B_0=z_1}\Big[Q^{n-\tau}(B_\tau,z_2)\mathbf{1}_{\{\tau<n\}}\Big]$$

for all $z_2 \leqslant X_0(n)$. Let us now denote

$$\widetilde{G}_{0,n}(z_1,z_2) := \mathbf{1}_{\{z_1>X_0(1)\}}\bar{Q}^{(n)}(z_1,z_2) + \mathbf{1}_{\{z_1\leqslant X_0(1)\}}\mathbb{E}_{B_0=z_1}\Big[\bar{Q}^{(n-\tau)}(B_\tau,z_2)\mathbf{1}_{\{\tau<n\}}\Big].$$

We claim that $\widetilde{G}_{0,n}(z_1,z_2) = G_{0,n}(z_1,z_2)$ for all $z_2 \leqslant X_0(n)$. To see this, note that

$$\chi_{X_0(1)}\bar{Q}^{(n)}\bar{\chi}_{X_0(n)} = \chi_{X_0(1)}Q^n\bar{\chi}_{X_0(n)},$$

thanks to the properties proved in Exercise 5.2.12, where the projections χ and $\bar{\chi}$ were defined in (4.0.2). For the other term, the last equality in (5.2.16) shows that we only need to check that $\chi_{X_0(k+1)}\bar{Q}^{(k+1)}\bar{\chi}_{X_0(n)} = \chi_{X_0(k+1)}Q^{k+1}\bar{\chi}_{X_0(n)}$ for $k = 0,\dots,n-1$, which follows again from the same fact proved in Exercise 5.2.12. To complete the proof, recall that, by Theorem 4.0.4, $K_t^{(n)}$ satisfies the following: for every fixed z_1, $2^{-z_2}K_t(z_1,z_2)$ is a polynomial of degree $n-1$ in z_2. It is easy to check that this implies that $G_{0,n} = Q^nR_tK_t^{(n)}R_t^{-1}$ satisfies the same. Since $\bar{Q}^{(k)}$ also satisfies this property for each $k = 0,\dots,n$, we deduce that $2^{-z_2}\widetilde{G}_{0,n}(z_1,z_2)$ is a polynomial in z_2. Since it coincides with $2^{-z_2}G_{0,n}(z_1,z_2)$ at infinitely many z_2's, we deduce that $\widetilde{G}_{0,n} = G_{0,n}$. □

In order to have a lighter notation, we define the following kernels

$$\mathcal{S}_{-t,-n}(z_1,z_2) := (e^{-\frac{t}{2}\nabla^-}Q^{-n})^*(z_1,z_2)$$

(5.2.17)
$$= \frac{1}{2\pi i}\oint_{\Gamma_0} dw\,\frac{(1-w)^n}{2^{z_2-z_1}w^{n+1+z_2-z_1}}e^{t(w-1/2)},$$

$$\bar{\mathcal{S}}_{-t,n}(z_1,z_2) := \bar{Q}^{(n)}e^{\frac{t}{2}\nabla^-}(z_1,z_2)$$

(5.2.18)
$$= \frac{1}{2\pi i}\oint_{\Gamma_0} dw\,\frac{(1-w)^{z_2-z_1+n-1}}{2^{z_1-z_2}w^n}e^{t(w-1/2)}.$$

Exercise 5.2.19. *Show that these operators indeed are given by the contour integrals.*

Furthermore, we define the following function

(5.2.20)
$$\bar{\mathcal{S}}^{\mathrm{epi}(X_0)}_{-t,n}(z_1,z_2) := \mathbb{E}_{B_0=z_1}\Big[\bar{\mathcal{S}}_{-t,n-\tau}(B_\tau,z_2)\mathbf{1}_{\{\tau<n\}}\Big],$$

where the superscript epi refers to the fact that τ (defined in (5.2.13)) is the hitting time of the epigraph of the curve $(X_0(k+1)+1)_{k=0,\dots,n-1}$ by the random walk B_k. With these operators at hand we have the following formula for TASEP with general right-finite initial data:

Theorem 5.2.21 (TASEP formula for right-finite initial data)**.** *Assume that initial values satisfy $X_0(j) = \infty$ for $j \leqslant 0$. Then for $1 \leqslant n_1 < n_2 < \cdots < n_M$ and $t > 0$ we have*

(5.2.22)
$$\mathbb{P}(X_t(n_j) > a_j,\ j = 1,\dots,M) = \det(I - \bar{\chi}_a K_t \bar{\chi}_a)_{\ell^2(\{n_1,\dots,n_M\}\times\mathbb{Z})},$$

where the kernel K_t *is given by*

$$K_t(n_i,\cdot;n_j,\cdot) = -Q^{n_j-n_i}\mathbf{1}_{\{n_i<n_j\}} + (\mathcal{S}_{-t,-n_i})^*\bar{\bar{\mathcal{S}}}^{\mathrm{epi}(X_0)}_{-t,n_j}. \tag{5.2.23}$$

Proof. If $X_0(1) < \infty$ then we are in the setting of the above sections. Formulas (5.2.22) and (5.2.23) follow directly from the above definitions together with (5.2.4) and Lemma 5.2.14. If $X_0(j) = \infty$ for $j = 1, \ldots, \ell$ and $X_0(\ell+1) < \infty$ then

$$\mathbb{P}_{X_0}\big(X_t(n_j) > a_j,\ j = 1, \ldots, M\big) = \det\big(I - \bar{\chi}_a \mathcal{K}^{(\ell)}_t \bar{\chi}_a\big)_{\ell^2(\{n_1,\ldots,n_M\}\times\mathbb{Z})}$$

with the correlation kernel

$$\mathcal{K}^{(\ell)}_t(n_i,\cdot;n_j,\cdot) = -Q^{n_j-n_i}\mathbf{1}_{\{n_i<n_j\}} + (\mathcal{S}_{-t,-n_i+\ell})^*\bar{\mathcal{S}}^{\mathrm{epi}(\theta_\ell X_0)}_{-t,n_j-\ell},$$

where $\theta_\ell X_0(j) = X_0(\ell+j)$. Using now the fact that $Q^\ell\bar{\mathcal{S}}^{\mathrm{epi}(\theta_\ell X_0)}_{-t,n_j-\ell} = \bar{\mathcal{S}}^{\mathrm{epi}(X_0)}_{-t,n_j}$ and (5.2.17) we conclude that (5.2.23) still holds in this case. □

Remark 5.2.24. One can write a formula for initial data which are not right finite [18], but they are a bit more cumbersome. In practice, one cuts of the data very far to the right, and uses the formula above. Since one has exact formulas, one can check that cutting off has a small effect.

For some special initial data one can get simpler expressions for the correlation kernel [18] and we can recover the formulas from [7], [13] and [6]. We leave these computations as exercises below.

Exercise 5.2.25. *Consider TASEP with step initial data,* $X_0(i) = -i$ *for* $i \geqslant 1$ *and show that*

$$\begin{aligned} K_t(n_i,z_1;n_j,z_2) = &-Q^{n_j-n_i}(z_1,z_2)\mathbf{1}_{n_i<n_j} \\ &+ \frac{1}{(2\pi i)^2}\oint_{\Gamma_0} dw \oint_{\Gamma_0} dv\, \frac{(1-w)^{n_i}(1-v)^{n_j+z_2}}{2^{z_1-z_2}w^{n_i+z_1+1}v^{n_j}}\frac{e^{t(w+v-1)}}{1-v-w}. \end{aligned}$$

Exercise 5.2.26.* *Consider TASEP with periodic initial data* $X_0(i) = 2i$, $i \in \mathbb{Z}$ *and show that*

$$K^{(n)}_t(z_1,z_2) = -\frac{1}{2\pi i}\oint_{1+\Gamma_0} dv\, \frac{v^{z_2+2n}}{2^{z_1-z_2}(1-v)^{z_1+2n+1}}\, e^{t(1-2v)}.$$

(Hint: approximate by finite periodic initial data $X_0(i) = 2(N-i)$ *for* $i = 1, \ldots, 2N$.*)*

5.3. Path integral formulas. In addition to the extended kernel formula (4.0.4), one has a *path integral formula*

$$\det\Big(I - K^{(n_m)}_t\big(I - Q^{n_1-n_m}\chi_{a_1}Q^{n_2-n_1}\chi_{a_2}\cdots Q^{n_m-n_{m-1}}\chi_{a_m}\big)\Big)_{\ell^2(\mathbb{Z})}, \tag{5.3.1}$$

where as before $K^{(n)}_t(z_1,z_2) = K_t(n,z_1;n,z_2)$. Such formulas were first obtained in [24] for the Airy$_2$ process (see [25] for the proof), and later was extended to the Airy$_1$ process in [26] and then to a very wide class of processes in [5]. We provide this result below in full generality.

For $t_1 < t_2 < \cdots < t_n$ we consider an extended kernel $\mathcal{K}^{\text{ext}}$ given as follows: for $1 \leqslant i, j \leqslant n$ and $x, y \in X$ (here (X, μ) is a given measure space),

$$\mathcal{K}^{\text{ext}}(t_i, x; t_j, y) := \begin{cases} \mathcal{W}_{t_i,t_j}\mathcal{K}_{t_j}(x,y), & \text{if } i \geqslant j, \\ -\mathcal{W}_{t_i,t_j}(I - \mathcal{K}_{t_j})(x,y), & \text{if } i < j. \end{cases} \tag{5.3.2}$$

Additionally, we are considering multiplication operators N_{t_i} acting on a measurable function f on X as $N_{t_i}f(x) = \varphi_{t_i}(x)f(x)$ for some measurable functions φ_{t_i} defined on X. M will denote the diagonal operator acting on functions f defined on $\{t_1, \ldots, t_n\} \times X$ as

$$Mf(t_i, \cdot) := N_{t_i} f(t_i, \cdot).$$

We provide below all the assumptions from [5] except their Assumption 2(iii) which has to be changed in our case.

Assumption 5.3.3. There are integral operators $\mathcal{Q}_t$ on $L^2(X)$ such that the following hold:

(i) The integral operators $\mathcal{Q}_{t_i}\mathcal{W}_{t_i,t_j}$, $\mathcal{Q}_{t_i}\mathcal{K}_{t_i}$, $\mathcal{Q}_{t_i}\mathcal{W}_{t_i,t_j}\mathcal{K}_{t_j}$ and $\mathcal{Q}_{t_j}\mathcal{W}_{t_j,t_i}\mathcal{K}_{t_i}$ for $1 \leqslant i < j \leqslant n$ are all bounded operators mapping $L^2(X)$ to itself.

(ii) If $\bar{\mathcal{Q}}_{t_i} := I - \mathcal{Q}_{t_i}$, then $\mathcal{K}_{t_1} - \bar{\mathcal{Q}}_{t_1}\mathcal{W}_{t_1,t_2}\bar{\mathcal{Q}}_{t_2}\cdots\mathcal{W}_{t_{n-1},t_n}\bar{\mathcal{Q}}_{t_n}\mathcal{W}_{t_n,t_1}\mathcal{K}_{t_1}$ is a bounded operator mapping $L^2(X)$ to itself.

Assumption 5.3.4. For each $i \leqslant j \leqslant k$ the following hold:

(i) *Right-invertibility*: $\mathcal{W}_{t_i,t_j}\mathcal{W}_{t_j,t_i}\mathcal{K}_{t_i} = \mathcal{K}_{t_i}$;

(ii) *Semigroup property*: $\mathcal{W}_{t_i,t_j}\mathcal{W}_{t_j,t_k} = \mathcal{W}_{t_i,t_k}$;

(iii) *Reversibility relation*: $\mathcal{W}_{t_i,t_j}\mathcal{K}_{t_j}\mathcal{W}_{t_j,t_i} = \mathcal{K}_{t_i}$, for all $t_i < t_j$.

Assumption 5.3.5. One can choose multiplication operators V_{t_i}, V'_{t_i}, U_{t_i} and U'_{t_i} acting on $\mathcal{M}(X)$, for $1 \leqslant i \leqslant n$, in such a way that:

(i) $V'_{t_i}V_{t_i}\mathcal{Q}_{t_i} = \mathcal{Q}_{t_i}$ and $\mathcal{K}_{t_i}U'_{t_i}U_{t_i} = \mathcal{K}_{t_i}$, for all $1 \leqslant i \leqslant n$.

(ii) The four operators $V_{t_i}\mathcal{Q}_{t_i}\mathcal{K}_{t_i}V'_{t_i}$, $V_{t_i}\mathcal{Q}_{t_i}\mathcal{W}_{t_i,t_j}V'_{t_j}$, $V_{t_i}\mathcal{Q}_{t_i}\mathcal{W}_{t_i,t_j}\mathcal{K}_{t_j}V'_{t_j}$ and $V_{t_j}\mathcal{Q}_{t_j}\mathcal{W}_{t_j,t_i}\mathcal{K}_{t_i}V'_{t_i}$ preserve $L^2(X)$ and are trace class in $L^2(X)$, for all $1 \leqslant i < j \leqslant n$.

(iii) The following operator preserves $L^2(X)$ and is trace class, for $1 \leqslant i \leqslant n$: $U_{t_i}\big[\mathcal{W}_{t_i,t_1}\mathcal{K}_{t_1} - \bar{\mathcal{Q}}_{t_i}\mathcal{W}_{t_i,t_{i+1}}\cdots\bar{\mathcal{Q}}_{t_{n-1}}\mathcal{W}_{t_{n-1},t_n}\bar{\mathcal{Q}}_{t_n}\mathcal{W}_{t_n,t_1}\mathcal{K}_{t_1}\big]U'_{t_1}$, where we define $\bar{\mathcal{Q}}_{t_i} := I - \mathcal{Q}_{t_i}$.

We are assuming here that $\mathcal{W}_{t_i,t_j}$ is invertible for all $t_i \leqslant t_j$, so that $\mathcal{W}_{t_j,t_i}$ is defined as a proper operator[3]. Moreover, we assume that it satisfies

$$\mathcal{W}_{t_j,t_i}\mathcal{K}_{t_i} = \mathcal{K}^{\text{ext}}(t_j, \cdot; t_i, \cdot) \tag{5.3.6}$$

for all $t_i \geqslant t_j$, and that the multiplication operators U_{t_i}, U'_{t_i} introduced in Assumption 5.3.5 satisfy Assumption 5.3.5(iii) with the operator in that assumption

[3] This is just for simplicity; it is possible to state a version of Theorem 5.3.7 asking instead that the product $\mathcal{K}_{t_j}\mathcal{W}_{t_j,t_i}$ be well defined.

replaced by

$$U_{t_i}\big[\mathcal{W}_{t_i,t_{i+1}}\overline{N}_{t_{i+1}}\cdots\mathcal{W}_{t_{n-1},t_n}\overline{N}_{t_n}\mathcal{K}_{t_n} - \mathcal{W}_{t_i,t_1}\overline{N}_{t_1}\mathcal{W}_{t_1,t_2}\overline{N}_{t_2}\cdots\mathcal{W}_{t_{n-1},t_n}\overline{N}_{t_n}\mathcal{K}_{t_n}\big]U'_{t_i}.$$

Theorem 5.3.7. *Under the assumptions above, we have the identity*

$$\det(I - N\mathcal{K}^{\mathrm{ext}})_{L^2(\{t_1,\dots,t_n\}\times X)} = \det(I - \mathcal{K}_{t_n} + \mathcal{K}_{t_n}\mathcal{W}_{t_n,t_1}\overline{N}_{t_1}\mathcal{W}_{t_1,t_2}\overline{N}_{t_2}\cdots\mathcal{W}_{t_{n-1},t_n}\overline{N}_{t_n})_{L^2(X)},$$

where $\overline{N}_{t_i} = I - N_{t_i}$.

Proof. The proof is a minor adaptation of the arguments in [5, Thm. 3.3], and we will use throughout it all the notation and conventions of that proof. We will just sketch the proof, skipping several technical details (in particular, we will completely omit the need to conjugate by the operators U_{t_i} and V_{t_i}, since this aspect of the proof can be adapted straightforwardly from [5]).

In order to simplify notation throughout the proof we will replace subscripts of the form t_i by i, so for example $\mathcal{W}_{i,j} = \mathcal{W}_{t_i,t_j}$. Let $K := N\mathcal{K}^{\mathrm{ext}}$. Then K can be written as

$$K = N(W^- K^d + W^+(K^d - I)) \qquad \text{with} \qquad K^d_{ij} := \mathcal{K}_i \mathbf{1}_{i=j}, \quad N_{i,j} := N_i \mathbf{1}_{i=j},$$

where W^-, W^+ are lower triangular, respectively strictly upper triangular, and defined by

$$W^-_{ij} = \mathcal{W}_{i,j}\mathbf{1}_{i\geqslant j}, \qquad W^+_{ij} = \mathcal{W}_{i,j}\mathbf{1}_{i<j}.$$

The key to the proof in [5] was $\big[(I+W^+)^{-1})\big]_{i,j} = \mathbf{1}_{j=i} - \mathcal{W}_{i,i+1}\mathbf{1}_{j=i+1}$, which then implies that $\big[(W^- + W^+)K^d(I+W^+)^{-1}\big]_{i,j} = \mathcal{W}_{i,1}\mathcal{K}_1\mathbf{1}_{j=1}$. The fact that only the first column of this matrix has non-zero entries is what ultimately allows one to turn the Fredholm determinant of an extended kernel into one of a kernel acting on $L^2(X)$. However, the derivation of this last identity uses $\mathcal{W}_{i,j-1}\mathcal{K}_{j-1}\mathcal{W}_{j-1,j} = \mathcal{W}_{i,j}\mathcal{K}_j$, which is a consequence of Assumptions 5.3.4(ii) and (iii), and thus is not available to us. In our case we may proceed similarly by observing that

$$\big[(W^-)^{-1}\big]_{i,j} = \mathbf{1}_{j=i} - \mathcal{W}_{i,i-1}\mathbf{1}_{j=i-1}, \tag{5.3.8}$$

as can be checked directly using Assumption 5.3.4(ii). Now using the identity

$$\mathcal{W}_{i,j+1}\mathcal{K}_{j+1}\mathcal{W}_{j+1,j} = \mathcal{W}_{i,j}\mathcal{K}_j,$$

which follows from Assumption 5.3.4(ii) and (iii), we get

$$\begin{aligned}\big[(W^- + W^+)K^d(W^-)^{-1}\big]_{i,j} &= \mathcal{W}_{i,j}\mathcal{K}_j - \mathcal{W}_{i,j+1}\mathcal{K}_{j+1}\mathcal{W}_{j+1,j}\mathbf{1}_{j<n} \\ &= \mathcal{W}_{i,n}\mathcal{K}_n\mathbf{1}_{j=n}.\end{aligned} \tag{5.3.9}$$

Note that now only the last column of this matrix has non-zero entries, which accounts for the difference between our result and that of [5]. To take advantage of (5.3.9) we write

$$I-K=(I+NW^+)\big[I-(I+NW^+)^{-1}N(W^-+W^+)K^d(W^-)^{-1}W^-\big].$$

Since NW^+ is strictly upper triangular, $\det(I+NW^+)=1$, which in particular shows that $I+NW^+$ is invertible. Thus by the cyclic property of the Fredholm determinant, $\det(I-K):=\det(I-\widetilde{K})$ with

$$\widetilde{K}=W^-(I+NW^+)^{-1}N(W^-+W^+)K^d(W^-)^{-1}.$$

Since only the last column of $(W^-+W^+)K^d(W^-)^{-1}$ is non-zero, the same holds for $\widetilde{K}$, and thus $\det(I-K)=\det(I-\widetilde{K}_{n,n})_{L^2(X)}$.

Our goal now is to compute $\widetilde{K}_{n,n}$. From (5.3.9) and Assumption 5.3.4(ii) we get, for $0\leqslant k\leqslant n-i$,

$$\Big[(NW^+)^kN(W^-+W^+)K^d(W^-)^{-1}\Big]_{i,n} = \sum_{i<\ell_1<\cdots<\ell_k\leqslant n} N_i\mathcal{W}_{i,\ell_1}N_{\ell_1}\mathcal{W}_{\ell_1,\ell_2}\cdots N_{\ell_{k-1}}\mathcal{W}_{\ell_{k-1},\ell_k}N_{\ell_k}\mathcal{W}_{\ell_k,n}\mathcal{K}_n,$$

while for $k>n-i$ the left-hand side above equals 0 (the case $k=0$ is interpreted as $N_i\mathcal{W}_{i,n}\mathcal{K}_n$). As in [5] this leads to

$$\widetilde{K}_{i,n}=\sum_{j=1}^{i}\sum_{k=0}^{n-j}(-1)^k\sum_{j=\ell_0<\ell_1<\cdots<\ell_k\leqslant n}\mathcal{W}_{i,j}N_j\mathcal{W}_{j,\ell_1}N_{\ell_1}\mathcal{W}_{\ell_1,\ell_2}N_{\ell_{k-1}}\mathcal{W}_{\ell_{k-1},\ell_k}N_{\ell_k}\mathcal{W}_{\ell_k,n}\mathcal{K}_n.$$

Replacing each N_ℓ except the first by $I-\overline{N}_\ell$ and simplifying as in [5] leads to

$$\widetilde{K}_{i,n}=\mathcal{W}_{i,i+1}\overline{N}_{i+1}\mathcal{W}_{i+1,i+2}\overline{N}_{i+2}\cdots\mathcal{W}_{n-1,n}\overline{N}_nK_n - \mathcal{W}_{i,1}\overline{N}_1\mathcal{W}_{1,2}\overline{N}_2\cdots\mathcal{W}_{n-1,n}\overline{N}_n\mathcal{K}_n.$$

Setting $i=n$ yields $\widetilde{K}_{n,n}=\mathcal{K}_n-\mathcal{W}_{n,1}\overline{N}_1\mathcal{W}_{1,2}\overline{N}_2\cdots\mathcal{W}_{n-1,n}\overline{N}_n\mathcal{K}_n$ and then an application of the cyclic property of the determinant gives the result. □

5.4. Proof of the TASEP path integral formula. To obtain the path integral version (5.3.1) of the TASEP formula we use Theorem 5.3.7. Recall from (4.0.12) that $Q^{n-m}\Psi^n_{n-k}=\Psi^m_{m-k}$. Then we can write

$$\begin{aligned} Q^{n_j-n_i}K_t^{(n_j)} &= \sum_{k=0}^{n_j-1}Q^{n_j-n_i}\Psi^{n_j}_k\otimes\Phi^{n_j}_k \\ &= \sum_{k=0}^{n_j-1}\Psi^{n_i}_{n_i-n_j+k}\otimes\Phi^{n_j}_k = K_t(n_i,\cdot;n_j,\cdot)+Q^{n_j-n_i}\mathbf{1}_{n_i<n_j}. \end{aligned} \tag{5.4.1}$$

Thus, the extended kernel K_t has exactly the structure specified in (5.3.2), taking $t_i=n_i$, $\mathcal{K}_{t_i}=K_t^{(n_i)}$, $\mathcal{W}_{t_i,t_j}=Q^{n_j-n_i}$ and $\mathcal{W}_{t_i,t_j}\mathcal{K}_{t_j}=K_t(n_i,\cdot;n_j,\cdot)$. It is not hard to check that Assumptions 1 and 3 of [5, Thm. 3.3] hold in our setting. The semigroup property (Assumption 2(ii)) is trivial in this case, while the right-invertibility condition (Assumption 2(i))

$$Q^{n_j-n_i}K_t(n_j,\cdot;n_i,\cdot)=K_t^{(n_i)}$$

for $n_i \leqslant n_j$ follows similarly to (5.4.1). However, Assumption 2(iii) of [5], which translates into $Q^{n_j - n_i} K_t^{(n_j)} = K_t^{(n_i)} Q^{n_j - n_i}$ for $n_i \leqslant n_j$, does not hold in our case (in fact, the right hand side does not even make sense as the product is divergent, as can be seen by noting that $\Phi_0^n(x) = 2^{x - X_0(n)}$; alternatively, note that the left hand side depends on the values of $X_0(n_{i+1}), \ldots, X_0(n_j)$ but the right hand side does not), which is why we need Theorem 5.3.7. To use it, we need to check that

$$Q^{n_j - n_i} K_t^{(n_j)} Q^{n_i - n_j} = K_t^{(n_i)}. \tag{5.4.2}$$

In fact, if $k \geqslant 0$ then (5.1.8a) and the easy fact that $h_k^n(\ell, z) = h_{k-1}^{n-1}(\ell - 1, z)$ imply that

$$(Q^*)^{n_i - n_j} h^{n_j}_{k + n_j - n_i}(0, z) = h^{n_j}_{k + n_j - n_i}(n_j - n_i, z) = h_k^{n_i}(0, z),$$

so that $(Q^*)^{n_i - n_j} \Phi^{n_j}_{k + n_j - n_i} = \Phi_k^{n_i}$. On the other hand, if $n_i - n_j \leqslant k < 0$ then we have

$$(Q^*)^{n_i - n_j} h^{n_j}_{k + n_j - n_i}(0, z) = (Q^*)^k h^{n_j}_{k + n_j - n_i}(k + n_j - n_i, z) = 0$$

thanks to (5.1.8b) and the fact that $2^z \in \ker(Q^*)^{-1}$, which gives the identity $(Q^*)^{n_i - n_j} \Phi^{n_j}_{k + n_j - n_i} = 0$. Therefore, proceeding as in (5.4.1), the left hand side of (5.4.2) equals

$$\begin{aligned}
&\sum_{k=0}^{n_j - 1} Q^{n_j - n_i} \Psi_k^{n_j} \otimes (Q^*)^{n_i - n_j} \Phi_k^{n_j} \\
&= \sum_{k=0}^{n_j - 1} \Psi^{n_i}_{n_i - n_j + k} \otimes (Q^*)^{n_i - n_j} \Phi_k^{n_j} \\
&= \sum_{k=0}^{n_i - 1} \Psi_k^{n_i} \otimes \Phi_k^{n_i}
\end{aligned}$$

as desired.

6. The KPZ fixed point

In this section we will take the KPZ scaling limit of the TASEP growth process, by using the formula from Theorem 5.2.21, and will get a complete characterisation of the limiting Markov process, called the *KPZ fixed point*. We start with introducing the topology and some operators.

6.1. State space and topology. The state space on which we define the KPZ fixed point is the following:

Definition 6.1.1 (UC functions)**.** We define UC as the space of upper semicontinuous functions $\mathfrak{h}\colon \mathbb{R} \to [-\infty, \infty)$ with $\mathfrak{h}(\mathbf{x}) \leqslant C(1 + |\mathbf{x}|)$ for some $C < \infty$.

Example 6.1.2. For $\mathbf{u} \in \mathbb{R}$ the function $\mathfrak{d}_{\mathbf{u}}(\mathbf{u}) := 0$, $\mathfrak{d}_{\mathbf{u}}(\mathbf{x}) := -\infty$ for $\mathbf{x} \neq \mathbf{u}$, is known as a *narrow wedge at* $\mathbf{u}$, and $\mathfrak{d}_{\mathbf{u}} \in \mathrm{UC}$. These arise naturally as $\mathfrak{d}_0$ is clearly the limit of the TASEP height function $h(x) = -|x|$ under the rescaling $\varepsilon^{1/2}h(\varepsilon^{-1}x)$.

We will endow this space with the topology of local UC convergence. This is the natural topology for lateral growth, and will allow us to compute the KPZ limit in all cases of interest[4]. In order to define this topology, recall that $\mathfrak{h}$ is upper semicontinuous (UC) if and only if its *hypograph*

$$\operatorname{hypo}(\mathfrak{h}) := \{(\mathbf{x},\mathbf{y}) : \mathbf{y} \leqslant \mathfrak{h}(\mathbf{x})\}$$

is closed in $\mathbb{R} \times [-\infty, \infty)$. Slightly informally, local UC convergence can be defined as follows:

Definition 6.1.3 (Local UC convergence). We say that $(\mathfrak{h}_\varepsilon)_\varepsilon \subseteq \mathrm{UC}$ *converges locally in* UC to $\mathfrak{h} \in \mathrm{UC}$ if there is a $C > 0$ such that $\mathfrak{h}_\varepsilon(\mathbf{x}) \leqslant C(1 + |\mathbf{x}|)$ for every $\varepsilon > 0$ and if for every $M \geqslant 1$ there is a $\delta = \delta(\varepsilon, M) > 0$ going to 0 as $\varepsilon \to 0$ such that the hypographs $\mathfrak{H}_{\varepsilon,M}$ and $\mathfrak{H}_M$ of $\mathfrak{h}_\varepsilon$ and $\mathfrak{h}$ restricted to $[-M, M]$ are δ-close in the sense that

$$\cup_{(\mathbf{t},\mathbf{x})\in\mathfrak{H}_{\varepsilon,M}} B_\delta((\mathbf{t},\mathbf{x})) \subseteq \mathfrak{H}_M \qquad \text{and} \qquad \cup_{(\mathbf{t},\mathbf{x})\in\mathfrak{H}_M} B_\delta((\mathbf{t},\mathbf{x})) \subseteq \mathfrak{H}_{\varepsilon,M}.$$

We will also use an analogous space LC, made of lower semicontinuous functions:

Definition 6.1.4 (LC functions and local convergence). Let $\mathrm{LC} := \{\mathfrak{g} \colon -\mathfrak{g} \in \mathrm{UC}\}$ and endow this space with the topology of *local* LC *convergence* which is defined analogously to local UC convergence, now in terms of *epigraphs*,

$$\operatorname{epi}(\mathfrak{g}) := \{(\mathbf{x},\mathbf{y}) : \mathbf{y} \geqslant \mathfrak{g}(\mathbf{x})\}.$$

Explicitly, $(\mathfrak{g}_\varepsilon)_\varepsilon \subset \mathrm{LC}$ converges locally in LC to $\mathfrak{g} \in \mathrm{LC}$ if and only if $-\mathfrak{g}_\varepsilon \to -\mathfrak{g}$ locally in UC.

Exercise 6.1.5. *Show that if $\mathfrak{h}$ is locally Hölder $\beta \in (0,1)$ then convergence in* UC *or* LC *implies uniform convergence on compact sets.*

Exercise 6.1.6.* *Show that if $\mathfrak{h}_0 \in \mathrm{LC}$ then the inviscid solution given by the Hopf-Lax formula*

$$h(t,x) = \sup_{y}\Big\{h_0(y) - t^{-1}(x-y)^2\Big\},$$

and for fixed $t > 0$ it is continuous in the UC *topology.*

6.2. Auxiliary operators. In order to state our main result we need to introduce several operators, which will appear in the explicit Fredholm determinant formula for the fixed point. Our basic building block is the following (almost) group of operators:

[4] Actually the bound $\mathfrak{h}(\mathbf{x}) \leqslant C(1 + |\mathbf{x}|)$ which we are imposing here and in [22] on UC functions is not as general as possible, but makes the arguments a bit simpler and it suffices for most cases of interest (see also [22, Foot. 9]).

Definition 6.2.1. For $\mathbf{x}, \mathbf{t} \in \mathbb{R}^2 \setminus \{\mathbf{x} < 0, \mathbf{t} = 0\}$ let us define the operator

$$\mathbf{S}_{\mathbf{t},\mathbf{x}} := \exp\{\mathbf{x}\partial^2 + \tfrac{\mathbf{t}}{3}\partial^3\}, \tag{6.2.2}$$

which satisfies the identity

$$\mathbf{S}_{\mathbf{s},\mathbf{x}}\mathbf{S}_{\mathbf{t},\mathbf{y}} = \mathbf{S}_{\mathbf{s}+\mathbf{t},\mathbf{x}+\mathbf{y}} \tag{6.2.3}$$

as long as all subscripts avoid the region $\{\mathbf{x} < 0, \mathbf{t} = 0\}$.

For $\mathbf{t} > 0$ the operator $\mathbf{S}_{\mathbf{t},\mathbf{x}}$ acts on nice functions by convolution with the kernel

$$\mathbf{S}_{\mathbf{t},\mathbf{x}}(z) = \frac{1}{2\pi \mathrm{i}} \int_{\langle} dw\, e^{\frac{\mathbf{t}}{3}w^3 + \mathbf{x}w^2 - zw} = \mathbf{t}^{-1/3} e^{\frac{2\mathbf{x}^3}{3\mathbf{t}^2} - \frac{z\mathbf{x}}{\mathbf{t}}} \operatorname{Ai}(-\mathbf{t}^{-1/3}z + \mathbf{t}^{-4/3}\mathbf{x}^2), \tag{6.2.4}$$

where $\langle$ is the positively oriented contour going in straight lines from $e^{-\mathrm{i}\pi/3}\infty$ to $e^{\mathrm{i}\pi/3}\infty$ through 0 and Ai is the Airy function

$$\operatorname{Ai}(z) = \frac{1}{2\pi \mathrm{i}} \int_{\langle} dw\, e^{\frac{1}{3}w^3 - zw}.$$

When $\mathbf{t} < 0$ we have the identity $\mathbf{S}_{\mathbf{t},\mathbf{x}} = (\mathbf{S}_{-\mathbf{t},\mathbf{x}})^*$, which in particular yields

$$(\mathbf{S}_{\mathbf{t},\mathbf{x}})^*\mathbf{S}_{\mathbf{t},-\mathbf{x}} = \mathbf{I}. \tag{6.2.5}$$

Exercise 6.2.6. *Prove that this identity indeed holds.*

Definition 6.2.7 (Hit operators). For $\mathfrak{g} \in \mathrm{LC}$ let us define

$$\begin{aligned} \bar{\mathbf{S}}^{\mathrm{epi}(\mathfrak{g})}_{\mathbf{t},\mathbf{x}}(v, u) &:= \mathbb{E}_{\mathbf{B}(0)=v}\left[\mathbf{S}_{\mathbf{t},\mathbf{x}-\boldsymbol{\tau}}(\mathbf{B}(\boldsymbol{\tau}), u)\mathbf{1}_{\{\boldsymbol{\tau}<\infty\}}\right] \\ &= \int_0^\infty \mathbb{P}_{\mathbf{B}(0)=v}(\boldsymbol{\tau} \in d\mathbf{s})\, \mathbf{S}_{\mathbf{t},\mathbf{x}-\mathbf{s}}(\mathbf{B}(\boldsymbol{\tau}), u) \end{aligned}$$

where $\mathbf{B}(x)$ is a Brownian motion with diffusion coefficient 2 and τ is the hitting time of the epigraph of the function $\mathfrak{g}$, $\boldsymbol{\tau} := \inf\{t \geqslant 0 : \mathbf{B}(t) \in \mathrm{epi}(\mathfrak{g})\}$.

Note that for $v \geqslant \mathfrak{g}(0)$ we trivially have the identity

$$\bar{\mathbf{S}}^{\mathrm{epi}(\mathfrak{g})}_{\mathbf{t},\mathbf{x}}(v, u) = \mathbf{S}_{\mathbf{t},\mathbf{x}}(v, u). \tag{6.2.8}$$

If $\mathfrak{h} \in \mathrm{UC}$, there is a similar operator $\bar{\mathbf{S}}^{\mathrm{hypo}(\mathfrak{h})}_{\mathbf{t},\mathbf{x}}$, except that now $\boldsymbol{\tau}$ is the hitting time of the hypograph of $\mathfrak{h}$ and $\bar{\mathbf{S}}^{\mathrm{hypo}(\mathfrak{h})}_{\mathbf{t},\mathbf{x}}(v, u) = \mathbf{S}_{\mathbf{t},\mathbf{x}}(v, u)$ for $v \leqslant \mathfrak{h}(0)$.

One way to think of $\bar{\mathbf{S}}^{\mathrm{epi}(\mathfrak{g})}_{\mathbf{t},\mathbf{x}}(v, u)$ is as a sort of asymptotic transformed transition 'probability' for the Brownian motion $\mathbf{B}$ to go from v to u hitting the epigraph of $\mathfrak{g}$ (note that $\mathfrak{g}$ is not necessarily continuous, so hitting $\mathfrak{g}$ is not the same as hitting $\mathrm{epi}(\mathfrak{g})$; in particular, $\mathbf{B}(\boldsymbol{\tau}) \geqslant \mathfrak{g}(\boldsymbol{\tau})$ and in general the equality need not hold). To see what we mean, write

$$\bar{\mathbf{S}}^{\mathrm{epi}(\mathfrak{g})}_{\mathbf{t},\mathbf{x}} = \lim_{\mathbf{T}\to\infty} \bar{\mathbf{S}}^{\mathrm{epi}(\mathfrak{g}),\mathbf{T}}\mathbf{S}_{\mathbf{t},\mathbf{x}-\mathbf{T}} \tag{6.2.9}$$

with

$$\bar{\mathbf{S}}^{\mathrm{epi}(\mathfrak{g}),\mathbf{T}}(v, u) := \mathbb{E}_{\mathbf{B}(0)=v}\left[\mathbf{S}_{0,\mathbf{T}-\boldsymbol{\tau}}(\mathbf{B}(\boldsymbol{\tau}), u)\mathbf{1}_{\{\boldsymbol{\tau}\leqslant\mathbf{T}\}}\right] \tag{6.2.10}$$

and note that $\bar{\mathbf{S}}^{\mathrm{epi}(\mathfrak{g}),\mathbf{T}}(v, u)$ is nothing but the transition probability for $\mathbf{B}$ to go from v at time 0 to u at time $\mathbf{T}$ hitting $\mathrm{epi}(\mathfrak{g})$ in $[0, \mathbf{T}]$.

Definition 6.2.11 (Brownian scattering operator)**.** For $\mathfrak{g} \in \mathrm{LC}$, $\mathbf{x} \in \mathbb{R}$ and $\mathbf{t} > 0$ we define

$$\mathbf{K}^{\mathrm{epi}(\mathfrak{g})}_{-\mathbf{t}} := \mathbf{I} - (\mathbf{S}_{-\mathbf{t},\mathbf{x}} - \bar{\mathbf{S}}^{\mathrm{epi}(\mathfrak{g}^-_{\mathbf{x}})}_{-\mathbf{t},\mathbf{x}})^* \bar{\chi}_{\mathfrak{g}(\mathbf{x})} (\mathbf{S}_{-\mathbf{t},-\mathbf{x}} - \bar{\mathbf{S}}^{\mathrm{epi}(\mathfrak{g}^+_{\mathbf{x}})}_{-\mathbf{t},-\mathbf{x}}), \tag{6.2.12}$$

where $\mathfrak{g}^+_{\mathbf{x}}(\mathbf{y}) := \mathfrak{g}(\mathbf{x}+\mathbf{y})$ and $\mathfrak{g}^-_{\mathbf{x}}(\mathbf{y}) := \mathfrak{g}(\mathbf{x}-\mathbf{y})$.

Exercise 6.2.13. *The projection $\bar{\chi}_{\mathfrak{g}(\mathbf{x})}$ can be removed from the formula without changing its meaning.*

Exercise 6.2.14.* *Show that the kernel $\mathbf{K}^{\mathrm{epi}(\mathfrak{g})}_{-\mathbf{t}}$ does not depend on $\mathbf{x}$.*

There is another operator which uses $\mathfrak{h} \in \mathrm{UC}$, and hits 'from above',

$$\mathbf{K}^{\mathrm{hypo}(\mathfrak{h})}_{\mathbf{t}} := \mathbf{I} - (\mathbf{S}_{\mathbf{t},\mathbf{x}} - \bar{\mathbf{S}}^{\mathrm{hypo}(\mathfrak{h}^-_{\mathbf{x}})}_{\mathbf{t},\mathbf{x}})^* \chi_{\mathfrak{h}(\mathbf{x})} (\mathbf{S}_{\mathbf{t},-\mathbf{x}} - \bar{\mathbf{S}}^{\mathrm{hypo}(\mathfrak{h}^+_{\mathbf{x}})}_{\mathbf{t},-\mathbf{x}}), \tag{6.2.15}$$

Exercise 6.2.16. *Show that the following identity holds*

$$\mathbf{K}^{\mathrm{hypo}(\mathfrak{h})}_{\mathbf{t}} = (\rho\, \mathbf{K}^{\mathrm{epi}(-\rho\mathfrak{h})}_{-\mathbf{t}}\, \rho)^*, \tag{6.2.17}$$

where $\rho\mathfrak{h}(\mathbf{x}) := \mathfrak{h}(-\mathbf{x})$.

As above, $\mathbf{S}_{\mathbf{t},\mathbf{x}} - \bar{\mathbf{S}}^{\mathrm{epi}(\mathfrak{g}_{\mathbf{x}})}_{\mathbf{t},\mathbf{x}}$ may be thought of as a sort of asymptotic transformed transition probability for a Brownian motion $\mathbf{B}$, started at time 0, not to hit $\mathrm{epi}(\mathfrak{g})$. Therefore $\mathbf{K}^{\mathrm{epi}(\mathfrak{g})}_{\mathbf{t}}$ may be thought of as the same sort of asymptotic transformed transition probability for $\mathbf{B}$, in this case hitting $\mathrm{epi}(\mathfrak{g})$, which is built out of the product of left and right 'no hit' operators.

6.3. The KPZ fixed point formula. At this stage we are ready to state our main result which we prove in Section 7.

Definition 6.3.1 (The KPZ fixed point formula)**.** The KPZ fixed point $\mathfrak{h}(\mathbf{t},\mathbf{x})$ is the Markov process on UC with transition probabilities

$$\mathbb{P}_{\mathfrak{h}_0}(\mathfrak{h}(\mathbf{t},\mathbf{x}) \leqslant \mathfrak{g}(\mathbf{x}),\, \mathbf{x} \in \mathbb{R}) = \det\left(\mathbf{I} - \mathbf{K}^{\mathrm{hypo}(\mathfrak{h}_0)}_{\mathbf{t}/2} \mathbf{K}^{\mathrm{epi}(\mathfrak{g})}_{-\mathbf{t}/2}\right)_{L^2(\mathbb{R})}, \tag{6.3.2}$$

where $\mathfrak{h}_0 \in \mathrm{UC}$ and $\mathfrak{g} \in \mathrm{LC}$. Here $\mathbb{P}_{\mathfrak{h}_0}$ means the process with initial data $\mathfrak{h}_0$.

Remark 6.3.3. The fact that the Fredholm determinant in the formula is finite is a consequence of the fact that there is a (multiplication) operator M such that the map $(\mathfrak{h}_0, \mathfrak{g}) \mapsto \mathrm{M}\mathbf{K}^{\mathrm{hypo}(\mathfrak{h}_0)}_{\mathbf{t}/2} \mathbf{K}^{\mathrm{epi}(\mathfrak{g})}_{-\mathbf{t}/2} \mathrm{M}^{-1}$ is continuous from $\mathrm{UC} \times \mathrm{LC}$ into the trace class (see [18]). We will not get into such issues in these notes.

Exercise 6.3.4 (Finite dimensional distributions)**.** *If $\mathfrak{h}_0 \in \mathrm{UC}$ and $\mathbf{x}_1 < \mathbf{x}_2 < \cdots < \mathbf{x}_M$, show that*

$$\begin{aligned} &\mathbb{P}_{\mathfrak{h}_0}(\mathfrak{h}(\mathbf{t},\mathbf{x}_1) \leqslant \mathbf{a}_1, \ldots, \mathfrak{h}(\mathbf{t},\mathbf{x}_M) \leqslant \mathbf{a}_M) \\ &= \det\Big(\mathbf{I} - \mathbf{K}^{\mathrm{hypo}(\mathfrak{h}_0)}_{\mathbf{t},\mathbf{x}_M} \\ &\qquad + \mathbf{K}^{\mathrm{hypo}(\mathfrak{h}_0)}_{\mathbf{t},\mathbf{x}_M} e^{(\mathbf{x}_1-\mathbf{x}_M)\partial^2} \bar{\chi}_{\mathbf{a}_1} e^{(\mathbf{x}_2-\mathbf{x}_1)\partial^2} \bar{\chi}_{\mathbf{a}_2} \cdots e^{(\mathbf{x}_M-\mathbf{x}_{M-1})\partial^2} \bar{\chi}_{\mathbf{a}_M}\Big) \end{aligned} \tag{6.3.5}$$

where the Fredholm determinant is over $L^2(\mathbb{R})$.

Remark 6.3.6 (Extended kernels). The formula in the preceding exercise can be rewritten as

$$\begin{aligned}\mathbb{P}_{\mathfrak{h}_0}\big(\mathfrak{h}(\mathbf{t},\mathbf{x}_1) \leqslant \mathbf{a}_1,\dots,\mathfrak{h}(\mathbf{t},\mathbf{x}_M) \leqslant \mathbf{a}_M\big)\\ = \det\Big(\mathbf{I}-\chi_{\mathbf{a}}\mathbf{K}^{\mathrm{hypo}(\mathfrak{h}_0)}_{\mathbf{t},\mathrm{ext}}\chi_{\mathbf{a}}\Big)_{L^2(\{x_1,\dots,x_M\}\times\mathbb{R})}\end{aligned} \tag{6.3.7}$$

where

$$\mathbf{K}^{\mathrm{hypo}(\mathfrak{h}_0)}_{\mathbf{t},\mathrm{ext}}(\mathbf{x}_i,\cdot;\mathbf{x}_j,\cdot) := -e^{(\mathbf{x}_j-\mathbf{x}_i)\partial^2}\mathbf{1}_{\mathbf{x}_i<\mathbf{x}_j} + e^{-\mathbf{x}_i\partial^2}\mathbf{K}^{\mathrm{hypo}(\mathfrak{h}_0)}_{\mathbf{t}}e^{\mathbf{x}_j\partial^2}. \tag{6.3.8}$$

The kernel in (6.3.8) is usually called an *extended kernel* (note that the Fredholm determinant is computed on the 'extended L^2 space' $L^2(\{\mathbf{x}_1,\dots,\mathbf{x}_M\}\times\mathbb{R})$). The kernel appearing after the second hypo operator in (6.3.5) is sometimes referred to as a *path integral kernel* [5], and should be thought of as a discrete, pre-asymptotic version of the epi operators (on a finite interval).

The fact that $e^{-x\partial^2}\mathbf{K}^{\mathrm{hypo}(\mathfrak{h}_0)}_{\mathbf{t}}e^{x\partial^2}$ makes sense is not entirely obvious, but follows from the fact that $\mathbf{K}^{\mathrm{hypo}(\mathfrak{h}_0)}_{\mathbf{t}}$ equals

$$\begin{aligned}(\mathbf{S}_{\mathbf{t},\mathbf{x}})^*\chi_{\mathfrak{g}(\mathbf{x})}\mathbf{S}_{\mathbf{t},-\mathbf{x}} + (\bar{\mathbf{S}}^{\mathrm{epi}(\mathfrak{g}_x^-)}_{\mathbf{t},\mathbf{x}})^*\bar{\chi}_{\mathfrak{g}(\mathbf{x})}\mathbf{S}_{\mathbf{t},-\mathbf{x}}\\ +(\mathbf{S}_{\mathbf{t},\mathbf{x}})^*\bar{\chi}_{\mathfrak{g}(\mathbf{x})}\bar{\mathbf{S}}^{\mathrm{epi}(\mathfrak{g}_x^+)}_{\mathbf{t},-\mathbf{x}} - (\bar{\mathbf{S}}^{\mathrm{epi}(\mathfrak{g}_x^-)}_{\mathbf{t},\mathbf{x}})^*\bar{\chi}_{\mathfrak{g}(\mathbf{x})}\bar{\mathbf{S}}^{\mathrm{epi}(\mathfrak{g}_x^+)}_{\mathbf{t},-\mathbf{x}},\end{aligned} \tag{6.3.9}$$

together with the group property (6.2.3) and the definition of the hit operators.

Example 6.3.10 (Airy processes). Let us write $\mathfrak{h}(\mathbf{t},\mathbf{x};\mathfrak{h}_0)$ for the KPZ fixed point $\mathfrak{h}(\mathbf{t},\mathbf{x})$ started at $\mathfrak{h}_0$. For special initial data, at time $\mathbf{t}=1$ we recover the known Airy$_1$, Airy$_2$ and Airy$_{2\to1}$ processes:

(1) Narrow wedge initial data leads to the *Airy$_2$ process* [16,24]:
$$\mathfrak{h}(1,\mathbf{x};\mathfrak{d}_{\mathbf{u}}) + (\mathbf{x}-\mathbf{u})^2 \;=\; \mathcal{A}_2(\mathbf{x}),$$
where $\mathfrak{d}_{\mathbf{u}}$ was defined in Example 6.1.2;

(2) *Flat* initial data $\mathfrak{h}_0\equiv 0$ leads to the *Airy$_1$ process* [7,28]:
$$\mathfrak{h}(1,\mathbf{x};0) \;=\; \mathcal{A}_1(\mathbf{x});$$

(3) *Wedge* or *half-flat* initial data $\mathfrak{h}_{\text{h-f}}(\mathbf{x}) = -\infty$ for $\mathbf{x}<0$ and $\mathfrak{h}_{\text{h-f}}(\mathbf{x})=0$ for $\mathbf{x}\geqslant 0$, leads to the *Airy$_{2\to1}$ process* [8]:
$$\mathfrak{h}(1,\mathbf{x};\mathfrak{h}_{\text{h-f}}) + \mathbf{x}^2\mathbf{1}_{\mathbf{x}<0} \;=\; \mathcal{A}_{2\to1}(\mathbf{x}).$$

We leave the derivation of the processes $\mathcal{A}_1$ and $\mathcal{A}_2$ as Exercise 6.3.18. To get the formula in the case (3) we need to show that the finite dimensional distributions match, by computing the kernel on the right hand side of (6.3.7) with $\mathfrak{h}_0=\mathfrak{h}_{\text{h-f}}$. An application of the reflection principle based on (6.2.9) yields the equality, for $v\geqslant 0$,

$$\bar{\mathbf{S}}^{\mathrm{hypo}(\mathfrak{h}_{\text{h-f}})}_{\mathbf{t},0}(v,u) = \int_0^\infty \mathbb{P}_v(\tau_0\in \mathbf{d}\mathbf{y})\mathbf{S}_{\mathbf{t},-\mathbf{y}}(0,u) = \mathbf{S}_{\mathbf{t},0}(-v,u),$$

which gives, with ρ the reflection operator $\rho f(x)=f(-x)$,

$$\mathbf{K}^{\mathrm{hypo}(\mathfrak{h}_{\text{h-f}})}_{\mathbf{t}} = \mathbf{I}-(\mathbf{S}_{\mathbf{t},0})^*\chi_0[\mathbf{S}_{\mathbf{t},0}-\rho\mathbf{S}_{\mathbf{t},0}] = (\mathbf{S}_{\mathbf{t},0})^*(\mathbf{I}+\rho)\bar{\chi}_0\mathbf{S}_{\mathbf{t},0}.$$

This yields, using (6.3.8),

$$\begin{aligned}\mathbf{K}^{\mathrm{hypo}(\mathfrak{h}_{\mathrm{h\text{-}f}})}_{\mathbf{t},\mathrm{ext}}(\mathbf{x}_i,\cdot;\mathbf{x}_j,\cdot) &= -e^{(\mathbf{x}_j-\mathbf{x}_i)\partial^2}\mathbf{1}_{\mathbf{x}_i<\mathbf{x}_j} + \mathbf{S}_{0,-\mathbf{x}_i}(\mathbf{S}_{\mathbf{t},0})^*(\mathbf{I}+\rho)\bar{\chi}_0\mathbf{S}_{\mathbf{t},0}\mathbf{S}_{0,\mathbf{x}_i}\\ &= -e^{(\mathbf{x}_j-\mathbf{x}_i)\partial^2}\mathbf{1}_{\mathbf{x}_i<\mathbf{x}_j} + (\mathbf{S}_{\mathbf{t},-\mathbf{x}_i})^*(\mathbf{I}+\rho)\bar{\chi}_0\mathbf{S}_{\mathbf{t}/2,\mathbf{x}_i}.\end{aligned}$$

Choosing $\mathbf{t}=1$ and using (6.2.4) yields that the second term on the right hand side equals

$$\begin{aligned}&\mathbf{K}^{\mathrm{hypo}(\mathfrak{h}_{\mathrm{h\text{-}f}})}_{\mathbf{t},\mathrm{ext}}(\mathbf{x}_i,u;\mathbf{x}_j,v)\\ &= \int_{-\infty}^{0} d\lambda\, e^{-2\mathbf{x}_i^3/3-\mathbf{x}_i(u-\lambda)}\,\mathrm{Ai}(u-\lambda+\mathbf{x}_i^2)\, e^{2\mathbf{x}_j^3/3+\mathbf{x}_j(v-\lambda)}\,\mathrm{Ai}(v-\lambda+\mathbf{x}_j^2)\\ &\quad+\int_{-\infty}^{0} d\lambda\, e^{-2\mathbf{x}_i^3/3-\mathbf{x}_i(u+\lambda)}\,\mathrm{Ai}(u+\lambda+\mathbf{x}_i^2)\, e^{2\mathbf{x}_j^3/3+\mathbf{x}_j(v-\lambda)}\,\mathrm{Ai}(v-\lambda+\mathbf{x}_j^2)\end{aligned}$$

which, after a simple conjugation, gives the kernel for the $\mathrm{Airy}_{2\to1}$ process.

Exercise 6.3.11. *Show that a conjugation gives the kernel for the $\mathrm{Airy}_{2\to1}$ process (see* [8]*).*

Example 6.3.12 (The Airy_2 process)**.** Given $\mathbf{x}_1<\cdots<\mathbf{x}_n$ in $\mathbb{R}$ and $\mathbf{a}_1,\ldots,\mathbf{a}_n\in\mathbb{R}$, we have

(6.3.13) $$\mathbb{P}\big(\mathcal{A}_2(\mathbf{x}_1)\leqslant \mathbf{a}_1,\ldots,\mathcal{A}_2(\mathbf{x}_n)\leqslant \mathbf{a}_n\big)=\det\big(I-\chi_{\mathbf{a}}K^{\mathrm{ext}}_{\mathrm{Ai}}\chi_{\mathbf{a}}\big)_{L^2(\{\mathbf{x}_1,\ldots,\mathbf{x}_n\}\times\mathbb{R})},$$

where the *extended Airy kernel* is defined by

(6.3.14)
$$K^{\mathrm{ext}}_{\mathrm{Ai}}(\mathbf{x},u;\mathbf{x}',u'):=\begin{cases}\int_0^\infty d\lambda\, e^{-\lambda(\mathbf{x}-\mathbf{x}')}\,\mathrm{Ai}(u+\lambda)\,\mathrm{Ai}(u'+\lambda), & \text{if } \mathbf{x}\geqslant\mathbf{x}',\\ -\int_{-\infty}^0 d\lambda\, e^{-\lambda(\mathbf{x}-\mathbf{x}')}\,\mathrm{Ai}(u+\lambda)\,\mathrm{Ai}(u'+\lambda), & \text{if } \mathbf{x}<\mathbf{x}'.\end{cases}$$

Example 6.3.15 (The Airy_1 process)**.** For $\mathbf{x}_i$ and $\mathbf{a}_i$ as in the previous example, we have the identity

(6.3.16) $$\mathbb{P}\big(\mathcal{A}_1(\mathbf{x}_1)\leqslant \mathbf{a}_1,\ldots,\mathcal{A}_1(\mathbf{x}_n)\leqslant \mathbf{a}_n\big)=\det\big(I-\chi_{\mathbf{a}}K^{\mathrm{ext}}_{1}\chi_{\mathbf{a}}\big)_{L^2(\{\mathbf{x}_1,\ldots,\mathbf{x}_n\}\times\mathbb{R})},$$

with the kernel

(6.3.17) $$\begin{aligned}K_1^{\mathrm{ext}}(\mathbf{x},u;\mathbf{x}',u') &:= -\frac{1}{\sqrt{4\pi(\mathbf{x}'-\mathbf{x})}}\exp\Big(-\frac{(u'-u)^2}{4(\mathbf{x}'-\mathbf{x})}\Big)\mathbf{1}_{\mathbf{x}'>\mathbf{x}}\\ &\quad+\mathrm{Ai}(u+u'+(\mathbf{x}'-\mathbf{x})^2)\exp\Big((\mathbf{x}'-\mathbf{x})(u+u')+\tfrac{2}{3}(\mathbf{x}'-\mathbf{x})^3\Big).\end{aligned}$$

Exercise 6.3.18. *Obtain the kernels for the Airy_2 and Airy_1 processes from the formula* (6.3.2).

6.4. Symmetries and invariance. The KPZ fixed point inherits several nice properties as a scaling limit of TASEP. We will write $\mathfrak{h}(\mathbf{t},\mathbf{x};\mathfrak{h}_0)$ for the KPZ fixed point $\mathfrak{h}(\mathbf{t},\mathbf{x})$ started at $\mathfrak{h}_0$.

Proposition 6.4.1 (Symmetries of $\mathfrak{h}$)**.** *The KPZ fixed point $\mathfrak{h}$ has the following properties:*

(i) *(1:2:3 scaling invariance)* $\alpha\mathfrak{h}(\alpha^{-3}\mathbf{t},\alpha^{-2}\mathbf{x};\alpha\mathfrak{h}_0(\alpha^{-2}\mathbf{x}))\overset{\mathrm{dist}}{=}\mathfrak{h}(\mathbf{t},\mathbf{x};\mathfrak{h}_0)$, $\alpha>0$;

(ii) *(Skew time reversal)* $\mathbb{P}\big(\mathfrak{h}(\mathbf{t},\mathbf{x};\mathfrak{g}) \leqslant -\mathfrak{f}(\mathbf{x})\big) = \mathbb{P}\big(\mathfrak{h}(\mathbf{t},\mathbf{x};\mathfrak{f}) \leqslant -\mathfrak{g}(\mathbf{x})\big)$, $\mathfrak{f},\mathfrak{g} \in \mathrm{UC}$;
(iii) *(Shift invariance)* $\mathfrak{h}(\mathbf{t},\mathbf{x}+\mathbf{u};\mathfrak{h}_0(\mathbf{x}+\mathbf{u})) \stackrel{\text{dist}}{=} \mathfrak{h}(\mathbf{t},\mathbf{x};\mathfrak{h}_0)$;
(iv) *(Reflection invariance)* $\mathfrak{h}(\mathbf{t},-\mathbf{x};\mathfrak{h}_0(-\mathbf{x})) \stackrel{\text{dist}}{=} \mathfrak{h}(\mathbf{t},\mathbf{x};\mathfrak{h}_0)$;
(v) *(Affine invariance)* $\mathfrak{h}(\mathbf{t},\mathbf{x};\mathfrak{f}(\mathbf{x})+\mathbf{a}+\mathbf{c}\mathbf{x}) \stackrel{\text{dist}}{=} \mathfrak{h}(\mathbf{t},\mathbf{x};\mathfrak{f}(\mathbf{x}+\frac{1}{2}\mathbf{c}\mathbf{t}))+\mathbf{a}+\mathbf{c}\mathbf{x}+\frac{1}{4}\mathbf{c}^2\mathbf{t}$;
(vi) *(Preservation of max)* $\mathfrak{h}(\mathbf{t},\mathbf{x};\mathfrak{f}_1 \vee \mathfrak{f}_2) = \mathfrak{h}(\mathbf{t},\mathbf{x};\mathfrak{f}_1) \vee \mathfrak{h}(\mathbf{t},\mathbf{x};\mathfrak{f}_2)$.

Exercise 6.4.2. *Prove property (i) from the fixed point formula in Definition* 6.3.1. *Prove that properties (ii), (iii),(iv) hold for TASEP, and therefore for the limiting fixed point.*

Properties (v) and (vi) require coupling and we provide their proofs in Section 6.7.

6.5. Markov property. The sets $A_{\mathfrak{g}} = \{\mathfrak{h} \in \mathrm{UC} : \mathfrak{h}(\mathbf{x}) \leqslant \mathfrak{g}(\mathbf{x}), \mathbf{x} \in \mathbb{R}\}$ with $\mathfrak{g} \in \mathrm{LC}$ form a generating family for the Borel sets $\mathcal{B}(\mathrm{UC})$. Hence from (6.3.2) we can define the fixed point transition probabilities $p_{\mathfrak{h}_0}(\mathbf{t}, A_{\mathfrak{g}})$, for which we have:

Lemma 6.5.1. *For fixed* $\mathfrak{h}_0 \in \mathrm{UC}$ *and* $\mathbf{t} > 0$, *the measure* $p_{\mathfrak{h}_0}(\mathbf{t},\cdot)$ *is a probability measure on* UC.

Sketch of the proof. It is clear by construction that $\mathbb{P}_{\mathfrak{h}_0}(\mathfrak{h}(\mathbf{t},\mathbf{x}_i) \leqslant \mathbf{a}_i, i = 1,\dots,n)$ is non-decreasing in each $\mathbf{a}_i$ and is in $[0,1]$. We need to show then that this quantity goes to 1 as all $\mathbf{a}_i$'s go to infinity, and to 0 if any $\mathbf{a}_i$ goes to $-\infty$. The first one is standard, and relies on the inequality $|\det(\mathbf{I}-\mathbf{K})-1| \leqslant \|\mathbf{K}\|_1 e^{\|\mathbf{K}\|_1+1}$ (with $\|\cdot\|_1$ denoting trace norm). The second limit is in general very hard to show for a formula given in terms of a Fredholm determinant, but it turns out to be rather easy in our case, because the multipoint probability is trivially bounded by $\mathbb{P}_{\mathfrak{h}_0}(\mathfrak{h}(\mathbf{t},\mathbf{x}_i) \leqslant \mathbf{a}_i)$, for any i. By the skew time reversal symmetry this becomes the probability that the Airy_2 process minus a parabola is bounded everywhere by $-\mathfrak{h}_0 + \mathbf{a}_i$, which clearly goes to 0 as $\mathbf{a}_i$ goes to $-\infty$ (see [16] and [11] for this property of the Airy_2 process). □

Theorem 6.5.2. *The KPZ fixed point* $\big(\mathfrak{h}(\mathbf{t},\cdot)\big)_{\mathbf{t}>0}$ *is a (Feller) Markov process taking values in* UC.

The proof is based on the fact that $\mathfrak{h}(\mathbf{t},\mathbf{x})$ is the limit of the rescaled TASEP height functions $\mathfrak{h}^{\varepsilon}(\mathbf{t},\mathbf{x})$, defined in (7.0.1) below, which is Markovian. To derive from this the Markov property of the limit requires some compactness, which in our case is provided by Theorem 6.6.3 below.

6.6. Regularity and local Brownian behavior. Up to this point we only know that the fixed point is in UC, but by the smoothing mechanism inherent to models in the KPZ class one should expect $\mathfrak{h}(\mathbf{t},\cdot)$ to at least be continuous for each fixed $\mathbf{t} > 0$. The next result shows that for every $M > 0$, $\mathfrak{h}(\mathbf{t},\cdot)\big|_{[-M,M]}$ is Hölder-β for any $\beta < 1/2$ with probability 1.

Definition 6.6.1 (Local Hölder spaces)**.** Let us define the space

$$\mathcal{C} := \big\{\mathfrak{h} \colon \mathbb{R} \to [-\infty,\infty) \text{ continuous with } \mathfrak{h}(\mathbf{x}) \leqslant C(1+|\mathbf{x}|) \text{ for some } C < \infty\big\}.$$

For $M > 0$ we define the local Hölder norm

$$\|\mathfrak{h}\|_{\beta,[-M,M]} := \sup_{x_1 \neq x_2 \in [-M,M]} \frac{|\mathfrak{h}(x_2) - \mathfrak{h}(x_1)|}{|x_2 - x_1|^\beta} \tag{6.6.2}$$

and the local Hölder spaces

$$\mathcal{C}^\beta := \bigcap_{M \in \mathbb{N}} \{\mathfrak{h} \in \mathcal{C} : \|\mathfrak{h}\|_{\beta,[-M,M]} < \infty\}.$$

The topology on UC, when restricted to $\mathcal{C}$, is the topology of uniform convergence on compact sets. UC is a Polish space and the spaces $\mathcal{C}^\beta$ are compact in UC.

Then we can get spatial regularity of the KPZ fixed point.

Theorem 6.6.3 (Space regularity). *Fix $t > 0$, $\mathfrak{h}_0 \in$ UC and initial data $\mathfrak{h}_0^\varepsilon$ for the TASEP height function such that $\mathfrak{h}_0^\varepsilon \to \mathfrak{h}_0$ locally in UC as $\varepsilon \to 0$. Then for each $\beta \in (0, 1/2)$ and $M > 0$ we have*

$$\lim_{A\to\infty} \limsup_{\varepsilon\to 0} \mathbb{P}\big(\|\mathfrak{h}^\varepsilon(t)\|_{\beta,[-M,M]} \geqslant A\big) = \lim_{A\to\infty} \mathbb{P}\big(\|\mathfrak{h}\|_{\beta,[-M,M]} \geqslant A\big) = 0, \tag{6.6.4}$$

where the rescaled height functions $\mathfrak{h}^\varepsilon$ are defined in (7.0.1).

The proof proceeds through an application of the Kolmogorov continuity theorem, which reduces regularity to two-point functions, and depends heavily on the representation (7.1.2) for the two-point function in terms of path integral kernels. We prefer to skip the details.

Remark 6.6.5. Since the theorem shows that this regularity holds uniformly (in $\varepsilon > 0$) for the approximating $\mathfrak{h}^\varepsilon(t,\cdot)$'s, we get the compactness needed for the proof of the Markov property.

Theorem 6.6.6 (Local Brownian behavior). *For any initial condition $\mathfrak{h}_0 \in$ UC the KPZ fixed point $\mathfrak{h}$ is locally Brownian in space in the sense that for each $y \in \mathbb{R}$, the finite dimensional distributions of*

$$\mathfrak{b}_\varepsilon^\pm(x) = \varepsilon^{-1/2}\big(\mathfrak{h}(t, y \pm \varepsilon x) - \mathfrak{h}(t, y)\big)$$

converge, as $\varepsilon \to 0$, to those of Brownian motions with diffusion coefficient 2.

A very brief sketch of the proof. The proof is based again on the arguments of [26]. One uses (7.1.2) and Brownian scale invariance to show that

$$\mathbb{P}\big(\mathfrak{h}(t, \varepsilon x_1) \leqslant u + \sqrt{\varepsilon} a_1, \dots, \mathfrak{h}(t, \varepsilon x_n) \leqslant u + \sqrt{\varepsilon} a_n \,\big|\, \mathfrak{h}(t, 0) = u\big)$$
$$= \mathbb{E}\Big(\mathbf{1}_{B(x_i) \leqslant a_i, i=1,\dots,n}\, \phi_{x,a}^\varepsilon(u, B(x_n))\Big),$$

for some explicit function $\phi_{x,a}^\varepsilon(u, b)$. The Brownian motion appears from the product of heat kernels in (7.1.2), while $\phi_{x,a}^\varepsilon$ contains the dependence on everything else in the formula (the Fredholm determinant structure and $\mathfrak{h}_0$ through the hypo operator $\mathbf{K}_t^{\mathrm{hypo}(\mathfrak{h}_0)}$). Then one shows that $\phi_{x,a}^\varepsilon(u, b)$ goes to 1 in a suitable sense as $\varepsilon \to 0$. □

Proposition 6.6.7 (Time regularity). *Fix* $\mathbf{x}_0 \in \mathbb{R}$ *and initial data* $\mathfrak{h}_0 \in \mathrm{UC}$. *For* $\mathbf{t} > 0$, $\mathfrak{h}(\mathbf{t}, \mathbf{x}_0)$ *is locally Hölder* α *in* $\mathbf{t}$ *for any* $\alpha < 1/3$.

The proof uses the variational formula for the fixed point, we sketch it in the next section.

6.7. Variational formulas and the Airy sheet

Definition 6.7.1 (Airy sheet). As before we write $\mathfrak{h}(\mathbf{t}, \mathbf{x}; \mathfrak{h}_0)$ for the KPZ fixed point $\mathfrak{h}(\mathbf{t}, \mathbf{x})$ started at $\mathfrak{h}_0$. The two parameter process

$$\mathcal{A}(\mathbf{x}, \mathbf{y}) \; := \; \mathfrak{h}(1, \mathbf{y}; \mathfrak{d}_{\mathbf{x}}) + (\mathbf{x} - \mathbf{y})^2$$

is called the *Airy sheet* [12]. Where to make this definition, we need to use the standard coupling for the KPZ fixed points described in Remark 6.7.2 below. Fixing either one of the variables, it is an Airy_2 process in the other. We also write

$$\hat{\mathcal{A}}(\mathbf{x}, \mathbf{y}) := \mathcal{A}(\mathbf{x}, \mathbf{y}) - (\mathbf{x} - \mathbf{y})^2.$$

Remark 6.7.2. The KPZ fixed point inherits from TASEP a canonical coupling between the processes started with different initial data (using the same 'noise'). It is this the property that allows us to define the two-parameter Airy sheet. An annoying difficulty is that we cannot prove that this process is unique. More precisely, the construction of the Airy sheet in [22] goes through using tightness of the coupled processes at the TASEP level and taking subsequential limits, and at the present time there seems to be no way to assure that the limit points are unique. This means that we have actually constructed 'an' Airy sheet, and the statements below should really be interpreted as about any such limit.

It is natural to wonder whether the fixed point formulas at our disposal determine the joint probabilities $\mathbb{P}(\mathcal{A}(\mathbf{x}_i, \mathbf{y}_i) \leqslant \mathbf{a}_i, i = 1, \ldots, M)$ for the Airy sheet. Unfortunately, this is not the case. In fact, the most we can compute using our formulas is

$$\mathbb{P}\big(\hat{\mathcal{A}}(\mathbf{x}, \mathbf{y}) \leqslant \mathfrak{f}(\mathbf{x}) + \mathfrak{g}(\mathbf{y}), \, \mathbf{x}, \mathbf{y} \in \mathbb{R}\big) = \det\Big(\mathbf{I} - \mathbf{K}_1^{\mathrm{hypo}(-\mathfrak{g})} \mathbf{K}_{-1}^{\mathrm{epi}(\mathfrak{f})}\Big).$$

To compute the two-point distribution of the Airy sheet $\mathbb{P}(\hat{\mathcal{A}}(\mathbf{x}_i, \mathbf{y}_i) \leqslant \mathbf{a}_i, \, i = 1, 2)$ from this, we would need to choose $\mathfrak{f}$ and $\mathfrak{g}$ taking two non-infinite values that yield a formula for $\mathbb{P}(\hat{\mathcal{A}}(\mathbf{x}_i, \mathbf{y}_j) \leqslant \mathfrak{f}(\mathbf{x}_i) + \mathfrak{g}(\mathbf{y}_j), \, i, j = 1, 2)$, and so would need $\mathfrak{f}(\mathbf{x}_1) + \mathfrak{g}(\mathbf{y}_1) = \mathbf{a}_1$, $\mathfrak{f}(\mathbf{x}_2) + \mathfrak{g}(\mathbf{y}_2) = \mathbf{a}_2$ and $\mathfrak{f}(\mathbf{x}_1) + \mathfrak{g}(\mathbf{y}_2) = \mathfrak{f}(\mathbf{x}_2) + \mathfrak{g}(\mathbf{y}_1) = L$ with $L \to \infty$. But $\{\mathfrak{f}(\mathbf{x}_i) + \mathfrak{g}(\mathbf{y}_j), \, i, j = 1, 2\}$ only spans a 3-dimensional linear subspace of $\mathbb{R}^4$, so this is not possible.

The preservation of max property allows us to write a variational formula for the KPZ fixed point in terms of the Airy sheet.

Theorem 6.7.3 (Airy sheet variational formula). *One has the identities*

$$\begin{aligned} \mathfrak{h}(\mathbf{t}, \mathbf{x}; \mathfrak{h}_0) &= \sup_{\mathbf{y}} \{\mathfrak{h}(\mathbf{t}, \mathbf{x}; \mathfrak{d}_{\mathbf{y}}) + \mathfrak{h}_0(\mathbf{y})\} \\ &\overset{\text{dist}}{=} \sup_{\mathbf{y}} \Big\{\mathbf{t}^{1/3} \hat{\mathcal{A}}(\mathbf{t}^{-2/3}\mathbf{x}, \mathbf{t}^{-2/3}\mathbf{y}) + \mathfrak{h}_0(\mathbf{y})\Big\}, \end{aligned} \tag{6.7.4}$$

where we need to use the standard coupling of the KPZ fixed points described in Remark 6.7.2. *In particular, the Airy sheet satisfies the* semi-group property: *If* $\hat{\mathcal{A}}^1$ *and* $\hat{\mathcal{A}}^2$ *are independent copies and* $\mathbf{t}_1 + \mathbf{t}_2 = \mathbf{t}$ *are all positive, then*

$$\sup_{\mathbf{z}}\left\{\mathbf{t}_1^{1/3}\hat{\mathcal{A}}^1(\mathbf{t}_1^{-2/3}\mathbf{x},\mathbf{t}_1^{-2/3}\mathbf{z}) + \mathbf{t}_2^{1/3}\hat{\mathcal{A}}^2(\mathbf{t}_2^{-2/3}\mathbf{z},\mathbf{t}_2^{-2/3}\mathbf{y})\right\} \stackrel{\text{dist}}{=} \mathbf{t}^{1/3}\hat{\mathcal{A}}^1(\mathbf{t}^{-2/3}\mathbf{x},\mathbf{t}^{-2/3}\mathbf{y}).$$

Proof. Let $\mathfrak{h}_0^n$ be a sequence of initial conditions taking finite values $\mathfrak{h}_0^n(\mathbf{y}_i^n)$ at $\mathbf{y}_i^n$, $i = 1,\dots,k_n$, and $-\infty$ everywhere else, which converges to $\mathfrak{h}_0$ in UC as $n \to \infty$. By repeated application of Proposition 6.4.1(v) (and the easy fact that $\mathfrak{h}(\mathbf{t},\mathbf{x};\mathfrak{h}_0 + \mathbf{a}) = \mathfrak{h}(\mathbf{t},\mathbf{x};\mathfrak{h}_0) + \mathbf{a}$ for $\mathbf{a} \in \mathbb{R}$) we get

$$\mathfrak{h}(\mathbf{t},\mathbf{x};\mathfrak{h}_0^n) = \sup_{i=1,\dots,k_n}\{\mathfrak{h}(\mathbf{t},\mathbf{x};\mathfrak{d}_{\mathbf{y}_i^n}) + \mathfrak{h}_0^n(\mathbf{y}_i^n)\},$$

and taking $n \to \infty$ yields the result (the second equality in (6.7.4) follows from scaling invariance, Proposition 6.4.1). □

One of the interests in this variational formula is that it leads to proofs of properties of the fixed point, as we already mentioned in earlier sections.

Proof of Proposition 6.4.1*(iv).* The fact that the fixed point is invariant under translations of the initial data is straightforward, so we may assume $\mathbf{a} = 0$. By Theorem 6.7.3 we have

$$\begin{aligned}
&\mathfrak{h}(\mathbf{t},\mathbf{x};\mathfrak{h}_0 + c\mathbf{x})\\
&\stackrel{\text{dist}}{=} \sup_{\mathbf{y}}\left\{\mathbf{t}^{1/3}\mathcal{A}(\mathbf{t}^{-2/3}\mathbf{x},\mathbf{t}^{-2/3}\mathbf{y}) - \mathbf{t}^{-1}(\mathbf{x}-\mathbf{y})^2 + \mathfrak{h}_0(\mathbf{y}) + c\mathbf{y}\right\}\\
&= \sup_{\mathbf{y}}\left\{\mathbf{t}^{1/3}\mathcal{A}(\mathbf{t}^{-2/3}\mathbf{x},\mathbf{t}^{-2/3}(\mathbf{y}+c\mathbf{t}/2)) - \mathbf{t}^{-1}(\mathbf{x}-\mathbf{y})^2 + \mathfrak{h}_0(\mathbf{y}+c\mathbf{t}/2) + c\mathbf{x} + c^2\mathbf{t}/4\right\}\\
&\stackrel{\text{dist}}{=} \sup_{\mathbf{y}}\left\{\mathbf{t}^{1/3}\hat{\mathcal{A}}(\mathbf{t}^{-2/3}\mathbf{x},\mathbf{t}^{-2/3}\mathbf{y}) + \mathfrak{h}_0(\mathbf{y}+c\mathbf{t}/2) + c\mathbf{x} + c^2\mathbf{t}/4\right\}\\
&= \mathfrak{h}(\mathbf{t},\mathbf{x};\mathfrak{h}_0(\mathbf{x}+c\mathbf{t}/2)) + c\mathbf{x} + c^2\mathbf{t}/4.
\end{aligned}$$

□

Sketch of the proof of Theorem 6.6.7. Fix $\alpha < 1/3$ and $\beta < 1/2$ so that $\beta/(2-\beta) = \alpha$. By the Markov property it is enough to assume that $\mathfrak{h}_0 \in \mathcal{C}^\beta$ and check the Hölder-α regularity at time 0. By space regularity of the Airy$_2$ process (proved in [26], but which also follows from Theorem 6.6.3) there is an $R < \infty$ a.s. such that $|\mathcal{A}_2(\mathbf{x})| \leqslant R(1 + |\mathbf{x}|^\beta)$, and making R larger if necessary we may also assume

$$|\mathfrak{h}_0(\mathbf{x}) - \mathfrak{h}_0(\mathbf{x}_0)| \leqslant R(|\mathbf{x}-\mathbf{x}_0|^\beta + |\mathbf{x}-\mathbf{x}_0|).$$

From the variational formula (6.7.4), $|\mathfrak{h}(\mathbf{t},\mathbf{x}_0) - \mathfrak{h}(0,\mathbf{x}_0)|$ is then bounded by

$$\sup_{\mathbf{x}\in\mathbb{R}}\left(R(|\mathbf{x}-\mathbf{x}_0|^\beta + |\mathbf{x}-\mathbf{x}_0| + \mathbf{t}^{1/3} + \mathbf{t}^{(1-2\beta)/3}|\mathbf{x}|^\beta) - \tfrac{1}{\mathbf{t}}(\mathbf{x}_0-\mathbf{x})^2\right).$$

The supremum above is attained roughly at $x - x_0 = \mathbf{t}^{-\eta}$ with η chosen so that $|\mathbf{x}-\mathbf{x}_0|^\beta \sim \frac{1}{\mathbf{t}}(\mathbf{x}_0-\mathbf{x})^2$. Then $\eta = 1/(2-\beta)$ and the supremum is bounded by a constant multiple of $\mathbf{t}^{\beta/(2-\beta)} = \mathbf{t}^\alpha$, as desired. □

7. The 1:2:3 scaling limit of TASEP

In this section we will prove that for a large class of initial data the growth process of TASEP converges to the KPZ fixed point defined in Section 6.3. To this end we consider the TASEP particles to be distributed with a density close to $\frac{1}{2}$, and take the following scaling of the height function h from Section 1.1:

$$\mathfrak{h}^{\varepsilon}(\mathbf{t},\mathbf{x}) := \varepsilon^{1/2}\Big[h_{2\varepsilon^{-3/2}\mathbf{t}}(2\varepsilon^{-1}\mathbf{x}) + \varepsilon^{-3/2}\mathbf{t}\Big]. \tag{7.0.1}$$

We will always consider the linear interpolation of $\mathfrak{h}^{\varepsilon}$ to make it a continuous function of $\mathbf{x} \in \mathbb{R}$. Suppose that we have initial data X_0^{ε} chosen to depend on ε in such a way that

$$\mathfrak{h}_0 = \lim_{\varepsilon \to 0} \mathfrak{h}^{\varepsilon}(0,\cdot) \tag{7.0.2}$$

in the UC topology. For fixed $\mathbf{t} > 0$, we will prove that the limit

$$\mathfrak{h}(\mathbf{t},\mathbf{x};\mathfrak{h}_0) = \lim_{\varepsilon \to 0} \mathfrak{h}^{\varepsilon}(\mathbf{t},\mathbf{x}) \tag{7.0.3}$$

exists, and coincides with the KPZ fixed point from Definition 6.3.1. We will often omit $\mathfrak{h}_0$ from the notation when it is clear from the context.

Exercise 7.0.4. *For any $\mathfrak{h}_0 \in$ UC, we can find initial data X_0^{ε} so that* (7.0.2) *holds.*

We have the following convergence result for TASEP:

Theorem 7.0.5. *For $\mathfrak{h}_0 \in$ UC, let X_0^{ε} be initial data for TASEP such that the corresponding rescaled height functions $\mathfrak{h}_0^{\varepsilon}$ converge to $\mathfrak{h}_0$ in the* UC *topology as $\varepsilon \to 0$. Then the limit* (7.0.3) *exists (in distribution) locally in* UC *and is the KPZ fixed point with initial value $\mathfrak{h}_0$, introduced in Definition* 6.3.1.

In other words, under the 1:2:3 scaling, as long as the initial data for TASEP converges in UC, the evolving TASEP height function converges to the KPZ fixed point.

We now sketch the proof. We must compute $\mathbb{P}_{\mathfrak{h}_0}\big(\mathfrak{h}(\mathbf{t},\mathbf{x}_i) \leqslant \mathbf{a}_i,\ i = 1,\dots,M\big)$. We chose for simplicity the frame of reference

$$X_0^{-1}(-1) = 1, \tag{7.0.6}$$

i.e. the particle labeled 1 is initially the rightmost in $\mathbb{Z}_{<0}$. Then it follows from (1.1.1), (7.0.1) and (7.0.3) that the required probability should coincide with the limit as $\varepsilon \to 0$ of

$$\mathbb{P}_{X_0}\Big(X_{2\varepsilon^{-3/2}\mathbf{t}}(\tfrac{1}{2}\varepsilon^{-3/2}\mathbf{t} - \varepsilon^{-1}\mathbf{x}_i - \tfrac{1}{2}\varepsilon^{-1/2}\mathbf{a}_i + 1) > 2\varepsilon^{-1}\mathbf{x}_i - 2,\ i = 1,\dots,M\Big). \tag{7.0.7}$$

We therefore want to consider Theorem 5.2.21 with

$$t = 2\varepsilon^{-3/2}\mathbf{t}, \qquad n_i = \tfrac{1}{2}\varepsilon^{-3/2}\mathbf{t} - \varepsilon^{-1}\mathbf{x}_i - \tfrac{1}{2}\varepsilon^{-1/2}\mathbf{a}_i + 1, \qquad a_i = 2\varepsilon^{-1}\mathbf{x}_i - 2. \tag{7.0.8}$$

Remark 7.0.9. One might worry that the initial data (5.2.23) is assumed to be right finite. In fact, one can obtain a formula without this condition, but it is awkward. On the other hand, one could always cut off the TASEP data far to the right, take the limit, and then remove the cutoff. If we call the macroscopic position of the cutoff L, this means the cutoff data is $X_0^{\varepsilon,\mathrm{L}}(n) = X_0^{\varepsilon}(n)$ if $n > -\lfloor \varepsilon^{-1}\mathrm{L}\rfloor$ and $X_0^{\varepsilon,\mathrm{L}}(n) = \infty$ if $n \leqslant -\lfloor \varepsilon^{-1}\mathrm{L}\rfloor$. This corresponds to replacing $\mathfrak{h}_0^{\varepsilon}(\mathbf{x})$ by $\mathfrak{h}_0^{\varepsilon,\mathrm{L}}(\mathbf{x})$ with a straight line with slope $-2\varepsilon^{-1/2}$ to the right of $\varepsilon X_0^{\varepsilon}(-\lfloor \varepsilon^{-1}\mathrm{L}\rfloor) \sim 2\mathrm{L}$. The question is whether one can justify the exchange of limits $\mathrm{L} \to \infty$ and $\varepsilon \to 0$. It turns out not to be a problem because one can use the exact formula to get uniform bound (in ε, and over initial data in UC with bound $C(1+|x|)$) that the difference of (7.0.7) computed with initial data X_0^{ε} and with initial data $X_0^{\varepsilon,\mathrm{L}}$ is bounded by $Ce^{-c\mathrm{L}^3}$.

Lemma 7.0.10. *Under the scaling* (7.0.8) *(dropping the* i *subscripts) and assuming that* (7.0.2) *holds, if we set* $z = 2\varepsilon^{-1}\mathbf{x} + \varepsilon^{-1/2}(u+\mathbf{a}) - 2$ *and* $y' = \varepsilon^{-1/2}v$*, then we have for* $\mathbf{t} > 0$*, as* $\varepsilon \to 0$*,*

$$\mathbf{S}^{\varepsilon}_{-\mathbf{t},\mathbf{x}}(v,u) := \varepsilon^{-1/2}\mathcal{S}_{-t,-n}(y',z) \longrightarrow \mathbf{S}_{-\mathbf{t},\mathbf{x}}(v,u), \tag{7.0.11}$$

$$\bar{\mathbf{S}}^{\varepsilon}_{-\mathbf{t},-\mathbf{x}}(v,u) := \varepsilon^{-1/2}\bar{\mathcal{S}}_{-t,n}(y',z) \longrightarrow \mathbf{S}_{-\mathbf{t},-\mathbf{x}}(v,u), \tag{7.0.12}$$

$$\bar{\mathbf{S}}^{\varepsilon,\mathrm{epi}(-\mathfrak{h}_0^-)}_{-\mathbf{t},-\mathbf{x}}(v,u) := \varepsilon^{-1/2}\bar{\mathcal{S}}^{\mathrm{epi}(X_0)}_{-t,n}(y',z) \longrightarrow \bar{\mathbf{S}}^{\mathrm{epi}(-\mathfrak{h}_0^-)}_{-\mathbf{t},-\mathbf{x}}(v,u) \tag{7.0.13}$$

pointwise, where $\mathfrak{h}_0^-(x) = \mathfrak{h}_0(-x)$ *for* $x \geqslant 0$*. Here* $\mathcal{S}_{-t,-n}$ *and* $\bar{\mathcal{S}}_{-t,n}$ *are defined in* (5.2.17) *and* (5.2.18).

Proof. Note that from (5.2.17),(5.2.18), $\bar{\mathcal{S}}_{-t,n}(z_1,z_2) = \mathcal{S}_{-t,-n+1-z_1+z_2}(z_2,z_1)$, so (7.0.12) follows from (7.0.11). By changing variables $w \mapsto \frac{1}{2}(1-\varepsilon^{1/2}\tilde{w})$ in (5.2.17), and using the scaling (7.0.8), we have

$$\mathbf{S}^{\varepsilon}_{-\mathbf{t},\mathbf{x}}(u) = \frac{1}{2\pi i}\oint_{C_\varepsilon} d\tilde{w}\, e^{\varepsilon^{-3/2}\mathbf{t}F_\varepsilon(\varepsilon^{1/2}\tilde{w},\varepsilon^{1/2}\mathbf{x}_\varepsilon/\mathbf{t},\varepsilon u_\varepsilon/\mathbf{t})}, \tag{7.0.14}$$

$$F(w,x,u) := (\operatorname{arctanh} w - w) - x\log(1-w^2) - u\operatorname{arctanh} w \tag{7.0.15}$$

where $\mathbf{x}_\varepsilon := \mathbf{x} + \varepsilon^{1/2}(\mathbf{a}-u)/2 + \varepsilon/2$ and $u_\varepsilon := u + \varepsilon^{1/2}$ and C_ε is a circle of radius $\varepsilon^{-1/2}$ centred at $\varepsilon^{-1/2}$ and $\operatorname{arctanh} w = \frac{1}{2}[\log(1+w) - \log(1-w)]$. Note that

$$\partial_w F(w,x,u) = (w-w_+)(w-w_-)(1-w^2)^{-1}, \qquad w_\pm = x \pm \sqrt{x^2+u}, \tag{7.0.16}$$

From (7.0.16) it is easy to see that as $\varepsilon \searrow 0$, $\varepsilon^{-3/2}\mathbf{t}F_\varepsilon(\varepsilon^{1/2}\tilde{w},\varepsilon^{1/2}\mathbf{x}_\varepsilon/\mathbf{t},\varepsilon u_\varepsilon/\mathbf{t})$ converges to the corresponding exponent in (6.2.4) (keeping in mind the fact that $\mathbf{S}_{-\mathbf{t},\mathbf{x}} = (\mathbf{S}_{\mathbf{t},\mathbf{x}})^*$). Alternatively, one can just use (7.0.15) and that for small w, $\operatorname{arctanh} w - w \sim w^3/3$, $-\log(1-w^2) \sim w^2$ and $\operatorname{arctanh} w \sim w$. Deform C_ε to the contour $\langle_\varepsilon \cup C_\varepsilon^{\pi/3}$ where $\langle_\varepsilon$ is the part of the Airy contour $\langle$ within the ball of radius $\varepsilon^{-1/2}$ centred at $\varepsilon^{-1/2}$, and $C_\varepsilon^{\pi/3}$ is the part of C_ε to the right of $\langle$. As $\varepsilon \searrow 0$, $\langle_\varepsilon \to \langle$, so it only remains to show that the integral over $C_\varepsilon^{\pi/3}$ converges to 0. To see this note that the real part of the exponent of the integral over Γ_0 in (5.2.17) is given by $\varepsilon^{-3/2}\frac{\mathbf{t}}{2}[\cos\theta - 1 + (\frac{1}{8} - c\varepsilon^{1/2})\log(1-4(\cos\theta-1)]$ where $w = \frac{1}{2}e^{i\theta}$and $c = \mathbf{x}/\mathbf{t} + \mathbf{a}/2\varepsilon^{1/2} + \varepsilon$. Using $\log(1+x) \leqslant x$ for $x \geqslant 0$, this is less than or equal to $\varepsilon^{-3/2}\frac{\mathbf{t}}{8}[\cos\theta - 1]$ for sufficiently small ε. The $\tilde{w} \in C_\varepsilon^{\pi/3}$ corresponds to $\theta \geqslant \pi/3$, so the exponent there is less than $-\varepsilon^{-3/2}\kappa\mathbf{t}$ for some $\kappa > 0$. Hence this part of the integral vanishes.

Now define the scaled walk $\mathbf{B}^\varepsilon(\mathbf{x}) = \varepsilon^{1/2}\big(B_{\varepsilon^{-1}\mathbf{x}} + 2\varepsilon^{-1}\mathbf{x} - 1\big)$ for $\mathbf{x} \in \varepsilon\mathbb{Z}_{\geqslant 0}$, interpolated linearly in between, and let $\boldsymbol{\tau}^\varepsilon$ be the hitting time by $\mathbf{B}^\varepsilon$ of the set $\operatorname{epi}(-\mathfrak{h}^\varepsilon(0,\cdot)^-)$. By Donsker's invariance principle [3], $\mathbf{B}^\varepsilon$ converges locally uniformly in distribution to a Brownian motion $\mathbf{B}(\mathbf{x})$ with diffusion coefficient 2, and therefore (using convergence of the initial values of TASEP) the hitting time $\boldsymbol{\tau}^\varepsilon$ converges to $\boldsymbol{\tau}$ as well. □

We will compute next the limit of (7.0.7) using (5.2.22) under the scaling (7.0.8). To this end we change variables in the kernel as in Lemma 7.0.10, so that for $z_i := 2\varepsilon^{-1}\mathbf{x}_i + \varepsilon^{-1/2}(u_i + \mathbf{a}_i) - 2$ we need to compute the limit of the kernel $\varepsilon^{-1/2}\big(\bar{\chi}_{2\varepsilon^{-1}\mathbf{x}-2} K_t \bar{\chi}_{2\varepsilon^{-1}\mathbf{x}-2}\big)(z_i, z_j)$. Note that the change of variables turns the projection $\bar{\chi}_{2\varepsilon^{-1}\mathbf{x}-2}(z)$ into $\bar{\chi}_{-\mathbf{a}}(u)$. We have $n_i < n_j$ for small ε if and only if $\mathbf{x}_j < \mathbf{x}_i$ and in this case we have, under our scaling,

$$\varepsilon^{-1/2} Q^{n_j - n_i}(z_i, z_j) \longrightarrow e^{(\mathbf{x}_i - \mathbf{x}_j)\partial^2}(u_i, u_j),$$

as $\varepsilon \to 0$. For the second term in (5.2.23) we have

$$\begin{aligned}
&\varepsilon^{-1/2}(\mathcal{S}_{t,-n_i})^* \bar{\mathcal{S}}_{t,n_j}(z_i, z_j) \\
&= \varepsilon^{-1}\int_{-\infty}^{\infty} dv\, (\mathbf{S}^\varepsilon_{-\mathbf{t},\mathbf{x}_i})^*(u_i, \varepsilon^{-1/2}v) \bar{\mathbf{S}}^{\varepsilon,\operatorname{epi}(-\mathfrak{h}_0^-)}_{-\mathbf{t},-\mathbf{x}_j}(\varepsilon^{-1/2}v, u_j) \\
&\xrightarrow[\varepsilon\to 0]{} (\mathbf{S}_{-\mathbf{t},\mathbf{x}_i})^* \mathbf{S}_{-\mathbf{t},-\mathbf{x}_j}(u_i, u_j)
\end{aligned}$$

(modulo suitable decay of the integrand). Thus we obtain that the rescaled kernel K, defined in (5.2.23), converges to the limiting kernel

(7.0.17)
$$\mathbf{K}_{\lim}(\mathbf{x}_i, u_i; \mathbf{x}_j, u_j) = -e^{(\mathbf{x}_i - \mathbf{x}_j)\partial^2}(u_i, u_j)\mathbf{1}_{\{\mathbf{x}_i > \mathbf{x}_j\}} + (\mathbf{S}_{-\mathbf{t},\mathbf{x}_i})^* \bar{\mathbf{S}}^{\operatorname{epi}(-\mathfrak{h}_0^-)}_{-\mathbf{t},-\mathbf{x}_j}(u_i, u_j),$$

surrounded by projections $\bar{\chi}_{-\mathbf{a}}$. Our computations here only give pointwise convergence of the kernels, but they can be upgraded to trace class convergence (see [18]), which thus yields convergence of the Fredholm determinants.

We prefer the projections $\bar{\chi}_{-\mathbf{a}}$ which surround (7.0.17) to read $\chi_{\mathbf{a}}$, so we change variables $u_i \longmapsto -u_i$ and replace the Fredholm determinant of the kernel by that of its adjoint to get

$$\det\Big(\mathbf{I} - \chi_a \mathbf{K}^{\operatorname{hypo}(\mathfrak{h}_0)}_{\mathbf{t},\mathrm{ext}} \chi_a\Big) \qquad \text{with} \qquad \mathbf{K}^{\operatorname{hypo}(\mathfrak{h}_0)}_{\mathbf{t},\mathrm{ext}}(u_i, u_j) = \mathbf{K}_{\lim}(\mathbf{x}_j, -u_j; \mathbf{x}_i, -u_i).$$

The choice of superscript $\operatorname{hypo}(\mathfrak{h}_0)$ in the resulting kernel comes from the fact

$$\bar{\mathbf{S}}^{\operatorname{epi}(-\mathfrak{h}_0^-)}_{-\mathbf{t},\mathbf{x}}(v, -u) = \big(\bar{\mathbf{S}}^{\operatorname{hypo}(\mathfrak{h}_0^-)}_{\mathbf{t},\mathbf{x}}\big)^*(u, -v),$$

which together with $\mathbf{S}_{-\mathbf{t},\mathbf{x}}(-u, v) = (\mathbf{S}_{\mathbf{t},\mathbf{x}})^*(-v, u)$ yield

$$\mathbf{K}^{\operatorname{hypo}(\mathfrak{h}_0)}_{\mathbf{t},\mathrm{ext}} = -e^{(\mathbf{x}_j - \mathbf{x}_i)\partial^2}\mathbf{1}_{\{\mathbf{x}_i < \mathbf{x}_j\}} + (\bar{\mathbf{S}}^{\operatorname{hypo}(\mathfrak{h}_0^-)}_{\mathbf{t},-\mathbf{x}_i})^* \mathbf{S}_{\mathbf{t},\mathbf{x}_j}. \tag{7.0.18}$$

This gives the following one-sided fixed point formula for the limit $\mathfrak{h}$.

Theorem 7.0.19 (One-sided fixed point formula). *Let* $\mathbf{x}$ *be defined in* (7.0.3) *for an initial state* $\mathfrak{h}_0 \in \mathrm{UC}$ *with* $\mathfrak{h}_0(\mathbf{x}) = -\infty$ *for* $\mathbf{x} > 0$. *Then given* $\mathbf{x}_1 < \mathbf{x}_2 < \cdots < \mathbf{x}_M$ *and*

$\mathbf{a}_1, \dots, \mathbf{a}_M \in \mathbb{R}$, *we have*

$$\begin{aligned}
&\mathbb{P}_{\mathfrak{h}_0}\big(\mathfrak{h}(\mathbf{t},\mathbf{x}_1) \leqslant \mathbf{a}_1, \dots, \mathfrak{h}(\mathbf{t},\mathbf{x}_M) \leqslant \mathbf{a}_M\big) \\
&= \det\Big(\mathbf{I} - \chi_{\mathbf{a}} \mathbf{K}^{\mathrm{hypo}(\mathfrak{h}_0)}_{\mathbf{t},\mathrm{ext}} \chi_{\mathbf{a}}\Big)_{L^2(\{x_1,\dots,x_M\}\times\mathbb{R})} \\
&= \det\Big(\mathbf{I} - \mathbf{K}^{\mathrm{hypo}(\mathfrak{h}_0)}_{\mathbf{t},\mathbf{x}_M} \\
&\quad + \mathbf{K}^{\mathrm{hypo}(\mathfrak{h}_0)}_{\mathbf{t},\mathbf{x}_M} e^{(x_1-x_M)\partial^2} \bar{\chi}_{\mathbf{a}_1} e^{(x_2-x_1)\partial^2} \bar{\chi}_{\mathbf{a}_2} \cdots e^{(x_M-x_{M-1})\partial^2} \bar{\chi}_{\mathbf{a}_M}\Big)
\end{aligned} \tag{7.0.20}$$

with the Fredholm determinant over $L^2(\mathbb{R})$ *and with the kernel*

$$\mathbf{K}^{\mathrm{hypo}(\mathfrak{h}_0)}_{\mathbf{t},\mathbf{x}}(\cdot,\cdot) := \mathbf{K}^{\mathrm{hypo}(\mathfrak{h}_0)}_{\mathbf{t},\mathrm{ext}}(\mathbf{x},\cdot;\mathbf{x},\cdot),$$

where the latter is defined in (7.0.18).

The assumption $\mathfrak{h}_0(\mathbf{x}) = -\infty$ for $\mathbf{x} > 0$ comes from the fact that we consider TASEP with right finite data. The second identity in (7.0.20) can be obtained similarly to (5.3.1) for the discrete kernels.

Our next goal is to take a continuum limit in the $\mathbf{a}_i$'s of the path-integral formula (7.0.20) on an interval $[-L, L]$ and then take $L \to \infty$. For this we take $\mathbf{x}_1, \dots, \mathbf{x}_M$ to be a partition of $[-L, L]$ and take $\mathbf{a}_i = \mathfrak{g}(\mathbf{x}_i)$. Then we get as in [27] (and actually dating back to [11])

$$\mathbf{S}_{t/2,-L} \bar{\chi}_{\mathfrak{g}(x_1)} e^{(x_2-x_1)\partial^2} \bar{\chi}_{\mathfrak{g}(x_2)} \cdots e^{(x_M-x_{M-1})\partial^2} \bar{\chi}_{\mathfrak{g}(x_M)} (\mathbf{S}_{t/2,-L})^*$$

whose limit as $M \to \infty$ is given by

$$\mathbf{S}_{t/2,-L} \widetilde{\Theta}^{\mathfrak{g}}_{-L,L} (\mathbf{S}_{t/2,-L})^*,$$

where

$$\widetilde{\Theta}^{\mathfrak{g}}_{\ell_1,\ell_2}(u_1,u_2) := \mathbb{P}_{\mathbf{B}(\ell_1)=u_1}\big(\mathbf{B}(s) \leqslant \mathfrak{g}(s)\ \forall s \in [\ell_1,\ell_2],\ \mathbf{B}(\ell_2) \in du_2\big)/du_2.$$

When we pass now to the limit $L \to \infty$, we see (at least roughly) that we obtain

$$\mathbf{S}_{t/2,-L} \widetilde{\Theta}^{\mathfrak{g}}_{-L,L} (\mathbf{S}_{t/2,-L})^* \longrightarrow \mathbf{I} - \mathbf{K}^{\mathrm{epi}(\mathfrak{g})}_{-t/2}.$$

One can find a rigorous proof of these results in [18]. After taking these limits, the Fredholm determinant from (7.0.20) thus converges to

$$\det\Big(\mathbf{I} - \mathbf{K}^{\mathrm{hypo}(\mathfrak{h}_0)}_{t/2} + \mathbf{K}^{\mathrm{hypo}(\mathfrak{h}_0)}_{t/2}\big(\mathbf{I} - \mathbf{K}^{\mathrm{epi}(\mathfrak{g})}_{-t/2}\big)\Big) = \det\Big(\mathbf{I} - \mathbf{K}^{\mathrm{hypo}(\mathfrak{h}_0)}_{t/2} \mathbf{K}^{\mathrm{epi}(\mathfrak{g})}_{-t/2}\Big),$$

which is exactly the content of Theorem 7.0.5.

As in the TASEP case, the kernel in (7.0.18) can be rewritten (thanks to the analog of (6.2.8)) as

$$\begin{aligned}
\mathbf{K}^{\mathrm{hypo}(\mathfrak{h}_0)}_{\mathbf{t},\mathrm{ext}}(x_i,\cdot;x_j,\cdot) = -e^{(x_j-x_i)\partial^2} \mathbf{1}_{\{x_i<x_j\}} + (\mathbf{S}_{\mathbf{t},-x_i})^* \bar{\chi}_{\mathfrak{h}_0(0)} \mathbf{S}_{\mathbf{t},x_j} \\
+ (\bar{\mathbf{S}}^{\mathrm{hypo}(\mathfrak{h}_0^-)}_{\mathbf{t},-x_i})^* \chi_{\mathfrak{h}_0(0)} \mathbf{S}_{\mathbf{t},x_j}.
\end{aligned}$$

7.1. From one-sided to two-sided formulas. Now we derive the formula for the process (7.0.3) with general initial data $\mathfrak{h}_0$. More precisely, we prove the following theorem.

Theorem 7.1.1 (Two-sided fixed point formula). *If* $\mathfrak{h}_0 \in \mathrm{UC}$ *and* $\mathbf{x}_1 < \mathbf{x}_2 < \cdots < \mathbf{x}_M$, *then for* $\mathfrak{h}(\mathbf{t}, \mathbf{x})$ *given as in* (7.0.3),

$$\begin{aligned} &\mathbb{P}_{\mathfrak{h}_0}\big(\mathfrak{h}(\mathbf{t}, \mathbf{x}_1) \leqslant \mathbf{a}_1, \ldots, \mathfrak{h}(\mathbf{t}, \mathbf{x}_M) \leqslant \mathbf{a}_M\big) \\ &= \det\Big(\mathbf{I} - \chi_{\mathbf{a}} \mathbf{K}^{\mathrm{hypo}(\mathfrak{h}_0)}_{\mathbf{t},\mathrm{ext}} \chi_{\mathbf{a}}\Big)_{L^2(\{\mathbf{x}_1,\ldots,\mathbf{x}_M\}\times\mathbb{R})} \\ &= \det\Big(\mathbf{I} - \mathbf{K}^{\mathrm{hypo}(\mathfrak{h}_0)}_{\mathbf{t},\mathbf{x}_M} \\ &\qquad + \mathbf{K}^{\mathrm{hypo}(\mathfrak{h}_0)}_{\mathbf{t},\mathbf{x}_M} e^{(\mathbf{x}_1-\mathbf{x}_M)\partial^2} \bar{\chi}_{\mathbf{a}_1} e^{(\mathbf{x}_2-\mathbf{x}_1)\partial^2} \bar{\chi}_{\mathbf{a}_2} \cdots e^{(\mathbf{x}_M-\mathbf{x}_{M-1})\partial^2} \bar{\chi}_{\mathbf{a}_M}\Big) \end{aligned} \tag{7.1.2}$$

where the kernels are as in Theorem 7.0.19, *and the last Fredholm determinant is over* $L^2(\mathbb{R})$.

Below we provide a sketch of the proof. We derive the formula (7.1.2) as the $L \to \infty$ limit of the formula with initial data

$$\mathfrak{h}_0^L(\mathbf{x}) = \mathfrak{h}_0(\mathbf{x}) \mathbf{1}_{\{\mathbf{x}\leqslant L\}} - \infty \cdot \mathbf{1}_{\{\mathbf{x}>L\}},$$

which can be obtained from Theorem 7.0.19 by translation invariance. We then take, in the next subsection, a continuum limit of the operator

$$e^{(\mathbf{x}_1-\mathbf{x}_M)\partial^2} \bar{\chi}_{a_1} e^{(\mathbf{x}_2-\mathbf{x}_1)\partial^2} \bar{\chi}_{\mathbf{a}_2} \cdots e^{(\mathbf{x}_M-\mathbf{x}_{M-1})\partial^2} \bar{\chi}_{\mathbf{a}_M}$$

on the right side of (7.0.20) to obtain a "hit" operator for the final data as well. The result of all this is the same as if we started with two-sided data for TASEP.

The shift invariance of TASEP, tells us that $\mathfrak{h}(\mathbf{t}, \mathbf{x}; \mathfrak{h}_0^L) \overset{\mathrm{dist}}{=} \mathfrak{h}(\mathbf{t}, \mathbf{x} - L; \theta_L \mathfrak{h}_0^L)$, where θ_L is the shift operator. Our goal then is to take $L \to \infty$ in the formula given in Theorem 7.0.19 for $\mathfrak{h}(\mathbf{t}, \mathbf{x} - L; \theta_L \mathfrak{h}_0^L)$. We get

$$\begin{aligned} &\mathbb{P}_{\theta_L \mathfrak{h}_0}\big(\mathfrak{h}(\mathbf{t}, \mathbf{x}_1 - L) \leqslant \mathbf{a}_1, \ldots, \mathfrak{h}(\mathbf{t}, \mathbf{x}_M - L) \leqslant \mathbf{a}_M\big) \\ &= \det\Big(\mathbf{I} - \chi_a \widetilde{\mathbf{K}}^{\theta_L \mathfrak{h}_0^L}_L \chi_a\Big)_{L^2(\{\mathbf{x}_1,\ldots,\mathbf{x}_M\}\times\mathbb{R})} \end{aligned}$$

with

$$\widetilde{\mathbf{K}}^{\theta_L \mathfrak{h}_0^L}_L(\mathbf{x}_i, \cdot; \mathbf{x}_j, \cdot) := e^{(\mathbf{x}_j-\mathbf{x}_i)\partial^2} \mathbf{1}_{\{\mathbf{x}_i<\mathbf{x}_j\}} + e^{(\mathbf{x}_j-\mathbf{x}_i)\partial^2} \big(\bar{\mathbf{S}}^{\mathrm{hypo}((\theta_L \mathfrak{h}_0^L)_0^-)}_{\mathbf{t},-\mathbf{x}_j+L}\big)^* \mathbf{S}_{\mathbf{t},\mathbf{x}_j-L}. \tag{7.1.3}$$

Since $(\theta_L \mathfrak{h}_0^L)_0^+ \equiv -\infty$, we may rewrite $\big(\bar{\mathbf{S}}^{\mathrm{hypo}((\theta_L \mathfrak{h}_0^L)_0^-)}_{\mathbf{t},-\mathbf{x}_j+L}\big)^* \mathbf{S}_{\mathbf{t},\mathbf{x}_j-L}$ as[5]

$$\mathbf{I} - \big(\mathbf{S}_{\mathbf{t},-\mathbf{x}_j+L} - \bar{\mathbf{S}}^{\mathrm{hypo}((\theta_L \mathfrak{h}_0^L)_0^-)}_{\mathbf{t},-\mathbf{x}_j+L}\big)^* \big(\mathbf{S}_{\mathbf{t},\mathbf{x}_j-L} - \bar{\mathbf{S}}^{\mathrm{hypo}((\theta_L \mathfrak{h}_0^L)_0^+)}_{\mathbf{t},\mathbf{x}_j-L}\big) = e^{-\mathbf{x}_j\partial^2} \mathbf{K}^{\mathrm{hypo}(\mathfrak{h}_0^L)}_{\mathbf{t}} e^{\mathbf{x}_j\partial^2},$$

where $\mathbf{K}^{\mathrm{hypo}(\mathfrak{h}_0)}_{\mathbf{t}}$ is the kernel defined in (6.2.15). Note the crucial fact that the right hand side depends on L only through $\mathfrak{h}_0^L$ (the various shifts by L on the left hand side of (7.1) play no role). It was shown in [27] that $\mathbf{K}^{\mathrm{hypo}(\mathfrak{h}_0^L)}_{\mathbf{t}} \longrightarrow \mathbf{K}^{\mathrm{hypo}(\mathfrak{h}_0)}_{\mathbf{t}}$ as $L \to \infty$. This tells us that the second term on the right hand side of (7.1.3)

[5] At first glance it may look as if the product $e^{-\mathbf{x}\partial^2} \mathbf{K}^{\mathrm{hypo}(\mathfrak{h})}_{\mathbf{t}} e^{\mathbf{x}\partial^2}$ makes no sense, because $\mathbf{K}^{\mathrm{hypo}(\mathfrak{h})}_{\mathbf{t}}$ is given in (6.2.15) as the identity minus a certain kernel, and applying $e^{\mathbf{x}\partial^2}$ to $\mathbf{I}$ is ill-defined for $\mathbf{x} < 0$. However, thanks to the analog of (6.3.9) below for $\mathbf{K}^{\mathrm{hypo}(\mathfrak{h})}_{\mathbf{t}}$, the action of $e^{\mathbf{x}\partial^2}$ on this kernel on the left and right is well defined for any $\mathbf{x} \in \mathbb{R}$. This also justifies the identity in (7.1).

equals $e^{-x_i\partial^2}\mathbf{K}_t^{\mathrm{hypo}(\mathfrak{h}_0^L)}e^{x_j\partial^2}$, which converges to $e^{-x_i\partial^2}\mathbf{K}_t^{\mathrm{hypo}(\mathfrak{h}_0)}e^{x_j\partial^2}$ as $L\to\infty$, and leads to (7.1.2).

7.2. Continuum limit. In order to complete the proof of Theorem 7.0.5, we need to go from the two-sided formula (7.1.2) to the KPZ fixed point formula (6.3.2). More precisely, we need to approximate any function $\mathfrak{g}\in\mathrm{LC}$ by a finite sequence of values $\mathbf{a}_i$'s and take a limit.

To do this, conjugate the kernel inside the determinant by $\mathbf{S}_{t/2,x_M}$, leading to

$$\widetilde{\mathbf{K}}^{\mathfrak{h}_0}_{t,x_M}-\widetilde{\mathbf{K}}^{\mathfrak{h}_0}_{t,x_M}\left[\mathbf{S}_{t/2,x_1}\bar{\chi}_{\mathbf{a}_1}e^{(x_2-x_1)\partial^2}\bar{\chi}_{\mathbf{a}_2}\cdots e^{(x_M-x_{M-1})\partial^2}\bar{\chi}_{\mathbf{a}_M}(\mathbf{S}_{t/2,-x_M})^*\right]$$

with $\widetilde{\mathbf{K}}^{\mathfrak{h}_0}_{t,x_M}:=\mathbf{S}_{t/2,x_M}\mathbf{K}^{\mathrm{hypo}(\mathfrak{h}_0)}_{t,x_M}(\mathbf{S}_{t/2,-x_M})^*=\mathbf{K}^{\mathrm{hypo}(\mathfrak{h}_0)}_{t/2}$ (the second equality follows from (6.2.15)). Now we take the limit of term in brackets, letting $x_1,\dots,x_M$ be a partition of $[-R,R]$ and taking $M\to\infty$ with $\mathbf{a}_i=\mathfrak{g}(x_i)$. As in [11] we have

$$\begin{aligned}\mathbf{S}_{t/2,-R}\,\bar{\chi}_{\mathfrak{g}(x_1)}e^{(x_2-x_1)\partial^2}\bar{\chi}_{\mathfrak{g}(x_2)}\cdots e^{(x_M-x_{M-1})\partial^2}\bar{\chi}_{\mathfrak{g}(x_M)}(\mathbf{S}_{t/2,-R})^*&\\ \longrightarrow \mathbf{S}_{t/2,-R}\,\widetilde{\Theta}^{\mathfrak{g}}_{-R,R}(\mathbf{S}_{t/2,-R})^*,&\end{aligned}$$

where $\widetilde{\Theta}^{\mathfrak{g}}_{\ell_1,\ell_2}(u_1,u_2):=\mathbb{P}_{\mathbf{B}(\ell_1)=u_1}(\mathbf{B}(s)\leqslant\mathfrak{g}(s)\;\forall s\in[\ell_1,\ell_2],\,\mathbf{B}(\ell_2)\in du_2)/du_2$,

Denote by $\mathbb{P}_{\ell_1,u_1;\ell_2,u_2}$ the distribution of a Brownian bridge $\mathbf{B}(s)$, $s\in[\ell_1,\ell_2]$, conditioned by $\mathbf{B}(\ell_1)=u_1$ and $\mathbf{B}(\ell_2)=u_2$. Then, for any $\alpha\in(\ell_1,\ell_2)$, we rewrite the probability as

$$\begin{aligned}\mathbb{P}_{\ell_1,u_1;\ell_2,u_2}(\mathbf{B}(s)\leqslant\mathfrak{g}(s)\;\forall s\in[\ell_1,\ell_2])&=\int_{-\infty}^{\mathfrak{g}(\alpha)}\mathbb{P}_{\ell_1,u_1;\ell_2,u_2}(\mathbf{B}(\alpha)\in dy)\\ \times\,\mathbb{P}_{\ell_1,u_1;\alpha,y}(\mathbf{B}(s)\leqslant\mathfrak{g}(s)\;\forall s\in[\ell_1,\alpha])&\mathbb{P}_{\alpha,y;\ell_2,u_2}(\mathbf{B}(s)\leqslant\mathfrak{g}(s)\;\forall s\in[\alpha,\ell_2]).\end{aligned}$$

Let us define the functions $\mathfrak{g}^{\pm}_{\alpha}(s):=\mathfrak{g}(\alpha\pm s)$, and let $\tau_{\mathfrak{g}}:=\inf\{t\geqslant 0:\mathbf{B}(t)>\mathfrak{g}(t)\}$ be the hitting time of the function $\mathfrak{g}$. Then the last probability in the above integral can be rewritten as

$$\begin{aligned}&\mathbb{P}_{0,y;\ell_2-\alpha,u_2}(\mathbf{B}(s)\leqslant\mathfrak{g}^+_{\alpha}(s)\;\forall s\in[0,\ell_2-\alpha])\\ &\qquad=1-\int_0^{\ell_2-\alpha}\mathbb{P}_{\mathbf{B}(0)=y}(\tau_{\mathfrak{g}^+_{\alpha}}\in dt)\frac{p(\ell_2-\alpha-t,u_2-\mathfrak{g}_{\alpha}(t))}{p(\ell_2-\alpha,u_2-y)},\end{aligned}$$

where $p(t,x)$ is the heat kernel. A similar identity can be written for the probability $\mathbb{P}_{\ell_1,u_1;\alpha,y}(\mathbf{B}(s)\leqslant\mathfrak{g}(s)\;\forall s\in[\ell_1,\alpha])$, now using $\tau_{\mathfrak{g}^-_{\alpha}}$, and going backwards from time α to time ℓ_1. Using this and writing $\mathbb{P}_{\ell_1,u_1;\ell_2,u_2}(\mathbf{B}(\alpha)\in dy)$ explicitly we find that

$$\begin{aligned}&\mathbb{P}_{\ell_1,u_1;\ell_2,u_2}(\mathbf{B}(s)<\mathfrak{g}(s)\;\forall s\in[\ell_1,\ell_2])\\ &\quad=\int_{-\infty}^{\mathfrak{g}(\alpha)}dy\sqrt{\frac{\ell_2-\ell_1}{4\pi(\alpha-\ell_1)(\ell_2-\alpha)}}\,e^{-\frac{((\ell_2-\alpha)u_1+(\alpha-\ell_1)u_2+(\ell_1-\ell_2)y)^2}{4(\alpha-\ell_1)(\ell_2-\alpha)(\ell_2-\ell_1)}}\\ &\qquad\times\left(1-\int_0^{\alpha-\ell_1}\mathbb{P}_{\mathbf{B}(0)=y}(\tau_{\mathfrak{g}^-_{\alpha}}\in dt_1)\frac{p(\alpha-\ell_1-t_1,u_1-\mathfrak{g}^-_{\alpha}(t_1))}{p(\alpha-\ell_1,u_1-y)}\right)\\ &\qquad\times\left(1-\int_0^{\ell_2-\alpha}\mathbb{P}_{\mathbf{B}(0)=y}(\tau_{\mathfrak{g}^+_{\alpha}}\in dt_2)\frac{p(\ell_2-\alpha-t_2,u_2-\mathfrak{g}^+_{\alpha}(t_2))}{p(\ell_2-\alpha,u_2-y)}\right).\end{aligned}$$

Recalling that in the formula for $\widetilde{\Theta}^{\mathfrak{g}}_{\ell_1,\ell_2}(u_1,u_2)$ this probability is premultiplied by $p(\ell_2-\ell_1,u_2-u_1+\ell_2^2-\ell_1^2)$ and observing that

$$\frac{p(\ell_2-\ell_1,u_2-u_1+\ell_2^2-\ell_1^2)}{p(\alpha-\ell_1,u_1-y)p(\ell_2-\alpha,u_2-y)}\sqrt{\frac{\ell_2-\ell_1}{4\pi(\alpha-\ell_1)(\ell_2-\alpha)}}e^{-\frac{((\ell_2-\alpha)u_1+(\alpha-\ell_1)u_2+(\ell_1-\ell_2)y)^2}{4(\alpha-\ell_1)(\ell_2-\alpha)(\ell_2-\ell_1)}}$$
$$= e^{\frac{1}{4}(\ell_1^2-\ell_2^2+2u_1-2u_2)(\ell_1+\ell_2)}$$

we deduce that

$$\widetilde{\Theta}^{\mathfrak{g}}_{\ell_1,\ell_2}(u_1,u_2) = e^{\frac{1}{4}(\ell_1^2-\ell_2^2+2u_1-2u_2)(\ell_1+\ell_2)}\int_{-\infty}^{\mathfrak{g}(\alpha)}dy$$
$$\times\left[p(\alpha-\ell_1,u_1-y)-\int_0^{\alpha-\ell_1}dt_1\,\mathbb{P}_{\mathbf{B}(0)=y}(\tau_{\mathfrak{g}_\alpha^-}\in dt_1)p(\alpha-\ell_1-t_1,u_1-\mathfrak{g}_\alpha^-(t_1))\right]$$
$$\times\left[p(\ell_2-\alpha,u_2-y)-\int_0^{\ell_2-\alpha}dt_2\,\mathbb{P}_{\mathbf{B}(0)=y}(\tau_{\mathfrak{g}_\alpha^+}\in dt_2)p(\ell_2-\alpha-t_2,u_2-\mathfrak{g}_\alpha^+(t_2))\right].$$

Let A be the *Airy transform*, defined by the integral kernel

$$A(x,\lambda) := \mathrm{Ai}(x-\lambda). \tag{7.2.1}$$

Taking then $-\ell_1=\ell_2=L$, for any $\alpha\in(-L,L)$ we have

$$A^*e^{-L\Delta}\widetilde{\Theta}^{\mathfrak{g}}_{-L,L}e^{-L\Delta}A(\lambda_1,\lambda_2) = \int_{-\infty}^{\mathfrak{g}(\alpha)}dy$$
$$\times\left[A^*e^{\alpha\Delta}(\lambda_1,y)-\int_0^{\alpha+L}dt_1\,\mathbb{P}_{\mathbf{B}(0)=y}(\tau_{\mathfrak{g}_\alpha^-}\in dt_1)A^*e^{(\alpha-t_1)\Delta}(\lambda_1,\mathfrak{g}_\alpha^-(t_1))\right]$$
$$\times\left[e^{-\alpha\Delta}A(y,\lambda_2)-\int_0^{L-\alpha}dt_2\,\mathbb{P}_{\mathbf{B}(0)=y}(\tau_{\mathfrak{g}_\alpha^+}\in dt_2)e^{-(\alpha+t_2)\Delta}A(\mathfrak{g}_\alpha^+(t_2),\lambda_2)\right].$$

Using the representation (6.2.4), we now have

$$\mathbf{S}_{\mathbf{t}/2,-R}\widetilde{\Theta}^{\mathfrak{g}}_{-R,R}(\mathbf{S}_{\mathbf{t}/2,-R})^* \longrightarrow \mathbf{I}-\mathbf{K}^{\mathrm{epi}(\mathfrak{g})}_{-\mathbf{t}/2}$$

as $R\to\infty$. Our Fredholm determinant (7.1.2) is thus now given by

$$\det\Big(\mathbf{I}-\mathbf{K}^{\mathrm{hypo}(\mathfrak{h}_0)}_{\mathbf{t}/2}+\mathbf{K}^{\mathrm{hypo}(\mathfrak{h}_0)}_{\mathbf{t}/2}(\mathbf{I}-\mathbf{K}^{\mathrm{epi}(\mathfrak{g})}_{-\mathbf{t}/2})\Big) = \det\Big(\mathbf{I}-\mathbf{K}^{\mathrm{hypo}(\mathfrak{h}_0)}_{\mathbf{t}/2}\mathbf{K}^{\mathrm{epi}(\mathfrak{g})}_{-\mathbf{t}/2}\Big),$$

which is the KPZ fixed point formula (6.3.2).

Remark 7.2.2. The Airy transform, defined in (7.2.1), satisfies $AA^*=I$, so that one has the identity $f(x)=\int_{-\infty}^{\infty}d\lambda\,\mathrm{Ai}(x-\lambda)Af(\lambda)$ for every function f from its domain. In other words, the shifted Airy functions $\{\mathrm{Ai}(x-\lambda)\}_{\lambda\in\mathbb{R}}$ (which are not in $L^2(\mathbb{R})$) form a generalized orthonormal basis of $L^2(\mathbb{R})$. Thus the *Airy kernel* $K_{\mathrm{Ai}}(x,y)=\int_{-\infty}^{0}d\lambda\,\mathrm{Ai}(x-\lambda)\mathrm{Ai}(y-\lambda)$ is the projection onto the subspace spanned by $\{\mathrm{Ai}(x-\lambda)\}_{\lambda\leq 0}$. Then $K_{\mathrm{Ai}}=A\bar{\chi}_0A^*$ and the distribution functions of the GUE and GOE Tracy-Widom distributions can be written as

$$F_{\mathrm{GUE}}(r) = \det(I-\chi_rK_{\mathrm{Ai}}\chi_r) = \det(I-K_{\mathrm{Ai}}\chi_rK_{\mathrm{Ai}}), \tag{7.2.3}$$

$$F_{\mathrm{GOE}}(4^{1/3}r) = \det(I-K_{\mathrm{Ai}}\rho_rK_{\mathrm{Ai}}), \tag{7.2.4}$$

where ρ_r is the reflection operator $\rho_rf(x):=f(2r-x)$.

References

[1] L. Bertini and G. Giacomin, *Stochastic Burgers and KPZ equations from particle systems*, Comm. Math. Phys. **183** (1997), no. 3, 571–607. ↑252

[2] H. A. Bethe, *On the theory of metals, i. eigenvalues and eigenfunctions of a linear chain of atoms*, Zeits. Phys. **74** (1931), 205–226. ↑254

[3] P. Billingsley, *Convergence of probability measures*, Second, Wiley Series in Probability and Statistics: Probability and Statistics, John Wiley & Sons, Inc., New York, 1999. A Wiley-Interscience Publication. ↑295

[4] A. Borodin, *Determinantal point processes*, The Oxford handbook of random matrix theory, 2011, pp. 231–249. ↑258

[5] A. Borodin, I. Corwin, and D. Remenik, *Multiplicative functionals on ensembles of non-intersecting paths*, Ann. Inst. H. Poincaré Probab. Statist. **51** (2015), no. 1, 28–58. ↑279, 280, 281, 282, 283, 287

[6] A. Borodin, P. L. Ferrari, and M. Prähofer, *Fluctuations in the discrete TASEP with periodic initial configurations and the* Airy_1 *process*, Int. Math. Res. Pap. IMRP (2007), Art. ID rpm002, 47. ↑253, 263, 274, 279

[7] A. Borodin, P. L. Ferrari, M. Prähofer, and T. Sasamoto, *Fluctuation properties of the TASEP with periodic initial configuration*, J. Stat. Phys. **129** (2007), no. 5-6, 1055–1080. ↑253, 263, 264, 274, 279, 287

[8] A. Borodin, P. L. Ferrari, and T. Sasamoto, *Transition between* $Airy_1$ *and* $Airy_2$ *processes and TASEP fluctuations*, Comm. Pure Appl. Math. **61** (2008), no. 11, 1603–1629. ↑274, 287, 288

[9] A. Borodin and G. Olshanski, *Distributions on partitions, point processes, and the hypergeometric kernel*, Comm. Math. Phys. **211** (2000), no. 2, 335–358. ↑258

[10] A. Borodin and E. M. Rains, *Eynard-Mehta theorem, Schur process, and their Pfaffian analogs*, J. Stat. Phys. **121** (2005), no. 3-4, 291–317. ↑258, 261, 262, 271

[11] I. Corwin, J. Quastel, and D. Remenik, *Continuum statistics of the* $Airy_2$ *process*, Comm. Math. Phys. **317** (2013), no. 2, 347–362. ↑289, 296, 298

[12] I. Corwin, J. Quastel, and D. Remenik, *Renormalization fixed point of the KPZ universality class* (English), J. Stat. Phys. **160** (2015), no. 4, 815–834. ↑291

[13] P. L. Ferrari, *Why random matrices share universal processes with interacting particle systems?*, 2013. ↑263, 264, 279

[14] P. L. Ferrari, *Dimers and orthogonal polynomials: connections with random matrices*, Dimer models and random tilings, 2015, pp. 47–79. ↑274

[15] J. B. Hough, M. Krishnapur, Y. Peres, and B. Virág, *Zeros of Gaussian analytic functions and determinantal point processes*, University Lecture Series, vol. 51, American Mathematical Society, Providence, RI, 2009. ↑258, 259

[16] K. Johansson, *Discrete polynuclear growth and determinantal processes*, Comm. Math. Phys. **242** (2003), no. 1-2, 277–329. ↑258, 259, 287, 289

[17] K. Johansson, *The arctic circle boundary and the Airy process*, Ann. Probab. **33** (2005), no. 1, 1–30. ↑260

[18] M. Kardar, G. Parisi, and Y.-C. Zhang, *Dynamical scaling of growing interfaces*, Phys. Rev. Lett. **56** (1986), no. 9, 889–892. ↑264, 279, 286, 295, 296

[19] S. Karlin and J. McGregor, *Coincidence probabilities*, Pacific J. Math. **9** (1959), 1141–1164. ↑259

[20] T. M. Liggett, *Interacting particle systems*, Grundlehren der Mathematischen Wissenschaften [Fundamental Principles of Mathematical Sciences], vol. 276, Springer-Verlag, New York, 1985. ↑252

[21] O. Macchi, *The coincidence approach to stochastic point processes*, Advances in Applied Probability **7** (1975), no. 01, 83–122, available at `https://doi.org/10.10172Fs0001867800040313`. ↑258, 262

[22] K. Matetski, J. Quastel, and D. Remenik, *The KPZ fixed point* (2016), available at `1701.00018`. ↑251, 274, 284, 291

[23] M. L. Mehta, *Random matrices*, Second, Academic Press Inc., Boston, MA, 1991. ↑260

[24] M. Prähofer and H. Spohn, *Scale invariance of the PNG droplet and the Airy process*, J. Stat. Phys. **108** (2002), no. 5-6, 1071–1106. ↑279, 287

[25] S. Prolhac and H. Spohn, *The one-dimensional KPZ equation and the Airy process*, J. Stat. Mech. Theor. Exp. **2011** (2011), no. 03, P03020. ↑279

[26] J. Quastel and D. Remenik, *Local behavior and hitting probabilities of the* $Airy_1$ *process* (English), Probability Theory and Related Fields **157** (2013), no. 3-4, 605–634. ↑279, 290, 292

[27] J. Quastel and D. Remenik, *How flat is flat in random interface growth?*, 2016. ↑296, 297

[28] T. Sasamoto, *Spatial correlations of the 1D KPZ surface on a flat substrate*, Journal of Physics A: Mathematical and General **38** (2005), no. 33, L549. ↑263, 287

[29] G. M. Schütz, *Exact solution of the master equation for the asymmetric exclusion process*, J. Statist. Phys. **88** (1997), no. 1-2, 427–445. ↑253

[30] F. Spitzer, *Interaction of Markov processes*, Advances in Math. **5** (1970), 246–290 (1970). ↑252

[31] C. A. Tracy and H. Widom, *Formulas and asymptotics for the asymmetric simple exclusion process*, Math. Phys. Anal. Geom. **14** (2011), no. 3, 211–235. ↑253, 257

University of Toronto, 40 St. George Street, Toronto, Ontario, Canada M5S 2E4
Email address: quastel@math.toronto.edu, matetski@math.toronto.edu

IAS/Park City Mathematics Series
Volume 26, Pages 303–340
https://doi.org/10.1090/pcms/026/00848

Delocalization of eigenvectors of random matrices

Mark Rudelson

Abstract. Let $x \in S^{n-1}$ be a unit eigenvector of an $n \times n$ random matrix. This vector is delocalized if it is distributed roughly uniformly over the real or complex sphere. This intuitive notion can be quantified in various ways. In these lectures, we will concentrate on the *no-gaps delocalization*. This type of delocalization means that with high probability, any non-negligible subset of the support of x carries a non-negligible mass. Proving the no-gaps delocalization requires establishing small ball probability bounds for the projections of random vector. Using Fourier transform, we will prove such bounds in a simpler case of a random vector having independent coordinates of a bounded density. This will allow us to derive the no-gaps delocalization for matrices with random entries having a bounded density. In the last section, we will discuss the applications of delocalization to the spectral properties of Erdős-Rényi random graphs.

1. Introduction

Let G be a symmetric random matrix with independent above the diagonal normal random entries having expectation 0 and variance 1 ($N(0,1)$ random variables). The distribution of such matrices is invariant under the action of the orthogonal group $O(n)$. Consider a unit eigenvector $v \in S^{n-1}$ of this matrix. The distribution of the eigenvector should share the invariance of the distribution of the matrix itself, so v is uniformly distributed over the real unit sphere $S_{\mathbb{R}}^{n-1}$. Similarly, if Γ is an $n \times n$ complex random matrix with independent entries whose real and imaginary part are independent $N(0,1)$ random variables, then the distribution of Γ is invariant under the action of the unitary group $U(n)$. This means that any unit eigenvector of Γ is uniformly distributed over the complex unit sphere $S_{\mathbb{C}}^{n-1}$. For a general distribution of entries, we cannot expect such strong invariance properties. Indeed, if the entries of the matrix are random variables taking finitely many values, the eigenvectors will take finitely many values as well, so the invariance is impossible. Nevertheless, as n increases, a central limit phenomenon should kick in, so the distribution of an eigenvector should be approximately uniform. This vague idea called delocalization can be made mathematically precise in a number of ways. Some of these formalizations use

2010 *Mathematics Subject Classification.* Primary 60B20; Secondary 05C80.
Key words and phrases. random matrices, eigenvectors, random graphs.
Partially supported by NSF grant, DMS-1464514.

the local structure of a vector. One can fix in advance several coordinates of the eigenvector and show that the joint distribution of these coordinates approaches the distribution of a properly normalized gaussian vector, see [6].

In these notes, we adopt a different approach to delocalization coming from the non-asymptotic random matrix theory. The asymptotic theory is concerned with establishing limit distributions of various spectral characteristics of a family of random matrices when the sizes of these matrices tend to infinity. In contrast to it, the non-asymptotic theory strives to obtain explicit, valid with high probability bounds for the matrices of a large fixed size. This approach is motivated by applications primarily to convex geometry, combinatorics, and computer science. For example, while analyzing performance of an algorithm solving a noisy linear system, one cannot let the size of the system go to infinity. An interested reader can find an introduction to the non-asymptotic theory in [21,22,27]. In this type of problems, strong probabilistic guarantees are highly desirable, since one typically wants to show that many "good" events occur at the same time. This will be the case in our analysis of the delocalization behavior as well.

We will consider the global structure of the eigenvector of a random matrix controlling all coordinates of it at once. The most classical type of such delocalization is the ℓ_∞ norm bound. If $v \in S^{n-1}$ is a random vector uniformly distributed over the unit sphere, then with high probability, all its coordinates are small. This is easy to check using the concentration of measure. Indeed, the vector v has the same distribution as $g/\|g\|_2$, where $g \in \mathbb{R}^n$ or $\mathbb{C}^n$ is the standard Gaussian vector, i.e., a vector with the independent $N(0,1)$ coordinates. By the concentration of measure, $\|g\|_2 = \sqrt{n}(1+o(1))$ with high probability. Also, since the coordinates of g are independent,

$$\mathbb{E}\,\|g\|_\infty = \mathbb{E} \max_{j \in [n]} |g_j| \leqslant C\sqrt{\log n},$$

and the measure concentration yields that $\|g\|_\infty \leqslant C'\sqrt{\log n}$ with high probability. Therefore, with high probability,

$$\|v\|_\infty \leqslant C\frac{\sqrt{\log n}}{\sqrt{n}}.$$

Here and below, $C, \bar{C}, C', c$, etc. denote absolute constants which can change from line to line, or even within the same line.

One would expect to have a similar ℓ_∞ delocalization for a general random matrix. The bound

$$\|v\|_\infty \leqslant C\frac{\log^c n}{\sqrt{n}}$$

for unit eigenvectors was proved in [12,13] for Hermitian random matrices and in [24] for random matrices all whose entries are independent. Moreover, in the case of the Hermitian random matrix with entries having more than four moments, the previous estimate has been established with the optimal power of the logarithm $c = 1/2$, see [14,28]. We will not discuss the detailed history and the methods of obtaining the ℓ_∞ delocalization in these notes, and refer a reader to a comprehensive recent survey [20].

Instead, we are going to concentrate on a different manifestation of the delocalization phenomenon. The ℓ_∞ delocalization rules out peaks in the distribution of mass among the coordinates of a unit eigenvector. In particular, it means that with high probability, most of the mass, i.e., ℓ_2 norm of a unit eigenvector cannot be localized on a few coordinates. We will consider a complementary phenomenon, namely ruling out chasms in the mass distribution. More precisely, we aim at showing that with high probability, any non-negligible set of the coordinates of a unit eigenvector carries a relatively large mass. We call this property of lack of almost empty zones in the support of the eigenvector the *no-gaps delocalization* property.

No-gaps delocalization property holds for the eigenvectors of many natural classes of random matrices. This includes matrices, whose all entries are independent, random real symmetric and skew-symmetric matrices, random complex hermitian matrices with independent real and imaginary parts of the entries, etc. We formulate the explicit assumption on the dependencies of the entries below.

Assumption 1.0.1 (Dependencies of entries)**.** Let A be an $n \times n$ random matrix. Assume that for any $i, j \in [n]$, the entry A_{ij} is independent of the rest of the entries except possibly A_{ji}. We also assume that the real part of A is random and the imaginary part is fixed.

Fixing the imaginary part in Assumption 1.0.1 allows us to handle real random matrices. This assumption can also be arranged for complex matrices with independent real and imaginary parts, once we condition on the imaginary part. One can even consider a more general situation where the real parts of the entries conditioned on the imaginary parts have variances bounded below.

We also assume that the operator norm of the matrix A satisfies $\|A\| = O(\sqrt{n})$ with high probability. This natural condition holds, in particular, if the entries of A have mean zero and bounded fourth moments (see, e.g., [26]). To make this rigorous, we fix a number $M \geqslant 1$ and introduce the boundedness event

$$\mathcal{B}_{A,M} := \left\{\|A\| \leqslant M\sqrt{n}\right\}. \tag{1.0.2}$$

We will give two versions of the no-gaps delocalization theorem, for absolutely continuous entries with bounded density and for general entries. Although the second case includes the first, the results assuming bounded density are stronger and the proofs significantly easier. We formulate the first assumption explicitly.

Assumption 1.0.3 (Continuous distributions)**.** We assume that the real parts of the matrix entries have densities bounded by some number $K \geqslant 1$.

Under Assumptions 1.0.1 and 1.0.3, we show that every subset of at least eight coordinates carries a non-negligible part of the mass of any eigenvector. This is summarized in the following theorem [25].

Theorem 1.0.4 (Delocalization: continuous distributions). *Let* A *be an* $n \times n$ *random matrix which satisfies Assumptions* 1.0.1 *and* 1.0.3. *Choose* $M \geqslant 1$. *Let* $\varepsilon \in [8/n, 1)$ *and* $s > 0$. *Then, the following event holds with probability at least*

$$1 - (Cs)^{\varepsilon n} - \mathbb{P}\,(\mathcal{B}_{A,M}^{c}).$$

Every eigenvector v *of* A *satisfies*

$$\|v_I\|_2 \geqslant (\varepsilon s)^6 \|v\|_2 \quad \textit{for all } I \subset [n],\ |I| \geqslant \varepsilon n.$$

Here $C = C(K, M) \geqslant 1$, *and* $[n]$ *denotes the set of all natural numbers from* 1 *to* n.

Note that we do not require any moments for the matrix entries, so heavy-tailed distributions are allowed. However, the boundedness assumption formalized by (1.0.2) implicitly yields some upper bound on the tails.

Further, we do not require that the entries of A have mean zero. Therefore, adding to A any fixed matrix of the operator norm $O(\sqrt{n})$ does not affect our results.

Extending Theorem 1.0.4 to general, possibly discrete distributions, is a challenging task. We are able to do this for matrices with identically distributed entries and under the mild assumption that the distributions of entries are not too concentrated near a single number.

Assumption 1.0.5 (General distribution of entries). We assume that the real parts of the matrix entries are distributed identically with a random variable ξ that satisfies

$$\sup_{u \in \mathbb{R}} \mathbb{P}\left\{|\xi - u| \leqslant 1\right\} \leqslant 1 - p, \quad \mathbb{P}\left\{|\xi| > K\right\} \leqslant p/2 \quad \text{for some } K, p > 0. \tag{1.0.6}$$

Assumption 1.0.5 holds for any non-constant random variable with some p, K after a proper scaling. Its meaning therefore is not to restrict the class of random variables, but to introduce parameters p and K which will be used in the formulation of Theorem 1.0.7 below.

With Assumption 1.0.3 replaced by Assumption 1.0.5, we can prove a general no-gaps delocalization result [25].

Theorem 1.0.7 (Delocalization: general distributions). *Let* A *be an* $n \times n$ *random matrix which satisfies Assumptions* 1.0.1 *and* 1.0.5. *Let* $M \geqslant 1$. *Let* $\varepsilon \geqslant 1/n$ *and* $s \geqslant c_1 \varepsilon^{-7/6} n^{-1/6} + e^{-c_2/\sqrt{\varepsilon}}$. *Then, the following event holds with probability at least*

$$1 - (Cs)^{\varepsilon n} - \mathbb{P}\,(\mathcal{B}_{A,M}^{c}).$$

Every eigenvector v *of* A *satisfies*

$$\|v_I\|_2 \geqslant (\varepsilon s)^6 \|v\|_2 \quad \textit{for all } I \subset [n],\ |I| \geqslant \varepsilon n.$$

Here $c_k = c_k(p, K, M) > 0$ *for* $k = 1, 2$ *and* $C = C(p, K, M) \geqslant 1$.

Remark 1.0.8. The assumption on s appearing in Theorem 1.0.7 forces us to consider only $\varepsilon \geqslant Cn^{-1/7}$ in contrast with Theorem 1.0.4 which yields non-trivial results as long as $\varepsilon \geqslant 8/n$. This assumption can be probably relaxed if one replaces using Berry-Esseen Theorem in the proof of Theorem 1.0.7 in [25] by a more complicated argument based on the least common denominator.

Remark 1.0.9. The proof of Theorem 1.0.7 presented in [25] can be modified to allow an extension to random matrices shifted by a constant multiple of the all ones matrix $\mathbf{1}_n$. More precisely, for a given $\mu \in \mathbb{C}$, the event described in the theorem holds with probability at least $1 - (Cs)^{\varepsilon n} - \mathbb{P}(\mathcal{B}^c_{A-\mu\mathbf{1}_n,M})$. This allows to consider random matrices with identically distributed entries having a non-zero expectation, in particular, with Bernoulli(p) entries for p being a constant. Moreover, tracing the proof appearing in [25], one can see that the constants c_k and C depend polynomially on p, which allows to extend no-gaps delocalization to matrices with i.i.d. Bernoulli entries for $p = \Omega(n^{-c'})$ for some absolute constant $c' \in (0,1)$.

Remark 1.0.10. The no-gaps delocalization phenomenon holds also for any unit vector which is a linear combination of eigenvectors whose eigenvalues are not too far apart, see Remark 2.1.8 for the details.

Acknowledgement These notes are based on the mini-courses given at Hebrew University of Jerusalem and at PCMI Summer School on Random Matrices. The author is grateful to Alex Samorodnitsky, Alexey Borodin, Ivan Corwin, and Alice Guionnet for their hospitality and an opportunity to present this material. The author is grateful to Feng Wei for running problem sessions at PCMI which were an integral part of the mini-course. He would also like to thank Anirban Basak, Konstantin Tikhomirov, and Feng Wei for careful reading of the manuscript and several suggestions on improving the presentation.

2. Reduction of no-gaps delocalization to invertibility of submatrices

2.1. From no-gaps delocalization to the smallest singular value bounds The first step in proving no-gaps delocalization is pretty straightforward. Let us consider the toy case when there exists a unit eigenvector u of the matrix A with $u_j = 0$ for all $j \in J$, where J is some subset of $[n]$. If we denote the corresponding eigenvalue by λ and the submatrix of A with columns from the set J^c by A_{J^c}, then we have that $(A_{J^c} - \lambda I_{J^c})u_{J^c} = 0$ so the kernel of $A_{J^c} - \lambda I_{J^c}$ is non-trivial. Here, $A_{J^c} - \lambda I_{J^c}$ is a "tall" matrix with the number of rows larger than the number of columns. A linear operator defined by a tall rectangular random matrix with sufficiently many independent entries is an injection with high probability. This means that the event that the probability of this "toy" case should be small. This idea is not directly applicable since the random eigenvalue λ depends on all entries of the matrix A, but this obstacle is easy to circumvent by discretizing the set of plausible values of λ and considering a deterministic λ from this discretization. If the probability that $A_{J^c} - \lambda I_{J^c}$ is close to a singular matrix is small for any fixed

λ, we can use the union bound over the dicretization along with approximation to show that, with high probability, the matrix $A_{J^c} - \lambda I_{J^c}$ has a trivial kernel for all λ from this plausible set simultaneously. This would imply the same statement for a random λ allowing us to avoid using hard to obtain information about its distribution except for a very rough bound defining the plausible set.

To implement this idea for a real setup, recall the definition of the singular values of a matrix. Let B be a real or complex $N \times n$ matrix, $N \geqslant n$. The singular values of B are defined as the square roots of eigenvalues of B^*B arranged in the decreasing order:

$$s_1(B) \geqslant s_2(B) \geqslant \ldots \geqslant s_n(B) \geqslant 0.$$

If B is real, and we consider this matrix as a linear operator $B : \mathbb{R}^n \to \mathbb{R}^N$, then the image of the Euclidean unit ball will be an ellipsoid whose semi-axes have lengths $s_1(B), \ldots, s_n(B)$. The extreme singular values have also an analytic meaning with

$$s_1(B) = \max_{x \in S^{n-1}} \|Bx\|_2 \quad \text{and}$$

$$s_n(B) = \min_{x \in S^{n-1}} \|Bx\|_2,$$

so $s_1(B) = \|B\|$, the operator norm of B, and $s_n(B)$ is the distance from B to the set of matrices of a rank smaller than n in the operator norm. Throughout these notes, we will also denote the smallest singular value by $s_{\min}(B)$. We will also abbreviate $A - \lambda I$ to $A - \lambda$.

Let us introduce the event that one of the eigenvectors is localized. Define the *localization event* by

$$\mathrm{Loc}(A, \varepsilon, \delta) := \left\{ \exists \text{ eigenvector } v \in S_{\mathbb{C}}^{n-1}, \exists I \subset [n], |I| = \varepsilon n : \|v_I\|_2 < \delta \right\}.$$

Since we assume in Theorem 1.0.4 that the boundedness event $\mathcal{B}_{A,M}$ holds with probability at least $1/2$, the conclusion of that theorem can be stated as follows:

$$\mathbb{P}\left\{ \mathrm{Loc}(A, \varepsilon, (\varepsilon s)^6) \text{ and } \mathcal{B}_{A,M} \right\} \leqslant (cs)^{\varepsilon n}. \tag{2.1.1}$$

The following proposition reduces proving a delocalization result like (2.1.1) to an invertibility bound.

Proposition 2.1.2 (Reduction of delocalization to invertibility). *Let A be an $n \times n$ random matrix with arbitrary distribution. Let $M \geqslant 1$ and $\varepsilon, p_0, \delta \in (0, 1/2)$. Assume that for any number $\lambda_0 \in \mathbb{C}$, $|\lambda_0| \leqslant M\sqrt{n}$, and for any set $I \subset [n]$, $|I| = \varepsilon n$, we have*

$$\mathbb{P}\left\{ s_{\min}\big((A - \lambda_0)_{I^c}\big) \leqslant 8\delta M\sqrt{n} \text{ and } \mathcal{B}_{A,M} \right\} \leqslant p_0. \tag{2.1.3}$$

Then

$$\mathbb{P}\left\{ \mathrm{Loc}(A, \varepsilon, \delta) \text{ and } \mathcal{B}_{A,M} \right\} \leqslant 5\delta^{-2}(e/\varepsilon)^{\varepsilon n} p_0.$$

Proof. Assume both the localization event and the boundedness event $\mathcal{B}_{A,M}$ occur. Use the definition of $\mathrm{Loc}(A, \varepsilon, \delta)$ to choose a localized eigenvalue-eigenvector pair (λ, v) and an index subset I. Decomposing the eigenvector as

$$v = v_I + v_{I^c}$$

and multiplying it by $A - \lambda$, we obtain

$$0 = (A-\lambda)v = (A-\lambda)_I v_I + (A-\lambda)_{I^c} v_{I^c}. \tag{2.1.4}$$

By triangle inequality, this yields

$$\|(A-\lambda)_{I^c} v_{I^c}\|_2 = \|(A-\lambda)_I v_I\|_2 \leqslant (\|A\| + |\lambda|)\|v_I\|_2.$$

By the localization event $\mathrm{Loc}(A, \varepsilon, \delta)$, we have $\|v_I\|_2 \leqslant \delta$. By the boundedness event $\mathcal{B}_{A,M}$ and since λ is an eigenvalue of A, we have $|\lambda| \leqslant \|A\| \leqslant M\sqrt{n}$. Therefore

$$\|(A-\lambda)_{I^c} v_{I^c}\|_2 \leqslant 2M\delta\sqrt{n}. \tag{2.1.5}$$

This happens for some λ in the disc $\{z \in \mathbb{C} : |z| \leqslant M\sqrt{n}\}$. We will now run a covering argument in order to fix λ. Let $\mathcal{N}$ be a $(2M\delta\sqrt{n})$-net of that disc. One can construct $\mathcal{N}$ so that

$$|\mathcal{N}| \leqslant \frac{5}{\delta^2}.$$

Choose $\lambda_0 \in \mathcal{N}$ so that $|\lambda_0 - \lambda| \leqslant 2M\delta\sqrt{n}$. By (2.1.5), we have

$$\|(A-\lambda_0)_{I^c} v_{I^c}\|_2 \leqslant 4M\delta\sqrt{n}. \tag{2.1.6}$$

Since $\|v_I\|_2 \leqslant \delta \leqslant 1/2$, we have $\|v_{I^c}\|_2 \geqslant \|v\|_2 - \|v_I\|_2 \geqslant 1/2$. Therefore, (2.1.6) implies that

$$s_{\min}((A-\lambda_0)_{I^c}) \leqslant 8M\delta\sqrt{n}. \tag{2.1.7}$$

Summarizing, we have shown that the events $\mathrm{Loc}(A, \varepsilon, \delta)$ and $\mathcal{B}_{A,M}$ imply the existence of a subset $I \subset [n]$, $|I| = \varepsilon n$, and a number $\lambda_0 \in \mathcal{N}$, such that (2.1.7) holds. Furthermore, for fixed I and λ_0, assumption (2.1.3) states that (2.1.7) together with $\mathcal{B}_{A,M}$ hold with probability at most p_0. So by the union bound we conclude that

$$\mathbb{P}\left\{\mathrm{Loc}(A, \varepsilon, \delta) \text{ and } \mathcal{B}_{A,M}\right\} \leqslant \binom{n}{\varepsilon n} \cdot |\mathcal{N}| \cdot p_0 \leqslant \left(\frac{e}{\varepsilon}\right)^{\varepsilon n} \cdot \frac{5}{\delta^2} \cdot p_0.$$

This completes the proof of the proposition. □

Remark 2.1.8. A simple analysis of the proof of Proposition 2.1.2 shows that it holds not only for eigenvectors of the matrix A, but for its *approximate* eigenvectors as well. Namely, instead of the event $\mathrm{Loc}(A, \varepsilon, \delta)$ one can consider the following event

$$\widetilde{\mathrm{Loc}}(A, \varepsilon, \delta) := \Big\{\exists v \in S_{\mathbb{C}}^{n-1},\ \exists \lambda \in \mathbb{C}\ |\lambda| \leqslant M\sqrt{n}\ \exists I \subset [n],\ |I| = \varepsilon n : \\ \|(A - \lambda I)v\|_2 \leqslant M\delta\sqrt{n} \text{ and } \|v_I\|_2 < \delta\Big\}.$$

This event obeys the same conclusion as $\mathrm{Loc}(A, \varepsilon, \delta)$:

$$\mathbb{P}\left\{\widetilde{\mathrm{Loc}}(A, \varepsilon, \delta) \text{ and } \mathcal{B}_{A,M}\right\} \leqslant 5\delta^{-2}(e/\varepsilon)^{\varepsilon n} p_0.$$

Indeed, equation (2.1.4) is replaced by

$$w = (A-\lambda)v = (A-\lambda)_I v_I + (A-\lambda)_{I^c} v_{I^c},$$

where w is a vector of a norm not exceeding $M\delta\sqrt{n}$. This in turn results in replacing $2M\delta\sqrt{n}$ by $3M\delta\sqrt{n}$ in (2.1.5) and $3M\delta\sqrt{n}$ by $4M\delta\sqrt{n}$ in (2.1.6). This observation shows, in particular, that the no-gaps delocalization phenomenon holds for any unit vector which is a linear combination of eigenvectors whose eigenvalues are at most $M\delta\sqrt{n}$ apart.

2.2. The ε-net argument. We have reduced the proof of the no-gaps delocalization to establishing quantitative invertibility of a matrix whose number of rows is larger than number of columns. This problem has been extensively studied, so before embarking on the real proof, let us check whether we can apply an elementary bound based on the discretization of the sphere. Assume for simplicity that all entries of the matrix A are real and independent, and the entries are centered and of the unit variance. We will formulate the result in a bigger generality than we need at this moment.

Lemma 2.2.1. *Let $M > 0$ and let A be an $m \times n$ matrix with real independent entries $A_{i,j}$ satisfying*

$$\mathbb{E}a_{i,j} = 0, \quad \mathbb{E}a_{i,j}^2 = 1, \quad \textit{and} \quad \mathbb{E}a_{i,j}^4 \leqslant C.$$

Let E be a linear subspace of $\mathbb{R}^n$ of dimension

$$k = \mathit{dim}(E) < c\frac{m}{\log(2+n/m)}.$$

Then with probability at least $1-\exp(-c'm) - \mathbb{P}\left\{\mathcal{B}_{A,M}^c\right\}$, all vectors $x \in E$ satisfy

$$\|Ax\|_2 \geqslant c\sqrt{m}.$$

The parameters c, c' here may depend on C.

Recall the definition of the ε-net. Let (X, d) be a metric space, and let $\varepsilon > 0$. A set $\mathcal{N} \subset X$ is called an ε-net for a set $V \subset X$ if for any $x \in V$, there exists $y \in \mathcal{N}$ with $d(x,y) < \varepsilon$. We will consider ε-nets for various subsets of $\mathbb{R}^n$ in the Euclidean metric below. These nets are useful in discretization of continuous structures. For instance, it is easy to show that if $\mathcal{N} \subset S^{n-1}$ is a $(1/2)$-net in the unit sphere S^{n-1}, then the operator norm of an $m \times n$ matrix A and the maximum of the Euclidean norm of Ax over the net are commensurate:

$$\max_{x\in\mathcal{N}} \|Ax\|_2 \leqslant \|A\| \leqslant 2\max_{x\in\mathcal{N}} \|Ax\|_2 .$$

The proof of Lemma 2.2.1 is based on the ε-net argument. To implement it, we need an elementary lemma.

Lemma 2.2.2. *Let $\varepsilon \in (0,1]$ and let $V \subset S_{\mathbb{R}}^{k-1}$ be any set. The set V contains an ε-net of cardinality at most $(1+2/\varepsilon)^k$.*

The proof of Lemma 2.2.2 relies on a simple volume comparison. Notice that the balls of radius $\varepsilon/2$ centered at the points of the net are disjoint. On the other hand, the union of these balls is contained in the ball of radius $(1+\varepsilon/2)$ centered at 0. We leave the details to a reader.

Proof of Lemma 2.2.1. Let $\varepsilon > 0$. It is enough to prove the norm bound for all vectors of $V := E \cap S^{n-1}$. Since the dimension of E is k, this set admits an ε-net $\mathcal{N}$ of cardinality $(1+2/\varepsilon)^k$. Let $y \in \mathcal{N}$, and let $z_j = (Ay)_j$ be the j-th coordinate of the vector Ay.

The Paley–Zygmund inequality asserts that a random variable $Y \geqslant 0$ satisfies

$$\mathbb{P}\{Y > t\} \geqslant \frac{(\mathbb{E}Y - t)^2}{\mathbb{E}Y^2} \quad \text{for any } t \in (0, \mathbb{E}Y).$$

If $Y = z_j^2$, the assumptions on $a_{i,j}$ imply $\mathbb{E}Y = 1$ and $\mathbb{E}Y^2 \leqslant C'$. Applying the Paley–Zygmund inequality with $t = 1/2$, we conclude that $\mathbb{P}\{|z_j| \geqslant 1/2\} \geqslant c$. Using Chernoff's inequality, we derive that

$$\begin{aligned}\mathbb{P}\left\{\|Ay\|_2 \leqslant \frac{1}{4}\sqrt{m}\right\} &= \mathbb{P}\left\{\sum_{j=1}^{m} |z_j|^2 \leqslant \frac{1}{16}m\right\} \\ &\leqslant \mathbb{P}\left\{|\{j : |z_j| \leqslant 1/2\}| \geqslant \frac{m}{2}\right\} \leqslant \exp(-c_2 m).\end{aligned} \tag{2.2.3}$$

In combination with the union bound, this yields

$$\mathbb{P}\left\{\exists y \in \mathcal{N} \ \|Ay\|_2 \leqslant (1/4)\sqrt{m}\right\} \leqslant (1+2/\varepsilon)^k \exp(-c_2 m). \tag{2.2.4}$$

Let Ω be the event that $\|Ay\|_2 > (1/4)\sqrt{m}$ for all $y \in \mathcal{N}$ intersected with $\mathcal{B}_{A,M}$. Assuming that Ω occurs, we will show that the matrix is invertible on the whole V. To this end, take any $x \in V$, and find $y \in \mathcal{N}$ such that $\|x - y\|_2 < \varepsilon$. Then

$$\|Ax\|_2 \geqslant \|Ay\|_2 - \|A\| \cdot \|x-y\|_2 \geqslant \frac{1}{4}\sqrt{m} - M\sqrt{n}\cdot\varepsilon \geqslant \frac{1}{8}\sqrt{m}$$

if we set

$$\varepsilon = \frac{1}{8M} \cdot \sqrt{\frac{m}{n}} \wedge 1.$$

It remains to estimate the probability that Ω does not occur. By (2.2.4),

$$\mathbb{P}\{\Omega^c\} \leqslant \exp(k\log(1+2/\varepsilon) - c_2 m) + \mathbb{P}\{\mathcal{B}^c_{A,M}\} \leqslant \exp\left(-\frac{c_2}{2}m\right) + \mathbb{P}\{\mathcal{B}^c_{A,M}\}$$

if we choose

$$k \leqslant c\frac{m}{\log(2+n/m)}.$$

□

Comparing the bound (2.1.3) needed to establish delocalization with the smallest singular value estimate of lemma 2.2.1, we see several obstacles preventing the direct use of the ε-net argument.

Lack of independence As we recall from Assumption 1.0.1, we are looking for ways to control symmetric and non-symmetric matrices simultaneously. This forces us to consider random matrices with dependent entries making Chernoff's inequality inapplicable.

Small exceptional probability required Lemma 2.2.1 provides the smallest singular value bound for rectangular matrices whose number of rows is significantly greater than the number of columns. If we are to apply it in combination with Proposition 2.1.2, we would have to assume in addition that $\varepsilon > 1 - \varepsilon_0$ for some small $\varepsilon_0 < 1$. Considering smaller values of ε would require a small ball probability bound better than (2.2.3) that we used in the proof. We will show that such bound is possible to obtain in the case when the entries have a bounded density. In the general case, however, such bound is unavailable. Indeed, if the entries of the matrix may take the value 0 with a positive probability, then $\mathbb{P}(Ae_1 = 0) = \exp(-cm)$, which shows that the bound (2.2.3) is, in general, optimal. Overcoming this problem for a general distribution would require a delicate stratification of the unit sphere according to the number-theoretic structure of the coordinates of a vector governing the small ball probability bound.

A closer look at Proposition 2.1.2 demonstrates that the demands for a small ball probability bound are even higher. We need that the delocalization result, and thus the invertibility bound (2.1.6), hold uniformly over all index subsets I of size εn. Since there are $\binom{n}{\varepsilon n} \sim \varepsilon^{-\varepsilon n}$ such sets, we would need the probability in (2.1.3) to be at most $\varepsilon^{\varepsilon n}$. Such small exceptional probabilities (smaller than $e^{-\varepsilon n}$) are hard to achieve in the general case.

Complex entries Even if the original matrix is real, its eigenvalues may be complex. This observation forces us to work with complex random matrices. Extending the known invertibility results to complex matrices poses two additional challenges. First, in order to preserve the matrix-vector multiplication, we replace a complex $n \times m$ random matrix $B = R + iT$ by the real $2m \times 2n$ random matrix $\left[\begin{smallmatrix} R & -T \\ T & R \end{smallmatrix}\right]$. The real and imaginary parts R and T each appear twice in this matrix, which causes extra dependencies of the entries. Besides that, we encounter a major problem while trying to apply the ε-net argument to prove the smallest singular value bound. Indeed, since we have to consider a real $2m \times 2n$ matrix, we will have to construct a net in a subset of the real sphere of dimension $2n$. The size of such net is exponential in the dimension. On the other hand, the number of independent rows of R is only m, so the small ball probability will be exponential in terms of m. If $m < 2n$, the union bound would not be applicable.

Each of these obstacles requires a set of rather advanced tools to deal with in general case, i.e. under Assumption 1.0.5. Fortunately, under Assumption 1.0.3, these problems can be addressed in a much easier way allowing a short and rather non-technical proof. For this reason, we are going to concentrate on the continuous density case below.

3. Small ball probability for the projections of random vectors

3.1. Density of a marginal of a random vector. The proof of the no-gaps delocalization theorem requires a result on the distribution of the marginals of a random vector which is of an independent interest. To simplify the presentation, we will consider a vector with independent coordinates having a bounded density. Let $X = (X_1, \dots, X_n)$ be independent real valued random variables with densities $f_{X_1}, \dots, f_{X_n}$ satisfying

$$f_{X_j}(t) \leqslant K \quad \text{for all } j \in [n],\ t \in \mathbb{R}.$$

The independence implies that the density of the vector is the product of the densities of the coordinates, and so, $f_X(x) \leqslant K^n$ for all $x \in \mathbb{R}^n$. Obviously, we can extend the previous observation to the coordinate projections of X showing that $f_{P_J X}(y) \leqslant K^{|J|}$ for any set $J \subset [n]$ and any $y \in \mathbb{R}^J$ with P_J standing for the coordinate projection of $\mathbb{R}^n$ to $\mathbb{R}^J$. It seems plausible that the same property should hold for the densities of all orthogonal projections to subspaces $E \subset \mathbb{R}^n$ with the dimension of E playing the role of $|J|$. Yet, a simple example shows that this statement fails even in dimension 2. Let X_1, X_2 be random variables uniformly distributed on the interval $[-1/2, 1/2]$, and consider the projection on the subspace $E \subset \mathbb{R}^2$ spanned by the vector $(1, 1)$. Then $Y = P_E X$ is the normalized sum of the coordinates of X:

$$P_Y = \frac{\sqrt{2}}{2}(X_1 + X_2).$$

A direct calculation shows that $f_Y(0) = \sqrt{2} > 1$. A delicate result of Ball [2] shows that this is the worst case for the uniform distribution. More precisely, consider a vector $X \in \mathbb{R}^n$ with i.i.d. coordinates uniformly distributed in the interval $[-1/2, 1/2]$. Then the projection of X onto any one-dimensional subspace $E = \operatorname{span}(a)$ with $a = (a_1, \dots, a_n) \in S^{n-1}$ is a weighted linear combination of the coordinates: $P_E(X) = \sum_{j=1}^n a_j X_j$. The theorem of Ball asserts that the density of such linear combination does not exceed $\sqrt{2}$ making $a = (\sqrt{2}/2, \sqrt{2}/2, 0, \dots, 0)$ the worst sequence of weights. This result can be combined with a theorem of Rogozin claiming that the density of a linear combination of independent random variables increases the most if these variables are uniformly distributed. This shows that if the coordinates of X are independent absolutely continuous random variables having densities uniformly bounded by K, then the density of $Y = \sum_{j=1}^n a_j X_j$ does not exceed $\sqrt{2}K$ for any $a = (a_1, \dots, a_n) \in S^{n-1}$.

Instead of discussing the proofs of the theorems of Ball and Rogozin, we will present here a simpler argument due to Ball and Nazarov [4] showing that the density of Y is bounded by CK for some unspecified absolute constant C. Moreover, we will show that this fact allows a multidimensional extension which we formulate in the following theorem [23].

Theorem 3.1.1 (Densities of projections). *Let $X = (X_1, \dots, X_n)$ where X_i are real-valued independent random variables. Assume that the densities of X_i are bounded by K almost everywhere. Let P be the orthogonal projection in $\mathbb{R}^n$ onto a d-dimensional subspace. Then the density of the random vector PX is bounded by $(CK)^d$ almost everywhere.*

To avoid ambiguity, let us mention that we consider the density in the range $P\mathbb{R}^n$, as the density in $\mathbb{R}^n$ does not exist.

This theorem shows that the density bound K^d for coordinate projections holds also for general ones if we include a multiplicative factor depending only on the dimension. Recently, Livshyts et al. [18] proved a multidimensional version of Rogozin's theorem. Combining it with the multidimensional version of Ball's theorem [3], one can show that the optimal value of the constant C is $\sqrt{2}$ as in the one-dimensional case.

Proof. We will start the proof from the one-dimensional case. The proof in this case is a nice illustration of the power of characteristic functions approach in deriving the small ball and density estimates. As before, we restate the one-dimensional version of the theorem as a statement about the density of a linear combination.

Step 1. Linear combination of independent random variables.

Fix $X_1, \dots, X_n$ be real-valued independent random variables with densities bounded by K almost everywhere and $\alpha_1, \dots, a_n$ real numbers with

$$\sum_{j=1}^{n} a_j^2 = 1.$$

Then the density of $\sum_{j=1}^n a_j X_j$ is bounded by CK almost everywhere.

We begin with a few easy reductions. By replacing X_j with KX_j we can assume that $K = 1$. By replacing X_j with $-X_j$ when necessary we can assume that all $a_j \geqslant 0$. We can further assume that $a_j > 0$ by dropping all zero terms from the sum. If there exists j_0 with $a_{j_0} > 1/2$, then the conclusion follows by conditioning on all X_j except X_{j_0}. Thus we can assume that

$$0 < a_j < \frac{1}{2} \quad \text{for all j.}$$

Finally, by translating X_j if necessary we reduce the problem to bounding the density of $S = \sum_j a_j X_j$ at the origin.

After these reductions, we proceed to bounding $f_S(0)$ in terms of the characteristic function

$$\phi_S(t) = \mathbb{E}e^{itS}.$$

We intend to use the Fourier inversion formula

$$f_S(0) = \frac{1}{2\pi} \int_{\mathbb{R}} \phi_S(x)\, dx.$$

This formula requires the assumption that $\phi_S \in L_1(\mathbb{R})$, while we only know that $\|\phi_S\|_\infty \leqslant 1$. This, however, is not a problem. We can add an independent $N(0, \sigma^2)$ random variable to each coordinate of X. In terms of the characteristic functions, this amounts to multiplying each $\phi_{X_j} \in L_\infty(\mathbb{R})$ by a scaled Gaussian density making it an L_1-function. The bound on the density we are going to obtain will not depend on σ which would allow taking $\sigma \to 0$.

By independence of the coordinates of X, $\phi_S(x) = \prod_j \phi_{X_j}(a_j t)$. Combining it with the Fourier inversion formula, we obtain

$$(3.1.2)\qquad f_S(0) = \frac{1}{2\pi}\int_{\mathbb{R}} \prod_j \phi_{X_j}(a_j x)\,dx \leqslant \frac{1}{2\pi}\prod_j \left(\int_{\mathbb{R}} |\phi_{X_j}(a_j x)|^{1/a_j^2}\,dx\right)^{a_j^2},$$

where we used Holder's inequality with exponents $1/a_j^2$ whose reciprocals sum up to 1.

We will estimate each integral appearing in the right hand side of (3.1.2) separately. Denote by λ the Lebesgue measure on $\mathbb{R}$. Using the Fubini theorem, we can rewrite each integral as

$$(3.1.3)\qquad \frac{1}{a_j}\cdot\int_{\mathbb{R}} |\phi_{X_j}(x)|^{1/a_j^2}\,dx = \int_0^1 \frac{1}{a_j^3}\cdot t^{1/a_j^2-1}\lambda\{x : |\phi_{X_j}(x)| > t\}\,dt.$$

To estimate the last integral, we need a bound on the measure of points where the characteristic function is large. Such bound is provided in the lemma below.

Lemma 3.1.4 (Decay of a characteristic function). *Let X be a random variable whose density is bounded by 1. Then the characteristic function of X satisfies*

$$\lambda\{x : |\phi_X(x)| > t\} \leqslant \begin{cases} \frac{2\pi}{t^2}, & t \in (0, 3/4) \\ C\sqrt{1-t^2}, & t \in [3/4, 1]. \end{cases}$$

The value 3/4 in Lemma 3.1.4 was chosen arbitrarily. It can be replaced by any other number $t_0 \in (0,1)$ at the price of changing the constant C.

Let us postpone the proof of the lemma for a moment and finish the proof of the one-dimensional case of Theorem 3.1.1. Fix $j \in [n]$ and denote for shortness $p = 1/a_j^2 \geqslant 4$. Combining Lemma 3.1.4 and (3.1.3), we obtain

$$\begin{aligned} &\frac{1}{a_j}\cdot\int_{\mathbb{R}} |\phi_{X_j}(x)|^{1/a_j^2}\,dx \\ &\leqslant p^{3/2}\cdot\left(\int_0^{3/4} t^{p-1}\cdot\frac{2\pi}{t^2}\,dt + \int_{3/4}^1 t^{p-1}\cdot C\sqrt{1-t^2}\,dt\right) \\ &\leqslant p^{3/2}\cdot\left(\frac{2\pi}{p-2}\cdot(3/4)^{p-2} + C\int_0^{\sqrt{7}/4} (1-s^2)^{(p-2)/2}\cdot s^2\,ds\right), \end{aligned}$$

where we used the substitution $s^2 = 1 - t^2$ in the second term. The function

$$u(p) = p^{3/2}\cdot\frac{2\pi}{p-2}\cdot(3/4)^{p-2}$$

is uniformly bounded for $p \in [4, \infty)$. To estimate the second term, we can use the inequality $1 - s^2 \leqslant \exp(-s^2)$, which yields

$$p^{3/2}\int_0^{\sqrt{7}/4} (1-s^2)^{(p-2)/2}\cdot s^2\,ds \leqslant p^{3/2}\int_0^{\infty} \exp\left(-\frac{p-2}{2}s^2\right) s^2\,ds.$$

The last expression is also uniformly bounded for $p \in [4, \infty)$. This proves that

$$\frac{1}{a_j}\cdot\int_{\mathbb{R}} |\phi_{X_j}(x)|^{1/a_j^2}\,dx \leqslant C$$

for all j, where C is an absolute constant. Substituting this into (3.1.2) and using that $\sum_{j=1}^{n} a_j^2 = 1$ yields $f_s(0) \leqslant C'$ completing the proof of Step 1 modulo Lemma 3.1.4. □

Let us prove the lemma now.

Proof of Lemma 3.1.4. The first bound in the lemma follows from Markov's inequality

$$\lambda\{x : |\phi_X(x)| > t\} \leqslant \frac{\|\phi_X\|_2^2}{t^2}.$$

To estimate the L_2-norm, we apply the Plancherel identity:

$$\|\phi_X\|_2^2 = 2\pi \|f_X\|_2^2 \leqslant 2\pi \|f_X\|_\infty \cdot \|f_X\|_1 \leqslant 2\pi. \tag{3.1.5}$$

The estimate for $t \in [3/4, 1]$ will be based on a regularity argument going back to Halasz [15].

We will start with the symmetrization. Let X' denote an independent copy of X. Then

$$\begin{aligned} |\phi_X(t)|^2 &= \mathbb{E}e^{itX}\, \overline{\mathbb{E}e^{itX}} = \mathbb{E}e^{itX}\, \mathbb{E}e^{-itX'} = \mathbb{E}e^{it(X-X')} \\ &= \phi_{\tilde{X}}(t), \quad \text{where } \tilde{X} := X - X'. \end{aligned}$$

Further, by symmetry of the distribution of $\tilde{X}$, we have

$$\phi_{\tilde{X}}(t) = \mathbb{E}\cos(t\tilde{X}) = 1 - 2\mathbb{E}\sin^2\Big(\frac{1}{2}t\tilde{X}\Big) =: 1 - \psi(t).$$

Denoting $s^2 = 1 - t^2$, we see that to prove that

$$\lambda\{x : \ |\phi_X(x)| > t\} \leqslant C\sqrt{1-t^2} \quad \text{for } t \in [3/4, 1],$$

it is enough to show that

$$\lambda\{\tau : \psi(\tau) \leqslant s^2\} \leqslant Cs, \quad \text{for } 0 < s \leqslant 1/2. \tag{3.1.6}$$

Observe that (3.1.6) holds for some fixed constant value of s. This follows from the identity $|\phi_X(\tau)|^2 = 1 - \psi(\tau)$ and inequality (3.1.5):

$$\lambda\Big\{\tau : \psi(\tau) \leqslant \frac{1}{4}\Big\} = \lambda\{\tau : |\phi_X(\tau)| \geqslant \sqrt{3/4}\} \leqslant 8\pi/3 \leqslant 9. \tag{3.1.7}$$

Next, the definition of $\psi(\cdot)$ and the inequality $|\sin(mx)| \leqslant m|\sin x|$ valid for $x \in \mathbb{R}$ and $m \in \mathbb{N}$ imply that

$$\psi(mt) \leqslant m^2\psi(t), \quad t > 0,\ m \in \mathbb{N}.$$

Therefore

$$\lambda\Big\{\tau : \psi(\tau) \leqslant \frac{1}{4m^2}\Big\} \leqslant \lambda\Big\{\tau : \psi(m\tau) \leqslant \frac{1}{4}\Big\} = \frac{1}{m}\lambda\Big\{\tau : \psi(\tau) \leqslant \frac{1}{4}\Big\} \leqslant \frac{9}{m}, \tag{3.1.8}$$

where in the last step we used (3.1.7). This establishes (3.1.6) for the discrete set of values $s = \frac{1}{2m}$, $m \in \mathbb{N}$. We can extend this to arbitrary $s > 0$ in a standard way, by applying (3.1.8) for $m \in \mathbb{N}$ such that $s \in (\frac{1}{4m}, \frac{1}{2m}]$. This proves (3.1.6) and completes the proof of Lemma 3.1.4. □

We now pass to the multidimensional case. As for one dimension, our strategy will depend on whether all vectors Pe_j are small or some Pe_j are large. In the first case, we proceed with a high-dimensional version of the argument from Step 1, where Hölder's inequality will be replaced by Brascamp-Lieb's inequality. In the second case, we will remove the large vectors Pe_j one by one, using the induction over the dimension.

Step 2. Let X be a random vector and P be a projection which satisfy the assumptions of Theorem 3.1.1. Assume that

$$\|Pe_j\|_2 \leqslant 1/2 \quad \text{for all } j = 1, \dots, n.$$

Then the density of the random vector PX is bounded by $(CK)^d$ almost everywhere.

The proof will be based on Brascamp-Lieb's inequality.

Theorem 3.1.9 (Brascamp-Lieb [7], see also [3]). *Let $u_1, \dots, u_n \in \mathbb{R}^d$ be unit vectors and $c_1, \dots, c_n > 0$ be real numbers satisfying*

$$\sum_{i=1}^{n} c_j u_j u_j^\top = I_d.$$

Let $f_1, \dots, f_n : \mathbb{R} \to [0, \infty)$ be integrable functions. Then

$$\int_{\mathbb{R}^n} \prod_{j=1}^{n} f_j(\langle x, u_j \rangle)^{c_j} \, dx \leqslant \prod_{j=1}^{n} \left(\int_{\mathbb{R}} f_j(t) \, dt \right)^{c_j}.$$

A short and very elegant proof of the Brascamp-Lieb inequality based on the measure transportation ideas can be found in [5].

The singular value decomposition of P yields the existence of a $d \times n$ matrix R satisfying

$$P = R^\top R, \quad RR^\top = I_d.$$

It follows that $\|Px\|_2 = \|Rx\|_2$ for all $x \in \mathbb{R}^n$. This allows us to work with the matrix R instead of P. As before, replacing each X_j by KX_j, we may assume that $K = 1$. Finally, translating X if necessary we reduce the problem to bounding the density of RX at the origin.

As in the previous step, Fourier inversion formula associated with the Fourier transform in n dimensions yields that the density of RX at the origin can be reconstructed from its Fourier transform as

$$f_{RX}(0) = (2\pi)^{-d} \int_{\mathbb{R}^d} \phi_{RX}(x) \, dx \leqslant (2\pi)^{-d} \int_{\mathbb{R}^d} |\phi_{RX}(x)| \, dx, \tag{3.1.10}$$

where

$$\phi_{RX}(x) = \mathbb{E} \exp\left(i \langle x, RX \rangle\right) \tag{3.1.11}$$

is the characteristic function of RX. Therefore, to complete the proof, it suffices to bound the integral in the right hand side of (3.1.10) by C^d.

In order to represent $\phi_{RX}(x)$ more conveniently for application of Brascamp-Lieb inequality, we denote

$$a_j := \|Re_j\|_2, \quad u_j := \frac{Re_j}{\|Re_j\|_2}.$$

Then $R = \sum_{j=1}^{n} a_j u_j e_j^\top$, so the identity $RR^\top = I_d$ can be written as

$$\sum_{j=1}^{n} a_j^2 u_j u_j^\top = I_d. \tag{3.1.12}$$

Moreover, we have $\langle x, RX\rangle = \sum_{i=1}^{n} a_j \langle x, u_j\rangle X_j$. Substituting this into (3.1.11) and using independence, we obtain

$$\phi_{RX}(x) = \prod_{j=1}^{n} \mathbb{E}\exp\left(ia_j \langle x, u_j\rangle X_j\right).$$

Define the functions $f_1, \ldots, f_n : \mathbb{R} \to [0, \infty)$ as

$$f_j(t) := \left|\mathbb{E}\exp(ia_j t X_j)\right|^{1/a_j^2} = \left|\phi_{X_j}(a_j t)\right|^{1/a_j^2}.$$

Recalling (3.1.12), we apply Brascamp-Lieb inequality for these functions and obtain

$$\begin{aligned}\int_{\mathbb{R}^d} |\phi_{RX}(x)|\,dx &= \int_{\mathbb{R}^d} \prod_{j=1}^{n} f_j\big(\langle x, u_j\rangle\big)^{a_j^2}\,dx \\ &\leqslant \prod_{j=1}^{n}\Big(\int_{\mathbb{R}} f_j(t)\,dt\Big)^{a_j^2} = \prod_{j=1}^{n}\Big(\int_{\mathbb{R}} \left|\phi_{X_j}(a_j t)\right|^{1/a_j^2}\,dt\Big)^{a_j^2}.\end{aligned} \tag{3.1.13}$$

We arrived at the same quantity as we encountered in one-dimensional argument in (3.1.2). Following that argument, which uses the assumption that all $a_j \leqslant 1/2$, we bound the product above by

$$(2C)^{\sum_{j=1}^{n} a_j^2}.$$

Recalling that $a_j = \|Re_j\|_2$ and , we find that

$$\sum_{j=1}^{n} a_j^2 = \sum_{j=1}^{n} \|Re_j\|_2^2 = \mathrm{Tr}(RR^\top) = \mathrm{Tr}(I_d) = d.$$

Thus the right hand side of (3.1.13) is bounded by $(2C)^d$. The proof of Theorem 3.1.1 in the case where all $\|Pe_j\|_2$ are small is complete.

Step 3. Inductive argument.

We will prove Theorem 3.1.1 by induction on the rank of the projection. The case $\mathrm{rank}(P) = 1$ has been already established. We have also proved the Theorem when $\|Pe_j\|_2 < 1/2$ for all j. Assume that the theorem holds for all projections Q with $\mathrm{rank}(Q) = d - 1$ and $\|Pe_1\|_2 \geqslant 1/2$.

The density function is not a convenient tool to run the inductive argument since the density of P_X does not usually splits into a product of densities related to the individual coordinates. Let us consider the *Lévy concentration function* of a random vector which would replace the density in our argument.

Definition 3.1.14. Let $r > 0$. For a random vector $Y \in \mathbb{R}^n$, define its Lévy concentration function by

$$\mathcal{L}(Y, r) := \sup_{y \in \mathbb{R}^n} \mathbb{P}\left\{ \|Y - y\|_2 \leqslant r \right\}.$$

Note that the condition that the density of Y is bounded is equivalent to

$$\mathcal{L}(Y, r\sqrt{n}) \leqslant (Cr)^n \quad \text{for any } r > 0.$$

One direction of this equivalence follows from the integration of the density function over the ball of radius $t\sqrt{n}$ with the center at any $y \in \mathbb{R}^n$; another one from the Lebesgue differentiation theorem.

In terms of the Lévy concentration function, the statement of the theorem is equivalent to the claim that for any $y \in P\mathbb{R}^n$ and any $t > 0$,

$$\mathbb{P}\left\{ \|PX - y\|_2 \leqslant t\sqrt{d} \right\} \leqslant (Mt)^d \tag{3.1.15}$$

for some absolute constant M and with $d = \operatorname{rank}(P)$. The induction assumption then reads: for all projections Q of rank $d-1$, $z \in Q\mathbb{R}^n$, and $t > 0$, we have

$$\mathbb{P}\left\{ \|QX - z\|_2 \leqslant t\sqrt{d-1} \right\} \leqslant (Mt)^{d-1}. \tag{3.1.16}$$

Comparison of (3.1.16) and (3.1.15) immediately shows the difficulties we are facing: the change from $d-1$ to d in the left hand side of these inequalities indicates that we have to work accurately to preserve the constant M while deriving (3.1.15) from (3.1.16). This is achieved by a delicate tensorization argument. By considering an appropriate shift of X, we can assume without loss of generality that $y = 0$. Let us formulate the induction step as a separate proposition.

Proposition 3.1.17 (Removal of large Pe_i). *Let X be a random vector satisfying the assumptions of Theorem* 3.1.1 *with* $K = 1$, *and let* P *be an orthogonal projection in* $\mathbb{R}^n$ *onto a* d*-dimensional subspace. Assume that*

$$\|Pe_1\|_2 \geqslant 1/2.$$

Define Q *to be the orthogonal projection in* $\mathbb{R}^n$ *such that*

$$\ker(Q) = \operatorname{span}\{\ker(P), Pe_1\}.$$

Let $M \geqslant C_0$ *where* C_0 *is an absolute constant. If*

$$\mathbb{P}\left\{ \|QX\|_2 \leqslant t\sqrt{d-1} \right\} \leqslant (Mt)^{d-1} \quad \textit{for all } t \geqslant 0, \tag{3.1.18}$$

then

$$\mathbb{P}\left\{ \|PX\|_2 \leqslant t\sqrt{d} \right\} \leqslant (Mt)^d \quad \textit{for all } t \geqslant 0.$$

Proof. Let us record a few basic properties of Q. It is straightforward to see that

$$P - Q \text{ is the orthogonal projection onto } \operatorname{span}(Pe_1). \tag{3.1.19}$$

Then $(P-Q)e_1 = Pe_1$, since the orthogonal projection of e_1 onto $\mathrm{span}(Pe_1)$ equals Pe_1. Canceling Pe_1 on both sides, we have

$$Qe_1 = 0. \tag{3.1.20}$$

It follows from (3.1.19) that P has the form

$$Px = \Big(\sum_{j=1}^{n} a_j x_j\Big) Pe_1 + Qx \quad \text{for } x = (x_1, \dots, x_n) \in \mathbb{R}^n, \tag{3.1.21}$$

where a_j are fixed numbers (independent of x). Substituting $x = e_1$, we obtain using (3.1.20) that $Pe_1 = a_1 Pe_1 + Qe_1 = a_1 Pe_1$. Thus

$$a_1 = 1. \tag{3.1.22}$$

Furthermore, we note that

$$Qx \text{ does not depend on } x_1 \tag{3.1.23}$$

since $Qx = Q(\sum_{i=1}^{n} x_j e_j) = \sum_{i=1}^{n} x_j Qe_j$ and $Qe_1 = 0$ by (3.1.20). Finally, since Pe_1 is orthogonal to the image of Q, the two vectors in the right side of (3.1.21) are orthogonal. Thus

$$\|Px\|_2^2 = \Big(\sum_{j=1}^{n} a_j x_j\Big)^2 \|Pe_1\|_2^2 + \|Qx\|_2^2. \tag{3.1.24}$$

Now let us estimate $\|PX\|_2$ for a random vector X. We express $\|PX\|_2^2$ using (3.1.24) and (3.1.22) as

$$\|PX\|_2^2 = \Big(X_1 + \sum_{j=2}^{n} a_j X_j\Big)^2 \|Pe_1\|_2^2 + \|QX\|_2^2 =: Z_1^2 + Z_2^2.$$

Since by (3.1.23) Z_2 is determined by $X_2, \dots, X_n$ (and is independent of X_1), and $\|Pe_i\|_2 \geqslant 1/2$ by a hypothesis of the proposition, we have

$$\begin{aligned} \mathbb{P}\left\{Z_1 \leqslant t \mid Z_2\right\} &\leqslant \max_{X_2,\dots,X_n} \mathbb{P}\left\{ \Big|X_1 + \sum_{j=2}^{n} a_j X_j\Big| \leqslant t/\|Pe_1\|_2 \;\Big|\; X_2, \dots, X_n \right\} \\ &\leqslant \max_{u \in \mathbb{R}} \mathbb{P}\left\{|X_1 - u| \leqslant 2t\right\} \leqslant 2t. \end{aligned}$$

The proof of the inductive step thus reduces to a two-dimensional statement, which we formulate as a separate lemma.

Lemma 3.1.25 (Tensorization). *Let $Z_1, Z_2 \geqslant 0$ be random variables and $K_1, K_2 \geqslant 0$, $d > 1$ be real numbers. Assume that*

(1) $\mathbb{P}\left\{Z_1 \leqslant t \mid Z_2\right\} \leqslant 2t$ *almost surely in Z_2 for all $t \geqslant 0$;*
(2) $\mathbb{P}\left\{Z_2 \leqslant t\sqrt{d-1}\right\} \leqslant (Mt)^{d-1}$ *for all $t \geqslant 0$.*

for a sufficiently large absolute constant M. Then

$$\mathbb{P}\left\{\sqrt{Z_1^2 + Z_2^2} \leqslant t\sqrt{d}\right\} \leqslant (Mt)^d \quad \textit{for all } t \geqslant 0.$$

The proof of the tensorization lemma requires an accurate though straightforward calculation. We write

$$\mathbb{P}\left\{\sqrt{Z_1^2+Z_2^2}\leqslant t\sqrt{d}\right\}=\int_0^{t^2 d}\mathbb{P}\left\{Z_1\leqslant (t^2 d-x)^{1/2}\mid Z_2^2=x\right\}\,dF_2(x)$$

where $F_2(x)=\mathbb{P}\left\{Z_2^2\leqslant x\right\}$ is the cumulative distribution function of Z_2^2. Using hypothesis (1) of the lemma, we can bound the right hand side by

$$2\int_0^{t^2 d}(t^2 d-x)^{1/2}\,dF_2(x)=\int_0^{t^2 d}F_2(x)(t^2 d-x)^{-1/2}\,dx,$$

where the last equation follows by integration by parts. Hypothesis (2) of the lemma states that

$$F_2(x)\leqslant M^{d-1}\left(\frac{x}{d-1}\right)^{(d-1)/2}.$$

Substituting this into the equality above and estimating the resulting integral explicitly, we obtain

$$\mathbb{P}\left\{\sqrt{Z_1^2+Z_2^2}\leqslant t\sqrt{d}\right\}\leqslant\int_0^{t^2 d}M^{d-1}\left(\frac{x}{d-1}\right)^{(d-1)/2}(t^2 d-x)^{-1/2}\,dx$$

$$=t^d\cdot M^{d-1}\frac{d^{d/2}}{(d-1)^{(d-1)/2}}\int_0^1 y^{(d-1)/2}(1-y)^{-1/2}\,dy\leqslant t^d\cdot M^{d-1}\cdot C,$$

where the last inequality follows with an absolute constant C from the known asymptotic of the beta-function. Alternatively, notice that

$$\frac{d^{d/2}}{(d-1)^{(d-1)/2}}\leqslant\sqrt{ed},$$

and

$$\int_0^1 y^{(d-1)/2}(1-y)^{-1/2}\,dy\leqslant\int_0^{1-1/d}y^{(d-1)/2}\sqrt{d}\,dy+\int_{1-1/d}^1(1-y)^{-1/2}\,dy$$

$$\leqslant\frac{2}{\sqrt{ed}}+\frac{1}{2\sqrt{d}}.$$

This completes the proof of the lemma if we assume that $M\geqslant C$. □

3.2. Small ball probability for the image of a vector. Let us derive an application of Theorem 3.1.1 which will be important for us in the proof of the no-gaps delocalization theorem. We will prove a small ball probability estimate for the image of a fixed vector under the action of a random matrix with independent entries of bounded density.

Lemma 3.2.1 (Lower bound for a fixed vector)**.** *Let G be an $l\times m$ matrix with independent complex random entries. Assume that the real parts of the entries have uniformly bounded densities, and the imaginary parts are fixed. For each $x\in S_{\mathbb{C}}^{m-1}$ and $\theta>0$, we have*

$$\mathbb{P}\left\{\|Gx\|_2\leqslant\theta\sqrt{l}\right\}\leqslant(C_0\theta)^l.$$

To prove this lemma, let us derive the small ball probability bound for a fixed coordinate of Gx first.

Lemma 3.2.2 (Lower bound for a fixed row and vector). *Let G_j denote the j-th row of G. Then for each j, $z \in S_{\mathbb{C}}^{n-1}$, and $\theta \geqslant 0$, we have*

$$\mathbb{P}\left\{|\langle G_j, z\rangle| \leqslant \theta\right\} \leqslant C_0 K\theta. \tag{3.2.3}$$

Proof. Fix j and consider the random vector $Z = G_j$. Expressing Z and z in terms of their real and imaginary parts as

$$Z = X + iY, \quad z = x + iy,$$

we can write the inner product as

$$\langle Z, z\rangle = [\langle X, x\rangle - \langle Y, y\rangle] + i\,[\langle X, y\rangle + \langle Y, x\rangle].$$

Since z is a unit vector, either x or y has norm at least $1/2$. Assume without loss of generality that $\|x\|_2 \geqslant 1/2$. Dropping the imaginary part, we obtain

$$|\langle Z, z\rangle| \geqslant |\langle X, x\rangle - \langle Y, y\rangle|.$$

The imaginary part Y is fixed. Thus

$$\mathbb{P}\left\{|\langle Z, z\rangle| \leqslant \theta\right\} \leqslant \mathcal{L}(\langle X, x\rangle, \theta). \tag{3.2.4}$$

We can express $\langle X, x\rangle$ in terms of the coordinates of X and x as the sum

$$\langle X, x\rangle = \sum_{k=1}^{n} X_k x_k.$$

Here X_k are independent random variables with densities bounded by K. Recalling that $\sum_{k=1}^{m} x_k^2 \geqslant 1/2$, we can apply Theorem 3.1.1 for a rank one projection. It yields

$$\mathcal{L}(\langle X, x\rangle, \theta) \leqslant CK\theta. \tag{3.2.5}$$

Substituting this into (3.2.4) completes the proof of Lemma 3.2.2. □

Now we can complete the proof of Lemma 3.2.1 We can represent $\|Gx\|_2^2$ as a sum of independent non-negative random variables $\sum_{j=1}^{l} |\langle G_j, x\rangle|^2$. Each of the terms $\langle G_j, x\rangle$ satisfies (3.2.3). Then the conclusion follows from the following Tensorization Lemma applied to $V_j = |\langle G_j, x\rangle|$.

Lemma 3.2.6. *Let $V_1, \dots, V_l$ be independent non-negative random variables satisfying*

$$\mathbb{P}\left\{V_j < t\right\} \leqslant Ct$$

for any $t > 0$. Then

$$\mathbb{P}\left\{\sum_{j=1}^{l} V_j^2 < t^2 l\right\} \leqslant (ct)^l.$$

Proof. Since the random variables $V_1^2, \dots, V_l^2$ are independent as well, the Laplace transform becomes a method of choice in handling this probability. By Markov's inequality, we have

$$\mathbb{P}\left\{\sum_{j=1}^{l} V_j^2 < t^2 l\right\} = \mathbb{P}\left\{l - \frac{1}{t^2}\sum_{j=1}^{l} V_j^2 > 0\right\} \leqslant \mathbb{E}\exp\left(l - \frac{1}{t^2}\sum_{j=1}^{l} V_j^2\right)$$
$$= e^l \prod_{j=1}^{l} \mathbb{E}\exp(-V_j^2/t^2).$$

To bound the expectations in the right hand side, we use the Fubini theorem:

$$\mathbb{E}\exp(-V_j^2/t^2) = \int_0^\infty 2xe^{-x^2}\mathbb{P}\left\{V_j < tx\right\}\, dx \leqslant Ct,$$

where the last inequality follows from the assumption on the small ball probability of V_j. Combining the previous two inequalities completes the proof. □

4. No-gaps delocalization for matrices with absolutely continuous entries.

In this section, we prove Theorem 1.0.4. To this end, we combine all the tools we discussed above: the bound on the density of a projection of a random vector obtained in Theorem 3.1.1, the ε-net argument, and the small ball probability bound of Lemma 3.2.1.

4.1. Decomposition of the matrix Let us recall that in Proposition 2.1.2 we have reduced the claim of delocalization Theorem 1.0.4 to the following quantitative invertibility problem:

- Let A be an $n \times n$ matrix satisfying Assumptions 1.0.1 and 1.0.3. Let $\varepsilon > 0$, $t > 0$, $M > 1$, and let $\lambda \in \mathbb{C}$, $|\lambda| \leqslant M\sqrt{n}$. Let $I \subset [n]$ be a fixed set of cardinality $|I| = \varepsilon n$. Estimate
$$p_0 := \mathbb{P}\left(s_{\min}((A-\lambda)_{I^c}) < t\sqrt{n} \text{ and } \|A\| \leqslant M\sqrt{n}\right).$$

Since the set I is fixed, we can assume without loss of generality that I consists of the last εn coordinates.

Let us decompose $(A-\lambda)_{I^c}$ as follows:

$$(A-\lambda)_{I^c} = \begin{bmatrix} B \\ G \end{bmatrix}, \tag{4.1.1}$$

where B and G are rectangular matrices of respective sizes $(1-\varepsilon/2)n \times (1-\varepsilon)n$ and $(\varepsilon/2)n \times (1-\varepsilon)n$. By Assumption 1.0.1, the random matrices B and G are independent, and moreover all entries of G are independent. At the same time, the matrix B is still rectangular, and the ratio of its number of rows and columns is similar to that of the matrix $(A-\lambda)_{I^c}$. This would allow us to prove a weaker statement for the matrix B. Namely, instead of bounding the smallest singular value, which is the minimum of $\|Bx\|_2$ over all unit vectors x, we will obtain the desired lower bound for all vectors which are far away from a certain low-dimensional subspace depending on B. The independence of B and G would make it possible to condition on B fixing this subspace and apply Lemma 2.2.1 to the matrix G restricted to this subspace to ensure that the matrix $(A-\lambda)_{I^c}$ is well invertible on this space as well.

Following this road map, we are going to show that either $\|Bx\|_2$ or $\|Gx\|_2$ is nicely bounded below for every vector $x \in S_{\mathbb{C}}^{n-1}$. To control B, we use the second negative moment identity to bound the Hilbert-Schmidt norm of the pseudo-inverse of B. We deduce from it that most singular values of B are not too small – namely, all but $0.01\varepsilon n$ singular values are bounded below by $\Omega(\sqrt{\varepsilon n})$. It follows that $\|Bx\|_2$ is nicely bounded below when x restricted to a subspace of codimension $0.01\varepsilon n$. (This subspace is formed by the corresponding singular vectors.) Next, we condition on B and we use G to control the remaining $0.01\varepsilon n$ dimensions. Therefore, either $\|Bx\|_2$ or $\|Gx\|_2$ is nicely bounded below on the entire space, and thus $\|(A-\lambda)_{I^c}x\|_2$ is nicely bounded below on the entire space as well.

We will now pass to the implementation of this plan. To simplify the notation, *assume that the maximal density of the entries is bounded by 1.*
The general case can be reduced to this by scaling the entries.

4.2. The negative second moment identity Let $k \geqslant m$. The Hilbert-Schmidt norm of a $k \times m$ matrix V is just the Euclidean norm of the km-dimensional vector consisting of its entries. Like the operator norm, the Hilbert-Schmidt norm is invariant under unitary or orthogonal transformations of the matrix V. This allows to rewrite it in two ways:

$$\|V\|_{HS}^2 = \sum_{j=1}^{m} \|V_j\|_2^2 = \sum_{j=1}^{m} s_j(V_j)^2,$$

where $V_1, \ldots, V_m$ are the columns of V, and $s_1(V) \geqslant s_2(V) \geqslant \ldots \geqslant s_m(V) \geqslant 0$ are its singular values. Applying this observation to the inverse of the linear operator defined by V considered as an operator from $V\mathbb{C}^m$ to $\mathbb{C}^m$, we obtain *the negative second moment identity*, see [26]:

$$\sum_{j=1}^{m} s_j(B)^{-2} = \sum_{i=1}^{m} \operatorname{dist}(B_j, H_j)^{-2}.$$

Here B_j denote the columns of B, and $H_j = \operatorname{span}(B_l)_{l \neq j}$.

Returning to the matrix B, denote for shortness $m = (1-\varepsilon)n$ and $\varepsilon' = \frac{\varepsilon}{2(1-\varepsilon)}$. In this notation, B is a $(1+\varepsilon')m \times m$ matrix. To bound the sum above, we have to establish a lower bound on the distance between the random vector $B_j \in \mathbb{C}^{(1+\varepsilon')m}$ and random subspace $H_j \subseteq \mathbb{C}^{(1+\varepsilon')m}$ of complex dimension $m-1$.

Enforcing independence of vectors and subspaces Let us fix j. If all entries of B are independent, then B_j and H_j are independent. However, Assumption 1.0.1 leaves a possibility for B_j to be correlated with j-th row of B. This means that B_j and H_j may be dependent, which would complicate the distance computation.

There is a simple way to remove the dependence by projecting out the j-th coordinate. Namely, let $B'_j \in \mathbb{C}^{(1+\varepsilon')m-1}$ denote the vector B_j with j-th coordinate removed, and let $H'_j = \operatorname{span}(B'_k)_{k \neq j}$. We note the two key facts. First, B'_j and H'_j are independent by Assumption 1.0.1. Second,

$$\operatorname{dist}(B_j, H_j) \geqslant \operatorname{dist}(B'_j, H'_j), \tag{4.2.1}$$

since the distance between two vectors can only decrease after removing a coordinate.

Summarizing, we have

$$\sum_{j=1}^{m} s_j(B)^{-2} \leqslant \sum_{j=1}^{m} \operatorname{dist}(B'_j, H'_j)^{-2}. \tag{4.2.2}$$

We are looking for a lower bound for the distances $\operatorname{dist}(B'_j, H'_j)$. It is convenient to represent them via the orthogonal projection of B'_j onto $(H'_j)^{\perp}$:

$$\operatorname{dist}(B'_j, H'_j) = \|P_{E_j} B'_j\|_2, \quad \text{where} \quad E_j = (H'_j)^{\perp}. \tag{4.2.3}$$

Recall that $B'_j \in \mathbb{C}^{(1+\varepsilon')m-1}$ is a random vector with independent entries whose real parts have densities bounded by 1 (by Assumptions 1.0.1 and 1.0.3); and H'_j is an independent subspace of $\mathbb{C}^{(1+\varepsilon')m-1}$ of complex dimension $m-1$. This puts us on a familiar ground as we have already proved Theorem 3.1.1. Now, the main strength of this result becomes clear. The bound of Theorem 3.1.1 is uniform over the possible subspaces E_j meaning that we do not need any information about the specific position of this subspace in $\mathbb{C}^{(1+\varepsilon')m-1}$. This is a major source of simplifications in the proof of Theorem 1.0.4 compare to Theorem 1.0.7. Under Assumption 1.0.5, a bound on the small ball probability for $\|P_{E_j} B'_j\|_2$ depends on the arithmetic structure of the vectors contained in the space E_j. Identifying subspaces of $\mathbb{C}^{(1+\varepsilon')m-1}$ containing vectors having exceptional arithmetic structure and showing that, with high probability, the space E_j avoids such positions, takes a lot of effort. Fortunately, under Assumption 1.0.3, this problem does not arise thanks to the uniformity mentioned above.

Transferring the problem from $\mathbb{C}$ to $\mathbb{R}$ If the real and the imaginary part of each entry of A are random variables of bounded density, one can apply Theorem 3.1.1 directly. However, this case does not cover many matrices satisfying Assumption 1.0.1, most importantly, the matrices with real entries and complex spectrum.

The general case, when only the real parts of the vector $B'_j \in \mathbb{C}^{(1+\varepsilon')m-1}$ are random, requires an additional symmetrization step. Indeed, if we transfer the problem from the complex vector space to a real one of the double dimension, only a half of the coordinates will be random. Such vector would not be absolutely continuous, so we cannot operate in terms of the densities. As in the previous section, the *Lévy concentration function* of a random vector would replace the density in our argument.

Let us formally transfer the problem from the complex to the real field. To this end, we define the operation $z \mapsto \operatorname{Real}(z)$ that makes complex vectors real in the obvious way: for $z = x + iy \in \mathbb{C}^N$, set $\text{Real}(z) := \left[\begin{smallmatrix} x \\ y \end{smallmatrix}\right] \in \mathbb{R}^{2N}$. Similarly, to make a complex subspace $E \subset \mathbb{C}^N$ real, we set $\text{Real}(E) := \{\text{Real}(z) : z \in E\} \subset \mathbb{R}^{2N}$. Note that this operation doubles the dimension of E.

We begin by recording two properties of this operation that follow straight from this definition.

Lemma 4.2.4. *(Elementary properties of operation $x \mapsto \text{Real}(x)$)*

(1) *For a complex subspace E and a vector z, one has*

$$\text{Real}(P_E z) = P_{\text{Real}(E)} \text{Real}(z).$$

(2) *For a complex-valued random vector X and $r \geqslant 0$, one has*

$$\mathcal{L}(\text{Real}(X), r) = \mathcal{L}(X, r).$$

The next symmetrization lemma allows randomizing all coordinates.

Lemma 4.2.5 (Randomizing all coordinates). *Let $Z = X + iY \in \mathbb{C}^N$ be a random vector whose imaginary part $Y \in \mathbb{R}^N$ is fixed and then set $\widehat{Z} = \left[\begin{smallmatrix} X_1 \\ X_2 \end{smallmatrix}\right] \in \mathbb{R}^{2N}$ with X_1 and X_2 independent copies of X. Let E be a subspace of $\mathbb{C}^N$. Then*

$$\mathcal{L}(P_E Z, r) \leqslant \left(\mathcal{L}(P_{\text{Real}(E)} \widehat{Z}, 2r)\right)^{1/2}, \quad r \geqslant 0.$$

Proof. Recalling the definition of the concentration function, in order to bound $\mathcal{L}(P_E Z, r)$ we need to choose arbitrary $a \in \mathbb{C}^N$ and find a uniform bound on the probability

$$p := \mathbb{P}\left\{\|P_E Z - a\|_2 \leqslant r\right\}.$$

By assumption, the random vector $Z = X + iY$ has fixed imaginary part Y. So it is convenient to express the probability as

$$p = \mathbb{P}\left\{\|P_E X - b\|_2 \leqslant r\right\}$$

where $b = a - P_E(iY)$ is fixed. Let us rewrite this identity using independent copies X_1 and X_2 of X as follows:

$$p = \mathbb{P}\left\{\|P_E X_1 - b\|_2 \leqslant r\right\} = \mathbb{P}\left\{\|P_E(iX_2) - ib\|_2 \leqslant r\right\}.$$

(The last equality follows trivially by multiplying by i inside the norm.) Using the independence of X_1 and X_2 and the triangle inequality, we obtain

$$\begin{aligned} p^2 &= \mathbb{P}\left\{\|P_E X_1 - b\|_2 \leqslant r \text{ and } \|P_E(iX_2) - ib\|_2 \leqslant r\right\} \\ &\leqslant \mathbb{P}\left\{\|P_E(X_1 + iX_2) - b - ib\|_2 \leqslant 2r\right\} \\ &\leqslant \mathcal{L}(P_E(X_1 + iX_2), 2r). \end{aligned}$$

Further, using part 2 and then part 1 of Lemma 4.2.4, we see that

$$\begin{aligned} \mathcal{L}(P_E(X_1 + iX_2), 2r) &= \mathcal{L}(P_{\text{Real}(E)}(\text{Real}(X_1 + iX_2)), 2r) \\ &= \mathcal{L}(P_{\text{Real}(E)} \widehat{Z}, 2r). \end{aligned}$$

Thus we showed that $p^2 \leqslant \mathcal{L}(P_{\text{Real}(E)} \widehat{Z}, 2r)$ uniformly in a. By definition of the Lévy concentration function, this completes the proof. □

Bounding the distances below We are ready to control the distances appearing in (4.2.3).

Lemma 4.2.6 (Distance between random vectors and subspaces). *For every* $j \in [n]$ *and* $\tau > 0$*, we have*

$$\mathbb{P}\left\{\operatorname{dist}(B_j', H_j') < \tau\sqrt{\varepsilon' m}\right\} \leqslant (C\tau)^{\varepsilon' m}. \tag{4.2.7}$$

Proof. Representing these distances via the projections of B_j' onto the subspaces $E_j = (H_j')^{\perp}$ as in (4.2.3), and using the definition of the Lévy concentration function, we have

$$p_j := \mathbb{P}\left\{\operatorname{dist}(B_j', H_j') < \tau\sqrt{\varepsilon' m}\right\} \leqslant \mathcal{L}(P_{E_j} B_j', \tau\sqrt{\varepsilon' m}).$$

Recall that B_j' and E_j are independent, and let us condition on E_j. Lemma 4.2.5 implies that

$$p_j \leqslant \left(\mathcal{L}(P_{\text{Real}(E_j)}\widehat{Z}, 2\tau\sqrt{\varepsilon' m})\right)^{1/2}$$

where $\widehat{Z}$ is a random vector with independent coordinates that have densities bounded by 1.

The space H_j' has codimension $\varepsilon' m$; thus E_j has dimension $\varepsilon' m$ and $\text{Real}(E_j)$ has dimension $2\varepsilon' m$. By Theorem 3.1.1, the density of $P_{\text{Real}(E_j)}\widehat{Z}$ is bounded by $C^{2\varepsilon' m}$. Integrating the density over a ball of radius $2\tau\sqrt{\varepsilon' m}$ in the subspace $\text{Real}(E_j)$ which has volume $(C\tau)^{2\varepsilon' m}$, we conclude that

$$\mathcal{L}(P_{\text{Real}(E_j)}\widehat{Z}, 2\tau\sqrt{\varepsilon n}) \leqslant (C\tau)^{2\varepsilon' m}.$$

It follows that

$$p_j \leqslant (C\tau)^{\varepsilon' m},$$

as claimed. The proof of Lemma 4.2.6 is complete. □

4.3. B is bounded below on a large subspace E^+

Using the second moment inequality Denote $p = \varepsilon' m/4$, and let

$$Y_j = \varepsilon' m \cdot \operatorname{dist}^{-2}(B_j', H_j') \quad \text{for } j \in [m].$$

By Lemma 4.2.6, for any $s > 0$,

$$\mathbb{P}\{Y_j > s\} \leqslant \left(\frac{C}{s}\right)^{2p}.$$

Using Fubini's theorem, we conclude that

$$\mathbb{E}Y_j^p \leqslant 1 + p\int_1^\infty s^{p-1} \cdot \mathbb{P}(Y_j > s)\, ds \leqslant 1 + \bar{C}^p,$$

so $\|Y_j\|_p = \left(\mathbb{E}Y_j^p\right)^{1/p} \leqslant C$. Here, once again the assumption of the bounded density of the entries simplifies the proof. For a general distribution of entries, the event $\operatorname{dist}(B_j', H_j') = 0$ may have a positive probability, so $\|Y_j\|_p$ may be infinite.

The bound on $\|Y_j\|_p$ yields $\left\|\sum_{j=1}^m Y_j\right\|_p \leqslant Cm$. By Markov's inequality, we get

$$\mathbb{P}\left(\sum_{j=1}^m \operatorname{dist}^{-2}(B_j', H_j') \geqslant \frac{1}{\varepsilon' t}\right) = \mathbb{P}\left(\sum_{j=1}^m Y_j \geqslant \frac{m}{t}\right)$$
$$\leqslant \frac{\mathbb{E}(\sum_{j=1}^m Y_j)^p}{(m/t)^p} \leqslant (Ct)^p$$

for any $t > 0$.

This estimate for $t = \tau^2$ combined with inequality (4.2.2) shows that the event

(4.3.1) $$\mathcal{E}_1 := \left\{\sum_{i=1}^m s_i(B)^{-2} \leqslant \frac{1}{\tau^2 \varepsilon'}\right\}$$

is likely: $\mathbb{P}((\mathcal{E}_1)^c) \leqslant (C'\tau)^{\varepsilon' m/2}$.

A large subspace E^+ on which B is bounded below Fix a parameter $\tau > 0$ for now, and assume that the event (4.3.1) occurs. By Markov's inequality, for any $\delta > 0$ we have

$$\left|\{i : s_i(B) \leqslant \delta\sqrt{m}\}\right| = \left|\{i : s_i(B)^{-2} \geqslant \frac{1}{\delta^2 m}\}\right| \leqslant \frac{\delta^2 m}{\tau^2 \varepsilon'}.$$

Setting $\delta = \tau\varepsilon'/10$, we have

(4.3.2) $$\left|\{i : s_i(B) \leqslant \frac{\tau\varepsilon'}{10}\sqrt{m}\}\right| \leqslant \frac{\varepsilon' m}{100}.$$

Let $v_i(B)$ be the right singular vectors of B, and consider the (random) orthogonal decomposition $\mathbb{C}^n = E^- \oplus E^+$, where

$$E^- = \operatorname{span}\{v_i(B) : s_i(B) \leqslant \frac{\tau\varepsilon'}{10}\sqrt{m}\},$$
$$E^+ = \operatorname{span}\{v_i(B) : s_i(B) > \frac{\tau\varepsilon'}{10}\sqrt{m}\}.$$

Inequality (4.3.2) means that $\dim_{\mathbb{C}}(E^-) \leqslant \frac{\varepsilon' m}{100}$.

Let us summarize. Recall that $\varepsilon' m = \varepsilon n/2$ and set $\tau = (\varepsilon s)^2$ for some $s \in (0,1)$. We proved that the event

$$\mathcal{D}_{E^-} := \left\{\dim(E^-) \leqslant \frac{\varepsilon' m}{100}\right\}$$

satisfies

(4.3.3) $$\mathbb{P}((\mathcal{D}_{E^-})^c) \leqslant (C_2\tau)^{\varepsilon' m} = (C_3 \varepsilon s)^{\varepsilon n},$$

so E^- is likely to be a small subspace and E^+ a large subspace. The choice of τ was made to create the factor $\varepsilon^{\varepsilon n}$ in the probability bound above ensuring that we can suppress the factor $\binom{n}{\varepsilon n}$ arising from the union bound. Moreover, by definition, B is nicely bounded below on $S_{E^+} = S^{n-1} \cap E^+$:

(4.3.4) $$\inf_{x \in S_{E^+}} \|Bx\|_2 \geqslant \frac{\tau\varepsilon'}{10}\sqrt{m} \geqslant \frac{s^2\varepsilon^3}{80}\sqrt{n}.$$

4.4. G **is bounded below on the small complementary subspace** E^- The previous argument allowed us to handle the subspace E_+ whose dimension is only slightly lower than m. Yet, it provided no information about the behavior of the infimum of $\|Bx\|_2$ over the unit vectors from the complementary subspace E_-. To get such a lower bound, we will use the submatrix G we have put aside. Recall that although the space E_- is random, it depends only on B, and thus is independent of G. Conditioning on the matrix B, we can regard this space as fixed. Our task therefore, is to establish a lower bound on $\|Gx\|_2$ over the unit vectors from E_-. To this end, we can use the Lemma 2.2.1. However, this lemma establishes the desired bound with probability at least $1-\exp(-c'\varepsilon' m)$. This probability is insufficient for our purposes (remember, the probability for a fixed set $I \subset [n]$ is multiplied by $\binom{n}{\varepsilon n} \sim (e/\varepsilon)^{\varepsilon n}$), but is easy to improve in case of the bounded densities. Replacing the small ball probability estimate for a fixed vector used in the proof of Lemma 2.2.1 with Lemma 3.2.1, we derive the following lemma.

Lemma 4.4.1 (Lower bound on a subspace)**.** *Let* $M \geqslant 1$ *and* $\mu \in (0,1)$. *Let* E *be a fixed subspace of* $\mathbb{C}^m$ *of dimension at most* $\varepsilon' m/100$. *Then, for every* $\rho > 0$, *we have*

$$\mathbb{P}\left\{\inf_{x\in S_E} \|Gx\|_2 < \rho\sqrt{\varepsilon' m} \text{ and } \mathcal{B}_{G,M}\right\} \leqslant \left(\frac{CM\rho^{0.98}}{\varepsilon'^{0.01}}\right)^{\varepsilon' m}. \tag{4.4.2}$$

The proof of this lemma follows the same lines as that of Lemma 2.2.1 and is left to a reader.

Lemma 4.4.1 provides the desired bound for the space E_-. Recall that we have $m = (1-\varepsilon)n$ and $\varepsilon' = \varepsilon/2(1-\varepsilon)$. Namely, if the events $\mathcal{B}_{G,M}$ and $\mathcal{D}_{E_-}$ occur, then the event

$$\mathcal{L}_{E_-} := \left\{\inf_{x\in S^{m-1}\cap E_-} \|Gx\|_2 \geqslant \rho\sqrt{\varepsilon' m}\right\}$$

holds with probability at least

$$1-\left(\frac{CM\rho^{0.98}}{\varepsilon'^{0.01}}\right)^{\varepsilon' m}.$$

This is already sufficient since choosing a sufficiently small ρ, say $\rho = (s\varepsilon')^3$ with any $s \in (0,1)$, we see that

$$\mathbb{P}\left(\mathcal{L}_{E_-}^c\right) \leqslant (CMs^3\varepsilon^{2.9})^{\varepsilon n/2},$$

so again we can suppress the factor $\binom{n}{\varepsilon n}$ arising from the union bound.

4.5. Extending invertibility from subspaces to the whole space. Assume that the events $\mathcal{D}_{E_-}$ and $\mathcal{L}_{E_-}$ occur. We know that if $\mathcal{B}_{A,M}$ occurs, then this is likely:

$$\mathbb{P}\left(\mathcal{B}_{A,M} \cap \mathcal{D}_{E_-} \cap \mathcal{L}_{E_-}\right) \geqslant \mathbb{P}\left(\mathcal{B}_{A,M}\right) - (Cs)^{\varepsilon n}.$$

Under this assumption, we have uniform lower bounds on $\|Ax\|_2$ on the unit spheres of both E_+ and E_-. The extension of these bounds to the whole unit sphere of $\mathbb{C}^m$ is now deterministic. It relies on the following lemma from linear algebra.

Lemma 4.5.1 (Decomposition). *Let A be an $m \times n$ matrix. Let us decompose A as*

$$A = \begin{bmatrix} B \\ G \end{bmatrix}, \quad B \in \mathbb{C}^{m_1 \times n},\ G \in \mathbb{C}^{m_2 \times n},\ m = m_1 + m_2.$$

Consider an orthogonal decomposition $\mathbb{C}^n = E^- \oplus E^+$ where E^- and E^+ are eigenspaces[1] *of B^*B. Denote*

$$\begin{aligned} s_A &= s_{\min}(A), \\ s_B &= s_{\min}(B|_{E^+}) = \min_{x \in S^{n-1} \cap E^+} \|Bx\|_2, \\ s_G &= s_{\min}(G|_{E^-}) = \min_{x \in S^{n-1} \cap E^-} \|Gx\|_2. \end{aligned}$$

Then

$$s_A \geqslant \frac{s_B s_G}{4\|A\|}. \tag{4.5.2}$$

Proof. Let $x \in S^{n-1}$. We consider the orthogonal decomposition

$$x = x^- + x^+, \quad x^- \in E^-, x^+ \in E^+.$$

We can also decompose Ax as

$$\|Ax\|_2^2 = \|Bx\|_2^2 + \|Gx\|_2^2.$$

Let us fix a parameter $\theta \in (0, 1/2)$ and consider two cases.

Case 1: $\|x^+\|_2 \geqslant \theta$. Since Bx^+ and Bx^- are orthogonal,

$$\|Ax\|_2 \geqslant \|Bx\|_2 \geqslant \|Bx^+\|_2 \geqslant s_B \cdot \theta.$$

Case 2: $\|x^+\|_2 < \theta$. In this case, $\|x^-\|_2 = \sqrt{1 - \|x^+\|_2^2} \geqslant 1/2$. Thus

$$\begin{aligned} \|Ax\|_2 &\geqslant \|Gx\|_2 \geqslant \|Gx^-\|_2 - \|Gx^+\|_2 \\ &\geqslant \|Gx^-\|_2 - \|G\| \cdot \|x^+\|_2 \geqslant s_G \cdot \frac{1}{2} - \|G\| \cdot \theta. \end{aligned}$$

Using that $\|G\| \leqslant \|A\|$, we conclude that

$$s_A = \inf_{x \in S^{n-1}} \|Ax\|_2 \geqslant \min\left(s_B \cdot \theta,\ s_G \cdot \frac{1}{2} - \|A\| \cdot \theta\right).$$

Optimizing the parameter θ, we conclude that

$$s_A \geqslant \frac{s_B s_G}{2(s_B + \|A\|)}.$$

Using that s_B is bounded by $\|A\|$, we complete the proof. □

Combining Lemma 4.5.1 with the bounds (4.3.4) and (4.4.2), we complete the proof of Proposition 2.1.2, and thus, the no-gaps delocalization Theorem 1.0.4.

5. Applications of the no-gaps delocalization

5.1. Erdős-Rényi graphs and their adjacency matrices In this section we consider two applications of the no-gaps delocalization to the spectral properties of

[1] In other words, E^- and E^+ are the spans of two disjoint subsets of right singular vectors of B.

the Erdős-Rényi random graphs. Let $p \in (0,1)$. Consider a graph $G = (V, E)$ with n vertices such that any pair of vertices is connected by an edge with probability p, and these events are independent for different edges. This model of a random graph is called an Erdős-Rènyi or $G(n,p)$ graph. Let A_G be the adjacency matrix of a graph G, i.e., the matrix of zeros and ones with 1 appearing on the spot (i,j) whenever the vertices i and j are connected. We will need several standard facts about the Erdős-Rényi graphs listed in the following proposition [10].

Proposition 5.1.1. *Let $p \geqslant C_0 \frac{\log n}{n}$ for some $C_0 > 1$. Let $G(V,E)$ be a $G(n,p)$ graph. Then G has the following properties with probability $1 - o(1)$.*

(1) *Let $R \subset V$ be an independent set, i.e., no two vertices from R are connected by an edge. Then*
$$|R| \leqslant C \frac{\log n}{p}.$$
(2) *Let $P, Q \subset V$ be disjoint sets of vertices with*
$$|P|, |Q| \geqslant C \frac{\log n}{p}.$$
Then there is an edge connecting a vertex from P and a vertex from Q.

(3) *The degree of any vertex $v \in V$ is close to its expectation:*
$$np - \log n \cdot \sqrt{np} \leqslant d_v \leqslant np + \log n \cdot \sqrt{np}$$
(4) *Define the normalized adjacency matrix of G to be $\hat{A} := D_G^{-1/2} A_G D_G^{-1/2}$ where D_G is the diagonal matrix $D_G = \operatorname{diag}(d_v, v \in V)$ and et $\hat{\lambda}_1 \geqslant, \dots, \geqslant \hat{\lambda}_n$ be eigenvalues of $\hat{A}$. Then*
$$\hat{\lambda}_1 = 1, \quad \text{and} \quad |\hat{\lambda}_j| \leqslant \frac{C}{\sqrt{np}} \text{ for } j \geqslant 1.$$
(5) *For every subset of vertices $J \subset V$, let Non-edges(J) be the set of all pairs of vertices $v, w \in J$ which are not connected by an edge. Then*
$$(1-p)\binom{|J|}{2} - n^{3/2} \leqslant |\text{Non-edges}(J)| \leqslant (1-p)\binom{|J|}{2} + n^{3/2}.$$

We leave the proof of these properties to a reader.

Considering the vector of all ones, we realize that $\|A_G\| = \Omega(np)$ with high probability. Hence, when p is fixed, and $n \to \infty$, this makes the event $\mathcal{B}_{A_G,M}$ unlikely. However, Remark 1.0.9 shows that we can replace this event by the event $\mathcal{B}_{A_G - p\mathbf{1}_n, M}$ which holds with probability close to 1. Indeed,
$$A_G - p\mathbf{1}_n = B - \Delta,$$
where B is a symmetric random matrix with centered Bernoulli(p) entries which are independent on and above the diagonal, and Δ is the diagonal matrix with i.i.d. Bernoulli(p) entries. Here, $\|\Delta\| \leqslant 1$ and $\|B\| \leqslant C\sqrt{np}$ with probability close to 1, by a simple ε-net argument.

This decomposition is reflected in the structure of the spectrum of A_G. Let us arrange the eigenvalues of A_G in the decreasing order: $\lambda_1(G) \geqslant \ldots \geqslant \lambda_n(G)$. Then with high probability, $\lambda_1(G) = \Omega(np)$ and $|\lambda_j(G)| = O(\sqrt{np})$, where the last equality follows from $\|A_G - p\mathbf{1}_n\| = O(\sqrt{np})$ and the interlacing property of the eigenvalues.

Remark 1.0.9 shows that no-gaps delocalization can be extended to the matrix A_G as well. We will use this result in combination with the ℓ_∞ delocalization which was established for the $G(n,p)$ graphs by Erdős et. al. [11]. They proved that with probability at least $1 - \exp(-c \log^2 n)$, any unit eigenvector x of A_G satisfies

$$\|x\|_\infty \leqslant \frac{\log^C n}{\sqrt{n}}. \tag{5.1.2}$$

5.2. Nodal domains of the eigenvectors of the adjacency matrix Let f be an eigenfunction of a self-adjoint linear operator. Define the (strong) nodal domains of f as connected components of the sets where f is positive or negative. Nodal domains of the Laplacian on a compact smooth manifold is a classical object in analysis. If the eigenvalues are arranged in the increasing order, the number of nodal domains of the eigenfunction corresponding to the k-th eigenvalue does not exceed k and tends to infinity as $k \to \infty$.

If we consider a finite-dimensional setup, the eigenfunctions of self-adjoint linear operators are replaced by the eigenvectors of symmetric matrices. In 2008, Dekel, Lee, and Linial [9] discovered that the nodal domains of the adjacency matrices of $G(n,p)$ graphs behave strikingly different from the eigenfunctions of the Laplacian on a manifold. Namely, they proved that with high probability, the number of nodal domains of any non-first eigenvector of a $G(n,p)$ graph is bounded by a constant depending only on p. Later, their result was improved by Arora and Bhaskara [1], who showed that with high probability, the number of nodal domains is 2 for all non-first eigenvectors. Also, Nguyen, Tao, and Vu [19] showed that the eigenvector of a $G(n,p)$ graph cannot have zero coordinates with probability close to 1. These two results in combination mean that for each non-first eigenvector, the set of vertices of a $G(n,p)$ graph splits into the set of positive and negative coordinates both of which are connected.

Let us derive Dekel-Lee-Linial-Arora-Bhaskara theorem from the delocalization properties of an eigenvector. Assume that p is fixed to make the presentation easier. Let $x \in S^{n-1}$ be a non-first eigenvector of A_G, and denote its coordinates by x_v, $v \subset V$. Let P and N be the largest nodal domains of positive and negative coordinates. Since x is orthogonal to the first eigenvector having all positive coordinates, both P and N are non-empty. Denote $W = V \setminus (P \cup N)$. Our aim is to prove that with high probability, $W = \varnothing$. We start with proving a weaker statement that the cardinality of W is small.

Proposition 5.2.1.

$$|W| \leqslant C \frac{\log^2 n}{p^2}$$

with probability $1 - o(1)$.

Proof. Pick a vertex from each positive nodal domain. These vertices cannot be connected by edges as they belong to different connected components, so they form an independent set. Using Proposition 5.1.1 (1), we derive that, with high probability, the number of such domains does not exceed $C\frac{\log n}{p}$. The same bound holds for the number of negative nodal domains.

Consider a nodal domain $W_0 \subset W$ and assume that $|W_0| \geqslant C\frac{\log n}{p}$. If this domain is positive, $|P| \geqslant C\frac{\log n}{p}$ as well, since P is the largest nodal domain. This contradicts Proposition 5.1.1 (2) as two nodal domains of the same sign cannot be connected. Combining this with the previous argument, we complete the proof of the proposition. □

Now, we are ready to prove that $W = \varnothing$ with probability $1 - o(1)$. Assume to the contrary that there is a vertex $v \in W$, and assume that $x_v < 0$. Let $\Gamma(v)$ be the set of its neighbors in G. Then $\Gamma(v) \cap N = \varnothing$ as otherwise v would be an element of N. Since x is an eigenvector,

$$\lambda x_v = \sum_{u \in \Gamma(v)} x_u = \sum_{u \in \Gamma(v) \cap P} x_u + \sum_{u \in \Gamma(v) \cap W} x_u.$$

Here $|\lambda| \leqslant \sqrt{np}$ because λ is a non-first eigenvalue. Then

$$\begin{aligned} \left\| x|_{\Gamma(v)} \right\|_1 &\leqslant \sum_{u \in \Gamma(v) \cap P} x_u + \sum_{u \in \Gamma(v) \cap W} |x_u| \leqslant 2 \sum_{u \in \Gamma(v) \cap W} |x_u| + |\lambda| \cdot |x_v| \\ &\leqslant (2|\Gamma(v) \cap W| + |\lambda|) \cdot \|x\|_\infty . \end{aligned}$$

By Proposition 5.2.1 and (5.1.2), this quantity does not exceed $\log^C n$ (recall that we assumed that $p \in (0,1)$ is fixed). Applying (5.1.2) another time, we conclude that

$$\left\| x|_{\Gamma(v)} \right\|_2 \leqslant \sqrt{\left\| x|_{\Gamma(v)} \right\|_1 \cdot \|x\|_\infty} \leqslant n^{-1/4} \log^C n.$$

In combination with Proposition 5.1.1 (3), this shows that a large set $\Gamma(v)$ carries a small mass, which contradicts the no-gaps delocalization. This completes the proof of Dekel-Lee-Linial-Arora-Bhaskara theorem.

The same argument shows that, with high probability, any vertex of the positive nodal domain is connected to the negative domain and vice versa, answering positively an old question of Linial. More precisely, we have the following stronger statement.

Lemma 5.2.2. *Let* $p \in (0,1)$. *Let* $x \in S^{n-1}$ *be a non-first eigenvector of* A_G. *Let* $V = P \cup N$ *be the decomposition of* V *into the positive and negative nodal domains corresponding to* x. *Then with probability greater than* $1 - \exp(-c' \log^2 n)$, *any vertex in* P *has at least* $\frac{n}{\log^C n}$ *neighbors in* N, *and any vertex in* N *has at least* $\frac{n}{\log^C n}$ *neighbors in* P.

Proof. Since λ is a non-first eigenvalue, $|\lambda| \leqslant c\sqrt{n}$ with high probability. Assume that the vector x is delocalized in both ℓ_∞ and no-gaps sense. Let $w \in P$, and assume that

$$|\Gamma(w) \cap N| \leqslant \frac{n}{\log^{4C} n},$$

where $\Gamma(w)$ denotes the set of neighbors of w. We have

$$\lambda x_w = \sum_{v \in \Gamma(w) \cap P} x_v + \sum_{v \in \Gamma(w) \cap N} x_v,$$

and as before,

$$\begin{aligned} \left\| x|_{\Gamma(w)} \right\|_1 &= \sum_{v \in \Gamma(w) \cap P} x_v + \sum_{v \in \Gamma \cap N} |x_v| \leqslant 2 \sum_{v \in \Gamma(w) \cap N} |x_v| + |\lambda| \cdot |x_w| \\ &\leqslant 2 \frac{n}{\log^{4C} n} \cdot \frac{\log^C n}{\sqrt{n}} + c\sqrt{n} \cdot \frac{\log^{4C} n}{\sqrt{n}}. \end{aligned}$$

Hence,

$$\left\| x|_{\Gamma(w)} \right\|_2 \leqslant \sqrt{\|x\|_\infty \cdot \left\| x|_{\Gamma(w)} \right\|_1} \leqslant \frac{\sqrt{2}}{\log^C n},$$

which contradicts the no-gaps delocalization, as $|\Gamma(w)| \geqslant cnp$ with high probability. The proof finishes by application of the union bound over w. □

5.3. Spectral gap of the normalized Laplacian and Braess's paradox In some cases, the addition of a new highway to an existing highway system may increase the traffic congestion. This phenomenon discovered in 1968 by Braess became known as Braess's paradox. Since its discovery, a number of mathematical models have been suggested to explain this paradox. We will consider one such model suggested by Chung et. al. [8].

We will model the highway system by an Erdős-Rènyi graph $G(n,p)$. The congestion of the graph will be measured in terms of its *normalized Laplacian* which we will define in a moment. Let A_G be the adjacency matrix of the graph G, and let $D_G = (d_v,\ v \in V)$ be $n \times n$ the diagonal matrix whose diagonal entries are the degrees of the vertices. The normalized Laplacian of G is defined as

$$\mathcal{L}_G := I_n - D_G^{-1/2} A_G D_G^{-1/2}.$$

The normalized Laplacian is a positive semidefinite matrix, so it has a real non-negative spectrum that we write in increasing order: $0 = \lambda_1(\mathcal{L}_G) \leqslant \ldots \leqslant \lambda_n(\mathcal{L}_G)$. The eigenvalue $\lambda_1(\mathcal{L}_G) = 0$ corresponds to the eigenvector Y, whose coordinates are $Y_v = d_v^{1/2}$, $v \in V$. The quantity $\lambda_2(\mathcal{L}_G)$ is called *the spectral gap* of G. The spectral gap appears in the Poincare inequality, so it is instrumental in establishing measure concentration properties of various functionals, see, e.g. [16]. Also, the reciprocal of the spectral gap defines the relaxation time for a random walk on a graph, [17]. In this quality, it can be used to measure the congestion of the graph considered as a traffic network: the smaller spectral gap corresponds to a bigger congestion.

For a graph G, and let $a_-(G)$ be the fraction of non-edges $(u,v) \notin E$ such that the addition of (u,v) to the set of edges decreases the spectral gap. Intuitively, the addition of an edge should increase the spectral gap as it brings the graph closer to the complete one, for which the spectral gap is maximal. However, the numerical experiments showed that the addition of an edge to a random graph frequently yields an opposite effect. This numerical data led to the following conjecture, which is a variant of the original conjecture of Chung.

Conjecture 5.3.1. For $p \in (0,1)$ fixed, there exists a constant $c(p)$ such that

$$\lim_{n \to \infty} \mathbb{P}\left\{a_-(G) \geqslant c(p)\right\} = 1.$$

This conjecture has been proved by Eldan, Ràsz, and Shramm [10]. Their proof is based on the following deterministic condition on the eigenvectors which ensures that the spectral gap decreases after adding an edge.

Proposition 5.3.2. *Let* G *be a graph such that* $(1/2)np \leqslant d_v \leqslant (3/2)np$ *for all vertices* $v \in V$*. Let* $x \in S^{n-1}$ *be the eigenvector of* $\mathcal{L}_G$ *corresponding to* $\lambda_2(G)$*. If* $(u,w) \notin E$ *is a non-edge, and*

$$\frac{1}{\sqrt{np}}\left(x_u^2 + x_w^2\right) + c_1(np)^{-2} < c_2 x_u x_v,$$

then the addition of the edge (u,w) *to* G *decreases the spectral gap.*

The proof of proposition 5.3.2 requires a tedious, although a rather straightforward calculation. Denote by $y \in S^{n-1}$ the *first* eigenvector of the Laplacian of graph G_+ obtained from G by adding the edge (u,w), and let $Q : \mathbb{R}^n \to \mathbb{R}^n$ be the orthogonal projection on the space $y^\perp$. By the variational definition of the second eigenvalue,

$$\lambda_2(G_+) = \inf_{z \in y^\perp \setminus \{0\}} \frac{\langle z, \mathcal{L}_{G_+} z\rangle}{\|z\|_2^2} \leqslant \frac{\langle Qx, \mathcal{L}_{G_+} Qx\rangle}{\|Qx\|_2^2} = \frac{\langle x, \mathcal{L}_{G_+} x\rangle}{1 - \langle x, y\rangle^2},$$

where the last equality follows since $\mathcal{L}_{G_+} y = 0$. In the last formula, $y = \Delta / \|\Delta\|_2$, where Δ is the vector with coordinates $\Delta_v = \sqrt{d_v}$ for $v \notin \{u,w\}$ and $\Delta_v = \sqrt{d_v + 1}$ for $v \in \{u,w\}$. The matrix $\mathcal{L}_{G_+}$ can be represented in a similar way:

$$\mathcal{L}_{G_+} = I_n - D_{G_+}^{-1/2} A_{G_+} D_{G_+}^{-1/2},$$

where $A_G + (e_u e_w^T + e_w e_u^T)$ and D_{G_+} is defined as D_G above. The proposition follows by substituting these formulas in the previous estimate of $\lambda_2(G_+)$ and simplifying the resulting expression. A reader can find the detailed calculation in [10].

Proposition 5.3.2 allows us to lower bound $a_-(G)$. The main technical tool in obtaining such a bound is delocalization. We will need both the ℓ_∞ and no-gaps delocalization of the second eigenvector of $\mathcal{L}_G$. Both properties hold for the eigenvectors of A_G, so our task is to extend them to the normalized Laplacian.

To derive the ℓ_∞ delocalization, we need some information on the distribution of the eigenvalues of A_G. The classical Wigner semi-circular law states that as $n \to \infty$, the percentage of eigenvalues of $n^{-1/2} A_G$ lying in any fixed interval

$(b, b+\theta) \subset \mathbb{R}$ approaches

$$\int_b^{b+\theta} \phi_{sc}(x)\,dx, \quad \text{where } \phi_{sc}(x) = \frac{1}{2\pi}\sqrt{(4-x^2)_+}.$$

Moreover, the *local* semi-circular law asserts that the same phenomenon holds on a short scale, namely when $\theta = \theta(n) = \Omega\left(\frac{\log^C n}{n}\right)$ with some absolute constant $C > 0$. Since ϕ_{sc} is a bounded function, this implies that as $n \to \infty$, an interval $(b, b+\rho)$ with $\rho = \Omega(1)$ should contain $O(\rho\sqrt{n})$ eigenvalues. Since we work with a fixed n instead of $n \to \infty$, we need a non-asymptotic version of the local semi-circular law proved by Erdős et. al. ([11], Theorem 2.10). Their result implies the following upper estimate for the number of eigenvalues in a fixed interval.

Theorem 5.3.3. *Let* $b \in \mathbb{R}$ *and* $\rho \geqslant 1$. *With probability greater than* $1 - \exp(-c\log^2 n)$, *the interval* $[b, b+\rho]$ *contains at most*

$$N(\rho) := c'\rho\sqrt{n}$$

eigenvalues of A_G. *The constants* $c, c' > 0$ *are absolute.*

Theorem 5.3.3 together with the ℓ_∞ delocalization for the eigenvectors of the adjacency matrix allows to prove a similar delocalization for the normalized Laplacian.

Lemma 5.3.4. *Let* $p \in (0,1)$. *Let* $f \in S^{n-1}$ *be the second eigenvector of* $\mathcal{L}_G$. *Then with probability at least* $1 - \exp(-c\log^2 n)$,

$$\|f\|_\infty \leqslant n^{-1/4}\log^C n$$

and

$$\left|\left\{v \in V : |f_v| \leqslant n^{-5/8}\right\}\right| \leqslant c' n^{1-1/48}.$$

Here, C, c, c' *are positive constants whose value may depend on* p.

Proof. Let us start with the ℓ_∞ delocalization. Let $d = np$ be the expected degree of a vertex, and set

$$x = d^{1/2} D_G^{-1/2} f.$$

By Proposition 5.1.1 (3), $d^{1/2}D_G^{-1/2} = \mathrm{diag}(s_v,\ v \in V)$, where $s_v = 1 + o(1)$ for all $v \in V$, and $\|x\|_2 = 1 + o(1)$ with probability close to 1. Hence, it is enough to bound $\|x\|_\infty$. Let us check that x is an approximate eigenvector of A_G corresponding to the approximate eigenvalue $\hat\lambda_2 d$, where $\hat\lambda_2$ is the second eigenvalue of the normalized adjacency matrix $D_G^{-1/2}A_G D_G^{-1/2}$. By Proposition 5.1.1 (4), $\hat\lambda_2 \leqslant c/\sqrt{np}$ with high probability, hence

$$\begin{aligned}
\left\|A_G D_G^{-1/2} f - \hat\lambda_2 d D_G^{-1/2} f\right\|_2 &= |\hat\lambda_2| \cdot \left\|D_G^{1/2} f - d D_G^{-1/2} f\right\|_2 \\
&\leqslant \frac{c}{\sqrt{np}} \cdot \max_{v\in V} d_v^{-1/2} \cdot \max_{v\in V} |d_v - d| \\
&\leqslant \frac{c}{np} \cdot \max_{v\in V} |d_v - d| \leqslant \frac{C\log n}{\sqrt{np}},
\end{aligned}$$

and so

$$\|A_G x - \hat{\lambda}_2 d x\|_2 \leqslant C \log n. \tag{5.3.5}$$

Denote the eigenvalues of A_G by $\mu_1, \dots, \mu_n$ and the corresponding eigenvectors by $u_1, \dots, u_n \in S^{n-1}$, and let $\alpha_j = \langle x, u_j \rangle$. Set $\mu = \hat{\lambda}_2 d$ and let P_τ be the orthogonal projection on the span of the eigenvectors corresponding to the eigenvalues of A_G in the interval $[\mu - \tau, \mu + \tau]$. Then

$$\begin{aligned}\tau \|(I - P_\tau)x\|_2 = \tau \left(\sum_{|\mu_j - \mu| > \tau} \alpha_j^2 \right)^{1/2} &\leqslant \left(\sum_{|\mu_j - \mu| > \tau} (\mu_j - \mu)^2 \alpha_j^2 \right)^{1/2} \\ &\leqslant \|(A_G - \mu)x\|_2 \leqslant C \log n.\end{aligned}$$

and so,

$$\|(I - P_\tau)x\|_2 \leqslant \left(C \frac{\log n}{\tau} \wedge 1 \right). \tag{5.3.6}$$

For any $\tau \geqslant 0$ and any $\rho \geqslant 1$,

$$\begin{aligned}\|(P_{\tau+\rho} - P_\tau)x\|_\infty &= \left\| \sum_{|\mu_j - \mu| \in [\tau, \tau+\rho]} \alpha_j u_j \right\|_\infty = \max_{\nu \in V} \left| \sum_{|\mu_j - \mu| \in [\tau, \tau+\rho]} \alpha_j u_{j,\nu} \right| \\ &\leqslant \left(\sum_{|\mu_j - \mu| \in [\tau, \tau+\rho]} \alpha_j^2 \right)^{1/2} \cdot \max_{\nu \in V} \left(\sum_{|\mu_j - \mu| \in [\tau, \tau+\rho]} u_{j,\nu}^2 \right)^{1/2} \\ &\leqslant \|(P_{\tau+\rho} - P_\tau)x\|_2 \cdot N^{1/2}(\rho) \cdot \max_{j \in [n]} \|u_j\|_\infty \\ &\leqslant \|(I - P_\tau)x\|_2 \cdot \sqrt{\rho n^{1/2}} \cdot \frac{\log^C n}{\sqrt{n}}\end{aligned}$$

with probability greater than $1 - \exp(-c \log^2 n)$, where we used (5.1.2) and Theorem 5.3.3 in the last inequality. Combining this with (5.3.6), we get

$$\|(P_{\tau+\rho} - P_\tau)x\|_\infty \leqslant C \sqrt{\rho} \frac{\log^C n}{n^{1/4}} \cdot \left(\tau^{-1} \wedge 1 \right).$$

By the union bound, with probability greater than $1 - \exp(-c \log^2 n)$, the same inequality holds for all $\tau = \rho = 2^k$, such that $1 \leqslant 2^k \leqslant 2n$. Also, with probability at least $1 - \exp(-cn)$, $I = P_{2n}$, i.e., there are no eigenvalues outside of the interval $[-2n, 2n]$. Therefore,

$$\begin{aligned}\|x\|_\infty &\leqslant \|P_1 x\|_\infty + \sum_{k=0}^{\log_2 2n} \|(P_{2^{k+1}} - P_{2^k})x\|_\infty \\ &\leqslant C \frac{\log^C n}{n^{1/4}} + \sum_{k=0}^{\log_2 2n} C 2^{-k/2} \frac{\log^C n}{n^{1/4}} \leqslant C n^{-1/4} \log^C n\end{aligned}$$

with the required probability. By the discussion above, $\|f\|_\infty \leqslant 2\|x\|_\infty$ which finishes the proof of the first part of the lemma.

Now, let us prove the lower bound on the absolute values of most of the coordinates of f. As before, it is enough to prove a similar bound on the coordinates of x. Assume to the contrary that there is a set $U \subset V$ with $|U| > cn^{1-1/48}$ such that for any $v \in U$, $|x_v| \leqslant n^{-5/8}$. Then

$$\|x_U\|_2 \leqslant \sqrt{n} \cdot n^{-5/8} = n^{-1/8}.$$

Inequality (5.3.5) shows that x is an approximate eigenvector of A_G. Since, by Remarks 1.0.9 and 2.1.8, $n^{-1/8} \gg Cn^{-1/2} \log^C n$, we can apply Theorem 1.0.7 to x with s being an appropriately small constant and $\varepsilon = (1/s)n^{-1/48}$, so $(\varepsilon s)^6 = n^{-1/8}$. This theorem shows that such set U exists with probability at most $\exp(-\varepsilon n) \ll \exp(-c \log^2 n)$. The proof of the lemma is complete. □

Equipped with Proposition 5.3.2 and Lemma 5.3.4, we can prove a stronger form of the conjecture showing that $c \geqslant 1/2 - o(1)$. Let us formulate it as a theorem.

Theorem 5.3.7. *Let* $p \in (0,1)$, *and let* G *be a* $G(n,p)$ *graph. Then with probability* $1 - o(1)$,

$$a_-(G) \geqslant \frac{1}{2} - O(n^{-c}).$$

Proof. Let $f \in S^{n-1}$ be the eigenvector of $\mathcal{L}_G$ corresponding to the second eigenvalue, and assume that the event described in Lemma 5.3.4 occurs. Let

$$W = \left\{ v \in V \colon \ |f_v| \geqslant n^{-5/8} \right\},$$

and set

$$W_+ = \{v \in W \colon \ f_v > 0\}, \quad \text{and} \quad W_- = \{v \in W \colon \ f_v < 0\}.$$

For any $v, w \in W_+$,

$$\frac{f_v^2 + f_w^2}{f_v f_w} \leqslant 2 \max_{v,w \in W_+} \frac{f_v}{f_w} \leqslant Cn^{3/8} \log^C n \ll \sqrt{n}.$$

Hence, if (v, w) is a non-edge, then Proposition 5.3.2 implies that adding it to G decreases the spectral gap. Similarly, we can show that adding any non-edge whose vertices belong to W_-, decreases the spectral gap as well. Let us count the number of the non-edges in W_+ and W_- and compare it to the total number of the non-edges. Using Property (5), and the bound $|W^c| \leqslant cn^{1-1/48}$, we obtain

$$\begin{aligned} a_-(G) &\geqslant \frac{|\text{Non-edges}(W_+)| + |\text{Non-edges}(W_-)|}{|\text{Non-edges}(V)|} \\ &\geqslant \frac{(1-p)\left[\binom{|W_+|}{2} + \binom{|W_-|}{2}\right] - 2n^{-3/2}}{(1-p)\binom{n}{2} + n^{3/2}} \\ &\geqslant \frac{(1-p)\left[\left(\frac{|W_+|+|W_-|}{2}\right)^2 - |W_+| - |W_-|\right] - 2n^{3/2}}{(1-p)\binom{n}{2} + n^{3/2}} \geqslant \frac{1}{2} - O(n^{-c}), \end{aligned}$$

as claimed. □

References

[1] S. Arora and A. Bhaskara, *Eigenvectors of random graphs: delocalization and nodal domains*, 2011. Manuscript, http://www.cs.princeton.edu/ bhaskara/files/deloc.pdf. ↑332

[2] K. Ball, *Cube slicing in* $\mathbb{R}^n$, Proc. Amer. Math. Soc. **97** (1986), 465–473. ↑313

[3] K. Ball, *Volumes of sections of cubes and related problems*, Geometric aspects of functional analysis (1987–88), 1989, pp. 251–260. ↑314, 317

[4] K. Ball and F. Nazarov, *Little level theorem and zero-Khinchin inequality*, 1996. Manuscript, http://www.math.msu.edu/ fedja/prepr.html. ↑313

[5] F. Barthe, *Inégalités de Brascamp-Lieb et convexité*, C. R. Acad. Sci. Paris Sér. I Math. **324** (1997), no. 8, 885–888. ↑317

[6] P. Bourgade and H.-T. Yau, *The eigenvector moment flow and local quantum unique ergodicity*, Comm. Math. Phys. **350** (2017), no. 1, 231–278. ↑304

[7] H. J. Brascamp and E. H. Lieb, *Best constants in Young's inequality, its converse, and its generalization to more than three functions*, Advances in Math. **20** (1976), 151–173. ↑317

[8] F. Chung, S. Young, and W. Zhao, *Braess's paradox in expanders*, Random Structures and Algorithms **41** (2012), no. 4, 451–468. ↑334

[9] Y. Dekel, J. R. Lee, and N. Linial, *Eigenvectors of random graphs: nodal domains*, Random Structures and Algorithms **39** (2011), no. 1, 39–58. ↑332

[10] R. Eldan, M. Rász, and T. Schramm, *Braess's paradox for the spectral gap in random graphs and delocalization of eigenvectors*, Random Structures and Algorithms **50** (2017), no. 4, 584–611. ↑331, 335

[11] L. Erdős, A. Knowles, H.-T. Yau, and J. Yin, *Spectral statistics of Erdős-Rényi graphs I: local semicircle law*, Annals of Probability **41** (2013), no. 3B, 2279–2375. ↑332, 336

[12] L. Erdős, B. Schlein, and H.-T. Yau, *Semicircle law on short scales and delocalization of eigenvectors for Wigner random matrices*, Annals of Probability **37** (2009), 815–852. ↑304

[13] L. Erdős, B. Schlein, and H.-T. Yau, *Local semicircle law and complete delocalization for Wigner random matrices*, Comm. Math. Phys. **287** (2009), 641–655. ↑304

[14] F. Götze, A. Naumov, A. Tikhomirov, and D. Timushev, *On the local semicircular law for Wigner ensembles*, 2016. arXiv:1602.03073v2. ↑304

[15] G. Halász, *Estimates for the concentration function of combinatorial number theory and probability*, Periodica Mathematica Hungarica **8** (1977), 197–211. ↑316

[16] M. Ledoux, *The concentration of measure phenomenon*, Mathematical Surveys and Monographs, vol. 89, American Mathematical Society, Providence, RI, 2001. ↑334

[17] Y. Peres D. A. Levin E. L. Wilmer, *Markov chains and mixing times*, American Mathematical Society, Providence, RI, 2017. ↑334

[18] G. Livshyts, G. Paouris, and P. Pivovarov, *On sharp bounds for marginal densities of product measures*, Israel J. Math. **216** (2016), no. 2, 877–889. ↑314

[19] H. Nguyen, T. Tao, and V. Vu, *Random matrices: tail bounds for gaps between eigenvalues*, Probability Theory Related Fields **167** (2017), no. 3–4, 777–816. ↑332

[20] S. O'Rourke, V. Vu, and K. Wang, *Eigenvectors of random matrices: a survey*, J. Combin. Theory Ser. A **144** (2016), 361–442. ↑305

[21] M. Rudelson, *Recent developments in non-asymptotic theory of random matrices*, Modern aspects of random matrix theory, 2014, pp. 1576–1602. ↑304

[22] M. Rudelson and R. Vershynin, *Non-asymptotic theory of random matrices: extreme singular values*, Proceedings of the International Congress of Mathematicians, 2010, pp. 83–120. ↑304

[23] M. Rudelson and R. Vershynin, *Small ball probabilities for linear images of high dimensional distributions*, Int. Math. Res. Not. **19** (2015), 9594–9617. ↑313

[24] M. Rudelson and R. Vershynin, *Delocalization of eigenvectors of random matrices with independent entries*, Duke Math. J. **164** (2015), no. 13, 2507–2538. ↑304

[25] M. Rudelson and R. Vershynin, *No-gaps delocalization for general random matrices*, Geom. Funct. Anal. **26** (2016), no. 6, 1716–1776. ↑305, 306, 307

[26] T. Tao, *Topics in random matrix theory*, Graduate Studies in Mathematics, vol. 132, American Mathematical Society, Providence, RI, 2012. ↑305, 324

[27] R. Vershynin, *Introduction to the non-asymptotic analysis of random matrices*, Compressed sensing, 2012, pp. 210–268. ↑304

[28] V. Vu and K. Wang, *Random weighted projections, random quadratic forms and random eigenvectors*, Random Structures and Algorithms **47** (2015), no. 4, 792–821. ↑304

Department of Mathematics, University of Michigan.
Email address: rudelson@umich.edu

IAS/Park City Mathematics Series
Volume 26, Pages 341–387
S 1079-5634(XX)0000-0

Microscopic description of Log and Coulomb gases

Sylvia Serfaty

Abstract. These notes are intended to review some recent results, obtained in large part with Thomas Leblé, on the statistical mechanics of systems of points with logarithmic or Coulomb interactions. After listing some motivations, we describe the "electric approach" which allows to get concentration results, Central Limit Theorems for fluctuations, and a Large Deviations Principle expressed in terms of the microscopic state of the system.

Contents

1 Introduction and motivations 342
1.1 Fekete points and approximation theory 343
1.2 Statistical mechanics 345
1.3 Two component plasmas 346
1.4 Random matrix theory 346
1.5 Complex geometry and theoretical physics 347
1.6 Vortices in condensed matter physics 348
2 Equilibrium measure and leading order behavior 348
2.1 The macroscopic behavior: empirical measure 348
2.2 Large Deviations Principle at leading order 350
2.3 Further questions 352
3 Splitting of the Hamiltonian and electric approach 354
3.1 The splitting formula 354
3.2 Electric interpretation 355
3.3 The case d = 1 358
3.4 The electric energy controls the fluctuations 359
3.5 Consequences for the energy and partition function 360
3.6 Consequence: concentration bounds 362
4 CLT for fluctuations in the logarithmic cases 363
4.1 Reexpressing the fluctuations as a ratio of partition functions 364
4.2 Transport and change of variables 364
4.3 Energy comparison 366

2010 *Mathematics Subject Classification.* 60F05, 60K35, 60B20, 82B05, 60G15, 82B21, 82B26, 15B52.
Key words and phrases. Coulomb gases, log gases, random matrices, jellium, large deviations, point processes.

4.4 Computing the ratio of partition functions 367
4.5 Conclusion in the one-dimensional one-cut regular case 368
4.6 Conclusion in the two-dimensional case or in the general one-cut case 369
5 The renormalized energy 371
5.1 Definitions 372
5.2 Scaling properties 373
5.3 Partial results on the minimization of $\mathcal{W}$, crystallization conjecture 374
5.4 Renormalized energy for point processes 374
5.5 Lower bound for the energy in terms of the empirical field 375
6 Large Deviations Principle for empirical fields 376
6.1 Specific relative entropy 376
6.2 Statement of the main result 378
6.3 Proof structure 380
6.4 Screening and consequences 381
6.5 Generating microstates and conclusion 383

1. Introduction and motivations

We are interested in the following class of energies

$$\mathcal{H}_N(x_1, \ldots, x_N) := \sum_{1 \leqslant i \neq j \leqslant N} g(x_i - x_j) + \sum_{i=1}^{N} NV(x_i). \tag{1.0.1}$$

where $x_1, \ldots, x_N$ are N points (or particles) in the Euclidean space $\mathbb{R}^d$ ($d \geqslant 1$), and N is large. The function V is a confining potential, growing fast enough at infinity, on which we shall make assumptions later. The pair interaction potential g is given by either

$$\text{(Log1 case)} \quad g(x) = -\log|x|, \quad \text{in dimension } d = 1, \tag{1.0.2}$$

$$\text{(Log2 case)} \quad g(x) = -\log|x|, \quad \text{in dimension } d = 2, \tag{1.0.3}$$

$$\text{(Coul case)} \quad g(x) = |x|^{2-d} \text{ in dimension } d \geqslant 3. \tag{1.0.4}$$

We will also say a few things about the more general case

$$\text{(Riesz case)} \quad g(x) = |x|^{-s}, \quad \text{with } \max(0, d-2) \leqslant s < d, \text{ in dimension } d \geqslant 1. \tag{1.0.5}$$

The interaction as in (1.0.2) (resp. (1.0.3)) corresponds to a one-dimensional (resp. two-dimensional) logarithmic interaction, we will call Log1, Log2 the *logarithmic cases*. The Log2 interaction is also the Coulomb interaction in dimension 2. For $d \geqslant 3$, Coul corresponds to the Coulomb interaction in higher dimension. In both instances Log2 and Coul we have

$$-\Delta g = c_d \delta_0 \tag{1.0.6}$$

where δ_0 is the Dirac mass at 0 and c_d is a constant given by

$$c_d = 2\pi \quad \text{if } d = 2 \qquad c_d = (d-2)|\mathbb{S}^{d-1}| \quad \text{for } d \geqslant 3. \tag{1.0.7}$$

Finally, the Riesz cases $\max(d-2,0) < s < d$ in (1.0.4) correspond to more general Riesz interactions, in what is called the *potential case*. The situation where $s > d$, i.e. that for which g is not integrable near 0 is called the *hypersingular case*. Some generalizations of what we discuss here to that case are provided in [57], see references therein for further background and results.

Whenever the parameter s appears, it will be with the convention that $s = 0$ in the logarithmic cases.

We will use the notational shortcut $\vec{X}_N$ for $(x_1, \dots, x_N)$ and $d\vec{X}_N$ for $dx_1 \dots dx_N$.

Consider a system of points with energy $\mathcal{H}_N$. In the "canonical formalism" where the number of particles is fixed, statistical mechanics postulates that if the inverse of the temperature is denoted $\beta > 0$, the (density of) probability of observing the system in the state $(x_1, \dots, x_N)$ is given by the Gibbs measure [1]

$$d\mathbb{P}_{N,\beta}(\vec{X}_N) = \frac{1}{Z_{N,\beta}} \exp\left(-\frac{\beta}{2}\mathcal{H}_N(\vec{X}_N)\right) d\vec{X}_N \tag{1.0.8}$$

where

$$Z_{N,\beta} = \int \exp\left(-\frac{\beta}{2}\mathcal{H}_N(\vec{X}_N)\right) d\vec{X}_N \tag{1.0.9}$$

is the normalizing constant that makes $\mathbb{P}_{N,\beta}$ a probability measure, called the *partition function*. It plays an important role in understanding the physics of the system. It is a priori not clear how β and N should be related when letting $N \to \infty$, in principle one should let β depend on N, as discussed further below.

The Gibbs measure can be characterized (this is a relatively easy calculation) by the fact that it is the unique minimizer of the *free energy*

$$\int_{(\mathbb{R}^d)^N} \mathcal{H}_N(\vec{X}_N) d\mathbb{P}(\vec{X}_N) + \frac{2}{\beta} \int_{(\mathbb{R}^d)^N} \mathbb{P}(\vec{X}_N) \log \mathbb{P}(\vec{X}_N), \tag{1.0.10}$$

among (symmetric) probability measures $\mathbb{P}$ on $(\mathbb{R}^d)^N$. Here, we are minimizing the sum of the energy and temperature times (minus) the entropy. The entropy part is a way of measuring the volume in phase-space of microscopic configurations that are underlying any given macroscopic configuration, and its presence in the minimization problem reflects the basic postulate of statistical mechanics that the most likely macroscopic configuration is the one which corresponds to the largest set of microscopic configurations. For these basic aspects, we refer to the textbooks, such as [60].

We now review various motivations for studying such systems.

1.1. Fekete points and approximation theory Fekete points arise in interpolation theory as the points minimizing interpolation errors for numerical integration [93]. Indeed, if one is looking for N interpolation points $\{x_1, \dots, x_N\}$ in some

[1] The choice of the $\beta/2$ normalization instead of β is to fit the usual convention in the literature.

compact set K and coefficients w_j such that the relation

$$\int_K f(x)dx = \sum_{j=1}^{N} w_j f(x_j)$$

is exact for the polynomials of degree $\leqslant N-1$, one sees that one needs to compute the coefficients w_j such that $\int_K x^k = \sum_{j=1}^{N} w_j x_j^k$ for $0 \leqslant k \leqslant N-1$, and this computation is easy if one knows to invert the Vandermonde matrix of the $\{x_j\}_{j=1\ldots N}$. The numerical stability of this operation is the same as the *condition number* of the matrix, i.e. the inverse Vandermonde determinant of the $(x_1, \ldots, x_N)$. The points that minimize the maximal interpolation error for general functions are easily shown to be the Fekete points, defined as those that maximize again the Vandermonde determinant

$$\prod_{i \neq j} |x_i - x_j|$$

or equivalently minimize

$$-\sum_{i \neq j} \log |x_i - x_j|.$$

They are often studied on manifolds, such as the d-dimensional sphere. In Euclidean space, one also considers "weighted Fekete points" which maximize

$$\left(\mathrm{prod}_{i \neq j} |x_i - x_j|\right) e^{-N \sum_i V(x_i)}$$

or equivalently minimize

$$-\sum_{i \neq j} \log |x_i - x_j| + N \sum_{i=1}^{N} V(x_i)$$

which in dimension 2 corresponds exactly to the minimization of our Hamiltonian $\mathcal{H}_N$ in the particular case Log2. They also happen to be zeroes of orthogonal polynomials, see [102].

Since $-\log|x|$ can be obtained as $\lim_{s \to 0} \frac{1}{s}(|x|^{-s} - 1)$, there is also interest in studying "Riesz s-energies" (for these aspects, we refer to the [22, 29, 92] and references therein, in that context such systems are mostly studied on the d-dimensional sphere or torus), i.e. the minimization of

$$\sum_{i \neq j} \frac{1}{|x_i - x_j|^s} \tag{1.1.1}$$

for all possible s, hence a motivation for (1.0.4). Varying s from 0 to ∞ connects *Fekete points* to optimal packing problems. The optimal packing problem has been solved in 1, 2 and 3 dimensions, as well as in dimensions 8 and 24 in the recent breakthrough [41, 109]. The solution in dimension 2 is the triangular lattice, in dimension 3 it is the FCC (face-centered cubic) lattice, in dimension 8 the E_8 lattice and in dimension 24 the Leech lattice. In other dimensions, the solution is not known and it is expected that in high dimension, where the problem is important for error-correcting codes, it is *not* a lattice (in dimension 10 already, a non-lattice competitor is known to beat the lattices), see [43] for these aspects.

The triangular lattice in dimension 2, the E_8 lattice in dimension 8 and the Leech lattice in dimension 24 were conjectured in [40] to be "universally minimizing", i.e. to be the minimizer for a broad class of interactions, and this has very recently been announced to be proven in the cases of dimensions 8 and 24 [42] thanks to the breakthrough of [41, 109]. On the other hand, the local universal minimality (i.e. minimality among periodic configurations) of the triangular lattice in dimension 2, the D_4 lattice in dimension 4, E_8 lattice in dimension 8, and Leech lattice in dimension 24, was shown in [44].

1.2. Statistical mechanics The ensemble given by (1.0.8) in the $\mathsf{Log2}$ case is called in physics a two-dimensional *Coulomb gas* or *one-component plasma* (see e.g. [4, 61, 62, 98] for a physical treatment). The Gibbs measure of the two-dimensional Coulomb gas happens to be directly related to exact solutions of several a priori unrelated models of quantum mechanics : the free electron gas and bosonic Tonks-Girardeau gas in harmonic traps [45] and the Laughlin wave-function in the fractional quantum Hall effect [56,106]. Ginzburg-Landau vortices [94] and vortices in superfluids and Bose-Einstein condensates also interact like two-dimensional Coulomb particles, cf. below. The Coul case with $d = 3$ can be seen as a toy model for (classical) matter (see e.g. [62,75,77,82]).

The general Riesz case can be seen as a generalization of the Coulomb case, motivations for studying Riesz gases are numerous in the physics literature (in solid state physics, ferrofluids, elasticity), see for instance [8, 34, 79, 107], they can also correspond to systems with Coulomb interaction constrained to a lower-dimensional subspace : for instance in the quantum Hall effect, electrons confined to a two-dimensional plane interact via the three-dimension Coulomb kernel.

In all cases of interactions, the systems governed by the Gibbs measure $\mathbb{P}_{N,\beta}$ are considered as difficult systems in statistical mechanics because the interactions are long-range (decay slower than $\frac{1}{|x|^d}$), singular, and the points are not constrained to live on a lattice.

As always in statistical mechanics [60], one would like to understand if there are phase-transitions for particular values of the (inverse) temperature β. For the systems studied here, one may expect what physicists call a liquid for small β, and a crystal for large β. The meaning of crystal in this instance is not to be taken literally as a lattice, but rather as a system of points whose "2-point correlation function" $\rho_2(x, y)$, defined[2] as the probability to have jointly one point at x and one point at y decays at most polynomially as $|x - y| \to \infty$ while retaining some orientational order.

This phase-transition at finite β has been discussed in the physics literature for the $\mathsf{Log2}$ case (see e.g. [30,33,37,105]) but its precise nature is still unclear. In view

[2]for instance the 2-point correlation associated to the Gibbs measure $\mathbb{P}_{N,\beta}$ is its second marginal, obtained by integrating the measure in all but its first two variables.

of the recent progress in computational physics concerning such phenomenon in two-dimensional systems (see e.g. [64]), which suggest a possibly very subtle transition between the liquid and solid phase, the question seems yet out of reach for a rigorous treatment.

1.3. Two component plasmas The two-dimensional "one-component plasma", consisting of positively charged particles, has a "two-component" counterpart which consists in N particles $x_1, \ldots, x_N$ of charge $+1$ and N particles $y_1, \ldots, y_N$ of charge -1 interacting logarithmically, with energy

$$\mathcal{H}_N(\vec{X}_N, \vec{Y}_N) = -\sum_{i \neq j} \log|x_i - x_j| - \sum_{i \neq j} \log|y_i - y_j| + \sum_{i,j} \log|x_i - y_j|$$

and the Gibbs measure

$$\frac{1}{Z_{N,\beta}} e^{-\beta \mathcal{H}_N(\vec{X}_N, \vec{Y}_N)} d\vec{X}_N \, d\vec{Y}_N.$$

Although the energy is unbounded below (positive and negative points attract), the Gibbs measure is well defined for β small enough, more precisely the partition function converges for $\beta < 2$. The system then forms dipoles which do not collapse due to the thermal agitation. The two-component plasma is interesting due to its close relation to two important theoretical physics models: the XY model and the sine-Gordon model (cf. the review [104]), which exhibit a Berezinski-Kosterlitz-Thouless phase transition (made famous by the 2016 physics Nobel prize, cf. [20]) consisting in the binding of these "vortex-antivortex" dipoles.

1.4. Random matrix theory The study of (1.0.8) has attracted a lot of attention due to its connection with random matrix theory (we refer to [50] for a comprehensive treatment). Random matrix theory (RMT) is a relatively old theory, pionereed by statisticians and physicists such as Wishart, Wigner and Dyson, and originally motivated by the understanding of the spectrum of heavy atoms, see [80]. For more recent mathematical references see [7, 46, 50]. The main question asked by RMT is : what is the law of the spectrum of a large random matrix? As first noticed in the foundational papers of [49, 110], in the particular cases (1.0.2)–(1.0.3) the Gibbs measure (1.0.8) corresponds in some particular instances to the joint law of the eigenvalues (which can be computed algebraically) of some famous random matrix ensembles:

- for `Log2`, $\beta = 2$ and $V(x) = |x|^2$, (1.0.8) is the law of the (complex) eigenvalues of an $N \times N$ matrix where the entries are chosen to be normal Gaussian i.i.d. This is called the *Ginibre ensemble* [54].
- for `Log1`, $\beta = 2$ and $V(x) = x^2/2$, (1.0.8) is the law of the (real) eigenvalues of an $N \times N$ Hermitian matrix with complex normal Gaussian iid entries. This is called the Gaussian Unitary Ensemble (GUE).
- for `Log1`, $\beta = 1$ and $V(x) = x^2/2$, (1.0.8) is the law of the (real) eigenvalues of an $N \times N$ real symmetric matrix with normal Gaussian iid entries. This is called the Gaussian Orthogonal Ensemble (GOE).

- for `Log1`, $\beta = 4$ and $V(x) = x^2/2$, (1.0.8) is the law of the eigenvalues of an $N \times N$ quaternionic symmetric matrix with normal Gaussian iid entries. This is called the Gaussian Symplectic Ensemble.

One thus observes in these ensembles the phenomenon of "repulsion of eigenvalues": they repel each other logarithmically, i.e. like two-dimensional Coulomb particles.

For the `Log1` and `Log2` cases, at the specific temperature $\beta = 2$, the law (1.0.8) acquires a special algebraic feature : it becomes a *determinantal* process, part of a wider class of processes (see [23, 59]) for which the correlation functions are explicitly given by certain determinants. This allows for many explicit algebraic computations, on which there is a large literature. One can also compute an expansion of $\log Z_{N,\beta}$ as $N \to \infty$ (see [80]) and the limiting processes at the microscopic scale: in the `Log1` case it is the sine process (and this can be generalized to any β, see [7, Sec. 3.1, 3.9]), and in the `Log2` case it is the Ginibre point process (see [59, Sec. 6.4]). Many relevant quantities that can be computed explicitly for $\beta = 2$ are not exactly known for the $\beta \neq 2$ case, even in the case of the potential $V(x) = |x|^2$. The particular case of (1.0.2) for all β and general V, also called β-ensembles, has however been well-understood. In particular, thanks to the works of [12, 13, 25, 26, 63, 99], one has expansions of $\log Z_{N,\beta}$, Central Limit Theorems for linear statistics, and *universality* (after suitable rescaling) of the microscopic behavior and local statistics of the points, i.e. the fact that they are essentially independent of V.

Considering the coincidence between a statistical mechanics model and the law of the spectrum of a random matrix model for several values of the inverse temperature, it is also natural to ask whether such a correspondence exists for any value of β, i.e. whether $\mathbb{P}_{N,\beta}$ as defined in (1.0.8) can be seen as a law of eigenvalues for some random matrix model. The answer is positive in dimension 1 for any β: a somehow complicated model of tridiagonal matrices can be associated to the Gibbs measure of the one-dimensional log-gas at inverse temperature β, see [48, 66]. This and other methods allow again to compute a lot explicitly, and to derive that the microscopic laws of the eigenvalues are those of a so called *sine-β process* [108].

1.5. Complex geometry and theoretical physics Two-dimensional Coulomb systems (in the determinantal case $\beta = 2$) are of interest to geometers because they serve to construct Kähler-Einstein metrics with positive Ricci curvature on complex manifolds, cf. [17]. Another important motivation is the construction of Laughlin states for the Fractional Quantum Hall effect, which naturally leads to analyzing a two-dimensional Coulomb gas (cf. [56, 91, 106]). When studying the Fractional Quantum Hall effect on a complex manifold, the coefficients in the expansion of the (logarithm of the) partition function have interpretations as geometric invariants, and it is thus of interest to be able to compute them, cf [67].

1.6. Vortices in condensed matter physics In both superconductors with applied magnetic fields (see [94] for general and mathematical reference) and rotating superfluids and Bose-Einstein condensates (see [3] for general and mathematical reference), one observes the occurrence of quantized "vortices" (which are local point defects of superconductivity or superfluidity, surrounded by a current loop). The vortices repel each other, while being confined together by the effect of the magnetic field or rotation, and the result of the competition between these two effects is that, as predicted by Abrikosov [1], they arrange themselves in a particular *triangular lattice* pattern, called *Abrikosov lattice*, cf. Fig. 1.6.1 (for more pictures, see `www.fys.uio.no/super/vortex/`).

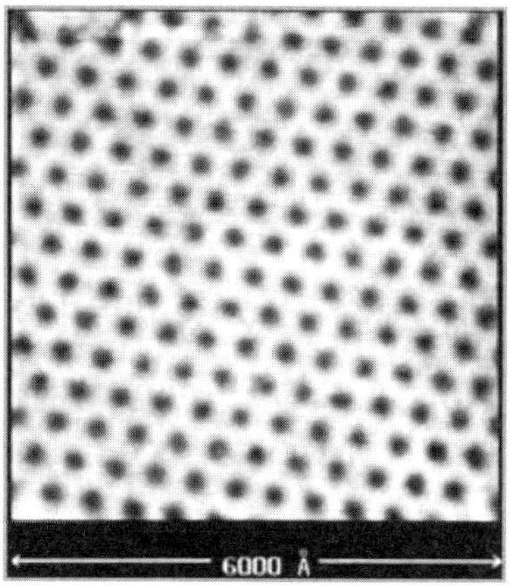

FIGURE 1.6.1. Abrikosov lattice, H. F. Hess et al. Bell Labs *Phys. Rev. Lett.* 62, 214 (1989)

When restricting to a two-dimensional situation, it can be shown formally, cf. [100, Chap. 1] for a formal derivation, but also rigorously [38, 94, 95], that the minimization problem governing the behavior of such systems can be reduced, in terms of the vortices, to the minimization of an energy of the form (1.0.1) in the case Log2. This naturally leads to the question of understanding the connection between minimizers of $\mathcal{H}_N$ in the Log2 case and the Abrikosov triangular lattice.

2. Equilibrium measure and leading order behavior

2.1. The macroscopic behavior: empirical measure It is well-known since [39] (see e.g. [93] for the logarithmic cases), that under suitable assumptions on V, the leading order behavior of $\mathcal{H}_N$ is governed by the minimization of the functional

$$\mathcal{I}_V(\mu) := \iint_{\mathbb{R}^d \times \mathbb{R}^d} g(x-y)\,d\mu(x)\,d\mu(y) + \int_{\mathbb{R}^d} V(x)\,d\mu(x) \tag{2.1.1}$$

defined over the space $\mathcal{P}(\mathbb{R}^d)$ of probability measures on $\mathbb{R}^d$, which may also take the value $+\infty$.

Note that $\mathcal{I}_V(\mu)$ is simply a continuum version of the discrete Hamiltonian $\mathcal{H}_N$. From the point of view of statistical mechanics, $\mathcal{I}_V$ is the "mean-field" limit (or also the Γ-limit) energy of $\frac{1}{N^2}\mathcal{H}_N$, while we will see that from the point of view of probability, $\mathcal{I}_V$ plays the role of a *rate function*.

Under suitable assumptions, there is a unique minimizer of $\mathcal{I}_V$ on the space $\mathcal{P}(\mathbb{R}^d)$ of probability measures on $\mathbb{R}^d$, it is called the *equilibrium measure* and we denote it by μ_V. Its uniqueness follows from the strict convexity of $\mathcal{I}_V$. For existence, one assumes

(A1): V is finite, lower semi-continuous, and bounded below

(A2): (growth assumption)[3]

$$\lim_{|x|\to+\infty}\left(\frac{V(x)}{2}+g(x)\right)=+\infty.$$

One then proceeds in a standard fashion, taking a minimizing sequence for $\mathcal{I}_V$ and using that $\mathcal{I}_V$ is coercive and lower semi-continuous. Finally, one has

Theorem 2.1.2 (Frostman [52], existence and characterization of the equilibrium measure). *Under the assumptions* ***(A1)-(A2)****, the minimum of* $\mathcal{I}_V$ *over* $\mathcal{P}(\mathbb{R}^d)$ *exists, is finite and is achieved by a unique* μ_V*, which has a compact support of positive capacity. In addition* μ_V *is uniquely characterized by the fact that there exists a constant* c *such that*

$$\begin{cases} h^{\mu_V}+\dfrac{V}{2}\geqslant c & \text{quasi-everywhere in } \mathbb{R}^d\\[2ex] h^{\mu_V}+\dfrac{V}{2}=c & \text{quasi-everywhere in the support of } \mu_V\end{cases}\tag{2.1.3}$$

where

$$h^{\mu_V}(x):=\int_{\mathbb{R}^d} g(x-y)\,d\mu_V(y)\tag{2.1.4}$$

is the electrostatic potential generated by μ_V*; and then the constant* c *must be equal to*

$$c=\mathcal{I}_V(\mu_V)-\frac{1}{2}\int_{\mathbb{R}^d}V(x)\,d\mu_V(x).\tag{2.1.5}$$

Quasi-everywhere means except on a set of zero capacity. The capacity of a set (see [2,93] or [76, Sec. 11.15]) is an appropriate notion of size, suffice it to say that a set of null capacity has zero Lebesgue measure (but the converse is not true). The proof of this theorem can easily be adapted from [93, Chap. 1] or [100, Chap. 2]. The relations (2.1.3) can be seen as the Euler-Lagrange equations associated to the minimization of $\mathcal{I}_V$, they are obtained by making variations of the form $(1-t)\mu_V+t\nu$ where μ_V is the minimizer, ν is an arbitrary probability measure of finite energy, and $t\in[0,1]$.

Remark 2.1.6. Note that by (1.0.6), in all Coulomb cases in dimension $d\geqslant 2$, the function h^{μ_V} solves :

$$-\Delta h^{\mu_V}=c_d\mu_V$$

[3]This assumption can be slightly weakened, and this is important for studying Fekete points on the two-dimensional sphere, see [18,58].

where c_d is the constant defined in (1.0.7).

Example 2.1.7 (C^2 potentials and RMT examples). In general, the relations (2.1.3) say that the total potential $h^{\mu_V} + \frac{V}{2}$ is constant on the support of the charges. Moreover, in dimension $d \geqslant 2$, applying the Laplacian on both sides of (2.1.3) and using Remark 2.1.6 gives that, on the interior of the support of the equilibrium measure, if $V \in C^2$,

$$c_d \mu_V = \frac{\Delta V}{2} \tag{2.1.8}$$

i.e. the density of the measure on the interior of its support is given by $\frac{\Delta V}{2c_d}$. For example if V is quadratic, then the density $\frac{\Delta V}{2c_d}$ is constant on the interior of its support, and if $V(x) = c|x|^2$ the equilibrium measure is a (multiple of) the characteristic function of a ball (by uniqueness, μ_V should be radially symmetric). This corresponds to the important example of the Ginibre ensemble: for the Log2 case with $V(x) = |x|^2$, one has $\mu_V = \frac{1}{\pi}\mathbf{1}_{B_1}$ where $\mathbf{1}$ denotes a characteristic function and B_1 is the unit ball (the combination of (2.1.8) with the constraint of being a probability measure imposes the support to be B_1). This is known as the *circular or circle law* for the Ginibre ensemble in the context of Random Matrix Theory (RMT). Its derivation is attributed to Ginibre [54], Mehta [80], an unpublished paper of Silverstein and Girko [55].

Note that in the Log1 case with $V(x) = x^2$, we obtain the equilibrium measure $\mu_V(x) = \frac{1}{2\pi}\sqrt{4-x^2}\mathbf{1}_{|x|\leqslant 2}$, which corresponds in the context of RMT (GUE and GOE ensembles) to *Wigner's semi-circle law*, cf. [80, 110, 111].

The equilibrium measure μ_V can also be interpreted in terms of the solution to a classical *obstacle problem*, which is essentially dual to the minimization of $\mathcal{I}_V$, and better studied from the PDE point of view (in particular the regularity of μ_V and of the boundary of its support). For this aspect, see [100, Chap. 2].

Definition 2.1.9. From now on, we denote by ζ the function

$$\zeta = h^{\mu_V} + \frac{V}{2} - c. \tag{2.1.10}$$

We note that in view of (2.1.3), $\zeta \geqslant 0$ and $\zeta = 0$ μ_V-a.e.

In the rest of the course, we will always assume that the support of μ_V, denoted Σ, is a set with a nice boundary (say in the Hölder class $C^{1,\alpha}$), and that the density $\mu_V(x)$ of μ_V is a Hölder continuous function in Σ.

2.2. Large Deviations Principle at leading order Here we want to study the typical behavior of the configurations under $\mathbb{P}_{N,\beta}$ in terms of the equilibrium measure. For us, a convenient macroscopic observable is given by the empirical measure of the particles: if $\vec{X}_N \in (\mathbb{R}^d)^N$ we form

$$\mu_N^{\mathrm{emp}}[\vec{X}_N] := \frac{1}{N}\sum_{i=1}^N \delta_{x_i}, \tag{2.2.1}$$

which is a probability measure on $\mathbb{R}^d$. The minimisation of $\mathcal{I}_V$ determines the macroscopic (or global) behavior of the system in the following sense:

- Minimizers of $\mathcal{H}_N$ are such that $\mu_N^{\mathrm{emp}}[\vec{X}_N]$ converges to μ_V as $N \to \infty$ (cf. [100, Chap. 2])
- In fact, for *fixed* β, $\mu_N^{\mathrm{emp}}[\vec{X}_N]$ converges weakly to μ_V as $N \to \infty$ *almost surely under the canonical Gibbs measures* $\mathbb{P}_{N,\beta}$, see below.

In other words, not only the minimisers of the energy, but almost every (under the sequence of Gibbs measures) sequence of particles is such that the empirical measure converges to the equilibrium measure, if β is taken to be independent of N. Since μ_V does not depend on the temperature, in particular *the asymptotic macroscopic behavior of the system is independent of* β.

The result is phrased as a Large Deviations Principle in a sense that we now recall (for reference see [47]).

Definition 2.2.2 (Rate function). Let X be a metric space (or a topological space). A rate function is a lower semi-continuous function $I : X \to [0, +\infty]$, it is called a *good rate function* if its sub-level sets $\{x, I(x) \leqslant \lambda\}$ are compact for all $\lambda \in \mathbb{R}$.

Definition 2.2.3 (Large deviations). Let $\{P_N\}_N$ be a sequence of Borel probability measures on X and $\{a_N\}_N$ a sequence of positive real numbers diverging to $+\infty$. Let also I be a (good) rate function on X. The sequence $\{P_N\}_N$ is said to satisfy a large deviation principle (LDP) at speed a_N with (good) rate function I if for every Borel set $E \subset X$ the following inequalities hold :

$$-\inf_{\overset{\circ}{E}} I \leqslant \liminf_{N\to+\infty} \frac{1}{a_N} \log P_N(E) \leqslant \limsup_{N\to+\infty} \frac{1}{a_N} \log P_N(E) \leqslant -\inf_{\bar{E}} I \tag{2.2.4}$$

where $\overset{\circ}{E}$ (resp. $\bar{E}$) denotes the interior (resp. the closure) of E for the topology of X.

Formally, it means that $P_N(E)$ should behave roughly like $e^{-a_N \inf_E I}$. The rate function I is the rate of exponential decay of the probability of rare events, and the events with larger probability are the ones on which I is smaller.

The formal result is the following Large Deviations Principle. Proofs of this result (or very close ones) can be found in [85] (in dimension 2), [15] (in dimension 1), [16] (in dimension 2) for the particular case of a quadratic potential (and $\beta = 2$), [35], [100] and [53], in more general setting. One can also see [17] for results in a more general (still determinantal) setting of multidimensional complex manifolds.

We need an additional assumption, which can be seen as a strengthening of **(A2)**:

(A3) Given β, for N large enough, we have

$$\int \exp\left(-\beta N \left(\frac{V(x)}{2} - \log|x|\right)\right) dx < \infty \tag{Log1, Log2}$$

$$\text{(Coul)} \int \exp\left(-\frac{\beta}{2}NV(x)\right)\, dx < +\infty.$$

Note in particular that **(A3)** ensures that the integral in (1.0.9) is convergent, hence $Z_{N,\beta}$ is well-defined, and so is $\mathbb{P}_{N,\beta}$.

Theorem 2.2.5 (Large deviations principle for the Coulomb gas at speed N^2). *Let $\beta > 0$ be given and assume that V is continuous, satisfies* **(A3)** *and that $(1-\alpha_0)V$ satisfies* **(A2)** *for some $\alpha_0 > 0$. Then the law of $\mu_N^{\text{emp}}[\vec{X}_N]$ under $\{\mathbb{P}_{N,\beta}\}_N$ satisfies a large deviations principle at speed N^2 with good rate function $\frac{\beta}{2}\hat{\mathcal{I}}_V$ where $\hat{\mathcal{I}}_V = \mathcal{I}_V - \min_{\mathcal{P}(\mathbb{R}^d)} \mathcal{I}_V = \mathcal{I}_V - \mathcal{I}_V(\mu_V)$. Moreover*

$$\lim_{N\to+\infty} \frac{1}{N^2} \log Z_{N,\beta} = -\frac{\beta}{2}\mathcal{I}_V(\mu_V) = -\frac{\beta}{2} \min_{\mathcal{P}(\mathbb{R}^d)} \mathcal{I}_V. \tag{2.2.6}$$

Here the underlying topology is that of weak convergence on $\mathcal{P}(\mathbb{R}^d)$.

The heuristic reading of the LDP is that

$$\mathbb{P}_{N,\beta}(E) \approx e^{-\frac{\beta}{2}N^2(\min_E \mathcal{I}_V - \min \mathcal{I}_V)}, \tag{2.2.7}$$

which in view of the uniqueness of the minimizer of $\mathcal{I}_V$ implies as stated above (informally) that configurations whose empirical measure does not converge to μ_V as $N \to \infty$ have exponentially decaying probability.

When β is allowed to depend on N, and in particular if β tends to 0 as $N \to \infty$ (i.e. temperature is rather large), then the macroscopic behavior of the system may no longer be independent of β. More specifically, if β is chosen to be $\frac{\beta_0}{N}$ for some fixed β_0, the temperature is felt at leading order and brings an entropy term: there is a temperature-dependent equilibrium measure $\mu_{V,\beta}$ which is the unique minimizer of

$$I_{V,\beta_0}(\mu) = \mathcal{I}_V(\mu) + \frac{2}{\beta_0}\int \mu \log \mu. \tag{2.2.8}$$

Contrarily to the equilibrium measure, μ_{V,β_0} is not compactly supported, but decays exponentially fast at infinity. This mean-field behavior and convergence of marginals was first established for logarithmic interactions [32,65] (see [81] for the case of regular interactions) using an approach based on de Finetti's theorem. In the language of Large Deviations, the same LDP as above then holds with rate function $I_{V,\beta_0} - \min I_{V,\beta_0}$, and the Gibbs measure now concentrates as $N \to \infty$ on a neighborhood of μ_{V,β_0}, for a proof see [53]. One can also refer to [88,89] for the mean-field and chaos aspects with a particular focus on their adaptation to the quantum setting.

The Coulomb nature of the interaction is not really needed for all these results on the leading order behavior and they can easily be generalized to a broader class of interactions (see [100, Chap. 2], [53]).

2.3. Further questions In contrast to the macroscopic result, several observations (e.g. by numerical simulation, see Figure 2.3.1) suggest that even in the scaling of β fixed for which the macroscopic behavior of the system is independent of β, the behavior of the system at the microscopic scale[4] depends heavily on β. The questions that one would like to answer are then to describe the system beyond the macroscopic scale, at the mesoscopic (i.e. $N^{-\alpha}$ for $0 < \alpha < 1/d$) scale, and at the microscopic ($N^{-1/d}$) scale.

FIGURE 2.3.1. Case Log2 with $N = 100$ and $V(x) = |x|^2$, for $\beta = 400$ (left) and $\beta = 5$ (right).

Since one already knows that $\sum_{i=1}^N \delta_{x_i} - N\mu_V$ is small (more precisely much smaller than N in some suitable sense), one knows that the so-called *discrepancy*

$$D(x,r) := \int_{B(x,r)} \sum_{i=1}^N \delta_{x_i} - N\, d\mu_V$$

is typically $o(N)$ as long as $r > 0$ is fixed)here $B(x,r)$ denotes the ball of center x and radius r). Is this still true at the mesoscopic scale for r of the order $N^{-\alpha}$ with $\alpha < 1/d$? Is it true down to the microscopic scale, i.e. for $r = RN^{-1/d}$ with $R \gg 1$? Does it hold regardless of the temperature? This would correspond to a *rigidity result*. Note that point processes of intensity 1 (the average number of points per unit volume) with discrepancies growing at most like the perimeter of the ball have been called *hyperuniform* and are of interest to physicists for a variety of applications (cf. [107]).

Once one proves rigidity down to the microscopic scale, one may also like to characterize the fluctuations of linear statistics of the form

$$\sum_{i=1}^N f(x_i) - N \int f d\mu_V,$$

where f is a regular enough test-function. In the logarithmic cases, they are proven to converge to a Gaussian distribution whose variance depends on the temperature, as will be seen below.

Another point of view is that of large deviations theory. Then, one wishes to study a microscopic observable, the microscopic point process obtained after blow-up by $N^{1/d}$, and characterize it as minimizing a certain (free) energy or rate function, thus connecting to a crystallization question.

In the course of the analysis we will see how a natural choice of scaling of β with respect to N emerges if one wishes to see the interaction energy and the entropy at the microscopic scale balance out.

[4]Since the N particles are typically confined in a set of order $O(1)$, the microscopic, inter-particle scale is $O(N^{-1/d})$.

In all the cases, one wants to understand precisely how the behavior depends on β, but also on V. It is believed that at the macroscopic and microscopic levels, in the logarithmic cases the behavior is independent on V, a universality feature.

3. Splitting of the Hamiltonian and electric approach

We now start to present the approach to these problems initiated in [96] and continued in [71,84,90,97]. It relies on a splitting of the energy into a fixed leading order term and a next order term expressed in terms of the charge fluctuations, and on a rewriting of this next order term via the "electric potential" generated by the points. As we will see in Corollary 3.6.4, it leads to "concentration bounds" that better quantify the probability that μ_N^{emp} can deviate from μ_V.

3.1. The splitting formula The splitting consists in an exact formula that separates the leading (N^2) order term in $\mathcal{H}_N$ from next order terms.

Since we expect $\mu_N^{\mathrm{emp}}[\vec{X}_N]$ to converge to μ_V, we may try to "expand" around μ_V. In all the sequel, we denote for any probability measure μ,

$$\mathrm{fluct}_N^{\mu}[\vec{X}_N] = N(\mu_N^{\mathrm{emp}}[\vec{X}_N] - \mu) = \sum_{i=1}^{N} \delta_{x_i} - N\mu. \tag{3.1.1}$$

Unless ambiguous, we will drop the $\vec{X}_N$ dependence.

Lemma 3.1.2 (Splitting formula). *Assume μ_V is absolutely continuous with respect to the Lebesgue measure. For any N and any $\vec{X}_N \in (\mathbb{R}^d)^N$ we have*

$$\mathcal{H}_N(\vec{X}_N) = N^2 \mathcal{I}_V(\mu_V) + 2N \sum_{i=1}^{N} \zeta(x_i) + F_N^{\mu_V}(\vec{X}_N) \tag{3.1.3}$$

where ζ was defined in (2.1.10) *and where we define for any probability measure μ*

$$F_N^{\mu}(\vec{X}_N) = \iint_{\mathbb{R}^d \times \mathbb{R}^d \setminus \triangle} g(x-y)\, d\mathrm{fluct}_N^{\mu}(x)\, d\mathrm{fluct}_N^{\mu}(y), \tag{3.1.4}$$

with $\triangle$ denoting the diagonal of $\mathbb{R}^d \times \mathbb{R}^d$.

Proof. We may write

$$\begin{aligned}
\mathcal{H}_N(\vec{X}_N) &= \sum_{i \neq j} g(x_i - x_j) + N \sum_{i=1}^{N} V(x_i) \\
&= N^2 \iint_{\triangle^c} g(x-y)\, d\mu_N^{\mathrm{emp}}(x)\, d\mu_N^{\mathrm{emp}}(y) + N^2 \int_{\mathbb{R}^d} V\, d\mu_N^{\mathrm{emp}}(x) \\
&= N^2 \iint_{\triangle^c} g(x-y)\, d\mu_V(x)\, d\mu_V(y) + N^2 \int_{\mathbb{R}^d} V\, d\mu_V \\
&\quad + 2N \iint_{\triangle^c} g(x-y)\, d\mu_V(x)\, d\mathrm{fluct}_N^{\mu_V}(y) + N \int_{\mathbb{R}^d} V\, d\mathrm{fluct}_N^{\mu_V} \\
&\quad + \iint_{\triangle^c} g(x-y)\, d\mathrm{fluct}_N^{\mu_V}(x)\, d\mathrm{fluct}_N^{\mu_V}(y).
\end{aligned} \tag{3.1.5}$$

We now recall that ζ was defined in (2.1.10) by

$$\zeta := h^{\mu_V} + \frac{V}{2} - c \qquad h^{\mu_V} := \int_{\mathbb{R}^d} g(\cdot - y)\, d\mu_V(y)$$

and that $\zeta = 0$ in Σ. Using this we may rewrite the second last line of (3.1.5) as

$$\begin{aligned}
&2N \iint_{\triangle^c} g(x-y)\, d\mu_V(x)\, d\mathrm{fluct}_N^{\mu_V}(y) + N \int_{\mathbb{R}^d} V d\mathrm{fluct}_N^{\mu_V} \\
&= 2N \int_{\mathbb{R}^d} (h^{\mu_V} + \frac{V}{2}) d\mathrm{fluct}_N^{\mu_V} = 2N \int_{\mathbb{R}^d} (\zeta + c) d\mathrm{fluct}_N^{\mu_V} \\
&= 2N^2 \int_{\mathbb{R}^d} \zeta d\mu_N^{\mathrm{emp}} - 2N^2 \int_{\mathbb{R}^d} \zeta d\mu_V + 2Nc \int_{\mathbb{R}^d} d\mathrm{fluct}_N^{\mu_V} = 2N^2 \int_{\mathbb{R}^d} \zeta d\mu_N^{\mathrm{emp}}.
\end{aligned}$$

The last equality is due to the facts that $\zeta \equiv 0$ on the support of μ_V and that μ_N^{emp} and μ_V are both probability measures so that $\int d\mathrm{fluct}_N^{\mu_V} = 0$. We also have to notice that since μ_V is absolutely continuous with respect to the Lebesgue measure, we may include the diagonal back into the domain of integration. By that same argument, one may recognize in the third last line of (3.1.5), the quantity $N^2 \mathcal{I}_V(\mu_V)$, cf. (2.1.1). □

The function ζ can be seen as an effective potential, whose sole role is to confine the points to the set Σ. We have thus reduced to studying $F_N^{\mu_V}$, which represents the total interaction of the neutral system formed by the singular charges at the x_i's and the negative background charge $-N\mu_V$. It is a priori not clear of which order this term is, and whether it is bounded below (because of the removal of the diagonal, F_N^{μ} is in general not positive)!

3.2. Electric interpretation To go further, we apply an electric interpretation of the energy $F_N^{\mu_V}$ first used in [96], and rewrite the energy via truncation as in [90] and [84]. Such a computation allows us to replace the sum of pairwise interactions of all the charges and "background" by an integral (extensive) quantity, which is easier to handle. This will be the first time that the Coulomb (or Riesz) nature of the interaction is really used. For simplicity of exposition, we will present it in the Coulomb cases only.

3.2.1. Electric potential and truncation *Electric potential.* For any N-tuple $\vec{X}_N$ of points in the space $\mathbb{R}^d$, and any probability density μ, we define the (electric) potential generated by $\vec{X}_N$ and μ as

$$H_N^{\mu}(x) := \int_{\mathbb{R}^d} -g(x-y) \left(\sum_{i=1}^N \delta_{x_i} - N d\mu \right)(y). \tag{3.2.1}$$

Note that in principle it should be denoted $H_N^{\mu}[\vec{X}_N](x)$, but we omit the $\vec{X}_N$ dependence for the sake of lightness of notation. In the Coulomb cases, the potential H_N^{μ} satisfies

$$-\Delta H_N^{\mu} = c_d \Big(\sum_{i=1}^N \delta_{x_i} - N d\mu \Big) \quad \text{in } \mathbb{R}^d. \tag{3.2.2}$$

Note that H_N^μ decays at infinity, because the charge distribution fluct_N is compactly supported and has zero total charge, hence, when seen from infinity behaves like a dipole. More precisely, H_N^μ decays like ∇g at infinity, that is $O(\frac{1}{|x|^{d-1}})$ and its gradient ∇H_N^μ decays like the second derivative $\nabla^2 g$, that is $O(\frac{1}{|x|^d})$.

Truncated potential. Let $\vec{X}_N \in (\mathbb{R}^d)^N$ be fixed. For any $\vec{\eta} = (\eta_1, \dots, \eta_N)$ we define the truncated potential

$$H_{N,\vec{\eta}}^\mu(x) = H_N^\mu(x) - \sum_{i=1}^N (g(x - x_i) - g(\eta_i))_+ \tag{3.2.3}$$

where $(\cdot)_+$ denotes the positive part of a number. Note that in view of the singular behavior of g at the origin, H_N^μ diverges at each x_i, and here we "chop off" these infinite peaks at distance η_i from x_i. We will also denote

$$f_\eta(x) = (g(x) - g(\eta))_+, \tag{3.2.4}$$

and point out that f_η is supported in $B(0, \eta)$. We note that by radial symmetry, letting $\delta_{x_i}^{(\eta_i)}$ denote the uniform measure of mass 1 on $\partial B(x_i, \eta_i)$, we have

$$f_\eta = g * \left(\delta_0 - \delta_0^{(\eta)}\right),$$

so that

$$H_{N,\vec{\eta}}^\mu(x) := \int_{\mathbb{R}^d} g(x-y) \left(\sum_{i=1}^N \delta_{x_i}^{(\eta_i)} - N d\mu \right)(y). \tag{3.2.5}$$

By $H_{N,\eta}$ we simply denote $H_{N,\vec{\eta}}$ when all the η_i are chosen equal to η.

3.2.2. Re-expressing the interaction term Formally, using Green's formula (or Stokes' theorem) and the definitions, one would like to write that in the Coulomb cases

$$F_N^\mu(\vec{X}_N) = \int H_N^\mu \, d\mathrm{fluct}_N = \int H_N^\mu \left(-\frac{1}{c_d} \Delta H_N^\mu \right) \approx \frac{1}{c_d} \int |\nabla H_N^\mu|^2 \tag{3.2.6}$$

This is the place where we really use for the first time in a crucial manner the Coulombic nature of the interaction kernel g. Such a computation allows to replace the sum of pairwise interactions of all the charges and "background" by an integral (extensive) quantity, which is easier to handle in some sense. However, (3.2.6) does not make sense because ∇H_N^μ fails to be in L^2 due to the presence of Dirac masses. Indeed, near each atom x_i, the vector-field ∇H_N^μ behaves like ∇g and the integrals $\int_{B(0,\eta)} |\nabla g|^2$ are divergent in all dimensions. Another way to see this is that the Dirac masses charge the diagonal $\triangle$ and so the restriction to its complement cannot be ignored. The point of the truncation above is precisely to remedy this and give a way of computing $\int |\nabla H_N^\mu|^2$ in a "renormalized" fashion.

We now give a proper meaning to the statement.

Lemma 3.2.7 (Electric rewriting of the next order energy). *Given $\vec{X}_N$, for any $\vec{\eta}$ with $\eta_i \leq \frac{1}{2}$ such that the $B(x_i, \eta_i)$ are disjoint, for any absolutely continuous probability* *measure* μ *we have*

$$F_N^\mu(\vec{X}_N) = \frac{1}{c_d}\left(\int_{\mathbb{R}^d} |\nabla H^\mu_{N,\vec{\eta}}|^2 - c_d \sum_{i=1}^N g(\eta_i)\right) + O\left(N\|\mu\|_{L^\infty} \sum_{i=1}^N \eta_i^2\right). \tag{3.2.8}$$

If the balls $B(x_i,\eta_i)$ *are not assumed to be disjoint, then we still have that the left-hand side is larger than the right-hand side.*

The last term in the right-hand side of (3.2.8) should be thought of as a small error. In practice, we will take $\eta_i \leqslant N^{-1/d}\eta$ and let $\eta \to 0$, which corresponds to truncating the charges at a lengthscale barely smaller than the typical interparticle distance. The error is then $O(\eta^2 N^{2-2/d})$. Choosing $\eta_i = N^{-1/d}$ we thus find that F_N^μ is bounded below as follows

$$F_N^\mu(\vec{X}_N) \geqslant \left(\frac{N}{2}\log N\right)_{\mathbf{1}_{\mathrm{Log2}}} - CN^{2-\frac{2}{d}},$$

where C depends only on μ and d.

Proof. For the proof, we drop the superscripts μ. First we notice that $\int_{\mathbb{R}^d} |\nabla H_{N,\vec{\eta}}|^2$ is a convergent integral and that

$$\int_{\mathbb{R}^d} |\nabla H_{N,\vec{\eta}}|^2 = c_d \iint g(x-y)\, d\left(\sum_{i=1}^N \delta_{x_i}^{(\eta_i)} - N\mu\right)(x)\, d\left(\sum_{i=1}^N \delta_{x_i}^{(\eta_i)} - N\mu\right)(y). \tag{3.2.9}$$

Indeed, we may choose R large enough so that all the points of $\vec{X}_N$ are contained in the ball $B_R = B(0,R)$. By Green's formula and (3.2.5), we have

$$\int_{B_R} |\nabla H_{N,\vec{\eta}}|^2 = \int_{\partial B_R} H_{N,\vec{\eta}} \frac{\partial H_N}{\partial \nu} - c_d \int_{B_R} H_{N,\vec{\eta}} \left(\sum_{i=1}^N \delta_{x_i}^{(\eta_i)} - N\mu\right). \tag{3.2.10}$$

In view of the decay of H_N and ∇H_N mentioned above, the boundary integral tends to 0 as $R \to \infty$, and so we may write

$$\int_{\mathbb{R}^d} |\nabla H_{N,\vec{\eta}}|^2 = c_d \int_{\mathbb{R}^d} H_{N,\vec{\eta}} \left(\sum_{i=1}^N \delta_{x_i}^{(\eta_i)} - N\mu\right)$$

and thus (3.2.9) holds in view of (3.2.1). We may next write

$$\begin{aligned}
&\iint g(x-y)\, d\left(\sum_{i=1}^N \delta_{x_i}^{(\eta_i)} - N\mu\right)(x)\, d\left(\sum_{i=1}^N \delta_{x_i}^{(\eta_i)} - N\mu\right)(y) \\
&\qquad - \iint_{\triangle^c} g(x-y)\, d\mathrm{fluct}_N(x)\, d\mathrm{fluct}_N(y) \\
&= \sum_{i=1}^N g(\eta_i) + \sum_{i\neq j} \iint g(x-y)\left(\delta_{x_i}^{(\eta_i)}\delta_{x_j}^{(\eta_j)} - \delta_{x_i}\delta_{x_j}\right) \\
&\qquad + 2N \sum_{i=1}^N \iint g(x-y)\left(\delta_{x_i} - \delta_{x_i}^{(\eta_i)}\right)\mu.
\end{aligned} \tag{3.2.11}$$

Let us now observe that $\int g(x-y)\delta_{x_i}^{(\eta_i)}(y)$, the potential generated by $\delta_{x_i}^{(\eta_i)}$ is equal to $\int g(x-y)\delta_{x_i}$ outside of $B(x_i,\eta_i)$, and smaller otherwise, this is "Newton's theorem", see for instance [76]. Since its Laplacian is $-c_d\delta_{x_i}^{(\eta_i)}$, a negative measure, this is also a superharmonic function, so by the maximum principle, its value at a point x_j is larger or equal to its average on a sphere centered at x_j. Moreover, outside $B(x_i,\eta_i)$ it is a harmonic function, so its values are equal to its averages. We deduce from these considerations, and reversing the roles of i and j, that for each $i \neq j$,

$$\int g(x-y)\delta_{x_i}^{(\eta_i)}\delta_{x_j}^{(\eta_j)} \leqslant \int g(x-y)\delta_{x_i}\delta_{x_j}^{(\eta_j)} \leqslant \int g(x-y)\delta_{x_i}\delta_{x_j}$$

with equality if $B(x_i,\eta_i)\cap B(x_j,\eta_j) = \emptyset$. We conclude that the second term in the right-hand side of (3.2.11) is nonnegative and equal to 0 if all the balls are disjoint. Finally, by the above considerations, since $\int g(x-y)\delta_{x_i}^{(\eta_i)}$ coincides with $\int g(x-y)\delta_{x_i}$ outside $B(x_i,\eta_i)$, we may rewrite the last term in the right-hand side of (3.2.11) as

$$2N\sum_{i=1}^{N}\int_{B(x_i,\eta_i)}(g(x-x_i)-g(\eta_i))d\mu.$$

On the other hand, recalling (3.2.4), if $\mu \in L^\infty$ then we have

$$\left|\sum_{i=1}^{N}\int_{\mathbb{R}^d} f_{\eta_i}\,d\mu\right| \leqslant C_d\|\mu\|_{L^\infty}\sum_{i=1}^{N}\eta_i^2. \tag{3.2.12}$$

Indeed, it suffices to observe that

$$\int_{B(0,\eta)} f_\eta = \int_0^\eta (g(r)-g(\eta))r^{d-1}\,dr = -\int_0^r g'(r)r^d\,dr, \tag{3.2.13}$$

with an integration by parts (and an abuse of notation, viewing g as a function on $\mathbb{R}$) and using the explicit form of g it follows that

$$\int_{B(0,\eta)} |f_\eta| \leqslant C_d\eta^2. \tag{3.2.14}$$

By the definition (3.2.4), we thus have obtained the result. □

3.3. The case $d = 1$ In the case Log1, or in the Riesz cases Riesz, g is no longer the Coulomb kernel, so the formal computation (3.2.6) does not work. However g is in the case Log1 the kernel of the half-Laplacian, and it is known that the half-Laplacian can be made to correspond to the Laplacian by adding one extra space dimension. In the same way, in the case Riesz, g is the kernel of a second order local operator, after adding one extra space dimension. In other words, in the case Log1 we should imbed the space $\mathbb{R}$ into the two-dimensional space $\mathbb{R}^2$ and consider the harmonic extension of $H_N^{\mu_V}$, defined in (3.2.1), to the whole plane. That extension will solve an appropriate Laplace equation, and we will reduce dimension 1 to a special case of dimension 2. An analogous procedure, popularized by Caffarelli-Silvestre [31] applies to the case Riesz. This is the approach that was proposed in [97] for the Log1 case and in [84] for the Riesz case.

Let us now get more specific about the extension procedure in the case `Log1`. We view $\mathbb{R}$ as identified with $\mathbb{R}\times\{0\}\subset\mathbb{R}^2=\{(x,y),x\in\mathbb{R},y\in\mathbb{R}\}$. Let us denote by $\delta_{\mathbb{R}}$ the uniform measure on $\mathbb{R}\times\{0\}$, i.e. such that for any smooth $\varphi(x,y)$ (with $x\in\mathbb{R},y\in\mathbb{R}$) we have

$$\int_{\mathbb{R}^2}\varphi\delta_{\mathbb{R}}=\int_{\mathbb{R}}\varphi(x,0)\,dx.$$

Let us still consider μ a measure on $\mathbb{R}$ (such as the equilibrium measure μ_V on $\mathbb{R}$, associated to $\mathcal{I}_V$ as in Theorem 2.1.2).

Given $x_1,\dots,x_N\in\mathbb{R}$, as explained above we identify them with the points $(x_1,0),\dots,(x_N,0)$ in $\mathbb{R}^2$, and we may then define the potentials H_N^μ and truncated potentials $H_{N,\vec\eta}^\mu$ in $\mathbb{R}^2$ by

$$H_N^\mu=g*\left(\sum_{i=1}^N\delta_{(x_i,0)}-N\mu\delta_{\mathbb{R}}\right)\qquad H_{N,\vec\eta}^\mu=g*\left(\sum_{i=1}^N\delta_{(x_i,0)}^{(\eta)}-N\mu\delta_{\mathbb{R}}\right).$$

Since g is naturally extended to a function in $\mathbb{R}^2$, these potentials make sense as functions in $\mathbb{R}^2$ and H_N^μ solves

$$-\Delta H_N^\mu=2\pi\left(\sum_{i=1}^N\delta_{(x_i,0)}-N\mu\delta_{\mathbb{R}}\right). \tag{3.3.1}$$

H_N^μ is nothing else than the harmonic extension to $\mathbb{R}^2$, away from the real axis, of the potential defined in dimension 1 by the analogue of (3.2.1). This is closely related to the *Stieltjes transform* of the empirical measure, a commonly used object in Random Matrix Theory: the Stieltjes transform S_μ of a probability measure μ is the function defined in the upper-half complex plane by

$$S_\mu(z)=\int_{\mathbb{R}}\frac{d\mu(x)}{x-z}.$$

Here the gradient of H_N^μ is essentially the Stieltjes transform of fluct_N^μ.

The proof of Lemma 3.2.7 then goes through without change, if one replaces $\int_{\mathbb{R}^d}|\nabla H_{N,\vec\eta}^\mu|^2$ with $\int_{\mathbb{R}^2}|\nabla H_{N,\vec\eta}^\mu|^2$.

3.4. The electric energy controls the fluctuations Since (3.2.2) (resp. (3.3.1)) holds, we can immediately relate the fluctuations to H_N^μ, and via the Cauchy-Schwarz inequality, control them by the electric energy.

Proposition 3.4.1 (Control of fluctuations by the electric energy). *Let φ be a compactly supported Lipschitz function in $\mathbb{R}^d$ supported in U, and μ be a bounded probability density on $\mathbb{R}^d$. Let $\vec\eta$ be a N-tuple of distances such that $\eta_i\leqslant N^{-1/d}$, for each $i=1,\dots,N$.*

For each configuration $\vec X_N$, we have

$$\left|\int_{\mathbb{R}^d}\varphi\,\mathrm{fluct}_N^\mu\right|\leqslant C(\|\varphi\|_{L^\infty}+\|\nabla\varphi\|_{L^\infty})\left(|U|^{\frac12}\|\nabla H_{N,\vec\eta}^\mu\|_{L^2(U)}+N^{1-\frac1d}\right) \tag{3.4.2}$$

where C depends only on d, and $|U|$ denotes the volume of U.

Proof. In the Log1 case, we first extend φ to a smooth compactly supported test function in $\mathbb{R}^2$ coinciding with $\varphi(x)$ for any (x,y) such that $|y| \leqslant 1$ and equal to 0 for $|y| \geqslant 2$. This is what will involve the term in $\|\varphi\|_{L^\infty}$ in the estimate.

In view of (3.2.2) (resp. (3.3.1)) and applying Cauchy-Schwarz, we have

$$\left|\int \Big(\sum_{i=1}^N \delta_{x_i}^{(\eta_i)} - N\mu_V\Big)\varphi\right| = \frac{1}{c_d}\left|\int_{\mathbb{R}^d} \nabla H_{N,\vec{\eta}}^{\mu_V} \cdot \nabla\varphi\right| \leqslant C|U|^{\frac12}\|\nabla\varphi\|_{L^\infty}\|\nabla H_{N,\vec{\eta}}^{\mu_V}\|_{L^2(U)}, \tag{3.4.3}$$

(resp. with an integral over $U \times \mathbb{R}$ in the 1D log case). Moreover, since $\eta_i \leqslant N^{-1/d}$, we have

$$\left|\int \left(\mathrm{fluct}_N^\mu - \Big(\sum_{i=1}^N \delta_{x_i}^{(\eta_i)} - N\mu\Big)\right)\varphi\right| = \left|\int \Big(\sum_{i=1}^N (\delta_{x_i} - \delta_{x_i}^{(\eta_i)})\Big)\varphi\right| \leqslant N^{1-\frac1d}\|\nabla\varphi\|_{L^\infty}. \tag{3.4.4}$$

The result follows. □

Corollary 3.4.5. *Applying this with* $\eta_i = N^{-1/d}$, *we deduce in view of Lemma* 3.2.7 *that*

$$\left|\int_{\mathbb{R}^d} \varphi\, \mathrm{fluct}_N^\mu\right| \leqslant C\|\nabla\varphi\|_{L^\infty}$$
$$\times \begin{cases} |\mathrm{Supp}\,\varphi|^{\frac12}\left(F_N^\mu(\vec{X}_N) + \frac{N}{d}\log N + CN\|\mu\|_{L^\infty}\right)^{\frac12} + N^{1-\frac1d} & \text{for } \mathit{Log1}, \mathit{Log2} \\ |\mathrm{Supp}\,\varphi|^{\frac12}\left(F_N^\mu(\vec{X}_N) + N^{2-\frac2d}\|\mu\|_{L^\infty}\right)^{\frac12} + N^{1-\frac1d} & \text{for } \mathit{Coul}, \end{cases}$$

where C *depends only on* d *and* $\|\mu\|_{L^\infty}$.

3.5. Consequences for the energy and partition function Thanks to the splitting, we can expand the energy to next order. For instance, combining (3.1.3) and Lemma 3.2.7, choosing again $\eta_i = N^{-1/d}$, we easily obtain

Corollary 3.5.1 (First energy lower bound). *If* $\mu_V \in L^\infty$, *then for any* $\vec{X}_N$

$$\mathcal{H}_N(\vec{X}_N) \geqslant N^2\mathcal{I}_V(\mu_V) + 2N\sum_{i=1}^N \zeta(x_i) - \left(\frac{N}{d}\log N\right)\mathbf{1}_{\mathit{Log1},\mathit{Log2}} - C\|\mu_V\|_{L^\infty}N^{2-2/d},$$

where C *depends only on* d.

Let us show how this lower bound easily translates into an upper bound for the partition function $Z_{N,\beta}$ in the case with temperature. In the cases Coul it is better to normalize the energy differently and define

$$\mathbb{P}_{N,\beta} = \frac{1}{Z_{N,\beta}} e^{-\frac{\beta}{2}N^{\min(\frac2d-1,0)}\mathcal{H}_N(\vec{X}_N)}d\vec{X}_N \tag{3.5.2}$$

and define $Z_{N,\beta}$ accordingly. Note that in the case Log1, Log2, this does not change anything.

Corollary 3.5.3 (An easy upper bound for the partition function). *Assume that* V *is continuous, satisfies* **(A1)–(A2)–(A3)** *and has an* L^∞ *density. Then for all* $\beta > 0$*, and for* N *large enough, we have*

$$\log Z_{N,\beta} \leqslant -\frac{\beta}{2} N^{\min(2,\frac{2}{d}+1)} \mathcal{I}_V(\mu_V) + \left(\frac{\beta}{4} N \log N\right) \mathbf{1}_{Log1,Log2} + C(1+\beta)N$$

where C *depends only on* μ_V *and the dimension.*

To prove this, let us state a lemma that we will use repeatedly and that exploits assumption **(A3)**.

Lemma 3.5.4. *Assume that* V *is continuous, such that* μ_V *exists, and satisfies* **(A3)**. *We have*

$$\lim_{N\to+\infty} \left(\int_{(\mathbb{R}^d)^N} e^{-\beta N \sum_{i=1}^N \zeta(x_i)} d\vec{X}_N \right)^{\frac{1}{N}} = |\omega| \tag{3.5.5}$$

where $\omega = \{\zeta = 0\}$.

Proof. First, by separation of variables, we have

$$\left(\int_{(\mathbb{R}^d)^N} e^{-\beta N \sum_{i=1}^N \zeta(x_i)} d\vec{X}_N \right)^{\frac{1}{N}} = \int_{\mathbb{R}^d} e^{-\beta N \zeta(x)}\, dx.$$

Second, we recall that since μ_V is a compactly supported probability measure, h^{μ_V} must asymptotically behave like $g(x)$ as $|x| \to \infty$, thus $\zeta = h^{\mu_V} + \frac{V}{2} - c$ grows like $g(x) + \frac{1}{2}V - c$. The assumption **(A3)** thus ensures that for N large enough, $\int e^{-\beta N \zeta(x)}\, dx < +\infty$. Moreover, by definition of ω,

$$e^{-\beta N \zeta} \to \mathbf{1}_\omega \quad \text{as } N \to +\infty$$

pointwise and monotonically, and ω has finite measure in view of the growth of h^{μ_V} and thus of ζ. The monotone convergence theorem allows to conclude that (3.5.5) holds. □

Proof of the corollary. Inserting the equation in Corollary 3.5.1 into the definition of $Z_{N,\beta}$ we are led to

$$\log Z_{N,\beta} \leqslant N^{\min(\frac{2}{d}-1,0)} \left(-\frac{\beta}{2} N^2 \mathcal{I}_V(\mu_V) + \left(\frac{\beta}{2d} N \log N\right) \mathbf{1}_{\mathrm{Log1,Log2}} + C\beta N^{2-\frac{2}{d}} \right)$$
$$+ \log \left(\int e^{-N\beta \sum_{i=1}^N \zeta(x_i)} d\vec{X}_N \right).$$

Using Lemma 3.5.4 to handle the last term, we deduce that

$$\log Z_{N,\beta} \leqslant N^{\min(\frac{2}{d}-1,0)} \left(-\frac{\beta}{2} N^2 \mathcal{I}_N(\mu_V) + \left(\frac{\beta}{2d} N \log N\right) \mathbf{1}_{\mathrm{Log1,Log2}} + C\beta N^{2-\frac{2}{d}} \right)$$
$$+ N(\log|\omega| + o_N(1))$$

which gives the conclusion. □

The converse inequality to Corollary 3.5.3 actually holds and we have

Proposition 3.5.6 (First expansion of the free energy). *Assume that V is continuous, such that μ_V exists, satisfies **(A3)** and has an L^∞ density. Then for all $\beta > 0$, and for N large enough, we have*

$$\log Z_{N,\beta} = -\frac{\beta}{2} N^{\min(2,\frac{2}{d}+1)} \mathcal{I}_V(\mu_V) + \left(\frac{\beta}{2d} N \log N\right) \mathbf{1}_{Log1,Log2} + O(N). \tag{3.5.7}$$

where $O(N)$ depends only on β, μ_V and the dimension.

One of the final goals on this course will be to present a more precise result with the constant in the order N term identified and characterized variationally. This will also provide information on the behavior of the configurations at the microscale. Anticipating the result, we will be able to show that in the logarithmic cases, $\log Z_{N,\beta}$ has an expansion where the V dependence can be decoupled:

$$\begin{aligned} \log Z_{N,\beta} &= -\frac{\beta}{2} N^2 \mathcal{I}_V(\mu_V) + \frac{\beta}{2} \frac{N \log N}{d} - NC(\beta, d) \\ &\quad - N\left(1 - \frac{\beta}{2d}\right) \int_\Sigma \mu_V(x) \log \mu_V(x)\, dx + o((\beta+1)N), \end{aligned} \tag{3.5.8}$$

where $C(\beta, d)$ is a constant depending only on β and the dimension (1 or 2). Obtaining (3.5.7) alone without an explicit constant is easier, and we will assume it for now.

In the sequel, it is convenient to use the splitting to rewrite

$$d\mathbb{P}_{N,\beta}(\vec{X}_N) = \frac{1}{K_{N,\beta}(\mu_V, \zeta)} e^{-\frac{\beta}{2} N^{\min(\frac{2}{d}-1,0)} \left(F_N^{\mu_V}(\vec{X}_N) + 2N \sum_{i=1}^N \zeta(x_i)\right)} d\vec{X}_N \tag{3.5.9}$$

where for ξ growing sufficiently fast at infinity, $K_{N,\beta}(\mu, \xi)$ is defined as

$$K_{N,\beta}(\mu, \xi) = \int e^{-\frac{\beta}{2} N^{\min(\frac{2}{d}-1,0)} \left(F_N^{\mu}(\vec{X}_N) + 2N \sum_{i=1}^N \xi(x_i)\right)} d\vec{X}_N. \tag{3.5.10}$$

In view of what precedes, we have

$$K_{N,\beta}(\mu_V, \zeta) = Z_{N,\beta} e^{\frac{\beta}{2} N^{\min(\frac{2}{d}+1,2)} \mathcal{I}_V(\mu_V)}. \tag{3.5.11}$$

3.6. Consequence: concentration bounds With Proposition 3.5.6, we deduce an upper bound on the exponential moments of the electric energy F_N:

Corollary 3.6.1 (Control of exponential moments of the electric energy). *We have, for some constant C depending on β and V,*

$$\left| \log \mathbf{E}_{\mathbb{P}_{N,\beta}} \exp\left(\frac{\beta}{4} \left(N^{\min(\frac{2}{d}-1,0)} F_N^{\mu_V}(\vec{X}_N) + (\frac{N}{d} \log N) \mathbf{1}_{Log1,Log2} \right)\right) \right| \leqslant CN. \tag{3.6.2}$$

Proof. We may write

$$\begin{aligned} &\mathbf{E}_{\mathbb{P}_{N,\beta}} \left[\exp\left(\frac{\beta}{4} N^{\min(\frac{2}{d}-1,0)} F_N^{\mu_V}(\vec{X}_N) \right) \right] \\ &= \frac{1}{K_{N,\beta}(\mu_V, \zeta)} \int \exp\left(-\frac{\beta}{4} \left(N^{\min(\frac{2}{d}-1,0)} (F_N^{\mu_V}(\vec{X}_N) + 2N \sum_{i=1}^N 2\zeta(x_i)) \right) \right) d\vec{X}_N \\ &= \frac{K_{N,\frac{\beta}{2}}(\mu_V, 2\zeta)}{K_{N,\beta}(\mu_V, \zeta)}. \end{aligned}$$

Taking the log and using (3.5.11), (3.5.7) to expand both terms up to order N yields the result. □

In [70] and [9], it is proven by a delicate bootstrap on the scales and in the case Log2 only (but this should really hold in higher dimensions and in the case Log1 as well) that this control can also be improved into a local control at all *mesoscales*: by this we mean the control

$$\left|\log \mathbf{E}_{\mathbb{P}_{N,\beta}}\left[\exp\left(\frac{\beta}{4}\int_{U}|\nabla H_{N,\vec{\eta}}|^2\right)\right]\right| \leqslant C|U|N, \tag{3.6.3}$$

for suitable $\vec{\eta}$ and cubes U of sidelength $N^{-\alpha}$, $\alpha < \frac{1}{2}$.

Combining (3.6.2) with Corollary 3.4.5, we may then easily obtain concentration bounds for the fluctuations:

Corollary 3.6.4 (Concentration of fluctuations). *For any Lipschitz function φ, we have*

$$\log \mathbf{E}_{\mathbb{P}_{N,\beta}}\left(e^{t(\int \varphi \mathrm{fluct}_N^{\mu_V})^2}\right) \leqslant CN \tag{3.6.5}$$

where the constant C depends on t, β and φ.

This already gives us a good control on the fluctuations: by Markov's inequality we may for instance deduce that the probability that $\left|\int \mathrm{fluct}_N^{\mu_V}\varphi\right| \gg \sqrt{N}$ is exponentially small (the control on the exponential moment is a stronger information though), or other moderate deviations bounds (compare with the total mass of $\mathrm{fluct}_N^{\mu_V}$ which is order N). This also improves on the a-priori concentration bound in $N\log N$ in the right-hand side of (3.6.5) generally obtained in the one-dimensional logarithmic case by the non-electric approach [27,78]. We also refer to [36] for a recent improved result in the same spirit valid in all Coulomb cases.

This is however not the best control one can obtain. The rest of these notes will be devoted to two things: obtaining a better control on the fluctuations when φ is assumed to be more regular, and obtaining a full LDP at next order for empirical fields, providing the exact constant in the order N term of (3.5.7).

4. CLT for fluctuations in the logarithmic cases

In this section, we restrict our attention to the logarithmic cases Log1, Log2 which are the only settings where the results that follow have been established. We expect that similar results may hold true in higher dimension but this is the object of future study. Let us define the fluctuations of the linear statistics associated to ξ as the random variable

$$\mathrm{Fluct}_N(\xi) := \int_{\mathbb{R}^2} \xi \, d\mathrm{fluct}_N^{\mu_V}, \tag{4.0.1}$$

where $\mathrm{fluct}_N^{\mu_V}$ is as in (3.1.1). The goal of this section is to present a result that $\mathrm{Fluct}_N(\xi)$ converges in law to a Gaussian random variable with explicit mean and variance. This is achieved by proving that the Laplace transform of $\mathrm{Fluct}_N(\xi)$ converges to that of the Gaussian.

This approach was pioneered in [63] and immediately leads to dealing with a Coulomb gas with potential V replaced by $V_t = V + t\xi$, with t small. We then follow here the approach of [11,72] where we use a simple change of variables which is a transport map between the equilibrium measure μ_V and the equilibrium measure μ_{V_t} for the perturbed potential. Note that the use of changes of variables in this context is not new, cf. [13,27,63,99]. In our approach, it essentially replaces the use of the so-called "loop" or Dyson-Schwinger equations.

4.1. Reexpressing the fluctuations as a ratio of partition functions We start by introducing the notation related to the perturbed potential and equilibrium measure.

Definition 4.1.1. For any $t \in \mathbb{R}$, we define

- The perturbed potential V_t as $V_t := V + t\xi$, and the perturbed equilibrium measure μ_{V_t}.
- The next-order confinement term $\zeta_t := \zeta_{V_t}$, as in (2.1.10).
- The next-order energy $F_N^{\mu_{V_t}}(\vec{X}_N)$ as in (3.1.4).
- The next-order partition function $K_{N,\beta}(\mu_{V_t}, \zeta_t)$ as in (3.5.10).

Lemma 4.1.2 (Reexpressing the Laplace transform of fluctuations). *For any* $t \in \mathbb{R}$ *we have*

$$\begin{aligned}(4.1.3)\quad &\mathbf{E}_{\mathbb{P}_{N,\beta}}\left[\exp\left(-\frac{\beta}{2} Nt\mathrm{Fluct}_N(\xi)\right)\right] \\ &= \frac{K_{N,\beta}(\mu_{V_t}, \zeta_t)}{K_{N,\beta}(\mu_V, \zeta)} \exp\left(-\frac{\beta}{2} N^2 \left(\mathcal{I}_{V_t}(\mu_{V_t}) - \mathcal{I}_V(\mu_V) - t\int \xi d\mu_V\right)\right).\end{aligned}$$

Proof. First, we notice that, for any t in $\mathbb{R}$

$$(4.1.4)\qquad \mathbf{E}_{\mathbb{P}_{N,\beta}}\left[\exp\left(-\frac{\beta}{2} Nt\mathrm{Fluct}_N(\xi)\right)\right] = \frac{Z_{N,\beta}^{V_t}}{Z_{N,\beta}} \exp\left(\frac{\beta}{2} N^2 t \int \xi \, d\mu_V\right).$$

Using the splitting formula (3.1.3) and the definition of $K_{N,\beta}$ as in (3.5.10) we see that for any t

$$(4.1.5)\qquad K_{N,\beta}(\mu_{V_t}, \zeta_t) = Z_{N,\beta}^{V_t} \exp\left(\frac{\beta}{2} N^2 \mathcal{I}_{V_t}(\mu_{V_t})\right),$$

thus combining (4.1.4) and (4.1.5), we obtain (4.1.3). □

To compute the limit of the Laplace transform of $\mathrm{Fluct}_N(\xi)$ we will just need to apply the formula (4.1.3) with $t = -\frac{2s}{\beta N}$ for some arbitrary number s, hence t will indeed be very small. We will see below that the term in the exponential in (4.1.3) is computable in terms of ξ and then there will remain to study the ratio of partition functions $\frac{K_{N,\beta}(\mu_{V_t}, \zeta_t)}{K_{N,\beta}(\mu_V, \zeta)}$.

4.2. Transport and change of variables Our method, introduced in [11,72] is based on the construction of a transport ϕ_t such that $\phi_t \# \mu_V = \mu_{V_t}$ (where #

denotes the push-forward of a measure) and with $\phi_t = \mathrm{Id} + t\psi$, i.e. a perturbation of identity. In fact the transport chosen will not satisfy exactly $\phi_t \# \mu_V = \mu_{V_t}$ but will satisfy it approximately as $t \to 0$, and in these notes we will neglect the error for simplicity.

In order to construct the correct transport map, one first needs to understand the perturbed equilibrium measure μ_{V_t}, or in other words, to understand how μ_V varies when V varies. In the case `Log1`, the dependence of μ_V on V is indirect, but well understood. At least in the "one-cut case" (i.e. when the support of μ_V is a single interval), the right perturbation map ψ is found by inverting with respect to ξ a so-called "master operator" Ξ, which already arose in the Dyson-Schwinger approach [13,27,28,99].

In dimension 2 (and in fact in all Coulomb cases), the perturbed equilibrium measure is easy to compute when ξ is supported in Σ (the support of μ_V) : one can check that it is $\mu_{V_t} = \mu_V - \frac{t}{c_d \beta}\Delta\xi$, and a correct transport map is $\mathrm{Id} + t\psi$ with $\psi = -\frac{\nabla \xi}{c_d \mu_V}$. The case where ξ is not supported in Σ i.e. has a support intersecting $\partial\Sigma$ is much more delicate: one needs to understand how $\partial\Sigma$ is displaced under the perturbation. This is described precisely (in all dimensions) in [101]. The PDE approach used there replaces Sakai's theory used in [6], which is restricted to the two-dimensional analytic setting.

In the interior (two-dimensional) Coulomb case, it follows from a direct computation based on the exact formula for μ_{V_t} that

$$\lim_{N \to \infty, t = -\frac{2s}{\beta N}} -\frac{\beta}{2} N^2 \left(\mathcal{I}_{V_t}(\mu_{V_t}) - \mathcal{I}_V(\mu_V) - t \int \xi d\mu_V \right) = \frac{s^2}{2\pi\beta} \int_{\mathbb{R}^2} |\nabla \xi|^2. \tag{4.2.1}$$

In the general case, thanks to the analysis of [101], we find that $\int |\nabla \xi|^2$ gets replaced by $\int_{\mathbb{R}^2} |\nabla \xi^\Sigma|^2$, where ξ^Σ denotes the harmonic extension of ξ outside Σ.

In the one-dimensional case, a more delicate computation based on the above facts reveals that

$$\lim_{N \to \infty, t = -\frac{2s}{\beta N}} -\frac{\beta}{2} N^2 \left(\mathcal{I}_{V_t}(\mu_{V_t}) - \mathcal{I}_V(\mu_V) - t \int \xi d\mu_V \right) = \frac{s^2}{2} \int \xi' \psi d\mu_V, \tag{4.2.2}$$

where ψ is defined above. We have thus identified the limit of the exponential term in (4.1.3) and now turn to the ratio of partition functions.

Once the (approximate) transport has been constructed and proven to be regular enough (it is typically once less differentiable than ξ) and such that we approximately have $\zeta_t \circ \phi_t = \zeta$, using the change of variables $y_i = \phi_t(x_i)$ we find

$$\begin{aligned} &K_{N,\beta}(\mu_{V_t}, \zeta_t) \\ &= \int \exp\left(-\frac{\beta}{2} \left(F_N^{\mu_{V_t}}(\Phi_t(\vec{X}_N)) + 2N \sum_{i=1}^N \zeta_t \circ \phi_t(x_i) \right) + \sum_{i=1}^N \log|\det D\phi_t|(x_i) \right) d\vec{X}_N \\ &= \int \exp\left(-\frac{\beta}{2} \left(F_N^{\mu_{V_t}}(\Phi_t(\vec{X}_N)) + 2N \sum_{i=1}^N \zeta(x_i) \right) + \sum_{i=1}^N \log|\det D\phi_t|(x_i) \right) d\vec{X}_N. \end{aligned}$$

Thus

$$(4.2.3)\qquad \frac{K_{N,\beta}(\mu_{V_t},\zeta_t)}{K_{N,\beta}(\mu_V,\zeta)} = \mathbf{E}_{\mathbb{P}_{N,\beta}}\left(e^{-\frac{\beta}{2}(F_N^{\mu_{V_t}}(\Phi_t(\vec{X}_N))-F_N^{\mu_V}(\vec{X}_N))+\sum_{i=1}^N \log|\det D\Phi_t|(x_i)}\right)$$

hence we need to evaluate the difference of energies before and after transport, $F_N^{\mu_{V_t}}(\Phi_t(\vec{X}_N)) - F_N^{\mu_V}(\vec{X}_N)$.

4.3. Energy comparison Let us present that computation in the Log1 case for simplicity, knowing that it has a more complex analogue in the Log2 case.

Lemma 4.3.1 (Energy linearization along a transport). *For any probability density μ, any $\vec{X}_N \in \mathbb{R}^N$, and any $\psi \in C_c^2(\mathbb{R})$, defining*

$$(4.3.2)\qquad A[\vec{X}_N,\psi,\mu] = \iint \frac{\psi(x)-\psi(y)}{x-y}\, d\mathrm{fluct}_N^{\mu}(x)\, d\mathrm{fluct}_N^{\mu}(y)$$

and letting $\Phi_t(\vec{X}_N) = (\phi_t(x_1),\cdots,\phi_t(x_N))$ with $\phi_t = \mathrm{Id} + t\psi$, we have

$$(4.3.3)\qquad \left|F_N^{\phi_t\#\mu}(\Phi_t(\vec{X}_N)) - F_N^{\mu}(\vec{X}_N) - \sum_{i=1}^N \log\phi_t'(x_i) + \frac{t}{2}A[\vec{X}_N,\psi,\mu]\right| \leqslant Ct^2\left(F_N^{\mu}(\vec{X}_N) + N\log N\right),$$

with C depending only on V,ψ,d.

This is the point that essentially replaces the loop equations.

Proof. We may write

$$\begin{aligned}
&F_N^{\phi_t\#\mu}(\Phi_t(\vec{X}_N)) - F_N^{\mu}(\vec{X}_N)\\
&= -\int_{\mathbb{R}\times\mathbb{R}\setminus\triangle} \log|x-y|\Big(\sum_{i=1}^N \delta_{\phi_t(x_i)} - N\phi_t\#\mu\Big)(x)\Big(\sum_{i=1}^N \delta_{\phi_t(x_i)} - N\phi_t\#\mu\Big)(y)\\
&\qquad + \int_{\mathbb{R}\times\mathbb{R}\setminus\triangle} \log|x-y|\, d\mathrm{fluct}_N^{\mu}(x)\, d\mathrm{fluct}_N^{\mu}(y)\\
&= -\int_{\mathbb{R}\times\mathbb{R}\setminus\triangle} \log\frac{|\phi_t(x)-\phi_t(y)|}{|x-y|}\, d\mathrm{fluct}_N^{\mu}(x)\, d\mathrm{fluct}_N^{\mu}(y)\\
&= -\int_{\mathbb{R}\times\mathbb{R}} \log\frac{|\phi_t(x)-\phi_t(y)|}{|x-y|}\, d\mathrm{fluct}_N^{\mu}(x)\, d\mathrm{fluct}_N^{\mu}(y) + \sum_{i=1}^N \log|\phi_t'(x_i)|.
\end{aligned}$$

Using that by definition $\phi_t = \mathrm{Id} + t\psi$ where ψ is in $C_c^2(\mathbb{R})$, and the fact that

$$\left\|\frac{\log(1+tx)-tx}{t^2}\right\|_{C^2(K)} \leqslant C_K$$

for all compact sets K, some constant C_K depending on ψ and t small enough, we get by the chain rule

$$\log\frac{|\phi_t(x)-\phi_t(y)|}{|x-y|} = t\,\frac{\psi(x)-\psi(y)}{x-y} + t^2\,\varepsilon_t(x,y),$$

with $\|\varepsilon_t\|_{C^2(\mathbb{R}^2)}$ uniformly bounded in t. Using Proposition 3.4.1 twice gives

$$\left|\iint \varepsilon_t(x,y)\,d\mathrm{fluct}_N^\mu(x)\,d\mathrm{fluct}_N^\mu(y)\right| \leqslant Ct^2\left(F_N^\mu(\vec{X}_N)+N\log N\right).$$

This yields the result in the Log1 case. □

4.4. Computing the ratio of partition functions Combining Lemma 4.3.1 (or its two-dimensional analogue) with (4.2.3) we obtain

$$\begin{aligned}
&\frac{K_{N,\beta}(\mu_{V_t},\zeta_t)}{K_{N,\beta}(\mu_V,\zeta)}\\
&= \mathbf{E}_{\mathbb{P}_{N,\beta}}\Big(\exp\Big(\Big(1-\frac{\beta}{2d}\Big)\sum_{i=1}^N \log|\det D\phi_t|(x_i)\\
&\qquad + \frac{\beta t}{4}A[\vec{X}_N,\psi,\mu_V] + O(t^2(F_N^{\mu_V}(\vec{X}_N)+N\log N))\Big)\Big)\\
&= \mathbf{E}_{\mathbb{P}_{N,\beta}}\Big(\exp\Big(\Big(1-\frac{\beta}{2d}\Big)\mathrm{Fluct}_N[\log|\det D\phi_t|]\\
&\qquad + \frac{\beta t}{4}A[\vec{X}_N,\psi,\mu_V] + O(t^2(F_N^{\mu_V}(\vec{X}_N)+N\log N))\Big)\Big)\\
&\qquad \times \exp\Big(\Big(1-\frac{\beta}{2d}\Big)N\int \log|\det D\phi_t|d\mu_V\Big).
\end{aligned}\tag{4.4.1}$$

Let us now examine the terms in this right-hand side.

First, since $\phi_t\#\mu_V=\mu_{V_t}$ (approximately), we have $\det D\phi_t = \frac{\mu_V}{\mu_{V_t}\circ\phi_t}$ and thus, by definition of the push-forward,

$$\int \log|\det D\phi_t|d\mu_V = \int \log\mu_V d\mu_V - \int \log \mu_{V_t} d\mu_{V_t},\tag{4.4.2}$$

which is of order t i.e. $O(\frac{s}{N})$, as $N\to\infty$. More precisely, we may compute that in the Log1 case

$$\lim_{N\to\infty, t=-\frac{2s}{\beta N}} (1-\frac{\beta}{2})N\left(\int \mu_V\log\mu_V - \int \mu_{V_t}\log\mu_{V_t}\right) = (1-\frac{2}{\beta})\int \psi' d\mu_V\tag{4.4.3}$$

and in the Log2 interior case the analogue is $\frac{s}{2\pi}\left(\frac{1}{\beta}-\frac{1}{4}\right)\int \Delta\xi\log\Delta V$.

There now remains to evaluate all the terms in the expectation. We note that by the Cauchy-Schwarz inequality, we have for instance

$$\log \mathbf{E}(e^{a+b+c}) \leqslant \log\mathbf{E}(e^{4a}) + \log\mathbf{E}(e^{4b}) + \log\mathbf{E}(e^{4c}).\tag{4.4.4}$$

First, by (3.6.2) and Hölder's inequality we have the control

$$\log \mathbf{E}_{\mathbb{P}_{N,\beta}}\left(\exp(O(t^2(F_N^{\mu_V}(\vec{X}_N)+N\log N)))\right) = O(\frac{s^2}{N^2}N) = O(\frac{s^2}{N})\tag{4.4.5}$$

(up to changing β in the formula), with the choice $t = O(s/N)$.

Since $\phi_t = \mathrm{Id} + t\psi$ with ψ regular enough, we find by combining Corollary 3.4.5 and (3.6.2) that

$$]\log\mathbf{E}_{\mathbb{P}_{N,\beta}}(\mathrm{Fluct}_N[\log|\det D\phi_t|]) = O(t\sqrt{N}) = O(\frac{s}{\sqrt{N}}).$$

Next, we turn to the term A (which we call anisotropy), which is the most delicate one. There are two ways to handle this term and conclude. The first is direct and works in the one-dimensional, one-cut regular case, in which case the operator Ξ that finds the right transport map is always invertible. This is described in the appendix of [11] and ressembles the method of [27]. The second way works in all logarithmic (one-dimensional possibly multi-cut or critical, and two-dimensional) and calls instead the information on $\log Z_{N,\beta}$ obtained in (3.5.8). It is described in [11] in the one-dimensional case and [72] in the two-dimensional case. Let us now present each.

4.5. Conclusion in the one-dimensional one-cut regular case We note that the term A can be seen as being essentially also a fluctuation term with

$$A[\vec{X}_N, \psi, \mu_V] = \int f(x)\, d\mathrm{fluct}_N(x) \qquad f(x) := \left(\int \hat{\psi}(x,y)\, d\mathrm{fluct}_N(y)\right) \tag{4.5.1}$$

where

$$\hat{\psi}(x,y) := \frac{\psi(x)-\psi(y)}{x-y}$$

is as regular as the regularity of ξ allows (note that this fact is only true in dimension 1 and so the argument breaks in dimension 2!). Using Proposition 3.4.1 twice, we obtain

$$\begin{aligned} \|\nabla f\|_{L^\infty} &\leqslant \left|\int \nabla_x \hat{\psi}(x,y)\, d\mathrm{fluct}_N(y)\right| \\ &\leqslant C\|\nabla_x \hat{\psi}\|_{C^1}\left(F_N^{\mu_V}(\vec{X}_N) + N\log N + CN\right)^{\frac{1}{2}} \end{aligned} \tag{4.5.2}$$

and

$$\begin{aligned} |A[\vec{X}_N, \psi, \mu_V]| &= \left|\int f(x)\, d\mathrm{fluct}_N(x)\right| \\ &\leqslant C\|f\|_{C^1}\left(F_N^{\mu_V}(\vec{X}_N) + N\log N + CN\right)^{\frac{1}{2}} \\ &\leqslant C\|\hat{\psi}\|_{C^2}\left(F_N^{\mu_V}(\vec{X}_N) + N\log N + CN\right). \end{aligned}$$

In view of (3.6.2), we deduce that

$$]\left|\mathbf{E}_{\mathbb{P}_{N,\beta}}\left[-\frac{\beta}{2} tA[\vec{X}_N, \psi, \mu_V]\right]\right| \leqslant O(\frac{s}{N}N) = O(s).$$

Combining all these elements and inserting them into (4.4.1), we deduce that $\frac{K_{N,\beta}(\mu_{V_t},\zeta_t)}{K_{N,\beta}(\mu_V,\zeta)}$ is of order 1 as $N \to \infty$, and inserting into (4.1.3) together with (4.2.1)-(4.2.2) we obtain that $\mathbf{E}_{\mathbb{P}_{N,\beta}}(e^{s\mathrm{Fluct}_N(\xi)})$ is bounded (in terms of s and ξ) as $N \to \infty$. We may then bootstrap this information by applying it to f, first improving (4.5.2) into

$$\begin{aligned} \|D^k f\|_{L^\infty} &\leqslant \left|\int D_x^k \hat{\psi}(x,y)\, d\mathrm{fluct}_N(y)\right| \\ &\leqslant C\|D_x^k \nabla_y \hat{\psi}\|_{L^\infty}\left(F_N^{\mu_V}(\vec{X}_N) + N\log N + CN\right)^{\frac{1}{2}} \end{aligned}$$

obtaining (with Hölder's inequality) if k is large enough, that

$$\log \mathbf{E}_{\mathbb{P}_{N,\beta}}\left(e^{t\mathrm{Fluct}_N(f)}\right) \leqslant Ct\sqrt{N} = O\left(\frac{s}{\sqrt{N}}\right).$$

We then conclude that in fact

$$\left|\log \mathbf{E}_{\mathbb{P}_{N,\beta}}\left[-\frac{\beta}{2}tA[\vec{X}_N,\psi,\mu_V]\right]\right| \leqslant o_N(1). \tag{4.5.3}$$

Combining all the previous relations and (4.4.4), (4.2.1)–(4.2.2), and inserting into (4.1.3), we conclude that

$$\lim_{N\to\infty} \log \mathbf{E}_{\mathbb{P}_{N,\beta}}\left(e^{s\mathrm{Fluct}_N(\xi)}\right) = m_\xi s + \frac{1}{2}v_\xi s^2$$

where

$$m_\xi = \left(1-\frac{2}{\beta}\right)\int_{\mathbb{R}} \psi' d\mu_V \qquad v_\xi = -\frac{2}{\beta}\int_{\mathbb{R}} \xi'\psi\, d\mu_V.$$

We have thus obtained that the Laplace transform of $\mathrm{Fluct}_N(\xi)$ converges to that of a Gaussian of mean m_ξ and variance v_ξ, concluding the proof in the one-dimensional one-cut regular case. Note that reinserting into (4.4.1) we have in fact obtained a precise expansion of $\frac{K_{N,\beta}(\mu_{V_t},\zeta_t)}{K_{N,\beta}(\mu_V,\zeta)}$, which also allows to directly evaluate $\frac{d}{dt}_{t=0}\log K_{N,\beta}(\mu_{V_t},\zeta_t)$ up to order $o(N)$. But this is valid for all variations of the type $V+t\xi$ of the potential, hence by interpolating between two potentials V_0 and V_1 which are such that their equilibrium measures are both one-cut regular, we can compute $\log K_{N,\beta}(\mu_{V_1},\zeta_1) - \log K_{N,\beta}(\mu_{V_0},\zeta_0)$ up to an error $o(N)$ and obtain effectively the formula (3.5.8). Moreover, since all the terms arising in the exponent in (4.4.1) are essentially fluctuations, the reasoning can be bootstrapped by inserting the CLT result and obtaining an expansion to higher order of $\frac{d}{dt}_{t=0}\log K_{N,\beta}(\mu_{V_t},\zeta_t)$. This is possible because the operator Ξ is always invertible in the one-cut case, allowing to build the transport map associated to the fluctuation of any function, provided that function is regular enough. This way, one may obtain a relative expansion of $\log Z_{N,\beta}$ with an error of arbitrary order, provided the potentials are regular enough, which is what is found in [27].

4.6. Conclusion in the two-dimensional case or in the general one-cut case

In these cases, we control the term A differently: instead we use the important relation (3.5.8) which we have assumed so far, and which provides another way of computing the left-hand side of (4.4.1). Comparing these two ways applied to a small, fixed t, we find on the one-hand

$$\log \frac{K_{N,\beta}(\mu_{V_t},\zeta_t)}{K_{N,\beta}(\mu_V,\zeta)} = \left(1-\frac{\beta}{2d}\right)N\left(\int \mu_V\log\mu_V - \int \mu_{V_t}\log\mu_{V_t}\right) + o(N)$$

and on the other hand combining (4.4.1), (4.4.2), (4.4.5),

$$\log \frac{K_{N,\beta}(\mu_{V_t},\zeta_t)}{K_{N,\beta}(\mu_V,\zeta)} = \left(1-\frac{\beta}{2d}\right)N\left(\int \mu_V\log\mu_V - \int \mu_{V_t}\log\mu_{V_t}\right)$$
$$+ \log \mathbf{E}_{\mathbb{P}_{N,\beta}}\left(\exp\frac{\beta t}{4}A[\vec{X}_N,\psi,\mu_V]\right) + o(N)$$

This implies that

$$\log \mathbf{E}_{\mathbb{P}_{N,\beta}}\left(\exp \frac{\beta t}{4} A[\vec{X}_N, \psi, \mu_V]\right) = o(N),$$

and this is true for any t small enough, say $t = -\varepsilon$ for some small enough ε. We then apply Hölder's inequality with exponent $p = \frac{\beta \varepsilon N}{4s}$ which is > 1 for N large enough. This yields

$$\mathbf{E}_{\mathbb{P}_{N,\beta}}\left(\exp \frac{-s}{2N} A[\vec{X}_N, \psi, \mu_V]\right) \leqslant \mathbf{E}_{\mathbb{P}_{N,\beta}}\left(\exp \frac{-ps}{2N} A[\vec{X}_N, \psi, \mu_V]\right)^{\frac{1}{p}}$$

hence

$$\log \mathbf{E}_{\mathbb{P}_{N,\beta}}\left(\exp \frac{-s}{2N} A[\vec{X}_N, \psi, \mu_V]\right) \leqslant \frac{o(N)}{\varepsilon N} = o(1).$$

Changing then ε into $-\varepsilon$ and s into $-s$ allows to conclude that

$$\log \mathbf{E}_{\mathbb{P}_{N,\beta}}\left(\exp \frac{-s}{2N} A[\vec{X}_N, \psi, \mu_V]\right) = o(1).$$

This replaces (4.5.3) and allows to finish the proof. The mean and variance are given by

(4.6.1)
$$m_\xi = \begin{cases} \left(1 - \frac{2}{\beta}\right) \int_{\mathbb{R}} \psi' d\mu_V & \text{if } d = 1 \\ \frac{1}{2\pi}\left(\frac{1}{\beta} - \frac{1}{4}\right) \int_{\mathbb{R}^2} \Delta\xi \left(\mathbf{1}_\Sigma + (\log \Delta V)^\Sigma\right) & \text{if } d = 2 \end{cases}$$

and

(4.6.2)
$$v_\xi = \begin{cases} -\frac{2}{\beta} \int_{\mathbb{R}} \xi' \psi \, d\mu_V & \text{if } d = 1 \\ \frac{1}{2\pi\beta} \int_{\mathbb{R}^2} |\nabla \xi^\Sigma|^2 & \text{if } d = 2. \end{cases}$$

In both cases, we have obtained the following

Theorem 4.6.3 (Central Limit Theorem for fluctuations in the logarithmic cases). *Assume ξ is regular enough. In the one-dimensional multi-cut or critical cases, assume in addition that ξ is in the range of Ξ, and an analogous condition in the two-dimension case with several connected components. Then* $\mathrm{Fluct}_N(\xi)$ *converges in law to a Gaussian random variable of mean m_ξ and variance v_ξ.*

Let us make several additional comments:

- In dimension 1, this theorem was first proven in [63] for polynomial V and ξ analytic. It was later generalized in [12, 27, 28, 68, 99], still with strong assumptions on ξ, and to the critical case in [11].
- If the extra conditions do not hold, then the CLT is not expected to hold. Rather, the limit should be a Gaussian convoled with a discrete Gaussian variable, as shown in the `Log1` case in [28].
- In dimension 2, the precise form of the variance as $C \int |\nabla \xi|^2$ (or more generally $C \int |\nabla \xi^\Sigma|^2$), means that fluct_N converges to a so-called Gaussian Free Field. This result was proven for the determinantal case $\beta = 2$ in [86] (for V quadratic) and [6] under analyticity assumptions. It was then proven for all β simultaneously in [72] and [10], with ξ roughly assumed to be C^4. Note that the result also applies in the case where ξ is supported

at a mesoscale, i.e. $\xi_N(x) = \xi(xN^\alpha)$ with $\alpha < \frac{1}{2}$ (the analogous result in dimension 1 is proven in [14]). The results of [6,72] are the only ones that also apply to the case where ξ is not supported inside Σ.
- This theorem is to be compared to Corollary 3.6.4. It shows that if ξ is smooth enough, the fluctuations $\mathrm{Fluct}_N(\xi)$ are in fact much smaller than could be expected. They are typically of order 1, to be compared with the sum of N iid random variables which is typically or order $\sqrt{N}$. This indicates strong correlations at all scales.
- Moderate deviation bounds similar to those of [9] can also easily be obtained as a by-product of the method presented above.

5. The renormalized energy

The goal of this section is to define a "renormalized energy" or jellium energy, that will arise as a rate function for the next order Large Deviation Principle that will be presented below.

It allows to compute a total Coulomb interaction for an infinite system of discrete point "charges" in a constant neutralizing background of fixed density $m > 0$. Such a system is often called a (classical) *jellium* in physics.

Starting again from the equations defining the electric potentials, (3.2.2) and (3.3.1), we first perform a blow-up at scale $N^{1/d}$ (the typical lengthscale of the systems). Denoting $x_i' = N^{1/d}x_i$ the blown-up points of the configuration, and the blown-up equilibrium measure $\mu_V{}'(x) = \mu_V(xN^{-1/d})$, we then define for shortcut the "electric fields"

$$E_N = \nabla H_N^{\mu_V{}'} \qquad H_N^{\mu_V{}'} = g * \left(\sum_{i=1}^{N} \delta_{x_i'} - \mu_V{}' \delta_{\mathbb{R}^d} \right), \tag{5.0.1}$$

where the measure $\delta_{\mathbb{R}^d}$ is needed only in the one-dimensional case, cf (3.3.1).

This way, E_N solves the linear equation

$$-\mathrm{div}\, E_N = c_d \Big(\sum_{i=1}^{N} \delta_{x_i'} - \mu_V{}' \delta_{\mathbb{R}^d} \Big), \tag{5.0.2}$$

in which it is easy, at least formally, to pass to the limit $N \to \infty$. Note that this is written when centering the blow-up at the origin. When the density μ_V is continuous, the blown-up measure $\mu_V{}'$ then converges to the constant $\mu_V(0)$ in the weak topology. If another origin x_0 was chosen and we defined $\mu_V'(x) = \mu_V((x - x_0)N^{1/d})$, the constant would be $\mu_V(x_0)$, the local density of the neutralizing background charge near x_0.

On the other hand, as $N \to +\infty$, the number of points becomes infinite and they fill up the whole space, their local density remaining bounded (by control on the fluctuations). This way $\sum_{i=1}^N \delta_{x_i}$ is expected to converge to a distribution $\mathcal{C} \in \mathrm{Config}(\mathbb{R}^d)$, where for A a Borel set of $\mathbb{R}^d$ we denote by $\mathrm{Config}(A)$ the set of locally finite point configurations in A or equivalently the set of non-negative, purely atomic Radon measures on A giving an integer mass to singletons. We mostly use $\mathcal{C}$ for denoting a point configuration and we will write $\mathcal{C}$ for $\sum_{p\in\mathcal{C}}\delta_p$.

We thus wish to define the energy associated to an electric field E solving an equation of the form

$$-\mathrm{div}\, E = c_d\left(\mathcal{C} - m\delta_{\mathbb{R}^d}\right) \quad \text{in } \mathbb{R}^{d+k} \tag{5.0.3}$$

where $\mathcal{C}\in \mathrm{Config}(\mathbb{R}^d)$ and m is a nonnegative constant (representing the density of points), and $k=1$ in the case `Log1` where we need to add one space dimension, and 0 otherwise. We say that a vector field E is an electric field compatible with $(\mathcal{C}, m)$ if it satisfies (5.0.3). By formulating in terms of electric fields, we have not retained the information that E_N was a gradient.

5.1. Definitions Let $\mathcal{C}$ be a point configuration, $m \geqslant 0$ and let E be compatible with $(\mathcal{C}, m)$. For any $\eta\in(0,1)$ we define the truncation of E as

$$E_\eta(x) := E(x) - \sum_{p\in\mathcal{C}} \nabla f_\eta(x-p), \tag{5.1.1}$$

where f_η is as in (3.2.4). Let us observe that

$$-\mathrm{div}\, E_\eta = c_d\left(\sum_{p\in\mathcal{C}} \delta_p^{(\eta)} - m\delta_{\mathbb{R}^d}\right). \tag{5.1.2}$$

This procedure is exactly the same, at the level of the electric fields, as the truncation procedure described in the previous sections. A change of variables in (3.2.8) allows to directly relate $F_N^\mu(\vec{X}_N)$ and $\int_{\mathbb{R}^d}|\nabla H^\mu_{N,\vec{\eta}}|^2$ hence to $\int_{\mathbb{R}^d}|E_{N,\eta}|^2$.

In the sequel, $\square_R$ still denotes the d-dimensional cubes $[-R/2, R/2]^d$.

Definition 5.1.3 (The renormalized energy)**.** The (Coulomb) renormalized energy of E with background m is

$$\mathcal{W}(E,m) := \lim_{\eta\to 0} \mathcal{W}_\eta(E,m) \tag{5.1.4}$$

where we let

$$\mathcal{W}_\eta(E,m) := \limsup_{R\to\infty} \frac{1}{R^d}\int_{\square_R\times\mathbb{R}^k} |E_\eta|^2 - mc_d g(\eta). \tag{5.1.5}$$

Definition 5.1.6. Let $\mathcal{C}$ be a point configuration and $m\geqslant 0$. We define the renormalized energy of $\mathcal{C}$ with background m as

$$\mathbb{W}(\mathcal{C},m) := \inf\{\mathcal{W}(E,m) \mid E \text{ compatible with } (\mathcal{C},m)\} \tag{5.1.7}$$

with the convention $\inf(\emptyset) = +\infty$.

The name *renormalized energy* (originating in Bethuel-Brezis-Hélein [19] in the context of two-dimensional Ginzburg-Landau vortices) reflects the fact that the integral of $|E|^2$ is infinite, and is computed in a renormalized way by first applying a truncation and then removing the appropriate divergent part $mc_d g(\eta)$.

It is not a priori clear how to define a total Coulomb interaction of such a jellium system, first because of the infinite size of the system as we just saw, second because of the lack of local charge neutrality of the system. The definitions we presented avoid having to go through computing the sum of pairwise interactions between particles (it would not even be clear how to sum them), but instead replace it with (renormalized variants of) the extensive quantity $\int |E|^2$ (see (3.2.6) and the comments following it).

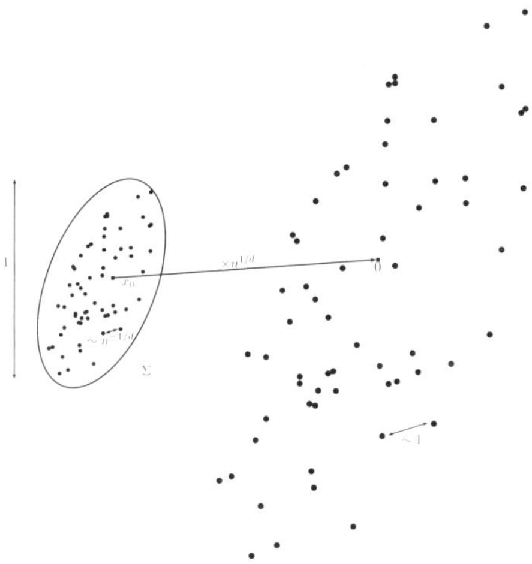

FIGURE 5.1.8. An arbitrary blown-up configuration

5.2. Scaling properties The dependence in m can be scaled out as follows: If $E \in \mathrm{Elec}_m$, we define $\sigma_m E$ by

$$\sigma_m E := m^{\frac{1}{d}-1} E\left(\frac{\cdot}{m^{\frac{1}{d}}}\right).$$

We have $\sigma_m E \in \mathrm{Elec}_1$ and in the case Coul,

$$\begin{aligned} \mathcal{W}(E, m) &= m^{2-\frac{2}{d}} \mathcal{W}(\sigma_m E, 1) \\ \mathcal{W}_\eta(E, m) &= m^{2-\frac{2}{d}} \mathcal{W}_{\eta m^{\frac{1}{d}}}(\sigma_m E, 1) \end{aligned} \tag{5.2.1}$$

and in the cases Log1, Log2

$$\begin{aligned} \mathcal{W}(E, m) &= m\mathcal{W}(\sigma_m E, 1) - \frac{2\pi}{d} m \log m \\ \mathcal{W}_\eta(E, m) &= m\mathcal{W}_{\eta m^{1/d}}(\sigma_m E, 1) - \frac{2\pi}{d} m \log m. \end{aligned} \tag{5.2.2}$$

It also follows that

$$\mathbb{W}(\mathcal{C}, m) = \begin{cases} m\mathbb{W}(\sigma_m \mathcal{C}, 1) - \frac{2\pi}{d} m \log mr & \text{for Log1, Log2} \\ m^{2-2/d}\mathbb{W}(\sigma_m \mathcal{C}, 1) & \text{for Coul.} \end{cases} \tag{5.2.3}$$

In view of the above scaling relations, it suffices to study $\mathbb{W}(\cdot, 1)$. It is proven in [90] that $\min \mathbb{W}(\cdot, 1)$ is finite and achieved. Moreover, the minimum coincides with the limit as $R \to +\infty$ of the minima of $\mathbb{W}(\cdot, 1)$ on configurations that are $(R\mathbb{Z})^d$-periodic (i.e. that live on the torus $\mathbb{T}_R = \mathbb{R}^d/(R\mathbb{Z})^d$) with $R \in \mathbb{N}$.

5.3. Partial results on the minimization of $\mathcal{W}$, crystallization conjecture We have seen that the minima of $\mathbb{W}$ can be achieved as limits of the minima over periodic configurations (with respect to larger and larger tori). In the `Log1` case, a convexity argument (for which we refer to [97, Prop. 2.3]) shows that the minimum is achieved when the points are equally spaced, in other words for the lattice or crystalline distribution $\mathbb{Z}$.

In higher dimension, determining the value of $\min \mathbb{W}$ is a difficult open question. One question that we can answer so far is that of the minimization over the restricted class of pure lattice configurations, in dimension $\mathsf{d} = 2$, i.e. vector fields which are gradient of functions that are periodic with respect to a lattice $\mathbb{Z}\vec{u} + \mathbb{Z}\vec{v}$ with $\det(\vec{u}, \vec{v}) = 1$, corresponding to configurations of points that can be identified with $\mathbb{Z}\vec{u} + \mathbb{Z}\vec{v}$. In this case, we have :

Theorem 5.3.1 (The triangular lattice is the minimizer over lattices in 2D)**.** *The minimum of $\mathbb{W}$ over this class of vector fields is achieved uniquely by the one corresponding to the triangular "Abrikosov" lattice.*

Here the triangular lattice means $\mathbb{Z} + \mathbb{Z}e^{i\pi/3}$, properly scaled, i.e. what is called the Abrikosov lattice in the context of superconductivity. This result is essentially a result of number theory, proven in the 50's.

One may ask whether this triangular lattice does achieve the global minimum of $\mathbb{W}$. The fact that the Abrikosov lattice is observed in superconductors, combined with the fact that $\mathbb{W}$ can be derived as the limiting minimization problem of Ginzburg-Landau [95], justify conjecturing this. It was recently proven in [18] that this conjecture is equivalent to a conjecture of Brauchart-Hardin-Saff [29] on the next order term in the asymptotic expansion of the minimal logarithmic energy on the sphere (an important problem in approximation theory, also related to Smale's "7th problem for the 21st century" [103]), which is obtained by formal analytic continuation, hence by very different arguments. In addition, the result of [44] essentially yields the local minimality of the triangular lattice within all periodic (with possibly large period) configurations.

In general dimension $\mathsf{d} \geqslant 3$ the minimization of $\mathbb{W}$ even restricted to the class of lattices is an open question except in dimensions 4, 8 and 24, and in dimensions 8 and 24 the global minimization of $\mathbb{W}$ should follow from [42].

One may expect that in general low dimensions, the minimum of $\mathbb{W}$ is achieved by some particular lattice, this is an instance of a *crystallization question*, scientifically fundamental for understanding the structure of matter (why do atoms arrange themselves periodically?), see for instance [21] for a review. A folklore conjecture is that lattices are not minimizing in large enough dimensions, as indicated by the situation for the sphere packing problem mentioned above.

5.4. Renormalized energy for point processes We wish to describe the limit of the *tagged empirical field* (according to the terminology of type-III large deviations) which we now define. The point x varying in Σ will represent the "tag", and objects integrated with respect to the tag will be recognizable by bars.

Definition 5.4.1 (Tagged empirical field). Given $\vec{X}_N = (x_1, \dots, x_N)$ in $(\mathbb{R}^d)^N$, denoting $\vec{X}'_N$ as the finite configuration rescaled by a factor $N^{1/d}$

(5.4.2)
$$\vec{X}'_N := \sum_{i=1}^{N} \delta_{N^{1/d} x_i},$$

we define the tagged empirical field associated to the configuration as the probability measure on $\Sigma \times \mathrm{Config}$ given by

(5.4.3)
$$\overline{\mathrm{Emp}}_N[\vec{X}_N] := \frac{1}{|\Sigma|} \int_\Sigma \delta_{\left(x, \theta_{N^{1/d}x} \cdot \vec{X}'_N\right)} \, dx,$$

where θ_x denotes the translation by $-x$.

For any $x \in \Sigma$, the term $\theta_{N^{1/d}x} \cdot \vec{X}'_N$ is an element of Config which represents the N-tuple of particles $\vec{X}_N$ centered at x and seen at microscopic scale (or, equivalently, seen at microscopic scale and then centered at $N^{1/d}x$). In particular any information about this point configuration in a given ball (around the origin) translates to an information about $\vec{X}'_N$ around x. We may thus think of $\theta_{N^{1/d}x} \cdot \vec{X}'_N$ as encoding the microscopic behavior of $\vec{X}'_N$ around x.

The limit of $\overline{\mathrm{Emp}}_N[\vec{X}_N]$ will be a probability measure on $\Sigma \times \mathrm{Config}$, typically denoted $\bar{P}$, whose first marginal is the (normalized) Lebesgue measure on Σ, and which is stationary with respect to the second variable. The space of such probability measures is denoted as $\mathcal{P}_s(\Sigma \times \mathrm{Config})$. We may then define the renormalized energy for $\bar{P} \in \mathcal{P}_s(\Sigma \times \mathrm{Config})$ as

(5.4.4)
$$\overline{\mathbb{W}}(\bar{P}, \mu_V) = \int_\Sigma \mathbf{E}_{\bar{P}^x}[\mathbb{W}(\cdot, \mu_V(x))] \, dx,$$

where $\bar{P}^x$ is the disintegration with respect to the variable x of the measure $\bar{P}$ (informally it is $\bar{P}(x, \cdot)$).

5.5. Lower bound for the energy in terms of the empirical field We may now relate rigorously the energy $F_N^{\mu_V}$ to $\overline{\mathbb{W}}$ by the following

Proposition 5.5.1 (Precise lower bound for the electric energy). *Assume $\partial\Sigma$ is C^1 and μ has a continuous density on its support. Let $\{\vec{X}_N\}_N$ be a sequence of N-tuples of points in $\mathbb{R}^d$. If the left-hand side below is bounded independently of N, then up to extraction the sequence $\overline{\mathrm{Emp}}_N[\vec{X}_N]$ converges to some $\bar{P}$ in $\mathcal{P}_s(\Sigma \times \mathrm{Config})$, and*

(5.5.2)
$$\liminf_{N\to\infty} F_N^{\mu_V}(\vec{X}_N) + \frac{N}{d} \log N \geq \overline{\mathbb{W}}(\bar{P}, \mu_V) \quad \text{in the cases Log1, Log2}$$
$$\liminf_{N\to\infty} N^{\min(\frac{2}{d}-2,-1)} F_N^{\mu_V}(\vec{X}_N) \geq \overline{\mathbb{W}}(\bar{P}, \mu_V) \quad \text{in the case Coul.}$$

Proof. For each configuration $\vec{X}_N$ let us define the tagged electric field process $P_N[\vec{X}_N]$ (then we will drop the $\vec{X}_N$) by

(5.5.3)
$$P_N[\vec{X}_N] := \frac{1}{|\Sigma|} \int_\Sigma \delta_{(x, \theta_{N^{1/d}x} \cdot E_N)} \, dx,$$

where E_N is as in (5.0.1). The result follows from relatively standard considerations, after noting that by Fubini's theorem we may write, for any $R > 1$, denoting by $\square_R$ the hypercube of sidelength R,

$$\begin{aligned}\int_{\mathbb{R}^d} |\nabla H_{N,\eta}^{\mu_V'}|^2 &\geqslant N \int_{\mathbb{R}^d} \frac{1}{|\square_R|} \mathbf{1}_{y\in\square_R} \int_\Sigma |E_{N,\eta}(N^{1/d}x + y)|^2 \, dx\, dy \\ &= N|\Sigma| \int f_{R,\eta}(E) \, dP_N(x, E)\end{aligned} \tag{5.5.4}$$

where $f_{R,\eta}$ is defined by

$$f_{R,\eta}(E) = \frac{1}{|\square_R|} \int_{\square_R} |E_\eta|^2$$

if E is of the form ∇H_N^μ for some N-point configuration, and $+\infty$ otherwise. It then suffices to show that $\{P_N\}_N$ is tight, and that up to extraction P_N converges to some stationary tagged electric field process. Taking the divergence of an electric field allows to recover the underlying point configuration via (5.0.3). By this mapping we can "project" the limiting tagged electric field process to some $\bar{P} \in \mathcal{P}_s(\Sigma \times \text{Config})$, and take the limit $N \to \infty$, $R \to \infty$, then $\eta \to 0$ in (5.5.4) to obtain the result. □

6. Large Deviations Principle for empirical fields

We now wish to state a large deviations result at the level of the point processes for the Gibbs measure (3.5.2), proven in [71].

6.1. Specific relative entropy We first need to define the analogue of the entropy at the level of point processes: the specific relative entropy with respect to the Poisson point process. The Poisson point process with intensity m is the point process characterized by the fact that for any bounded Borel set B in $\mathbb{R}^d$

$$P\left(N(B) = n\right) = \frac{(m|B|)^n}{n!} e^{-m|B|}$$

where $N(B)$ denotes the number of points in B. The expectation of the number of points in B can then be computed to be $m|B|$, and the numbers of points in two disjoint sets are independent, thus the points "don't interact".

For any $m \geqslant 0$, we denote by Π^m the (law of the) Poisson point process of intensity m in $\mathbb{R}^d$ (it is an element of $\mathcal{P}_s(\text{Config})$). Let P be in $\mathcal{P}_s(\text{Config})$.

Definition 6.1.1 (Specific relative entropy). The specific relative entropy of P with respect to Π^1 is defined as

$$\text{ent}[P|\Pi^1] := \lim_{R\to\infty} \frac{1}{R^d} \text{ent}[P_{\square_R}|\Pi^1_{\square_R}], \tag{6.1.2}$$

where $P_{\square_R}, \Pi^1_{\square_R}$ denote the restrictions of the processes to the hypercube $\square_R$. Here, $\text{ent}[\cdot|\cdot]$ denotes the *usual* relative entropy (or Kullback-Leibler divergence) of two probability measures defined on the same probability space, namely

$$\text{ent}[\mu|\nu] := \int \log \frac{d\mu}{d\nu} \, d\mu$$

if μ is absolutely continuous with respect to ν, and $+\infty$ otherwise.

Lemma 6.1.3. *The following properties are known (see* [51, Sec. 6.9.2]*) :*

(1) *The limit in* (6.1.2) *exists for* P *stationary.*
(2) *The map* $P \mapsto \mathsf{ent}[P|\Pi^1]$ *is affine and lower semi-continuous on* $\mathcal{P}_s(\text{Config})$.
(3) *The sub-level sets of* $\mathsf{ent}[\cdot|\Pi^1]$ *are compact in* $\mathcal{P}_s(\text{Config})$ *(it is a good rate function).*
(4) *We have* $\mathsf{ent}[P|\Pi^1] \geqslant 0$ *and it vanishes only for* $P = \Pi^1$.

Next, if $\bar{P}$ is in $\mathcal{P}_s(\Sigma \times \text{Config})$, we define the tagged specific relative entropy as

$$\overline{\mathsf{ent}}[\bar{P}|\Pi^1] := \int_\Sigma \mathsf{ent}[\bar{P}^x|\Pi^1]dx. \tag{6.1.4}$$

For any $N \geqslant 1$ we define $\mathbb{Q}_{N,\beta}$ as

$$d\mathbb{Q}_{N,\beta}(x_1, \dots, x_N) := \prod_{i=1}^N \frac{\exp\left(-N^{1-\frac{s}{d}}\beta\zeta(x_i)\right) dx_i}{\int_{\mathbb{R}^d} \exp\left(-N^{1-\frac{s}{d}}\beta\zeta(x)\right) dx}, \tag{6.1.5}$$

(with $s = 0$ by convention in the logarithmic cases). The integral converges by assumption, and we let $\bar{\mathfrak{Q}}_{N,\beta}$ be the push-forward of $\mathbb{Q}_{N,\beta}$ by the map $\overline{\mathrm{Emp}}_N$, as defined in (5.4.3).

We also introduce the following constant:

$$c_{\omega,\Sigma} := \log|\omega| - |\Sigma| + 1, \tag{6.1.6}$$

where ω is the zero-set of ζ (see (2.1.10)).

The following result expresses how the specific relative entropy evaluates the volume of "microstates" $\vec{X}_N$ near a reference tagged point process $\bar{P}$ (one may think of ζ as 0 in (6.1.5), then $\mathbb{Q}_{N,\beta}$ measures the volume in Σ^N). It is a Large Deviations Principle for the tagged empirical field, when the points are distributed according to a reference measure on $(\mathbb{R}^d)^N$ where there is no interaction. This is a microscopic or so-called "type III" analogue of Sanov's theorem for the empirical measures.

Proposition 6.1.7 (Sanov type result for the tagged empirical fields). *Given any* $A \subset \mathcal{P}_s(\Sigma \times \text{Config})$, *we have*

$$\begin{aligned} -\inf_{\mathring{A}\cap\mathcal{P}_{s,1}} \overline{\mathsf{ent}}[\bar{P}|\Pi^1] - c_{\omega,\Sigma} &\leqslant \liminf_{N\to\infty} \frac{1}{N}\log\bar{\mathfrak{Q}}_{N,\beta}(A) \\ \leqslant \limsup_{N\to\infty}\frac{1}{N}\log\bar{\mathfrak{Q}}_{N,\beta}(A) &\leqslant -\inf_{\bar{P}\in\bar{A}} \overline{\mathsf{ent}}[\bar{P}|\Pi^1] - c_{\omega,\Sigma}. \end{aligned} \tag{6.1.8}$$

With this result at hand we may expect the LDP result we are searching for (or at least an LDP upper bound): indeed, the splitting (3.1.3) and the rewriting (3.5.9) may be combined to Proposition 5.5.1, which bounds from above the exponent in $\mathbb{P}_{N,\beta}$, while Proposition 6.1.7 allows to evaluate the volume via the $\mathbb{Q}_{N,\beta}$ contribution. The hard part is to obtain the LDP lower bound, which involves instead an upper bound for the energy, together with a lower bound for the volume of configurations.

In view of (5.5.2), we also see that if we want the (specific) relative entropy term to intervene at the same order as the energy term, we have to take an inverse temperature of the form

$$\beta N^{\min(\frac{2}{d}-1,0)} \tag{6.1.9}$$

with β independent of N, or more generally $\beta N^{-\frac{s}{d}}$ in all cases (with the convention $s = 0$ in the cases `Log1`, `Log2`). Tis is, to consider as already mentioned in (3.5.2),

$$d\mathbb{P}_{N,\beta}(\vec{X}_N) = \frac{1}{Z_{N,\beta}} \exp\left(-\frac{\beta N^{-\frac{s}{d}}}{2}\mathcal{H}_N\right) d\vec{X}_N. \tag{6.1.10}$$

6.2. Statement of the main result For any $\beta > 0$, we define a free energy functional $\overline{\mathcal{F}}_\beta$ as

$$\overline{\mathcal{F}}_\beta(\bar{P}) := \frac{\beta}{2}\overline{\mathbb{W}}(\bar{P}, \mu_V) + \overline{\mathrm{ent}}[\bar{P}|\Pi^1]. \tag{6.2.1}$$

For any N, β we let $\overline{\mathfrak{P}}_{N,\beta}$ be push-forward of the canonical Gibbs measure $\mathbb{P}_{N,\beta}$ by the *tagged empirical field map* $\overline{\mathrm{Emp}}_N$ as in (5.4.3) (in other words, $\overline{\mathfrak{P}}_{N,\beta}$ is the law of the tagged empirical field when the particles are distributed according to $\mathbb{P}_{N,\beta}$).

Theorem 6.2.2 (Large Deviation Principle for the tagged empirical fields [71]). *In all `Log1`, `Log2` and `Riesz` cases, under suitable assumptions on the regularity of $\partial\Sigma$ and μ_V, for any $\beta > 0$ the sequence $\{\overline{\mathfrak{P}}_{N,\beta}\}_N$ satisfies a large deviation principle at speed N with good rate function $\overline{\mathcal{F}}_\beta - \inf \overline{\mathcal{F}}_\beta$.*

In particular, in the limit as $N \to \infty$, the law $\overline{\mathfrak{P}}_{N,\beta}$ concentrates on minimizers of $\overline{\mathcal{F}}_\beta$.

- One readily sees the effect of the temperature: in the minimization there is a competition between the term $\overline{\mathbb{W}}$ based on the renormalized energy, which is expected to favor very ordered configurations, and the entropy term which in contrast favors disorder (the entropy is minimal for a Poisson point process). As $\beta \to \infty$, we expect configurations to concentrate on minimizers of $\overline{\mathbb{W}}$ hence to crystallize on lattices (in dimension 1, there is a complete proof [69]), in low dimensions. As $\beta \to 0$, the limiting point processes (if they exist) converge to the Poisson point process. This was already known in the `Log1` case, see again [69] or [5].
- In the case `Log2`, it is proven in [70] that the similar result holds down to the mesoscales (see also [9] for results without a full LDP).
- The corresponding, simpler, result for minimizers of $\mathcal{H}_N$ (which are of interest on their own, as seen in Section 1) can be proven as well. More precisely, the lower bound of Proposition 5.5.1 is sharp, and the scaling

property of $\overline{\mathbb{W}}$ allow to obtain

$$\min \mathcal{H}_N(\vec{X}_N) = N^2 \mathcal{I}_V(\mu_V) + \tag{6.2.3}$$
$$\begin{cases} -\frac{N}{d}\log N + N\left(-\frac{1}{2}\int_\Sigma \mu_V \log \mu_V + \frac{1}{c_d}\min \mathbb{W}(\cdot,1)\right) & \text{for } \mathsf{Log1}, \mathsf{Log2} \\ N^{1+\frac{s}{d}} \frac{1}{c_{d,s}} \min \mathbb{W}(\cdot,1) \int_\Sigma \mu_V^{1+\frac{s}{d}} & \text{for } \mathsf{Coul}, \mathsf{Riesz}. \end{cases}$$

Moreover, if $\vec{X}_N$ minimizes $\mathcal{H}_N$ then, up to extraction $\overline{\mathrm{Emp}_N[\vec{X}_N]}$ converges to a minimizer of $\overline{\mathbb{W}}(\cdot, \mu_V)$. These are results of [84,90,96,97]. More precise results including rigidity down to the microscopic scales and separation of points can be found in [83,87].

As is usual and as alluded to before, as a by-product of the large deviation principle we obtain the order N term in the expansion of the partition function.

Corollary 6.2.4 (Next-order expansion and thermodynamic limit). *Under the same assumptions, we have, as* $N \to \infty$*:*

- *In the logarithmic cases* Log1, Log2

$$\log Z_{N,\beta} = -\frac{\beta}{2} N^2 \mathcal{I}_V(\mu_V) + \frac{\beta}{2}\frac{N\log N}{d} - N \min \overline{\mathcal{F}}_\beta + N(|\Sigma| - 1) + o((\beta+1)N). \tag{6.2.5}$$

- *In the* Riesz *cases*

$$\log Z_{N,\beta} = -\frac{\beta}{2} N^{2-\frac{s}{d}} \mathcal{I}_V(\mu_V) - N \min \overline{\mathcal{F}}_\beta + N(|\Sigma| - 1) + o((\beta+1)N). \tag{6.2.6}$$

The scaling properties allow to rewrite the previous expansion in the cases Log1, Log2 *as*

$$\log Z_{N,\beta} = -\frac{\beta}{2} N^2 \mathcal{I}_V(\mu_V) + \frac{\beta}{2}\frac{N\log N}{d} - NC(\beta, d) - N\left(1 - \frac{\beta}{2d}\right)\int_\Sigma \mu_V(x)\log \mu_V(x)\, dx + o((\beta+1)N), \tag{6.2.7}$$

where $C(\beta, d)$ *is a constant depending only on* β *and the dimension* d*, but independent of the potential.*

In the particular case of Log1 with a quadratic potential $V(x) = x^2$, the equilibrium measure is known to be Wigner's semi-circular law and the limiting process at the microscopic scale when centered around a point $x \in (-2, 2)$ (let us emphasize that here there is *no averaging*) has been identified for any $\beta > 0$ in [66,108]. It is called the *sine-*β *point process* and we denote it by $\mathrm{Sine}_\beta(x)$ (so that $\mathrm{Sine}_\beta(x)$ has intensity $\frac{1}{2\pi}\sqrt{4-x^2}$). For $\beta > 0$ fixed, the law of these processes do not depend on x up to rescaling and we denote by Sine_β the corresponding process with intensity 1.

A corollary of our main result is then a new variational property of Sine_β: it minimizes the sum of $\beta/2$ times the renormalized energy (with background 1) and the specific relative entropy.

6.3. Proof structure To prove the LDP, the standard method is to evaluate the logarithm of $\overline{\mathfrak{P}}_{N,\beta}(B(\bar{P},\varepsilon))$, where $\bar{P}$ is a given tagged point process and $B(\bar{P},\varepsilon)$ is a ball of small radius ε around it, for a distance that metrizes the weak topology.

In view of the splitting formula, we may write as in (3.5.9),

$$\overline{\mathfrak{P}}_{N,\beta}(B(\bar{P},\varepsilon)) = \frac{1}{K_{N,\beta}}\left(\int_{\mathbb{R}^d}\exp\left(-N^{1-\frac{s}{d}}\beta\zeta\right)\right)^{-N} \times \int_{\overline{\mathrm{Emp}}_N(\vec{X}_N)\in B(\bar{P},\varepsilon)}\exp\Big(-\frac{\beta N^{-\frac{s}{d}}}{2}F_N^{\mu_V}(\vec{X}_N)\Big)dQ_{N,\beta}(x_1,\dots,x_N).$$

Recalling that $\zeta \geqslant 0$ and $\zeta(x)=0$ if and only if $x\in\Sigma$, and if we believe in equality in (5.5.2) we obtain formally

$$\lim_{\varepsilon\to 0}\log\overline{\mathfrak{P}}_{N,\beta}(B(\bar{P},\varepsilon)) \simeq -\log K_{N,\beta} - N\log|\Sigma| - \frac{\beta N}{2}\overline{\mathbb{W}}(\bar{P},\mu_V) + \lim_{\varepsilon\to 0}\log Q_{N,\beta}\left(\{\overline{\mathrm{Emp}}_N(\vec{X}_N)\in B(\bar{P},\varepsilon)\}\right). \tag{6.3.1}$$

Extracting this way the exponential of a function is the idea of Varadhan's integral lemma (cf. [47, Theorem 4.3.1]).

Using the definition of $\bar{\mathfrak{Q}}_{N,\beta}$ as the push-forward of $Q_{N,\beta}$ by $\overline{\mathrm{Emp}}_N$), Proposition 6.1.7 yields that

$$\frac{1}{N}\log Q_{N,\beta}\left(\{\overline{\mathrm{Emp}}_N\in B(\bar{P},\delta_1)\}\right) \simeq -\overline{\mathrm{ent}}[\bar{P}|\Pi^1] - c_{\omega,\Sigma}, \tag{6.3.2}$$

giving rise to the relative entropy term. When looking to bound from above $\log\overline{\mathfrak{P}}_{N,\beta}(B(\bar{P},\varepsilon))$, combining Propositions 5.5.1 and 6.1.7 actually suffices.

Thus, there remains to prove the corresponding lower bound by justifying the replacement of $N^{-\frac{s}{d}}F_N^{\mu_V}(\vec{X}_N)$ by $N\overline{\mathbb{W}}(\bar{P},\mu_V)$. This is significantly harder and relies on proving the following *quasi-continuity* of the interaction.

For any integer N, and $\delta>0$ and any $\bar{P}\in\mathcal{P}_s(\Sigma\times\mathrm{Config})$, let us define

$$T_N(\bar{P},\delta) := \left\{\vec{X}_N\in(\mathbb{R}^d)^N\right\} \text{ such that } N^{-1-\frac{s}{d}}\left(F_N^{\mu_V}(\vec{X}_N) + (\frac{N}{d}\log N)\mathbf{1}_{\mathrm{Log1,Log2}}\right) \leqslant \overline{\mathbb{W}}(\bar{P},\mu_V)+\delta. \tag{6.3.3}$$

This is the set of configurations whose energy is not much larger than the limiting energy. This certainly excludes some configurations, typically those with possibly rare pairs of points getting very close to each other, for which $N^{-1-\frac{s}{d}}F_N^{\mu_V}(\vec{X}_N)$ (which tends to $+\infty$) is not well approximated by $\overline{\mathbb{W}}(\bar{P},\mu_V)$ (which is finite for typical $\bar{P}$'s). We need to show that $T_N(\bar{P},\delta)$ contains a large enough volume of "good" configurations, in the following sense.

Proposition 6.3.4. *Let* $\bar{P}\in\mathcal{P}_{s,1}(\Sigma\times\mathrm{Config})$. *For all* $\delta_1,\delta_2>0$ *we have*

$$\liminf_{N\to\infty}\frac{1}{N}\log Q_{N,\beta}\left(\{\overline{\mathrm{Emp}}_N\in B(\bar{P},\delta_1)\}\cap T_N(\bar{P},\delta_2)\right) \geqslant -\overline{\mathrm{ent}}[\bar{P}|\Pi^1] - c_{\omega,\Sigma}. \tag{6.3.5}$$

In view of (6.3.2), to obtain Proposition 6.3.4 we need to show that the event $T_N(\bar{P},\delta_2)$ has enough volume in phase-space near $\bar{P}$. This relies on the screening construction which we now describe.

6.4. Screening and consequences The screening procedure allows, starting from a given configuration or set of configurations, to build configurations whose energy is well controlled and can be computed additively in blocks of large microscopic size.

Let us zoom everything by $N^{1/d}$ and work in the rescaled set $\Sigma' = N^{1/d}\Sigma$. It will be split into cubes K of size R with R independent of N but large. This can be done except in a thin boundary layer which needs to be treated separately (for simplicity we will not discuss this).

In a cube K (or in $K \times \mathbb{R}$ in the `Log1` case), the screening procedure takes as input a given configuration whose electric energy in the cube is not too large, and replaces it by a better configuration which is neutral (the number of points is equal to $\int_K \mu_V{}'$), and for which there is a compatible electric field in K whose energy is controlled by the initial one plus a negligible error as $R \to \infty$. The point configuration is itself only modified in a thin layer near ∂K, and the electric field is constructed in such a way as to obtain $E \cdot \vec{\nu} = 0$ on ∂K. The configuration is then said to be screened, because in some sense it then generates only a negligible electric field outside of K.

Starting from a given vector field, we not only construct a modified, screened, electric field, but a whole family of electric fields and configurations (which are small perturbations of each other). This is important for the LDP lower bound where we need to exhibit a large enough volume of configurations having well-controlled energy.

Let us now get into more detail, restricting to the Coulomb cases.

6.4.1. Compatible and screened electric fields Let K be a compact subset of $\mathbb{R}^d$ with piecewise C^1 boundary (typically, K will be an hyperrectangle or the support Σ), $\mathcal{C}$ be a finite point configuration in K, μ be a nonnegative density in $L^\infty(K)$, and E be a vector field.

- We say that E is compatible with $(\mathcal{C}, \mu)$ in K provided

$$-\operatorname{div} E = c_d\,(\mathcal{C} - \mu) \quad \text{in } K. \tag{6.4.1}$$

- We say that E *is compatible with* $(\mathcal{C}, \mu)$ *and screened in* K when (6.4.1) holds and moreover

$$E \cdot \vec{\nu} = 0 \quad \text{on } \partial K, \tag{6.4.2}$$

  where $\vec{\nu}$ is the outer unit normal vector.

We note that integrating (6.4.1) over K and using (6.4.2), we find that $\int_K \mathcal{C} = \int_K \mu$. In other words, screened electric fields are necessarily associated with *neutral* systems, in which the positive charge $\int_K \mathcal{C}$ is equal to the negative charge $\int_K \mu$. In particular $\int_K \mu$ has to be an integer.

6.4.2. Computing additively the energy Vector fields that are screened can be naturally "pasted" together: we may construct point configurations by elementary blocks (hyperrectangles) and compute their energy additively in these blocks.

More precisely, assume that space is partitioned into a family of hyperrectangles $K \in \mathcal{K}$ such that $\int_K d\mu_V{}'$ are integers. We would like to construct a vector field $E^{(K)}$ in each K such that

$$-\mathrm{div}\, E^{(K)} = c_d \left(\mathcal{C}_K - \mu_V{}'\right) \text{ in } K \text{ and } E^{(K)} \cdot \vec{\nu} = 0 \text{ on } \partial K \tag{6.4.3}$$

(where $\vec{\nu}$ is the outer unit normal to K) for some discrete set of points $\mathcal{C}_K \subset K$, and with

$$\int_K |E_\eta^{(K)}|^2$$

well-controlled.

When solving (6.4.3), we could take $E^{(K)}$ to be a gradient, but we do not require it, in fact relaxing this constraint is what allows to paste together the vector fields. Once the relations (6.4.3) are satisfied on each K, we may paste together the vector fields $E^{(K)}$ into a unique vector field E^{tot}, and the discrete sets of points $\mathcal{C}_K$ into a configuration $\mathcal{C}^{\mathrm{tot}}$. The cardinality of $\mathcal{C}^{\mathrm{tot}}$ will be equal to $\int_{\mathbb{R}^d} d\mu_V{}'$, which is exactly N.

We thus obtain a configuration of N points, whose interaction energy $F_N^{\mu_V}$ we may evaluate. The important fact is that the enforcement of the boundary condition $E^{(K)} \cdot \vec{\nu} = 0$ on each boundary ensures that

$$-\mathrm{div}\, E^{\mathrm{tot}} = c_d \left(\mathcal{C}^{\mathrm{tot}} - \mu_V{}'\right) \quad \text{in } \mathbb{R}^d \tag{6.4.4}$$

holds globally. In general, a vector field which is discontinuous across an interface has a distributional divergence concentrated on the interface equal to the jump of the normal derivative, but here by construction the normal components coincide hence there is no divergence created across these interfaces.

Even if the $E^{(K)}$'s were gradients, the global E^{tot} is in general no longer a gradient. This does not matter however, since the energy of the *local electric field* $\nabla H_N'$ generated by the finite configuration $\mathcal{C}^{\mathrm{tot}}$ (and the background $\mu_V{}'$) is always smaller, as stated in

Lemma 6.4.5 (The local energy minimizes the energy among compatible electric fields). *Let Ω be a compact subset of $\mathbb{R}^d$ with piecewise C^1 boundary, let $N \geqslant 1$, let $\vec{X}_N$ be in $(\mathbb{R}^d)^N$. We assume that all the points of $\vec{X}_N$ belong to K and that $\Omega \subset \Sigma$. Let E be a vector field such that*

$$-\mathrm{div}\, E = c_d \left(\sum_{i=1}^{N} \delta_{x_i'} - \mu_V{}'\right) \text{ in } \Omega \quad \text{and} \quad E \cdot \vec{\nu} = 0 \text{ on } \partial\Omega. \tag{6.4.6}$$

Then, for any $\eta \in (0,1)$ we have

$$\int_{\mathbb{R}^d} |\nabla H_{N,\eta}^{\mu_V}{}'|^2 \leqslant \int_K |E_\eta|^2. \tag{6.4.7}$$

Proof. The point is that $\nabla H_{N,\eta}^{\mu_V}{}'$ is the L^2 projection of any compatible E onto gradients, and that the projection decreases the L^2 norm. The truncation procedure can be checked not to affect that property. □

We thus have

$$\int_{\mathbb{R}^d} |\nabla H_{N,\eta}^{\mu_V}{}'|^2 \leqslant \sum_{K\in\mathcal{K}} \int_K |E_\eta^{(K)}|^2,$$

and after a change of variables this bounds $\int_{\mathbb{R}^d} |\nabla H_{N,\vec{\eta}}^{\mu_V}|^2$, so that the energy $F_N^{\mu_V}(\mathcal{C}^{tot})$ can indeed be computed (or at least bounded above, which is precisely what we care about at this stage). This is the core of the method: by passing to the electric representation, we transform a sum of pairwise interactions into an additive (extensive) energy.

The screening theorem (see for instance [71, Section 5]) states that this program can be achieved while adding only a negligible error to the energy, modifying only a small fraction of the points, and preserving the good separation. It also allows, when starting from a family of configurations, to build families of configurations which occupy enough volume in configuration space (i.e. losing only a negligible logarithmic volume).

6.5. Generating microstates and conclusion To prove Proposition 6.3.4, we need to examine configurations sampled "at random" in $B(\bar{P}, \alpha)$ according to the law $Q_{N,\beta}$, and then modify them via the screening procedure.

A way to generate such configurations approximating (after averaging over translations) the tagged point process $\bar{P}$ is to draw a point configuration in each hypercube $K \in \mathcal{K}$ jointly at random according to a Bernoulli point process (indeed $Q_{N,\beta}$ coincides with the uniform measure in Σ), then Sanov's theorem imply that the discrete averages of most of these configurations are close to $\bar{P}$ as desired.

Using Proposition 6.1.7 such configurations will still occupy a volume bounded below by $-\overline{\mathrm{ent}}[\bar{P}_{\underline{m}}|\Pi^1] - c_{\omega,\Sigma}$.

These configurations then need to be modified as follows:

- we apply to them the screening procedure, which allows to compute their energy additively over the hypercubes
- we apply to them an additional "regularization procedure", which separates pairs of points that are too close while decreasing the energy, thus ensuring that the error between $\overline{\mathbb{W}}_\eta$ and $\overline{\mathbb{W}}$ can be made small as $\eta \to 0$ uniformly among the configurations and ensuring that the resulting configurations are in $T_N(\bar{P}, \delta_2)$.

At the same time, we check that these procedure

- do not increase the distance to $\bar{P}$ by too much, so that the configurations remain in $B(\bar{P}, \delta_1)$, provided that α was chosen small enough
- do not make us lose too much volume of configurations so that we keep a lower bound of the same form as that of Proposition 6.1.7.

This concludes the (sketch of the) proof of Proposition 6.3.4. The proof of the LDP lower bound in Theorem 6.2.2 follows after letting $N \to \infty$ and the other parameters tend to 0.

References

[1] A. Abrikosov, On the magnetic properties of superconductors of the second type. Soviet Phys. JETP **5** (1957), 1174–1182. ← 348

[2] D. R. Adams and L. I. Hedberg, *Function Spaces and Potential Theory*, Springer, 1999. ← 349

[3] A. Aftalion, *Vortices in Bose-Einstein Condensates*, Birkhaüser, 2006. ← 348

[4] A. Alastuey and B. Jancovici, On the classical two-dimensional one-component Coulomb plasma. *J. Physique* **42** (1981), no. 1, 1–12. ← 345

[5] R. Allez and L. Dumaz, From Sine Kernel to Poisson statistics. *Electron. J. Probab.* **19** (2014) no. 114, 1–25. ← 378

[6] Y. Ameur, H. Hedenmalm and N. Makarov, Fluctuations of eigenvalues of random normal matrices, *Duke Math. J.* **159** (2011), no. 1, 31–81. ← 365, 370, 371

[7] G. W. Anderson, A. Guionnet and O. Zeitouni, *An introduction to random matrices.* Cambridge University Press, 2010. ← 346, 347

[8] J. Barré, F. Bouchet, T. Dauxois, and S. Ruffo. Large deviation techniques applied to systems with long-range interactions. *J. Stat. Phys*, 119(3-4):677–713, 2005. ← 345

[9] R. Bauerschmidt, P. Bourgade, M. Nikula and H. T. Yau, Local density for two-dimensional one-component plasma, *Comm. Math. Phys.* **356** (2017), no. 1, 189–230. ← 363, 371, 378

[10] R. Bauerschmidt, P. Bourgade, M. Nikula and H. T. Yau, The two-dimensional Coulomb plasma: quasi-free approximation and central limit theorem, 1609.08582. ← 370

[11] F. Bekerman, T. Leblé and S. Serfaty, CLT for Fluctuations of β-ensembles with general potential, 1706.09663., to appear in *Electron. J. Probab.* 364, 368, 370

[12] F. Bekerman and A. Lodhia, Mesoscopic central limit theorem for general β-ensembles, 1605.05206. 347, 370

[13] F. Bekerman, A. Figalli and A. Guionnet, Transport maps for Beta-matrix models and Universality, *Comm. Math. Phys.* **338** (2015), no. 2, 589–619. ← 347, 364, 365

[14] F. Bekerman and A. Lodhia, Mesoscopic central limit theorem for general β-ensembles, 1605.05206. ← 371

[15] G. Ben Arous and A. Guionnet, Large deviations for Wigner's law and Voiculescu's non-commutative entropy, *Probab. Theory Related Fields* **108** (1997), no. 4, 517–542. ← 351

[16] G. Ben Arous and O. Zeitouni, Large deviations from the circular law. *ESAIM Probab. Statist.* **2** (1998), 123–134. ← 351

[17] R. J. Berman, Determinantal point processes and fermions on complex manifolds: large deviations and bosonization, 0812.4224. ← 347, 351

[18] L. Bétermin and E. Sandier, Renormalized energy and asymptotic expansion of optimal logarithmic energy on the sphere, to appear in *Constr. Approx.* 349, 374

[19] F. Bethuel, H. Brezis and F. Hélein, *Ginzburg-Landau Vortices*, Birkhäuser, 1994. ← 372

[20] W. Bietenholz and U. Gerber, Berezinskii-Kosterlitz-Thouless Transition and the Haldane Conjecture: Highlights of the Physics Nobel Prize 2016 - *Rev. Cub. Fis.* **33** (2016) 156-168, 1612.06132. ← 346

[21] X. Blanc and M. Lewin, The Crystallization Conjecture: A Review, *EMS surveys* **2** (2015), 255–306. ← 374

[22] S. Borodachov, D. Hardin and E. Saff, *Minimal Discrete Energy on Rectifiable Sets*, to appear. ← 344

[23] A. Borodin, Determinantal point processes, in *Oxford Handbook of Random Matrix Theory*, G. Akemann, J. Baik, and P. Di Francesco, eds. Oxford, 2011. 347

[24] A. Borodin and C. D. Sinclair, The Ginibre ensemble of real random matrices and its scaling limits. *Comm. Math. Phys.* **291** (2009), no. 1, 177–224. ←

[25] P. Bourgade, L. Erdős and H.-T. Yau, Universality of general β-ensembles, *Duke Math. J.* **163** (2014), no. 6, 1127–1190. ← 347

[26] P. Bourgade, L. Erdős and H. T. Yau, Bulk Universality of General β-ensembles with non-convex potential. *J. Math. Phys.* **53**, No. 9, (2012), 095221. ← 347

[27] G. Borot and A. Guionnet, Asymptotic expansion of β matrix models in the one-cut regime, *Comm. Math. Phys.* **317** (2013), 447–483. ← 363, 364, 365, 368, 369, 370

[28] G. Borot and A. Guionnet, Asymptotic expansion of beta matrix models in the multi-cut regime, 1303.1045. ← 365, 370

[29] S. Brauchart, D. Hardin and E. Saff, The next-order term for optimal Riesz and logarithmic energy asymptotics on the sphere. Recent advances in orthogonal polynomials, special functions, and their applications, 31?61, Contemp. Math., 578, Amer. Math. Soc., Providence, RI, 2012. ← 344, 374
[30] S. G. Brush, H. L. Sahlin and E. Teller. Monte-Carlo study of a one-component plasma. *J. Chem. Phys*, 45:2102–2118, 1966. ← 345
[31] L. Caffarelli and L. Silvestre, An extension problem related to the fractional Laplacian. *Comm. PDE* **32**, No. 8, (2007), 1245–1260. ← 358
[32] E. Caglioti, P. L. Lions, C. Marchioro and M. Pulvirenti, A Special Class of Stationary Flows for Two-Dimensional Euler Equations: A Statistical Mechanics Description. *Comm. Math. Phys.* **143**, 501–525 (1992). ← 352
[33] J.-M. Caillol, D. Levesque, J.-J. Weis and J.-P. Hansen. A monte carlo study of the classical two-dimensional one-component plasma. *Journal of Statistical Physics*, 28(2):325–349, 1982. ← 345
[34] A. Campa, T. Dauxois and S. Ruffo, Statistical mechanics and dynamics of solvable models with long-range interactions, *Physics Reports* **480** (2009), 57–159. ← 345
[35] D. Chafaï, N. Gozlan and P.-A. Zitt, First-order global asymptotics for confined particles with singular pair repulsion. *Ann. Appl. Probab.* **24** (2014), no. 6, 2371–2413. ← 351
[36] D. Chafai, A. Hardy and M. Maida, Concentration for Coulomb gases and Coulomb transport inequalities, 1610.00980. ← 363
[37] P. Choquard and J. Clerouin. Cooperative phenomena below melting of the one-component two-dimensional plasma. *Physical review letters*, 50(26):2086, 1983. ← 345
[38] M. Correggi and N. Rougerie, Inhomogeneous vortex patterns in rotating Bose-Einstein condensates, *Comm. Math. Phys.* **321** (2013), 817–860. ← 348
[39] G. Choquet. Diamètre transfini et comparaison de diverses capacités. Technical report, Faculté des Sciences de Paris, 1958. ← 348
[40] H. Cohn and A. Kumar, Universally optimal distribution of points on spheres. *J. Amer. Math. Soc.* **20** (2007), no. 1, 99–148. ← 345
[41] H. Cohn, A. Kumar, S. D. Miller, D. Radchenko and M. Viazovska, The sphere packing problem in dimension 24, *Ann. of Math.* (2) **185** (2017), no. 3, 1017–1033. ← 344, 345
[42] H. Cohn, A. Kumar, S. D. Miller, D. Radchenko and M. Viazovska, to appear. ← 345, 374
[43] J. Conway and J. Sloane, *Sphere Packings, Lattices and Groups*, Springer, 1999. ← 345
[44] R. Coulangeon and A. Schürmann, Energy minimization, periodic sets and spherical designs, *Intern. Math. Research Notices*, (2012), no. 4, 829–848. ← 345, 374
[45] D. S. Dean, P. Le Doussal, S. N. Majumdar and G. Schehr, Non-interacting fermions at finite temperature in a d-dimensional trap: universal correlations, *Phys. Rev. A* **94**, 063622, (2016). ← 345
[46] P. Deift, *Orthogonal Polynomials and Random Matrices: A Riemann-Hilbert Approach.* Courant Lecture Notes in Mathematics, AMS, 1999. ← 346
[47] A. Dembo and O. Zeitouni, *Large deviations techniques and applications*, Springer-Verlag, 2010. ← 351, 380
[48] I. Dumitriu and A. Edelman, Matrix models for beta ensembles, *J. Math. Phys.* **43** (2002), 5830–5847. ← 347
[49] F. Dyson, Statistical theory of the energy levels of a complex system. Part I, *J. Math. Phys.* **3**, 140–156 (1962); Part II, ibid. 157–185; Part III, ibid. 166–175 346
[50] P. J. Forrester, *Log-gases and random matrices*. London Mathematical Society Monographs Series, 34. Princeton University Press, 2010. ← 346
[51] S. Friedli and Y. Velenik, *Statistical Mechanics of Lattice Systems: a Concrete Mathematical Introduction*, Cambridge University Press , 2017. ← 377
[52] O. Frostman, Potentiel d'équilibre et capacité des ensembles avec quelques applications à la théorie des fonctions. *Meddelanden Mat. Sem. Univ. Lund 3*, **115 s** (1935). ← 349
[53] C. García-Zelada, A large deviation principle for empirical measures on Polish spaces: Application to singular Gibbs measures on manifolds, 1703.02680. ← 351, 352
[54] J. Ginibre, Statistical ensembles of complex, quaternion, and real matrices. *J. Math. Phys.* **6** (1965), 440–449. ← 346, 350
[55] V. L. Girko, Circle law. *Theory Probab. Appl.* **29** (1984), 694-706. ← 350
[56] S. Girvin, Introduction to the fractional quantum Hall effect, *Séminaire Poincaré* **2**, 54–74 (2004). ← 345, 347

[57] D. Hardin, T. Leblé, E. B. Saff and S. Serfaty, Large Deviations Principle for Hypersingular Riesz Gases, 1702.02894. To appear in *Constr. Approx.* 343

[58] A. Hardy, A Note on Large Deviations for 2D Coulomb Gas with Weakly Confining Potential, *Elec. Comm. Proba.* **17** (2012), 1–12. ← 349

[59] J. B. Hough, M. Krishnapur, Y. Peres and B. Virág, *Zeros of Gaussian analytic functions and determinantal point processes,* University Lecture Series, 51. AMS, 2009. ← 347

[60] K. Huang, *Statistical Mechanics*, Wiley, 1987. ← 313, 345

[61] B. Jancovici, Classical Coulomb systems: screening and correlations revisited. *J. Statist. Phys.* **80** (1995), no. 1-2, 445–459. ← 345

[62] B. Jancovici, J. Lebowitz and G. Manificat, Large charge fluctuations in classical Coulomb systems. *J. Statist. Phys.* **72** (1993), no. 3-4, 773–7. ← 345

[63] K. Johansson, On fluctuations of eigenvalues of random Hermitian matrices, *Duke Math. J.* **91** (1998), no. 1, 151–204. ← 347, 364, 370

[64] S. Kapfer and W. Krauth. Two-dimensional melting: From liquid-hexatic coexistence to continuous transitions. *Physical review letters*, 114(3):035702, 2015. ← 346

[65] M. K. Kiessling, Statistical mechanics of classical particles with logarithmic interactions. *Comm. Pure Appl. Math.* **46** (1993), no. 1, 27–56. ← 352

[66] R. Killip and I. Nenciu, Matrix models for circular ensembles. *Int. Math. Res. Not.* **50**, (2004), 2665–2701. ← 347, 379

[67] S. Klevtsov, Geometry and large N limits in Laughlin states, Lecture notes from the School on Geometry and Quantization, ICMAT, Madrid, September 7-11, 2015, 1608.02928. ← 347

[68] G. Lambert, M. Ledoux and C. Webb, Stein's method for normal approximation of linear statistics of beta-ensembles, 1706.10251. ← 370

[69] T. Leblé, Logarithmic, Coulomb and Riesz energy of point processes, *J. Stat. Phys,* **162** (4), (2016) 887–923. ← 378

[70] T. Leblé, Local microscopic behavior for 2D Coulomb gases, *Probab. Theory Relat. Fields* (2016). ← 363, 378

[71] T. Leblé and S. Serfaty, Large Deviation Principle for Empirical Fields of Log and Riesz gases, *Inventiones Math.* **210**, No. 3, 645–757. ← 354, 376, 378, 383

[72] T. Leblé and S. Serfaty, Fluctuations of Two-Dimensional Coulomb Gases, *Geom. Funct. Anal.* **28** (2018), no. 2, 443–508 1609.08088. ← 364, 368, 370, 371

[73] T. Leblé, S. Serfaty and O. Zeitouni, Large deviations for the two-dimensional two-component plasma. *Comm. Math. Phys.* **350** (2017), no. 1, 301–360. ←

[74] E. H. Lieb and M. Loss, *Analysis*, American Mathematical Society, 2001. ← 349, 358

[75] E. H. Lieb and J. Lebowitz. The constitution of matter: Existence of thermodynamics for systems composed of electrons and nuclei. *Advances in Math.* **9** (1972), 316–398. ← 345

[76] E.H. Lieb and M. Loss, Analysis, Graduate Studies in Mathematics **14**, AMS, Providence, 1997. ← 349, 358

[77] E. H. Lieb and H. Narnhofer, The thermodynamic limit for jellium. *J. Statist. Phys.* **12** (1975), 291–310. ← 345

[78] M. Maida and E. Maurel-Segala, Free transport-entropy inequalities for non-convex potentials and application to concentration for random matrices, *Probab. Theory Relat. Fields*, **159** (2014) No. 1, 329–356. ← 363

[79] M. Mazars, Long ranged interactions in computer simulations and for quasi-2D systems, *Physics Reports* **500**, (2011), 43–116. ← 345

[80] M. L. Mehta, *Random matrices. Third edition.* Elsevier/Academic Press, 2004. ← 346, 347, 350

[81] J. Messer and H. Spohn, Statistical mechanics of the isothermal Lane-Emden equation, *J. Stat. Phys.* **29** (1982), no. 3, 561–578. ← 352

[82] O. Penrose and E.R. Smith, Thermodynamic Limit for Classical Systems with Coulomb Interactions in a Constant External Field, *Comm. Math. Phys.* **26**, 53–77 (1972). ← 345

[83] M. Petrache and S. Rota Nodari, Equidistribution of jellium energy for Coulomb and Riesz Interactions, 1609.03849. To appear in *Constr. Approx.* 379

[84] M. Petrache and S. Serfaty, Next Order Asymptotics and Renormalized Energy for Riesz Interactions, *J. Inst. Math. Jussieu* **16** (2017) No. 3, 501–569. ← 354, 355, 358, 379

[85] D. Petz and F. Hiai, Logarithmic energy as an entropy functional, *Advances in differential equations and mathematical physics*, 205–221, Contemp. Math., 217, Amer. Math. Soc., Providence, RI, 1998. ← 351

[86] B. Rider and B. Virag, The noise in the circular law and the Gaussian free field, *Int. Math. Res. Not* **2**, (2007). 370
[87] S. Rota Nodari and S. Serfaty, Renormalized energy equidistribution and local charge balance in 2D Coulomb systems, *Inter. Math. Research Notices* **11** (2015), 3035–3093. ← 379
[88] N. Rougerie, *De Finetti theorems, mean-field limits and Bose-Einstein condensation*, LMU lecture notes, 2014, 1506.05263. ← 352
[89] N. Rougerie, *Théorèmes de De Finetti, limites de champ moyen et condensation de Bose-Einstein*, Les cours Peccot, Spartacus IHD, Paris, 2016. ← 352
[90] N. Rougerie and S. Serfaty, Higher Dimensional Coulomb Gases and Renormalized Energy Functionals, *Comm. Pure Appl. Math* **69** (2016), 519–605. ← 354, 355, 373, 379
[91] N. Rougerie and J. Yngvason, Incompressibility estimates for the Laughlin phase, *Comm. Math. Phys.* **336**, (2015), 1109–1140. ← 347
[92] E. Saff and A. Kuijlaars, Distributing many points on a sphere. *Math. Intelligencer* **19** (1997), no. 1, 5–11. ← 344
[93] E. Saff and V. Totik, *Logarithmic potentials with external fields*, Springer-Verlag, 1997. 343, 348, 349
[94] E. Sandier and S. Serfaty, *Vortices in the Magnetic Ginzburg-Landau Model*, Birkhäuser, 2007. ← 345, 348
[95] E. Sandier and S. Serfaty, From the Ginzburg-Landau model to vortex lattice problems, *Comm. Math. Phys.* **313** (2012), 635–743. ← 348, 374
[96] E. Sandier and S. Serfaty, 2D Coulomb Gases and the Renormalized Energy, *Annals of Proba*, **43**, no 4, (2015), 2026–2083. ← 354, 355, 379
[97] E. Sandier and S. Serfaty, 1D Log Gases and the Renormalized Energy: Crystallization at Vanishing Temperature, *Proba. Theor. Rel. Fields*, **162**, no 3, (2015), 795–846. 354, 358, 374, 379
[98] R. Sari and D. Merlini, On the ν-dimensional one-component classical plasma: the thermodynamic limit problem revisited. *J. Statist. Phys.* **14** (1976), no. 2, 91–100. ← 345
[99] M. Shcherbina, Fluctuations of linear eigenvalue statistics of β matrix models in the multi-cut regime, J. Stat. Phys. **151**, (2013), 1004–1034 ← 347, 364, 365, 370
[100] S. Serfaty, *Coulomb gases and Ginzburg-Landau vortices*, Zurich Lectures in Advanced Mathematics, 70, Eur. Math. Soc., 2015. ← 348, 349, 350, 351, 352
[101] S. Serfaty and J. Serra, Quantitative stability of the free boundary in the obstacle problem *Anal. PDE* **11** (2018), no. 7, 1803–1839, 1708.01490. ← 365
[102] B. Simon, The Christoffel-Darboux kernel, in "Perspectives in PDE, Harmonic Analysis and Applications," a volume in honor of V.G. Maz'ya's 70th birthday, *Proc. Symp. Pure Math.* **79** (2008), 295–335. ← 344
[103] S. Smale, Mathematical problems for the next century. *Mathematics: Frontiers and Perspectives*, Amer. Math. Soc., pages 271–294, 2000. ← 374
[104] T. Spencer, Scaling, the free field and statistical mechanics, in The Legacy of Norbert Wiener: A. Centennial Symposium, *Proc. Sympos. Pure Math* **60**, AMS, (1997). ← 346
[105] S. M. Stishov. Does the phase transition exist in the one-component plasma model? *Jour. Exp. Theor. Phys. Lett*, 67(1):90–94, 1998. ← 345
[106] H. Stormer, D. Tsui and A. Gossard, The fractional quantum Hall effect, *Reviews of Modern Physics* **71**, (1999), S298. ← 345, 347
[107] S. Torquato, Hyperuniformity and its Generalizations, *Phys. Rev. E* **94** (2016) 022122. ← 345, 353
[108] B. Valkó and B. Virág, Continuum limits of random matrices and the Brownian carousel. *Invent. Math.* **177** (2009), no. 3, 463–508. ← 347, 379
[109] M. Viazovska, The sphere packing problem in dimension 8, *Ann. of Math.* **185** (2) (2017), no. 3, 991–1015. ← 344, 345
[110] E. Wigner, Characteristic vectors of bordered matrices with infinite dimensions, *Ann. Math.* **62**, (1955), 548–564. ← 346, 350
[111] E. Wigner, On the distributions of the roots of certain symmetric matrices. *Ann. of Math.* **67**, (1958), 325–327. ← 350

Courant Institute of Mathematical Sciences, New York University, 251 Mercer Street, New York, NY 10012
Email address: `serfaty@cims.nyu.edu`

IAS/Park City Mathematics Series
Volume 26, Pages 389–459
https://doi.org/10.1090/pcms/026/00850

Random matrices and free probability

Dimitri Shlyakhtenko

Abstract. Voiculescu invented his free probability theory to approach problems in von Neumann algebras. A key feature of his theory is the treatment of free independence — based on the notion of free products, such as free products of groups — as a surprisingly close parallel to classical independence. Rather unexpectedly it turned out that there are deep connections between his theory and the theory of random matrices: very roughly, free probability describes certain aspects of asymptotic behavior of random matrix models. In this course, we will start with an introduction to free probability theory, discuss connections with random matrix theory, and finally describe some applications of results from free probability in random matrices.

Contents

Introduction. 390
Lecture 0: Non-commutative probability spaces. 391
  0.1 Executive summary. 391
  0.2 Non-commutative measure spaces. 392
  0.3 Non-commutative probability spaces. 397
  0.4 Summary: non-commutative measure spaces. 399
  0.5 Exercises. 399
Lecture 1: Non-commutative Laws. Classical and Free Independence. 401
  1.1 Executive summary. 401
  1.2 Non-commutative laws. 402
  1.3 Examples of non-commutative probability spaces and laws. 404
  1.4 Notions of independence. 405
  1.5 The free Gaussian Functor. 408
  1.6 Exercises 416
Lecture 2: R-transform and Free Harmonic Analysis. 418
  2.1 Executive summary. 418
  2.2 Additive and multiplicative free convolutions. 418
  2.3 Computing $\boxplus$: R-transform. 419

2010 *Mathematics Subject Classification.* Primary 60B20; Secondary 46L54.
Key words and phrases. Random matrix theory, free probability theory.

2.4 Combinatorial interpretation of the R-transform. 422
2.5 Properties of free convolution. 423
2.6 Free subordination. 424
2.7 Multiplicative convolution $\boxtimes$. 425
2.8 Other operations. 425
2.9 Multivariable and matrix-valued results. 426
2.10 Exercises. 427
Lecture 3: Free Probability Theory and Random Matrices. 428
3.1 Executive summary. 428
3.2 Non-commutative laws of random matrices. 428
3.3 Random matrix models. 429
3.4 GUE matrices: bound on eigenvalues. 434
3.5 GUE matrices and Stein's method: proof of Theorem 3.3.5(1′). 435
3.6 On the proof of Theorem 3.3.5(2) and (3). 439
3.7 Exercises. 440
Lecture 4: Free Entropy Theory and Applications. 441
4.1 Executive summary. 441
4.2 More on free difference quotient derivations. 441
4.3 Free difference quotients and free subordination. 443
4.4 Free Fisher information and non-microstates free entropy. 444
4.5 Free entropy and large deviations: microstates free entropy χ. 445
4.6 χ vs χ^*. 447
4.7 Lack of atoms. 448
4.8 Exercises 450
Addendum: Free Analogs of Monotone Transport Theory. 451
5.1 Classical transport maps. 451
5.2 Non-commutative transport. 452
5.3 The Monge-Ampère equation. 452
5.4 The Free Monge-Ampère equation. 453
5.5 Random Matrix Applications. 454

Introduction

The goal of these notes is to describe the deep connection between Voiculescu's free probability theory and random matrix theory. Voiculescu's free probability theory has its origins in operator algebras, a study of operators acting on a Hilbert space. Voiculescu's deep insight was his discovery of an analogy between tensor products of von Neumann algebras — an operation key to the notion of independence in classical and non-commutative probability — and free products, a purely non-commutative operation. The resulting free probability theory, built around the notion of free independence, has many parallels with classical probability. For example, it turned out that in free probability the semicircle law plays the role of the Gaussian law from classical probability theory. This very same semicircle law is a centerpiece of random matrix theory: a celebrated result going back to Wigner's work in the 1950s shows that the empirical measure of many random matrices converges to the semicircle law as the matrix size goes to infinity. Voiculescu's work in the 1990s showed that the appearance of the semicircle law in both results is no coincidence: the underlying fact is that many $N \times N$ random matrix models give rise to freely independent families of non-commuting variables in the large N limit.

The connection between random matrices and free probability has been an extremely fruitful one. Voiculescu's introduction of free entropy — intimately connected to large deviations for random matrix models — has seen a number of spectacular applications to operator algebras. Conversely, a number of free probability techniques, such as free convolution and the resulting theory of subordination functions, or the theory of free monotone transportation, have found applications in random matrices (sometimes surprising ones, e.g. in connection to questions about outlier eigenvalues).

It is not the aim of these notes to summarize either random matrix theory or free probability theory – each of these fields deserves (and got!) books devoted to it — but rather to indicate a few of the methods and ideas that have been influential; for this same reason the bibliography is not meant to be comprehensive. While we did not aim to provide shortest possible proofs, our goal was to chose ones that illuminate the connections between the subjects and point out some of the techniques.

Acknowledgements. The author would like to thank Ian Charlesworth and the anonymous referees of these notes for their careful proofreading of these notes and for several helpful suggestions. I am also grateful to the organizers of the PCMI summer school for their support and encouragement. Research supported by NSF grant DMS-1500035.

Lecture 0: Non-commutative probability spaces

0.1. Executive summary. Classical measure spaces can be characterized as *abelian von Neumann algebras endowed with normal states.* Dropping the abelian assumption leads to the notion of a *non-commutative probability space* in which random variables are realized as operators on some Hilbert space. **This section can be skipped during the first reading, and serves only as motivation and background reference.** Its main purpose is to motivate viewing an algebra of operators acting on Hilbert space H endowed with a distinguished unit vector ξ as an analog of a classical probability space.

0.2. Non-commutative measure spaces. Classical standard measure spaces can be described in terms of the σ-algebra of measurable subsets of some measure space (X, μ). An equivalent description replaces sets by their indicator functions. Without recalling all of the axioms of a σ-algebra, note that the set of indicator functions forms a lattice $\Sigma(X)$ with order and max/min operations given by

$$1_A \leqslant 1_B \iff A \subset B$$
$$1_A \vee 1_B = 1_{A\cup B}$$
$$1_A \wedge 1_B = 1_{A\cap B}$$

and with maximum element $1 := 1_X$ and minimum element $0 := 1_\emptyset$. A *random variable* can then be expressed as a combination of these indicator functions. For example, a discrete $\{-1,1\}$-valued random variable $f : X \to \{-1,1\}$ can be written as the sum $f = (-1) \cdot 1_{\{x:f(x)=-1\}} + 1_{\{x:f(x)=1\}}$, an integer-valued random variable $g : X \to \mathbb{Z}$ is an infinite sum

$$g(x) = \sum_{n\in\mathbb{Z}} n \cdot 1_{\{x:g(x)=n\}}$$

and a real-valued random variable $f : X \to \mathbb{R}$ can be written as an integral

$$f = \int_{\mathbb{R}} t \cdot dP(t)$$

where $P(t)$ is a measure on $\mathbb{R}$ valued in $\Sigma(X)$ and given by $P(B) = 1_{\{x:f(x)\in B\}}$.

It is well-known that standard measure spaces can be completely classified in terms of their lattices of indicator functions. One says that $0 \neq p \in \Sigma(X)$ is *minimal* if $0 \neq q \leqslant p \implies q = p$. It then turns out that any standard measure space X is isomorphic to the union $X_c \cup X_d$ where X_c is empty or $X_c = ([0,1], \lambda)$ with its standard σ-algebra of Lebesgue-measurable subsets (with no minimal elements with non-zero measure) and X_d is one of the spaces $\emptyset$, $\{1\}$, $\{1,2\}$, ... or $\mathbb{N}$, with their standard σ-algebras of measurable subsets (which have minimal elements of non-zero measure).

Motivated by the goal of a mathematical formulation of quantum mechanics, von Neumann proposed in 1930s to investigate non-commutative analogs of classical measure spaces. His deep insight was to attempt to replace in the description above the lattice of indicator functions by some lattice of projections on a Hilbert space H. For a closed Hilbert subspace $K \subset H$, let us denote by P_K the orthogonal projection onto K. Such projections form a lattice:

$$P_K \subset P_L \iff K \subset L$$
$$P_K \vee P_L = P_{\overline{K+L}}$$
$$P_K \wedge P_L = P_{K\cap L}.$$

To see the connection with the classical picture, we can start with a measure space X with a probability measure μ and let $H = L^2(X, \mu)$. Each measurable subset $A \subset X$ gives rise to the subspace $L^2(A) = \{f \in L^2(X,\mu) : f(x) = 0\ \forall x \notin A\}$. It is easy to see that $1_A \mapsto P_{L^2(A)}$ is a mapping of lattices (whose kernel consists precisely of indicator functions of measure zero sets). Of course, the projection lattice is different in that the analog of $f_A \vee f_B = f_A + f_B - f_A \wedge f_B$ does not hold for projections (see Exercise 0.5.1).

Tantalizingly, the spectral theorem associates to every self-adjoint operator T on H a projection-valued measure E defined on $\mathbb{R}$ so that

$$T = \int_{\mathbb{R}} \lambda dE(\lambda).$$

(As an easy example, if T is a finite-rank operator, then $T = \sum_i \lambda_i E_i$ where E_i is the λ_i -eigenspace of T; in this case the measure is atomic with atoms λ_i, assigning to the atom λ_i the projection E_i.) Thus, if we accept that the lattice of projections on a Hilbert space is the lattice of indicator function on "some probability space", then self-adjoint operators are the "random variables on that space". Of course, no such probability space can exist, simply because the lattice of projections on a Hilbert space does not have all the properties of the lattice of indicator functions (this is related to the fact that operators on a Hilbert space do not commute). This non-commutativity is often attached to names of various objects in this context: projections can be called "non-commutative indicator functions", operators — "non-commutative random variables", and the whole set-up "non-commutative probability".

This set of observations led von Neumann to try to formulate an abstract definition of what a "non-commutative lattice of projections" would mean.

One very convenient property of the classical lattice of indicator functions is that the min operation $\wedge$ coincides with the operation of product of functions: $1_A \cdot 1_B = 1_{A \cap B} = 1_A \wedge 1_B$. This does not at all hold for projections (Exercise 0.5.1) except in rare circumstances. For this reason, lattices of non-commutative projections are more awkward to deal with than their commutative analogs. A better object of study is what has become known as a *von Neumann algebra*. It is the non-commutative analog of the space of essentially bounded functions $L^\infty(X, \mu)$ (which contains the indicator functions themselves). Amazingly enough, the set of axioms defining a von Neumann algebra is extremely short!

Definition 0.2.1. A *von Neumann algebra* is normed $*$-algebra $(M, \|\cdot\|)$ satisfying the following properties:

- $\|x\| = \|x^*\| = \|x^*x\|^{1/2}$ for all $x \in M$ and M is complete in this norm (this means that M is a C^**-algebra*);
- As a Banach space, M is the dual of some separable Banach space M_*; moreover, the multiplication maps $x \mapsto xy$, $x \mapsto yx$ and conjugation map $x \mapsto x^*$ are weak-$*$ continuous in x.

The second property automatically implies that M comes equipped with a topology, called the weak-$*$ topology: a net $x_\iota \in M$ converges to $x \in M$ if and only if $\phi(x_i) \to \phi(x)$ for all $\phi \in M_*$.

A norm satisfying the first property above, $\|x\| = \|x^*\| = \|x^*x\|^{1/2}$, is the *operator norm* on the space $B(H)$ of all bounded operators on a Hilbert space H. This norm is defied by

$$\|T\|_\infty = \sup_{\|\xi\|=1} \|T\xi\|.$$

In fact, it turns out that this is a universal example, in the sense that any Banach $*$-algebra equipped with such a norm is (isometrically) a subalgebra of $B(H)$ for some H.

If M is a von Neumann algebra, an element $p \in M$ satisfying $p = p^* = p^2$ is called a *projection*. We write $P(M)$ for the set of projections on M.

For $x \in M$ we say that $x \geqslant 0$ iff $x = y^*y$ for some $y \in M$. We say that $x \geqslant y$ iff $x - y \geqslant 0$. With this notation we obtain a partial order on the set of projections on M: $p \geqslant q \iff p - q \geqslant 0$. (See Exercise 0.5.2). It turns out that the operations of sup and inf for a pair of projections are also well-defined (see Exercise 0.5.4).

Theorem 0.2.2. *Let M be a von Neumann algebra and assume that M is abelian. Then $M = L^\infty(X, \nu)$ for some standard measure space (X, ν). Furthermore, in this identification the lattice of projections of M becomes the lattice of indicator functions $\Sigma(X)$.*

It is quite striking that Theorem 0.2.2 can be used to characterize standard measure spaces (or at least algebras of random variables); at least formally the definition is much shorter than of a standard measure space. Exercise 0.5.3 explains why the algebra $B(H)$ of all bounded operators on a Hilbert space is a von Neumann algebra: $B(H)$ is the Banach space dual of the space $\mathcal{L}^1(H)$ of trace-class operators on H. This duality induces a topology on $B(H)$ called the σ-WOT topology (WOT stands for weak operator topology). In this topology, a net of operator T_ι converges to an operator T if and only if $\mathrm{Tr}(T_\iota S)$ converges to $\mathrm{Tr}(TS)$ for any trace-class operator S.

Definition 0.2.3. A linear map between two von Neumann algebras is called *normal* if it is continuous with respect to the weak topology induced by the preduals. In particular, elements of the predual are exactly the normal linear functionals.

The lattice of projections of $B(H)$ is much more complicated than the lattice of indicator functions on any commutative measure space, but if anything it resembles most closely that of a purely atomic space (i.e., $\{1, \ldots, n\}$ or $\mathbb{N}$). Indeed, it is easily seen that a rank-one projection P is minimal, in that no non-zero projection Q satisfies $Q \leqslant P$ without Q begin equal to P. Similarly, the indicator function of single point of nonzero mass is a minimal non-zero element of the lattice $\Sigma(X)$. It turns out that more general von Neumann algebras can have more interesting lattices. Before studying these, let us formulate a fundamental theorem showing that $B(H)$ is in a sense the universal von Neumann algebra.

Theorem 0.2.4. *Let M be a von Neumann algebra, so that M is the dual of a Banach space M_*. Assume that M_* is separable. Then there exists a separable Hilbert space* *and a unital isometric $*$-homomorphism $\pi : M \to B(H)$ so that π is a homeomorphism from M (with the weak topology induced by M_*) to $B(H)$ with the σ-WOT topology. In particular, for any $\phi \in M_*$ there exists a $T \in \mathcal{L}^1(H)$ so that $\phi(x) = \mathrm{Tr}(T\pi(x))$ for any $x \in M$.*

Thus every von Neumann algebra is a σ-WOT closed von Neumann subalgebra of $B(H)$. Conversely, it is not hard to see that any σ-WOT closed subalgebra of $B(H)$ has a pre-dual (because $B(H)$ has a predual, which induces the σ-WOT topology). There is an alternative characterization of such subalgebras, usually called the von Neumann bicommutant theorem:

Theorem 0.2.5. *Let $M \subset B(H)$ be a unital $*$-subalgebra. Let*

$$M' = \{T \in B(H) : Tx = xT \ \forall x \in M\}$$

be the commutant of M. Then M is σ-WOT closed iff $M = (M')'$.

In particular, if X is any set of operators in $B(H)$ so that $x \in X \iff x^* \in X$, then X' is a von Neumann algebra; and X'' is the smallest von Neumann subalgebra of $B(H)$ containing X (often denoted $W^*(X)$; it's of course the same as the σ-WOT closure of the unital algebra generated by X).

It is not hard to come up with an example of a von Neumann algebra whose projection lattice has no nonzero minimal elements – indeed, $L^\infty(X,\mu)$ with non-atomic μ would do. Remarkably, there are such examples which are purely non-commutative.

Definition 0.2.6. A von Neumann algebra M is called a *factor* if the center of M is trivial: if $xy = yx$ for all $y \in M$, then $x \in \mathbb{C}1$.

Let us give the first example of a factor whose projection lattice has no non-zero minimal projections.

Example 0.2.7. Let $R_n = M_{2^n \times 2^n}$ and denote by $i_n : R_n \to R_{n+1}$ the map

$$x \mapsto \begin{bmatrix} x & 0 \\ 0 & x \end{bmatrix}.$$

If $\phi_n : R_n \to \mathbb{C}$ is the normalized trace $\phi_n(x) = 2^{-n}\mathrm{Tr}(x)$, then $\phi_{n+1} \circ i_n = \phi_n$. Denoting $\|x\|_1 = \phi_n((x^*x)^{1/2})$, and by $\|x\|$ the operator norm, we note that

$$(R_n, \|\cdot\|) = (R_n, \|\cdot\|_1)^*$$

if we identify $x \in M_n$ with the linear functional $y \mapsto \phi_n(xy)$. In this identification, $\phi_n : M_n \to \mathbb{C}$ corresponds to $1 \in (R_n, \|\cdot\|_1)$.

The map i_n gives an isometric inclusion of Banach spaces

$$i_n : (R_n, \|\cdot\|_1) \to (R_{n+1}, \|\cdot\|_1).$$

Denote by R_* the inductive limit of these spaces, i.e., the completion in the norm $\|\cdot\|_1$ of the union $\cup_n R_n$. Finally, let $R = (R_*)^*$ with the dual Banach norm. It is not hard to see that R is an algebra and thus a von Neumann algebra; moreover

$1 \in R_*$ gives rise a linear function $\phi : R \to \mathbb{C}$ satisfying $\phi(xy) = \phi(yx)$ for all $x, y \in R$.

The algebra R is called the *hyperfinite II_1 factor.* Its construction is reminiscent of the Cantor set, if one identifies $R_n \cong M_{2\times 2}^{\otimes n}$.

Note: There are some obvious modifications of this construction, in which one uses some other inclusion, e.g. $M_{3^n \times 3^n} \to M_{3^{n+1}\times 3^{n+1}}$. It is a somewhat challenging exercise to prove that one actually gets the same von Neumann algebra R in the end.

Let us see what can be said about the projection lattice of R. First of all, note that any element of R_n is clearly in R; thus the inductive limit of algebras R_n belongs to R. A little thought shows that in fact this inductive limit is dense in R (for the weak topology induced by the predual). In particular, any $x \in R$ can be approximated in the weak topology by an element of R_n for a sufficiently large n. If $p \in R$ is a projection one can in fact choose a sequence $p_n \in R_n$ consisting of *projections* with $p_n \to p$ weakly. Now, since R_n is a matrix algebra, if $p, p' \in R_n$ are two projections so that $\phi_n(p) \leqslant \phi_n(p')$ then there exists an element $v \in R_n$ so that $v_n^* v_n = p$ and $v_n v_n^* \leqslant p'$. Using this one can show that if $p, p' \in R$ are any two projections with $\phi(p) \leqslant \phi(p')$, then there exists some $v \in R$ so that $v^*v = p$ and $vv^* \leqslant p'$ (note that vv^* is also a projection). By the trace property, $\phi(vv^*) = \phi(v^*v)$; thus if $\phi(p) < \phi(p')$ then for some v, $v^*v = p$, $vv^* \leqslant p'$ and moreover $\phi(p') - \phi(vv^*) = \phi(p') - \phi(p) > 0$, so that $vv^* \neq p'$. Assume for contradiction that $q \in R$ is a minimal projection and $q \neq 0$. It can be shown that then $\phi(q) > 0$. Choosing n large enough we can then find $q' \in R_n$ with $\phi(q') = 2^{-n} < \phi(q)$, which then implies that there is a projection $0 \neq q'' < q$ and $q'' \neq q$, so that q is not minimal.

In this example we have see that projections in R can be classified by the value of ϕ, in the following sense: $\phi(p) = \phi(q)$ iff for some v, $p = v^*v$ and $q = vv^*$. This is similar to the situation for projections in finite-dimensional matrices: they are in a similar way classified by their trace (i.e., the dimension of their image). The difference is that the values of ϕ on projections in R can be arbitrary numbers in the interval $[0, 1]$, a phenomenon von Neumann referred to as *continuous dimension.* This situation is much more general and applies to so-called *finite* von Neumann algebras (in fact, it turns out that any such von Neumann algebra that is infinite dimensional and has trivial center has to contain a copy of R).

Definition 0.2.8. Let M be a factor. We say that M is *finite* if there exists a linear functional $\phi : M \to \mathbb{C}$ satisfying $\phi(1) = 1$ and $\phi(xy) = \phi(yx)$ for all $x, y \in M$ and which is *faithful*, i.e., $\phi(x^*x) = 0$ iff $x = 0$. Such ϕ is called a *trace.*

Theorem 0.2.9. *Let M be a finite factor.*

(1) *Any $x \in M$ admits a decomposition*

$$x = x_0 + \sum_{i=1}^{8} x_i$$

with $x_0 \in \mathbb{C}1$ *and* $x_i = y_i z_i - z_i y_i$ *for some* $y_i, z_i \in M$. *In particular, the trace on* M *is unique.*

(2) *Denote the trace by* τ. *Then* $\tau \in M_*$, $\|\tau\|_{M_*} = 1$, $\tau(x^*x) \geqslant 0$ *for any* $x \in M$ *and* $\tau(x^*x) = 0$ *iff* $x = 0$.

(3) *If* $p \in M$ *is a projection, then* $\tau(p) \in [0,1]$ *and moreover* $\tau(p) \geqslant \tau(p')$ *iff for some* $v \in M$, $p = vv^*$ *and* $v^*v \geqslant p'$.

Many properties of projections in finite dimensions extend to projections in finite von Neumann algebras. One of the most useful is the analog of the statement that large-dimensional subspaces of a finite-dimensional vector space must intersect:

Theorem 0.2.10. *Let* $p, q \in M$ *be two projections, for which* $\tau(p) + \tau(q) - 1 = a > 0$. *Then* $\tau(p \wedge q) \geqslant a$; *in particular,* $p \wedge q \neq 0$.

There is also an analog of the rank-nullity theorem:

Theorem 0.2.11. *Let* $x \in M$. *Define the kernel and image projections of* x *by*

$$p_{\ker x} = \sup\{p : xp = 0\}, \qquad p_{\operatorname{im} x} = \inf\{q : qx = x\}.$$

Then $\tau(p_{\ker x}) + \tau(p_{\operatorname{im} x}) = 1$.

0.3. Non-commutative probability spaces. Let M be a von Neumann algebra. As before, we say that an element $x \in M$ is positive if $x = y^*y$ for some $y \in M$.

Definition 0.3.1. A linear functional $\phi : M \to \mathbb{C}$ is called *positive* if $\phi(x) \geqslant 0$ whenever $x \geqslant 0$. A functional is called *normal* if it is weakly continuous for the topology induced on M by the predual M_* (i.e., $\phi \in M_*$). A positive normal functional satisfying $\phi(1) = 1$ is called a (normal) *state*.

The terminology *state* comes from quantum mechanics.

Theorem 0.3.2. *Let* $M = L^\infty(X, \mu)$ *be an abelian von Neumann algebra. Every normal state* $\phi : M \to \mathbb{C}$ *has the form*

$$\phi(f) = \int f d\nu$$

for some probability measure ν *which is* μ*-absolutely continuous.*

The previous theorem justifies the following definition:

Definition 0.3.3. A W^**-non-commutative probability space* is a pair (M, ϕ) where M is a von Neumann algebra and $\phi : M \to \mathbb{C}$ is a normal state. We say that the space is *tracial* if ϕ is a trace: $\phi(xy) = \phi(yx)$ for all $x, y \in M$.

Often we just say "non-commutative probability space".

Just as every von Neumann algebra can be represented on a Hilbert space, so can every non-commutative probability space. In fact, there is a canonical choice of the Hilbert space: the state ϕ determines a pre-Hilbert space structure on M by

$$\langle x, y \rangle = \phi(x^*y), \qquad x, y \in M.$$

The completion of M with respect to this inner product is denoted $L^2(M, \phi)$ (to be more precise, one needs to complete the quotient of M by vectors of length zero). If ϕ is tracial, both the left and right multiplication representations of M on itself

$$\rho(x) : y \mapsto yx, \qquad \lambda(x) : y \mapsto xy, \qquad x, y \in M$$

extend to bounded operators $y \mapsto yx$ and $y \mapsto xy$ for $y \in L^2(M, \phi)$ (in the non-tracial case, the right multiplication action may fail to be bounded). Thus we get a representation of M on $L^2(M, \phi)$ (say, by left multiplication λ). As a bonus, we see that ϕ is also represented as a "matrix coefficient" of the representation:

$$\phi(x) = \langle 1, \lambda(x)1 \rangle, \qquad \forall x \in M.$$

This construction is often called the GNS representation (after Gelfand, Naimark and Segal). If we require ϕ to be *faithful*, i.e., $\phi(x^*x) = 0$ iff $x = 0$, then M injects into $L^2(M, \phi)$ and in particular the representation λ is faithful. A finite factor with a separable predual always has a faithful representation on a separable Hilbert space.

Remaining in the faithful tracial case, one can attach to M a number of other Banach spaces: $L^p(M, \phi)$ defined as completions with respect to the norms

$$\|x\|_p = \phi(|x|^p)^{1/p}, \qquad x \in M$$

where we put $|x| = (x^*x)^{1/2}$. Just as in the case of a classical probability space one has the usual dualities $L^p(M, \phi)^* = L^q(M, \phi)$ with $1/p + 1/q = 1$; $L^1(M, \phi) = M_*$ and $L^\infty(M, \phi) = (L^1(M, \phi))^* = M$.

It remains to understand the non-commutative analog of one more function space: that of measurable functions. Classically, we may view any measurable function $f : X \to \mathbb{C}$ as an unbounded multiplication operator on $L^2(X, \mu)$.

Definition 0.3.4. Let $M \subset B(H)$ be a von Neumann algebra, and assume that M has a faithful normal trace (e.g., M is a finite factor). An unbounded operator $T : H \to H$ with domain D is said to be affiliated to M if (i) T is closed and (ii) for any $S \in M'$, $TS = ST$ (in particular, this means that $SD \subset D$).

The main fact is the following beautiful theorem due to von Neumann:

Theorem 0.3.5. *If M is a finite von Neumann algebra, the set $\mathcal{M}(M)$ of affiliated operators to M forms a $*$-algebra.*

This is of course highly unexpected: generally, the sum or product of two unbounded operators may have no domain of definition except for the zero vector. The proof is sketched in Exercise 0.5.5.

Theorem 0.3.5 fails for non-finite von Neumann algebras (e.g. it fails for $B(H)$ if H is infinite dimensional).

A final aspect of measure theory that is nicely generalized in the tracial context has to do with conditioning. Recall that if $\Sigma(X)$ is the σ-algebra of measurable sets on X and Σ' is some σ-subalgebra, the measure μ on X defines a *conditional expectation* from $\Sigma(X)$ to Σ'. In the setting of measure spaces this corresponds to a measure-preserving onto map $\pi : X \to Y$, so that Σ' consists precisely of the sets $\pi^{-1}(B)$, $B \in \Sigma(Y)$. The measure μ disintegrates into a family of probability measures ν_y on the preimages $\pi^{-1}(y)$, $y \in Y$. Then the conditional expectation of $f : X \to \mathbb{C}$ is determined by

$$E(f)(y) = \int_{\pi^{-1}(y)} f(t) d\nu_y(t).$$

Theorem 0.3.6. *Let* (M, ϕ) *be a non-commutative probability space, and assume that* ϕ *is a normal faithful trace. If* $N \subset M$ *is a von Neumann subalgebra, then the orthogonal projection* $P : L^2(M, \phi) \to L^2(N, \phi|_N) \subset L^2(M, \phi)$ *restricts to a map ("conditional expectation")* $E = E_N^M : M \to N$ *satisfying:*

(1) $E(nmn') = nE(m)n', \forall n, n' \in N,\ m \in M;$
(2) $\tau \circ E = \tau;$
(3) $E(x^*x) \geqslant 0$, $E(x^*x) = 0 \iff x = 0$, *for any* $x \in M$;
(4) E *is normal, i.e., continuous with respect to the topology induced on* M *and* N *by their preduals.*

Furthermore, the map E *is characterized by properties 1, 2, 3, 4 listed above.*

The simplest case of the preceding theorem corresponds to $N = \mathbb{C}1 \subset M$. In that case E_N^M is the trace on M. If one drops the traciality assumption on ϕ, there may fail to be a conditional expectation onto a subalgebra $N \subset M$. This is the case, for example, if $M = B(L^2[0,1])$ and $N = L^\infty[0,1]$ identified with multiplication operators on $L^2[0,1]$.

0.4. Summary: non-commutative measure spaces. In summary, we have the following properties of a tracial non-commutative probability space (M, τ) under the assumptions that τ is faithful:

- M is a von Neumann algebra with pre-dual $L^1(M, \tau)$
- M can be viewed as a von Neumann subalgebra of the algebra of bounded operators on a Hilbert space, with unit vector $\xi \in H$ so that $\tau(x) = \langle \xi, x\xi \rangle$ for all $x \in M \subset B(H)$; in fact, we can take $H = L^2(M, \tau)$ and $\xi = 1$
- $M \subset \mathcal{M}(M)$, the algebra of unbounded operators affiliated to M
- In this set-up, if M is assumed to be abelian, then $M = L^\infty(X, \mu)$ so that $\tau(f) = \int f(x) d\mu(x)$; $L^p(M, \tau) = L^p(X, \mu)$ and $\mathcal{M}(M)$ is the algebra of all measurable functions on X.

0.5. Exercises.

Exercise 0.5.1. Let H be a Hilbert space.

(1) Give an example of two projections P, Q onto subspaces of H for which $P \vee Q \neq P + Q - P \wedge Q$.
(2) Give an example of two projections P, Q onto subspaces of H for which $PQ \neq P \wedge Q$.

(3) Show that if P, Q are projections onto subspaces of H, then for any $\xi \in H$,

$$(P \wedge Q)\xi = \lim_{n\to\infty} (PQP)^n \xi = \lim_{n\to\infty} (QPQ)^n \xi$$

(both limits taken in the norm topology on H).

Note: the case of finite-dimensional H is interesting enough for these questions.

Exercise 0.5.2. Let P, Q be operators on a Hilbert space H so that P, Q are projections (i.e., $P = P^* = P^2$, $Q = Q^* = Q^2$). Show that

(1) P is the orthogonal projection onto PH;
(2) $P - Q \geqslant 0$ (i.e., $P - Q = T^*T$ for some operator T) iff $QH \subset PH$.
(3) Define a topology on the set of bounded operators on H by $T_i \to T$ iff $\|T_i\xi - T\xi\| \to 0$ for all $\xi \in H$ (this is called the Strong Operator Topology, SOT). Using part (3) of Exercise 0.5.1, show that given P, Q there exist projections $S, T \in \overline{\mathrm{Alg}(P, Q)}^{SOT}$ so that S is the smallest projection with $S \geqslant P$ and $S \geqslant Q$ and T is the largest projection with $T \leqslant P$, $T \leqslant Q$.

Exercise 0.5.3. Let H be a Hilbert space and let $B(H)$ be the algebra of all bounded operators on it. For $T \in H$ finite-rank, define $|T| = (T^*T)^{1/2}$ and let

$$\|T\|_1 = \mathrm{Tr}(|T|) = \sum \mu_j$$

where μ_j are the singular values of T, i.e., eigenvalues of $|T|$.

(1) Show that $\|TS\|_1 \leqslant \|T\|_1\|S\|$, where $\|S\| = \sup_{\|\xi\|=1,\ \xi\in H} \|S\xi\|$ is the operator norm of S. Conclude that the closure of finite-rank operators in the norm $\|\cdot\|_1$ is a Banach space. Denote this Banach space $\mathcal{L}^1(H)$. Show that it is an ideal in $B(H)$, i.e., that $TS \in \mathcal{L}^1(H)$ whenever $S \in \mathcal{L}^1(H)$, $T \in B(H)$.
(2) Show that $B(H) = \mathcal{L}^1(H)^*$ with, for $T \in \mathcal{L}^1(H)$, $S \in B(H)$, the duality is implemented by taking $\langle T, S\rangle = \mathrm{Tr}(TS)$.
(3) Show that the weak topology induced by duality on $B(H)$ is the σ-Weak Operator Topology. By definition, a net $T_i \to T$ in σ-WOT if and only if $\sum_j \langle T_i\xi_j, \eta_j\rangle \to \sum_j \langle T\xi_j, \eta_j\rangle$ for any sequences $\xi_j, \eta_j \in H$ satisfying $\sum_j \|\xi_j\|^2 + \sum_j \|\eta_j\|^2 < \infty$.
(4) Fix an orthonormal basis ξ_i for H. For $d_i \in \mathbb{C}$, let $T = \mathrm{diag}(d_1, d_2, \dots)$ be the operator determined by $T\xi_j = d_j\xi_j$. Show that $\|T\| = \|(d_i)\|_{\ell^\infty}$ and $\|T\|_1 = \|(d_i)\|_{\ell^1}$.

Remark: it is also possible to define the p-norm $\|T\|_p = \mathrm{Tr}(|T|^p)^{1/p}$. These norms lead to Banach spaces of so-called Schatten p-class operators; the dualities between these Banach spaces extend dualities between classical ℓ^p spaces identified with spaces of diagonal operators.

Exercise 0.5.4. Let $M \subset B(H)$ be a $*$-subalgebra closed in σ-WOT (see Exercise 0.5.3). Show that σ-WOT is weaker than SOT and then use Exercise 0.5.2 to conclude that if $P, Q \in M$ are projections, then $P \vee Q$ and $P \wedge Q$ are both in M. Alternatively, use the von Neumann bicommutant theorem to obtain the same conclusion (note that $P \in M$ iff PH is M'-invariant).

Exercise 0.5.5. Give a proof of Theorem 0.3.5 along the following lines.

(1) Prove that if $T \in \mathcal{M}(M)$, then for any $\epsilon > 0$ there exists a projection $p \in M$ so that $\tau(1-p) < \epsilon$ and the operators pT and Tp are *bounded* and belong to M. To do so use the polar decomposition theorem saying that $T = U|T|$, with U bounded and $|T|$ self-adjoint; the fact that T is affiliated to M implies that both U and $|T|$ (and hence the spectral projections of T) commute with M'. It follows that U and the spectral projections of T are in M. Let q be the spectral projection of $|T|$ corresponding to the set $[0, R]$. Set $q' = UqU^*$. Then Tq and $q'T$ are both bounded. Thus if $p = q \wedge q'$, both Tp and pT are bounded. But why isn't $p = 0$? As $R \to \infty$, $q \uparrow 1$, so we can choose q so that $\tau(q) > 1 - \epsilon/2$. Then also $\tau(q') > 1 - \epsilon/2$ and so $\tau(p) = \tau(q \wedge q') > 1 - \epsilon$ by Theorem 0.2.10.

(2) Now use part (1) to deduce that if $T, T' \in \mathcal{M}(M)$, we can choose p, p' so that $\tau(1-p) > 1 - \epsilon/2$, $\tau(1-p') > 1 - \epsilon/2$ and pT, Tp, $p'T'$, $T'p'$ are all bounded. Deduce that $T + T'$ is bounded on the range of $p \wedge p'$ which has trace at least $1 - \epsilon$. Deduce similarly that if $T' = U'|T'|$, then TT' is bounded on the range of $(U')^* p' U' \wedge p$, a projection of trace at least $1 - \epsilon$. Conclude that $T + T'$ and TT' are both densely defined, and prove that they are closable with closures affiliated to M.

Exercise 0.5.6. Let $M \subset B(H)$ be a von Neumann algebra.

(1) Suppose that $x \in M$. Show that if γ is any path in the complex plane so that γ surrounds a domain D containing the spectrum $\sigma(x)$ and f is some function analytic on D, then the contour integral $\frac{1}{2\pi i} \int_\gamma \frac{f(z)}{z-x} dz$ defines an element in M. Let us call this element $f(x)$. Show that if g is analytic on some domain containing $f(\sigma(x))$, then $(g \circ f)(x) = g(f(x))$. Thus M is closed under *analytic functional calculus.*

(2) Suppose that $x \in M$ is a *normal* element, i.e., $xx^* = x^*x$. Use the spectral theorem to identify $W^*(x)$ with the algebra of functions $L^\infty(\sigma(x), dP)$, where dP is in the absolute continuity class of the spectral measure of x. Conclude that if f is any essentially bounded (for this measure) function on $L^\infty(\sigma(x))$, then $f(x) \in M$ is well-defined. Thus M is closed under *measurable functional calculus applied to normal elements.*

Lecture 1: Non-commutative Laws. Classical and Free Independence

1.1. Executive summary. We introduce the notion of a non-commutative law and of equivalence in law. We discuss Voiculescu's discovery that in the non-commutative case, there exists a new notion of independence in addition to the straightforward generalization of classical independence. The main reference for much of this section is the book [37].

1.2. Non-commutative laws. Let (M, τ) be a non-commutative probability space; thus we are assuming that $M \subset B(H)$ is a $*$-subalgebra of the algebra of bounded operators on a Hilbert space and that $\tau(x) = \langle \xi, x\xi \rangle$ for some fixed unit vector $\xi \in H$. We will usually assume that M is a von Neumann algebra (i.e., M is closed in the σ-weak operator topology), but this assumption will not matter very much in this lecture. We will assume at the very least that if $x \in M$, then the resolvent $(z1 - x)^{-1}$ also belongs to M for all z in some domain in $\mathbb{C}$ (see Exercise 0.5.6).

Our first task is to formulate the notion of a *non-commutative law* of an n-tuple of variables $x_1, \ldots, x_n \in M$.

1.2.1. Classical laws. Classically, the law $\mu_{Z_1,\ldots,Z_n}$ of an n-tuple of real-valued random variables $Z_1, \ldots, Z_n$ (viewed as functions on some probability space (X, μ)) is the measure on $\mathbb{R}^n$ defined as the push-forward of the measure μ by the map $x \mapsto (Z_1(x), \ldots, Z_n(x)) \in \mathbb{R}^n$. We say that two n-tuples $Z_1, \ldots, Z_n$ and $Z_1', \ldots, Z_n'$ *have the same law*, or are *equal in law* if $\mu_{Z_1,\ldots,Z_n} = \mu_{Z_1',\ldots,Z_n'}$.

This law $\mu_{Z_1,\ldots,Z_n}$ has the property that

$$\mathbb{E}[f(Z_1, \ldots, Z_n)] = \int f(t_1, \ldots, t_n) d\mu_{Z_1,\ldots,Z_n}(t_1, \ldots, t_n);$$

indeed, both of these quantities are equal to $\int f(Z_1(x), \ldots, Z_n(x)) d\mu(x)$. In other words, the map

$$f \mapsto \int f d\mu_{Z_1,\ldots,Z_n} = \mathbb{E}[f(Z_1, \ldots, Z_n)]$$

is a linear functional on a certain space S of test functions f. There are many natural choices for the space S. For example, we can take $S = L^1(\mathbb{R}^n, \mu_{Z_1,\ldots,Z_n})$, but this choice is inconvenient since the space depends on the law itself. A choice independent of μ is to take S to be all bounded Borel functions on $\mathbb{R}^n$. However, all we really care about is to be able to distinguish different laws; thus we can make a much smaller choice of S as long as it contains enough test functions to separate points in the space of laws, i.e., the space of probability measures on $\mathbb{R}^n$. If one *a priori* restricts to laws that have some regularity (e.g. growth conditions at ∞), there are even more choices of S available. Here are some popular classical choices for S (we take $n = 1$ in these examples for simplicity; we assume that Z is a real-valued random variable with law μ).

(1) Take S to be the linear span of functions $\phi_t(x) = \exp(itx)$. Then

$$\mathbb{E}[\phi_t(Z)] = \mathbb{E}[\exp(itZ)] = \hat{\mu}(t)$$

is (up to normalization) the Fourier transform of μ.

(2) Take S to be the linear span of functions $\theta_t(x) = \exp(-tx)$. Then

$$\mathbb{E}[\theta_t(Z)] = \mathbb{E}[\exp(-tZ)]$$

is the Laplace transform of μ (note that this is defined only if we require some growth conditions on μ; under such growth conditions the space of test functions θ_t is sufficient to determine μ).

(3) Take S to be the linear span of functions $\zeta_z(x) = (x-z)^{-1}$ for z in the upper-half-plane $\mathbb{C}_+ := \{z \in \mathbb{C} : \operatorname{Im} z > 0\}$. Then

$$\mathbb{E}[\zeta_z(Z)] = \int \frac{1}{x-z} d\mu(x)$$

is the Cauchy (or Stieltjes) transform of μ.

(4) Take S to be the space of all polynomial functions. For $f_m = x^m \in S$, the numbers $m_n = \mathbb{E}[f_n(Z)] = \mathbb{E}[Z^n]$ are the *moments* of μ. Under growth conditions on μ (such as Carleman's condition: $\sum_n m_{2n}^{-1/2n} = +\infty$) the moments are well-defined and encode μ uniquely. (Note that this is the case for any compactly supported measure).

The advantage of the first two choices is that the test functions $\phi_t(x)$ have some magic properties relating differentiation in x to multiplication in t.

We shall see later (see Lemma 4.3.1) that the third choice also has a magic property but with respect to a different kind of differentiation in x; this makes it particularly suitable to free probability.

The fourth choice is the simplest one to make algebraically and it is a suitable one in many cases. As we shall see, a large number of interesting laws occurring in free probability theory (such as, for example, analogs of the Gaussian and of the Poisson laws) are compactly supported, and thus the fourth approach is available in their case as well.

1.2.2. Non-commutative laws. Let us again return to the set-up of $X_1, \ldots, X_n$ self-adjoint operators in a non-commutative probability space (M, τ). Arguing by analogy, we define the notion of a non-commutative law to be a linear functional $\mu_{X_1,\ldots,X_n}$ defined on a suitable space of test functions S by

$$\mu_{X_1,\ldots,X_n}(f) = \tau[f(X_1, \ldots, X_n)], \qquad f \in S.$$

Since in our case the elements $X_1, \ldots, X_n \in M$ are assumed to be bounded, the easiest approach to take is the analog of 4 of §1.2.1 and to take S to be space of all non-commutative polynomial functions, i.e., the vector space with a basis consisting of all non-commutative monomials: expressions of the form $t_{i_1} \ldots t_{i_j}$ where $i_1, \ldots, i_j \in \{1, \ldots, n\}$ and $t_1, \ldots, t_n$ are formal variables. We say that $X_1, \ldots, X_n$ and $X_1', \ldots, X_n'$ have *the same law* or are *equal in law*, if $\mu_{X_1,\ldots,X_n} = \mu_{X_1',\ldots,X_n'}$.

It's worth noting that the notion of equivalence in law remains unchanged if we replace our choice of S by the analog of any of spaces 1, 2 or 3 described in §1.2.1; for example,

$$C_{i_1,\ldots,i_k}(z_1, \ldots, z_k) = \tau\left((X_{i_1} - z_1)^{-1} \cdots (X_{i_k} - z_k)^{-1}\right)$$

is the analog of 3. In this case the notion of a law is perfectly defined for $X_1, \ldots, X_n$ self-adjoint operators affiliated with M, so that the boundedness restriction on $X_1, \ldots, X_n$ is completely artificial. We shall nonetheless stick to our definition (which imposes the boundedness restriction) leaving it to the reader to modify various statements to avoid it.

The following Lemma shows that equivalence in law has consequences for von Neumann algebras. Denote by $W^*(X_1,\dots,X_n)$ the von Neumann algebra generated by $X_1,\dots,X_n$ acting on the Hilbert space obtained by completing all non-commutative polynomials in $X_1,\dots,X_n$ in the norm $\|p\|_2 = \tau(p^*p)^{1/2}$.

Lemma 1.2.3. *Suppose that $X_1,\dots,X_n$ and $Y_1,\dots,Y_n$ are equal in law and the traces on $W^*(X_1,\dots,X_n)$ and $W^*(Y_1,\dots,Y_n)$ are faithful. Then the map $X_j \mapsto Y_j$ extends to an isomorphism $W^*(X_1,\dots,X_n) \cong W^*(Y_1,\dots,Y_n)$.*

Proof. The map $X_j \mapsto Y_j$ extends to an isometric isomorphism between the Hilbert space completions of non-commutative polynomials in $X_1,\dots,X_n$ and $Y_1,\dots,Y_n$: this is a direct consequence of equivalence in law. This isomorphism takes the left multiplication action of X_j to that of Y_j, so that the von Neumann algebras are isomorphic. □

In the case that $W^*(X_1,\dots,X_n)$ and $W^*(Y_1,\dots,Y_n)$ are factors faithfulness turns out to be automatic, due to the uniqueness of the trace.

It is worth noting that in the case of a single variable, the notions of non-commutative and classical laws are exactly the same; thus a (non-commutative) law of a single self-adjoint variable $X \in M$ has an interpretation as a measure on $\mathbb{R}$. The measure we thus get is also called the *spectral measure* of X because of the following connection with the spectral theorem. Let P be the projection valued measure on $\sigma(X) \subset \mathbb{R}$ so that

$$X = \int_{\lambda\in\mathbb{R}} \lambda dP(\lambda).$$

Then $\mu = \tau \circ P$ is a scalar-valued measure on $\mathbb{R}$. Moreover,

$$\int f(\lambda)d\mu(\lambda) = \int f(\lambda)\tau(dP(\lambda)) = \tau\left(\int f(\lambda)dP(\lambda)\right) = \tau(f(X))$$

so that μ is exactly the law of X in our sense.

1.3. Examples of non-commutative probability spaces and laws. Let Γ be a discrete group (e.g. $\Gamma = \mathbb{Z}$ or $\Gamma = \mathbb{F}_n = \mathbb{Z} * \mathbb{Z} * \cdots * \mathbb{Z}$). Let $\ell^2(\Gamma)$ be the Hilbert space with orthonormal basis $\{\delta_g : g \in \Gamma\}$. For $g \in \Gamma$ define the unitary operator $\lambda(g) : \ell^2(\Gamma) \to \ell^2(\Gamma)$ by $\lambda(g)\delta_h = \delta_{gh}$. This gives a unitary representation of Γ, called the left regular representation. Let $L(\Gamma)$ be the von Neumann algebra generated by $\lambda(\Gamma)$, and denote by $\tau : L(\Gamma) \to \mathbb{C}$ the linear functional $\tau(x) = \langle\delta_e, x\delta_e\rangle$.

Lemma 1.3.1. *τ is a faithful trace on $L(\Gamma)$.*

Proof. Because τ is σ-WOT continuous, it is sufficient to check that $\tau(xy) = \tau(yx)$ for x, y in a σ-WOT dense set. We can thus assume that x, y are finite linear combinations of operators $\lambda(g_i)$ for some $g_i \in \Gamma$. By linearity, it is therefore sufficient to check that $\tau(\lambda(g_1)\lambda(g_2)) = \tau(\lambda(g_2)\lambda(g_1))$, i.e., that $\tau(\lambda(g_1g_2)) = \tau(\lambda(g_2g_1))$. Now, $\tau(\lambda(g)) = 1$ if $g = e$ and $\tau(\lambda(g)) = 0$ if $g \neq e$; it's easy to check that $g_1g_2 = e$ iff $g_2g_1 = e$, which completes the proof of the trace property.

We now claim that τ is faithful, i.e., for all $x \in L(\Gamma)$, $\tau(x^*x) = 0$ iff $x = 0$.

For $g \in \Gamma$ denote by $\rho(g)$ the unitary map $\rho(g)\delta_h = \delta_{hg^{-1}}$. Then ρ is the right regular representation of Γ; moreover, $\lambda(g)\rho(h) = \rho(h)\lambda(g)$ for all $g, h \in \Gamma$. Taking linear combinations and σ-WOT closures gives us that $x\rho(h) = \rho(h)x$ for all $x \in L(\Gamma)$.

Suppose now that $\tau(x^*x) = 0$; by definition this means that $\langle x\delta_e, x\delta_e\rangle = 0$ so that $x\delta_e = 0$. But then for any $h \in \Gamma$, $\rho(h)x\delta_e = x\rho(h)\delta_e = x\delta_h$ so that the equality $x\delta_e = 0$ implies that $x\delta_h = 0$ for all h, and thus $x = 0$. □

It follows that $(L(\Gamma), \tau)$ is a tracial non-commutative probability space. (If Γ is abelian, $L(\Gamma)$ is of course commutative; in that case $L(\Gamma) = L^\infty(\hat{\Gamma})$ and τ is integration with respect to the Haar measure on the Pontrjagin dual $\hat{\Gamma}$).

Let now $X_1, \ldots, X_n$ be self-adjoint elements in M; such elements can be viewed as non-commutative random variables. A particular example is the often-studied Laplace operator related to random walks on Cayley graphs. Let us assume that $S \subset \Gamma$ is a finite generating set, and assume that $S^{-1} = S$ (i.e., $g \in S$ iff $g^{-1} \in S$). The set S allows us to define a random walk on Γ as follows. Having arrived at some element $h \in \Gamma$, we choose uniformly at random an element $g \in S$ and then jump from h to gh.

Define

$$L = \frac{1}{|S|} \sum_{g \in S} \lambda(g) \in L(\Gamma).$$

Then $L = L^*$ and the associated law is a probability measure μ_L on $\mathbb{R}$ (supported on $[-1, 1]$ since $\|L\| \leqslant |S|^{-1} \sum_{g \in S} \|\lambda(g)\| = |S|^{-1}|S| = 1$).

Note that if $h \in \Gamma$, then $L\delta_h = |S|^{-1} \sum_{g \in S} \lambda(g)\delta_h = |S|^{-1} \sum_{g \in S} \delta_{gh}$ has a simple probabilistic interpretation: the dot product $\langle \delta_g, L\delta_h\rangle$ is the probability of jumping to g starting at h. It is not hard to see that a similar interpretation exists for higher powers: $\langle \delta_g, L^n\delta_h\rangle$ is the probability of arriving at g after n jumps, having started at h. It follows that the moments of μ_L are given by

$$\int t^n \, d\mu_L(t) = \tau(L^n) = \langle \delta_e, L^n \delta_e\rangle = \text{Probability of return to } e \text{ after } n \text{ jumps.}$$

It is instructive to carry out some computations of μ_L (see Exercises below).

1.4. Notions of independence. By definition, two classical random variables X and Y are independent if $\mathbb{E}[f(X)g(Y)] = \mathbb{E}[f(X)]\mathbb{E}[g(Y)]$ for all test functions f, g (with enough growth conditions so that the expected values are defined). It should be noted right away that independence is a property of the algebras generated by the variables X and Y and not of the variables themselves; this is apparent from the definition. It is not hard to prove that if $X_1, \ldots, X_n$ are an n-tuple independent from the m-tuple $Y_1, \ldots, Y_m$ then one can realize the joint law of $X_1, \ldots, X_n, Y_1, \ldots, Y_m$ by some random variables $X_1', \ldots, X_n', Y_1', \ldots, Y_m'$ on a probability space (Z, μ) which has a special property that $(Z, \mu) = (X, \nu) \times (Y, \eta)$ and X_j' (resp. Y_j') are functions that depend only on the X (resp. only on the Y) coordinates. From the point of view of von Neumann algebras, this means that $W^*(X_1',\dots,X_n',Y_1',\dots,Y_m') \cong W^*(X_1',\dots,X_n')\bar{\otimes}W^*(Y_1',\dots,Y_m')$. (Here $\bar{\otimes}$ is the von Neumann algebra tensor product. If M_1, M_2 are two von Neumann algebras, one chooses realizations $M_i \subset B(H_i)$ and defines the tensor product to be the von Neumann algebra generated by $M_1 \otimes 1$ and $1 \otimes M_2$ inside bounded operators on the tensor product Hilbert space $H_1 \otimes H_2$.) Note that because of Lemma 1.2.3, we actually have

$$W^*(X_1,\dots,X_n,Y_1,\dots,Y_m) \cong W^*(X_1,\dots,X_n)\bar{\otimes}W^*(Y_1,\dots,Y_m).$$

1.4.1. Classical independence in the non-commutative setting. All of this can be readily generalized to the non-commutative setting:

Definition 1.4.2. Let $X_1,\dots,X_n,Y_1,\dots,Y_m \in (M,\tau)$ be non-commutative random variables. Then the families $(X_1,\dots,X_n)$ and $(Y_1,\dots,Y_m)$ are *classically independent* if

(1) $X_iY_j = Y_jX_i$ for all i,j
(2) $\tau(f(X_1,\dots,X_n)g(Y_1,\dots,Y_m)) = \tau(f(X_1,\dots,X_n))\tau(g(Y_1,\dots,Y_m))\ \forall f,g \in S$.

Here S is the suitable set of test functions (e.g., non-commutative polynomials). The following two statements are straightforward:

Proposition 1.4.3. *Assume that $(X_1,\dots,X_n)$ and $(Y_1,\dots,Y_m)$ are classically independent. Then*

$$W^*(X_1,\dots,X_n,Y_1,\dots,Y_m) \cong W^*(X_1,\dots,X_n)\bar{\otimes}W^*(Y_1,\dots,Y_m).$$

Moreover, the joint law of $X_1,\dots,X_n,Y_1,\dots,Y_m$ is completely determined by the joint laws of $X_1,\dots,X_n$ and $Y_1,\dots,Y_m$.

Note that to compute the joint law of $X_1,\dots,X_n,Y_1,\dots,Y_m$ we in principle need to know the value of expressions of the form $\tau(X_{i_1}Y_{j_1}X_{i_2}Y_{j_2})$; but because of commutativity and the independence assumption, we have

$$\tau(X_{i_1}Y_{j_1}X_{i_2}Y_{j_2}) = \tau(X_{i_1}X_{i_2}Y_{j_1}Y_{j_2}) = \tau(X_{i_1}X_{i_2})\tau(Y_{j_1}Y_{j_2}).$$

One example of non-commutative classical independence comes from groups:

Proposition 1.4.4. *Let $\Gamma = \Gamma_1 \times \Gamma_2$ be a product group, let $g_1,\dots,g_n \in \Gamma_1$ and let $h_1,\dots,h_m \in \Gamma_2$. Then $(\lambda(g_1),\dots,\lambda(g_n))$ and $(\lambda(h_1),\dots,\lambda(h_m))$ are classically independent in $(L(\Gamma),\tau)$. In fact, $(L(\Gamma),\tau) = (L(\Gamma_1)\bar{\otimes}L(\Gamma_2),\tau\otimes\tau)$.*

1.4.5. Free independence. Voiculescu discovered [25] that the non-commutative setting allows for a new notion of independence, having to do with free products. This new notion of independence has the property that elements $g_1,\dots,g_n \in \Gamma_1$ and $h_1,\dots,h_m \in \Gamma_2$ lead to *freely independent* families of variables in the free product $\Gamma_1 * \Gamma_2$. Before proceeding further, let us review the notion of the free product of groups.

Let Γ_1, Γ_2 be two groups. Their free product is by definition the group generated by Γ_1 and Γ_2 with no relations. In other words, elements of the free product are formal words of the form $h_1h_2\cdots h_k$ where h_j are elements of Γ_1

or Γ_2; we are assuming that the word is reduced, i.e., that all possible multiplications within Γ_1 and Γ_2 have been carried out. This means that consecutive letters come from different groups (i.e., if $h_j \in \Gamma_{i(j)}$, then $i(1) \neq i(2)$, $i(2) \neq i(3)$ and so on) and that each h_j is not the neutral element. Multiplication of two such words is defined as the *reduction* of $h_1 \cdots h_k h_1' \cdots h_{k'}'$. This reduction is defined to be $h_1 \cdots h_k h_1' \cdots h_{k'}'$ if that is a reduced word. Otherwise, it must be that $h_k h_1' \in \Gamma_{i(k)}$ and the reduction is defined to be the reduction of the shorter word $h_1 \cdots (h_k h_1') \cdot h_2' \cdots h_{k'}'$.

Lemma 1.4.6. *Let* $\Gamma = \Gamma_1 * \Gamma_2$. *Assume that* $X_1, \ldots, X_n$ *are elements of* $L(\Gamma)$ *so that* $X_j \in L(\Gamma_{i(j)})$. *Assume that* $i(1) \neq i(2)$, $i(2) \neq i(3)$, ..., $i(n-1) \neq i(n)$ *and that* $\tau(X_1) = \tau(X_2) = \cdots = \tau(X_n) = 0$. *Then* $\tau(X_1 \cdots X_n) = 0$.

Proof. Note that $\tau(\lambda(g)) = 0$ iff $g \neq e$. By density and linearity of τ we may assume that $X_j = \lambda(g_j)$ with $g_j \in \Gamma_{i(j)}$ and $g_j \neq e$ (because $\tau(X_j) = 0$). Thus

$$X_1 \cdots X_n = \lambda(g_1 \cdots g_n) \neq \lambda(e)$$

since $g_1 \cdots g_n$ is a reduced word. Thus $\tau(X_1 \cdots X_n) = 0$. □

This example motivates the definition of free independence:

Definition 1.4.7. [25,37] Let $A_1, \ldots, A_n \subset (M, \tau)$ be unital subalgebras of a non-commutative probability space (M, τ). We say that $A_1, \ldots, A_n$ are *freely independent* if

$$\tau(X_1 \cdots X_n) = 0$$

whenever

$$X_j \in A_{i(j)}, \qquad \tau(X_1) = \tau(X_2) = \cdots \tau(X_n) = 0$$
$$\text{and } i(1) \neq i(2), i(2) \neq i(3), \ldots, i(n-1) \neq i(n).$$

Given families $F_1 = (X_1^{(1)}, \ldots, X_{n_1}^{(1)}), \ldots, F_k = (X_1^{(k)}, \ldots, X_{n_k}^{(k)})$, we say that they are freely independent if the algebras $A_i = \mathrm{Alg}(1, F_i)$ are freely independent.

Note that the definition does not require (M, τ) to be a tracial non-commutative probability space; it even makes sense for any unital algebra M with a linear functional τ that sends 1 to 1.

Lemma 1.4.8. *Suppose that* $A_1, \ldots, A_n$ *are freely independent in* (M, τ), *and let* A *be the algebra generated by* $A_1, \ldots, A_n$. *Then the restriction of* τ *to* A *is uniquely determined by its restrictions to the algebras* $A_1, \ldots, A_n$.

Proof. Suppose $x_1 \cdots x_n \in A$ with $x_j \in A_{i(j)}$. By grouping variables together, we may assume that $i(1) \neq i(2), i(2) \neq i(3), \ldots, i(n-1) \neq i(n)$. Then, rewriting $x_j = x_j' + \tau(x_j)1$ with $x_j' = x_j - \tau(x_j)1$ allows us to write $x_1 \cdots x_n$ as the sum $x_1' \cdots x_n' + w'$ where w' is a product of shorter length. Since $\tau(x_j') = 0$ we see from the freeness condition that $\tau(x_1' \cdots x_n') = 0$, so that $\tau(x_1 \cdots x_n) = \tau(w')$. Proceeding by induction we see that τ is completely determined by its values on products of length 1, i.e., by its restriction to the algebras $A_1, \ldots, A_n$. □

Corollary 1.4.9. *Suppose that* $X_1, \ldots, X_n$ *and* $Y_1, \ldots, Y_m$ *are two families which are freely independent. Then the joint law* $\mu_{X_1,\ldots,X_n,Y_1,\ldots,Y_m}$ *is completely determined by the two joint laws* $\mu_{X_1,\ldots,X_n}$ *and* $\mu_{Y_1,\ldots,Y_m}$.

Given two non-commutative probability spaces (M_i, ϕ_i), $i = 1, 2$, there exists a construction of a larger non-commutative probability space (M, ϕ) and maps $\alpha_i : M_i \to M$ so that $\phi \circ \alpha_i = \phi_i$ and the images $\alpha_i(M_i)$ are free inside M and generate M as a von Neumann algebra (see Exercise 1.6.6). If ϕ_i are faithful, α_i are injections and ϕ is also faithful. If ϕ_i are traces, then ϕ is also a trace. The functional ϕ is called the *free product* of the functionals ϕ_i, $\phi = \phi_1 * \phi_2$. The von Neumann algebra generated by the freely independent copies of M_1 and M_2 is called the *free product* of the von Neumann algebras M_1 and M_2, and we write $(M, \phi) = (M_1, \phi_1) * (M_2, \phi_2)$. Unlike tensor products, the free product is highly sensitive to the choice of states ϕ_i on the algebras M_i.

1.5. The free Gaussian Functor. Every real Hilbert space H comes with a canonical Gaussian measure; it's determined by the condition that if $\xi_1, \ldots, \xi_n \in H$, then the linear functions $\phi_j = \langle \xi_j, \cdot \rangle$ form a family of centered real Gaussian random variables satisfying

$$\mathbb{E}[\phi_i \phi_j] = \langle \xi_i, \xi_j \rangle.$$

(This condition determines the joint law of all linear functions on H and thus a probability measure on H, at least for H finite-dimensional). In particular, variables associated to perpendicular vectors are independent. There is also a complex analog in which the variables are complex Gaussians satisfying

$$\mathbb{E}[\bar{\phi}_i \psi_j] = \langle \xi_i, \xi_j \rangle .$$

Below we describe the analog of this functor in free probability [25, 28, 37].

Let H be a real Hilbert space and let $H_{\mathbb{C}} = H \otimes_{\mathbb{R}} \mathbb{C}$ be its complexification. Define

$$\mathcal{F}(H) = \mathbb{C}\Omega \oplus H_{\mathbb{C}} \oplus (H_{\mathbb{C}} \otimes H_{\mathbb{C}}) \oplus \cdots \oplus H_{\mathbb{C}}^{\otimes n} \oplus \cdots$$

to be the *full Fock space* over $H_{\mathbb{C}}$. For $h \in H_{\mathbb{C}}$ define the *free creation operator* $\ell(h) : \mathcal{F}(H) \to \mathcal{F}(H)$ by

$$\ell(h)\xi_1 \otimes \cdots \otimes \xi_n = h \otimes \xi_1 \otimes \cdots \otimes \xi_n.$$

(By convention we treat Ω as a zeroth tensor power, so that $\ell(h)\Omega = h$). The adjoint of $\ell(h)$ is given by

$$\ell(h)^* \xi_1 \otimes \cdots \otimes \xi_n = \langle h, \xi_1 \rangle \xi_2 \otimes \cdots \otimes \xi_n$$

(and $\ell(h)^*\Omega = 0$). It is not hard to verify that $\ell(h)^*\ell(g) = \langle h, g \rangle 1$ and so $\ell(h)$ is a bounded operator of norm $\|h\|$.

For $h \in H \subset H_{\mathbb{C}}$ define

$$s(h) = \ell(h) + \ell(h)^*.$$

Proposition 1.5.1. *Let $\Phi(H) = W^*(s(h) : h \in H)$. Then the linear functional $\tau(x) = \langle \Omega, x\Omega\rangle$ defines a faithful trace on $\Phi(H)$.*

To prove traciality, we note that τ is a trace iff for all x, y self-adjoint, $\tau(xy) \in \mathbb{R}$. Indeed, $\overline{\tau(xy)} = \tau((xy)^*) = \tau(yx)$. If a is a product $s(h_1)\cdots s(h_n)$ and b is a product $s(g_1)\cdots s(g_m)$ then it's clear that $\tau(ab) \in \mathbb{R}$, being a linear combination of inner products between the vectors $h_1, \ldots, h_n, g_1, \ldots, g_m \in H \subset H_{\mathbb{C}}$ (and all of these inner products are real valued). Thus $\tau((a + a^*)(b + b^*)) \in \mathbb{R}$, since a^* and b^* are of the same form. Since elements of the form $a + a^*$ densely span the set of self-adjoint elements in $\Phi(H)$, traciality follows. Faithfulness of τ is proved in Exercise 1.6.4.

It is not hard to see that $h \mapsto s(h)$ is $\mathbb{R}$-linear; furthermore, $\tau(s(h)s(g)) = \langle h, g\rangle$.

Proposition 1.5.2. *Let $h \in H$ be a unit vector and let $x = s(h)$. Then the law of x is the semicircular law*

$$\frac{1}{2\pi}\sqrt{4 - t^2}\chi_{[-2,2]}(t).$$

In particular, $\|x\| = 2$.

We will give three proofs of this statement.

1.5.3. Combinatorial proof. Since x is bounded, μ_x is determined by its moments. Let $m_n = \int t^n d\mu_x(t) = \tau(x^n) = \langle \Omega, s(h)^n\Omega\rangle$. Let us write $e_n = h^{\otimes n}$, $e_0 = \Omega$. Then $s(h)e_n = e_{n-1} + e_{n+1}$ if $n \geqslant 1$ and $s(h)e_0 = e_1$. Recall that a Dyck path of length n is a function $f : \{0, \ldots, n\} \to \mathbb{N} \cup \{0\}$ satisfying $f(0) = 0$ and $f(k+1) = f(k) \pm 1$ (the word path comes from identification with paths in the integer lattice that start at the origin, have slope ± 1 and which never go below the x-axis). Let $D(n)$ be the set of all Dyck paths and $D'(n)$ be the subset of $D(n)$ consisting of paths f satisfying the condition $f(n) = 0$. We then claim that

$$x^n\Omega = \sum_{f \in D(n)} e_{f(n)}.$$

This formula is easily proved recursively: we have $\Omega = e_0$ and

$$x \cdot \sum_{f \in D(n)} e_{f(n)} = \sum_{f \in D(n+1)} e_{f(n+1)}$$

since each path in $D(n + 1)$ comes from a path in $D(n)$ by appending a single segment of slope 1 or -1.

Thus

$$m_n = \langle \Omega, x^n\Omega\rangle = \sum_{f \in D(n)} \langle e_0, e_{f(n)}\rangle = \#\{f \in D(n) : f(n) = 0\}.$$

In particular, note that $m_0 = 1$ and $m_1 = 0$.

We now claim that

$$m_n = \sum_{a+b=n-2} m_a m_b. \tag{1.5.4}$$

We allow a or b to be zero in this sum.

To see this, let $D'(n) = \{f \in D(n) : f(n) = 0\}$ so that $m_n = \#D'(n)$ (we define $D'(0)$ to consist of single path of length zero). It is thus sufficient to provide a bijection between $D'(a) \times D'(b)$ and $D'(a+b+2)$.

To pair of paths $(f_1, f_2) \in D'(a) \times D'(b)$ let us associate a path $f \in D'(a+b+2)$ as follows. Starting at zero, we go up by 1. Next, we follow the ups/downs of $f_1 \in D'(a)$. By definition, we are now 1 unit above the x-axis. Next, we go down by 1 to the x-axis and finally follow f_2. The result is clearly a path in $D'(a+b+2)$; it's defined by

$$f(k) = \begin{cases} 0 & k = 0 \\ 1 + f_1(k-1) & 1 \leqslant k \leqslant a+1 \\ 0 & k = a+2 \\ g(k-a-2) & a+2 < k \leqslant a+b+2. \end{cases}$$

The map $(f_1, f_2) \mapsto f$ is clearly injective, since by construction $a+2$ is the smallest $k > 0$ at which $f(k) = 0$, and once we know a we can easily reconstruct f_1 and f_2. The same argument shows that the map is surjective: given an arbitrary path f we can define a by saying that it is smallest such that $f(a+2) = 0$, recover f_1 and f_2 and see that f is built from them.

Together with $m_0 = 1$, $m_1 = 0$ equation (1.5.4) defines m_n recursively.

Let us now define

$$m'_n = \frac{1}{2\pi} \int_{-2}^{2} t^n \sqrt{4 - t^2} dt.$$

Let $t = 2 \sin u$ so that $dt = 2 \cos u du$:

$$\begin{aligned} m'_n &= \frac{1}{2\pi} \int_{-\pi}^{\pi} 2^n \sin^n u \cdot 2^2 \cos^2 u \, du \\ &= \frac{2^{n+1}}{2\pi} \int_{-\pi}^{\pi} \sin^n u (1 - \sin^2 u) du \\ &= \frac{2^{n+1}}{2\pi} \left[\int_{-\pi}^{\pi} \sin^n u du - \frac{n+1}{n+2} \int_{-\pi}^{\pi} \sin^n u du \right] \\ &= \frac{2^{n+1}}{2\pi(n+2)} \int_{-\pi}^{\pi} \sin^n u du \\ &= 4 \cdot \frac{n-1}{(n+2)} \cdot \frac{2^{n-1}}{2\pi n} \int_{-\pi}^{\pi} \sin^{n-2} u du \\ &= 4 \frac{n-1}{n+2} m'_{n-2}. \end{aligned}$$

Coupled with $m'_0 = 1$ this defines m'_n recursively. Note that $m'_n = 0$ for n odd by parity.

Let C_k be the k-th Catalan number:

$$\begin{aligned} C_k &= \frac{1}{k+1} \binom{2k}{k} = \frac{(2k) \cdots (k+2)}{k \cdots 1} = \frac{4(2k-1)}{2(k+1)} \cdot \frac{(2k-2) \cdots (k+2)(k+1)}{(k-1) \cdots 1} \\ &= 4 \frac{n-1}{n+2} \cdot \frac{1}{k} \binom{2(k-1)}{k-1} = 4 \frac{n-1}{n+2} C_{k-1}. \end{aligned}$$

Since $C_0 = 1$ we see that $m'_{2k} = C_k$. Let us show that Catalan numbers satisfy the analog of (1.5.4); this will end the proof showing that $m_n = m'_n$ (since we clearly have $m_n = 0 = m'_n$ for n odd).

Note that

$$\sqrt{1+y} = \sum_{n\geqslant 0} \binom{\frac{1}{2}}{n} y^n = \sum_{n\geqslant 0} \frac{(-1)^{n+1}}{4^n(2n-1)} \binom{2n}{n} y^n$$

so that if we set

$$c(z) = \frac{1-\sqrt{1-4z}}{2z} = \sum_{n\geqslant 0} \binom{2n}{n} \frac{z^n}{n+1} = \sum_{n\geqslant 0} C_n z^n$$

then $c(x)$ is a generating function for the Catalan numbers. Then

$$zc(x)^2 = \frac{2-4z-2\sqrt{1-4z}}{4z} = \frac{1-\sqrt{1-4z}}{2z} - 1 = c(z) - 1$$

so that

$$c(z) = 1 + zc(z)^2.$$

But this means precisely that

$$C_n = \sum_{a+b=n-1} C_a C_b.$$

This completes the proof.

For future use, recall that a *partition* of $\{1, \ldots, n\}$ is a covering of the set $\{1, \ldots, n\}$ by disjoint subsets called *classes* of the partition. A partition is called *non-crossing* if whenever $i < j < k < l$, the fact that i, k and j, l are in the same class implies that all four i, j, k, l are in the same class. Drawing the numbers $\{1, \ldots, n\}$ cyclically around the circle and for each class drawing a convex polygon with vertices from that class justifies the name "non-crossing": the resulting diagram can be drawn with no crossings precisely when the original partition is non-crossing. A partition is called a *pairing* if all classes contain exactly two elements. We write $NC(n)$ for the set of all non-crossing partitions of $\{1, \ldots, n\}$ and $NC_2(n)$ for the set of all non-crossing pairings.

Let us record that C_n is also the number of non-crossing pairings. Indeed, there is a simple bijection between Dyck paths and non-crossing pairings: to a Dyck path we associate a pairing on the segments of the paths in which we pair each upwards segment with the closest downward segment to the right of it which is at the same level. Thus:

$$m_n = \#NC_2(n).$$

1.5.5. Analytic proof. Let P be the orthogonal projection onto the vector Ω. Then for any operator y,

$$\tau(y) = \operatorname{Tr}(PyP)$$

and

$$\tau(y)^2 = \operatorname{Tr}(PyPyP),$$

where Tr denotes the usual trace on finite rank operators given by

$$\mathrm{Tr}(T) = \sum \langle \zeta_n, T\zeta_n \rangle$$

where ζ_n is any orthonormal basis (a convenient one to use for our purposes is one where the first vector is Ω, which makes the identity trivial).

Let now $r(h)$ be the operator $r(h)\xi_1 \otimes \cdots \otimes \xi_n = \xi_1 \otimes \cdots \otimes \xi_n \otimes h$. Then we easily have:

$$\ell(h)r(g) = r(g)\ell(h)$$
$$\ell^*(h)r(g) = r(g)\ell^*(h) + \langle h, g \rangle P.$$

Denote by $[A, B] = AB - BA$, the commutator of two operators. Then setting $r = r(h)$ we have

$$P = [\ell^*(h), r(h)] = [s(h), r(h)] = [x, r].$$

We now return to our computation. Let

$$G(z) = \tau((z - x)^{-1}) = \int \frac{d\mu_x(t)}{z - t}$$

be the Cauchy transform of the measure μ_x. Since x is a bounded operator, its spectrum is contained in some interval. Thus for z sufficiently large, we can expand $(z - x)^{-1}$ as a power series in z^{-1}:

$$\frac{1}{z - x} = \frac{1}{z}\frac{1}{1 - x/z} = z^{-1} \sum_{n \geqslant 0} (x/z)^n.$$

Let us now compute $[(z - x)^{-1}, r]$:

$$\begin{aligned}
[(z - x)^{-1}, r] &= z^{-1}[\sum_{n \geqslant 0} (x/z)^n, r] \\
&= z^{-1} \sum_{n \geqslant 0} [x^n, r] z^{-n} \\
&= z^{-1} \sum_{n \geqslant 0} \sum_{k=0}^{n-1} x^k [x, r] x^{n-k-1} z^{-n} \\
&= z^{-2} \sum_{n \geqslant 0} \sum_{k=0}^{n-1} z^{-k} x^k P z^{-(n-k-1)} x^{n-k-1} \\
&= (z - x)^{-1} P (z - x)^{-1}.
\end{aligned}$$

By analytic continuation, this formula is still true for all z with positive imaginary part. Thus

$$\begin{aligned}
\mathrm{Tr}(P[(z - x)^{-1}, r]P) &= \mathrm{Tr}(P(z - x)^{-1} P (z - x)^{-1} P) \\
&= G(z)^2.
\end{aligned}$$

On the other hand, $Pr = 0$ since the range of r is perpendicular to Ω, and $rP = xP$ since the range of P consists of multiples of Ω and $x\Omega = r\Omega = h$. Thus

$$P[(z - x)^{-1}, r]P = P(z - x)^{-1} rP - Pr(z - x)^{-1} P = P(z - x)^{-1} xP$$

so that

$$\mathrm{Tr}(P[(z-x)^{-1}, r]P) = \mathrm{Tr}(P(z-x)^{-1}xP) = \tau((z-x)^{-1}x).$$

Since

$$\frac{x}{z-x} = \frac{x-z}{z-x} + \frac{z}{z-x} = -1 + \frac{z}{z-x}$$

we conclude that

$$\begin{aligned} G(z)^2 &= \mathrm{Tr}(P[(z-x)^{-1}, r]P) = \tau((z-x)^{-1}x) \\ &= -1 + z\tau((z-x)^{-1}) = -1 + zG(z). \end{aligned}$$

We can now solve this equation for $G(z)$:

$$G(z)^2 - zG(z) + 1 = 0 \implies G(z) = \frac{z \pm \sqrt{z^2-4}}{2}.$$

Since $G(z) \sim z^{-1}$ near $i\infty$, $G(z) \to 0$ as $z \to i\infty$, which implies that

$$G(z) = \frac{z - \sqrt{z^2-4}}{2}$$

where we choose the branch of the square root so that $\sqrt{z^2} = z$ for z in the upper half plane.

We now recall how the measure μ can be recovered from the knowledge of its Cauchy transform.

For this we need the following Lemma:

Lemma 1.5.6. *Let ν be a probability measure on $\mathbb{R}$ and denote by $G_\nu(z)$ the Cauchy transform*

$$G_\nu(z) = \int \frac{d\nu(t)}{z-t}.$$

Then for any interval $I = [a, b] \subset \mathbb{R}$,

$$\lim_{y\downarrow 0} \int_I -\frac{1}{\pi}\Im[G_\nu(x+iy)]dx = \nu((a,b)) + \frac{1}{2}(\nu(a) + \nu(b)).$$

In particular, if ν has a continuous density ρ at a point $x \in \mathbb{R}$, then

$$\rho(x) = \lim_{y\downarrow 0} -\frac{1}{\pi}\Im[G_\nu(x+iy)].$$

Proof. Note that if $z = x + iy$, then

$$\begin{aligned} -\frac{1}{\pi}\Im G_\nu(x+iy) &= -\frac{1}{\pi}\int \Im[(iy + x - t)^{-1}]d\nu(t) \\ &= \frac{1}{\pi}\int \frac{y}{(x-t)^2 + y^2} d\nu(t) = \nu * \alpha_y \end{aligned}$$

where

$$\alpha_y(x) = \frac{y}{x^2 + y^2}$$

is the Cauchy law with scale parameter y. If X and Y are independent random variables so that $X \sim \nu$ and $Y \sim \alpha_1$ then

$$\nu * \alpha_y \sim X + yY$$

so that

$$\int_I -\frac{1}{\pi}\Im(G_\nu(x+iy))dx = \mathbb{E}[1_{X+yY\in[a,b]}].$$

Now as $y \downarrow 0$,

$$1_{X+yY\in[a,b]} \to 1_{X\in(a,b)} + 1_{X=a\ \&\ Y\geqslant 0} + 1_{X=b\ \&\ Y\leqslant 0},$$

which gives the advertised formula for the limit (noting that X and Y are independent and the probability that $Y \geqslant 0$ is $1/2$). □

We are now ready to apply the lemma to our formula for G:

$$\begin{aligned}-\frac{1}{\pi}\Im G(x+iy) &= \frac{1}{2\pi}\Im[\sqrt{(x+iy)^2-4}-x-iy] \\ &= \frac{1}{2\pi}(\Im[\sqrt{x^2-4-y^2+2ixy}]-y).\end{aligned}$$

We can now take the limit as $y \downarrow 0$ to obtain

$$\frac{1}{2\pi}\Im[\sqrt{x^2-4}].$$

If $|x| \geqslant 4$ this is zero. If $|x| < 4$ then we get $\sqrt{x^2-4} = i\sqrt{4-x^2}$ so that the limit becomes $(2\pi)^{-1}\sqrt{4-x^2}$. Since this is the density of a probability measure, we conclude that $G(z) = G_\mu(z)$ where μ is the measure

$$d\mu(x) = \frac{1}{2\pi}\sqrt{4-x^2}\chi_{[-2,2]}(x)dx.$$

The equation $G(z)^2 = zG(z) - 1$ is the direct analog of 1.5.4; indeed, note that for z near $i\infty$, $G(z) = z^{-1}\sum_{n\geqslant 0}\tau(x/z)^n = \sum_{n\geqslant 0} m_n z^{n-1}$; we have just arrived at this equation in a different way.

1.5.7. Remarks. At the heart of our analytic proof was the use of the linear map $\delta : z \mapsto [z, r]$ which is a derivation having the special property that $\delta(x) = P$. It's worth pointing out that if $z = p(x)$ is a polynomial in x, then $\delta(z)$ is a finite-rank operator; in fact, if we denote by K the integral operator with kernel

$$K(s,t) = \frac{p(s)-p(t)}{s-t}$$

then

$$\delta(p)(f)(s) = \int K(s,t)f(t)d\mu(t) \qquad \forall f \in L^2(\mu).$$

The map

$$\partial : p \mapsto \frac{p(s)-p(t)}{s-t} \in L^2(\mu \times \mu)$$

is also a derivation, called the *difference quotient*. This derivation plays a special role in free probability theory; we shall see later that it is a kind of replacement for the usual derivative d/dx. The resolvent has a special property when differentiated with respect to ∂ (and which we used in the proof):

$$\partial\frac{1}{z-x} = \frac{\frac{1}{z-s}-\frac{1}{z-t}}{s-t} = \frac{1}{z-s}\cdot\frac{1}{z-t}.$$

In particular

$$\iint (\partial \frac{1}{z-x}) d\mu(s) d\mu(t) = G_\mu(z)^2.$$

1.5.8. Another analytic proof. Let $e_n = h^{\otimes n}$, $e_0 = \Omega$, $\ell = \ell(h)$. For $z \in \mathbb{C}$, $|z| < 1$, let

$$\omega_z = (1 - z\ell)^{-1}\Omega = \sum_{n \geqslant 0} z^n e_n.$$

Then

$$\ell\omega_z = \frac{1}{z}(\omega_z - \Omega), \qquad 0 < |z| < 1$$

and similarly

$$\ell^* \omega_z = z\omega_z, \qquad |z| < 1.$$

Thus

$$x\omega_z = (\ell + \ell^*)\omega_z = \frac{1}{z}(\omega_z - \Omega) + z\omega_z = (z + \frac{1}{z})\omega_z - \frac{1}{z}\Omega.$$

Thus

$$\frac{1}{z}\Omega = (z + \frac{1}{z} - x)\omega_z.$$

Let $\zeta = z + z^{-1}$. Since $\zeta \to \infty$ as $z \to 0$ we may choose some $\delta > 0$ so that for $0 < |z| < \delta$, $|\zeta| > \|x\|$ and so $\zeta - x$ is invertible. We then get:

$$(\zeta - x)^{-1}\Omega = (z + \frac{1}{z} - x)^{-1}\Omega = z\omega_z$$

whence

$$G(\zeta) = \tau((\zeta - x)^{-1}) = \langle \Omega, (\zeta - x)^{-1}\Omega \rangle = \langle \Omega, z\omega_z \rangle = z.$$

We have thus proved that

$$G(z + z^{-1}) = z, \qquad 0 < |z| < \delta,$$

so that G is the functional inverse of the map $z \mapsto z + z^{-1}$. From this (and the fact that $G(z) \sim z^{-1}$ near $i\infty$) one can easily recover that $G(z) = \frac{z - \sqrt{z^2-4}}{2}$ and then proceed as in the previous proof to recover the measure μ.

We conclude with a proof that semicircular operators $s(h)$ associated to orthogonal vectors are freely independent:

Proposition 1.5.9. *Let K_1 and K_2 in H be a pair of subspaces for which $K_1 \perp K_2$. If $M_i = \Phi(K_i) = W^*(s(h) : h \in K_i)$, then M_1 and M_2 are freely independent in $(\Phi(H), \tau)$.*

Proof. We will prove a stronger statement: letting $N = W^*(\ell(H))$ with (non-tracial) linear functional $\phi = \langle \Omega, \cdot\Omega \rangle$, we will prove that $N_i = W^*(\ell(K_i))$ are freely independent in (N, ϕ). To do so, we will assume that $y_i \in N_1$, $z_j \in N_2$, $\phi(z_j) = \phi(y_i) = 0$; we need to prove that $\phi(y_1 z_1 \cdots y_k z_k) = 0$ (actually we need to also prove $\phi(z_1 y_1 \cdots y_k z_k) = 0$ and two more similar identities, but the proofs are identical).

By a standard operator algebras argument (relying on the Kaplansky density theorem, which we will not discuss here), we may approximate in the weak topology each y_j by elements $y_j^{(n)}$ which are non-commutative polynomials in $\{\ell(h_j), \ell^*(h_j)\}_{j\in J}$ for $h_j \in K_1$, and similarly for z_k. Next, replacing $y_j^{(n)}$ by $y_j^{(n)} - \phi(y_j^{(n)})$ (which still converges to y_j) we may assume that $\phi(y_j^{(n)}) = 0$. Thus we may reduce to the case that y_j are polynomials in $\{\ell(h_j), \ell(h_j)^*\}$, $j \in K_1$ and z_j are polynomials in $\{\ell(g_j), \ell(g_j)^*\}$ for $g_j \in K_2$.

Assume that $w - \ell(h_1)^{\epsilon_1}\cdots\ell(h_n)^{\epsilon_n}$ is a monomial, with $\epsilon_j \in \{\cdot, *\}$. Using the relation $\ell(h)^*\ell(g) = \langle h, g\rangle 1$, we may assume there is a j such that $k \leqslant j$ implies $\epsilon_j = \cdot$ and $k > j$ implies that $\epsilon_k = *$. But then $\phi(w) = 0$ unless $n = 0$ (i.e., $w \in \mathbb{C}1$). As a corollary, we see that the *space of non-commutative polynomials in* $\ell(h), \ell(h)^*$ *that belong to the kernel of* ϕ *is spanned by the non-constant irreducible monomials* $w = \ell(h_1)^{\epsilon_1}\cdots\ell(h_n)^{\epsilon_n}$ *with* $n > 0$ *and such that* $\epsilon_j = *,\ k > j \implies \epsilon_k = *$.

By linearity, we thus assume that each y_i, z_j is such a monomial.

But then $X = y_1z_1\cdots y_kz_k$ is either an irreducible monomial (in which case $\phi(X) = 0$) or the reduction comes from the fact that some y_j ends in $\ell(g)^*$, $g \in K_2$ and y_{j+1} starts with $\ell(h)$ with $h \in K_1$ (or a similar situation with the roles of y, z switched). But then $X = 0$ because $\ell(g)^*\ell(h) = \langle g, h\rangle = 0$ since $K_1 \perp K_2$. □

1.6. Exercises

Exercise 1.6.1. Let $\Gamma = \mathbb{Z}$ with generating set $S = \{\pm 1\}$. Identify $\lambda(\pm 1)$ with the operator of multiplication by $\exp(\pm it)$ on the unit circle $\mathbb{T} = \{\exp(it) : t \in [0, 2\pi]\}$ by using Fourier transform isomorphism of $L^2(\mathbb{T})$ with $\ell^2(\mathbb{Z})$. Show that in this identification $\tau(x) = \langle 1, x1\rangle$ (here 1 refers to the constant function 1). Use this to compute the spectral measure of $L = \frac{1}{2}(\lambda(1) + \lambda(-1))$.

Exercise 1.6.2. Show that $L(\Gamma)$ is a factor (i.e., has trivial center) iff the group Γ is ICC (Infinite Conjugacy Classes), i.e., for any $h \neq e$, the set $\{ghg^{-1} : g \in \Gamma\}$ is infinite. *Hint:* the map $x \mapsto x\delta_e$ is an injective map from $L(\Gamma)$ into $\ell^2(\Gamma)$. Use this to show that if $x \in L(\Gamma)$ satisfies $x\lambda(h) = \lambda(h)x$ for all h, then $x\delta_e = \sum_{g\in\Gamma} a_g\delta_g$ must satisfy $a_{hgh^{-1}} = a_g$ for all h; now use that $a_g \in \ell^2$.

Exercise 1.6.3. If $M \subset B(H)$ is a von Neumann algebra, $\xi \in H$ is a unit vector, $\phi : M \to \mathbb{C}$ is given by $\phi(x) = \langle \xi, x\xi\rangle$, define $M' = \{T \in B(H) : Tx = xT\ \forall x \in M\}$. Show that ϕ is faithful iff $M'\xi$ is dense in H. *Hint*: $\phi(x^*x) = 0 \iff x\xi = 0$. Assuming $x\xi = 0$ use density of $M'\xi$ and commutation of elements of M' with x to conclude that $x = 0$. In the opposite direction, show that if $K = \overline{M\xi}$ then the orthogonal projection p onto $K^\perp$ belongs to M (use the double commutant theorem) and satisfies $\phi(p) = 0$.

Exercise 1.6.4. Use the result of the previous exercise to prove that τ is faithful on $\Phi(H)$ by showing that for $h \in H$, $d(h) = r(h) + r(h)^*$ where

$$r(h)\xi_1 \otimes \cdots \otimes \xi_n = \xi_1 \otimes \cdots \otimes \xi_n \otimes h$$

belong to $\Phi(H)'$ and conclude that $\Phi(H)'\Omega$ is dense in $\mathcal{F}(H)$.

Exercise 1.6.5. Let $A \subset (M, \tau)$ be a subalgebra, and assume that $u \in M$ is a unitary (so $u^*u = uu^* = 1$) so that $\tau(u^k) = 0$ for all $k \neq 0$. Assume that A and $\{u, u^*\}$ are freely independent. Show that the algebras A, uAu^*, $u^2A(u^*)^2$ are freely independent.

Exercise 1.6.6. Let (M_i, ϕ_i) be non-commutative probability spaces, and assume that $\pi_i : M_i \to B(H_i)$ are representations on Hilbert spaces H_i, and that $\xi_i \in H_i$ are unit vectors so that $\phi_i(x) = \langle \xi_i, \pi_i(x)\xi_i \rangle$, $i = 1, 2$. Set

$$H_i^o = (\mathbb{C}\xi_i)^\perp \cap H_i = H_i \ominus \mathbb{C}\xi_i.$$

The purpose of this exercise is to construct the free product of the representations π_i and to define the free product of the non-commutative probability spaces $(M, \phi) = (M_1, \phi_1) * (M_2, \phi_2)$.

Define H to be the direct sum of a one-dimensional space with unit basis vector ξ and alternating tensor products of H_i^o:

$$H = \mathbb{C}\xi \oplus \bigoplus_{k \geqslant 1} \bigoplus_{i_1 \neq i_2, i_2 \neq i_3, \cdots, i_{k-1} \neq i_k} H_{i_1}^o \otimes \cdots \otimes H_{i_k}^o.$$

(1) Let

$$H(i) = H_i \otimes \left[\xi_i \oplus \bigoplus_{k \geqslant 1} \bigoplus_{i \neq i_2, i_2 \neq i_3, \cdots, i_{k-1} \neq i_k} H_{i_2}^o \otimes \cdots \otimes H_{i_k}^o \right]$$

For any $h = \alpha\xi_i + h_1 \in H_i$ and any $h_2, \ldots, h_n$ satisfying $h_j \in H_{i_j}^o$ and $i \neq i_2, \ldots, i_{k-1} \neq i_k$, define the map V_i by

$$V_i[(\alpha\xi_i + h_1) \otimes (h_2 \otimes \cdots \otimes h_k)] = \alpha(h_2 \otimes \cdots \otimes h_k) + h_1 \otimes \cdots \otimes h_k, \qquad k \geqslant 1$$

$$V_i(h \otimes \xi_i) = h, \qquad h \in H_i.$$

Show that V_i extends to an isometric isomorphism of $H(i)$ with H.

(2) For $T \in B(H_i)$, view $T \otimes 1 \in B(H(i))$ using the tensor product decomposition

$$H(i) = H_i \otimes \bigoplus_{k \geqslant 1} \bigoplus_{i \neq i_2, i_2 \neq i_3, \cdots, i_{k-1} \neq i_k} H_{i_2}^o \otimes \cdots \otimes H_{i_k}^o.$$

For $x \in M_i$, define $\lambda_i(x) \in B(H)$ by

$$\lambda_i(x) = V_i(\pi_i(x) \otimes 1)V_i^*.$$

Show that $\lambda_i : M_i \to B(H)$ satisfy $\langle \xi, \lambda_i(x)\xi \rangle = \phi_i(x)$, for all $x \in M_i$.

(3) Let $M = W^*(\lambda_i(M_i) : i = 1, 2) \subset B(H)$, and let $\phi(x) = \langle \xi, x\xi \rangle$ for all $x \in M$. Show that $\lambda_i(M_i) \subset M$ are freely independent with respect to ϕ.

(4) Show that if ϕ_i are traces, then so is ϕ.

(5) Let Γ_i, $i = 1, 2$ be discrete groups, let $H_i = \ell^2(\Gamma_i)$, $\xi_i = \delta_e$, $M_i = L(\Gamma_i)$, $\phi_i(x) = \langle \xi_i, \cdot \xi_i \rangle$. Show that H can be naturally identified with $\ell^2(\Gamma_1 * \Gamma_2)$ and that the representations $\lambda_i : \Gamma_i \to B(H)$ are exactly the restrictions of the left regular representation $\lambda : \Gamma \to B(\ell^2(\Gamma))$ to $\Gamma_i \subset \Gamma$.

Exercise 1.6.7. Let $B \subset M$ be a von Neumann subalgebra, and assume that there exists a unital map $E : M \to B$ so that $E(x^*x) \geqslant 0$ for all $x \in M$ and so that $E(b_1 x b_2) = b_1 E(x) b_2$ for all $x \in M$, $b_1, b_2 \in B$. (By definition this means that $E_i : M \to B$ is a conditional expectation.)

(1) Define two subalgebras $M_1, M_2 \subset M$ to be *free with amalgamation over* B if $B \subset M_1 \cap M_2$ and

$$E(x_1 \cdots x_n) = 0$$

whenever $x_j \in M_{i(j)}$, $E(x_j) = 0$ and $i(1) \neq i(2)$, $i(2) \neq i(3)$, etc. Show that if M_1 and M_2 are free with amalgamation over B, then the restrictions of E to M_1 and to M_2 determine the restriction of E to $W^*(M_1, M_2)$.

(2) Let Γ_i be discrete groups with a common subgroup Λ. Give a definition of the amalgamated free product $\Gamma = \Gamma_1 *_\Lambda \Gamma_2$. Define a conditional expectation $E : L(\Gamma) \to L(\Lambda)$ by $E(\lambda(g)) = \lambda(g)$ if $g \in \Lambda$ and $E(\lambda(g)) = 0$ if $g \in \Gamma \setminus \Lambda$ (extended by linearity and continuity). Show that $L(\Gamma_1)$ and $L(\Gamma_2)$ are free with amalgamation over $L(\Lambda)$ inside $(L(\Gamma), E)$.

(3) Carry out the analog of Exercise 1.6.6 for amalgamated free products over B (to avoid technical difficulties, assume that B is a finite dimensional von Neumann algebra, e.g. a matrix algebra).

Lecture 2: R-transform and Free Harmonic Analysis

2.1. Executive summary. We discuss free multiplicative and additive convolutions of probability measures as well as the associated theory of the R-transform. A good reference for this lecture is Voiculescu's book [37].

2.2. Additive and multiplicative free convolutions. Let μ, ν be two probability measures, and let X, Y be two random variables so that $X \sim \mu$, $Y \sim \nu$. If we assume that X and Y are (classically) independent, then the law of $X+Y$ is the convolution $\mu * \nu$. This could be used as a definition of the convolution.

Definition 2.2.1. [26] Let μ, ν be two probability measures on $\mathbb{R}$. Their *free (additive) convolution* $\mu \boxplus \nu$ is defined to be the law of the variable $X+Y$ where X and Y are freely independent self-adjoint variables and $X \sim \mu$, $Y \sim \nu$.

One can also consider multiplication of variables, rather than their addition. Classically, there isn't a separate theory since to compute the law of XY we may as well compute the law of $\log(XY) = \log X + \log Y$. This is not the case in the free case.

Definition 2.2.2. [27] Let μ, ν be probability measures on $[0, +\infty)$. Their *multiplicative free convolution* $\mu \boxtimes \nu$ is defined to be the law of the variable $X^{1/2} Y X^{1/2}$ where X and Y are freely independent self-adjoint variables and $X \sim \mu$, $Y \sim \nu$ (that is, they have laws μ and ν, respectively).

Because the restriction of the expectation functional to $W^*(X)$ and $W^*(Y)$ is tracial (these algebras are commutative), the free product functional is also tracial. For this reason the law of $X^{1/2}YX^{1/2}$ is the same as that of $Y^{1/2}XY^{1/2}$. This is easily seen on the level of moments:

$$\tau(X^{1/2}YX^{1/2}\cdots X^{1/2}YX^{1/2}) = \tau(X^{1/2}YXYX\cdots YX^{1/2})$$
$$= \tau(YXYX\cdots YX) = \tau(Y^{1/2}XY^{1/2}\cdots Y^{1/2}XY^{1/2}).$$

It is also possible to define free convolution for measures supported on the unit circle: in this case X and Y can be chosen to be unitary operators and the free convolution is by definition the law of the unitary operator XY.

2.3. Computing $\boxplus$: R-transform. Let μ be a probability measure, and let

$$G_\mu(z) = \int \frac{d\mu(t)}{z-t}$$

be its Cauchy transform. Then G_μ is a holomorphic function defined on the upper half plane $\mathbb{C}_+$, and takes values in $\mathbb{C}_-$, the lower half-plane. It is a little easier to work with the function

$$F_\mu(z) = 1/G_\mu(z),$$

since F_μ takes $\mathbb{C}_+$ to itself; since $G_\mu(z) \sim z^{-1}$ as $z \to i\infty$, $F_\mu(z) \sim z$ as $z \to i\infty$. It turns out that $F_\mu(z)$ always has a functional inverse F_μ^{-1} defined on a set of the form

$$\Gamma_{\eta,M} = \{z \in \mathbb{C}_+ : |\Re(z)| < \eta\Im(z),\ \Im z > M\}.$$

Let now

$$\phi_\mu(z) = F_\mu^{-1}(z) - z,$$

defined on the same set. Finally, let

$$R_\mu(z) = \phi_\mu(z^{-1})$$

defined on the set of z for which $z^{-1} \in \Gamma_{\eta,M}$, in particular on a wedge-shaped region whose boundary includes 0.

If μ is compactly supported, G_μ is analytic on the complement of the support of μ and thus is analytic near ∞. It then follows that R_μ is analytic in some disk around 0.

Returning to the general situation, if we denote by K_μ the functional inverse to G_μ,

$$G_\mu(K_\mu(\zeta)) = \zeta, \qquad K_\mu(G_\mu(z)) = z$$

then

$$R_\mu(\zeta) = K_\mu(\zeta) - \frac{1}{\zeta}.$$

By definition we have

$$G_\mu(\zeta^{-1} + R_\mu(\zeta)) = \zeta, \qquad \frac{1}{G_\mu(z)} + R_\mu(G_\mu(z)) = z.$$

The main theorem is that R_μ linearizes $\boxplus$:

Theorem 2.3.1. [16, 26] *Let μ, ν be two probability measures on $\mathbb{R}$. Then on a wedge-shaped domain including* 0,

$$R_{\mu\boxplus\nu} = R_\mu + R_\nu.$$

Proof. We give the proof for the case that the measures are compactly supported. In that case, $G_\mu(z)$ is analytic near ∞ and thus also $R_\mu(\zeta)$ is analytic near 0. We follow the proof of Haagerup [17].

Let $\ell = \ell(h)$ be the creation operator as described in §1.5 associated to a unit vector h, let $e_n = h^{\otimes n}$, $e_0 = \Omega$ and let $\omega_z = \sum_{n\geqslant 0} z^n e_n$. In other words, our set up is very similar to that of §1.5.8. For $z \in \mathbb{C}$, $|z| < 1$ let

$$\omega_z = (1 - z\ell)^{-1}\Omega = \sum_{n\geqslant 0} z^n e_n.$$

Then

$$\ell\omega_z = \frac{1}{z}(\omega_z - \Omega), \qquad 0 < |z| < 1$$

and similarly

$$\ell^*\omega_z = z\omega_z, \qquad |z| < 1.$$

Let now $\alpha > 0$ be a fixed number so that $\alpha = \|\alpha\ell\|$ is inside some disk around 0 on which both R_μ and R_ν are analytic. Let

$$X = \alpha^{-1}\ell + R_\mu(\alpha\ell^*).$$

Then

$$X\omega_z = (\frac{1}{\alpha z} + R_\mu(\alpha z))\omega_z - \frac{1}{\alpha z}\Omega$$

so that

$$\frac{1}{\alpha z}\Omega = (\frac{1}{\alpha z} + R_\mu(\alpha z) - X)\omega_z.$$

Substituting z for αz we get

$$\frac{1}{z}\Omega = (\frac{1}{z} + R_\mu(z) - X))\omega_{z/\alpha} = (K_\mu(z) - X)\omega_{z/\alpha}.$$

We can now choose $\delta > 0$ so that $0 < |z| < \delta$ implies that $|K_\mu(z)| \geqslant \|X\|$ so that $(K_\mu(z) - X)$ is an invertible operator. Then

$$(K_\mu(z) - X)^{-1}\Omega = z\omega_{z/\alpha}$$

so that, substituting $\zeta = K_\mu(z)$,

$$\langle\Omega, (\zeta - X)^{-1}\Omega\rangle = z,$$

which implies that

$$G_\mu(\zeta) = \langle\Omega, (\zeta - X)^{-1}\Omega\rangle.$$

Let now $Y = \alpha^{-1}\ell(g) + R_\nu(\alpha\ell^*(g))$ for some unit vector $g \perp h$. Arguing in exactly the same way, we get that

$$G_\nu(\zeta) = \langle\Omega, (\zeta - Y)^{-1}\Omega\rangle.$$

By Proposition 1.5.9, Y and X are free. Thus

$$G_{\mu\boxplus\nu}(\zeta) = \langle\Omega, (\zeta - (X+Y))^{-1}\Omega\rangle.$$

Now, if we write $R_\mu(z) = \sum a_n z^n$, $R_\nu(z) = \sum b_n z^n$ then

$$X + Y = \alpha^{-1}\ell(g+h) + \sum_n (a_n[\ell(h)^*]^n + b_n[\ell(g)^*]^n)\alpha^n.$$

Denote by S the linear subspace spanned by vectors Ω and $(g+h)^{\otimes n}$, $n = 1, 2, \ldots$. Then S is invariant under $X+Y$, as it is invariant under $\ell(g+h)$ and each operator $Z_n = [\ell(h)^*]^n + [\ell(g)^*]^n$. In fact, $Z_n \cdot (g+h)^{\otimes m} = (\ell(h+g)^*)^n (g+h)^{\otimes m})$. Therefore, if we set

$$Z = (\alpha/\sqrt{2})^{-1} l + \sum_n (\alpha/\sqrt{2})^n (a_n + b_n)(l^*)^n = (\alpha')^{-1} l + (R_\mu + R_\nu)(\alpha' l^*),$$

where $l = \ell((g+h)/\sqrt{2})$ and $\alpha' = \alpha/\sqrt{2}$, then

$$(\zeta - Z)^{-1}\Omega = (\zeta - (X+Y))^{-1}\Omega$$

for sufficiently large ζ. Thus

$$G_{\mu\boxplus\nu}(\zeta) = \langle\Omega, (\zeta - Z)^{-1}\Omega\rangle.$$

But the proof above can be applied to Z to show that

$$\langle\Omega, (\frac{1}{z} + R_\mu(z) + R_\nu(z) - Z)^{-1}\Omega\rangle = z$$

which implies that

$$G_{\mu\boxplus\nu}(\frac{1}{z} + R_\mu(z) + R_\nu(z)) = z$$

Applying $K_{\mu\boxplus\nu}$ to this identity and noting that it is the functional inverse of $G_{\mu\boxplus\nu}$ we conclude:

$$K_{\mu\boxplus\nu}(z) = \frac{1}{z} + R_\mu(z) + R_\nu(z)$$

so that

$$R_{\mu\boxplus\nu}(z) = R_\mu(z) + R_\nu(z)$$

as claimed. □

We refer the reader to several exercises that demonstrate the use of R-transform.

We are now ready to state the free central limit theorem [37]:

Theorem 2.3.2. *Let $X_i \in (M, \tau)$ be non-commutative random variables, so that X_i are freely independent and each X_i has the same compactly supported law, μ_X. Assume further that $\tau(X_i) = 0$, $\tau(X_i^2) = a^2$. Let*

$$Y_N = \frac{X_1 + \cdots + X_N}{\sqrt{N}}.$$

Then $\mu_{Y_N} \to \mu$ weakly, where μ is the semicircle law

$$\mu(dx) = \frac{1}{2\pi a^2}\sqrt{4a^2 - x^2}\,\chi_{[-2a,2a]}(x)dx.$$

Proof. It is an easy exercise to verify that the R-transform of the semicircle law $(2\pi a^2)^{-1}\sqrt{4a^2 - x^2}\chi_{[-2a,2a]}(x)dx$ is az^2 (for example, one can use the formula for the Cauchy transform $G(z)$ derived in §1.5.5.

Let $\mu_N = \mu_{Y_N}$. Then, denoting by $D_s\nu$ the push-forward of a measure ν by the map $x \mapsto sx$, we have that

$$\mu_N = D_{1/\sqrt{N}}(\mu_X \boxplus \cdots \boxplus \mu_X)$$

(N times). Using Exercise 2.10.1 and additivity of R-transform, we see that

$$R_{\mu_N}(z) = N \cdot N^{-1/2} R_{\mu_X}(N^{-1/2}z) = N^{1/2} R_{\mu_X}(N^{-1/2}z).$$

Since μ_X is compactly supported, $R_{\mu_X}(z)$ is a power series in z. We have

$$G_{\mu_X}(z) = \int \frac{d\mu_X(t)}{z-t} = \frac{1}{z}\int \frac{d\mu_X(t)}{1-(t/z)} = z^{-1}\sum_n z^{-n}\int t^n d\mu(t)$$
$$= z^{-1} + z^{-2}\int t d\mu_X(t) + z^{-3}\int t^2 d\mu_X(t) + O(z^{-4}).$$

Since $\tau(X_i) = 0$ and $\tau(X_i^2) = a^2$, we see that

$$G_{\mu_X}(z) = z^{-1} + a^2 z^{-3} + O(z^{-4}).$$

It follows that $K_{\mu_X}(z) = z^{-1} + a^2 z + O(z^2)$ and so $R_{\mu_X}(z) = a^2 z + O(z^2)$. Let $f(z) = z^{-1}R_{\mu_X}(z)$. Then $f(z)$ is an analytic function. Moreover,

$$z^{-1}R_{\mu_N}(z) = z^{-1}N^{1/2}R_{\mu_X}(N^{-1/2}z) = f(N^{-1/2}z) \to f(0) = a^2$$

uniformly on compact sets. It follows that

$$R_{\mu_N}(z) \to a^2 z$$

as $N \to \infty$ uniformly on compact sets. This implies in fact that $G_{\mu_N}(z) \to G_\mu(z)$ on compact sets, which turns out to imply weak convergence. □

2.4. Combinatorial interpretation of the R-transform. As we just saw, for a compactly supported probability measure μ, $G_\mu(z) = \sum_{n\geqslant 0} z^{-(n+1)} m_n$ where $m_n = \int t^n d\mu(t)$ are the moments of μ. At the same time, $R_\mu(z) = \sum_{n\geqslant 1}\alpha_n z^{n-1}$. General theory of inversion of power series implies that each coefficient α_n is a polynomial in the moments $m_0, \ldots, m_n$. The numbers α_n are called *free cumulants.* There is a very nice combinatorial formula linking the two:

Theorem 2.4.1. [23] *With the notations of the preceding paragraph,*

$$m_n = \sum_{\pi\in NC(n)} \prod_{B \text{ block of } \pi} \alpha_{|B|}.$$

Furthermore, this equation determines the sequence $\{m_n\}_{n\geqslant 1}$ in terms of the sequence $\{\alpha_m\}_{m\geqslant 1}$ and vice-versa.

Before giving the proof, let us explain the notation. As before, $NC(n)$ stands for the set of all non-crossing partitions of $\{1, \ldots, n\}$. We call each equivalence class of the partition a *block.* Thus if the partition is $\pi = \{\{1,3\},\{2\},\{4,5,6\}\}$ then it has three blocks: $\{1,3\}$, $\{2\}$ and $\{4,5,6\}$. The corresponding term in the summation above would be $\alpha_2\alpha_1\alpha_3$, since the blocks have cardinalities 2, 1 and 3 respectively.

Proof. We have seen in the proof of Theorem 2.3.1 that if

$$X = \ell + f(\ell^*)$$

with $f(z) = \sum_{n=1}^{N} \alpha_n z^{n-1}$, then

$$\langle \Omega, (K(z) - X)^{-1}\Omega\rangle = z,$$

where we set $K(z) = z^{-1} + f(z)$. Denoting by G the function inverse of $K(z)$ and substituting $z = G(\zeta)$ gives us

$$\langle \Omega, (\zeta - X)^{-1}\Omega\rangle = G(\zeta).$$

Since by general theory of inverting power series, the coefficients of order at most ζ^{-M} of G depend only on the coefficients of order at most z^M for f, it follows that if we want to find a formula for a specific coefficient of G, we only need to make sure to take N sufficiently large, and the formula expressing a specific coefficient of G in terms of coefficients of f will then not depend on N. In particular, if μ has R-transform $\sum \alpha_n z^{n-1}$, then the first N terms of the power series expansion at ∞ of $G_\mu(\zeta)$ and $G(\zeta)$ are the same; thus

$$m_n = \langle \Omega, X^n \Omega\rangle$$

for any $n < N$.

It is therefore sufficient to prove that

$$\langle \Omega, X^n\Omega\rangle = \langle (X^*)^n\Omega, \Omega\rangle = \sum_{\pi\in NC(n)} \prod_{B} \alpha_{|B|}.$$

This proof is left as an exercise, along the lines of our combinatorial computation of the law of $\ell + \ell^*$.

It is clear that the sequence $\{\alpha_m\}_{m\geqslant 1}$ determines the sequence $\{m_n\}_{n\geqslant 1}$. Conversely, the expression for m_n involves a single term equal to α_n as well as other terms involving products of α_m for $m < n$. Thus the equation can be solved for α_n. □

2.5. Properties of free convolution. Amazingly enough, free convolution has a number of properties in common with classical convolution. In fact, if ν is the Cauchy law, then for any μ, $\mu \boxplus \nu$ is exactly the same as the classical convolution $\mu * \nu$.

Just as classical convolution, free convolution is a regularizing operation. For example, $\mu \boxplus \nu$ cannot have large atoms:

Proposition 2.5.1. [6] *Let* $t \in \mathbb{R}$. *Then* $(\mu \boxplus \nu)(\{t\}) > 0$ *implies that for some* $a, b \in \mathbb{R}$, $a + b = t$ *and* $\mu(\{a\}) + \nu(\{b\}) - 1 > 0$.

This theorem is perhaps not very surprising. In the case that μ and ν are atomic measures considered in Exercise 2.10.4, this statement corresponds to the fact that two freely independent projections P, Q have a non-zero intersection

$P \wedge Q$ iff $\tau(P) + \tau(Q) - 1 > 0$; in other words, free projections intersect "only when they are forced to". We refer the reader to [6, 37] for more regularization properties of $\boxplus$; for example, in many cases $\mu \boxplus \nu$ has a smooth density, and so on.

A well-understood topic concerns *free convolution powers.* Let $\mu^{\boxplus n}$ denote the n-fold free convolution of μ with itself. Then

$$R_{\mu^{\boxplus n}} = nR_\mu.$$

We can then try to define $\mu^{\boxplus \alpha}$ by the requirement that $R_{\mu^{\boxplus \alpha}} = \alpha R_\mu$ for any $\alpha > 0$. Of course, the issue is that αR_μ need not be equal to R_ν for any positive probability measure ν and thus $\mu^{\boxplus \alpha}$ may fail to exist. In fact, classical intuition would suggest that there should be measures μ so that only integer convolution powers exist (classically, this is true for many atomic measures). Strikingly,

Proposition 2.5.2. [20] *For any $\alpha \geqslant 1$ and any probability measure μ, $\mu^{\boxplus \alpha}$ exists. There are measures for which $\mu^{\boxplus \alpha}$ does not exist for any $\alpha < 1$.*

A measure μ for which $\mu^{\boxplus \alpha}$ exists for all $\alpha > 0$ is called *$\boxplus$-infinitely divisible.* Surprisingly, there is a natural 1-1 correspondence between classical and $\boxplus$-infinitely divisible measures. We refer the reader to [5] for more details.

2.6. Free subordination. One of the tools in the study of free convolution is the following *analytic subordination* result, which we will prove later once we develop a few more techniques (see §4.3 for the proofs). For now we just state it and point out a few consequences.

Theorem 2.6.1. [30] *Let X_1, X_2 be two free random variables. Then there exists a pair of holomorphic maps $\omega_1, \omega_2 : \mathbb{C}_+ \to \mathbb{C}_+$ so that*

$$G_{\mu_{X_1+X_2}}(z) = G_{\mu_{X_j}}(\omega_j(z)), \qquad j = 1, 2.$$

Moreover, if each μ_j is compactly supported, then $\lim_{z\to\infty} \Im(\omega_j(z) - z) = 0$.

The following Lemma (see [30, Lemma 4.1]) is not hard to prove using standard complex analysis techniques:

Lemma 2.6.2. *Let $f \in L^p(\mathbb{R})$, $1 \leqslant p \leqslant \infty$ and let $F(x + iy) = P_y * f$ be the harmonic extension of f to the upper half-plane $\mathbb{C}_+$ (here P_y is the Poisson kernel). If $\omega : \mathbb{C}_+ \to \mathbb{C}_+$ is analytic,* $\lim_{z\to\infty} |\omega(z)| = \infty$ *and* $\lim_{z\to\infty} \Im(\omega(z) - z) = 0$, *then for any $\epsilon > 0$,*

$$\|x \mapsto F(x + i\epsilon)\|_{L^p(\mathbb{R})} \geqslant \|x \mapsto F(\omega(x + i\epsilon))\|_{L^p(\mathbb{R})}.$$

Corollary 2.6.3. *Assume that X_1, X_2 are free and μ_{X_1} has density ρ_1 with respect to Lebesgue measure and that $\rho_1 \in L^p(\mathbb{R})$ for some $p \in (1, \infty]$. Then $\mu = \mu_1 \boxplus \mu_2$ is also Lebesgue absolutely continuous and its density ρ satisfies $\rho \in L^p(\mathbb{R})$.*

The idea of the proof is to use the analytic subordination theorem, which states that $G_\mu(z) = G_{\mu_1}(\omega(z))$. From this one can deduce that for every ϵ,

$$\|x \mapsto \Im G_\mu(x + i\epsilon)\|_{L^p} \leqslant \|x \mapsto \Im G_{\mu_1}(x + i\epsilon)\|_{L^p}$$

which implies the stated property of ρ once we send ϵ to zero.

2.7. Multiplicative convolution $\boxtimes$. Similarly to the R-transform, there is an S-transform that can be used to compute $\boxtimes$. We will not give proofs, but just supply the relevant formulas. For a probability measure μ on $[0, +\infty)$ such that $\mu(\{0\}) < 1$, define

$$\Psi_\mu(z) = \int_0^\infty \frac{zx}{1 - zx} d\mu(x).$$

This function turns out to be univalent on the left half-plane $i\mathbb{C}_+$. Let χ_μ be the functional inverse to Ψ_μ. Let

$$S_\mu(z) = \chi(z) \cdot \frac{1+z}{z}.$$

Theorem 2.7.1. [27] $S_{\mu \boxtimes \nu} = S_\mu \cdot S_\nu$.

2.8. Other operations. The definition of $\boxplus$ and $\boxtimes$ relied on simple operations on a pair of random variables: $(X, Y) \to (X + Y)$ and $(X, Y) \to (XY)$ (or, more precisely, $X^{1/2} Y X^{1/2}$). Clearly, more complicated operations are possible; one can, for example, associate to X, Y the law of an arbitrary non-commutative polynomial $Z = P(X, Y)$ (perhaps taking care that P produces a self-adjoint variable for any self-adjoint X, Y). Two such examples are the anti-commutator $P(X, Y) = XY + YX$ or the commutator $P(X, Y) = i(XY - YX)$. One does not have as well-developed a theory for these operations, but there are nonetheless a few results that can be established:

Proposition 2.8.1. *Let $X_1, \ldots, X_n$ be freely independent self-adjoint random variables, and let P be a self-adjoint non-commutative polynomial. Let $Z = P(X_1, \ldots, X_n)$. Then:*

(1) *if each $G_{\mu_{X_i}}$ is an algebraic function of z, then so is G_{μ_Z}. In particular, μ_Z has a smooth density except possibly at a finite number of points* [1];

(2) *if each $G_{\mu_{X_i}}$ is non-atomic and P is not the constant polynomial, then μ_Z is non-atomic* [9, 19].

(3) *For each i, let*

$$S_i = \{0\} + \{\mu_{X_j}(J) : J \text{ union of connected components of the support of } \mu_{X_j}\}.$$

Let

$$S = [0, 1] \cap (S_1 + \cdots + S_n).$$

Then for any connected component J of the support of μ_Z, $\mu_Z(J) \in S$. In particular, if for each i, the support of μ_{X_i} is connected, then the support of μ_Z is connected as well [12, 13].

We explain the proof of 3 here; the proof of 1 will not be given, and the proof of 2 will be postponed until a later section (see §4.7).

Let A_i be the algebra of continuous functions on the support of μ_i. Let $p \in A_i$ be a projection, i.e., a real-valued function satisfying $p^2 = p$. Then p is valued in $\{0, 1\}$ and is thus constant on the connected components of μ_i. In particular, if we denote by τ_i the trace on A_i corresponding to integration against μ_i, then $S_i = \{\tau(p) : p \in A_i \text{ is a projection}\}$. The element Z belongs to the C^*-algebra free product $(A,\tau) = *_i(A_i,\tau_i)$. If J is a connected component of the spectrum of Z, then the characteristic function χ_J of J is a continuous function on the spectrum of $Z \in A$ and thus $\chi_J(Z) \in A$ is a projection. Remarkably, it is possible to "classify" projections in A up to unitary conjugation in matrix algebras over A (i.e., $p \sim q$ if for some n, $p' = \text{diag}(p,0,\dots,0)$ and $q' = \text{diag}(q,0,\dots,0)$, regarded as elements of $M_{n\times n}(A)$, satisfy $p' = uq'u^*$ for some unitary $u \in M_{n\times n}(A)$). Such a classification is equivalent to the computation of the so-called K_0 group of A and it is a deep result of C^*-algebraic K-theory that the K-groups of free products can be computed (see [12, 13]). Essentially, the classification states that any projection $p \in A$ must be equivalent to $\text{diag}(p_1,\dots,p_n) \in M_{n\times n}(A)$ for projections $p_j \in A_j$. But if $p \sim q$ then $\tau(p) = \tau(q)$; thus $\tau(p)$ must be in the set S. In particular, if the support of each μ_{X_i} is connected, then any $p \in A$ is equivalent to some $q = \text{diag}(p_1,\dots,p_n) \in M_{n\times n}(A)$; but since each p_i must have integer trace, so does q and hence $\tau(p) = 0$ or $\tau(p) = 1$, so that $\mu_Z(J) \in \{0,1\}$ and thus μ has connected support.

2.9. Multivariable and matrix-valued results. Much of what we discussed in this lecture concerns the "non-commutative" law of single variable. Such a law can be identified with a measure; the Cauchy transform, analytic subordination, etc. then become familiar objects and subjects from complex analysis. Of course, if one is given an n-tuple of self-adjoint elements $x_1,\dots,x_n$ such an interpretation is no longer possible.

There are two ways to proceed. The first is via a combinatorial study of joint laws of such n-tuples. This way, for example, one can generalize notions of free cumulants to such laws. A number of combinatorial results can be proved [23].

The second way involves a matrix trick. Let (M,τ) be a non-commutative probability space. Let $M_n = M_{n\times n}(M)$ be the algebra of $n\times n$ matrices over M. The scalar $n\times n$ matrices $B_n = M_{n\times n}(\mathbb{C}) \subset M_n$ is a unital subalgebra; the conditional expectation $E : M_n \to B_n$ is given by $E([x_{ij}]_{i,j}) = [\tau(x_{ij})]_{i.j}$. Thus the values of expectations of $E(b_1X_1b_2X_2\cdots b_nX_nb_{n+1})$ and $b_j \in B_n$ are completely determined by the joint laws of the entries of the matrices $X_1,\dots,X_n$. These *operator-valued* expectations lead naturally to a generalization of free probability in which the linear functional $\tau : M \to \mathbb{C}$ is replaced by a B-bilinear map $E : M \to B$ for a unital subalgebra $B \subset M$. We only briefly touch upon this subject and refer the reader to [32, 37] for further development of these ideas.

Lemma 2.9.1. *Let* $x_1,\dots,x_n$ *be an* n*-tuple of random variables in* (M,τ)*, and then let* $X = \text{diag}(x_1,\dots,x_n) \in M_n$*. For each* m*, consider the matrix-valued Cauchy transform*

$$G_m(b) = E_{nm}([b - X\otimes 1_m]^{-1}), \qquad b \in B_{nm},\ \Im b > 0.$$

Then the functions $G_m(b)$ *completely encode the joint law of* $x_1,\dots,x_n$.

Proof. Expanding as power series for b sufficiently large, we have

$$G_m(b) = \sum E[b^{-1}X\otimes 1_m b^{-1}\cdots b^{-1}X\otimes 1_m^{-1} b^{-1}],$$

which can be viewed as a formal power series in b; thus $G_m(b)$ encodes the values of all expressions of the form

$$E[bX \otimes 1_m b^{-1} \cdots b^{-1} X \otimes 1_m b^{-1}].$$

With an appropriate choice of m and b, we can make the $1,1$-entry of this matrix equal to the trace of an arbitrary monomial in $x_1, \dots, x_n$. □

This Lemma motivates the study of matrix-valued Cauchy transforms, a topic of active research. What is important is to keep track of the functions $G_m(b)$ for all values of m. We refer the reader to [18, 36, 38] for some basic results, including a characterization of which such "hierarchies" of functions $G_m(b)$ come from non-commutative probability laws. Amazingly, a number of single-variable results, including for example results on analytic subordination, have matricial analogs.

2.10. Exercises.

Exercise 2.10.1. Let μ be a probability measure on $\mathbb{R}$, and let $D_\alpha\mu$ be the dilation of μ, i.e., the push-forward of μ by the map $x \mapsto \alpha x$. In other words, if $X \sim \mu$ then $\alpha X \sim D_\alpha \mu$. Show that

$$G_{D_\alpha\mu}(z) = \alpha^{-1} G_\mu(z/\alpha)$$

and

$$R_{D_\alpha\mu}(z) = \alpha R_\mu(\alpha z).$$

Exercise 2.10.2. Compute G_μ for the following probability measures:

(1) $\mu = \sum \alpha_j \delta_{x_j}$, $x_j \in \mathbb{R}$, $\alpha_j \in (0,1]$, $\sum_j \alpha_j = 1$.
(2) The spectral measure you computed in Exercise 1.6.1.

Exercise 2.10.3. Let $\mathbb{F}_n$ be the free group on n generators $g_1, \dots, g_n$, and denote by u_j the unitary $\lambda(g_j)$ acting in the left regular representation $\lambda : \mathbb{F}_n \to \ell^2(\mathbb{F}_n)$. As in §1.3, let $L^{(n)} = \frac{1}{2n} \sum_j u_j + u_j^* = \frac{1}{n} \sum_j X_j$ where $X_j = \frac{1}{2}(u_j + u_j^*)$. Using Exercise 1.6.1 and part (2) of the previous exercise, compute

$$\mu_{L^{(n)}} = D_{1/n}(\mu_X \boxplus \mu_X \boxplus \cdots \boxplus \mu_X) \quad (n \text{ times}).$$

The computation of $\mu_{L^{(n)}}$ was first carried out by Kesten in 1950s, by a different method.

Explicitly show that $D_{\sqrt{n}}\mu_{L^{(n)}}$ converges to the semicircle law.

Exercise 2.10.4. Let $\mu = \alpha\delta_0 + (1-\alpha)\delta_1$ and $\nu = \beta\delta_0 + (1-\beta)\delta_1$ with $\alpha, \beta \in [0,1]$. Compute $\mu \boxplus \nu$ and $\mu^{\boxplus\gamma}$ for all $\gamma \geqslant 1$, while showing that $\mu^{\boxplus\gamma}$ does not exist for $0 < \gamma < 1$.

Exercise 2.10.5. Let $X = \ell + f(\ell^*)$ with $f(z) = \sum_{n=0}^N z^n \alpha_{n+1}$. Show that for any $n < N$,

$$\langle (X^*)^n \Omega, \Omega \rangle = \sum_{\pi \in NC(n)} \prod_B \alpha_{|B|}.$$

To do this, first show that $\langle (X^*)^n\Omega, \Omega\rangle$ is equal to the summation over functions $f : \{0, \dots, n\} \to \mathbb{N} \cup \{0\}$ satisfying $f(0) = 0$, $f(n) = 0$ and for each k, either $f(k) \geqslant f(k-1)$ or $f(k) = f(k-1) - 1$ (i.e., these are "Dyck paths", only we allow an arbitrary non-negative slope), of products $\alpha_f = \prod_{k:f(k+1)\geqslant f(k)} \alpha_{f(k+1)-f(k)}$. Then produce a bijection between such functions f and non-crossing partitions, by associating to each f a partition that has a class $\{k_0 < k_1 < \dots < k_r\}$ whenever $f(k_0+1) - f(k_0) = r$ and $k_1, \dots, k_r$ are the starting points of the first descending segments at height in the interval $[f(k_0), f(k_0+1)]$ and use it to prove the stated formula.

Exercise 2.10.6. (Free Poisson laws). Fix $\lambda \geqslant 0$ and define

$$\nu_N = ((1-\lambda/N)\,\delta_0 + (\lambda/N)\delta_\alpha)$$

and let $\mu_N = \nu_N \boxplus \cdots \boxplus \nu_N$ (N-times) be its N-fold free additive convolution. Likewise, let $\psi_N = \nu_N * \cdots * \nu_N$ be the N-fold classical convolution.

(1) Show that ψ_N converges to the classical Poisson distribution of intensity λ (this is the Poisson limit theorem) and jump size α.

(2) Show that $\mu_N \to \mu$ where

$$\mu = \begin{cases} (1-\lambda)\delta_0 + \lambda\nu, & 0 < \lambda < 1 \\ \nu, & \lambda \geqslant 1 \end{cases}$$

where

$$d\nu(t) = \frac{1}{2\pi\alpha} t^{-1} \sqrt{4\lambda\alpha^2 - (t - \alpha(1+\lambda))^2}\chi_{[\alpha(1-\sqrt{\lambda})^2, \alpha(1+\sqrt{\lambda})^2]}\,dt.$$

This *free Poisson law* occurs in random matrix theory under a different name, the Marčenko-Pastur distribution.

(3) Let X be a semicircular variable of variance α and let p be a self-adjoint element free from X so that $\mu_p = \lambda\delta_1 + (1-\lambda)\delta_0$ (i.e., p is a projection with $\tau(p) = \lambda$). Show that $\mu_{(pXp)^2|}$ is the free Poisson law with parameters α and λ. (*Hint:* $\tau([(pXp)^2]^n) = \tau((pX)^{2n})$ so $\mu_{(pXp)^2}$ can be computed in terms of $\mu_{pX} = \mu_X \boxtimes \mu_p$).

Lecture 3: Free Probability Theory and Random Matrices

3.1. Executive summary. The link between random matrices and free probability theory comes from Voiculescu's amazing insight that certain classes of random matrices become asymptotically free as their size goes to infinity. We review several examples of this behavior, giving both combinatorial and analytic proofs. An excellent reference is the book [2].

3.2. Non-commutative laws of random matrices. A family of random matrices is an example of a family of non-commuting random variables. Indeed, if $X = (X_{ij})_{ij=1}^N$ is a random matrix, then we can regard the random entries X_{ij} as functions on of some commutative probability space (Ω, ν) and view X as an element of the non-commutative probability space $M_{N\times N}(L^\infty(\Omega,\nu))$ with the tracial state

$$\tau(\cdot) = \mathbb{E}_\nu[\frac{1}{N}\mathrm{Tr}(\cdot)].$$

(Strictly speaking, this disallows random matrices with unbounded entries, such as Gaussian random matrices; very little below is changed if we allow entries to live in the space $L^{\infty-}(\Omega,\nu) := \bigcap_{p\geqslant 1} L^p(\Omega,\nu)$; from the von Neumann algebra point of view such elements are affiliated operators to the von Neumann algebra $M_{N\times N}(L^\infty(\Omega,\nu))$ satisfying growth conditions.)

If X is self-adjoint, its non-commutative law μ_X has an interpretation in terms of eigenvalues of X. Indeed, for any test function f, we have

$$\int f d\mu_X = \tau(f(X)) = \mathbb{E}_\nu[\frac{1}{N}\mathrm{Tr}(f(X)] = \mathbb{E}_\nu[\frac{1}{N}\sum_{j=1}^N f(\lambda_j)],$$

where $\lambda_1 < \lambda_2 < \cdots < \lambda_N$ are the eigenvalues of X (viewed as random variables). Alternatively, we can rewrite this as

$$\mu_X = \mathbb{E}[\frac{1}{N}\sum_{j=1}^N \delta_{\lambda_j}],$$

i.e., μ_X is the expected value of the random measure that puts an atom of size $1/N$ at each eigenvalue of X.

In particular,

$$\mu_X[a,b] = \frac{1}{N}\mathbb{E}[\#\{\text{eigenvalues of X which lie in } [a,b]\}].$$

3.3. Random matrix models. Free probability is able to provide information about asymptotics of certain types of random matrices. We will state several theorems and then discuss their proofs and improvements.

3.3.1. Haar unitary matrices. In this model we are given a family of deterministic matrices $\{B_j^{(N)}\}_{j=1}^d$ of size $N \times N$; we are assuming that their joint laws $\mu_{B_1^{(N)},\dots,B_d^{(N)}}$ converge (let us say, on all polynomial test functions) to some non-commutative law $\mu_{b_1,\dots,b_d}$; often we also assume that the operator norms $\|B_j^{(N)}\|$ are uniformly bounded in N.

To give a concrete example of such a set-up, we could take $d = 1$ and set $B^{(N)} = \mathrm{diag}(1,\dots,1,0,\dots,0)$ to be a projection matrix of rank $N/2$.

Theorem 3.3.2. [29,34] *Let* $\{B_j^{(N)}\}_{j=1}^d$ *be deterministic matrices so that*

$$\mu_{B_1^{(N)},\dots,B_d^{(N)}} \to \mu_{b_1,\dots,b_d} \textit{ and } \sup_{j,N}\|B_j^{(N)}\| < \infty.$$

If $U_1^{(N)},\dots,U_q^{(N)}$ *are unitary matrices drawn independently at random with respect to Haar measure, then*

$$\mu_{U_1^{(N)},\dots,U_q^{(N)},[U_1^{(N)}]^*,\dots,[U_q^{(N)}]^*,B_1^{(N)},\dots,B_d^{(N)}} \to \mu_{u_1,\dots,u_q,u_1^*,\dots,u_q^*,b_1,\dots,b_d}$$

where u_j are unitaries satisfying $\tau(u_j^k) = 0$ for all integer $k \neq 0$ and so that the families $(b_1, \ldots, b_d)$ and the families $(u_j, u_j^*), j = 1, \ldots, q$ are freely independent from each other.

In particular,

$$\mu_{B_1^{(N)},\ldots,B_d^{(N)},U_1^{(N)}B_1^{(N)}[U_1^{(N)}]^*,\ldots,U_1^{(N)}B_d^{(N)}[U_1^{(N)}]^*} \to \mu_{b_1,\ldots,b_d,b_1',\ldots,b_d'}$$

where $\mu_{b_1,\ldots,b_d} = \mu_{b_1',\ldots,b_d'}$ and $(b_1, \ldots, b_d)$ and $(b_1', \ldots, b_d')$ are free.

The second part of the statement of the theorem is an consequence of Exercise 1.6.5, since we can take $b_j' = u_1 b_j u_1^*$.

Here is an immediate corollary of this theorem. Suppose that we have taken as before $d = q = 1$, $A^{(N)}$ a deterministic diagonal rank $\lfloor N/2 \rfloor$ projection and $U^{(N)}$ a random Haar unitary. Then $\mu_{A^{(N)}+U^{(N)}A^{(N)}[U^{(N)}]^*} \to \nu \boxplus \nu$ (free additive convolution), where $\nu = \frac{1}{2}\delta_0 + \frac{1}{2}\delta_1$ is the limiting distribution of $A^{(N)}$.

We can also use other regularity results associated with freeness. For example, we can use Proposition 2.8.1 to conclude:

Corollary 3.3.3. *Suppose that $A^{(N)}$ are deterministic matrices so that $\mu_{A^{(N)}} \to \mu$ and μ is a measure with connected support. Let P be a self-adjoint polynomial in 3 variables, and let $Z^{(N)} = P(A^{(N)}, U^{(N)}, [U^{(N)}]^*)$, where $U^{(N)}$ is a Haar unitary random matrix. Then $\mu_{Z^{(N)}} \to \mu_Z$, a measure with connected support. If μ is non-atomic and P is non-constant, then μ_Z is non-atomic.*

3.3.4. Mulimatrix models with a potential. Assume now that V is a fixed non-commutative polynomial, and assume that for each N the integral

$$Z_N = \int \exp(-N\mathrm{Tr}(V(A_1, \ldots, A_d)))dA_1 \ldots dA_d$$

is finite (here $A_1, \ldots, A_d$ are $N \times N$ self-adjoint matrices and $dA_1 \ldots dA_d$ stands for Lebesgue measure associated to the real Hilbert space inner product given by $\langle A, B\rangle = \mathrm{Tr}(AB)$). Let

$$\nu_N^{(V)} = \frac{1}{Z_N} \exp(-N\mathrm{Tr}(V(A_1, \ldots, A_d)))dA_1 \ldots dA_d$$

be a probability measure on d-tuples of $N \times N$ self-adjoint random matrices.

An example of this situation is $V(A_1, \ldots, A_d) = \sum A_j^2$; in this case $\nu_N^{(V)}$ is just the Gaussian measure.

For more general V we may introduce a cutoff parameter R and define

$$Z_{N,R} = \int_{\|A_j\| \leqslant R} \exp(-N\mathrm{Tr}(V(A_1, \ldots, A_d)))dA_1 \ldots dA_d$$

and

$$\nu_{N,R}^{(V)} = \frac{1}{Z_{N,R}} \exp(-N\mathrm{Tr}(V(A_1, \ldots, A_d))) \prod_j \chi_{\|A_j\| \leqslant R} dA_1 \ldots dA_d.$$

This defines random matrices $\{A_j^{(N,R)}\}_{j=1}^d$.

It is a challenging and unsolved question to find out exactly for which functions V the joint laws $\mu_{A_1^{(N)},\ldots,A_d^{(N)}}$ have a limit. However, there is the following result:

Theorem 3.3.5. (1) *If* $V(A_1,\ldots,A_d) = \sum A_j^2$, *then* $\mu_{A_1^{(N)},\ldots,A_d^{(N)}} \to \mu_{a_1,\ldots,a_d}$ *where* $a_1,\ldots,a_d$ *are free semicircular variables* [29].

(1′) *Let* $\{B_j^{(N)}\}_{j=1}^k$ *be fixed deterministic matrices so that* $\sup_{N,j} \|B_j^{(N)}\| < \infty$, *and suppose that* $\mu_{B_1^{(N)},\ldots,B_k^{(N)}} \to \mu_{b_1,\ldots,b_k}$. *Then*

$$\mu_{A_1^{(N)},\ldots,A_d^{(N)},B_1^{(N)},\ldots,B_k^{(N)}} \to \mu_{a_1,\ldots,a_d,b_1,\ldots,b_k}$$

where $a_1,\ldots,a_d$ *are free semicircular variables, free from the family* $(b_1,\ldots,b_k)$ [29,37].

(2) *Suppose that* $V(A_1,\ldots,A_d) = \sum A_j^2 + \epsilon W(A_1,\ldots,A_d)$ *where W is a polynomial. Then there exists* $\epsilon_0 > 0$ *and* $0 < R_0 < R_1$ *so that for all* $|\epsilon| < \epsilon_0$ *and all* $R_0 < R < R_1$,

$$\mu_{A_1^{(N,R)},\ldots,A_d^{(N,R)}} \to \mu^{\epsilon}_{a_1,\ldots,a_d},$$

which does not depend on R [14].

(3) *In the setting of (2), if* $W(A_1,\ldots,A_d) = W_1(A_1,\ldots,A_r) + W_2(A_{r+1},\ldots,A_d)$, *then the families* $(a_1,\ldots,a_r)$ *and* $(a_{r+1},\ldots,a_d)$ *are free under* $\mu^{\epsilon}_{a_1,\ldots,a_d}$.

In any of the cases (1), (2) and (3) the convergence is almost sure.

Sketch of proof of Theorem 3.3.2 *given Theorem* 3.3.5*(1′).* Assume that

$$A_1^{(N)}, A_2^{(N)}, \ldots, A_{2r}^{(N)}$$

are as in Theorem 3.3.5(1′) and let

$$C_j^{(N)} = A_{2j}^{(N)} + iA_{2j-1}^{(N)}, \qquad j = 1,\ldots,r.$$

Then $C_j^{(N)}$ has as entries iid complex Gaussian variables; in particular, for any fixed $N \times N$ unitary matrix W, $(WC_j^{(N)} : j = 1,\ldots,r\} \sim \{C_j^{(N)} : j = 1,\ldots,r\}$.

Let $a_1,\ldots,a_{2r}$ be a semicircular family, and let $c_j = a_{2j} + ia_{2j-1}$, $j = 1,\ldots,r$.

Let $U_j^{(N)}$ be the polar part of $C_j^{(N)}$ in its polar decomposition

$$C_j^{(N)} = U_j^{(N)} |C_j^{(N)}|;$$

similarly, let u_j be the polar part of c_j. Because of unitary invariance of $C_j^{(N)}$, it follows that the law of $U_j^{(N)}$ is unitarily invariant and thus $U_j^{(N)}$ is distributed with respect to Haar measure. Since the entries $C_j^{(N)}$ are independent for different values of j, it follows that $U_j^{(N)}$ are independent for different j.

Let now $B_s^{(N)}$, $s = 1,\ldots,q$ be deterministic matrices as in the hypothesis of Theorem 3.3.2.

A restatement of Theorem 3.3.5(1′) that we will often find useful is by saying that $(C_j^{(N)}, [C_j^{(N)}]^*, B_s^{(N)} : j = 1,\ldots,r, s = 1,\ldots,q)$ converge in law to the family $(c_j, c_j^*, b_s : j = 1,\ldots,r, s = 1,\ldots,q)$, where the families $(c_j, c_j^* : j = 1,\ldots,r)$ and $(b_s : s = 1,\ldots,q)$ are free. A small argument then shows that this implies that

$$\mu_{(U_1^{(N)},\ldots,U_r^{(N)},[U_1^{(N)}]^*,\ldots,[U_r^{(N)}]^*,B_1^{(N)},\ldots,B_q^{(N)})} \to \mu_{(u_1,\ldots,u_r,u_1^*,\ldots,u_r^*,b_1,\ldots,b_q)}.$$

Since $u_j \in W^*(c_j, c_j^*)$, it follows that the families (u_j, u_j^*), $j = 1,\ldots,s$ are free from each other and from $(b_1,\ldots,b_q)$, as claimed. □

3.3.6. Theorem 3.3.5(1): combinatorial proof [14]. Assume that

$$(A_1^{(N)}, \ldots, A_d^{(N)}) \sim \nu_N = Z_N^{-1} \exp(-N\mathrm{Tr} \sum A_j^2) \prod dA_j$$

are Gaussian, i.e., that the entries $\{A_k^{(N)}(i,j) : 1 \leqslant i \leqslant j \leqslant N, k = 1, \ldots, d\}$ are independent Gaussian random variables, which are complex when $i \neq j$, real when $i = j$ and have variance $\mathbb{E}[|A_k^{(N)}(i,j)|^2] = N^{-1}$. Of course, since the matrices $A_k^{(N)}$ are self-adjoint,

$$A_k^{(N)}(j,i) = \overline{A_k^{(N)}(i,j)}.$$

These are sometimes called GUE matrices.

We will directly compute

$$\int \frac{1}{N}\mathrm{Tr}(p(A_1, \ldots, A_d))d\nu_N$$

for a monomial p. We will proceed with the proof in the case that $d = 1$ for simplicity of notation, but the idea is exactly the same in the general case.

We first start with a Lemma, which is standard:

Lemma 3.3.7. *Suppose that $g_1, \ldots, g_r$ are jointly complex Gaussian random variables. Then*

$$\mathbb{E}[g_{i_1} \cdots g_{i_k}] = \sum_{\pi} \prod_{B=\{a,b\}} \mathbb{E}[g_{i_a} g_{i_b}]$$

where the summation is over all partitions π of $\{1, \ldots, k\}$ into pairs and the product is over these pairs.

Note that if we assume that $g_1, \ldots, g_r$ are independent, then $\mathbb{E}[g_{i_a} g_{i_b}] = 0$ unless $g_{i_b} = \overline{g_{i_a}}$.

Let us now return to the question of computing

$$\mathbb{E}_{\nu_N}\left[\frac{1}{N}\mathrm{Tr}(A^p)\right] = \frac{1}{N} \sum_{j_1,\ldots,j_p} \mathbb{E}[A^{(N)}(j_1, j_2) \cdots A^{(N)}(j_{p-1}, j_p) A^{(N)}(j_p, j_1)].$$

Since all the entries $A^{(N)}(i,j)$ are Gaussian, we may use our Lemma to conclude that

$$\mathbb{E}_{\nu_N}\left[\frac{1}{N}\mathrm{Tr}(A^p)\right] = \frac{1}{N} \sum_{\pi} \prod_{B=\{a,b\}} \mathbb{E}[A^{(N)}(j_a, j_{a+1}) A^{(N)}(j_b, j_{b+1})]$$

(we set $j_{b+1} = j_1$ if $b = p$). Because of independence, the resulting product is nonzero iff $A^{(N)}(j_a, j_{a+1}) = \overline{A^{(N)}(j_b, j_{b+1})} = A^{(N)}(j_{b+1}, j_b)$ so that we end up with

$$\mathbb{E}_{\nu_N}\left[\frac{1}{N}\mathrm{Tr}(A^p)\right] = \frac{1}{N} \sum_{\pi} \prod_{B=\{a,b\}} \left[\frac{1}{N} \delta_{j_a = j_{b+1}} \delta_{j_{a+1} = j_b}\right].$$

Now, given a pairing π we associate to it a diagram in the following way. At each vertex of a p-gon we place two lines, one outgoing and one incoming. At the k-th vertex, we label the lines j_k and j_{k+1} (we put a $*$ to indicate which vertex is the first). Next, if the partition π contains the block $\{a, b\}$, we connect the pairs of lines associated to vertex a and vertex b by a pair of dashed lines. Such a diagram is called a *map*; see Figure 3.3.8.

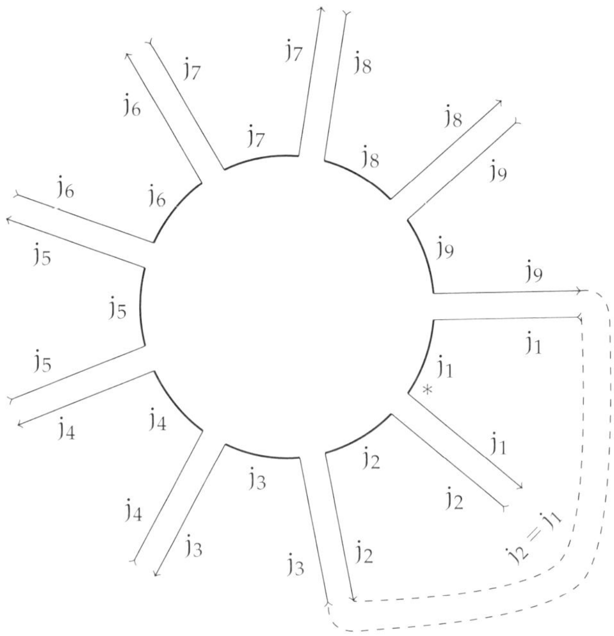

FIGURE 3.3.8. A diagram associated to a pairing.

It is not hard to see that for some given numbers $j_1, \ldots, j_k$, the condition

$$\prod_{B=\{a,b\}} \delta_{j_a = j_b} \delta_{j_{a+1} = j_{b+1}} \neq 0$$

is equivalent to the ability to consistently extend the labeling of solid lines by $j_1, \ldots, j_p$ to the dashed lines. For example, in the Figure 3.3.8 it must be that $j_9 = j_3$ and $j_1 = j_2$.

Suppose that $j_1, \ldots, j_p$ is such a labeling. Starting from some vertex of the polygon, follow the outbound line from this vertex; next, follow the dashed line and the inbound line of some other vertex. Next, continue following the side of the polygon, then an output vertex, etc. until one gets a closed curve. For each such closed curve, glue in a topological disk with that curve as the boundary. Our condition on $j_1, \ldots, j_p$ implies that each such topological disk receives a consistent labeling by $j_1, \ldots, j_p$; therefore, the number of labelings compatible with a given partition π is simply N^d, where d is the number of disks that we have glued in.

Contract the interior of the p-gon to single point P, and glue pairs of parallel dashed and solid lines into single lines. The result can be interpreted as a 2-skeleton of an orientable compact topological space, in which we have d two-cells (the disks we glued in), $p/2$ one-cells (formed by the pairs of lines we glued together) and 1 zero-cell (the point P). We have thus obtained a CW-complex representing a curve of genus g, where by Euler's formula,

$$2-2g=1-p/2+d$$

so that

$$d=1+p/2-2g.$$

We have thus obtained the so-called *genus expansion*:

$$\mathbb{E}_{\nu_N}[\frac{1}{N}\mathrm{Tr}(A^p)] = N^{-1-p/2}\sum_g \sum_{\pi\in\Pi_g(p)} N^d = \sum_g N^{-2g} \sum_{\pi\in\Pi_g(p)} 1, \tag{3.3.9}$$

where we are summing over the set Π_g of all pairings whose associated CW-complex has genus g.

Since $N\to\infty$ the terms corresponding to $g>0$ disappear in the limit; thus

$$\lim_{N\to\infty} \mathbb{E}_{\nu_N}[\frac{1}{N}\mathrm{Tr}(A^p)] = \#\Pi_0(p).$$

It remains to remark that a pairing whose resulting CW-complex has genus 0 (i.e., a sphere) is one that can be drawn on a sphere with no crossings; in other words, $\#\Pi_0(p) = \#NC_2(p)$. Since the even moments of the semicircle law are given by cardinalities of $NC_2(p)$, the result follows.

3.4. GUE matrices: bound on eigenvalues. We follow [2]:

Corollary 3.4.1. *Let $A_k^{(N)}, k\in\mathbb{N}$ be independent GUE matrices*

(1) For all $\ell \leqslant N/\log N$,

$$\mathbb{E}[\frac{1}{N}\mathrm{Tr}([A_k^{(N)}]^{2\ell})] \leqslant 2^{2\ell}\sum_g \left(\frac{(2\ell)^2}{N}\right)^{2g} \leqslant 3^{2\ell}.$$

(2) For all monomials q with total degree 2ℓ, $\ell\leqslant\sqrt{N}$

$$\left|\mathbb{E}[\frac{1}{N}\mathrm{Tr}(q(A_k^{(N)}, k\in\mathbb{N}))]\right| \leqslant 3^{2\ell}.$$

Proof. We use (3.3.9); note that the cardinality of $\Pi_g(2\ell)$ is at most $(2\ell)^{4g}2^{2\ell}$, the first term corresponding to the choice of $2g$ half-edges to force the genus to be g. This gives the first inequality. The second inequality can be obtained by Holder's inequality. □

Corollary 3.4.2. *Let $\lambda_1(N)$ and $\lambda_N(N)$ be the smallest and largest eigenvalues of a GUE matrix $A^{(N)}$. Then almost surely $\lambda_1(N)\to -2$ and $\lambda_N(N)\to 2$.*

Proof. From Corollary 3.4.1 for any even number d we get the estimate:

$$\left|\mathbb{E}\left[\frac{1}{N}\mathrm{Tr}([A^{(N)}]^d)\right]\right| \leqslant 2^d\sum_g \left(\frac{d^2}{N}\right)^{2g}.$$

In particular, if $d^2/N<1$ then we get

$$\left|\mathbb{E}\left[\frac{1}{N}\mathrm{Tr}([A^{(N)}]^d)\right]\right| \leqslant \frac{2^d}{1-(d^2/N)^2}.$$

Now choose a sequence of even numbers $d(N)$. As long as $(d(N))^2/N < 1$,

$$\left|\mathbb{E}\left[\frac{1}{N}\mathrm{Tr}([A^{(N)}]^{d(N)})\right]\right| \leqslant \frac{2^{d(N)}}{1-(d^2/N)^2}.$$

Thus the probability that the largest eigenvalue λ_N exceeds $2+\delta$ or the lowest eigenvalue is less than $-(2+\delta)$ can be estimated by:

$$\begin{aligned}
\mathbb{P}(\lambda_N > 2) &\leqslant \mathbb{P}(\mathrm{Tr}([A^{(N)}]^{d(N)}) > (2+\delta)^{d(N)}) \\
&\leqslant \frac{\mathbb{E}(\mathrm{Tr}([A^{(N)}]^{d(N)}))}{(2+\delta)^{d(N)}} \\
&\leqslant N\left(\frac{2}{2+\delta}\right)^{d(N)} \frac{1}{1-(d^2(N)/N)^2} \\
&\leqslant CN\left(\frac{2}{2+\delta}\right)^{d(N)}.
\end{aligned}$$

Now choose $d(N)$ so that $d(N)^2/N \to 0$ but $d(N)/\log N \to \infty$. Taking logarithms gives us

$$\log N\left(\frac{2}{2+\delta}\right)^{d(N)} = d(N)\log\frac{2}{2+\delta} + \log N \to -\infty$$

since $\log(2/(2+\delta))$ is negative and $d(N)/\log N \to \infty$.

A little more work (Chebyshev inequality) actually gives an almost sure result that no eigenvalue above 2 or below -2 exist. □

3.5. GUE matrices and Stein's method: proof of Theorem 3.3.5(1′). In this section we develop a more analytical approach to asymptotic freeness which is inspired by Stein's method in classical probability: the key idea is to show that $*$-moments of GUE matrices satisfy equations, known as Schwinger-Dyson or loop equations, that describe uniquely their limits. We start by recalling basics of Stein's method and then generalize the ideas to random matrix theory. We will see further that this approach generalizes to dependent random matrices.

3.5.1. Stein's method in classical probability Stein's method is useful for proving weak convergence of a sequence of probability measures to a target measure μ. The idea is that given a probability measure P, one finds an operator $\mathcal{L}$ acting on a dense set $\mathcal{F}$ of test functions (such as C^1 or C^2 functions) such that

$$\nu(\mathcal{L}f) = 0 \quad \forall f \in \mathcal{F} \quad \Leftrightarrow \quad \nu = \mu. \tag{3.5.2}$$

An important example is where μ is the standard Gaussian law, $\mathcal{F} = C_b^1(\mathbb{R})$ and $\mathcal{L}f(x) = xf(x) - f'(x)$. There are several ways to see that this example satisfies (3.5.2). The first is to observe that

$$xf(x) - f'(x) = h(x) - \int h(y)\,d\mu(y)$$

where $f(x) = e^{x^2/2}\int_{-\infty}^{x}[h(u) - \int h(y)dP(y)]e^{-u^2/2}du$.

Another approach is to take $\mathcal{F}$ to be the set of polynomials, to observe that (3.5.2) defines uniquely the moments under ν, and that the measure μ is uniquely defined by its moments as they do not grow too fast. This latter approach will generalize to the non-commutative set-up.

One can then prove weak convergence of a sequence $(\nu_n)_{n\geqslant 0}$ of probability measures towards μ by showing that it is tight and moreover that for all $f \in \mathcal{F}$

$$\lim_{n\to\infty} \nu_n(\mathcal{L}f) = 0.$$

If the topology for which tightness is proved allows one to deduce that any limit point ν satisfies (3.5.2), then any such any limit point ν must equal μ thus giving the desired conclusion that $\nu_n \to \mu$ weakly.

3.5.3. Stein's method in free probability To describe a free probability analog of Stein's method, we start by introducing the natural derivatives in functions of several non-commutative variables $X_1, \ldots, X_k$. It will be convenient to consider also an algebra D of "constants", which will be in the kernel of the derivations. We thus assume that D is a unital algebra and let $D\langle X_1, \ldots, X_n\rangle$ be the free algebra generated by D and n indeterminates $X_1, \ldots, X_n$ (in other words, this algebra is the free product with amalgamation over the unit of D and the algebra of non-commutative polynomials in $X_1, \ldots, X_n$; it is linearly spanned by words of the form $d_0 X_{i_1} d_1 \cdots X_{i_k} d_k$ with $d_j \in D$ and $i_1, \ldots, i_k \in \{1, \ldots, n\}$).

Definition 3.5.4. [33] The free difference quotient ∂_{X_i} (sometimes abbreviated ∂_i) with respect to the variable X_i is the unique derivation

$$\partial_{X_i} : D\langle X_1, \ldots, X_k\rangle \to D\langle X_1, \ldots, X_k\rangle \otimes D\langle X_1, \ldots, X_k\rangle$$

given by

$$\partial_{X_i}(d_0 X_{j_1} d_1 X_{j_2} \cdots X_{j_m} d_m) = \sum_{j_\ell = i} d_0 X_{j_1} \cdots X_{j_{\ell-1}} d_\ell \otimes d_{\ell+1} X_{j_{\ell+1}} \cdots X_{j_m} d_m$$

for any $j_1, \ldots, j_m \in \{1, \ldots, k\}$ and $d_0, \ldots, d_m \in D$. In other words ∂_{X_i} is the unique linear map which satisfies the Leibniz rule

$$\partial_i PQ = \partial_i P\, Q + P\, \partial_i Q \tag{3.5.5}$$

and for which $\partial_i X_j = \mathbf{1}_{i=j} 1 \otimes 1$, $\partial_i d = 0 \otimes 0 \quad \forall d \in D$. (Here we use the notation $P\,(a \otimes b) = (Pa) \otimes b$, $(a \otimes b)\, Q = a \otimes (bQ)$).

The analog of Stein's equation is the following. Let τ_D be a fixed trace on D. We say that τ satisfies the free Stein's equation relative to (D, τ_D) iff for all $q \in D\langle X_i : 1 \leqslant i \leqslant m\rangle$, we have

$$\tau(X_i q) = \tau \otimes \tau(\partial_i q), \tag{3.5.6}$$

and

$$\tau|_D = \tau_D, \ \tau(PQ) = \tau(QP), \tag{3.5.7}$$

where $P, Q \in D\langle X_i : 1 \leqslant i \leqslant m\rangle$.

Lemma 3.5.8. [33] *There exists a unique solution τ to (3.5.6) and (3.5.7). This is the law of m free semicircular variables, free from* D.

Proof. Uniqueness of the solution is clear. If $p \in D\langle X_1, \ldots, X_m \rangle$ is a monomial of degree zero in $X_1, \ldots, X_m$, then $p \in D$ and (3.5.7) determines $\tau(p)$. Next, if $p \in D\langle X_1, \ldots, X_m \rangle$ is a monomial of degree $d+1$ in the variables $X_1, \ldots, X_m$, then by the trace property, $\tau(p) = \tau(X_j q)$ for some j and a degree d monomial q. Then the equation

$$\tau(p) = \tau(X_j q) = \tau \otimes \tau(\partial_j q)$$

expresses the value of $\tau(p)$ in terms of products of values of τ on monomials of degree at most $d-1$ entering in the expression for $\partial_j q$. Thus the values of τ are completely determined.

It remains to prove that the law of a semicircular m-tuple free from D satisfies (3.5.6) and (3.5.7). The proof for (3.5.7) is easy, and we concentrate on (3.5.6).

The algebra $D\langle X_1, \ldots, X_m \rangle$ is linearly spanned by elements

$$W = d_0 Q_1 d_1 Q_2 \cdots d_{n-1} Q_n d_n$$

where $d_j \in D$, $Q_j \in \mathbb{C}\langle X_1, \ldots, X_m \rangle$ and $\tau(d_j) = \tau(Q_j) = 0$ for all j (except we may allow d_0 and/or d_n to be equal to 1. Note that $\tau(W) = 0$.

If $\tau(d_0) = 0$, then by freeness $\tau(X_j W) = 0$. If $d_0 = 1$, then, again by freeness,

$$\tau(X_j W) = \tau(X_j Q_1 d_1 \cdots Q_n d_n) = \tau(X_j Q_1)\tau(d_1 Q_1 \cdots Q_n d_n.$$

The latter term is zero unless $n = 1$ and $d_1 = 1$.

We now compute the right hand side of (3.5.6).

$$\partial_j W = \sum_{k=1}^{n} d_0 Q_1 \cdots d_{k-1} \partial_j(Q_k) d_k Q_{k+1} d_k \cdots Q_n d_n.$$

Note that $\partial_j(Q_k) = \sum_s A_s \otimes B_s$ where A_s, B_s are (possibly degree zero) monomials in $\mathbb{C}\langle X_1, \ldots, X_m \rangle$. From this it follows that if $k \neq 1$,

$$\tau \otimes \tau(d_0 Q_1 \cdots d_{k-1} \partial_j(Q_k) d_k Q_{k+1} d_k \cdots Q_n d_n) = 0.$$

We consider the remaining term corresponding to $k = 1$. If $\tau(d_0) = 0$ we get (writing again $\partial_j Q_1 = \sum_s A_s \otimes B_s$)

$$\sum_s \tau(d_0 A_s)\tau(B_s d_1 Q_2 \cdots d_n)$$

which is zero by freeness. If $d_0 = 1$ we get

$$\sum_s \tau(A_s)\tau(B_s d_1 Q_2 \cdots d_n).$$

This term is also zero by freeness unless $n = 1$ and $d_1 = 1$.

It follows that it remains to prove (3.5.6) in the case that $W = Q_1$ for some $Q_1 \in \mathbb{C}\langle X_1, \ldots, X_m \rangle$; in other words, we can reduce to the case that $D = \mathbb{C}$.

To see this, we imitate the proof in §1.5.5 (we use the notation of that section). We write $X_j = \ell(h_j) + \ell(h_j)^*$ where $h_1, \ldots, h_m$ are an orthonormal family of vectors and $\ell(\cdot)$ are free creation operators on a free Fock space. Then as in §1.5.5,

$$[X_i, r(h_j)] = \delta_{i=j} P$$

where P denotes the projection onto the vector Ω. It follows from the Leibnitz rule that if $Q \in \mathbb{C}\langle X_1, \dots, X_m \rangle$, and $\partial_j Q = \sum_s A_s \otimes B_s$, then

$$[Q, r(h_j)] = \sum_s A_s P B_s$$

and thus

$$\mathrm{Tr}(P[Q, r(h_j)]P) = \sum_s \mathrm{Tr}(PA_s PB_s).$$

Since $\tau(x) = \mathrm{Tr}(PxP)$ and $\tau(x)\tau(y) = \mathrm{Tr}(PxPyP)$ for $x, y \in \mathbb{C}\langle X_1, \dots, X_m \rangle$, it follows that

$$\tau \otimes \tau(\partial_j Q) = \sum_s \tau(A_s)\tau(B_s) = \sum_s \mathrm{Tr}(PA_s PB_s P) = \mathrm{Tr}(P[Q, r(h_j)]P).$$

On the other hand, $r(h_j)P = X_j P$ and $Pr(h_j) = 0$ giving us finally

$$\tau \otimes \tau(\partial_j Q) = \mathrm{Tr}(PQr(h_j)P) - \mathrm{Tr}(Pr(h_j)Q) = \mathrm{Tr}(PQX_j P) = \tau(X_j Q).$$

This concludes the proof. □

It is perhaps worth to add a small note of explanation for the end of the proof. If one identifies $\mathbb{C}\langle X_1, \dots, X_m \rangle \otimes \mathbb{C}\langle X_1, \dots, X_m \rangle$ with a set of finite-rank operators on $L^2(\mathbb{C}\langle X_1, \dots, X_m \rangle)$ via the map

$$x \otimes y \mapsto xPy,$$

then, as we have shown, the derivation ∂_j is simply given as a commutator:

$$\partial_j = [\cdot, r(h_j)].$$

The free difference quotient satisfies the following universal property. Let us write $(a \otimes b)\#c = acb$. Let $P \in D\langle X_1, \dots, X_d \rangle$. Consider, for $\epsilon \in \mathbb{R}$,

$$Q(X_1, \dots, X_d, Y) = P(X_1, \dots, X_{j-1}, X_j + \epsilon Y, X_{j+1}, \dots, X_d) \in D\langle X_1, \dots, X_d, Y \rangle.$$

Then

$$Q = P + \epsilon \partial_j P \# Y + O(\epsilon^2). \tag{3.5.9}$$

Let $A_k^{(N)}(i, j)$ be the entries of the random matrices $A_k^{(N)}$ as described in Theorem 3.3.5(1′). These random variables are functions on the space of d-tuples of $N \times N$ self-adjoint matrices endowed with the Gaussian measure.

As functions, $A_k^{(N)}(i, j)$ evaluated at some self-adjoint matrices $(\omega_1, \dots, \omega_d)$ is just the i, j-th entry $\omega_k(i, j)$ of ω_k. Under the Gaussian measure, we have the Stein identity

$$\mathbb{E}[\omega_k(j, i) f(\omega_1, \dots, \omega_d)] = \frac{1}{N} \mathbb{E}[\partial_{\omega_k(i,j)} f]$$

(here $\partial_{\omega_k(i,j)}$ refers to the usual partial derivative, not the difference quotient. The switch in i, j has to do with the fact that $\omega_k(i, j)$ is complex valued and $\omega_k(j, i) = \overline{\omega_k(i, j)}$).

Let $P \in D\langle X_1, \dots, X_d \rangle$ be a fixed polynomial; we assume that D is generated by $(b_1, \dots, b_k)$. For each N, we can substitute $A_1^{(N)}, \dots, A_d^{(N)}$ for $X_1, \dots, X_d$ and

$B_1^{(N)},\dots,B_k^{(N)}$ for $b_1,\dots,b_k$. Thus if we set

$$Z = P(A_1^{(N)},\dots,A_d^{(N)})$$

then Z is a certain function of the entries $(\omega_k(i,j))_{i,j,k}$; a simple computation involving (3.5.9) shows that

$$\partial_{\omega_k(i,j)} P(A_1^{(N)},\dots,A_d^{(N)}) = \partial_k P \# E_{ij}$$

where E_{ij} is the matrix with all entries zero, except that the i,j-th entry is 1.

Therefore, using the fact that $\sum_j E_{jj}$ is the identity matrix,

$$\begin{aligned}
&\frac{1}{N}\mathbb{E}_{\nu_N}[\mathrm{Tr}(A_k^{(N)}P(A_1^{(N)},\dots,A_d^{(N)})]\\
&\quad= \frac{1}{N}\mathbb{E}_{\nu_N}[\mathrm{Tr}(\sum_{i,j} E_{jj}A_k^{(N)}E_{ii}P(A_1^{(N)},\dots,A_d^{(N)}))]\\
&\quad= \frac{1}{N}\mathbb{E}_{\nu_N}[\mathrm{Tr}(\sum_{i,j} \omega_k(j,i)E_{ji}E_{ii}P(A_1^{(N)},\dots,A_d^{(N)})E_{jj})]\\
&\quad= \frac{1}{N^2}\mathbb{E}_{\nu_N}[\mathrm{Tr}(\sum_{i,j} \partial_{\omega_k(i,j)}E_{ji}E_{ii}P(A_1^{(N)},\dots,A_d^{(N)})E_{jj})]\\
&\quad= \frac{1}{N^2}\mathbb{E}_{\nu_N}[\mathrm{Tr}(\sum_{i,j} E_{ji}(\partial_k P\#E_{ij})E_{jj})]\\
&\quad= \mathbb{E}_{\nu_N}[\frac{1}{N}\mathrm{Tr}\otimes\frac{1}{N}\mathrm{Tr}(\partial_k P)].
\end{aligned}$$

One can now use Gaussian concentration to conclude that

$$\mathbb{E}_{\nu_N}[\frac{1}{N}\mathrm{Tr}\otimes\frac{1}{N}\mathrm{Tr}(\partial_k P)] \approx (\mathbb{E}_{\nu_N}\circ\frac{1}{N}\mathrm{Tr})\otimes(E_{\nu_N}\circ\frac{1}{N}\mathrm{Tr})(\partial_k P)$$

so that in the limit the hypothesis of Lemma 3.5.8 is satisfied.

3.6. On the proof of Theorem 3.3.5(2) and (3). In the case of part (2) one integrates with respect to a measure different from the Gaussian measure. The relevant formula replacing Stein's equation is

$$\frac{1}{N}\mathbb{E}_{\nu_{N,R}^{(V)}}[\omega_k(i,j)f] = \mathbb{E}_{\nu_{N,R}^{(V)}}[\partial_{\omega_k(i,j)} f\mathrm{Tr}(V)].$$

Let $\mathcal{D}_j$ be the linear operator defined on $\mathbb{C}\langle X_1,\dots,X_d\rangle$ by

$$\mathcal{D}_j X_{i_1}\cdots X_{i_r} = \sum_{q:i_q=j}(X_{i_{q+1}}\cdots X_{i_r})\cdot(X_{i_1}\cdots X_{i_{q-1}}).$$

If one denotes by m the map $m(a\otimes b)=ab$ and by σ the flip $\sigma(a\otimes b)=b\otimes a$, then $\mathcal{D}_j = m\circ\sigma\circ\partial_j$. This operator is called the j-th *cyclic derivative*. It is not hard to show, along the lines of (3.5.9), that for any trace τ,

$$\tau(P(X_1,\dots,X_{j-1},X_j+\epsilon Y,X_{j+1},\dots,X_d) = \tau(P(X_1,\dots,X_n)) + \tau(\mathcal{D}_j P\cdot Y)\epsilon + O(\epsilon^2).$$

Arguing as in the Gaussian case, we obtain the identity

$$\begin{aligned}
&\frac{1}{N}\mathbb{E}_{\nu_{N,R}^{(V)}}[\mathrm{Tr}((\mathcal{D}_j V)(A_1^{(N)},\dots,A_d^{(N)})\cdot P(A_1^{(N)},\dots,A_d^{(N)})]\\
&\quad= \mathbb{E}_{\nu_{N,R}^{(V)}}[\frac{1}{N}\mathrm{Tr}\otimes\frac{1}{N}\mathrm{Tr}(\partial_j P)].
\end{aligned}$$

Together with concentration (which can be proved under the assumptions that $\mathrm{Tr}(V)$ is a strictly convex function on the space of d-tuples of self-adjoint matrices of bounded operator norm; this convexity is ensured by making a suitable choice of ϵ_0), this implies that the limit state τ satisfies

$$\tau(\mathcal{D}_j V \cdot P) = \tau \otimes \tau(\partial_j P) \tag{3.6.1}$$

for all P. This equation is often called the *Schwinger-Dyson equation.* A supplementary argument (requiring a choice of ϵ_0) is required to show that under the assumptions of Theorem 3.3.5(2) and boundedness of our matrices, this completely characterizes τ.

Part (3) follows because of unitary invariance: for any unitary U,

$$(A_1^{(N)},\ldots,A_r^{(N)},A_{r+1}^{(N)},\ldots,A_d^{(N)}) \sim (A_1^{(N)},\ldots,A_r^{(N)},UA_{r+1}^{(N)}U^*,\ldots,UA_d^{(N)}U^*).$$

3.7. Exercises.

Exercise 3.7.1. Fix $0 < a, b < 1$, and let P_N and Q_N be $N \times N$ complex projection matrices of rank $\lfloor aN \rfloor$ and $\lfloor bN \rfloor$, respectively, chosen at random with respect to the Haar measure on the appropriate Grassmanian. (Here $\lfloor x \rfloor$ denotes the largest integer not exceeding x.) Show that $\mu_{P+Q} \to \nu_a \boxplus \nu_b$ and $\mu_{PQ} \to \nu_a \boxtimes \nu_b$, where $\nu_c = c\delta_1 + (1-c)\delta_0$.

Exercise 3.7.2. Let X_N be an $N \times M$ random matrix with iid complex Gaussian entries of variance $1/N$. Assume that $N/M \to \lambda \in (0, +\infty)$. Let $Y_N = X^*X$. The nonzero eigenvalues of Y_N are squares of the singular values of X_N. Show that $\mu_{Y_N} \to \mu$, the Marčenko-Pastur (a.k.a. free Poisson) law of Exercise 2.10.6. This is the celebrated Marčenko-Pastur theorem.

Exercise 3.7.3. Prove the analog of Theorem 3.3.5(1) for matrices with *real* Gaussian entries. A significant difference is that, unlike in the complex case, the analog of (3.3.9) will have terms of order $1/N$.

Exercise 3.7.4. Use the previous exercise to prove an analog of Theorem 3.3.2 for real orthogonal matrices.

Exercise 3.7.5. Let $\Sigma_1^{(N)},\ldots,\Sigma_k^{(N)}$ be random permutation matrices. Show that

$$\mu_{\Sigma_1^{(N)},\ldots,\Sigma_k^{(N)},[\Sigma_1^{(N)}]^*,\ldots[\Sigma_k^{(N)}]^*} \to \mu_{u_1,\ldots,u_k,u_1^*,\ldots,u_k^*}$$

where $u_1,\ldots,u_k$ are unitaries which are freely independent and satisfy $\tau(u_j^p) = 0$ for all $p \neq 0$. Show that the analog of Theorem 3.3.2 for random permutation matrices does *not* hold (*hint*: the conjugation of a diagonal matrix by a permutation matrix stays diagonal).

Exercise 3.7.6. Suppose that the law of an n-tuple $X_1,\ldots,X_n \in (M, \tau_\epsilon)$ satisfies the Schwinger-Dyson equation (3.6.1) where

$$V(X_1,\ldots,X_n) = \frac{1}{2}\sum X_j^2 + \epsilon W(X_1,\ldots,X_n)$$

with W a fixed non-commutative polynomial. Prove that τ_ϵ admits a formal expansion

$$\tau_\epsilon = \sum \tau^{(k)} \epsilon^k$$

with $\tau^{(0)}$ the semicircle law and $\tau^{(k)}$ determined recursively. *Hint:* equation (3.6.1) becomes

$$\tau_\epsilon((X_j + \epsilon W_j)P) = \tau_\epsilon \otimes \tau_\epsilon(\partial_j P)$$

which can be rewritten as

$$\tau_\epsilon(X_j P) = \tau_\epsilon \otimes \tau_\epsilon(\partial_j P) - \epsilon\tau_\epsilon(W_j P).$$

Conclude that $\tau_k(X_j P)$ can be evaluated in terms of τ_ϵ evaluated on polynomials of smaller degree as well as $\tau^{(j)}$ for $j < k$ evaluated on higher degree polynomials.

Using this idea one can argue that τ_ϵ is an analytic function of ϵ (see [14] for more detail on this).

Lecture 4: Free Entropy Theory and Applications

4.1. Executive summary. Difference quotient derivations, which made their appearance in the previous lectures, are fundamental objects in free probability theory. We show how they can be used to define an entropy theory and also to prove subordination and regularity results for free convolution.

4.2. More on free difference quotient derivations. Let $A = \mathbb{C}\langle t_1, \ldots, t_n\rangle$ be the algebra of non-commutative polynomials in n indeterminates $t_1 \ldots, t_n$. We have already encountered the difference quotient derivations $\partial_j : A \to A \otimes A$.

In the case $n = 1$, this definition is equivalent to taking a difference quotient (see Exercise 4.8.1).

Let $X_1, \ldots, X_n \in (M, \tau)$ be non-commutative random variables. Then we have a canonical evaluation homomorphism $ev : A \to M$ given by $p \mapsto p(X_1, \ldots, X_n)$.

Definition 4.2.1. We say that $X_1, \ldots, X_n$ are *freely differentiable* if there exist closed (possibly unbounded densely defined) operators $\bar{\partial}_i : D_i \to L^2(M, \tau)\bar{\otimes}L^2(M, \tau)$ so that $ev(A) \subset D_i$ and $\bar{\partial}_i(ev(p)) = (ev \otimes ev)(\bar{\partial}_i(p))$.

Note that the differentiability assumption implies that ∂_i gives rise to a linear operator densely defined on the image of ev inside $L^2(M, \tau)$, and that this operator is closable (this explains the notation $\bar{\partial}_i$.)

In the case $n = 1$ we may assume that $M = L^\infty(\mathbb{R}, \mu_X)$ is abelian and $X = X_1$ is just multiplication by x. In that case the differentiability requirement is exactly that the map

$$f \mapsto \frac{f(s) - f(t)}{s - t} \in L^2(\mu_X \times \mu_X)$$

defines a closable operator. This is the case iff μ_X is non-atomic (see Exercise 4.8.2).

Recall that a densely defined operator $T : D \to H$ is closable if and only if its adjoint T^* is densely defined; the domain of T^* consists of all $h \in H$ for which $|\langle T(g), h\rangle| \leqslant C\|g\|$, and for such an h, T^*h is the unique Riesz vector representing the (bounded) functional $g \mapsto \langle T(g), h\rangle$, i.e., T^*h is the unique vector u so that $\langle g, u\rangle = \langle T(g), h\rangle$ for all $g \in D$.

Definition 4.2.2. [33] We say that $\xi_j \in L^2(W^*(X_1, \dots, X_n), \tau)$, $j = 1, \dots, n$ are *conjugate variables* to $X_1, \dots, X_n$ if for all polynomials $p \in A$,

$$\langle \xi_j, p(X_1, \dots, X_n)\rangle = \langle 1 \otimes 1, (ev \otimes ev)\partial_j(p)\rangle, \qquad j = 1, \dots, n.$$

We write $\xi_j = J(X_j : X_1, \dots, \hat{X}_j, \dots, X_n)$. If $n = 1$ we simply write $\xi = J(X)$.

Since the Hilbert space inner product between a conjugate variable ξ_j and a weakly dense subset of the Hilbert space $L^2(W^*(X_1, \dots, X_n), \tau)$ is determined, conjugate variables are unique, if they exist.

The conjugate variable has a classical interpretation (see Exercise 4.8.3) which we will set aside for now. As we have seen in Lemma 3.5.8, free semicircular n-tuples are characterized by $\xi_j = X_j$.

Existence of conjugate variables is useful as a condition that implies free differentiability:

Lemma 4.2.3. [33] *Suppose that the conjugate variable* $\xi_i = J(X_i : X_1, \dots, \hat{X}_i, \dots, X_n)$ *exist for all* i. *Then* $X_1, \dots, X_n$ *are freely differentiable.*

Proof. One can verify by a direct computation that for any polynomials $p, q, r \in A$, the element

$$\zeta = ev(p)\xi_i ev(q) - (ev \otimes \tau \circ ev)(\partial_i p)q - (\tau \circ ev \otimes ev)(\partial_i q)$$

satisfies

$$\langle \zeta, ev(r)\rangle = \langle ev(p) \otimes ev(q), (ev \otimes ev)\partial_i(r)\rangle.$$

This implies that ∂_j is well-defined as map on $ev(A)$ and that ∂_j^* is densely defined. The closure $\bar{\partial}_i$ is then the desired derivation, showing that $X_1, \dots, X_n$ are freely differentiable. □

Lemma 4.2.4. [33] *Suppose that* $J(X : Y_1, \dots, Y_n)$ *exists. Then:*

(1) *For any* $m < n$, $J(X : Y_1, \dots, Y_m)$ *also exists, and*

$$J(X : Y_1, \dots, Y_m) = E_{W^*(X, Y_1, \dots, Y_m)}(J(X : Y_1, \dots, Y_n)).$$

(2) *If* Z *is freely independent from* $(X, Y_1, \dots, Y_m)$, *then*

$$J(X : Y_1, \dots, Y_m, Z) = J(X : Y_1, \dots, Y_m).$$

(3) $J(X + Y_1 : Y_1, \dots, Y_m) = J(X : Y_1, \dots, Y_m)$. *In particular, if* $J(Z)$ *exists and* Z *is free from* $X, Y_1, \dots, Y_m$, *then* $J(X + Z : Y_1, \dots, Y_m)$ *exists and equals* $E_{W^*(X+Z, Y_1, \dots, Y_m)}(J(Z))$.

(4) $J(\lambda X : Y_1, \dots, Y_n) = \lambda^{-1} J(X : Y_1, \dots, Y_n)$ *for all* $\lambda \neq 0$.

Proof. The first and next to last statements follow from the fact that ∂_X restricted to the algebra generated by $X, Y_1, \ldots, Y_m$ (resp., restricted to the algebra generated by $X+Z, Y_1, \ldots, Y_m$) coincides with ∂_X (resp., ∂_{X+Z}). The proof of the second statement is identical to the proof of the second part of Lemma 3.5.8. The last statement follows from the identity $\partial_X(\lambda X) = \lambda 1 \otimes 1$ so $\partial_{\lambda X} = \lambda^{-1}\partial_X$. □

Corollary 4.2.5. [33] *Let $X_1, \ldots, X_n$ be arbitrary, and let $S_1, \ldots, S_n$ be a free semicircular system, free from $X_1, \ldots, X_n$ and of variance 1. Let $Z_j = X_j + \epsilon S_j$. Then*

$$J(Z_j : Z_1, \ldots, \hat{Z}_j, \ldots, Z_n) = \frac{1}{\epsilon} E_{W^*(Z_1,\ldots,Z_n)}(S_j).$$

Proof. This follows from the facts that S_j are free from $(X_1, \ldots, X_n)$ and that $J(\epsilon S_j) = \epsilon^{-1} J(S_j) = \epsilon^{-1} S_j$. □

A corollary is that the free semicircular perturbation $(X_1 + \epsilon S_1, \ldots, X_n + \epsilon S_n)$ is always differentiable for any $\epsilon > 0$.

Proposition 4.2.6. *Suppose that $X_1, \ldots, X_n$ are limits of a random matrix model and satisfy the Schwinger-Dyson equation (3.6.1). Then $J(X_i : X_1, \ldots, \hat{X}_i, \ldots, X_n)$ exists and equals $\mathcal{D}_i V$.*

Proof. Equation (3.6.1) immediately implies this. □

4.3. Free difference quotients and free subordination. In this section we prove a version of the free subordination theorem (Theorem 2.6.1) using the approach from [35]. The general case of the subordination theorem can be obtained from this statement by regularizing X (e.g., replacing X by $X + tS$ with S a free semicircular) and passing to limits. We start with the following Lemma, which characterizes resolvents as functions having a particular behavior under difference quotients:

Lemma 4.3.1. [35] *Let X be a differentiable variable. Then*

$$\bar{\partial}_X((z-X)^{-1}) = (z-X)^{-1} \otimes (z-X)^{-1}.$$

Conversely, if for some $f \in L^2(W^*(X))$, $\bar{\partial}_X f = f \otimes f$, then $f = (z-X)^{-1}$ for some $z \in \mathbb{C} \setminus \sigma(X)$.

Proof. The closure of ∂_X is given by $g \mapsto (g(s) - g(t))/(s-t)$. Thus

$$\bar{\partial}_X((z-X)^{-1}) = \frac{(z-s)^{-1} - (z-t)^{-1}}{s-t} = (z-s)^{-1}(z-t)^{-1}.$$

Conversely, suppose that for some f, $\bar{\partial}_X f = f \otimes f$. Thus

$$f(s)f(t) = (s-t)^{-1}(f(s) - f(t))$$

so that

$$f(s) - f(t) = (s-t)f(s)f(t)$$

and thus f is a continuous function on the spectrum $\sigma(X)$ (equal to the support of μ_X).

Thus for some t_0,

$$f(s)f(t_0) = (s-t_0)^{-1}(f(s)-f(t_0))$$

for μ_X-a.e. $s \in \sigma(X)$. By continuity, this is true for all s and thus

$$f(s) = f(t_0)(1-f(t_0)(s-t_0))^{-1} = ((f(t_0)^{-1}+t_0)-s)^{-1} = (\lambda - s)^{-1}$$

for $\lambda = f(t_0)^{-1} + t_0$. By continuity of f, $\lambda \in \mathbb{C} \setminus \sigma(X)$. □

Theorem 4.3.2. [35] *Suppose that X and Y are self-adjoint, freely independent and differentiable (i.e., each has non-atomic law; for example, we may assume that $J(X)$ exists). Then there exists an analytic function $\omega : \mathbb{C}_+ \to \mathbb{C}_+$ so that*

$$E_{W^*(X)}((z-(X+Y))^{-1}) = (\omega(z)-X)^{-1}.$$

Here $E_{W^*(X)} : W^*(X,Y) \to W^*(X)$ is the conditional expectation satisfying

$$\tau \circ E_{W^*(X)} = \tau.$$

In particular,

$$\tau((z-(X+Y))^{-1}) = \tau((\omega(z)-X)^{-1}).$$

Proof. It is clear that $E_{W^*(X)}((z-(X+Y))^{-1})$ depends analytically on z. Thus if we can show that $E_{W^*(X)}((z-(X+Y))^{-1}) = (\eta - X)^{-1}$ for some $\eta \in \mathbb{C}$, it follows that $\eta = \omega(z)$ depends on z analytically. Furthermore, since $X+Y$ and X are self-adjoint, it follows that if $z \in \mathbb{C}_+$ then also $\eta \in \mathbb{C}_+$.

It is easy to check that

$$E_{W^*(X)\bar{\otimes}W^*(X)} \circ \bar{\partial}_X = \bar{\partial}_X \circ E_{W^*(X)}.$$

Restricting this identity to the algebra generated by $X+Y$ and noting that on this algebra ∂_{X+Y} coincides with ∂_X allows us to conclude that

$$E_{W^*(X)\bar{\otimes}W^*(X)} \circ \bar{\partial}_{X+Y} = \bar{\partial}_X \circ E_{W^*(X)}$$

on the algebra generated by $X+Y$. Next, we combine this with the identity

$$\bar{\partial}_{X+Y}[(z-(X+Y)]^{-1} = (z-(X+Y))^{-1} \otimes (z-(X+Y))^{-1}$$

from Lemma 4.3.1 to conclude that, with $f_z = E_{W^*(X)}[(z-(X+Y)]^{-1}$,

$$\bar{\partial}_X f_z = f_z \otimes f_z.$$

Then Lemma 4.3.1 implies that $f_z = (\eta - X)^{-1}$ for some $\eta = \eta(z) \in \mathbb{C}_+ \setminus \sigma(X)$. □

4.4. Free Fisher information and non-microstates free entropy.

Definition 4.4.1. [33] Let $X_1, \dots, X_n$ be non-commutative random variables. The *free Fisher information* is defined by

$$\Phi^*(X_1,\dots,X_n) = \sum_j \|J(X_j : X_1,\dots,\hat{X}_j,\dots,X_n)\|_2^2.$$

The free Fisher information is the analog of the classical Fisher information, given (see Exercise 4.8.3), for a random variable with density ρ, by

$$F = \int_{\mathbb{R}} \frac{(\rho'(t))^2}{\rho(t)} dt.$$

If $X_1, \ldots, X_n$ are freely independent, $\Phi^*(X_1, \ldots, X_n) = \sum_j \Phi^*(X_j)$. For a single variable X with law ρdt, $\Phi^*(X) = \int f^2 d\mu$ where f is the Hilbert transform of ρ:

$$f(t) = \text{P.V.} \int \frac{\rho(s)}{s-t} dt.$$

Classical Fisher information has the property that it is a derivative of entropy with respect to Gaussian regularization: if $X \sim \rho dt$ and B is a Gaussian variable independent of X, then

$$F(X) = \frac{1}{2} \frac{d}{dt}\Big|_{t=0} H(X + t^{1/2} B)$$

where H stands for entropy: if $Z \sim \eta(t)dt$, $H(Z) = \int \eta(t) \log \eta(t) dt$. Motivated by this, Voiculescu gave the following definition of free entropy:

Definition 4.4.2. [33] The non-microstates free entropy χ^* is given by the formula

$$\chi^*(x_1, \ldots, x_n) = \frac{1}{2} \int_0^\infty \left[\frac{n}{1+t} - \Phi^*(x_1 + t^{1/2} S_1, \ldots, x_n + t^{1/2} S_n) \right] dt + \frac{n}{2} \log(2\pi e)$$

where $S_1, \ldots, S_n$ is a semicircular n-tuple free from $x_1, \ldots, x_n$.

The entropy χ^* inherits a number of properties from the Fisher information. For example, if $X_1, \ldots, X_n$ are freely independent, $\chi^*(X_1, \ldots, X_n) = \sum \chi(X_j)$. For a single variable X with law μ, $\chi^* = \iint \log|s-t| d\mu(s) d\mu(t) + \frac{1}{2} \log(2\pi e)$.

4.5. Free entropy and large deviations: microstates free entropy χ. Voiculescu gave an alternative definition of free entropy, which is closely connected to large deviations (though the explanation of this connection is beyond the scope of these notes [3]).

Let $x_1, \ldots, x_n$ be self-adjoint non-commutative random variables in (M, τ). Fix integers N, m and real numbers $R, \epsilon > 0$. Then define the set

$$\Gamma_R(x_1, \ldots, x_n, N; m, \epsilon) = \Big\{ (X_1, \ldots, X_n) \in M_{N \times N} : X_j = X_j^*, \|X_j\| < R$$

$$\text{such that for any monomial } p \text{ with } \deg p \leqslant m,$$

$$\left| \frac{1}{N} \mathrm{Tr}(p(X_1, \ldots, X_n)) - \tau(p(x_1, \ldots, x_n)) \right| < \epsilon \Big\}.$$

Then the free entropy is defined by

$$\chi(x_1, \ldots, x_n) = \sup_R \inf_{m,\epsilon} \limsup_{N \to \infty} \{ \frac{1}{N^2} \log \mathrm{Vol}(\Gamma_R(x_1, \ldots, x_n, N; m, \epsilon) + \frac{n}{2} \log N \}.$$

The volume is computed with respect to Lebesgue measure on self-adjoint matrices corresponding to the inner product $\langle A, B \rangle = \mathrm{Tr}(AB)$.

Theorem 4.5.1. [30, 31, 33] *For a single variable, $\chi(x) = \chi^*(x)$. If $x_1, \ldots, x_n$ are free, then $\chi(x_1, \ldots, x_n) = \sum \chi(x_j)$.*

We only sketch the proof. Let $X \in \Gamma_R(x, N; m, \epsilon)$. Choose a unitary U so that $X = UDU^*$, where D is a diagonal matrix with eigenvalues $\lambda_1 > \lambda_2 > \cdots$. The condition that X belongs to Γ_R is the condition that $\frac{1}{N} \sum \lambda_j^r \approx \tau(x^r)$ for all

$0 \leqslant r \leqslant m$. In other words, if we write $\mu_X = \frac{1}{N}\sum \delta_{\lambda_j}$, then $\mu_X \approx \mu_x$ in the weak topology.

The map $(U, D = \mathrm{diag}(\lambda_1, \lambda_2, \dots)) \mapsto UDU^*$ is a measure-space isomorphism between the product of the unitary group and ordered n-tuples of real numbers $U(N) \times \mathbb{R}^N_>$ and self-adjoint $N \times N$ matrices. Via this isomorphism, Lebesgue measure on self-adjoint matrices corresponds to the measure

$$\begin{aligned}
&(\text{Haar measure}) \times C_N \prod_{i \neq j} |\lambda_i - \lambda_j| d\lambda_1 \cdots d\lambda_N \\
&\quad = (\text{Haar measure}) \times C_N \exp\Big(\sum_{i \neq j} \log|\lambda_i - \lambda_j|\Big) d\lambda_1 \cdots d\lambda_N \\
&\quad = (\text{Haar measure}) \times C_N \exp\Big(\iint_{s \neq t} \log|s-t| d\mu_X(s) d\mu_X(t)\Big) d\lambda_1 \cdots d\lambda_N,
\end{aligned}$$

for a certain constant C_N. Ignoring technicalities associated with the singularity of the log (which are not important since the set on which λ_i are close together has small measure), it follows that the restriction of Lebesgue measure to $\Gamma_R(X, N)$ can be identified with the product of the Haar measure and a constant multiple of $\exp(\iint \log|s-t| d\mu_x(s) d\mu_x(t))$. From this identification one deduces the formula $\chi(x) = \iint \log|s-t| d\mu(s) d\mu(t) + \log 2\pi e$. This proves the first part.

If $x_1, \dots, x_n$ are free, then one uses concentration and Theorem 3.3.2 to prove that for arbitrary diagonal matrices $D_1, \dots, D_n$ which satisfy $\mu_{D_j} \approx \mu_{x_j}$ and "most" unitary matrices $U_1, \dots, U_n$,

$$\mu_{U_1 D_1 U_1^*, \dots, U_n D_n U_n^*} \approx \mu_{x_1, \dots, x_n}$$

(this step uses crucially freeness of $x_1, \dots, x_n$). Together with the proof of the first part (which shows that $\Gamma(x_j)$ is approximately the unitary orbit of diagonal matrices), it follows that the volumes of $\Gamma(x_1, \dots, x_n)$ and $\Gamma(x_1) \times \cdots \times \Gamma(x_n)$ are comparable.

Theorem 4.5.2 (Change of variables), cf. [31]). *Let $x_1, \dots, x_n, y_1, \dots, y_n \in (M, \tau)$ be such that for certain convergent non-commutative power series P_i, Q_i with*

$$P_i(Q_1(t_1, \dots, t_n), \dots, Q_n(t_1, \dots, t_n)) = t_i = Q_i(P_1(t_1, \dots, t_n), \dots, P_n(t_1, \dots, t_n))$$

one has

$$x_j = Q_j(y_1, \dots, y_n), \qquad y_j = P_j(x_1, \dots, x_n).$$

Then

$$\chi(y_1, \dots, y_n) = \chi(x_1, \dots, x_n) + \mathrm{Tr} \otimes (\tau \otimes \tau) \left(\log \left|[\partial_j P_i]_{ij}\right|\right).$$

The proof of this theorem relies on (3.5.9) to show that the linear operator $([\partial_j P_i]\#)_{ij}$ is the derivative of the non-linear transformation

$$(X_1, \dots, X_n) \mapsto (P_1(X_1, \dots, X_n), \dots, P_n(X_1, \dots, X_n))$$

from n-tuples of self-adjoint $N \times N$ matrices to themselves. The logarithm of its Jacobian is then given approximately by the formula $\mathrm{Tr} \otimes (\tau \otimes \tau)(\log|[\partial_j P_i]_{ij}|)$.

4.6. χ vs χ^*. The equality between χ and χ^* is not known. If it were to hold, it would resolve in the affirmative the famous Connes Approximate Embedding (CAE) Question (see [22] for more details about this conjecture and some of its equivalent formulations):

Conjecture 4.6.1 (Connes, [10]). *For $(x_1, \ldots, x_n) \in (M, \tau)$ is it true that for all $\epsilon > 0$, $m \in \mathbb{N}$ and $R > \max \|x_j\|$, there exists some N so that the set $\Gamma_R(x_1, \ldots, x_n, N; m, \epsilon)$ is non-empty?*

Indeed, given $x_1, \ldots, x_n$, let $s_1, \ldots, s_n$ be a semicircular n-tuple which is free from $(x_1, \ldots, x_n)$. Since $\chi^*(x_1 + ts_1, \ldots, x_n + ts_n)$ is finite for any $t > 0$, the equality $\chi = \chi^*$ would imply that the sets $\Gamma_R(x_1 + ts_1, \ldots, x_n + ts_n, N; m, \epsilon)$ are asymptotically non-empty. A diagonal argument then implies that $x_1, \ldots, x_n$ satisfy CAE.

A fundamental result of Biane, Capitaine and Guionnet states that the inequality $\chi \leqslant \chi^*$ always holds:

Theorem 4.6.2. [7] $\chi \leqslant \chi^*$.

We give an indication of the idea of the proof. Let $x_i^t = (1 - t^{1/2})x_i + t^{1/2}s_i$, where $s_1, \ldots, s_n$ is a semicircular n-tuple free from $(x_1, \ldots, x_n)$. Denote by γ_N^t the restriction of Lebesgue measure to $\Gamma_R(x_1^t, \ldots, x_n^t, N; m, \epsilon)$ and normalized to be a probability measure. It turns out that for all $t \in (0, 1]$,

$$\chi(x_1^t, \ldots, x_n^t) = \inf_{m,\epsilon} \limsup_{N \to \infty} \frac{1}{N^2} H(\gamma_N^t), \tag{4.6.3}$$

where H stands for classical entropy ($H(\rho dt) = \int \rho \log \rho dt$). Moreover,

$$\chi(x_1^t, \ldots, x_n^t) = \chi^*(x_1^t, \ldots, x_n^t)$$

when $t = 1$ since in that case $x_j^t = s_j$, which are free, so

$$\chi(s_1, \ldots, s_n) = \sum \chi(s_j) = \sum \chi^*(s_j) = \chi^*(s_1, \ldots, s_n).$$

It is therefore sufficient to show that the derivatives in t of $\chi(x_1^t, \ldots, x_n^t)$ and $\chi^*(x_1^t, \ldots, x_n^t)$ are related by an inequality. Because of (4.6.3), this amounts to comparing classical Fisher information of γ_N^t and free Fisher information of $x_1^t, \ldots, x_n^t$.

Both of these quantities have expressions as conditional expectations of some variable onto a certain von Neumann algebra.

Because sums of semicircular variables are semicircular variables (of a different variance), we find that $x_j + (t + \delta)^{1/2}s_j \sim (x_j + t^{1/2}s_j) + \delta^{1/2}s_j'$, where $s_1', \ldots, s_j'$ are semicircular variables free from $(x_1, \ldots, x_n, s_1, \ldots, s_n)$. Thus,

$$\Phi^*(x_1^{t+\delta}, \ldots, x_n^{t+\delta}) = \frac{1}{\delta} \sum_j \|E_{W^*(x_1^t, \ldots, x_n^t)}(s_j')\|_2^2$$

while

$$F(\gamma_N^{t+\delta}) = \frac{1}{\delta} \sum_j \|E_{\sigma-\text{algebra generated by entries of matrices } X_j + t^{1/2}B_j}(B_j')\|_2^2$$

where B_j, B_j' are Gaussian random matrices, independent from the matrices $X_1, \ldots, X_n \sim \gamma_N^0$. Because X_j, B_j, B_j' converge in non-commutative law to x_j, s_j, s_j', the length of any approximation to the vector $E_{W^*(x_1^t,\ldots,x_n^t)}(s_j')$ by polynomials in $x_1^t, \ldots, x_n^t$ when evaluated in $X_j + t^{1/2}B_j$ gives rise to a *lower* estimate on the length of the vector

$$E_{\sigma-\text{algebra generated by entries of matrices } X_j+t^{1/2}B_j}(B_j')$$

(roughly speaking because it gives an approximation to the vector

$$E_{\text{linear space spanned by entries of matrix polynomials in } X_j+t^{1/2}B_j}(B_j'),$$

which involves conditioning onto a *smaller* Hilbert space). This implies that $F(\gamma_N^t) \geqslant \Phi^*(x_1^t, \ldots, x_n^t)$ and thus $\chi \leqslant \chi^*$ (because the Fisher information enters into the integral formula for entropy with a negative sign).

4.7. Lack of atoms. We will prove that if $x_1, \ldots, x_n$ are a free semicircular family and p is a non-constant polynomial, then the law of $z = |p(x_1, \ldots, x_n)|$ is non-atomic. We first need a simple Lemma, which is a direct consequence of the fact that operators affiliated to a tracial von Neumann algebra form an algebra.

Lemma 4.7.1. *Let (M, τ) be a tracial non-commutative probability space.*

(1) *Let $y \in M$. Suppose that for some $y' \neq 0$, $yy' = 0$. Then for some $y'' \neq 0$, $y''y = 0$; moreover, we can arrange that the traces of the range projections of y' and y'' be the same.*
(2) *For an element $y \in M$ the following are equivalent: (i) $\mu_{|y|}$ has no atom at zero; (ii) if $py = 0$ or $yp = 0$ for some projection p, then $p = 0$; (iii) if $y'y = 0$ or $yy' = 0$ for some $y' \in M$, then $y' = 0$.*

Proof. (2) If $y' \in M$, let q be the range projection of y'. Then there exists an unbounded operator z affiliated with M so that $q = y'z$. Indeed, one can simply form the polar decomposition $y' = |y'|w$ and set $z = w^*f(|y'|)$ with $f(t) = t^{-1}$ for $t > 0$ and $f(0) = 0$.

Thus if $yy' = 0$ then $yp = 0$ where p is the range projection of y'. Applying this to y^* completes the proof that (iii) and (ii) are equivalent.

Let $y = |y|v$ be the polar decomposition of y. Then $|y|$ has no atom at zero iff $|y|^{-1}$ is a (possibly unbounded densely defined) operator affiliated with M, v is a unitary and the identity $v = |y|^{-1}y$ holds. Thus, defining $z = v^*|y|^{-1}$, we find that $zy = 1$. Hence if $yy' = 0$, then also $zyy' = y' = 0$.

Conversely, if $\mu_{|y|}$ has an atom at zero, letting q be the (nonzero!) kernel projection of $|y|$ gives us that $qy = 0$.

(1) By the first part of the proof, if $yy' = 0$ then $yp = 0$ where p is the range projection of y'. Let $y = |y|u$ be the polar decomposition of y; then $u^*u \leqslant p^\perp$. It follows that $uu^* \leqslant up^\perp u^*$, which means that the range projection of $y = uu^*|y|u$ is dominated by $up^\perp u^*$, so that $qy = 0$ with $q = 1 - up^\perp u^*$. Now note that $\tau(q) = \tau(uu^*) = \tau(u^*u) = \tau(p)$. □

Theorem 4.7.2. [9, 19] *Let $x_1, \dots, x_n$ be a free semicircular family, let p be a non-constant non-commutative polynomial and $z = p(x_1, \dots, x_n)$. Then the law of $\mu_{|z|}$ has no atoms. In particular, if $z = z^*$, μ_z has no atoms.*

Proof. We will prove that if p is non-zero and $yz = 0$, then $y = 0$. This will imply that the law of $|z|$ has no atom at zero, and by translation we can conclude that the law of z is non-atomic. The proof is by induction on the degree of p. If p has degree zero, the statement is clearly true.

Suppose by contradiction that $yz = 0$. Then by Lemma 4.7.1, also $zy' = 0$ for some y'; moreover, we may assume that the traces of range projections of y and y' are the same.

Thus $(y \otimes y')\#(z \otimes 1 - 1 \otimes z) = 0$. But an easy computation shows that

$$(z \otimes 1 - 1 \otimes z) = \sum_j \partial_j z\#(x_i \otimes 1 - 1 \otimes x_i)$$

a non-commutative form of the identity $[(f(s) - f(t))/(s - t)] \cdot (s - t) = f(s) - f(t)$. Thus

$$\sum_j [(y \otimes y')\#\partial_j z]\#(x_i \otimes 1 - 1 \otimes x_i) = 0. \tag{4.7.3}$$

Represent $x_1, \dots, x_n$ on the Fock space (see section §1.5, especially §1.5.5) so that $x_j = \ell(h_j) + \ell(h_j)^*$ and let P be the rank-one projection onto the vector Ω. Let also $r(h) : \xi_1 \otimes \cdots \otimes \xi_n \mapsto \xi_1 \otimes \cdots \otimes \xi_n \otimes h$. It's not hard to check that

$$[\ell(h), r(g)] = 0,$$
$$[\ell^*(h), r(g)] = \langle h, g \rangle P$$

where we use the commutator notation $[a, b] = ab - ba$.

Let

$$d_j = r(h_j) + r(h_j)^*.$$

Then d_j commutes with $x_1, \dots, x_n$:

$$[r(h_j) + r(h_j)^*, \ell(h_i) + \ell(h_i)^*] = \langle h_j, h_i \rangle P - \langle h_i, h_j \rangle P = 0.$$

Thus d_j also commutes with $y, y' \in W^*(x_1, \dots, x_n)$; moreover, $x_j P = d_j P$ and $P x_j = P d_j$.

We now use (4.7.3) to deduce that

$$\begin{aligned} 0 &= \sum_j (y \otimes y')\#(\partial_j z)\#(x_i \otimes 1 - 1 \otimes x_i)\#P \\ &= \sum_j (y \otimes y')\#(\partial_j z)\#(x_i P - P x_i) \\ &= \sum_j (y \otimes y')\#(\partial_j z)\#(d_i P - P d_i) \\ &= \sum_j [d_i, y(\partial_j z\#P)y']. \end{aligned}$$

Using the identity $\mathrm{Tr}([a,b]c) = \mathrm{Tr}(b[c,a])$ (valid whenever b is a finite rank operator) and noting that $\partial_j z\#P$ is finite rank, we have

$$0 = \mathrm{Tr}(\sum_j [d_i, y(\partial_j z\#P)y']\ell(h_k)) = \mathrm{Tr}(\sum_j y(\partial_j z\#P)y' \cdot [\ell(h_k), d_i])$$
$$= \mathrm{Tr}(\sum_j y(\partial_j z\#P)y' \cdot \delta_{ki}P) = \tau\otimes\tau((y\otimes y')\#\partial_k z).$$

Here we used that $\mathrm{Tr}(PaPb) = \tau(a)\tau(b)$ for $a, b \in W^*(x_1, \ldots, x_n)$ (see §1.5.5).

Replacing y with ay and y' with $y'b$ for arbitrary a, b allows us to deduce from faithfulness of τ that $(y \otimes y')\#\partial_k z = 0$. In particular,

$$(1\otimes\tau)((y\otimes y')\#\partial_k z) = 0$$

so that

$$y \cdot \{(1\otimes\tau)((1\otimes y')\#\partial_k z\} = 0.$$

We now note that $z' = (1\otimes\tau)((1\otimes y')\#\partial_k z$ is again a polynomial in $x_1, \ldots, x_n$ but of degree lower than p. By the inductive hypothesis, it follows that z' must be constant, which means that z is a linear polynomial. But then

$$z = \sum \alpha_i x_i + \beta = s(\sum \alpha_i h_i) + \beta$$

is (up to a shift and dilation) a semicircular variable, and $yz' = 0$ cannot hold unless $y = 0$. □

4.8. Exercises

Exercise 4.8.1. Let $A = \mathbb{C}[r]$ be the algebra of polynomials in one variable r, and let $\hat{A} = \mathbb{C}[s,t]$ be the algebra of (commutative) polynomials in two variables, s and t.

(1) Show that $A\otimes A \cong \hat{A}$, the map being given by $p(r)\otimes q(r) \mapsto p(s)q(t)$.
(2) Show that the difference quotient map $\partial : A \to \hat{A}$ defined by setting $\partial(q) = (q(s) - q(t))/(s-t)$ is a derivation and satisfies $\partial(r) = 1$. Show that, up to isomorphism $\hat{A} \cong A\otimes A$ ∂ is the difference quotient derivation of Definition 3.5.4 corresponding to the single variable case $n = 1$.

Exercise 4.8.2. Show that a single random variable X is freely differentiable iff its law is non-atomic.

Exercise 4.8.3. Let $x_1, \ldots, x_n$ be classical random variables, represented as coordinate functions on $\mathbb{R}^n$ endowed with the measure μ; assume that μ is Lebesgue-absolutely continuous with density ρ. Define derivations

$$d_i : f \mapsto \frac{\partial_i f}{\partial x_i}$$

viewed as densely defined operators on $L^2(\mu)$.

Show that if $\zeta_i = (\partial_i/\partial x_i)\log\rho = \frac{(\partial/\partial x_i)\rho}{\rho}$ belongs to $L^2(\mu)$, then $\zeta_i = d_i^*(1)$. This function is called the *score function* in statistics. Use this to show that the classical Fisher information $F = \int(\rho')^2/\rho\, dt = \|\zeta\|^2_{L^2(\mu)}$.

Exercise 4.8.4. (1) Prove the *free Cramer-Rao inequality*: if $\tau(X_j^* X_j) = 1$ for all j, then $\Phi^*(X_1, \ldots, X_n) \geqslant n$ with equality iff $X_1, \ldots, X_n$ are free semicircular variables. *Hint*: consider $\langle \xi_j, X_j \rangle$ and apply Cauchy-Schwartz.

(2) Use this to show that $\chi^*(X_1, \ldots, X_n)$ is maximal among all n-tuples with $\tau(X_j^* X_j) = 1$ for all j iff $X_1, \ldots, X_n$ are free semicircular variables.

Exercise 4.8.5. Let X be a random variable with non-atomic law, and let A be the algebra generated by 1, X and all resolvents $(z - X)^{-1}$. The difference quotient is then a map $\partial : A \to A \otimes A$. If A^* is the dual space to A, then the adjoint map $\partial^* : A^* \hat{\otimes} A^* \to A^*$ is a multiplication. Show that ∂ is *co-associative* (equivalently, ∂^* is an associative multiplication), meaning that $(1 \otimes \partial) \circ \partial = (\partial \otimes 1) \circ \partial$. In this way, the equation $\partial f = f \otimes f$ satisfied by the resolvents (see Lemma 4.3.1) shows that the resolvents are precisely the co-representations of A.

Exercise 4.8.6. The goal of this exercise is to show that if $X_1, \ldots, X_d$ are differentiable, τ is a faithful trace and $d \geqslant 2$, then the von Neumann algebra $M = W^*(X_1, \ldots, X_d)$ generated by $X_1, \ldots, X_d$ on $L^2(\tau)$ is a *factor*, i.e., its center is trivial (thus if $Z \in M$ satisfies $[Z, X_j] = ZX_j - X_j Z = 0$ for all j, then $Z \in \mathbb{C}1$).

(1) Let $\Delta = \bar{\partial}_1^* \partial_1$, and let $\zeta_\alpha = (\alpha/(\alpha + \Delta))^{1/2}$. It turns out that $\partial_1 \circ \zeta_\alpha$ is always bounded. For all $x \in M$, $\|\zeta_\alpha(x) - x\|_2 \to 0$ for as $\alpha \to \infty$. Moreover, if $\partial_1 Y = 0$ then $\zeta_\alpha(PY) = \zeta_\alpha(P)Y$ and $\zeta_\alpha(YP) = Y\zeta_\alpha(P)$ for all P. Let $\partial_1^{(\alpha)} = \bar{\partial}_1 \circ \zeta_\alpha$. Use this to show that since $[Z, X_2] = 0$ then

$$0 = \partial_1^{(\alpha)}([Z, X_2]) = [X_2, \partial_1^{(\alpha)}(Z)].$$

(2) Show that if P_1 is the orthogonal projection onto $1 \in L^2(M)$, the identification $a \otimes b \mapsto aPb$ extends to an isometry between $L^2(M) \bar{\otimes} L^2(M)$ and the space of Hilbert-Schmidt operators, i.e., compact operators T satisfying $\mathrm{Tr}(T^*T) < \infty$ with the inner product $\langle T, T' \rangle = \mathrm{Tr}(T^*T')$. Conclude that for $T \in L^2(M) \bar{\otimes} L^2(M)$ and $X \in M$ has diffuse spectrum, if $[X, T] = 0$ then $T = 0$ (use the fact that Hilbert-Schmidt operators are compact and that no compact operator can commute with an operator with diffuse spectrum).

(3) Returning to the setting of (1), conclude from differentiability that X_2 has diffuse spectrum. Apply (2) to the conclusion of (1) to conclude that $\partial_1^{(\alpha)}(Z) = 0$. Use closability of ∂_1 to conclude that Z belongs to the domain of $\bar{\partial}$ and $\bar{\partial}_1 Z = 0$. Now apply $\bar{\partial_1}$ to $0 = [Z, X_1]$ to conclude that $Z \otimes 1 - 1 \otimes Z = 0$. Now apply $P_1 \otimes 1$ to deduce that $Z = \tau(Z)1 \in \mathbb{C}1$.

Addendum: Free Analogs of Monotone Transport Theory

5.1. Classical transport maps. A map $f : \mathbb{R}^n \to \mathbb{R}^n$ is said to be *monotone* if $f = \nabla \phi$ for some convex function ϕ. This is obviously a generalization to n dimensions of the second-derivative test for monotonicity of a function of a single variable: indeed, convexity of ϕ amounts to the requirement that $\mathrm{Hess}(\phi) = Df$ is a positive matrix.

Let μ and ν be two probability measures supported on a ball in $\mathbb{R}^n$. If μ and ν are diffuse, there exist many maps f with the property that $f_*\mu = \nu$.

A remarkable theorem of Brenier states that, under non-degeneracy conditions on these measures (e.g. Lebesgue absolute continuity, although the actual requirements are substantially less), there exists a unique *monotone transport map* f (see [8, 24] for more details). The unique map f has a number of amazing properties; among them is that it is the minimizer of the *optimal transport problem*: among all transport maps g the map f minimizes the distance $\int \|f(x) - x\|^2 d\mu(x)$, a kind of energy required to transport μ into ν.

5.2. Non-commutative transport. It is tempting to look for analogs of transport maps in the non-commutative setting.

Definition 5.2.1. Let $X_1, \ldots, X_n \in (M, \tau)$, $Y_1, \ldots, Y_n \in (\hat{M}, \hat{\tau})$ be non-commutative random variables. A *transport map* F from $\mu_{X_1,\ldots,X_n}$ to $\mu_{Y_1,\ldots,Y_n}$ is an element $F = (F_1, \ldots, F_n) \in (W^*(X_1, \ldots, X_n))^n$ such that $\mu_{F_1,\ldots,F_n} = \mu_{Y_1,\ldots,Y_n}$.

Note that $W^*(X_1, \ldots, X_n)$ is the analog of the algebra of essentially bounded functions of $X_1, \ldots, X_n$ sor we can view F as a vector-valued map of n variables.

Unlike the classical situation, it is clear that there are many obstructions to the existence of the map F. For example, suppose that $X_1, \ldots, X_n$ commute with one another; then clearly so will $F_1, \ldots, F_n$. Thus not every law $Y_1, \ldots, Y_n$ can be obtained as the transport of $\mu_{X_1,\ldots,X_n}$. Actually, the problem is substantially deeper; going beyond easy counterexamples (such as the case when $X_1, \ldots, X_n$ generate a matrix algebra), a result of Ozawa [21] states no separable von Neumann algebra contains all other separable von Neumann algebras. Since $\mu_{F_1,\ldots,F_n} = \mu_{Y_1,\ldots,Y_n}$ implies that $W^*(Y_1, \ldots, Y_n) \cong W^*(F_1, \ldots, F_n) \subset W^*(X_1, \ldots, X_n)$, this shows that existence of non-commutative transport maps is quite delicate. This is hardly surprising, since there is absolutely no non-commutative analog of the classification of classical measure spaces.

5.3. The Monge-Ampère equation. If f is a (classical) transport map from μ to ν, and $d\mu = \rho(\vec{x})d\vec{x}$, $\nu = \eta(\vec{x})d\vec{x}$, then the condition $f_*\mu = \nu$ entails

$$|\det[Df](\vec{x})|\rho(\vec{x}) = \eta(f(\vec{x})).$$

Using $|\det A| = \exp \operatorname{Tr} \log |A|$, this can be rewritten as

$$\exp[\operatorname{Tr} \log |Df(\vec{x})|]\rho(\vec{x}) = \eta(f(\vec{x}));$$

taking log of both sides then gives

$$\operatorname{Tr} \log |Df(\vec{x})| + \log \rho(\vec{x}) = \log \eta(f(\vec{x})).$$

Suppose now that $\rho(\vec{x}) = \exp(-V(\vec{x}))$ and $\eta(\vec{x}) = \exp(-U(\vec{x}))$. Then we get

$$\operatorname{Tr} \log |Df(\vec{x})| = V(\vec{x}) - U(f(\vec{x})).$$

This equation is equivalent to the transport equation for f, assuming invertibility of the map f.

If we now assume that $f = \nabla\phi$ for a convex function ϕ, then $Df = \mathrm{Hess}(\phi) \geqslant 0$ and we get

$$\mathrm{Tr}\log \mathrm{Hess}(\phi) = V(\vec{x}) - U\big(\nabla\phi(\vec{x})\big). \tag{5.3.1}$$

This is the so-called *Monge-Ampère equation* (although it's usually written in a different form, equivalent to the exponential of the equation we introduced above).

Let us now assume that the measure ν is the Gaussian measure so that we have $U(\vec{x}) = \frac{1}{2}\|\vec{x}\|^2$ and that the measure μ is a perturbation of the Gaussian measure, so that $V(\vec{x}) = \frac{1}{2}\|\vec{x}\|^2 + \epsilon W(\vec{x})$, for some small ϵ. In that case we may expect that $f = \mathrm{id} + \epsilon g$, so that $\phi(\vec{x}) = \frac{1}{2}\|\vec{x}\|^2 + \epsilon\psi$. Substituting this into the equation (5.3.1) yields

$$\begin{aligned}\mathrm{Tr}\log(1+\epsilon D\nabla\psi) &= \frac{1}{2}\|\vec{x}\|^2 + \epsilon W(\vec{x}) - \frac{1}{2}\|\vec{x} + \epsilon\nabla\psi(\vec{x})\|^2 \\ &= \epsilon W(\vec{x}) - \epsilon\nabla\psi(\vec{x})\cdot\vec{x} - \frac{\epsilon^2}{2}\|\nabla\psi(\vec{x})\|^2.\end{aligned}$$

The operator $N : \psi \mapsto \nabla\psi(\vec{x})\cdot\vec{x}$ acts on monomials in $x_1,\dots,x_n$ by multiplying a monomial by its degree. We can thus rewrite the equation as

$$N\psi = W(\vec{x}) - \frac{\epsilon}{2}\|\nabla\psi(\vec{x})\|^2 - \frac{1}{\epsilon}\mathrm{Tr}\log(1+\epsilon D\nabla\psi) \stackrel{\mathrm{def}}{=} F(W, \nabla\psi, D\nabla\psi).$$

We can clearly change ψ by adding a constant, thus we can assume that $\psi(0) = 0$. Since N is invertible on power series vanishing at zero, we can then write

$$\psi = N^{-1}F(W, \nabla\psi, D\nabla\psi).$$

5.4. The Free Monge-Ampère equation. Let us return to the non-commutative situation. Let's say that we are given $X_1,\dots,X_n$ and a free semicircular system $Y_1,\dots,Y_n$; let's further assume that the conjugate variables to $X_1,\dots,X_n$ exist and satisfy (3.6.1) for some polynomial V, i.e.,

$$\tau(\mathcal{D}_j V \cdot P) = \tau\otimes\tau(\partial_j P)$$

holds for any polynomial P. Let's assume that

$$V(Z_1,\dots,Z_n) = \sum Z_j^2 + \epsilon W(Z_1,\dots,Z_n).$$

Proceeding by analogy, we write an analog of the classical equation, replacing gradients ∇ by cyclic gradients $\mathcal{D}$ and Jacobian derivatives $DF = ((\partial/\partial x_j)F_i)_{ij}$ by difference quotient Jacobian $JF = (\partial_j F_i)_{ij}$. We thus arrive at the *free Monge-Ampère equation*

$$(1\otimes\tau + \tau\otimes 1)\otimes \mathrm{Tr}(\log(1+\epsilon J\mathcal{D}g)) = \epsilon W - \epsilon\mathcal{D}g(X).X + \frac{1}{2}\epsilon^2\mathcal{D}g.\mathcal{D}g$$

where $X = (X_1,\dots,X_n)$ and we write $(a_1,\dots,a_n).(b_1,\dots,b_n) = \frac{1}{2}\sum a_i b_i + b_i a_i$. Once again, we get:

$$\mathcal{D}g.X = \mathcal{F}(W, \mathcal{D}g, J\mathcal{D}g).$$

Let us try to solve this equation assuming that $g(X_1,\dots,X_n)$ is a power series in $X_1,\dots,X_n$. We note that we can certainly modify g by a constant, and thus assume that $g(0,\cdots,0)=0$. Moreover, we can replace any monomial term in g by its cyclic permutation; indeed, $\mathcal{D}a=\mathcal{D}b$ if a and b are cyclic permutations of the same monomial. It follows that we may as well assume that g is a *cyclically symmetric* function. In that case the map

$$\mathcal{N}: g \mapsto \mathcal{D}g.X$$

acts on monomials by multiplying them by their degree. Thus we can once again write

$$g=\mathcal{N}^{-1}\mathcal{F}(W,\mathcal{D}g,J\mathcal{D}g).$$

We can then solve the resulting equation using Picard iterations in the space of absolutely convergent non-commutative power series; the inverse of $\mathcal{N}$ ensures that the map $\mathcal{N}^{-1}\mathcal{F}$ is uniformly Lipschitz. As a result, we obtain a solution g for which $F_i=X_i+\epsilon\mathcal{D}_i g(X)$ satisfy

$$\tau(F_iP(F))=\tau\otimes\tau(\partial_i(P)(F)),$$

an equation that uniquely determines the semicircle law (see Lemma 3.5.8). It follows that we constructed a transport map from the law of $X_1,\ldots,X_n$ to the semicircle law; we can argue that since F is sufficiently close to the identity map, it has an inverse, and the inverse transports the semicircle law back into the law of $X_1,\ldots,X_n$. We have thus proved:

Theorem 5.4.1. [15] *Let $X_1,\ldots,X_n$ have the law satisfying the Schwinger-Dyson equation (3.6.1) for $V(X)=\frac{1}{2}\sum X_j^2+\epsilon W(X)$ and $Y_1,\ldots,Y_n$ be a free semicircular n-tuple. Then for sufficiently small ϵ, $W^*(X_1,\ldots,X_n)\cong W^*(Y_1,\ldots,Y_n)$. In fact, holomorphic functional calculus closures of the algebras generated by $X_1,\ldots,X_n$ and $Y_1,\ldots,Y_n$ are isomorphic.*

5.5. Random Matrix Applications. Theorem 5.4.1 has immediate random matrix corollaries. Recall (see Theorem 3.3.5) that if $V(X)=\frac{1}{2}\sum X_j^2+\epsilon W(X)$ and ϵ is sufficiently small, then the $N\to\infty$ limit law of n-tuples of random matrices $A_1,\ldots,A_n$ drawn according to the probability measure

$$\mu_N=\frac{1}{Z_N}\exp(-N\mathrm{Tr}[V(A_1,\ldots,A_n)])dA_1\cdots dA_n$$

satisfies (3.6.1) and thus the hypothesis of 5.4.1. It follows that if $X_1,\ldots,X_n$ have this law, they can be written as convergent power series in semicircular variables $Y_1,\ldots,Y_n$. In particular:

Theorem 5.5.1. *Let P be a self-adjoint non-commutative polynomial, and keep the assumptions and notations of this subsection. Then the limit law of $P(A_1,\ldots,A_n)$ has connected support.*

Proof. The limit law of $Z_N=P(A_1,\ldots,A_n)$ is $Z=P(X_1,\ldots,X_n)\in C^*(Y_1,\ldots,Y_n)$ which has no projections (see proof of Theorem 2.8.1). □

In fact, random matrices provide a link between classical and free transport. Let ν_N be the Gaussian measure on n-tuples of $N\times N$ random matrices, and let

$F^{(N)}$ be the Brenier transport map from μ_N to ν_N. Let F be the free transport map we constructed in Theorem 5.4.1.

Theorem 5.5.2. [15] $F^{(N)}(A_1,\ldots,A_n) \to F(A_1,\ldots,A_n)$ *almost surely.*

We give a sketch of the proof.

The Gaussian measure ν_N is the unique minimizer of the functional

$$H_N(\rho dA_1\cdots dA_n) = \int \rho\log\rho dA_1\cdots dA_n - \int \frac{1}{N}\mathrm{Tr}(\frac{1}{2}\sum A_j^2)\rho dA_1\cdots dA_n$$

called the *relative entropy*. Thus the transport map F_N is *characterized* by the equality

$$H_N((F^{(N)})_*\mu_N) = H_N(\nu_N).$$

The change of variables formula for entropy allows us to write

$$\begin{aligned} H_N(\nu_N) = H_N((F^{(N)})_*\mu_N) &= H_N(\mu_N) + \mathbb{E}_{\mu_N}[\frac{1}{N}\mathrm{Tr}(\frac{1}{2}\sum A_j^2)] \\ &- \mathbb{E}_{\mu_N}[\frac{1}{N}\mathrm{Tr}(\frac{1}{2}\sum [F_N(A)]_j^2)] + \mathbb{E}_{\mu_N}[\mathrm{Tr}\log|DF^{(N)}|] \\ &= H_N(\mu_N) + \mathbb{E}_{\mu_N}[\frac{1}{N}\mathrm{Tr}(\frac{1}{2}\sum A_j^2)] - \Omega_N(F^{(N)}) \end{aligned}$$

where for $G = (G_1,\ldots,G_n)$ we set

$$\Omega_N(G) = -\mathbb{E}_{\mu_N}[\frac{1}{N}\mathrm{Tr}(\frac{1}{2}\sum [G_j(A)]^2)] + \mathbb{E}_{\mu_N}[\mathrm{Tr}\log(DG)].$$

Note that $F^{(N)} \mapsto \Omega_N(F^{(N)})$ is a (strictly) concave function of $F^{(N)}$ (under the assumption that $F^{(N)}$ is the gradient of a convex function) and thus $F^{(N)}$ is the maximizer of Ω_N.

Similarly, the semicircle law τ can be characterized as being the minimizer of the free entropy functional $\chi_s(x_1,\ldots,x_n) = \chi(x_1,\ldots,x_n) - \tau(\frac{1}{2}\sum x_j^2)$, where χ is Voiculescu's free entropy (see §4.5). Thus the transport map $F = (F_1,\ldots,F_n)$ satisfies

$$\chi_s(F_1,\ldots,F_n) - \tau(\frac{1}{2}\sum F_j^2) = \chi_s(Y_1,\ldots,Y_n)$$

where $Y_1,\ldots,Y_n$ is a free semicircular n-tuple. Since F is close to identity, we may use the change of variables formula (Theorem 4.5.2) to deduce that

$$\begin{aligned} &\chi_s(F_1(X_1,\ldots,X_n),\ldots,F_n(X_1,\ldots,X_n)) \\ =&\chi(X_1,\ldots,X_n) + \tau(\frac{1}{2}\sum X_j^2) \\ &\quad - \tau(\frac{1}{2}\sum F_j^2) + \tau\otimes\tau\otimes\mathrm{Tr}((\partial_j F_i)_{ij}) \\ =&\chi(X_1,\ldots,X_n) + \tau(\frac{1}{2}\sum X_j^2) - \Omega(F) \end{aligned}$$

with

$$\Omega(F) = -\tau(\frac{1}{2}\sum F_j^2) + \tau\otimes\tau\otimes\mathrm{Tr}((\partial_j F_i)_{ij}),$$

also a (strictly) concave function of F.

Using convergence of the random matrix model, we can show that $\Omega_N(G) \to \Omega(G)$ for any reasonably smooth function G close to identity. In particular,

$$\Omega_N(F) \to \Omega(F) \text{ as } N \to \infty.$$

One can further prove that $H_N(\mu_N) \to \chi_s(X_1, \dots, X_N)$, $\mathbb{E}_{\mu_N}[\frac{1}{N}\mathrm{Tr}(\frac{1}{2}\sum A_j^2)] \to \tau(\frac{1}{2}\sum X_j^2)$, $H_N(\nu_N) \to \chi_s(Y_1, \dots, Y_n)$. Thus

$$\begin{aligned} \Omega_N(F^{(N)}) &= H_N(\nu_N) - H_N(\mu_N) - \mathbb{E}_{\mu_N}[\frac{1}{N}\mathrm{Tr}(\frac{1}{2}\sum A_j^2)] \\ &\to \chi_s(X_1, \dots, X_n) - \chi_s(Y_1, \dots, Y_n) - \tau(\frac{1}{2}\sum X_j^2) \\ &= \Omega(F) = \lim_{N\to\infty} \Omega_N(F) \end{aligned}$$

It follows that

$$\Omega_N(F) - \Omega_N(F^{(N)}) \to 0.$$

But since $F^{(N)}$ is the unique minimizer of Ω_N we can use strict convexity of Ω_N to deduce that

$$\|F - F^{(N)}\|_{L^2} \to 0.$$

Concentration allows us to upgrade this to an almost sure convergence.

We refer the reader to [4, 11] for more applications of transportation ideas to random matrices.

Index

C^*-algebra, 393

affiliated operator, 398

Cauchy transform, 413, 426
Cayley graph, 405
$\boxplus, \boxtimes$, 418
commutant, 395
commutator, $[\ ,\ ]$, 412
conditional expectation, 399, 418
conjugate variable, 442
creation operator, 408
cyclic derivative, 439
cyclic gradient, 439
cyclically symmetric function, 454

$\mathcal{D}_i$, 439
∂_X, ∂_i, 436
difference quotient, 414, 436
Dyck path, 409

entropy
  free, 445
  relative, 455

factor, 395, 451
  finite, 396
faithful
  trace, 396
Fisher information, 444
Fock space, 408
free convolution, 418, 424
free Poisson law, 428, 440
free product, 408, 417
  of groups, 406
free subordination, 424, 443
freely differentiable, 441

genus expansion, 434

Hilbert-Schmidt operator, 451

independence
  classical, 406
  free, 406, 407
  free with amalgamation, 418

Laplace operator, 405
law, 402, 403

Marčenko-Pastur law, 428, 440
Monge-Ampère equation, 453
  free, 453
monotone map, 451

normal
  linear functional, 394
  map, 394

probability space, 397
projection, 394

R-transform, 419, 422
random matrix, 428
random variable, 392
random walk, 405

S-transform, 425
Schwinger-Dyson equation, 440
semicircular law, 409

Tr (trace on finite-rank operators), 412
trace, 396
  faithful, 398, 404
transport map, 452
  monotone, 452

von Neumann algebra, 391, 393
  finite, 396

References

[1] G. W. Anderson, *Preservation of algebraicity in free probability*, 2014. Preprint. ↑425

[2] G. W. Anderson, A. Guionnet, and O. Zeitouni, *An introduction to random matrices*, Cambridge Studies in Advanced Mathematics, vol. 118, Cambridge University Press, Cambridge, 2010. MR2760897 ↑428, 434

[3] G. B. Arous and A. Guionnet, *Large deviations for Wigner's law and Voiculescu's non-commutative entropy*, Probab. Theory Related Fields **108** (1997), no. 4, 517–542. MR1465640 ↑445

[4] F. Bekerman, A. Figalli, and A. Guionnet, *Transport maps for β-matrix models and universality*, Comm. Math. Phys. **338** (2015), no. 2, 589–619, DOI 10.1007/s00220-015-2384-y. ↑456

[5] H. Bercovici and V. Pata, *Stable laws and domains of attraction in free probability theory*, Ann. of Math. (2) **149** (1999), no. 3, 1023–1060. With an appendix by Philippe Biane. MR1709310 ↑424

[6] H. Bercovici and D.-V. Voiculescu, *Regularity questions for free convolution*, Nonselfadjoint operator algebras, operator theory, and related topics, 1998, pp. 37–47. MR1639647 ↑423, 424

[7] P. Biane, M. Capitaine, and A. Guionnet, *Large deviation bounds for matrix Brownian motion*, Invent. Math. **152** (2003), no. 2, 433–459. MR1975 007 ↑447

[8] Y. Brenier, *Polar factorization and monotone rearrangement of vector-valued functions*, Comm. Pure Appl. Math **44** (1991), 375–417. ↑452

[9] I. Charlesworth and D. Shlyakhtenko, *Free entropy dimension and regularity of non-commutative polynomials*, J. Funct. Anal. **271** (2016), no. 8, 2274–2292. MR3539353 ↑425, 449

[10] A. Connes, *Classification of injective factors. Cases* II_1, II_∞, III_λ, $\lambda \neq 1$, Ann. of Math. (2) **104** (1976), no. 1, 73–115. MR56#12908 ↑447

[11] A. Figalli and A. Guionnet, *Universality in several-matrix models via approximate transport maps*, Acta Math. **217** (2016), no. 1, 81–176, DOI 10.1007/s11511-016-0142-4. ↑456

[12] P. Fima and E. Germain, *The KK-theory of fundamental* C^**-algebras*, Trans. Amer. Math. Soc. **370** (2018), no. 10, 7051–7079, DOI 10.1090/tran/7211. ↑425, 426

[13] E. Germain, *KK-theory of reduced free-product* C^**-algebras*, Duke Math. J. **82** (1996), 707–723. ↑425, 426

[14] A. Guionnet and E. Maurel-Segala, *Combinatorial aspects of matrix models*, ALEA Lat. Am. J. Probab. Math. Stat. **1** (2006), 241–279. MR2249657 ↑431, 432, 441

[15] A. Guionnet and D. Shlyakhtenko, *Free monotone transport*, Invent. Math. **197** (2014), no. 3, 613–661. MR3251831 ↑454, 455

[16] Hari Bercovici and Dan Voiculescu, *Free Convolution of Measures with Unbounded Support*, Indiana University Mathematics Journal **42** (1993), no. 3, 733–773. ↑420

[17] U. Haagerup, *On Voiculescu's* R*- and* S*-transforms for free non-commuting random variables*, Free probability theory (Waterloo, ON, 1995), 1997, pp. 127–148. MR98c:46137 ↑420

[18] D. Kaliuzhnyi-Verbovetskyi and V. Vinnikov, *Foundations of free noncommutative function theory*, Mathematical Surveys and Monographs, vol. 199, American Mathematical Society, Providence, RI, 2014. MR3244229 ↑427

[19] T. Mai, R. Speicher, and M. Weber, *Absence of algebraic relations and of zero divisors under the assumption of full non-microstates free entropy dimension*, Adv. Math. **304** (2017), 1080–1107, DOI 10.1016/j.aim.2016.09.018. ↑425, 449

[20] A. Nica and R. Speicher, *On the multiplication of free* n*-tuples of non-commutative random variables*, American Journal of Mathematics **118** (1996). ↑424

[21] N. Ozawa, *There is no separable universal* II_1*-factor*, Proc. Amer. Math. Soc. **132** (2004), 487–490. ↑452

[22] N. Ozawa, *About the Connes embedding conjecture*, Japanese Journal of Mathematics **8** (2013), no. 1, 147–183. ↑447

[23] R. Speicher, *Multiplicative functions on the lattice of non-crossing partitions and free convolution*, Math. Annalen **298** (1994), 193–206. ↑422, 426

[24] C. Villani, *Topics in optimal transportation*, Graduate Studies in Mathematics, Vol. 58, AMS, Providence, RI, 2003. ↑452

[25] D.-V. Voiculescu, *Symmetries of some reduced free product* C^**-algebras*, Operator algebras and their connections with topology and ergodic theory, 1985, pp. 556–588. ↑406, 407, 408

[26] D.-V. Voiculescu, *Addition of certain non-commuting random variables*, J. Funct. Anal. **17** (1986), 323–345. ↑418, 420

[27] D.-V. Voiculescu, *Multiplication of certain non-commuting random variables*, J. Operator Theory **18** (1987), 223–235. ↑418, 425
[28] D.-V. Voiculescu, *Circular and semicircular systems and free product factors*, Operator algebras, unitary representations, enveloping algebras, and invariant theory, 1990, pp. 45–60. ↑408
[29] D.-V. Voiculescu, *Limit laws for random matrices and free products*, Invent. math **104** (1991), 201–220. ↑429, 431
[30] D.-V. Voiculescu, *The analogues of entropy and of Fisher's information measure in free probability theory I*, Commun. Math. Phys. **155** (1993), 71–92. ↑424, 445
[31] D.-V. Voiculescu, *The analogues of entropy and of Fisher's information measure in free probability theory II*, Invent. Math. **118** (1994), 411–440. ↑445, 446
[32] D.-V. Voiculescu, *Operations on certain non commutative operator-valued random variables*, Recent advances in operator algebras (Orléans, 1992), 1995, pp. 243–275. ↑426
[33] D.-V. Voiculescu, *The analogues of entropy and of Fisher's information measure in free probability, V*, Invent. Math. **132** (1998), 189–227. ↑436, 442, 443, 444, 445
[34] D.-V. Voiculescu, *A strengthened asymptotic freeness result for random matrices with applications to free entropy*, IMRN **1** (1998), 41 –64. ↑429
[35] D.-V. Voiculescu, *The coalgebra of the free difference quotient and free probability*, Internat. Math. Res. Notices **2** (2000), 79–106. MR1744647 ↑443, 444
[36] D.-V. Voiculescu, *Free analysis questions i: Duality transform for the coalgebra of* $\partial_{X:B}$, International Math. Res. Notices **16** (2004), 793–822. ↑427
[37] D.-V. Voiculescu, K. Dykema, and A. Nica, *Free random variables*, CRM monograph series, vol. 1, American Mathematical Society, 1992. ↑401, 407, 408, 418, 421, 424, 426, 431
[38] J. Williams, *Analytic function theory for operator-valued free probability*, J. Reine Angew. Math. **729** (2017), 119–149, DOI 10.1515/crelle-2014-0106. ↑427

Department of Mathematics, UCLA, Los Angeles, CA 90095
Email address: shlyakht@math.ucla.edu

IAS/Park City Mathematics Series
Volume 26, Pages 461–498
https://doi.org/10.1090/pcms/026/00851

Least singular value, circular law, and Lindeberg exchange

Terence Tao

Abstract. These lectures cover three loosely related topics in random matrix theory. First we discuss the techniques used to bound the least singular value of (non-Hermitian) random matrices, focusing particularly on the matrices with jointly independent entries. We then use these bounds to obtain the *circular law* for the spectrum of matrices with iid entries of finite variance. Finally, we discuss the *Lindeberg exchange method* which allows one to demonstrate universality of many spectral statistics of matrices (both Hermitian and non-Hermitian).

1. The least singular value

This section[1] of the lecture notes is concerned with the behaviour of the *least singular value* $\sigma_n(M)$ of an $n \times n$ matrix M (or, more generally, the least non-trivial singular value $\sigma_p(M)$ of a $n \times p$ matrix with $p \leqslant n$). This quantity controls the invertibility of M—indeed, M is invertible precisely when $\sigma_n(M)$ is non-zero, and the ℓ^2 operator norm $\|M^{-1}\|_{op}$ of M^{-1} is given by $1/\sigma_n(M)$—and is also related to the *condition number* $\sigma_1(M)/\sigma_n(M) = \|M\|_{op}\|M^{-1}\|_{op}$ of M, which is of importance in numerical linear algebra. As we shall see in Section 2, the least singular value of M (and more generally, of the shifts $\frac{1}{\sqrt{n}}M - zI$ for complex z) is important in rigorously establishing the *circular law* for iid random matrices M.

The least singular value[2]

$$\sigma_n(M) = \inf_{\|x\|=1} \|Mx\|,$$

which sits at the "hard edge" of the spectrum, bears a superficial similarity to the operator norm

$$\|M\|_{op} = \sigma_1(M) = \sup_{\|x\|=1} \|Mx\|$$

at the "soft edge" of the spectrum. For strongly rectangular matrices, the techniques that are useful to control the latter can also control the former, but the situation becomes more delicate for square matrices. For instance, the "epsilon net"

2010 *Mathematics Subject Classification.* Primary 60B20; Secondary 60F17.
Key words and phrases. Least singular value, circular law, Lindeberg exchange method, random matrix.
[1]The material in this section and the next is based on that in [56]. Thanks to Nick Cook and the anonymous referees for corrections and suggestions. The author is supported by NSF grant DMS-1266164, the James and Carol Collins Chair, and by a Simons Investigator Award.
[2]In these notes we use $\|\|$ to denote the ℓ^2 norm on $\mathbb{R}^n$ or $\mathbb{C}^n$.

method that is so useful for understanding the operator norm can control some "low entropy" portions of the infimum that arise from "structured" or "compressible" choices of x, but are not able to control the "generic" or "incompressible" choices of x, for which new arguments will be needed. Similarly, the moment method can give the coarse order of magnitude (for instance, for rectangular matrices with $p = yn$ for $0 < y < 1$, it gives an upper bound of $(1 - \sqrt{y} + o(1))n$ for the least singular value with high probability, thanks to the *Marcenko-Pastur law*), but again this method begins to break down for square matrices, although one can make some partial headway by considering *negative* moments such as $\operatorname{tr} M^{-2}$, though these are more difficult to compute than positive moments $\operatorname{tr} M^k$.

So one needs to supplement these existing methods with additional tools. It turns out that the key issue is to understand the distance between one of the n rows $X_1, \ldots, X_n \in \mathbf{C}^n$ of the matrix M, and the hyperplane spanned by the other $n-1$ rows. The reason for this is as follows. First suppose that $\sigma_n(M) = 0$, so that M is non-invertible, and there is a linear dependence between the rows $X_1, \ldots, X_n$. Thus, one of the X_i will lie in the hyperplane spanned by the other rows, and so one of the distances mentioned above will vanish; in fact, one expects many of the n distances to vanish. Conversely, whenever one of these distances vanishes, one has a linear dependence, and so $\sigma_n(M) = 0$.

More generally, if the least singular value $\sigma_n(M)$ is small, one generically expects many of these n distances to be small also, and conversely. Thus, control of the least singular value is morally equivalent to control of the distance between a row X_i and the hyperplane spanned by the other rows. This latter quantity is basically the dot product of X_i with a unit normal n_i of this hyperplane.

When working with random matrices with jointly independent coefficients, we have the crucial property that the unit normal n_i (which depends on all the rows other than X_i) is *independent* of X_i, so even after conditioning n_i to be fixed, the entries of X_i remain independent. As such, the dot product $X_i \cdot n_i$ is a familiar scalar random walk, and can be controlled by a number of tools, most notably Littlewood-Offord theorems and the Berry-Esséen central limit theorem. As it turns out, this type of control works well except in some rare cases in which the normal n_i is "compressible" or otherwise highly structured; but epsilon-net arguments can be used to dispose of these cases[3].

These methods rely quite strongly on the joint independence on all the entries; it remains a challenge to extend them to more general settings. Even for Wigner matrices, the methods run into difficulty because of the non-independence of some of the entries (although it turns out one can understand the least singular value in such cases by rather different methods).

To simplify the exposition, we shall focus primarily here on just one specific ensemble of random matrices, the *Bernoulli ensemble* $M = (\xi_{ij})_{1 \leqslant i \leqslant p; 1 \leqslant j \leqslant n}$ of

[3]This general strategy was first developed for the technically simpler singularity problem in [43], and then extended to the least singular value problem in [52].

random sign matrices, where $\xi_{ij} = \pm 1$ are independent Bernoulli signs. However, the results can extend to more general classes of random matrices, with the main requirement being that the coefficients are jointly independent.

Throughout these notes, we use $X \ll Y$, $Y \gg X$, or $X = O(Y)$ to denote a bound of the form $|X| \leqslant CY$ for some absolute constant C; we take n as an asymptotic parameter, and write $X = o(Y)$ to denote a bound of the form $|X| \leqslant c(n)Y$ for some quantity $c(n)$ that goes to zero as n goes to infinity (holding other parameters fixed). If C or $c(n)$ needs to depend on additional parameters, we will denote this by subscripts, e.g. $X \gg_\delta Y$ denotes a bound of the form $X \geqslant c_\delta |Y|$ for some $c_\delta > 0$.

1.1. The epsilon-net argument We begin by using the epsilon net argument to upper bound the operator norm:

Theorem 1.1.1 (Upper bound for operator norm). *Let* $M = (\xi_{ij})_{1 \leqslant i,j \leqslant n; 1 \leqslant j \leqslant n}$ *be an* $n \times n$ *Bernoulli matrix. Then with exponentially high probability (i.e.* $1 - O(e^{-cn})$ *for some* $c > 0$*), one has*

$$\|M\|_{op} = \sigma_1(M) \leqslant C\sqrt{n} \tag{1.1.2}$$

for some absolute constant C.

We remark that the above theorem in fact holds for any $C > 2$; this follows for instance from the work of Geman [32] combined with the Talagrand concentration inequality (see Exercise 1.4.5 below). We will not use this improvement to this theorem here.

Proof. We write

$$\|M\|_{op} = \sup_{x \in \mathbb{R}^n : \|x\| = 1} \|Mx\|,$$

thus the failure event $\|M\|_{op} > C\sqrt{n}$ is the union of the events $\|Mx\| > C\sqrt{n}$ as x ranges over the unit sphere. One cannot apply the union bound immediately because the number of points on the unit sphere is uncountable. Instead, we first discretise the unit sphere using the "epsilon net argument". Let Σ be a maximal $1/2$-net of the unit sphere in $\mathbb{R}^n$, that is to say a maximal $1/2$-separated subset of the sphere. Observe that if x attains the supremum in the above equation, and y is the nearest element of Σ to x, then $\|y - x\| \leqslant 1/2$, and hence by the triangle inequality

$$\|M\|_{op} = \|Mx\| \leqslant \|My\| + \|M\|_{op}\|y - x\|$$

and thus

$$\|M\|_{op} \leqslant \sup_{x \in \Sigma} \|Mx\| + \frac{1}{2}\|M\|_{op}$$

or equivalently

$$\|M\|_{op} \leqslant 2 \sup_{x \in \Sigma} \|Mx\|,$$

and so it suffices to show that

$$\mathbb{P}(\sup_{x\in\Sigma} \|Mx\| \geqslant \frac{C}{2}\sqrt{n})$$

is exponentially small in n. From the union bound, we can upper bound this by

$$\sum_{x\in\Sigma} \mathbb{P}(\|Mx\| \geqslant \frac{C}{2}\sqrt{n}).$$

The balls of radius $1/4$ around each point in Σ are disjoint, and lie in the $1/4$-neighbourhood of the sphere. From volume considerations we conclude that

$$|\Sigma| \leqslant O(1)^n \tag{1.1.3}$$

We set aside this bound as an "entropy cost" to be paid later, and focus on upper bounding, for each $x \in \Sigma$, the probability

$$\mathbb{P}(\|Mx\| \geqslant \frac{C}{2}\sqrt{n}).$$

If we let $Y_1, \ldots, Y_n \in \mathbb{R}^n$ be the rows of M, we can write this as

$$\mathbb{P}\left(\sum_{j=1}^{n} |Y_j \cdot x|^2 \geqslant \frac{C^2}{4} n\right).$$

To bound this expression, we use the same exponential moment method used to prove the Chernoff inequality. Indeed, for any parameter $c > 0$, we can use Markov's inequality to bound the above by

$$e^{-cC^2n/4}\mathbb{E}\exp(c\sum_{j=1}^{n}|Y_j \cdot x|^2).$$

As the random vectors $Y_1, \ldots, Y_n$ are iid, this is equal to

$$e^{-cC^2n/4}\left(\mathbb{E}\exp(c|Y \cdot x|^2)\right)^n.$$

Using the Chernoff inequality, the vectors $Y \cdot x$ are uniformly subgaussian, in the sense that there exist constants $C, c > 0$ such that

$$\mathbb{P}(|\xi_{ij}| \geqslant t) \leqslant C\exp(-ct^2)$$

for all $t > 0$. In particular one has

$$\mathbb{E}\exp(c|Y \cdot x|^2) \ll 1$$

if $c > 0$ is a sufficiently small absolute constant. This implies the bound

$$\mathbb{P}(\|Mx\| \geqslant \frac{C}{2}\sqrt{n}) \ll \exp(-cC^2n/8 + O(n)).$$

Taking C large enough, we obtain the claim. □

Now we use the same method to establish a lower bound in the rectangular case, first established in [4]:

Theorem 1.1.4 (Lower bound). *Let $M = (\xi_{ij})_{1\leqslant i\leqslant p; 1\leqslant j\leqslant n}$ be an $n \times p$ Bernoulli matrix, where $1 \leqslant p \leqslant (1-\delta)n$ for some $\delta > 0$ (independent of n). Then with exponentially high probability, one has $\sigma_p(M) \gg_\delta \sqrt{n}$.*

To prove this theorem, we again use the "epsilon net argument". We write

$$\sigma_p(M) = \inf_{x \in \mathbb{R}^p : \|x\|=1} \|Mx\|.$$

Let $\varepsilon > 0$ be a parameter to be chosen later. Let Σ be a maximal ε-net of the unit sphere in $\mathbb{R}^p$. Then we have

$$\sigma_p(M) \geqslant \inf_{x \in \Sigma} \|Mx\| - \varepsilon \|M\|_{op}$$

and thus by (1.1.2), we have with exponentially high probability that

$$\sigma_p(M) \geqslant \inf_{x \in \Sigma} \|Mx\| - C\varepsilon\sqrt{n},$$

and so it suffices to show that

$$\mathbb{P}(\inf_{x \in \Sigma} \|Mx\| \leqslant 2C\varepsilon\sqrt{n})$$

is exponentially small in n. From the union bound, we can upper bound this by

$$\sum_{x \in \Sigma} \mathbb{P}(\|Mx\| \leqslant 2C\varepsilon\sqrt{n}).$$

The balls of radius $\varepsilon/2$ around each point in Σ are disjoint, and lie in the $\varepsilon/2$-neighbourhood of the sphere. From volume considerations we conclude that

$$|\Sigma| \leqslant O(1/\varepsilon)^p \leqslant O(1/\varepsilon)^{(1-\delta)n}. \tag{1.1.5}$$

We again set aside this bound as an "entropy cost" to be paid later, and focus on upper bounding, for each $x \in \Sigma$, the probability

$$\mathbb{P}(\|Mx\| \leqslant 2C\varepsilon\sqrt{n}).$$

If we let $Y_1, \dots, Y_n \in \mathbb{R}^p$ be the rows of M, we can write this as

$$\mathbb{P}\left(\sum_{j=1}^{n} |Y_j \cdot x|^2 \leqslant 4C^2\varepsilon^2 n\right).$$

By Markov's inequality, the only way that this event can hold is if we have

$$|Y_j \cdot x|^2 \leqslant 8C^2\varepsilon^2/\delta$$

for at least $(1-\delta/2)n$ values of j. The number of such sets of values is at most 2^n. Applying the union bound (and paying the entropy cost of 2^n) and using symmetry, we may thus bound the above probability by

$$\leqslant 2^n \mathbb{P}(|Y_j \cdot x|^2 \leqslant 8C^2\varepsilon^2/\delta \text{ for } 1 \leqslant j \leqslant (1-\delta/2)n).$$

Now observe that the random variables $Y_j \cdot x$ are independent, and so we can bound this expression by

$$\leqslant 2^n \mathbb{P}(|Y \cdot x| \leqslant \sqrt{8}C\varepsilon/\delta^{1/2})^{(1-\delta/2)n}$$

where $Y = (\xi_1, \dots, \xi_n)$ is a random vector of iid Bernoulli signs.

We write $x = (x_1, \dots, x_n)$, so that $Y \cdot x$ is a random walk

$$Y \cdot x = \xi_1 x_1 + \cdots + \xi_n x_n.$$

To understand this walk, we apply (a slight variant) of the Berry-Esséen theorem:

Exercise 1.1.6. Show[4] that

$$\sup_t \mathbf{P}(|Y \cdot x - t| \leqslant r) \ll \frac{r}{\|x\|} + \frac{1}{\|x\|^3} \sum_{j=1}^{n} |x_j|^3$$

for any $r > 0$ and any non-zero x.

Conclude in particular that if

$$\sum_{j:|x_j| \leqslant \varepsilon} |x_j|^2 \geqslant \eta$$

for some $\eta > 0$, then

$$\sup_t \mathbf{P}(|Y \cdot x - t| \leqslant \sqrt{8} C\varepsilon) \ll_\eta \varepsilon.$$

(*Hint:* condition out all the x_j with $|x_j| > \varepsilon$.)

Let us temporarily call x *incompressible* if

$$\sum_{j:|x_j| \leqslant \varepsilon} |x_j|^2 \geqslant \eta$$

and *compressible* otherwise, where $\eta > 0$ is a parameter (independent of n) to be chosen later. If we only look at incompressible elements of Σ, we can now bound

$$\mathbf{P}(\|Mx\| \leqslant 2C\varepsilon\sqrt{n}) \ll O_\eta(\varepsilon)^{(1-\delta/2)n},$$

and comparing this against the entropy cost (1.1.5) we obtain an acceptable contribution for ε small enough (here we are crucially using the rectangular condition $p \leqslant (1-\delta)n$).

It remains to deal with the compressible vectors. Observe that such vectors lie within η of a *sparse* unit vector which is only supported in at most ε^{-2} positions. The η-entropy of these sparse vectors (i.e. the number of balls of radius η needed to cover this space) can easily be computed to be of polynomial size $O(n^{O_{\varepsilon,\eta}(1)})$ in n; thus, if η is chosen to be less than $\varepsilon/2$, the number of compressible vectors in Σ is also $O(n^{O_{\varepsilon,\eta}(1)})$. Meanwhile, we have the following crude bound:

Exercise 1.1.7. For any unit vector x, show that

$$\mathbf{P}(|Y \cdot x| \leqslant \kappa) \leqslant 1 - \kappa$$

for $\kappa > 0$ a small enough absolute constant. (*Hint:* Use the *Paley-Zygmund inequality* $\mathbf{P}(Z \geqslant \theta \mathbf{E} Z) \geqslant (1-\theta)^2 \frac{(\mathbf{E}Z)^2}{\mathbf{E}(Z^2)}$, valid for any non-negative random variable Z of finite non-zero variance, and any $0 \leqslant \theta \leqslant 1$. Bounds on higher moments on $|Y \cdot x|$ can be obtained for instance using Hoeffding's inequality, or by direct computation.) Use this to show that

$$\mathbf{P}(\|Mx\| \leqslant \alpha\sqrt{n}) \ll \exp(-cn)$$

for all such x and a sufficiently small absolute constant $\alpha > 0$, with $c > 0$ independent of α and n.

---

[4]Actually, for the purposes of this section, it would suffice to establish a weaker form of the Berry-Esséen theorem with $\sum_{j=1}^{n} |x_j|^3/\|x\|^3$ replaced by $(\sum_{j=1}^{3} |x_j|^3/\|x\|^3)^c$ for any fixed $c > 0$. This can for instance be done using the Lindeberg exchange method, discussed in Section 3.

Thus the net contribution of the compressible vectors is $O(n^{O_{\varepsilon,\eta}(1)}) \times \exp(-cn)$ which is acceptable. This concludes the proof of Theorem 1.1.4.

1.2. Singularity probability Now we turn to square iid matrices. Before we investigate the size of the least singular value of M, we first tackle the easier problem of bounding the *singularity probability*

$$\mathbb{P}(\sigma_n(M) = 0),$$

i.e. the probability that M is not invertible. The problem of computing this probability exactly is still not completely settled. Since M is singular whenever the first two rows (say) are identical, we obtain a lower bound

$$\mathbb{P}(\sigma_n(M) = 0) \geqslant \frac{1}{2^n},$$

and it is conjectured that this bound is essentially tight in the sense that

$$\mathbb{P}(\sigma_n(M) = 0) = \left(\frac{1}{2} + o(1)\right)^n,$$

but this remains open; the best bound currently is [18], and gives

$$\mathbb{P}(\sigma_n(M) = 0) \leqslant \left(\frac{1}{\sqrt{2}} + o(1)\right)^n.$$

We will not prove this bound here, but content ourselves with a weaker bound, essentially due to Komlós [43]:

Proposition 1.2.1. *We have* $\mathbb{P}(\sigma_n(M) = 0) \ll 1/n^{1/2}$.

To show this, we need the following combinatorial fact, due to Erdős [25]:

Proposition 1.2.2 (Erdős Littlewood-Offord theorem). *Let* $x = (x_1, \ldots, x_n)$ *be a vector with at least* k *nonzero entries, and let* $Y = (\xi_1, \ldots, \xi_n)$ *be a random vector of iid Bernoulli signs. Then* $\mathbb{P}(Y \cdot x = 0) \ll k^{-1/2}$.

Proof. By taking real and imaginary parts we may assume that x is real. By eliminating zero coefficients of x we may assume that $k = n$; reflecting we may then assume that all the x_i are positive. The set of $Y = (\xi_1, \ldots, \xi_n) \in \{-1, 1\}^n$ with $Y \cdot x = 0$ forms an *antichain*[5]. The product partial ordering on $\{-1, 1\}^n$ is defined by requiring $(x_1, \ldots, x_n) \leqslant (y_1, \ldots, y_n)$ iff $x_i \leqslant y_i$ for all i. On the other hand, *Sperner's theorem* asserts that all anti-chains in $\{-1, 1\}^n$ have cardinality at most $\binom{n}{\lfloor n/2 \rfloor}$. in $\{-1, 1\}^n$ with the product partial ordering. The claim now easily follows from this theorem and Stirling's formula. □

Note that we also have the obvious bound

$$\mathbb{P}(Y \cdot x = 0) \leqslant 1/2 \tag{1.2.3}$$

for any non-zero x.

Now we prove the proposition. Arguing in analogy with Section 1.1, we write

$$\mathbb{P}(\sigma_n(M) = 0) = \mathbb{P}(Mx = 0 \text{ for some nonzero } x \in \mathbb{C}^n)$$

[5] An *antichain* in a partially ordered set X is a subset S of X such that no two elements in S are comparable in the order.

(actually we can take $x \in \mathbb{R}^n$ or even $x \in \mathbb{Z}^n$ since M is integer-valued). We divide into compressible and incompressible vectors as before, but our definition of compressibility and incompressibility is slightly different now. Also, one has to do a certain amount of technical maneuvering in order to preserve the crucial independence between rows and columns.

Namely, we pick an $\varepsilon > 0$ and call x *compressible* (or more precisely *sparse*) if it is supported on at most εn coordinates, and *incompressible* otherwise.

Let us first consider the contribution of the event that $Mx = 0$ for some nonzero compressible x. Pick an x with this property which is as sparse as possible, say k-sparse for some $1 \leqslant k < \varepsilon n$. Let us temporarily fix k. By paying an entropy cost of $\lfloor \varepsilon n \rfloor \binom{n}{k}$, we may assume that it is the first k entries that are non-zero for some $1 \leqslant k \leqslant \varepsilon n$. This implies that the first k columns $Y_1, \ldots, Y_k$ of M have a linear dependence given by x; by minimality, $Y_1, \ldots, Y_{k-1}$ are linearly independent. Thus, x is uniquely determined (up to scalar multiples) by $Y_1, \ldots, Y_k$. Furthermore, as the $n \times k$ matrix formed by $Y_1, \ldots, Y_k$ has rank $k-1$, there is some $k \times k$ minor which already determines x up to constants; by paying another entropy cost of $\binom{n}{k}$, we may assume that it is the top left minor which does this. In particular, we can now use the first k rows $X_1, \ldots, X_k$ to determine x up to constants. But the remaining $n-k$ rows are independent of $X_1, \ldots, X_k$ and still need to be orthogonal to x; by Proposition 1.2.2and (1.2.3), this happens with probability at most $\min(1/2, O(1/\sqrt{k}))^{(n-k)}$, giving a total cost of

$$\sum_{1 \leqslant k \leqslant \varepsilon n} \lfloor \varepsilon n \rfloor \binom{n}{k}^2 \min(1/2, O(1/\sqrt{k}))^{-(n-k)},$$

which by Stirling's formula is acceptable (in fact this gives an exponentially small contribution).

The same argument gives that the event that $y^* M = 0$ for some nonzero compressible y also has exponentially small probability. The only remaining event to control is the event that $Mx = 0$ for some incompressible x, but that $Mz \neq 0$ and $y^* M \neq 0$ for all nonzero compressible z, y. Call this event E.

Since $Mx = 0$ for some incompressible x, we see that for at least εn values of $k \in \{1, \ldots, n\}$, the column Y_k lies in the vector space V_k spanned by the remaining $n-1$ rows of M. Let E_k denote the event that E holds, and that Y_k lies in V_k; then we see from double counting that

$$\mathbb{P}(E) \leqslant \frac{1}{\varepsilon n} \sum_{k=1}^{n} \mathbb{P}(E_k).$$

By symmetry, we thus have

$$\mathbb{P}(E) \leqslant \frac{1}{\varepsilon} \mathbb{P}(E_n).$$

To compute $\mathbb{P}(E_n)$, we freeze $Y_1, \ldots, Y_{n-1}$ consider a normal vector x to V_{n-1}; note that we can select x depending only on $Y_1, \ldots, Y_{n-1}$. We may assume that an incompressible normal vector exists, since otherwise the event E_n would be empty. We make the crucial observation that Y_n is still independent of x. By Proposition 1.2.2, we thus see that the conditional probability that $Y_n \cdot x = 0$, for fixed $Y_1, \dots, Y_{n-1}$, is $O_\varepsilon(n^{-1/2})$. We thus see that $\mathbb{P}(E) \ll_\varepsilon 1/n^{1/2}$, and the claim follows.

Remark 1.2.4. Further progress has been made on this problem by a finer analysis of the concentration probability $\mathbb{P}(Y \cdot x = 0)$, and in particular in classifying those x for which this concentration probability is large (this is known as the *inverse Littlewood-Offord problem*). Important breakthroughs in this direction were made by Halász [38] (introducing Fourier-analytic tools) and by Kahn, Komlós, and Szemerédi [40] (introducing an efficient "swapping" argument). In [57] tools from additive combinatorics (such as Freiman's theorem) were introduced to obtain further improvements, leading eventually to the results from [18] mentioned earlier.

1.3. Lower bound for the least singular value Now we return to the least singular value $\sigma_n(M)$ of an iid Bernoulli matrix, and establish a lower bound. Given that there are n singular values between 0 and $\sigma_1(M)$, which is typically of size $O(\sqrt{n})$, one expects the least singular value to be of size about $1/\sqrt{n}$ on the average. Another argument supporting this heuristic scomes from the following identity:

Exercise 1.3.1 (Negative second moment identity). Let M be an invertible $n \times n$ matrix, let $X_1, \dots, X_n$ be the rows of M, and let $C_1, \dots, C_n$ be the columns of M^{-1}. For each $1 \leqslant i \leqslant n$, let V_i be the hyperplane spanned by all the rows $X_1, \dots, X_n$ other than X_i. Show that $\|C_i\| = \mathrm{dist}(X_i, V_i)^{-1}$ and that

$$\sum_{i=1}^{n} \sigma_i(M)^{-2} = \sum_{i=1}^{n} \mathrm{dist}(X_i, V_i)^{-2}.$$

The expression $\mathrm{dist}(X_i, V_i)^2$ has a mean comparable to 1, which suggests that $\mathrm{dist}(X_i, V_i)^{-2}$ is typically of size $O(1)$, which from the negative second moment identity suggests that $\sum_{i=1}^{n} \sigma_i(M)^{-2} = O(n)$; this is consistent with the heuristic that the eigenvalues $\sigma_i(M)$ should be roughly evenly spaced in the interval $[0, 2\sqrt{n}]$ (so that $\sigma_{n-i}(M)$ should be about $(i+1)/\sqrt{n}$).

Now we give a rigorous lower bound:

Theorem 1.3.2 (Lower tail estimate for the least singular value). *For any $\lambda > 0$, one has*

$$\mathbb{P}(\sigma_n(M) \leqslant \lambda/\sqrt{n}) \ll o_{\lambda \to 0}(1) + o_{n \to \infty; \lambda}(1)$$

where $o_{\lambda \to 0}(1)$ goes to zero as $\lambda \to 0$ uniformly in n, and $o_{n \to \infty;\lambda}(1)$ goes to zero as $n \to \infty$ for each fixed λ.

This is a weaker form of a result of Rudelson and Vershynin [53] (which obtains a bound of the form $O(\lambda) + O(c^n)$ for some $c < 1$), which builds upon the earlier works [52], [59], which obtained variants of the above result.

The scale $1/\sqrt{n}$ that we are working at here is too fine to use epsilon net arguments (unless one has a *lot* of control on the entropy, which can be obtained in some cases thanks to powerful inverse Littlewood-Offord theorems, but is difficult to obtain in general.)

We can prove this theorem along similar lines to the arguments in the previous section; we sketch the method as follows. We can take λ to be small. We write the probability to be estimated as

$$\mathbb{P}(\|Mx\| \leqslant \lambda/\sqrt{n} \text{ for some unit vector } x \in \mathbb{C}^n).$$

We can assume that $\|M\|_{op} \leqslant C\sqrt{n}$ for some absolute constant C, as the event that this fails has exponentially small probability.

We pick an $\varepsilon > 0$ (not depending on λ) to be chosen later. We call a unit vector $x \in \mathbb{C}^n$ *compressible* if x lies within a distance ε of a εn-sparse vector. Let us first dispose of the case in which $\|Mx\| \leqslant \lambda/\sqrt{n}$ for some compressible x. In fact we can even dispose of the larger event that $\|Mx\| \leqslant \lambda\sqrt{n}$ for some compressible x. By paying an entropy cost of $\binom{n}{\lfloor \varepsilon n \rfloor}$, we may assume that x is within ε of a vector y supported in the first $\lfloor \varepsilon n \rfloor$ coordinates. Using the operator norm bound on M and the triangle inequality, we conclude that

$$\|My\| \leqslant (\lambda + C\varepsilon)\sqrt{n}.$$

Since y has norm comparable to 1, this implies that the least singular value of the first $\lfloor \varepsilon n \rfloor$ columns of M is $O((\lambda + \varepsilon)\sqrt{n})$. But by Theorem 1.1.4, this occurs with probability $O(\exp(-cn))$ (if λ, ε are small enough). So the total probability of the compressible event is at most

$$\binom{n}{\lfloor \varepsilon n \rfloor} O(\exp(-cn)),$$

which is acceptable if ε is small enough.

Thus we may assume now that $\|Mx\| > \lambda/\sqrt{n}$ for all compressible unit vectors x; we may similarly assume that $\|y^*M\| > \lambda/\sqrt{n}$ for all compressible unit vectors y. Indeed, we may also assume that $\|y^*M_i\| > \lambda/\sqrt{n}$ for every i, where M_i is M with the ith column removed.

The remaining case is if $\|Mx\| \leqslant \lambda/\sqrt{n}$ for some incompressible x. Let us call this event E. Write $x = (x_1, \ldots, x_n)$, and let $Y_1, \ldots, Y_n$ be the column of M, thus

$$\|x_1 Y_1 + \cdots + x_n Y_n\| \leqslant \lambda/\sqrt{n}.$$

Letting W_i be the subspace spanned by all the $Y_1, \ldots, Y_n$ except for Y_i, we conclude upon projecting to the orthogonal complement of W_i that

$$|x_i| \operatorname{dist}(Y_i, W_i) \leqslant \lambda/\sqrt{n}$$

for all i (compare with Exercise 1.3.1).

On the other hand, since x is incompressible, we see that $|x_i| \geqslant \varepsilon/\sqrt{n}$ for at least εn values of i, and thus

$$\operatorname{dist}(Y_i, W_i) \leqslant \lambda/\varepsilon. \tag{1.3.3}$$

for at least εn values of i. If we let E_i be the event that E and (1.3.3) both hold, we thus have from double-counting that

$$\mathbb{P}(E) \leqslant \frac{1}{\varepsilon n}\sum_{i=1}^{n}\mathbb{P}(E_i)$$

and thus by symmetry

$$\mathbb{P}(E) \leqslant \frac{1}{\varepsilon}\mathbb{P}(E_n)$$

(say). However, if E_n holds, then setting y to be a unit normal vector to W_i (which is necessarily incompressible, by the hypothesis on M_i), we have

$$|Y_i \cdot y| \leqslant \lambda/\varepsilon.$$

Again, the crucial point is that Y_i and y are independent. The incompressibility of y, combined with a Berry-Esséen type theorem, then gives

Exercise 1.3.4. Show that

$$\mathbb{P}(|Y_i \cdot y| \leqslant \lambda/\varepsilon) \ll \varepsilon^2$$

(say) if λ is sufficiently small and n is sufficiently large depending on ε.

This gives a bound of $O(\varepsilon)$ for $\mathbb{P}(E)$ if λ is small enough depending on ε, and n is large enough; this gives the claim.

Remark 1.3.5. By refining these arguments one can obtain an estimate of the form

$$\sigma_n\left(\frac{1}{\sqrt{n}}M_n - zI\right) \geqslant n^{-A} \tag{1.3.6}$$

for any given $A > 0$ and z of polynomial size in n, with a failure probability that is of the form $O(n^{-B})$ for some $B > 0$. By using inverse Littlewood-Offord theorems rather than the Berry-Esséen theorem, one can make B arbitrarily large (if A is sufficiently large depending on A), which is important for several applications. There are several results of this type, with overlapping ranges of generality (and various values of A) [36, 51, 58], and the exponent A is known to degrade if one has too few moment assumptions on the underlying random matrix M. This type of result (with an unspecified A) is important for the circular law, discussed in the next section.

1.4. Upper bound for the least singular value One can complement the lower tail estimate with an upper tail estimate:

Theorem 1.4.1 (Upper tail estimate for the least singular value)**.** *For any $\lambda > 0$, one has*

$$\mathbb{P}(\sigma_n(M) \geqslant \lambda/\sqrt{n}) \ll o_{\lambda\to\infty}(1) + o_{n\to\infty;\lambda}(1). \tag{1.4.2}$$

We prove this using an argument of Rudelson and Vershynin [54]. Suppose that $\sigma_n(M) > \lambda/\sqrt{n}$, then, for all y,

$$\|y^*M^{-1}\| \leqslant \sqrt{n}\|y\|/\lambda. \tag{1.4.3}$$

Next, let $X_1, \ldots, X_n$ be the rows of M, and let $C_1, \ldots, C_n$ be the columns of M^{-1}, thus $C_1, \ldots, C_n$ is a *dual basis* for $X_1, \ldots, X_n$. From (1.4.3) we have

$$\sum_{i=1}^{n} |y \cdot C_i|^2 \leqslant n\|y\|^2/\lambda^2.$$

We apply this with y equal to $X_n - \pi_n(X_n)$, where π_n is the orthogonal projection to the space V_{n-1} spanned by $X_1, \ldots, X_{n-1}$. On the one hand, we have

$$\|y\|^2 = \operatorname{dist}(X_n, V_{n-1})^2$$

and on the other hand we have for any $1 \leqslant i < n$ that

$$y \cdot C_i = -\pi_n(X_n) \cdot C_i = -X_n \cdot \pi_n(C_i)$$

and so

$$\sum_{i=1}^{n-1} |X_n \cdot \pi_n(C_i)|^2 \leqslant n \operatorname{dist}(X_n, V_{n-1})^2/\lambda^2. \tag{1.4.4}$$

If (1.4.4) holds, then $|X_n \cdot \pi_n(C_i)|^2 = O(\operatorname{dist}(X_n, V_{n-1})^2/\lambda^2)$ for at least half of the i, so the probability in (1.4.2) can be bounded by

$$\ll \frac{1}{n} \sum_{i=1}^{n-1} \mathbb{P}(|X_n \cdot \pi_n(C_i)|^2 = O(\operatorname{dist}(X_n, V_{n-1})^2/\lambda^2))$$

which by symmetry can be bounded by

$$\ll \mathbb{P}(|X_n \cdot \pi_n(C_1)|^2 = O(\operatorname{dist}(X_n, V_{n-1})^2/\lambda^2)).$$

Let $\varepsilon > 0$ be a small quantity to be chosen later. We now need an application of Talagrand's inequality:

Exercise 1.4.5. Talagrand's inequality [55] implies that if A is a convex subset of $\mathbb{R}^n$, A_t is the t-neighbourhood of A in the Euclidean metric for some $t > 0$, and X_n is a random Bernoulli vector, then

$$\mathbb{P}(X_n \in A)\mathbb{P}(X_n \notin A_t) \leqslant \exp(-ct^2)$$

for some absolute constant $c > 0$. Use this to show that for any d-dimensional subspace V of $\mathbb{R}^n$, we have

$$\mathbb{P}(|\operatorname{dist}(X_n, V) - \sqrt{n-d}| \geqslant t) \ll \exp(-ct^2)$$

for some absolute constant t. (*Hint:* one will need to combine Talagrand's inequality with a computation of $\mathbb{E} \operatorname{dist}(X_n, V)^2$.)

From Exercise 1.4.5 we know that $\operatorname{dist}(X_n, V_{n-1}) = O_\varepsilon(1)$ with probability $1 - O(\varepsilon)$, so we obtain a bound of

$$\ll \mathbb{P}(X_n \cdot \pi_n(C_1) = O_\varepsilon(1/\lambda)) + O(\varepsilon).$$

Now a key point is that the vectors $\pi_n(C_1), \ldots, \pi_n(C_{n-1})$ depend only on $X_1, \ldots, X_{n-1}$ and not on X_n; indeed, they are the dual basis for $X_1, \ldots, X_{n-1}$ in V_{n-1}. Thus, after conditioning $X_1, \ldots, X_{n-1}$ and thus $\pi_n(C_1)$ to be fixed, X_n is still a Bernoulli random vector. Applying a Berry-Esséen inequality, we obtain a bound of $O(\varepsilon)$ for the conditional probability that $X_n \cdot \pi_n(C_1) = O_\varepsilon(1/\lambda)$ for λ sufficiently large depending on ε, unless $\pi_n(C_1)$ is compressible (in the sense that, say, it is within ε of an εn-sparse vector). But this latter possibility can be controlled (with exponentially small probability) by the same type of arguments as before; we omit the details.

We remark that stronger bounds for the upper tail have recently been obtained in [48].

1.5. Asymptotic for the least singular value The distribution of singular values of a Gaussian random matrix can be computed explicitly. In particular, if M is a real Gaussian matrix (with all entries iid with distribution $N(0,1)_\mathbb{R}$), it was shown in [23] that, as $n \to \infty$, $\sqrt{n}\sigma_n(M)$ converges in distribution to the distribution $\mu_E := \frac{1+\sqrt{x}}{2\sqrt{x}} e^{-x/2-\sqrt{x}}\, dx$. It turns out that this result can be extended to other ensembles with the same mean and variance. In particular, we have the following result from [60]:

Theorem 1.5.1. *If M is an iid Bernoulli matrix, then $\sqrt{n}\sigma_n(M)$ also converges in distribution to μ_E as $n \to \infty$. (In fact there is a polynomial rate of convergence.)*

This should be compared with Theorems 1.3.2, 1.4.1 showing that $\sqrt{n}\sigma_n(M)$ have a tight sequence of distributions in $(0,+\infty)$. The arguments from [60] thus provide an alternate proof of these two theorems. The same result in fact holds for all iid ensembles obeying a finite moment condition.

The arguments used to prove Theorem 1.5.1 do not establish the limit μ_E directly, but instead use the result of [23] as a black box, focusing instead on establishing the *universality* of the limiting distribution of $\sqrt{n}\sigma_n(M)$, and in particular that this limiting distribution is the same whether one has a Bernoulli ensemble or a Gaussian ensemble.

The arguments are somewhat technical and we will not present them in full here, but instead give a sketch of the key ideas.

In previous sections we have already seen the close relationship between the least singular value $\sigma_n(M)$, and the distances $\mathrm{dist}(X_i, V_i)$ between a row X_i of M and the hyperplane V_i spanned by the other $n-1$ rows. It is not hard to use the above machinery to show that as $n \to \infty$, $\mathrm{dist}(X_i, V_i)$ converges in distribution to the absolute value $|N(0,1)_\mathbb{R}|$ of a Gaussian regardless of the underlying distribution of the coefficients of M (i.e. it is asymptotically universal). The basic point is that one can write $\mathrm{dist}(X_i, V_i)$ as $|X_i \cdot n_i|$ where n_i is a unit normal of V_i (we will assume here that M is non-singular, which by previous arguments is true asymptotically almost surely). The previous machinery lets us show that n_i is incompressible with high probability, and the claim then follows from the Berry-Esséen theorem.

Unfortunately, despite the presence of suggestive relationships such as Exercise 1.3.1, the asymptotic universality of the distances $\mathrm{dist}(X_i, V_i)$ does not directly imply asymptotic universality of the least singular value. However, it turns out that one can obtain a higher-dimensional version of the universality of the scalar quantities $\mathrm{dist}(X_i, V_i)$, as follows. For any small k (say, $1 \leqslant k \leqslant n^c$ for some small $c > 0$) and any distinct $i_1, \dots, i_k \in \{1, \dots, n\}$, a modification of the above argument shows that the covariance matrix

$$(\pi(X_{i_a}) \cdot \pi(X_{i_b}))_{1 \leqslant a,b \leqslant k} \tag{1.5.2}$$

of the orthogonal projections $\pi(X_{i_1}), \dots, \pi(X_{i_k})$ of the k rows $X_{i_1}, \dots, X_{i_k}$ to the complement $V_{i_1,\dots,i_k}^\perp$ of the space $V_{i_1,\dots,i_k}$ spanned by the other $n-k$ rows of M, is also universal, converging in distribution to the covariance[6] matrix

$$(G_a \cdot G_b)_{1 \leqslant a,b \leqslant k}$$

of k iid Gaussians $G_a \equiv N(0,1)_{\mathbb{R}}$. (Note that the convergence of $\mathrm{dist}(X_i, V_i)$ to $|N(0,1)_{\mathbb{R}}|$ is the $k = 1$ case of this claim.)

The key point is that one can show that the complement $V_{i_1,\dots,i_k}^\perp$ is usually "incompressible" in a certain technical sense, which implies that the projections $\pi(X_{i_a})$ behave like iid Gaussians on that projection thanks to a multidimensional Berry-Esséen theorem.

On the other hand, the covariance matrix (1.5.2) is closely related to the inverse matrix M^{-1}:

Exercise 1.5.3. Show that (1.5.2) is also equal to A^*A, where A is the $n \times k$ matrix formed from the $i_1, \dots, i_k$ columns of M^{-1}.

In particular, this shows that the set of singular values of k randomly selected columns of M^{-1} have a universal distribution.

Recall our goal is to show that $\sqrt{n}\sigma_n(M)$ has an asymptotically universal distribution, which is equivalent to asking that $\frac{1}{\sqrt{n}}\|M^{-1}\|_{op}$ has an asymptotically universal distribution. The goal is then to extract the operator norm of M^{-1} from looking at a random $n \times k$ minor B of this matrix. This comes from the following application of the second moment method:

Exercise 1.5.4. Let A be an $n \times n$ matrix with columns $R_1, \dots, R_n$, and let B be the $n \times k$ matrix formed by taking k of the columns $R_1, \dots, R_n$ at random. Show that

$$\mathbb{E}\|A^*A - \frac{n}{k}B^*B\|_F^2 \leqslant \frac{n}{k}\sum_{k=1}^{n} \|R_k\|^4,$$

where $\| \|_F$ is the Frobenius norm $\|A\|_F := \mathrm{tr}(A^*A)^{1/2}$.

Recall from Exercise 1.3.1 that $\|R_k\| = 1/\mathrm{dist}(X_k, V_k)$, so we expect each $\|R_k\|$ to have magnitude about $O(1)$. As such, we expect $\sigma_1((M^{-1})^*(M^{-1})) = \sigma_n(M)^{-2}$ to differ by $O(n^2/k)$ from $\frac{n}{k}\sigma_1(B^*B) = \frac{n}{k}\sigma_1(B)^2$. In principle, this gives us asymptotic universality on $\sqrt{n}\sigma_n(M)$ from the already established universality of B.

There is one technical obstacle remaining, however: while we know that each $\mathrm{dist}(X_k, V_k)$ is distributed like a Gaussian, so that each individual R_k is going to

[6]These covariance matrix distributions are also known as *Wishart distributions*.

be of size $O(1)$ with reasonably good probability, in order for the above exercise to be useful, one needs to bound *all* of the R_k *simultaneously* with high probability. A naive application of the union bound leads to terrible results here. Fortunately, there is a strong correlation between the R_k: they tend to be large together or small together, or equivalently that the distances $\operatorname{dist}(X_k, V_k)$ tend to be small together or large together. Here is one indication of this:

Lemma 1.5.5. *For any $1 \leqslant k < i \leqslant n$, one has*

$$\operatorname{dist}(X_i, V_i) \geqslant \frac{\|\pi_i(X_i)\|}{1 + \sum_{j=1}^{k} \frac{\|\pi_i(X_j)\|}{\|\pi_i(X_i)\| \operatorname{dist}(X_j, V_j)}},$$

where π_i is the orthogonal projection onto the space spanned by $X_1, \ldots, X_k, X_i$.

Proof. We may relabel so that $i = k+1$; then projecting everything by π_i we may assume that $n = k+1$. Our goal is now to show that

$$\operatorname{dist}(X_n, V_{n-1}) \geqslant \frac{\|X_n\|}{1 + \sum_{j=1}^{n-1} \frac{\|X_j\|}{\|X_n\| \operatorname{dist}(X_j, V_j)}}.$$

Recall that $R_1, \ldots, R_n$ is a dual basis to $X_1, \ldots, X_n$. This implies in particular that

$$x = \sum_{j=1}^{n} (x \cdot X_j) R_j$$

for any vector x; applying this to X_n we obtain

$$X_n = \|X_n\|^2 R_n + \sum_{j=1}^{n-1} (X_j \cdot X_n) R_j$$

and hence by the triangle inequality

$$\|X_n\|^2 \|R_n\| \leqslant \|X_n\| + \sum_{j=1}^{n-1} \|X_j\| \|X_n\| \|R_j\|.$$

Using the fact that $\|R_j\| = 1/\operatorname{dist}(X_j, R_j)$, the claim follows. □

In practice, once k gets moderately large (e.g. $k = n^c$ for some small $c > 0$), one can control the expressions $\|\pi_i(X_j)\|$ appearing here by Talagrand's inequality (Exercise 1.4.5), and so this inequality tells us that once $\operatorname{dist}(X_j, V_j)$ is bounded away from zero for $j = 1, \ldots, k$, it is bounded away from zero for all other k also. This turns out to be enough to get enough uniform control on the R_j to make Exercise 1.5.4 useful, and ultimately to complete the proof of Theorem 1.5.1.

2. The circular law

In this section, we leave the realm of self-adjoint matrix ensembles, such as Wigner random matrices, and consider instead the simplest examples of non-self-adjoint ensembles, namely the iid matrix ensembles.

The basic result in this area is

Theorem 2.0.1 (Circular law). *Let M_n be an $n \times n$ iid matrix, whose entries ξ_{ij}, $1 \leqslant i, j \leqslant n$ are iid with a fixed (complex) distribution $\xi_{ij} \equiv \xi$ of mean zero and variance one. Then the spectral measure*

$$\mu_{\frac{1}{\sqrt{n}}M_n}$$

converges (in the vague topology) both in probability and almost surely to the circular law

$$\mu_{\mathrm{circ}} := \frac{1}{\pi} 1_{|x|^2+|y|^2 \leqslant 1}\, dxdy,$$

where x, y are the real and imaginary coordinates of the complex plane.

This theorem has a long history; it is analogous to the semicircular law, but the non-Hermitian nature of the matrices makes the spectrum so unstable that key techniques that are used in the semicircular case, such as truncation and the moment method, no longer work; significant new ideas are required. In the case of random Gaussian matrices (and using the expected spectral measure rather than the actual spectral measure), this result was established by Ginibre [33] (in the complex case) and by Edelman [22] (in the real case), using the explicit formulae for the joint distribution of eigenvalues available in these cases. In 1984, Girko [34] laid out a general strategy for establishing the result for non-gaussian matrices, which formed the base of all future work on the subject; however, a key ingredient in the argument, namely a bound on the least singular value of shifts $\frac{1}{\sqrt{n}}M_n - zI$, was not fully justified at the time. A rigorous proof of the circular law was then established by Bai [9], assuming additional moment and boundedness conditions on the individual entries. These additional conditions were then slowly removed in a sequence of papers [35, 36, 51, 58], with the last moment condition being removed in [63]. There have since been several further works extending the circular law to other ensembles [1–3, 5, 10–13, 20, 21, 47, 50, 67], including several models in which the entries are no longer jointly independent; see also the surveys [14, 58]. The method also applies to various ensembles with a different limiting law than the circular law; see e.g. [49], [37]. More recently, local circular laws have also been established for various models [5, 16, 17, 65, 68].

2.1. Spectral instability One of the basic difficulties present in the non-Hermitian case is *spectral instability*: small perturbations in a large matrix can lead to large fluctuations in the spectrum. In order for any sort of analytic technique to be effective, this type of instability must somehow be precluded.

The canonical example of spectral instability comes from perturbing the right shift matrix

$$U_0 := \begin{pmatrix} 0 & 1 & 0 & \dots & 0 \\ 0 & 0 & 1 & \dots & 0 \\ \vdots & \vdots & \vdots & \ddots & \vdots \\ 0 & 0 & 0 & \dots & 0 \end{pmatrix}$$

to the matrix

$$U_\varepsilon := \begin{pmatrix} 0 & 1 & 0 & \dots & 0 \\ 0 & 0 & 1 & \dots & 0 \\ \vdots & \vdots & \vdots & \ddots & \vdots \\ \varepsilon & 0 & 0 & \dots & 0 \end{pmatrix}$$

for some $\varepsilon > 0$.

The matrix U_0 is nilpotent: $U_0^n = 0$. Its characteristic polynomial is $(-\lambda)^n$, and it thus has n repeated eigenvalues at the origin. In contrast, U_ε obeys the equation $U_\varepsilon^n = \varepsilon I$, its characteristic polynomial is $(-\lambda)^n - \varepsilon(-1)^n$, and it thus has n eigenvalues at the nth roots $\varepsilon^{1/n} e^{2\pi i j/n}$, $j = 0, \dots, n-1$ of ε. Thus, even for exponentially small values of ε, say $\varepsilon = 2^{-n}$, the eigenvalues for U_ε can be quite far from the eigenvalues of U_0, and can wander all over the unit disk. This is in sharp contrast with the Hermitian case, where eigenvalue inequalities such as the Weyl inequalities or Wielandt-Hoffman inequalities ensure stability of the spectrum.

One can explain the problem in terms of *pseudospectrum*[7]. The only spectrum of U is at the origin, so the resolvents $(U_0 - zI)^{-1}$ of U_0 are finite for all non-zero z. However, while these resolvents are finite, they can be extremely large. Indeed, from the nilpotent nature of U_0 we have the Neumann series

$$(U_0 - zI)^{-1} = -\frac{1}{z} - \frac{U_0}{z^2} - \dots - \frac{U_0^{n-1}}{z^n}$$

so for $|z| < 1$ we see that the resolvent has size roughly $|z|^{-n}$, which is exponentially large in the interior of the unit disk. This exponentially large size of resolvent is consistent with the exponential instability of the spectrum:

Exercise 2.1.1. Let M be a square matrix, and let z be a complex number. Show that $\|(M - zI)^{-1}\|_{op} \geqslant R$ if and only if there exists a perturbation $M + E$ of M with $\|E\|_{op} \leqslant 1/R$ such that $M + E$ has z as an eigenvalue.

This already hints strongly that if one wants to rigorously prove control on the spectrum of M near z, one needs some sort of upper bound on $\|(M - zI)^{-1}\|_{op}$, or equivalently one needs some sort of lower bound on the least singular value $\sigma_n(M - zI)$ of $M - zI$.

Without such a bound, though, the instability precludes the direct use of the *truncation method*, which was so useful in the Hermitian case. In particular, there is no obvious way to reduce the proof of the circular law to the case of bounded coefficients, in contrast to the semicircular law for Wigner matrices where this reduction follows easily from the Wielandt-Hoffman inequality. Instead, we must continue working with unbounded random variables throughout the argument (unless, of course, one makes an additional decay hypothesis, such as assuming

[7]The *pseudospectrum* of an operator T is the set of complex numbers z for which the operator norm $\|(T - zI)^{-1}\|_{op}$ is either infinite, or larger than a fixed threshold $1/\varepsilon$. See [66] for further discussion.

certain moments are finite; this helps explain the presence of such moment conditions in many papers on the circular law).

2.2. Incompleteness of the moment method In the Hermitian case, the moments

$$\frac{1}{n}\operatorname{tr}(\frac{1}{\sqrt{n}}M)^k = \int_{\mathbb{R}} x^k \, d\mu_{\frac{1}{\sqrt{n}}M_n}(x)$$

of a matrix can be used (in principle) to understand the distribution $\mu_{\frac{1}{\sqrt{n}}M_n}$ completely (at least, when the measure $\mu_{\frac{1}{\sqrt{n}}M_n}$ has sufficient decay at infinity). This is ultimately because the space of real polynomials $P(x)$ is dense in various function spaces (the Weierstrass approximation theorem).

In the non-Hermitian case, the spectral measure $\mu_{\frac{1}{\sqrt{n}}M_n}$ is now supported on the complex plane rather than the real line. One still has the formula

$$\frac{1}{n}\operatorname{tr}(\frac{1}{\sqrt{n}}M)^k = \int_{\mathbb{C}} z^k \, d\mu_{\frac{1}{\sqrt{n}}M_n}(z)$$

but it is much less useful now, because the space of complex polynomials $P(z)$ no longer has any good density properties[8]. In particular, the moments no longer uniquely determine the spectral measure.

This can be illustrated with the shift examples given above. It is easy to see that U and U_ε have vanishing moments up to $(n-1)^{\text{th}}$ order, i.e.

$$\frac{1}{n}\operatorname{tr}\left(\frac{1}{\sqrt{n}}U\right)^k = \frac{1}{n}\operatorname{tr}\left(\frac{1}{\sqrt{n}}U_\varepsilon\right)^k = 0$$

for $k = 1, \ldots, n-1$. Thus we have

$$\int_{\mathbb{C}} z^k \, d\mu_{\frac{1}{\sqrt{n}}U}(z) = \int_{\mathbb{C}} z^k \, d\mu_{\frac{1}{\sqrt{n}}U_\varepsilon}(z) = 0$$

for $k = 1, \ldots, n-1$. Despite this enormous number of matching moments, the spectral measures $\mu_{\frac{1}{\sqrt{n}}U}$ and $\mu_{\frac{1}{\sqrt{n}}U_\varepsilon}$ are dramatically different; the former is a Dirac mass at the origin, while the latter can be arbitrarily close to the unit circle. Indeed, even if we set *all* moments equal to zero,

$$\int_{\mathbb{C}} z^k \, d\mu = 0$$

for $k = 1, 2, \ldots$, then there are an uncountable number of possible (continuous) probability measures that could still be the (asymptotic) spectral measure μ: for instance, any measure which is rotationally symmetric around the origin would obey these conditions.

If one could somehow control the mixed moments

$$\int_{\mathbb{C}} z^k \bar{z}^l \, d\mu_{\frac{1}{\sqrt{n}}M_n}(z) = \frac{1}{n}\sum_{j=1}^n \left(\frac{1}{\sqrt{n}}\lambda_j(M_n)\right)^k \left(\frac{1}{\sqrt{n}}\overline{\lambda_j}(M_n)\right)^l$$

of the spectral measure, then this problem would be resolved, and one could use the moment method to reconstruct the spectral measure accurately. However,

[8]For instance, the uniform closure of the space of polynomials on the unit disk is not the space of continuous functions, but rather the space of holomorphic functions that are continuous on the closed unit disk.

there does not appear to be any easy way to compute this quantity; the obvious guess of $\frac{1}{n}\operatorname{tr}(\frac{1}{\sqrt{n}}M_n)^k(\frac{1}{\sqrt{n}}M_n^*)^l$ works when the matrix M_n is *normal*, as M_n and M_n^* then share the same basis of eigenvectors, but generically one does not expect these matrices to be normal.

Remark 2.2.1. The failure of the moment method to control the spectral measure is consistent with the instability of spectral measure with respect to perturbations, because moments are stable with respect to perturbations.

Exercise 2.2.2. Let $k \geqslant 1$ be an integer, and let M_n be an iid matrix whose entries have a fixed distribution ξ with mean zero, variance 1, and with k^{th} moment finite. Show that $\frac{1}{n}\operatorname{tr}(\frac{1}{\sqrt{n}}M_n)^k$ converges to zero as $n \to \infty$ in expectation, in probability, and in the almost sure sense. Thus we see that $\int_{\mathbb{C}} z^k \, d\mu_{\frac{1}{\sqrt{n}}M_n}(z)$ converges to zero in these three senses also. This is of course consistent with the circular law, but does not come close to establishing that law, for the reasons given above.

Remark 2.2.3. The failure of the moment method also shows that methods of free probability (see e.g. [46]) do not work directly. For instance, observe that for fixed ε, U_0 and U_ε (in the noncommutative probability space $(\mathrm{Mat}_n(\mathbb{C}), \frac{1}{n}\operatorname{tr})$) both converge in the sense of $*$-moments as $n \to \infty$ to that of the right shift operator on $\ell^2(\mathbb{Z})$ (with the trace $\tau(T) = \langle e_0, Te_0\rangle$, with e_0 being the Kronecker delta at 0); but the spectral measures of U_0 and U_ε are different. Thus the spectral measure cannot be read off directly from the free probability limit.

2.3. The logarithmic potential With the moment method out of consideration, attention naturally turns to the Stieltjes transform

$$s_n(z) = \frac{1}{n}\operatorname{tr}\left(\frac{1}{\sqrt{n}}M_n - zI\right)^{-1} = \int_{\mathbb{C}} \frac{d\mu_{\frac{1}{\sqrt{n}}M_n}(w)}{w - z}.$$

This is a rational function on the complex plane. Its relationship with the spectral measure is as follows:

Exercise 2.3.1. Show that

$$\mu_{\frac{1}{\sqrt{n}}M_n} = \frac{1}{\pi}\partial_{\bar{z}} s_n(z)$$

in the sense of distributions, where

$$\partial_{\bar{z}} := \frac{1}{2}\left(\frac{\partial}{\partial x} + i\frac{\partial}{\partial y}\right)$$

is the Cauchy-Riemann operator, and the Stieltjes transform is interpreted distributionally in a principal value sense.

One can control the Stieltjes transform quite effectively away from the origin. Indeed, for iid matrices with subgaussian entries, one can show that the spectral radius of $\frac{1}{\sqrt{n}}M_n$ is $1 + o(1)$ almost surely; this, combined with (2.2.2) and Laurent expansion, tells us that $s_n(z)$ almost surely converges to $-1/z$ locally uniformly in the region $\{z : |z| > 1\}$, and that the spectral measure $\mu_{\frac{1}{\sqrt{n}}M_n}$ converges almost surely to zero in this region (which can of course also be deduced directly from the spectral radius bound). This is of course consistent with the circular law, but is not sufficient to prove it (for instance, the above information is also consistent with the scenario in which the spectral measure collapses towards the origin). One also needs to control the Stieltjes transform inside the disk $\{z : |z| \leqslant 1\}$ in order to fully control the spectral measure.

For this, many existing methods for controlling the Stieltjes transform are not particularly effective in this non-Hermitian setting (mainly because of the spectral instability, and also because of the lack of analyticity in the interior of the spectrum). Instead, one proceeds by relating the Stieltjes transform to the *logarithmic potential*

$$f_n(z) := \int_{\mathbf{C}} \log |w - z| d\mu_{\frac{1}{\sqrt{n}} M_n}(w).$$

It is easy to see that $s_n(z)$ is essentially the (distributional) gradient of $f_n(z)$:

$$s_n(z) = \left(-\frac{\partial}{\partial x} + i\frac{\partial}{\partial y}\right) f_n(z),$$

and thus g_n is related to the spectral measure by the distributional formula[9]

$$\mu_{\frac{1}{\sqrt{n}} M_n} = \frac{1}{2\pi} \Delta f_n \tag{2.3.2}$$

where $\Delta := \frac{\partial^2}{\partial x^2} + \frac{\partial^2}{\partial y^2}$ is the Laplacian.

The following basic result relates the logarithmic potential to probabilistic notions of convergence.

Theorem 2.3.3 (Logarithmic potential continuity theorem). *Let M_n be a sequence of random matrices, and suppose that for almost every complex number z, $f_n(z)$ converges almost surely (resp. in probability) to*

$$f(z) := \int_{\mathbf{C}} \log |z - w| d\mu(w)$$

for some probability measure μ. Then $\mu_{\frac{1}{\sqrt{n}} M_n}$ converges almost surely (resp. in probability) to μ in the vague topology.

Proof. We prove the almost sure version of this theorem, and leave the convergence in probability version as an exercise.

On any bounded set K in the complex plane, the functions $\log |\cdot - w|$ lie in $L^2(K)$ uniformly in w. From Minkowski's integral inequality, we conclude that the f_n and f are uniformly bounded in $L^2(K)$. On the other hand, almost surely the f_n converge pointwise to f. From the dominated convergence theorem this implies that $\min(|f_n - f|, M)$ converges in $L^1(K)$ to zero for any M; using the uniform bound in $L^2(K)$ to compare $\min(|f_n - f|, M)$ with $|f_n - f|$ and then letting $M \to \infty$, we conclude that f_n converges to f in $L^1(K)$. In particular, f_n converges to f in the sense of distributions; taking distributional Laplacians using (2.3.2) we obtain the claim. □

[9]This formula just reflects the fact that $\frac{1}{2\pi} \log |z|$ is the *Newtonian potential* in two dimensions.

Exercise 2.3.4. Establish the convergence in probability version of Theorem 2.3.3.

Thus, the task of establishing the circular law then reduces to showing, for almost every z, that the logarithmic potential $f_n(z)$ converges (in probability or almost surely) to the right limit $f(z)$.

Observe that the logarithmic potential

$$f_n(z) = \frac{1}{n} \sum_{j=1}^{n} \log \left| \frac{\lambda_j(M_n)}{\sqrt{n}} - z \right|$$

can be rewritten as a log-determinant:

$$f_n(z) = \frac{1}{n} \log \left| \det(\frac{1}{\sqrt{n}} M_n - zI) \right|.$$

To compute this determinant, we recall that the determinant of a matrix A is not only the product of its eigenvalues, but also has a magnitude equal to the product of its *singular* values:

$$|\det A| = \prod_{j=1}^{n} \sigma_j(A) = \prod_{j=1}^{n} \lambda_j(A^* A)^{1/2}$$

and thus

$$f_n(z) = \frac{1}{2} \int_0^\infty \log x \, d\nu_{n,z}(x)$$

where $d\nu_{n,z}$ is the spectral measure of the matrix $\left(\frac{1}{\sqrt{n}} M_n - zI\right)^* \left(\frac{1}{\sqrt{n}} M_n - zI\right)$.

The advantage of working with this spectral measure, as opposed to the original spectral measure $\mu_{\frac{1}{\sqrt{n}} M_n}$, is that the matrix $(\frac{1}{\sqrt{n}} M_n - zI)^*(\frac{1}{\sqrt{n}} M_n - zI)$ is *self-adjoint*, and so methods such as the moment method or free probability can now be safely applied to compute the limiting spectral distribution. Indeed, Girko [34] established that for almost every z, $\nu_{n,z}$ converged both in probability and almost surely to an explicit (though slightly complicated) limiting measure ν_z in the vague topology. Formally, this implied that $f_n(z)$ would converge pointwise (almost surely and in probability) to

$$\frac{1}{2} \int_0^\infty \log x \, d\nu_z(x).$$

A lengthy but straightforward computation then showed that this expression was indeed the logarithmic potential $f(z)$ of the circular measure μ_{circ}, so that the circular law would then follow from the logarithmic potential continuity theorem.

Unfortunately, the vague convergence of $\nu_{n,z}$ to ν_z only allows one to deduce the convergence of $\int_0^\infty F(x) \, d\nu_{n,z}$ to $\int_0^\infty F(x) \, d\nu_z$ for F continuous and compactly supported. The logarithm function $\log x$ has singularities at zero and at infinity, and so the convergence

$$\int_0^\infty \log x \, d\nu_{n,z}(x) \to \int_0^\infty \log x \, d\nu_z(x)$$

can fail if the spectral measure $\nu_{n,z}$ sends too much of its mass to zero or to infinity.

The latter scenario can be easily excluded, either by using operator norm bounds on M_n (when one has enough moment conditions) or even just the Frobenius norm bounds (which require no moment conditions beyond the unit variance). The real difficulty is with preventing mass from going to the origin.

The approach of Bai [9] proceeded in two steps. Firstly, he established a polynomial lower bound

$$\sigma_n\left(\frac{1}{\sqrt{n}}M_n - zI\right) \geqslant n^{-C}$$

asymptotically almost surely for the least singular value of $\frac{1}{\sqrt{n}}M_n - zI$. This has the effect of capping off the $\log x$ integrand to be of size $O(\log n)$. Next, by using Stieltjes transform methods, the convergence of $\nu_{n,z}$ to ν_z in an appropriate metric (e.g. the Levi distance metric) was shown to be polynomially fast, so that the distance decayed like $O(n^{-c})$ for some $c > 0$. The $O(n^{-c})$ gain can safely absorb the $O(\log n)$ loss, and this leads to a proof of the circular law assuming enough boundedness and continuity hypotheses to ensure the least singular value bound and the convergence rate. This basic paradigm was also followed by later works [36,51,58], with the main new ingredient being the advances in the understanding of the least singular value (Section 1).

Unfortunately, to get the polynomial convergence rate, one needs some moment conditions beyond the zero mean and unit variance rate (e.g. finite $2+\eta^{\text{th}}$ moment for some $\eta > 0$). In [63] the additional tool of the Talagrand concentration inequality (see Exercise 1.4.5) was used to eliminate the need for the polynomial convergence. Intuitively, the point is that only a small fraction of the singular values of $\frac{1}{\sqrt{n}}M_n - zI$ are going to be as small as n^{-c}; most will be much larger than this, and so the $O(\log n)$ bound is only going to be needed for a small fraction of the measure. To make this rigorous, it turns out to be convenient to work with a slightly different formula for the determinant magnitude $|\det(A)|$ of a square matrix than the product of the eigenvalues, namely the base-times-height formula

$$|\det(A)| = \prod_{j=1}^{n} \operatorname{dist}(X_j, V_j)$$

where X_j is the jth row and V_j is the span of $X_1, \ldots, X_{j-1}$.

Exercise 2.3.5. Establish the inequality

$$\prod_{j=n+1-m}^{n} \sigma_j(A) \leqslant \prod_{j=1}^{m} \operatorname{dist}(X_j, V_j) \leqslant \prod_{j=1}^{m} \sigma_j(A)$$

for any $1 \leqslant m \leqslant n$. (*Hint:* the middle product is the product of the singular values of the first m rows of A, and so one should try to use the Cauchy interlacing inequality for singular values.) Thus we see that $\operatorname{dist}(X_j, V_j)$ is a variant of $\sigma_j(A)$.

The least singular value bounds above (and in Remark 1.3.5), translated in this language (with $A := \frac{1}{\sqrt{n}}M_n - zI$), tell us that $\operatorname{dist}(X_j, V_j) \geqslant n^{-C}$ with high probability; this lets us ignore the most dangerous values of j, namely those j that are equal to $n - O(n^{0.99})$ (say). For low values of j, say $j \leqslant (1-\delta)n$ for some small δ, one can use the moment method to get a good lower bound for the distances and the singular values, to the extent that the logarithmic singularity of $\log x$ no longer causes difficulty in this regime; the limit of this contribution can then be seen by moment method or Stieltjes transform techniques to be *universal* in the sense that it does not depend on the precise distribution of the components of M_n. In the medium regime $(1-\delta)n < j < n - n^{0.99}$, one can use Talagrand's inequality (Exercise 1.4.5) to show that $\operatorname{dist}(X_j, V_j)$ has magnitude about $\sqrt{n-j}$, giving rise to a net contribution to $f_n(z)$ of the form $\frac{1}{n}\sum_{(1-\delta)n<j<n-n^{0.99}} O(\log\sqrt{n-j})$, which is small. Putting all this together, one can show that $f_n(z)$ converges to a universal limit as $n \to \infty$ (independent of the component distributions); see [63] for details. As a consequence, once the circular law is established for one class of iid matrices, such as the complex Gaussian random matrix ensemble, it automatically holds for all other ensembles also.

Exercise 2.3.6. (i) If μ is the circular law measure $\frac{1}{\pi}1_{|z|\leqslant 1}dz$, show that the logarithmic potential $f(z)$ is equal to $\log|z|$ for $|z| \geqslant 1$ and $\frac{|z|^2-1}{2}$ for $|z| \leqslant 1$.

(ii) If M_n is a random Bernoulli matrix and z is a fixed complex number, show that the quantity $\det(\frac{1}{\sqrt{n}}M_n - zI)$ has mean $(-z)^n$ and variance $\sum_{j=0}^{n-1}\frac{n!|z|^{2j}}{n^{n-j}j!}$. Use this to obtain a heuristic prediction as to the size of the quantity $f_n(z)$ discussed above, and argue why this is consistent with the circular law and the calculation in (i).

3. The Lindeberg exchange method

The *central limit theorem* asserts that if $X_1, X_2, \ldots$ are a sequence of iid real random variables of mean zero and variance one, then the normalised means

$$\frac{X_1 + \cdots + X_n}{\sqrt{n}}$$

converge in distribution to the normal distribution $N(0,1)$ as $n \to \infty$, thus

$$\lim_{n\to\infty} \mathbb{P}\left(\frac{X_1 + \cdots + X_n}{\sqrt{n}} \leqslant t\right) = P(G \leqslant t)$$

for any $t \in \mathbb{R}$, where G is a normally distributed random variable of mean zero and variance one. Equivalently, if $F : \mathbb{R} \to \mathbb{R}$ is any smooth, compactly supported function, then one has

$$\mathbb{E}F\left(\frac{X_1 + \cdots + X_n}{\sqrt{n}}\right) = \mathbb{E}F(G) + o(1) \tag{3.0.1}$$

as $n \to \infty$, where $o(1)$ denotes a quantity that goes to zero as $n \to \infty$.

Exercise 3.0.2. Why are these two claims equivalent?

A consequence of the central limit theorem is that the statistics $\mathbb{E}F(\frac{X_1+\cdots+X_n}{\sqrt{n}})$ are asymptotically *universal* in the sense that they do not depend on the precise distribution of the individual random variables. In particular, if $Y_1, \ldots, Y_n$ are another sequence of iid random variables, with a different distribution than the $X_1, \ldots, X_n$, then we have the asymptotic universality property

$$\mathbb{E}F\left(\frac{X_1 + \cdots + X_n}{\sqrt{n}}\right) = \mathbb{E}F\left(\frac{Y_1 + \cdots + Y_n}{\sqrt{n}}\right) + o(1) \tag{3.0.3}$$

as $n \to \infty$.

The central limit theorem is traditionally proven by Fourier analytic methods, and then the universality (3.0.3) obtained as a corollary. However, one could also argue in the reverse direction, establishing the central limit theorem (3.0.1) as a consequence. Indeed, in 1922 Lindeberg [44] established the central limit theorem by establishing the following three claims:

(1) If $Y_1, Y_2, \ldots$ were an iid sequence of *gaussian* random variables of mean zero and variance one, then
$$\mathbb{E}F\left(\frac{Y_1 + \cdots + Y_n}{\sqrt{n}}\right) = \mathbb{E}F(G) + o(1).$$

(2) If $X_1, X_2, \ldots$ were an iid sequence of random variables mean zero and variance one, and finite third moment $\mathbb{E}|X_i|^3 < \infty$, then
$$\mathbb{E}F\left(\frac{X_1 + \cdots + X_n}{\sqrt{n}}\right) - \mathbb{E}F\left(\frac{Y_1 + \cdots + Y_n}{\sqrt{n}}\right) = o(1). \tag{3.0.4}$$

(3) If the central limit theorem (3.0.1) was true for iid random variables of mean zero, variance one, and finite third moment, it would also be true without the finite third moment condition.

Clearly, the central limit theorem is immediate from the above three claims.

The first claim is an easy computation: since the sum of any finite number of independent gaussian random variables is still gaussian, the variable $\frac{Y_1+\cdots+Y_n}{\sqrt{n}}$ is also gaussian. Since this variable has mean zero and variance one, the claim follows (indeed we don't even have the $o(1)$ error in this case).

The third claim is also an easy consequence of a standard *truncation argument*. Suppose that $X_1, X_2, \ldots$ were iid copies of a random variable X of mean zero and variance one, but possibly infinite third moment. For any cutoff parameter M, the truncation $X1_{|X| \leqslant M}$ will have finite third moment; it might not have mean zero and variance one, but from the dominated convergence theorem we see that the mean of this random variable goes to zero, and the variance goes to one, as $M \to \infty$. Similarly, the tail $X1_{|X|>M}$ has variance going to zero as $M \to \infty$. By adjusting the former random variable slightly to normalise the mean and variance, we thus see that for any $\varepsilon > 0$, one can obtain a decomposition
$$X = X'_\varepsilon + X''_\varepsilon$$
where X'_ε is a random variable of mean zero, variance one, and finite third moment, while X''_ε is a random variable of variance at most ε (and necessarily of mean zero, by linearity of expectation). The random variables X'_ε and X''_ε may be coupled to each other, but this will not concern us. We can therefore split each

X_i as $X_i = X'_{i,\varepsilon} + X''_{i,\varepsilon}$, where $X'_{1,\varepsilon}, \dots, X'_{n,\varepsilon}$ are iid copies of X'_ε, and $X''_{1,\varepsilon}, \dots, X''_{n,\varepsilon}$ are iid copies of X''_ε. We then have

$$\frac{X_1 + \cdots + X_n}{\sqrt{n}} = \frac{X'_{1,\varepsilon} + \cdots + X'_{n,\varepsilon}}{\sqrt{n}} + \frac{X''_{1,\varepsilon} + \cdots + X''_{n,\varepsilon}}{\sqrt{n}}.$$

If the central limit theorem holds in the case of finite third moment, then the random variable

$$\frac{X'_{1,\varepsilon} + \cdots + X'_{n,\varepsilon}}{\sqrt{n}}$$

converges in distribution to G. Meanwhile, the random variable

$$\frac{X''_{1,\varepsilon} + \cdots + X''_{n,\varepsilon}}{\sqrt{n}}$$

has mean zero and variance at most ε. From this, we see that (3.0.1) holds up to an error of $O(\varepsilon)$; sending ε to zero, we obtain the claim.

The heart of the Lindeberg argument is in the second claim. Without loss of generality we may take the tuple $(X_1, \dots, X_n)$ to be independent of the tuple $(Y_1, \dots, Y_n)$. The idea is not to swap the $X_1, \dots, X_n$ with the $Y_1, \dots, Y_n$ all at once, but instead to swap them one at a time. Indeed, one can write the difference

$$\mathbb{E}F\left(\frac{X_1 + \cdots + X_n}{\sqrt{n}}\right) - \mathbb{E}F\left(\frac{Y_1 + \cdots + Y_n}{\sqrt{n}}\right)$$

as a telescoping series formed by the sum of the n terms

$$\begin{aligned} &\mathbb{E}F\left(\frac{Y_1 + \cdots + Y_{i-1} + X_i + X_{i+1} + \cdots + X_n}{\sqrt{n}}\right) \\ &- \mathbb{E}F\left(\frac{Y_1 + \cdots + Y_{i-1} + Y_i + \cdots + Y_n}{\sqrt{n}}\right) \end{aligned} \tag{3.0.5}$$

for $i = 1, \dots, n$, each of which represents a single swap from X_i to Y_i. Thus, to prove (3.0.4), it will suffice to show that each of the terms (3.0.5) is of size $o(1/n)$ (uniformly in i).

We can write (3.0.5) as

$$\mathbb{E}F\left(Z_i + \frac{1}{\sqrt{n}}X_i\right) - \mathbb{E}F\left(Z_i + \frac{1}{\sqrt{n}}Y_i\right)$$

where Z_i is the random variable

$$Z_i := \frac{Y_1 + \cdots + Y_{i-1} + X_{i+1} + \cdots + X_n}{\sqrt{n}}.$$

The key point here is that Z_i is independent of both X_i and Y_i. To exploit this, we use the smoothness of F to perform a Taylor expansion

$$F(Z_i + \frac{1}{\sqrt{n}}X_i) = F(Z_i) + \frac{1}{\sqrt{n}}X_i F'(Z_i) + \frac{1}{2n}X_i^2 F''(Z_i) + O\left(\frac{1}{n^{3/2}}|X_i|^3\right)$$

and hence on taking expectations (and using the finite third moment hypothesis and independence of X_i from Z_i)

$$\mathbb{E}F(Z_i + \frac{1}{\sqrt{n}}X_i) = \mathbb{E}F(Z_i) + \frac{1}{\sqrt{n}}\mathbb{E}X_i \mathbb{E}F'(Z_i) + \frac{1}{2n}\mathbb{E}X_i^2 \mathbb{E}F''(Z_i) + O\left(\frac{1}{n^{3/2}}\right).$$

Similarly

$$\mathbb{E}F(Z_i + \frac{1}{\sqrt{n}}Y_i) = \mathbb{E}F(Z_i) + \frac{1}{\sqrt{n}}\mathbb{E}Y_i\mathbb{E}F'(Z_i) + \frac{1}{2n}\mathbb{E}Y_i^2\mathbb{E}F''(Z_i) + O\left(\frac{1}{n^{3/2}}\right).$$

Now observe that as X_i and Y_i both have mean zero and variance one, their first two moments match: $\mathbb{E}X_i = \mathbb{E}Y_i$ and $\mathbb{E}X_i^2 = \mathbb{E}Y_i^2$. As such, the first three terms of the above two right-hand sides match, and thus (3.0.5) is bounded by $O(n^{-3/2})$, which is $o(1/n)$ as required.

Note how important it was that we had two matching moments; if we only had one matching moment, then the bound for (3.0.5) would only be $O(1/n)$, which is not sufficient. (And of course the central limit theorem would fail as stated if we did not correctly normalise the variance.) One can therefore think of the central limit theorem as a *two moment theorem*, asserting that the asymptotic behaviour of the statistic $\mathbb{E}F(\frac{X_1+\cdots+X_n}{\sqrt{n}})$ for iid random variables $X_1, \ldots, X_n$ depends only on the first two moments of the X_i.

Exercise 3.0.6 (Lindeberg central limit theorem). Let $m_1, m_2, \ldots$ be a sequence of natural numbers. For each n, let $X_{n,1}, \ldots, X_{n,m_n}$ be a collection of independent real random variables of mean zero and total variance one, thus

$$\mathbb{E}X_{n,i} = 0 \; \forall 1 \leqslant i \leqslant m_n$$

and

$$\sum_{i=1}^{m_n} \mathbb{E}X_{n,i}^2 = 1.$$

Suppose also that for every fixed $\delta > 0$ (not depending on n), one has

$$\sum_{i=1}^{m_n} \mathbb{E}X_{n,i}^2 1_{|X_{n,i}| \geqslant \delta} = o(1)$$

as $n \to \infty$. Show that the random variables $\sum_{i=1}^{m_n} X_{n,i}$ converge in distribution as $n \to \infty$ to a normal variable of mean zero and variance one.

Exercise 3.0.7 (Martingale central limit theorem). Let $\mathcal{F}_0 \subset \mathcal{F}_1 \subset \mathcal{F}_2 \subset \ldots$ be an increasing collection of σ-algebras in the ambient sample space. For each n, let X_n be a real random variable that is measurable with respect to $\mathcal{F}_n$, with conditional mean and variance one with respect to $\mathcal{F}_0$, thus

$$\mathbb{E}(X_n | \mathcal{F}_{n-1}) = 0$$

and

$$\mathbb{E}(X_n^2 | \mathcal{F}_{n-1}) = 1$$

almost surely. Assume also the bounded third moment hypothesis

$$\mathbb{E}(|X_n|^3 | \mathcal{F}_{n-1}) \leqslant C$$

almost surely for all n and some finite C independent of n. Show that the random variables $\frac{X_1+\cdots+X_n}{\sqrt{n}}$ converge in distribution to a normal variable of mean zero and variance one. (One can relax the hypotheses on this martingale central limit theorem substantially, but we will not explore this here.)

Exercise 3.0.8 (Weak Berry-Esséen theorem). Let $X_1, \dots, X_n$ be an iid sequence of real random variables of mean zero and variance 1, and bounded third moment: $\mathbb{E}|X_i|^3 = O(1)$. Let G be a gaussian random variable of mean zero and variance one. Using the Lindeberg exchange method, show that

$$\mathbb{P}(\frac{X_1 + \cdots + X_n}{\sqrt{n}} \leqslant t) = \mathbb{P}(G \leqslant t) + O(n^{-1/8})$$

for any $t \in \mathbb{R}$. (The full Berry-Esséen theorem improves the error term to $O(n^{-1/2})$, but this is difficult to establish purely from the Lindeberg exchange method; one needs alternate methods, such as Fourier-based methods or Stein's method, to recover this improved gain.) What happens if one assumes more matching moments between the X_i and G, such as, for example, matching third moment $\mathbb{E}X_i^3 = \mathbb{E}G^3$ or matching fourth moment $\mathbb{E}X_i^4 = \mathbb{E}G^4$?

One can think of the Lindeberg method as having the schematic form of the telescoping identity

$$X^n - Y^n = \sum_{i=1}^n Y^{i-1}(X - Y)X^{n-i}$$

which is valid in any (possibly non-commutative) ring. It breaks the symmetry of the indices $1, \dots, n$ of the random variables $X_1, \dots, X_n$ and $Y_1, \dots, Y_n$, by performing the swaps in a specified order. A more symmetric variant of the Lindeberg method was introduced recently by Knowles and Yin [41], and has the schematic form of the fundamental theorem of calculus identity

$$X^n - Y^n = \int_0^1 \sum_{i=1}^n ((1-\theta)X + \theta Y)^{i-1}(X - Y)((1-\theta)X + \theta Y)^{n-i}\, d\theta,$$

which is valid in any (possibly non-commutative) real algebra, and can be established by computing the θ derivative of $((1-\theta)X + \theta Y)^n$. We illustrate this method by giving a slightly different proof of (3.0.4). We may again assume that the $X_1, \dots, X_n$ and $Y_1, \dots, Y_n$ are independent of each other. We introduce auxiliary random variables $t_1, \dots, t_n$, drawn uniformly at random from $[0, 1]$, independently of each other and of the $X_1, \dots, X_n$ and $Y_1, \dots, Y_n$. For any $0 \leqslant \theta \leqslant 1$ and $1 \leqslant i \leqslant n$, let $X_i^{(\theta)}$ denote the random variable

$$X_i^{(\theta)} = 1_{t_i \leqslant \theta} X_i + 1_{t_i > \theta} Y_i,$$

thus for instance $X_i^{(0)} = Y_i$ and $X_i^{(1)} = X_i$ almost surely. One then has the following key derivative computation:

Exercise 3.0.9. With the notation and assumptions as above, show that

$$\frac{d}{d\theta}\mathbb{E}F\left(\frac{X_1^{(\theta)} + \cdots + X_n^{(\theta)}}{\sqrt{n}}\right) = \sum_{i=1}^n \mathbb{E}F\left(Z_i^{(\theta)} + \frac{1}{\sqrt{n}}X_i\right) - \mathbb{E}F\left(Z_i^{(\theta)} + \frac{1}{\sqrt{n}}Y_i\right)$$

where

$$Z_i^{(\theta)} := \frac{X_1^{(\theta)} + \cdots + X_{i-1}^{(\theta)} + X_{i+1}^{(\theta)} + \cdots + X_n^{(\theta)}}{\sqrt{n}}.$$

In particular, the derivative on the left-hand side exists and depends continuously on θ.

From the above exercise and the fundamental theorem of calculus, we can write the left-hand side of (3.0.4) as

$$\int_0^1 \sum_{i=1}^n \mathbf{E}F(Z_i^{(\theta)} + \frac{1}{\sqrt{n}}X_i) - \mathbf{E}F(Z_i^{(\theta)} + \frac{1}{\sqrt{n}}Y_i)\, d\theta.$$

Repeating the Taylor expansion argument used in the original Lindeberg method, we see that

$$\mathbf{E}F\left(Z_i^{(\theta)} + \frac{1}{\sqrt{n}}X_i\right) - \mathbf{E}F\left(Z_i^{(\theta)} + \frac{1}{\sqrt{n}}Y_i\right) = O(n^{-3/2})$$

thus giving an alternate proof of (3.0.4).

Remark 3.0.10. In this particular instance, the Knowles-Yin version of the Lindeberg method did not offer any significant advantages over the original Lindeberg method. However, the more symmetric form of the former is useful in some random matrix theory applications due to the additional cancellations this induces. See [41] for an example of this.

The Lindeberg exchange method was first employed to study statistics of random matrices in [19]; the method was applied to local statistics first in [61] and then (with a simplified approach) in [42]. (See also [6–8] for another application of the Lindeberg method to random matrix theory.) To illustrate this method, we will (for simplicity of notation) restrict our attention to real Wigner matrices $M_n = \frac{1}{\sqrt{n}}(\xi_{ij})_{1\leqslant i,j\leqslant n}$, where the ξ_{ij} are real random variables of mean zero and variance either one (if $i \neq j$) or two (if $i = j$), with the symmetry condition $\xi_{ij} = \xi_{ji}$ for all $1 \leqslant i, j \leqslant n$, and with the upper-triangular entries ξ_{ij} for $1 \leqslant i \leqslant j \leqslant n$ jointly independent. For technical reasons we will also assume that the ξ_{ij} are uniformly subgaussian, thus there exist constants $C, c > 0$ such that

$$\mathbf{P}(|\xi_{ij}| \geqslant t) \leqslant C\exp(-ct^2)$$

for all $t > 0$. In particular, the kth moments of the ξ_{ij} will be bounded for any fixed natural number k. These hypotheses can be relaxed somewhat, but we will not aim for the most general results here. One could easily consider Hermitian Wigner matrices instead of real Wigner matrices, after some minor modifications to the notation and discussion below (e.g. replacing the Gaussian Orthogonal Ensemble (GOE) with the Gaussian Unitary Ensemble (GUE)). The $\frac{1}{\sqrt{n}}$ term appearing in the definition of M_n is a standard normalising factor to ensure that the spectrum of M_n (usually) stays bounded (indeed it will almost surely obey the famous *semicircle law* restricting the spectrum almost completely to the interval $[-2, 2]$).

The most well known example of a real Wigner matrix ensemble is the *Gaussian Orthogonal Ensemble* (GOE), in which the ξ_{ij} are all real gaussian random variables (with the mean and variance prescribed as above). This is a highly symmetric ensemble, being invariant with respect to conjugation by any element of the orthogonal group $O(n)$, and as such many of the sought-after spectral statistics of GOE matrices can be computed explicitly (or at least asymptotically) by direct computation of certain multidimensional integrals (which, in the case of GOE, eventually reduces to computing the integrals of certain Pfaffian kernels). We will not discuss these explicit computations further here, but view them as the analogue to the simple gaussian computation used to establish Claim 1 of Lindeberg's proof of the central limit theorem. Instead, we will focus more on the analogue of Claim 2 - using an exchange method to compare statistics for GOE to statistics for other real Wigner matrices.

We can write a Wigner matrix $\frac{1}{\sqrt{n}}(\xi_{ij})_{1\leqslant i,j\leqslant n}$ as a sum

$$M_n = \frac{1}{\sqrt{n}} \sum_{(i,j)\in\Delta} \xi_{ij} E_{ij}$$

where Δ is the upper triangle

$$\Delta := \{(i,j) : 1 \leqslant i \leqslant j \leqslant n\}$$

and the real symmetric matrix E_{ij} is defined to equal $E_{ij} = e_i^T e_j + e_j^T e_i$ for $i < j$, and $E_{ii} = e_i^T e_i$ in the diagonal case $i = j$, where $e_1, \dots, e_n$ are the standard basis of $\mathbb{R}^n$ (viewed as column vectors). Meanwhile, a GOE matrix G_n can be written

$$G_n = \sum_{(i,j)\in\Delta} \eta_{ij} E_{ij}$$

where $\eta_{ij}, (i,j) \in \Delta$ are jointly independent gaussian real random variables, of mean zero and variance either one (for $i \neq j$) or two (for $i = j$). For any natural number m, we say that M_n and G_n have m *matching moments* if we have

$$\mathbb{E}\xi_{ij}^k = \mathbb{E}\eta_{ij}^k$$

for all $k = 1, \dots, m$. Thus for instance, we always have two matching moments, by our definition of a Wigner matrix.

If $S(M_n)$ is any (deterministic) statistic of a Wigner matrix M_n, we say that we have a m *moment theorem* for the statistic $S(M_n)$ if one has

$$S(M_n) - S(G_n) = o(1) \tag{3.0.11}$$

whenever M_n and G_n have m matching moments. In most applications m will be very small, either equal to 2, 3, or 4.

One can prove moment theorems using the Lindeberg exchange method. For instance, consider statistics of the form $S(M_n) = \mathbb{E}F(M_n)$, where $F : V \to \mathbb{C}$ is a bounded measurable function on the space V of real symmetric matrices. Then the left-hand side of (3.0.11) is the sum of $|\Delta| = \frac{n(n+1)}{2}$ terms of the form

$$\mathbb{E}F\Big(M_{n,i,j} + \frac{1}{\sqrt{n}}\xi_{ij}E_{ij}\Big) - \mathbb{E}F\Big(M_{n,i,j} + \frac{1}{\sqrt{n}}\eta_{ij}E_{ij}\Big)$$

where for each $(i,j) \in \Delta$, $M_{n,i,j}$ is the real symmetric matrix

$$M_{n,i,j} := \sum_{(i',j')<(i,j)} \frac{1}{\sqrt{n}} \eta_{i'j'} E_{i'j'} + \sum_{(i',j')>(i,j)} \frac{1}{\sqrt{n}} \xi_{i'j'} E_{i'j'}$$

where one imposes some arbitrary ordering $<$ on Δ (e.g. the lexicographical ordering). Strictly speaking, the $M_{n,i,j}$ are not Wigner matrices, because their ij entry vanishes and thus has zero variance, but their behaviour turns out to be almost identical to that of a Wigner matrix (being a rank one or rank two perturbation of such a matrix). Thus, to prove an m moment theorem for $\mathbb{E}F(M_n)$, it thus suffices by the triangle inequality to establish a bound of the form

$$\mathbb{E}F\left(M_{n,i,j} + \frac{1}{\sqrt{n}}\xi_{ij}E_{ij}\right) - \mathbb{E}F\left(M_{n,i,j} + \frac{1}{\sqrt{n}}\eta_{ij}E_{ij}\right) = o(n^{-2}) \tag{3.0.12}$$

for all $(i,j) \in \Delta$, whenever ξ_{ij} and η_{ij} have m matching moments. (For the diagonal entries $i = j$, a bound of $o(n^{-1})$ will in fact suffice, as the number of such entries is n rather than $O(n^2)$.)

As with the Lindeberg proof of the central limit theorem, one can establish (3.0.12) for various statistics F by performing a Taylor expansion with remainder. To illustrate this method, we consider the expectation $\mathbb{E}s(M_n, z)$ of the *Stieltjes transform*

$$s(M_n, z) := \frac{1}{n}\operatorname{tr} R(M_n, z)$$

for some complex number $z = E + i\eta$ with $\eta > 0$, where $R(M_n, z) := (M_n - z)^{-1}$ denotes the resolvent (also known as the *Green's function*), and we identify z with the matrix zI_n. The application of the Lindeberg exchange method to quantities relating to Green's functions is also referred to as the *Green's function comparison method*. The Stieltjes transform is closely tied to the behavior of the eigenvalues $\lambda_1, \dots, \lambda_n$ of M_n (counting multiplicity), thanks to the spectral identity

$$s(M_n, z) = \frac{1}{n}\sum_{j=1}^{n} \frac{1}{\lambda_j - z}. \tag{3.0.13}$$

On the other hand, the resolvents $R(M_n, z)$ are particularly amenable to the Lindeberg exchange method, thanks to the *resolvent identity*

$$R(M_n + A_n, z) = R(M_n, z) - R(M_n, z)A_n R(M_n + A_n, z)$$

whenever M_n, A_n, z are such that both sides are well defined (e.g. if M_n, A_n are real symmetric and z has positive imaginary part); this identity is easily verified by multiplying both sides by $M_n - z$ on the left and $M_n + A_n - z$ on the right. One can iterate this to obtain the Neumann series

$$R(M_n + A_n, z) = \sum_{k=0}^{\infty} (-R(M_n, z)A_n)^k R(M_n, z), \tag{3.0.14}$$

assuming that the matrix $R(M_n, z)A_n$ has spectral radius less than one. Taking normalised traces and using the cyclic property of trace, we conclude in particular that

$$\begin{aligned} &s(M_n + A_n, z) \\ (3.0.15) \qquad &= s(M_n, z) + \sum_{k=1}^{\infty} \frac{(-1)^k}{n} \operatorname{tr}\left(R(M_n, z)^2 A_n (R(M_n, z) A_n)^{k-1}\right). \end{aligned}$$

To use these identities, we invoke the following useful facts about Wigner matrices:

Theorem 3.0.16. *Let M_n be a real Wigner matrix, let $\lambda_1, \dots, \lambda_n$ be the eigenvalues, and let $u_1, \dots, u_n$ be an orthonormal basis of eigenvectors. Let $A > 0$ be any constant. Then with probability $1 - O_A(n^{-A})$, the following statements hold:*

(i) *(Weak local semi-circular law) For any interval $I \subset \mathbb{R}$, the number of eigenvalues in I is at most $n^{o(1)}(1 + n|I|)$.*
(ii) *(Eigenvector delocalisation) All coefficients of all of the eigenvectors $u_1, \dots, u_n$ have magnitude $O(n^{-1/2+o(1)})$.*

The same claims hold if one replaces one of the entries of M_n, together with its transpose, by zero.

Proof. See for instance [61, Theorem 60, Proposition 62, Corollary 63]; related results are also given in the lectures of Erdős. Estimates of this form were introduced in the work of Erdős, Schlein, and Yau [27–29]. One can sharpen these estimates in various ways (e.g. one has improved estimates near the edge of the spectrum), but the form of the bounds listed here will suffice for the current discussion. □

This yields some control on resolvents:

Exercise 3.0.17. Let M_n be a real Wigner matrix, let $A > 0$ be a constant, and let $z = E + i\eta$ with $\eta > 0$. Show that with probability $1 - O_A(n^{-A})$, all coefficients of $R(M_n, z)$ are of magnitude $O(n^{o(1)}(1 + \frac{1}{n\eta}))$, and all coefficients of $R(M_n, z)^2$ are of magnitude $O(n^{o(1)}\eta^{-1}(1 + \frac{1}{n\eta}))$. (*Hint:* use the spectral theorem to express $R(M_n, z)$ in terms of the eigenvalues and eigenvectors of M_n. You may wish to treat the $\eta > 1/n$ and $\eta \leqslant 1/n$ cases separately.) The bounds here are not optimal regarding the off-diagonal terms of $R(M_n, z)$ or $R(M_n, z)^2$; see [31] for some stronger estimates.

On the exceptional event where the above exercise fails, we can use the crude estimate (from spectral theory) that $R(M_n, z)$ has operator norm at most $1/\eta$.

We are now ready to establish the bound (3.0.12) for the Stieltjes transform statistic $F(M_n) := s(M_n, z)$ for certain choices of Wigner ensemble M_n and certain choices of spectral parameter $z = E + i\eta$. Let (i, j) be an element of Δ. By Exercise 3.0.17, we have with probability $1 - O(n^{-100})$ (say) that all coefficients of $R(M_{n,i,j}, z)$ are of magnitude $O(n^{o(1)}(1 + \frac{1}{n\eta}))$, and all coefficients of $R(M_{n,i,j}, z)^2$ are of magnitude $O(\eta^{-1} n^{o(1)}(1 + \frac{1}{n\eta}))$; also from the subgaussian hypothesis we may assume that $\xi_{ij} = O(n^{o(1)})$ without significantly increasing the failure probability of the above event. Among other things, this implies that $R(M_{n,i,j}, z)E_{ij}$ has spectral radius $O(n^{o(1)}(1 + \frac{1}{n\eta}))$. Conditioning to this event, and assuming that $\eta \geqslant n^{-3/2+\varepsilon}$ for some fixed $\varepsilon > 0$ (to keep the spectral radius of $R(M_{n,i,j}, z)\frac{1}{\sqrt{n}}\xi_{ij}E_{ij}$ less than one), we then see from (3.0.15) that

$$F\Big(M_{n,i,j} + \frac{\xi_{ij}E_{ij}}{\sqrt{n}}\Big) = F(M_{n,i,j}) + \sum_{k=1}^{\infty} \frac{(-\xi_{ij})^k}{n^{1+k/2}} \operatorname{tr}\left(R(M_{n,i,j}, z)^2 E_{ij}(R(M_{n,i,j}, z)E_{ij})^{k-1}\right).$$

The coefficient bounds show that the trace here is of size $O(\eta^{-1}(n^{o(1)}(1 + \frac{1}{n\eta}))^k)$ (where the $n^{o(1)}$ expression, or the implied constant in the $O()$ notation, does not depend on k). Thus we may truncate the sum at any stage to obtain

$$F\Big(M_{n,i,j} + \frac{\xi_{ij}E_{ij}}{\sqrt{n}}\Big) = F(M_{n,i,j}) + \sum_{k=1}^{m} \frac{(-\xi_{ij})^k}{n^{1+k/2}} \operatorname{tr}\left(R(M_{n,i,j}, z)^2 E_{ij}(R(M_{n,i,j}, z)E_{ij})^{k-1}\right) + O_m\left(\eta^{-1} n^{-\frac{m+3}{2}+o(1)}(1 + \frac{1}{n\eta})^{m+1}\right)$$

with probability $1 - O(n^{-100})$. On the exceptional event, we can bound all terms on the left and right-hand side crudely by $O(n^{10})$ (say). Taking expectations, and using the independence of ξ_{ij} from $M_{n,i,j}$, we conclude that

$$\mathbb{E}F\Big(M_{n,i,j} + \frac{\xi_{ij}E_{ij}}{\sqrt{n}}\Big) = \mathbb{E}F(M_{n,i,j}) + \sum_{k=1}^{m} \frac{\mathbb{E}(-\xi_{ij})^k}{n^{1+k/2}} \mathbb{E}\operatorname{tr}\left(R(M_{n,i,j}, z)^2 E_{ij}(R(M_{n,i,j}, z)E_{ij})^{k-1}\right) + O\left(\eta^{-1} n^{-\frac{m+3}{2}+o(1)}(1 + \frac{1}{n\eta})^{m+1}\right)$$

for any $1 \leqslant m \leqslant 10$ (say). Similarly with ξ_{ij} replaced by η_{ij}. If we assume that ξ_{ij} and η_{ij} have m matching moments in the sense that

$$\mathbb{E}\xi_{ij}^k = \mathbb{E}\eta_{ij}^k$$

for all $k = 1, \ldots, m$, we conclude on subtracting that

$$\mathbb{E}F\left(M_{n,i,j} + \frac{1}{\sqrt{n}}\xi_{ij}E_{ij}\right) - \mathbb{E}F\left(M_{n,i,j} + \frac{1}{\sqrt{n}}\eta_{ij}E_{ij}\right) = O\left(\eta^{-1} n^{-\frac{m+3}{2}+o(1)}(1 + \frac{1}{n\eta})^{m+1}\right).$$

Comparing this with (3.0.12), we arrive at the following conclusions for the statistic $F(M_n) = s(M_n, z)$:

- By definition of a Wigner matrix, we already have 2 matching moments. Setting $m = 2$, we conclude that $\mathbb{E}s(M_n, z)$ enjoys a two moment theorem whenever $\eta \geqslant n^{-1/2+\varepsilon}$ for some fixed $\varepsilon > 0$.

- If we additionally assume a third matching moment $\mathbb{E}\xi_{ij}^3 = \mathbb{E}\eta_{ij}^3$, then we may set $m = 3$, and we conclude that $\mathbb{E}s(M_n, z)$ enjoys a three moment theorem whenever $\eta \geqslant n^{-1+\varepsilon}$ for a fixed $\varepsilon > 0$.
- If we assume both third and fourth matching moments $\mathbb{E}\xi_{ij}^3 = \mathbb{E}\eta_{ij}^3$ and $\mathbb{E}\xi_{ij}^4 = \mathbb{E}\eta_{ij}^4$, then we may set $m = 4$, and we conclude that $\mathbb{E}s(M_n, z)$ enjoys a four moment theorem whenever $\eta \geqslant n^{-\frac{13}{12}+\varepsilon}$ for some $\varepsilon > 0$.

Thus we see that the expected Stieltjes transform $\mathbb{E}s(M_n, z)$ has some universality, although the amount of universality provided by the Lindeberg exchange method degrades as z approaches the real axis. Additional matching moments beyond the fourth will allow one to approach the real axis even further, although with the bounds provided here, one cannot get closer than $n^{-3/2}$ due to the potential divergence of the Neumann series beyond this point. Some of the exponents in the above results can be improved by using more refined control on the Stieltjes transform, and by using the Knowles-Yin variant of the Lindeberg exchange method; see for instance [41].

One can generalise these arguments to more complicated statistics than the Stieltjes transform $s(M_n, z)$. For instance, one can establish a four moment theorem for multilinear averages of the Stieltjes transform:

Exercise 3.0.18. Let k be a fixed natural number, and let $\psi : \mathbb{R}^k \to \mathbb{C}$ be a smooth compactly supported function, both of which are independent of n. Let $z = E + i\eta$ be a complex number with $\eta \geqslant n^{-1-\frac{1}{100k}}$. Show that the statistic

$$\mathbb{E}\int_{\mathbb{R}^k} \psi(t_1, \dots, t_k) \prod_{l=1}^{k} s\left(M_n, z + \frac{t_l}{n}\right)\, dt_1 \dots dt_k$$

enjoys a four moment theorem. Similarly if one replaces one or more of the $s(M_n, z + \frac{t_l}{n})$ with their complex conjugates.

Using this, one can then obtain analogous four moment theorems for k-point correlation functions. Recall that for any fixed $1 \leqslant k \leqslant n$, the k-point correlation function $\rho^{(k)} : \mathbb{R}^k \to \mathbb{R}^+$ of a random real symmetric (or Hermitian) matrix M_n is defined via duality by requiring that $\rho^{(k)}$ be symmetric and obey the relation

$$\int_{\mathbb{R}^k} \rho^{(k)}(x_1, \dots, x_k) F(x_1, \dots, x_k)\, dx_1 \dots dx_k = \mathbb{E} \sum_{1 \leqslant i_1 < \dots < i_k \leqslant n} F(\lambda_{i_1}, \dots, \lambda_{i_k})$$

for all continuous compactly supported symmetric functions $F : \mathbb{R}^k \to \mathbb{C}$, where $\lambda_1 \leqslant \dots \leqslant \lambda_n$ denote the eigenvalues of M_n arranged in increasing order (and counting multiplicity). From the Riesz representation theorem we see that this defines $\rho^{(k)}$ as a Radon measure at least; if M_n has an absolutely continuous distribution (as is the case for instance with the GOE ensemble) then ρ will in fact be a locally integrable function. Setting $k = 1$ and $F(x) = \frac{1}{x-z}$, we see in particular that

$$\int_{\mathbb{R}} \rho^{(1)}(x) \frac{dx}{x - z} = n\mathbb{E}s(M_n, z). \tag{3.0.19}$$

Similarly, setting $k = 2$ and $F(x_1, x_2) = \frac{1}{x_1 - z_1}\frac{1}{x_2 - z_2} + \frac{1}{x_1 - z_2}\frac{1}{x_2 - z_1}$, then setting $k = 1$ and $F(x) = \frac{1}{(x - z_1)(x - z_2)}$, and adding, we see that

$$2\int_{\mathbb{R}^2} \rho^{(2)}(x_1, x_2)\, \frac{dx_1 dx_2}{(x_1 - z_1)(x_2 - z_2)} + \int_{\mathbb{R}^2} \rho^{(1)}(x)\, \frac{dx}{(x - z_1)(x - z_2)} = n^2 \mathbb{E} s(M_n, z_1) s(M_n, z_2).$$

By combining these sorts of identities with Exercise 3.0.18, one can obtain four moment theorems for correlation functions. For instance, we have

Proposition 3.0.20. *Let* E *be a real number, and let* $\psi : \mathbb{R} \to \mathbb{R}$ *be a smooth compactly supported function. Then the statistic*

$$\frac{1}{n}\int_{\mathbb{R}} \rho^{(1)}(E + \frac{s}{n})\psi(s)\, ds$$

enjoys a four moment theorem.

Proof. Set $\eta := n^{-1-\frac{1}{100}}$. From Exercise 3.0.18, the statistic

$$\mathbb{E}\int_{\mathbb{R}} \psi(t) s(M_n, E + \frac{t}{n} + i\eta)\, dt$$

enjoys a four moment theorem. Applying (3.0.19), we conclude that

$$\frac{1}{n}\int_{\mathbb{R}}\int_{\mathbb{R}} \rho^{(1)}(x)\psi(t)\, \frac{dx dt}{x - E - \frac{t}{n} - i\eta}$$

enjoys a four moment theorem. The change of variables $x = E + \frac{s}{n}$, yields

$$\frac{1}{n}\int_{\mathbb{R}} \rho^{(1)}(E + \frac{s}{n}) \left(\int_{\mathbb{R}} \frac{\psi(t)\, dt}{s - t - in\eta}\right) ds.$$

Taking imaginary parts and dividing by π, we conclude that

$$\frac{1}{n}\int_{\mathbb{R}} \rho^{(1)}(E + \frac{s}{n}) \left(\int_{\mathbb{R}} \frac{n\eta\psi(t)\, dt}{\pi((s - t)^2 + (n\eta)^2)}\right) ds$$

enjoys a four moment theorem. However, from the smoothness and compact support of ψ, and the fact that the Cauchy distribution $\frac{n\eta}{\pi((s-t)^2 + (n\eta)^2)}$ integrates to one, one can establish a bound of the form

$$\int_{\mathbb{R}} \frac{n\eta\psi(t)\, dt}{\pi((s - t)^2 + (n\eta)^2)} = \psi(s) + O\left(\frac{n^{-1/100}}{1 + s^2}\right)$$

(Exercise!). On the other hand, from (3.0.19) and Theorem 3.0.16 one can show that

$$\frac{1}{n}\int_{\mathbb{R}} \rho^{(1)}\left(E + \frac{s}{n}\right) \frac{ds}{1 + s^2} \ll n^{o(1)}$$

(Exercise!). The claim now follows from the triangle inequality. □

Exercise 3.0.21. Justify the two steps marked (Exercise!) in the above proof.

Exercise 3.0.22. If k is a fixed natural number, $E_1, \ldots, E_k$ are real numbers, and $\psi : \mathbb{R}^k \to \mathbb{R}$ is a smooth compactly supported function, show that the statistic

$$\frac{1}{n^k}\int_{\mathbb{R}} \rho^{(k)}\left(E_1 + \frac{s_1}{n}, \ldots, E_k + \frac{s_k}{n}\right) \psi(s_1, \ldots, s_k)\, ds_1 \ldots s_k$$

enjoys a four moment theorem.

With a bit more effort (using now the real part of the Stieltjes transform, in addition to the imaginary part), one can also use Exercise 3.0.18 to establish a four moment theorem for statistics involving the individual eigenvalues and eigenvectors of M_n; see [41] for details. Such theorems were also established by direct Taylor expansion of the eigenvalues and eigenvectors; see [61], [62].

Of course, one would like to also establish universality results for classes of random matrices with fewer than four matching moments. At the scale of the mean eigenvalue spacing of $1/n$, it appears (in the bulk, at least) that the Lindeberg exchange method is not powerful enough on its own to accomplish this task. However, the Lindeberg exchange method combines well with other universality results involving ensembles in which the third and fourth moments are allowed to vary. For instance, a breakthrough work of Johansson [39] obtained universality results for matrices that were *gauss divisible* in the sense that they were of the form

$$M_n^t := (1-t)^{1/2} M_n + t^{1/2} G_n.$$

where G_n was a GUE matrix, M_n was a (Hermitian) Wigner matrix (with the same first two moments as the GUE matrix), and $0 < t < 1$ was a fixed parameter. This combines well with the four moment theorem to obtain universality results for a relatively wide class of Hermitian Wigner matrices; see [61]. This approach has not yet been extended to the case of real Wigner matrices. However, a different way to obtain universality between a real Wigner matrix M_n and a GOE matrix G_n (which we take to be independent of M_n) is not to exchange the elements from M_n to G_n one at a time, but instead to consider an Ornstein-Uhlenbeck type process that flows continuously from M_n to G_n from time $t = 0$ to time $t = 1$, by the formula

$$M_n^t := (1-t)^{1/2} M_n + t^{1/2} G_n.$$

Clearly M_n^t equals M_n when $t = 0$ and G_n when $t = 1$. For any intermediate time t, M_n^t is a real Wigner matrix (of a special type known as a *Gauss divisible* Wigner matrix), and the third and fourth moments of the components of M_n^t vary in some explicit polynomial fashion from those of ξ_{ij} to those of η_{ij}. A key feature of this flow is that the eigenvalues of M_n^t evolve by the laws of *Dyson Brownian motion*. Using techniques such as the method of *local relaxation flow* as discussed in the lecture notes of Erdős, one can obtain good universality results[10] that compare M_n^t to G_n for t as low as $n^{-1+\varepsilon}$ for any fixed $\varepsilon > 0$. Meanwhile, the Lindeberg exchange method can be used to compare[11] M_n^t to M_n for $t = n^{-1+\varepsilon}$.

[10]Strictly speaking, the method of local relaxation flow requires an additional averaging in the energy parameter E in order to obtain this claim. However, more recent methods are now available to avoid this averaging in energy: see [15].

[11]There is a minor additional issue because the third and fourth moments of the coefficients of M_n and M_n^t are not quite identical, however it turns out that the additional error terms caused by this are negligible when $t = n^{-1+\varepsilon}$ for ε small enough.

Combining these two claims, one can obtain universality results for Wigner matrices that do not require additional matching moments beyond the second; see e.g. [26] or [31]. On the other hand, there are some spectral statistics, particularly those involving a fixed eigenvalue of a Wigner matrix, for which the matching of the fourth moment is in fact necessary: see [64], [24]. Also, at this present time, the Lindeberg exchange method is the only tool that has been effectively used to obtain local universality results for non-Hermitian matrices [65].

References

[1] R. Adamczak, *On the Marchenko-Pastur and circular laws for some classes of random matrices with dependent entries*, Electron. J. Probab. **16** (2011), no. 37, 1068–1095, DOI 10.1214/EJP.v16-899. MR2820070 ↑476

[2] R. Adamczak and D. Chafaï, *Circular law for random matrices with unconditional log-concave distribution*, Commun. Contemp. Math. **17** (2015), no. 4, 1550020, 22, DOI 10.1142/S0219199715500200. MR3359233 ↑476

[3] R. Adamczak, D. Chafaï, and P. Wolff, *Circular law for random matrices with exchangeable entries*, Random Structures Algorithms **48** (2016), no. 3, 454–479, DOI 10.1002/rsa.20599. MR3481269 ↑476

[4] R. Adamczak, O. Guédon, A. Litvak, A. Pajor, and N. Tomczak-Jaegermann, *Smallest singular value of random matrices with independent columns* (English, with English and French summaries), C. R. Math. Acad. Sci. Paris **346** (2008), no. 15-16, 853–856, DOI 10.1016/j.crma.2008.07.011. MR2441920 ↑464

[5] J. Alt, L. Erdős, and T. Krueger, *Local inhomogeneous circular law*, preprint. ↑476

[6] M. Banna, *Limiting spectral distribution of Gram matrices associated with functionals of β-mixing processes*, J. Math. Anal. Appl. **433** (2016), no. 1, 416–433, DOI 10.1016/j.jmaa.2015.07.064. MR3388800 ↑488

[7] M. Banna and F. Merlevède, *Limiting spectral distribution of large sample covariance matrices associated with a class of stationary processes*, J. Theoret. Probab. **28** (2015), no. 2, 745–783, DOI 10.1007/s10959-013-0508-x. MR3370674 ↑488

[8] M. Banna, F. Merlevède, and M. Peligrad, *On the limiting spectral distribution for a large class of symmetric random matrices with correlated entries*, Stochastic Process. Appl. **125** (2015), no. 7, 2700–2726, DOI 10.1016/j.spa.2015.01.010. MR3332852 ↑488

[9] Z. D. Bai, *Circular law*, Ann. Probab. **25** (1997), no. 1, 494–529, DOI 10.1214/aop/1024404298. MR1428519 ↑476, 482

[10] A. Basak, N. Cook, and O. Zeitouni, *Circular law for the sum of random permutation matrices*, preprint. ↑476

[11] A. Basak and M. Rudelson, *Circular law for sparse matrices*, preprint. ↑476

[12] A. a. D. Basak Amir, *Limiting spectral distribution of sums of unitary and orthogonal matrices*, Electron. Commun. Probab. **18** (2013), no. 69, 19. ↑476

[13] C. Bordenave, P. Caputo, and D. Chafaï, *Circular law theorem for random Markov matrices*, Probab. Theory Related Fields **152** (2012), no. 3-4, 751–779, DOI 10.1007/s00440-010-0336-1. MR2892961 ↑476

[14] C. Bordenave and D. Chafaï, *Around the circular law*, Probab. Surv. **9** (2012), 1–89, DOI 10.1214/11-PS183. MR2908617 ↑476

[15] P. Bourgade, L. Erdős, H.-T. Yau, and J. Yin, *Fixed energy universality for generalized Wigner matrices*, Comm. Pure Appl. Math. **69** (2016), no. 10, 1815–1881, DOI 10.1002/cpa.21624. MR3541852 ↑495

[16] P. a. Y. Bourgade Horng-Tzer and Yin, *Local circular law for random matrices*, Probab. Theory Related Fields **159** (2014), no. 3-4, 545–595. ↑476

[17] P. a. Y. Bourgade Horng-Tzer and Yin, *The local circular law II: the edge case*, Probab. Theory Related Fields **159** (2014), no. 3-4, 619–660. ↑476

[18] J. Bourgain, V. H. Vu, and P. M. Wood, *On the singularity probability of discrete random matrices*, J. Funct. Anal. **258** (2010), no. 2, 559–603, DOI 10.1016/j.jfa.2009.04.016. MR2557947 ↑467, 469

[19] S. Chatterjee, *A simple invariance theorem*, preprint. ↑488

[20] N. Cook, *The circular law for random regular digraphs*, preprint. ↑476

[21] N. Cook, W. Hachem, J. Najim, and D. Renfrew, *Limiting spectral distribution for non-hermitian random matrices with a variance profile*, preprint. ↑476

[22] A. Edelman, *The probability that a random real Gaussian matrix has* k *real eigenvalues, related distributions, and the circular law*, J. Multivariate Anal. **60** (1997), no. 2, 203–232, DOI 10.1006/jmva.1996.1653. MR1437734 ↑476

[23] A. Edelman, *Eigenvalues and condition numbers of random matrices*, SIAM J. Matrix Anal. Appl. **9** (1988), no. 4, 543–560, DOI 10.1137/0609045. MR964668 ↑473

[24] A. Edelman, A. Guionnet, and S. Péché, *Beyond universality in random matrix theory*, Ann. Appl. Probab. **26** (2016), no. 3, 1659–1697, DOI 10.1214/15-AAP1129. MR3513602 ↑496

[25] P. Erdős, *On a lemma of Littlewood and Offord*, Bull. Amer. Math. Soc. **51** (1945), 898–902, DOI 10.1090/S0002-9904-1945-08454-7. MR0014608 ↑467

[26] L. Erdős, J. Ramírez, B. Schlein, T. Tao, V. Vu, and H.-T. Yau, *Bulk universality for Wigner Hermitian matrices with subexponential decay*, Math. Res. Lett. **17** (2010), no. 4, 667–674, DOI 10.4310/MRL.2010.v17.n4.a7. MR2661171 ↑496

[27] L. Erdős, B. Schlein, and H.-T. Yau, *Semicircle law on short scales and delocalization of eigenvectors for Wigner random matrices*, Ann. Probab. **37** (2009), no. 3, 815–852, DOI 10.1214/08-AOP421. MR2537522 ↑491

[28] L. Erdős, B. Schlein, and H.-T. Yau, *Local semicircle law and complete delocalization for Wigner random matrices*, Comm. Math. Phys. **287** (2009), no. 2, 641–655, DOI 10.1007/s00220-008-0636-9. MR2481753 ↑491

[29] L. Erdős, B. Schlein, and H.-T. Yau, *Wegner estimate and level repulsion for Wigner random matrices*, Int. Math. Res. Not. IMRN **3** (2010), 436–479, DOI 10.1093/imrn/rnp136. MR2587574 ↑491

[30] L. Erdős, H.-T. Yau, and J. Yin, *Bulk universality for generalized Wigner matrices*, Probab. Theory Related Fields **154** (2012), no. 1-2, 341–407, DOI 10.1007/s00440-011-0390-3. MR2981427 ↑

[31] L. Erdős, H.-T. Yau, and J. Yin, *Rigidity of eigenvalues of generalized Wigner matrices*, Adv. Math. **229** (2012), no. 3, 1435–1515, DOI 10.1016/j.aim.2011.12.010. MR2871147 ↑491, 496

[32] S. Geman, *A limit theorem for the norm of random matrices*, Ann. Probab. **8** (1980), no. 2, 252–261. ↑463

[33] J. Ginibre, *Statistical ensembles of complex, quaternion, and real matrices*, J. Mathematical Phys. **6** (1965), 440–449. ↑476

[34] V. L. Girko, *The circular law* (Russian), Teor. Veroyatnost. i Primenen. **29** (1984), no. 4, 669–679. MR773436 ↑476, 481

[35] V. L. Girko, *The strong circular law. Twenty years later. II*, Random Oper. Stochastic Equations **12** (2004), no. 3, 255–312, DOI 10.1163/1569397042222477. MR2085255 ↑476

[36] F. Götze and A. Tikhomirov, *On the circular law*, preprint. ↑471, 476, 482

[37] A. Guionnet, M. Krishnapur, and O. Zeitouni, *The single ring theorem*, Ann. of Math. (2) **174** (2011), no. 2, 1189–1217, DOI 10.4007/annals.2011.174.2.10. MR2831116 ↑476

[38] G. Halász, *Estimates for the concentration function of combinatorial number theory and probability*, Period. Math. Hungar. **8** (1977), no. 3-4, 197–211, DOI 10.1007/BF02018403. MR0494478 ↑469

[39] K. Johansson, *Universality of the local spacing distribution in certain ensembles of Hermitian Wigner matrices*, Comm. Math. Phys. **215** (2001), no. 3, 683–705, DOI 10.1007/s002200000328. MR1810949 ↑495

[40] J. Kahn, J. Komlós, and E. Szemerédi, *On the probability that a random ± 1-matrix is singular*, J. Amer. Math. Soc. **8** (1995), no. 1, 223–240, DOI 10.2307/2152887. MR1260107 ↑469

[41] A. Knowles and J. Yin, *Anisotropic local laws for random matrices*, preprint. ↑487, 488, 493, 495

[42] A. Knowles and J. Yin, *Eigenvector distribution of Wigner matrices*, Probab. Theory Related Fields **155** (2013), no. 3-4, 543–582, DOI 10.1007/s00440-011-0407-y. MR3034787 ↑488

[43] J. Komlós, *On the determinant of $(0,1)$ matrices*, Studia Sci. Math. Hungar **2** (1967), 7–21. MR0221962 ↑462, 467

[44] J. W. Lindeberg, *Eine neue Herleitung des Exponentialgesetzes in der Wahrscheinlichkeitsrechnung* (German), Math. Z. **15** (1922), no. 1, 211–225, DOI 10.1007/BF01494395. MR1544569 ↑484

[45] M. L. Mehta, *Random matrices*, 3rd ed., Pure and Applied Mathematics (Amsterdam), vol. 142, Elsevier/Academic Press, Amsterdam, 2004. MR2129906 ↑

[46] J. A. a. S. Mingo Roland, *Free probability and random matrices*, Fields Institute Monographs, vol. 35, Springer, New York; Fields Institute for Research in Mathematical Sciences, Toronto, ON, 2017. ↑479

[47] H. H. Nguyen, *Random doubly stochastic matrices: the circular law*, Ann. Probab. **42** (2014), no. 3, 1161–1196, DOI 10.1214/13-AOP877. MR3189068 ↑476

[48] H. H. Nguyen and V. H. Vu, *Normal vector of a random hyperplane* **3** (2016), available at `1604.04897`. ↑473

[49] H. Nguyen and S. O'Rourke, *The elliptic law*, preprint. ↑476

[50] H. H. Nguyen and V. H. Vu, *Circular law for random discrete matrices of given row sum*, J. Comb. **4** (2013), no. 1, 1–30, DOI 10.4310/JOC.2013.v4.n1.a1. MR3064040 ↑476

[51] G. Pan and W. Zhou, *Circular law, extreme singular values and potential theory*, J. Multivariate Anal. **101** (2010), no. 3, 645–656, DOI 10.1016/j.jmva.2009.08.005. MR2575411 ↑471, 476, 482

[52] M. Rudelson, *Invertibility of random matrices: norm of the inverse*, Ann. of Math. (2) **168** (2008), no. 2, 575–600, DOI 10.4007/annals.2008.168.575. MR2434885 ↑462, 469

[53] M. Rudelson and R. Vershynin, *The Littlewood-Offord problem and invertibility of random matrices*, Adv. Math. **218** (2008), no. 2, 600–633, DOI 10.1016/j.aim.2008.01.010. MR2407948 ↑469

[54] M. Rudelson and R. Vershynin, *The least singular value of a random square matrix is* $O(n^{-1/2})$ (English, with English and French summaries), C. R. Math. Acad. Sci. Paris **346** (2008), no. 15-16, 893–896, DOI 10.1016/j.crma.2008.07.009. MR2441928 ↑471

[55] M. Talagrand, *Concentration of measure and isoperimetric inequalities in product spaces*, Inst. Hautes Études Sci. Publ. Math. **81** (1995), 73–205. MR1361756 ↑472

[56] T. Tao, *Topics in random matrix theory*, Graduate Studies in Mathematics, vol. 132, American Mathematical Society, Providence, RI, 2012. MR2906465 ↑461

[57] T. Tao and V. Vu, *On the singularity probability of random Bernoulli matrices*, J. Amer. Math. Soc. **20** (2007), no. 3, 603–628, DOI 10.1090/S0894-0347-07-00555-3. MR2291914 ↑469

[58] T. Tao and V. Vu, *Random matrices: the circular law*, Commun. Contemp. Math. **10** (2008), no. 2, 261–307, DOI 10.1142/S0219199708002788. MR2409368 ↑471, 476, 482

[59] T. Tao and V. H. Vu, *Inverse Littlewood-Offord theorems and the condition number of random discrete matrices*, Ann. of Math. (2) **169** (2009), no. 2, 595–632, DOI 10.4007/annals.2009.169.595. MR2480613 ↑469

[60] T. Tao and V. Vu, *Random matrices: the distribution of the smallest singular values*, Geom. Funct. Anal. **20** (2010), no. 1, 260–297, DOI 10.1007/s00039-010-0057-8. MR2647142 ↑473

[61] T. Tao and V. Vu, *Random matrices: universality of local eigenvalue statistics*, Acta Math. **206** (2011), no. 1, 127–204, DOI 10.1007/s11511-011-0061-3. MR2784665 ↑488, 491, 495

[62] T. Tao and V. Vu, *Random matrices: universal properties of eigenvectors*, Random Matrices Theory Appl. **1** (2012), no. 1, 1150001, 27, DOI 10.1142/S2010326311500018. MR2930379 ↑495

[63] T. Tao and V. Vu, *Random matrices: universality of ESDs and the circular law*, Ann. Probab. **38** (2010), no. 5, 2023–2065, DOI 10.1214/10-AOP534. With an appendix by Manjunath Krishnapur. MR2722794 ↑476, 482, 483

[64] T. Tao and V. Vu, *Random matrices: localization of the eigenvalues and the necessity of four moments*, Acta Math. Vietnam. **36** (2011), no. 2, 431–449. MR2908537 ↑496

[65] T. a. V. Tao Van, *Random matrices: universality of local spectral statistics of non-Hermitian matrices*, Ann. Probab. **43** (2015), no. 2, 782–874. ↑476, 496

[66] L. N. Trefethen, *Pseudospectra of matrices*, Numerical analysis 1991 (Dundee, 1991), Pitman Res. Notes Math. Ser., vol. 260, Longman Sci. Tech., Harlow, 1992, pp. 234–266. MR1177237 ↑477

[67] P. M. Wood, *Universality and the circular law for sparse random matrices*, Ann. Appl. Probab. **22** (2012), no. 3, 1266–1300, DOI 10.1214/11-AAP789. MR2977992 ↑476

[68] J. Yin, *The local circular law III: general case*, Probab. Theory Related Fields **160** (2014), no. 3-4, 679–732. ↑476

UCLA Department of Mathematics, Los Angeles, CA 90095-1555
Email address: `tao@math.ucla.edu`

Published Titles in This Series

26 **Alexei Borodin, Ivan Corwin, and Alice Guionnet, Editors,** Random Matrices, 2019
25 **Michael W. Mahoney, John C. Duchi, and Anna C. Gilbert, Editors,** The Mathematics of Data, 2018
24 **Roman Bezrukavnikov, Alexander Braverman, and Zhiwei Yun, Editors,** Geometry of Moduli Spaces and Representation Theory, 2017
23 **Mark J. Bowick, David Kinderlehrer, Govind Menon, and Charles Radin, Editors,** Mathematics and Materials, 2017
22 **Hubert L. Bray, Greg Galloway, Rafe Mazzeo, and Natasa Sesum, Editors,** Geometric Analysis, 2016
21 **Mladen Bestvina, Michah Sageev, and Karen Vogtmann, Editors,** Geometric Group Theory, 2014
20 **Benson Farb, Richard Hain, and Eduard Looijenga, Editors,** Moduli Spaces of Riemann Surfaces, 2013
19 **Hongkai Zhao, Editor,** Mathematics in Image Processing, 2013
18 **Cristian Popescu, Karl Rubin, and Alice Silverberg, Editors,** Arithmetic of L-functions, 2011
17 **Jeffery McNeal and Mircea Mustaţă, Editors,** Analytic and Algebraic Geometry, 2010
16 **Scott Sheffield and Thomas Spencer, Editors,** Statistical Mechanics, 2009
15 **Tomasz S. Mrowka and Peter S. Ozsváth, Editors,** Low Dimensional Topology, 2009
14 **Mark A. Lewis, Mark A. J. Chaplain, James P. Keener, and Philip K. Maini, Editors,** Mathematical Biology, 2009
13 **Ezra Miller, Victor Reiner, and Bernd Sturmfels, Editors,** Geometric Combinatorics, 2007
12 **Peter Sarnak and Freydoon Shahidi, Editors,** Automorphic Forms and Applications, 2007
11 **Daniel S. Freed, David R. Morrison, and Isadore Singer, Editors,** Quantum Field Theory, Supersymmetry, and Enumerative Geometry, 2006
10 **Steven Rudich and Avi Wigderson, Editors,** Computational Complexity Theory, 2004
9 **Brian Conrad and Karl Rubin, Editors,** Arithmetic Algebraic Geometry, 2001
8 **Jeffrey Adams and David Vogan, Editors,** Representation Theory of Lie Groups, 2000
7 **Yakov Eliashberg and Lisa Traynor, Editors,** Symplectic Geometry and Topology, 1999
6 **Elton P. Hsu and S. R. S. Varadhan, Editors,** Probability Theory and Applications, 1999
5 **Luis Caffarelli and Weinan E, Editors,** Hyperbolic Equations and Frequency Interactions, 1999
4 **Robert Friedman and John W. Morgan, Editors,** Gauge Theory and the Topology of Four-Manifolds, 1998
3 **János Kollár, Editor,** Complex Algebraic Geometry, 1997
2 **Robert Hardt and Michael Wolf, Editors,** Nonlinear partial differential equations in differential geometry, 1996
1 **Daniel S. Freed and Karen K. Uhlenbeck, Editors,** Geometry and Quantum Field Theory, 1995